Extracting the Science
A Century of Mining Research

Edited by Jürgen Brune

Published by

SME
Society for
Mining, Metallurgy
& Exploration

Society for Mining, Metallurgy, and Exploration, Inc. (SME)
8307 Shaffer Parkway
Littleton, Colorado, USA 80127
(303) 948-4200 / (800) 763-3132
www.smenet.org

SME advances the worldwide mining and minerals community through information exchange and professional development. SME is the world's largest association of mining and minerals professionals.

ISBN: 978-0-87335-322-9

Library of Congress Cataloging-in-Publication Data

Extracting the science : a century of mining research / edited by Jürgen Brune.
    p. cm.
  Includes bibliographical references and index.
  ISBN 978-0-87335-322-9
1. Mining engineering—History. I. Brune, Jürgen F.

TN19.E97 2010
622.09—dc22

                                                    2009051063

Contents

SECTION 3.   GROUND CONTROL

SECTION 4.   MINE VENTILATION

SECTION 5.   TRAINING

Preface

The year 2010 marks the 100th anniversary of the formation of the Bureau of Mines as the premier mining research organization in the United States. The Society for Mining, Metallurgy, and Exploration, Inc. (SME), is celebrating this anniversary with a special symposium, "Extracting the Science: 100 Years of Mining Research," held in conjunction with the 2010 Annual Meeting in Phoenix, Arizona, USA.

When the United States Congress founded the Bureau of Mines in 1910, it established the primary government agency to conduct scientific research and to disseminate information on the extraction, processing, use, and conservation of mineral resources. The initial focus of the bureau's work was to conduct research on the prevention of disastrous mine accidents. On December 6, 1907, the explosion of the Monongah mine in West Virginia killed 362 miners. This disaster, the worst in U.S. mining history, was followed 2 weeks later by a severe explosion that killed 239 miners at Jacobs Creek, Pennsylvania. One hundred fifty-four miners died at Marianna, Pennsylvania, in November of 1908, and 259 were killed at Cherry, Illinois, in November 1909. Mine disasters in European countries, specifically the catastrophic explosion at the French Courrières mine in 1906 that killed 1,099 miners, sparked similar organized research efforts in Europe.

During the past 100 years, the U.S. mining industry has made remarkable improvements in mine safety and health, and equally remarkable advances in mining technology, equipment, and mine productivity. Mine operators benefit from the economics of scale offered by larger mines with higher-capacity equipment and increasingly automated control, remote sensing, and diagnostic technology. Computers and digital communication systems are common in all mine types. Sophisticated mine planning and design tools are available to optimize cut-off grades and mining sequences.

Along with the above advances, the productivity of mining operations has vastly improved with the introduction of more powerful equipment and sophisticated technologies, more extensive mine workings, and more efficient production technology. These improvements have led to specialized mining research challenges, including the following:

- To prevent catastrophic events such as fires, explosions, inundations, and major roof or slope failures
- To provide adequate ventilation to dilute explosive or toxic gases
- To prevent respiratory illness caused by inhalation of harmful dusts, diesel particulates, blasting or welding fumes, and radon progeny
- To properly engineer and design ground support and prevent falls of roof or back
- To provide protection from moving equipment and machinery
- To prevent hearing loss caused by workplace noise
- To facilitate an ergonomic work environment and prevent musculoskeletal diseases
- To provide emergency communications and tracking technology as well as life support for miners trapped in underground mines
- To conduct surveillance studies in mine worker safety and health
- To develop training materials and industry guidelines for the improvement of safety, health, and productivity in mines

For this symposium, we invited more than 50 leading scientists from 11 mining countries around the world to present their research findings. Their papers, as compiled in this volume, provide an in-depth cross section of research and technological achievements in the mining industry aimed at improving the health and safety of mine workers as well as furthering mine productivity. We asked the authors to include a brief historical view but to mainly focus on today's knowledge and on directions and expectations for future development. Keynote presentations introduce the subjects and provide a framework for the more detailed technical papers.

We have divided the volume into five subject areas:

1. General Mining
2. Mine Productivity
3. Ground Control
4. Mine Ventilation
5. Training

The presentations cover a wide spectrum of research and technology in underground and surface mining for a variety of minerals. Aspects of safety and health are blended with considerations for economics and the environment. Prevention of mine disasters remains a key responsibility for the industry, along with advancements in emergency response, personal protective equipment, communications and tracking, and training. Innovative technical solutions include computer-controlled mining equipment and remote monitoring of air quality and ground control parameters, while virtual reality training systems provide realistic experience within the safety of a controlled environment.

The research papers represent the viewpoints of mine operators, researchers in academia, private and government institutions, as well as regulators and enforcement agencies.

We wish to thank all authors and presenters for their outstanding contributions. We also wish to thank SME for hosting the symposium and for providing its support with the organization and publication of the papers.

Section 1. General Mining

The Development of Mining and Minerals Research in Canada

Louise Laverdure
CANMET, Mining & Mineral Sciences Laboratories, NRC, Ottawa, Canada

John Udd
CANMET, Mining & Mineral Sciences Laboratories, NRC, Ottawa, Canada

Gilles Tremblay
CANMET, Mining & Mineral Sciences Laboratories, NRC, Ottawa, Canada

Thomas Hynes
CANMET, Mining & Mineral Sciences Laboratories, NRC, Ottawa, Canada

INTRODUCTION

Against the background that many mineral deposits have unique attributes in size, depth, location, tenor, and mineralogy, mining has always been regarded as being an art as well as a science. Mine designers often face unique challenges in order to obtain the optimal economic results from the orebody which is to be exploited. Further, health and safety, as well as environmental and societal considerations, are paramount for a successful operation.

Advances in the art and craft of mining have been at a measured pace, as mine operators have, of necessity, been conservative in their approach to the development of mining and mineral processing operations. Thus, the industrial revolution in mining followed the advances which had been made in the agriculture, transportation and manufacturing industries. Beginning in the late 18th century and continuing to the present at an ever-increasing pace, manual mining methods have been replaced by mechanization and then automation. Some key developments were the introduction of steam-powered machinery, the advent of the internal combustion engine and the introduction of dynamite and other explosives. In mining and mineral processing operations, some of the key innovations in the 19th century were the introduction of compressed air, the development of rock drills, crushers, stamp mills, mine hoists and pumping systems, and rail haulage underground.

In the latter half of the 19th century the Silver Islet silver mine, located across the bay from today's city of Thunder Bay, was renowned for its technological innovations which it introduced in the mining and milling processes. Burleigh drills were introduced in Silver Islet in 1872, followed by the first use of the Holt and Severance diamond drill in Canada, in 1873. Compressed air systems were introduced into the mining industry during the late 1860s and early 1870s. Mineral processing stamp mills were common in the 19th century. No less than 80 Ball stamps mills were used at Silver Islet—a very large number by the standards of the day. In mineral concentration, a Frue Vanner, a Canadian invention, was developed at Silver Islet during the period 1872–1874 (Newell, 1986).

During the 19th century underground coal and other stratified deposits, which, because of their tabular nature, were particularly amenable to mechanization, were the first to be mechanized. One early innovation was the use of the Holt and Severance drill used for tunnelling at the Spring Hill Collieries, in Nova Scotia, in 1873. By the middle of the 20th century underground hard-rock selective mining methods were being replaced by bulk mining methods and open pits. Larger underground openings became possible through having bigger equipment and having the means to stabilize larger spanned openings. Today, environmental concerns now drive the means of handling disposal of mine and mill wastes. Mine backfill, used in Canadian mines since the 1930s, was once considered to be a waste disposal issue, has now become an integral part of engineered mine support.

In 1933 reverberatory furnace slag and pyrrhotite tailings were used as mine backfill at Noranda's Horne mine, in Noranda, Québec. The Falconbridge Mine, in Falconbridge (near Sudbury, Ontario) used sand and gravel for backfilling in 1935 In 1959, Falconbridge experimented with cemented tailings for the floors of cut-and-fill stopes. In 1960, INCO added Portland cement to hydraulic fill (from Udd, 1989). Today, backfill is engineered to provide the support needed in large blasthole stopes. The most recent innovations have been the use of cemented rock fill (CRF) and paste fill. Much work today has focussed on finding a replacement for Portland cement as a binding agent to reduce the costs to the mining companies.

During the past two centuries the mining and mineral processing industries have become locally professionalized. It should be noted that prior to the formation of the first mining engineering and mineral processing programs at North American universities (at Columbia University, in 1864, and Canada's McGill University, in 1871) all of the mining engineers in North America had been "imported" from older European universities. That saying is true for miners as well, since most mines had people that were connected to the Cornish mining industry in the UK at one time of another. The growth of our mining industry drove the development of our own educational programs.

THE DEVELOPMENT OF MINERALS RESEARCH

Concurrent with the development of processing programs at Canadian universities came the beginnings of mining and minerals research. The sources of funding for these research programs may come from industry, government, consortia, or private funding organizations. Of these sources, the most important one in Canada today is the Natural Sciences and Engineering Research Council of Canada (NSERC). At present NSERC supports the research interests of about 11,000 professors per annum, of which only a few hundred are in mining. Grant applications are reviewed by specialized committees of peers who review and adjudicate the requests for funding. It should also be noted that academics in Canadian universities, as elsewhere are free to pursue their own research and other interests.

During the early days of Canadian minerals research (during the late 19th Century) the emphasis was very much on mineral processing technology as the recovery of ore minerals was very low by present day standards. Much of the ore was lost in stamp milling and hand sorting (cobbing) operations. An example that helped recovery from this period is how, Professor Haultain at the University of Toronto developed the Haultain Infra-Sizer and the Super-Panner. The former was (and is) an air elutriating device through which the variously sized particles in crushed minerals could be separated into fractions. The Super-Panner was developed to improve the separation of gold particles from the gangue. Both may be found in mineral processing laboratories today, and are still in use—the best possible recognition for noteworthy innovations!

The Canadian Mines Branch, formed in 1907, eventually became CANMET and other parts within the Department of Natural Resources Canada. During the growth of the Canadian mining industry during the early part of the 20th century many requests were made to the Mines Branch for assistance addressing the industry's mineral processing problems. During the First World War the Mines Branch Laboratories, as a part of its activities, operated a custom milling operation to concentrate the strategic minerals needed for the war operations—such as molybdenum. During 1917, the demand for the metal was so great that the Ore Dressing Laboratory was operated as a molybdenite mill on an around-the-clock basis! By March 31,1935, some fifty new gold mills had entered production in Canada, with most using processes developed in the Ore Dressing and Metallurgical Laboratories of the Mines Branch. It was later said that most of the ore milling plants in Canada have benefited from improvements that resulted from the use of CANMET's expertise.

In 1931, following the discovery of rich silver ore and pitchblendeat Echo Bay, on Great Bear Lake, a shipment of twenty tons of ore was sent to the Mines Branch for hydrometallurgical treatment. Subsequently, the carbonate leaching method for treating the radium ores from Great Bear Lake that was developed by the Mines Branch was used by Eldorado Gold Mines Limited at its plant and radium refinery, which opened in January, 1933, at Port Hope, Ontario. Thus began the Mines Branch's involvement in a process that would ultimately be connected to the Manhattan Project of World War II. This began a long period of work on radioactive ores, which resulted in the discovery of the pressure acid leaching process, which was first used at Port Radium, in the Northwest Territories, in 1952. Other sites followed including Beaverlodge, in northern Saskatchewan, in 1953, Gunnar, also in northern Saskatchewan, in 1955, and the process became the foundation for the treatment of the uranium ores of Bancroft and the Elliot Lake mining camps, in Ontario, during the late 1950s and onwards.

In academia the research performed has been driven by the interests of the staff at the universities. Through the years the Canadian university mining and mineral processing programs have evolved in parallel with the evolution of the various departments. As an example, at McGill the Mining Engineering Department became the Department of Mining Engineering and Applied Geophysics in the 1960s, then the Department of Mining and Metallurgical Engineering, and then the Department of Mining and Materials Engineering, and presently the Department of Mining, Metals and Materials Engineering. During this period the mineral processing staff found themselves in the Department of Metallurgical Engineering, then the Department of Mining Engineering and Applied Geophysics, and then into today's department which reflects the full span of the minerals industries.

Other Canadian university minerals-related departments have followed similar paths as their faculties have evolved to reflect the interests of the universities themselves. During the last fifty years a great many North American mining programs have been terminated, since these were thought not to be cost-effective from the point-of-view of the institutions. Where there had been twenty-seven university mining programs in the USA up to about 1990, by 2007 the number had shrunk to 13 (McCarter, 2007). Likewise, the number of professors, which had numbered about 118 in 1984, had shrunk to about 72, in 2007.

Canadian Mining Programs have been under pressure since the 1960s. All have survived, however, albeit through reinstatement (at the University of Toronto), or mergers with other university departments. In mineral processing, the work is carried out through: the Department of Chemicals and Materials Engineering at the University of Alberta (Edmonton, Alberta); the Department of Materials Engineering at the University of British Columbia (Vancouver, British Columbia); the Department of Mineral Resource Engineering at Dalhousie University (Halifax, Nova Scotia); the Department of Chemical Engineering at Lakehead University (Thunder Bay, Ontario); the School of Engineering at Laurentian University (Sudbury, Ontario); the Department of Mining, Metallurgical and Materials Engineering at Laval University (Quebec City, Quebec); the Department of Mining, Metals and Materials Engineering at McGill University (Montreal, Quebec); the INCO Innovation Centre—Process Engineering and Hydrometallurgy at Memorial University of Newfoundland (St. John's, Newfoundland and Labrador);the Department of Mining Engineering at Queen's University (Kingston, Ontario); the Department of Civil and Environmental Engineering at the University of Saskatchewan (Saskatoon, Saskatchewan); and the Department of Chemical Engineering and Applied Chemistry at the University of Toronto (Toronto, Ontario).

The details on all of these mining programs may be found on the MineCan website (http://www .minecan.org/). Minecan is a Canadian national database on minerals research being developed by CANMET. It will be used as the principal database for mining and minerals research by the Canada Mining Innovation Council (CMIC).

Of all the research programs mentioned above, those at the University of Alberta, the University of British Columbia, and McGill are probably the best known for the breadth, depth, and continuity of their research. Regarding the education of mining engineers, Queen's University, with many more mining engineering students than the others, has probably produced the greatest number of mining engineering graduates from any Canadian university.

In the Canadian industry substantial mineral processing research has been done by a number of organizations including, in alphabetical order: Cameco Research (Port Hope, Ontario); COREM, (Québec City, Québec); Québec Iron and Titanium (Trois Rivières Rivers, Québec); the Saskatchewan Research Council's Pipe Flow Technology Centre™ (Saskatoon, Saskatchewan); SGS Lakefield (formerly Lakefield Research) (at Lakefield, near Peterborough, Ontario); Sherritt International Corporation (Fort Saskatchewan, Alberta); Syncrude Research (Edmonton, Alberta); Teck-Cominco (Trail, British Columbia);Vale-INCO's J. Roy Gordon Centre (Sheridan Park, Mississauga, Ontario); Xstrata Process Support (Falconbridge, Ontario).

While more will be said of industrial research in the next section, it should be noted that INCO (now Vale-INCO) closed its mining research facilities in Copper Cliff, Ontario, in 2001, and, the following year Noranda closed its Technology Centre, at Point-Claire, Québec. In the West, Placer Dome disbanded its Vancouver research division, while, in 2003, the Cambior organization also left the field. Essentially, the industry withdrew from doing its own basic research and development and is concentrating its present efforts on process and process-enhancement research.

In the broad scheme of things it must be mentioned that mineral processors in government and academia have always had strong connections and good working relationships with their counterparts in the industry. In 1962 the Mines Branch (CANMET's predecessor) organized a conference of Eastern Canadian gold processors. By the following year, 1963, the Canadian Mineral Processors (CMP) was formed, and then became a constituent Society of the Canadian Institute of Mining, Metallurgy and Petroleum. The organization exists to this day, with the 42nd Annual Conference planned for Ottawa, Ontario, in January 2010. All of the annual conferences have been held in Ottawa, with the result being that very strong relationships exist between government, industry, consultants and academia. In essence, the researchers have strong connectivity with those in industry who make the decisions regarding research.

THE DEVELOPMENT OF MINING RESEARCH

In mining research the pattern of communications was very different, and did not feature ongoing discussions among industry, government, and academia. For a great many years mine operators and academics had not come together to discuss the research needed. The focus of mine operators has always been to mine the ore safely and at a profit. There was very little commonality of views between operators and researchers. This gap in communications has been detrimental to organizing mining research in Canada, and continues to a much-reduced extent to this day.

During the 1930s and the 1940s, the narrow vein gold mines of Ontario were hard-hit by rockbursts and fatal accidents as the result of miners "chasing the vein" and practicing narrow-vein mining on irregular patterns. In Canada, the rocks of the Canadian Shield can be highly stressed due to tectonic forces, The results were that stress concentrations, which exceeded the strengths of the rocks, developed around sharp corners and in narrow support pillars. Explosive failures of rocks ensued, with damage to the mining operations and their personnel.

In the ensuing years both geophysicists and mining engineers studied the situation and possible solutions. As a result, with the development of the "doming theory, mining practices were altered through the development of the concept of sequential mining and the avoidance of practices that would lead to excessively high stress concentrations. The earliest mining research in Canada took place in the mining departments at McGill and Queen's universities and the University of Toronto, and the federal Mines Branch. In the latter, small regional offices were established in Blairmore, Alberta, and Fernie, British Columbia, at the beginning of the 1950s, in order to study the ground control problems being encountered in the underground coal mines in the Crowsnest Pass area. In 1953, additional offices were established in Springhill, Nova Scotia, and Calgary, Alberta. Since then, other units were opened in Edmonton, Alberta (1956–1986); Devon, Alberta (1986–present); Glace Bay, Nova Scotia (1959); Elliot Lake, Ontario (1964–1998); Quebec City (1973–1975); Sydney, Nova Scotia (1981–1998); Sudbury, Ontario (1987–present); Varennes, Quebec (1992–present); Val-d'Or, Quebec (1993–present). In essence, the federal government, the key player in Canadian minerals and mining research, had responded to the needs of industry, by establishing regional laboratories.

In the years following the mid-1960s great strides have been made in mining research in Canada, the United States, Europe, South Africa, Australia and elsewhere in the world. Increasingly powerful computers were facilitating studies in numerical modelling and improved instrumentation was appearing for studies in geomechanics. "Stressmeters" had been developed for measuring the in-situ stresses in rock, as had load cells and extensometers. The Mines Branch staff had been very active in all of these aspects, and so had others. During 1965 a comparative study was made of the various stressmeters available and, as the result, the decision was made to examine the borehole deformation meter developed by the United States Bureau of Mines (USBM) and the "doorstopper" method developed by the Council of Scientific and Industrial Research of South Africa (CSIR). Both had been developed for the purpose of measuring stresses in hard rocks (Ignatieff, p. 291). Beginning in 1965, work with both instruments began at the Elliot Lake Laboratory and in Rio Algom's Nordic Mine. Work with the CSIR device and a hollow inclusion stressmeter subsequently developed by the Commonwealth Scientific and Industrial Research Organisation (CSIRO), of Australia, continues today at the Bells Corners Laboratory.

In 1968, in response to the needs of the Canadian Advisory Committee on Rock Mechanics, the Mines Branch commissioned a study by consultants and issued *"Tentative Design Guide for Mine Waste Embankments in Canada"* in 1972. With a view towards improving the stabilities of waste embankments and other high slopes at mining properties, this was the first of two major studies that would have a significant impact on open pit mining operations throughout the world. In April 1972, the Department announced the first phase of a five-year $3.3 million program to produce standard engineering procedures for the design and support of open pit walls, and reducing the size of open pits. At its conclusion, this would lead to the publication of the multi-volume "Pit Slope Manual," one of the most significant scientific and engineering contributions ever made by mining researchers in the Mines Branch. With tens of thousands of copies of the individual chapters having been sold since 1977, it is the Mine Branch's/CANMET's best-selling publication. It resulted in the elevation of the Canadian government's mining research to that of an international world-class player.

On April 22, 1981, CANMET and the United States Bureau of Mines (of the Department of the Interior) (USBM) signed a five-year "Memorandum of Understanding" (MOU) for cooperative activities in mineral science and technology. At the outset, it was visualized that the research would be into radiation detection equipment and health and safety in the mineral industry. Shortly afterwards, the Ontario Ministry of Labour (MOL), CANMET, and the United States Bureau of Mines signed a MOU for research into understanding and reducing the toxicity of diesel emissions in underground mines. A further similar agreement concerned research on the means of inspecting the wire hoisting ropes which are the lifeline of any underground mine. This relationship, which included annual meetings to discuss the mining research needs in both countries continued until the closure of the USBM in 1996. It continues through meetings with NIOSH. (The United States National Institute for Occupational safety and Health).

Mining education began in Canada at McGill University (Montreal, Québec), in 1871. This was followed by École Polytechnique (Montréal, Québec), in 1873; the University of Toronto (Toronto, Ontario), in 1878; Queen's University (Kingston, Ontario), in 1893; the Nova Scotia Technical College (later the Technical University of Nova Scotia, and now a part of Dalhousie University, Halifax, Nova Scotia), in 1907; the University of Alberta (Edmonton, Alberta), in 1914; the University of British Columbia (Vancouver, British Columbia), in 1915; Laval University (Québec City, Québec), in 1938; Laurentian University (Sudbury, Ontario), in 1960; and the University of Saskatchewan (Saskatoon, Saskatchewan), in 1963.

Inasmuch as mining engineering education has experienced cycles of popularity which follow the economic cycles of the minerals industry there have been pressures on all of these departments during recessionary times. The names of the university mining departments have changed often as the result of mergers with other university programs. At the Universities of Toronto and Saskatchewan, respectively, the programs were terminated in the mid 1960s and 1977. Most recently, some programs are being rejuvenated through major donations. Thus, the Lassonde Mineral Engineering Program, named after a gift from Dr. Pierre Lassonde, was implemented at the University of Toronto, in 2000; likewise, the Norman B. Keevil Institute of Mining Engineering, was opened at the University of British Columbia, in 2003; and the Queen's University Department of Mining was renamed the Buchan Department of Mining, after the receipt of a $10,000,000 gift from Robert M. Buchan (the largest single donation to Canadian mining education to date) in mid-2009.

Traditionally, the faculty involved in mining education at Canadian universities has been small—perhaps a half-dozen people. With each having their own particular specialization and interests there has been a lack of significant team research. Traditionally mining research grants have been able to fund the studies of a professor and one or two graduate students. Hopefully, major gifts, such as those mentioned above, will encourage major cutting-edge research projects.

As mentioned previously, the names of mining researchers and their technical interests may be found on the MineCan website (http://www.minecan.org/).

As mentioned previously, in the hard rock mines of the Canadian Shield, the creation of underground openings can result in the generation of zones of unusually high stress. During the 1930s and the 1940s, the narrow vein gold mines of Ontario were hard-hit by rockbursts and fatal accidents as the result of this. During the early 1980s, and particularly 1984, rockbursting (or, the explosive failure of highly-stressed rock) had re-emerged as a serious problem in Ontario mines. By mid-1985, the Canadian government, the Government of Ontario and its mining industry had come together to form the five-year Canada-Ontario-Industry Rockburst Research Project (COIRRP). Through this, CANMET agreed to provide its research expertise, while the government of Ontario and the industry, provided the equipment needed and field installations and support, respectively. Each of the three contributions were valued at $1.4 million, making it the largest tri-partite mining research project that had been established by that time. Because of the COIRRP, seismic monitoring systems were being improved, local networks expanded, and interpretative techniques mastered. By 1989, microseismic monitoring systems had been installed in all of the Ontario mines where these had been needed. Further, the Eastern Canada Seismic Network, operated by the Geological Survey of Canada, had been augmented through the installation of seismograph stations in the Sudbury, Elliot Lake, Kirkland Lake and Red Lake mining camps.

Further, the mining companies involved had started to take the measures needed to alleviate their rockbursting problems. These included better sequencing of extraction to minimize the development of excessive stress concentrations; the use of engineered mine backfills to provide improved and predictable support; the use of new support techniques including yieldable support; and destress blasting through which excessively stressed ground could be "softened."

Beyond doubt, the COIRRP was one of the most successful projects ever undertaken by the Mines Branch/CANMET. After the conclusion of the first phase, the accounting firm of Peat, Marwick, Stevenson and Kellogg stated that the annual losses to the Ontario industry, during 1982–1984, prior to the implementation of the project were about $200 million. The total benefits of the project, in addition to the lives saved, and which cost CANMET $4.2 million for its share over five years, were estimated at $245 million, an excellent return on investment.During 1990–1991 a second phase of rockburst research, known as the Canadian Rockburst Research Program, and scheduled to end in 1995, was begun. The new project, now managed by the Ontario Mining Research Directorate, in Sudbury, was aimed to transfer the technology which had been developed during the previous five years and to bring about improvements in mine design in order to reduce the risks of rockbursts. In this phase of the work, the industry was assuming a larger role in the management and direction of the project. CANMET had discharged its role of initiating new research when it was urgently needed and then, with new Canadian expertise in place, transferring the continuation of the work to the industry. By 1992, with the signing of the Canada/Québec Mineral Development Agreement, the Québec Government and mineral industry joined with CANMET to form the Québec Rockburst Research Program.

The roles of Canadian industry in mining research have been varied. Some organizations have opted to perform their own in-house research for proprietary reasons. One of the most outstanding examples was the five-year multimillion dollar Mining Automation Project initiated by INCO in 1995–1996, with the objectives of introducing a very high degree of automation into mines. In addition to INCO, Tamrock, Dyno Nobel, and CANMET participated in the consortium. The goals were to facilitate the mining of both high-grade and low-grade deposits at various depths in the Sudbury basin. Notable advances were made in the development of tele-remote operated mining vehicles and in blasting technology. Unfortunately, on the conclusion of the project, INCO made the decision to disband the research team and assigned the personnel to their operations.

Concurrently, the industry has supported the development of mining research centres of excellence on campuses—such as the example above—the Ontario Mining Research Directorate (MRD), which is located at Laurentian University, in Sudbury, Ontario. With the ambition of becoming the "Harvard of the Mining Industry," Laurentian University has also supported the development of: the Mining Innovation, Rehabilitation and Applied Research Corporation (MIRARCO), the Centre for Environmental Monitoring Information (CEM), the Centre for Excellence in Mining Innovation (CEMI),the Geomechanics Research Centre (GRC), the Centre in Mining Materials Research (CIMMR), the Mining Exploration Research Centre (MERC), and the Laurentian University Mining Automation Laboratory (LUMAL).

Likewise, at the University of British Columbia, in Vancouver the Bridge Program Research Centre, the Centre for Environmental Research in Minerals, Metals, and Material (CERM3), the Centre for Industrial Minerals' Innovations (CIMI), the Mine Automation/Environmental Simulation Laboraory (MAESL), and the Sustainability Working Group (SWG) Research Centre have all been created in recent years.

While all of these centres of excellence could bode well for the future of mining research it must be emphasised that the pool of talent that is available internationally is quite small—the effects of substantial reductions in mining education and research in Europe, South Africa, and North America.

THE DEVELOPMENT OF MINING ENVIRONMENTAL RESEARCH

The first mining which took place in Canada was by the aboriginals, several thousands of years ago, as they mined and traded: in native copper from the Lake Superior Region; in chert, from Ramah Bay, Labrador; and, in native silver from the Cobalt, Ontario, area. There are thousands of very old small pits which dot the landscape across Canada (Udd, 2000).

Later, after the arrival of the Europeans on Canada's east coast, primitive mining took place as the development of Canada proceeded inland along the nation's dendritic river systems. Early in the 19th century the nation's first railroads were developed to provide overland transportation from mine sites to the ports beside the Great Lakes and the St. Lawrence and Ottawa Rivers. Concurrently, thousands of small deposits were exploited by the settlers as they sought an additional source of income as they cleared the land. There may be as many, or more, than 100,000 small abandoned mine sites in the country, with about 40,000 in Ontario alone, and no less than about 5,000 in the area of Eastern Ontario and Western Québec, which surrounds the nation's capital, Ottawa (Udd, 2005).

During these period of early mining there was probably little thought of the environmental consequences of mining or of managing the "footprints" which were left behind. Today, major remediation efforts are being made in both Ontario and Québec, and other jurisdictions, as old pits and underground mines are either fenced off, or backfilled, or bull-dozed. The potential hazards to people walking in "the bush" are significant due to these historical abandoned and unprotected shafts and pits.

Most of these early operations would have been small, with the mining and "milling" (ie., hand cob-bing of ores) involving a few dozen people at each. With the advent of the use of cyanide to extract gold, the use of chemicals in the flotation process, and the development of major mining, mineral concentra-tion, smelting and refining plants, the possibilities of environmental pollution acquired a huge new dimen-sion. Today, we are aware of the long-term effects of tailings pond, waste dumps, slag heaps, and smelter emissions. One of the greatest concerns, because of the prevalence of sulphide ores in Canada, is acid mine drainage (AMD). The environmental liability which the industry faces is potentially in the multi-billions of dollars.

In CANMET emphasis was placed on under-ground mining environments beginning in the 1960s. The work continues to this day. In the mid-1970s , with a goal of developing more environmen-tally-friendly mineral treatment processes, the work on using hydrometallurgical methods to treat sul-phide ores began. In the late 1970s, with the goals of improving recoveries and reducing environmen-tal hazards, CANMET worked on the development of better and newer means of treating uranium ores. At Elliot Lake, trials were being made on the reveg-etation of tailings ponds and the use of microorgan-isms to reduce the environmental hazards caused by these ponds.

In 1988 the Department of Energy, Mines and Resources (NRCan's predecessor) launched its "*Action Plan on the Environment*." Some of the objec-tives were to ensure the use of science for the benefit of Canadians; to integrate environmental considerations into decision-making in the Department; to encourage the adoption of environmentally-sensitive practices by the industry; to work with the provinces as full partners in environmental protection; and, to review past practices. During the same year, the Department also initiated a review of the long-term manage-ment of nuclear fuel and waste in Canada. During 1988, there was also an awakening of concern on the possibilities of global warming and of the possible consequences.

Also in 1988, the $12.5-million Reactive Mine Tailings Stabilization (RATS) was born in CANMET. This program, involving 15 mining com-panies, several federal departments and the provin-cial governments of British Columbia, Manitoba, Ontario, Quebec and New Brunswick was working

on finding long-term solutions for AMD. By the middle of 1989, some 20 collaborative projects with industry, with a value of $3.1 million, had been signed. In 1990–1991, an additional$5.5 million was added to the research program and a new name was established to reflect the addition of waste rock to the issue. The $18.0 million dollar program was now known as Mine Environment Neutral Drainage (MEND). The goal was to develop techniques which could be used to enable the use and abandonment of acid-generating tailings disposal areas in a safer, more economical and environmentally-friendly manner. This program would have huge implications for the Canadian mining industry, which typically mined sulphide orebodies. As time progressed, it was recognized that the total national costs of reclaiming and treating non-ferrous metal mines wastes could be as high as $5 billion over the next 20 years.

By 1995, no fewer than 20 mining companies were involved in MEND. CANMET, which had just completed a major study on the mineralogi-cal characteristics and degradation of mill tailings, continued to provide its secretariat as well as pro-viding research expertise and funds for contracting. With the program entering its seventh year, the focus was shifting to the transfer of technologies which had been developed. Communications packages had been developed and workshops were being held across the country. One major company was reported to have stated that the transfer of these technologies had contributed to a $500 million reduction in the costs of future decommissioning. An evaluation of the MEND program in 1997 concluded that for five respondents, it was estimated that their savings in liabilities had been $400 million (Udd, 2009).

About 1990–1991, the concept of Sustainable Development (a term of Canadian origin) became prominent. The minerals industries were encouraged to adopt an attitude of environmental stewardship of minerals through the wise management of resources, the reduction of effluents, and integration of environ-mental thinking into normal business operations.

By 2000, there were now four programs in MMSL's Environment Laboratory: Mine Effluents; Tailings and Waste Rock; Metals and the Environment; and, MEND. In the first-mentioned, the focus was on providing R&D support to the industry to com-ply with both the regulatory and legal requirements regarding liquid effluents from mines, mills, and smelters. As a part of this, a Slag Leach Consortium had been formed to assess the potential impacts of slags from non-ferrous smelters. With the support of the MEND consortium, work was taking place on the methods that could be used for the disposal of acidic mine drainage (AMD) treatment sludge.

By 2005, CANMET-MMSL, was operating laboratories in Sudbury, Val-d'Or, and the Ottawa area—both downtown and at Bells Corners. With a focus on mining health and safety, the environmental challenges facing the industry, and improving both productivity and competitiveness, its research-based programs were: Mineralogy and Metallurgical Processing (Ottawa); Ground Control (Bells Corners, Sudbury, and Val-d'Or); Mine Mechanization (Val-d'Or); Underground Mine Environment (Bells Corners and Sudbury); Mine Waste Management (Ottawa); Mine Effluents (Ottawa); and, Metals in the Environment (Ottawa). Concurrently, through its Special Projects Program, it provided the Secretariat for both the Mine Environment Neutral Drainage (MEND) Program and the National Orphaned/Abandoned Mines Initiative (NAOMI). The former, begun in 1989, and now co-financed by Natural Resources Canada and The Mining Association of Canada, had been tremendously successful in reducing the industry's liabilities for acidic drainage by an estimated $400 million, while the latter, launched in 2002, was focused on the development of a large-scale national program for the rehabilitation of orphaned/abandoned mine sites in Canada.

Elsewhere in Canada, environmental research in connection with mining is being done at the University of British Columbia (Vancouver, British Columbia), Dalhousie University (Halifax, Nova Scotia), École Polytechnique (Montreal, Quebec), Lakehead University (Thunder Bay, Ontario), Laurentian University (Sudbury, Ontario), and Laval University (Quebec City, Quebec). The names of the researchers and their technical interests may be found on the MineCan website (http://www.minecan.org/).

In addition, many of the major consulting firms in Canada are deeply involved in environmental studies and research.

Current research themes in Canada now focus around two new research initiatives—The Canadian Mining innovation Council (CMIC), and the Green Mining initiative. Both are multi-stakeholder (industry, academia, government, and others) research programs that are intended to address the mining research needs of Canada in the years to come.

REFERENCES

McCarter, M. K., 2007, "Mining Faculty in the United States: Current Status and Sustainability,"

Mining Engineering, ISSN 0026-5187, Society for Mining, Metallurgy and Exploration, SME, Littleton, Colorado, Vol. 59, No. 9, September, 2007, pp. 28–33.

Newell, Diane, 1986, "Technology on the Frontier: Mining in Old Ontario," UBC Press, University of British Columbia, Vancouver, ISBN 0-7748-0240-5.

Udd, John E., 1989, "Backfill Research in Canadian Mines," in "Innovations in Mining Backfill Research," Proceedings of the 4th International Symposium on Mining with Backfill, Montreal, Quebec, October 2–5, 1989, Edited by F. P. Hassani, M. J. Scoble & T. R. Yu, A. A. Balkema, Rotterdam & Brookfield, ISBN 90 6191 985 1.

Udd, John E., 2000, "A Century of Achievement: The Development of Canada's Minerals Industries," CIM Special Volume 52, Canadian Institute of Mining, Metallurgy and Petroleum, Montreal, Quebec, ISBN 0-919086-99-3, 206 pp.

Udd, John E., "A Guide to the Mineral Deposits of Southeastern Ontario and Southwestern Quebec," CJ Multi-Media Inc., Ottawa, Ontario, ISBN 0-9685147-3-1, 634 pp.

Udd, John E., 2009, "A Century of Service: The Hundred-Year History of the Canadian Mines Branch/Bureau of Mines/CANMET, 1907–2007," CANMET MMSL-Report MMSL 08-010, Natural Resources Canada, Ottawa, Ontario, 390 pp.

Coal Mining Research in the United Kingdom: A Historical Review

Sevket Durucan
Department of Earth Science and Engineering, Royal School of Mines,
Imperial College London, United Kingdom

Robert Jozefowicz
Mines Rescue Service Limited, Mansfield, Nottinghamshire, United Kingdom

David Brenkley
Mines Rescue Service Limited, Mansfield, Nottinghamshire, United Kingdom

ABSTRACT: This paper reviews the history of coal mining research in the United Kingdom, referring to the development of the industry from its early days a few centuries ago, the subsurface environmental conditions which led to many explosions and loss of lives, and the development of mining research as a consequence. The decline of UK coal mining industry and the changes this brought about are discussed. Using methane prediction technology as an example, brief details of industrial and academic research carried out at the end of the 20th Century is presented.

INTRODUCTION

Coal has fuelled the industrial revolution and, over the years, had major economic, social and political impacts on communities worldwide. Throughout the history of coal production, the industry has continuously evolved and the focal points of production have moved around the globe. Development of new technologies and the geographical shifts in coal production worldwide meant that the science of coal has been re-invented many times over. The world currently is in a new economical and technological era where it is suggested that the internet and e-commerce is a revolution as great as the industrial revolution. As surprising as it may seem, coal still has its place in this revolution. As Mark P. Mills (1999) once wrote, "the internet begins with coal." In recent years, the internet has been responsible for a significant part of the growth in electricity demand and, although over-estimated by nearly an order of magnitude, Mills (1999) had argued that a kilogram of coal is being burned for every 4 megabytes of information moved over the internet. Until the 1960s, coal was the single most important source of the world's primary energy. In the late 1960s it was overtaken by oil; but it is forecast that coal will remain an extremely important part of the world's energy mix for many years to come. Currently, coal provides 26.5% of global primary energy needs and generates 41.5% of the world's electricity. Estimates of proven coal reserves worldwide are considered to be around 850 billion tonnes with total global hard coal and brown coal production in 2008 estimated at ~5.9Gt and ~0.95Gt respectively.

Presented in two parts, this paper reviews the historical development of coal mining research in the United Kingdom and, by using methane prediction technology as an example, demonstrates how complementary (and sometimes competing) strands of industrial and academic research has helped further our knowledge in this field.

PART I—THE DEVELOPMENT OF COAL MINING RESEARCH IN THE UK

Pre-1994 Industrial Research in the UK

Within a historical review of coal mining research it is useful to appraise the industrial and social developments which shaped the research direction and research resource which was made available at that time. As reported by Unwin (2007), one of the earliest deaths attributed to an explosion of firedamp occurred in 1621. With the increase in the demand for coal and, as a result, the depth and size of mines, the number and severity of these incidents rose subsequently. The urgent requirement for better control over the mining environment, and in particular the dire need to prevent or at least reduce the subsequent violence of methane (firedamp) and coal dust explosions led to the earliest mining research per se.

The explosion on 25 May 1812 at Felling Colliery near Gateshead-on-Tyne claimed the lives of ninety-two men and boys and led to the establishment of the "Sunderland Society" on 1 October 1813, under the patronage of the Duke of Northumberland. The explosion, indirectly, resulted in the invention by Sir Humphrey Davy of the flame safety lamp, an important milestone. The wire gauze safety lamp was tested underground at Hebburn Colliery on 9 January 1816 (Saxton 1986). Following the death of fifty-two men and boys in an explosion in June 1839 at St Hilda Colliery, County Durham the 'South Shields Committee' was formed to study the prevention of accidents in coal mines. The report produced by the Committee concluded that "Ventilation seems to be the only certain and secure means of safety in inflammable mines…. Whilst this imperfect ventilation is allowed to continue, the mining districts and the public must prepare themselves for the continual recurrence of these dreadful calamities." It was recommended that, to improve safety underground, barometers, thermometers and anemometers must be used (Unwin 2007).

The UK's historical lead in the intensive exploitation of its coal deposits led to early state intervention in the regulation and inspection of mines. For example, the first Coal Mines Act was passed in 1850 (Statham 1958), which created a Mines Inspectorate with the power to enter any workings and advise the manager of any dangerous conditions found therein. However, the progression towards effective health, safety, and environmental controls was reactive and disaster-led. A number of Acts and Supplementary Acts can be linked as a response to particular disasters over time; for example the Hartley Colliery Disaster in 1862, with 204 entombed, resulting in a Supplementary Act requiring two means of egress, and Aberfan, with 109 killed, leading to the Mines and Quarries (Tips) Act of 1969 and Mines and Quarries (Tips) Regulations 1971 (MQ&A 1972). The Senghenydd Colliery disaster, Glamorgan, South Wales in 1913 killed 439 miners. This explosion was attributed to possible electric sparking from equipment, such as electric bell signalling gear, and reinforced the need for spark-free electrical systems. It was, however, not until 1974 with the arrival of the Health and Safety at Work Act that this reactive cycle of legislative formulation was effectively broken. This has progressed to the present day with legislation that recognises the importance of risk assessment, where proper consideration of hazards and adoption of appropriate methods to reduce their severity and frequency is fundamental to reducing injury and asset loss.

It is interesting to reflect that the UK, which has arguably been at the forefront of mining research and development in the second half of the 20th Century, witnessed a significant decline in its coal mining industry during this period. The problems of the UK coal industry in the interwar years were related to a dependence on the exported coal trade together with delays in mechanisation compared with nearby European competitors. Exports fell sharply during this period, whilst UK domestic consumption remained static. This resulted in major unemployment problems in the coastal coalfields. The industry was traditional in its structure and included many small companies. It was also heavily dependent on manual labour, with labour costs accounting for between two thirds and three quarters of the cost of the coal mined. This situation was compounded in that the introduction of mechanisation was very slow. In 1913 less than ten percent of coal was cut by machine. Between 1913 and 1939, output per man shift rose by 81 percent in the Ruhr in Germany, and by 73 percent in Poland, but by only 10 percent in the UK. The decline in the UK's position as an international coal producer can therefore be traced back to fragmented ownership within the UK, coupled with increasing international competition compounded by delays in introducing mechanisation. The latter can be effectively considered to be a delay in exploiting the fruits of mining equipment development.

Arguably, the major thrust in mining research and development in the UK occurred after the nationalisation of the coal industry. However, it would be unfair to leave an impression that little or no Government-sponsored research was conducted prior to 1947. By way of example, in 1911 the UK Government agreed to fund an experimental station at Eskmeals in Cumberland for the investigation of explosions in coal mines. In 1924 the Harpur Hill, Buxton site was acquired for large scale mining safety work, following on from the establishment of the Safety in Mines Research Board in 1921. The UK government took control of the coal mining industry during and after World War I (1917–1921) and during World War II. On 1st January 1947 the coal industry became wholly nationalised under the body of the National Coal Board (NCB). The large reserves of coal, coupled with an anticipated increasing requirement for coal led to many modernisation projects associated with large capital investment. However, the increasing availability and low cost of production of Middle-East oil, partly coupled with an emerging nuclear power industry led to a rapid run-down in the coal mining industry. In 1961, more than 600,000 were employed at 698 collieries. By 1971, those figures were down to 300,000 and 292 respectively. In 1965, coinciding with the discovery of substantial reserves of North Sea oil and gas, the UK Government published its National Plan for Coal which endorsed an intensification of the industry contraction programme. A turning point was reached in 1973, when the UK's economy was affected significantly by oil price rises arising when the members of the Organization of Arab Petroleum Exporting Countries initiated an oil embargo in response to the

US decision to re-supply the Israeli military during the Yom Kippur war. This lasted until March 1974 and resulted in a fundamental reappraisal of national energy policies. Consistent with a view that coal should maintain a significant part in the primary energy mix considerable attention was given to coal related research on both sides of the Atlantic.

At the time of nationalisation in 1947 National Coal board established the Operational Research Executive (ORE), which played a major part in developing the longwall system of mining coal. Within the UK, coal related research was broadly split into two avenues; (1) initiatives to advance the efficiency of combustion and scope of coal utilisation, conducted at the Coal Research Establishment (CRE) located at Stoke Orchard, and (2) wide-ranging mining research focussing on all aspects of exploration, coal production and workplace environmental controls. The NCB's first research establishment at Stoke Orchard in Gloucestershire was founded in 1950 with Jacob Bronowski as Director of Research. The establishment at Stoke Orchard was cited, in the words of the UK Government's Parliamentary Office of Science and Technology, as a "world centre of excellence." The work undertaken there, particularly the work on clean coal technology and the coal Topping cycle, if applied to major power stations, was recognised as potentially increasing the thermal efficiency of generation from about 36 to 47 per cent. Global expenditure on coal combustion research has been generally maintained at a high level compared with mining related research.

In 1952 the structure and organisation were changed and the Mining Research Establishment (MRE) set up, which became the principal focus for mining research in the UK until 1994. MRE was located at Isleworth, near London. This was the former site of Isleworth Studios, where The African Queen was shot (with a few other noteworthy films). Desired improvements in the productivity of the British mining industry required reliable machinery. As a result, and to assist in its development, in 1955 the NCB set up the Central Engineering Establishment (CEE). This was sited at Stanhope-Bretby, near Burton on Trent, closer to the coalfields. In 1969 MRE and CEE were merged to form the Mining Research and Development Establishment (MRDE) at Stanhope-Bretby. It is noted that explosion and electrical safety research was undertaken at the Safety in Mines Research Establishment (SMRE) located jointly at Buxton and Sheffield and established in 1947 as part of the Ministry of Fuel and Power. A further notable mining industry related research development in the UK was the formation of the Institute of Occupational Medicine medical research centre in Edinburgh. The Institute of Occupational Medicine (IOM) was

founded in 1969 by the National Coal Board (NCB) as an independent charity, but with staff employed by the NCB. Its research was overseen by a Council of Management with representatives of the NCB, the mining trade unions and independent scientists. The early history of the IOM's research is inextricably linked to the NCB's Pneumoconiosis Field Research (PFR). The PFR started in the early 1950s with the objective of determining how much and what types of coal caused pneumoconiosis and what dust levels should be attained in order to prevent miners from becoming disabled by the air they breathed. These objectives were ahead of their time, implying a requirement to measure both exposure and outcome in a large cohort of miners over a prolonged period, and to use these quantitative data to set protective health standards in the industry. Thus started the largest occupational epidemiological study ever carried out. Early in the history of IOM, the NCB recognised the importance of ergonomic research to the coal industry, in terms of protecting men from adverse environments and in the design of mining equipment. This became an important additional area of medical research in addition to that of the mechanisms of respirable dust exposure and risks of pneumoconiosis.

The period 1973 until 1988 represents a period of significant research intensity in the UK. The fact that most of the energy industry (apart from petroleum) belonged to the public sector in the United Kingdom during this period resulted in extensive research and development related to energy being undertaken, with a fair balance between short-term and long-term objectives. Emphasis was given to improving the reliability and efficiency of existing machines and systems for the short term. For the longer term, the principal developments related to the effects of the depths at which mining was carried out in the UK. This included research into the feasibility of extending systems for remote operations and control, with overriding aims of reducing the hazards of mining and increasing productive capacity. One significant development, well in advance of its time, was early experiments to enhance the performance of the main power-loading equipment with ROLF (Remotely Operated Longwall Face). This demonstrated the possibility of remote control of coal face equipment, but also exposed the limitations of electro-hydraulic controls of roof supports and the crudeness of the prototype power-loader steering. Subsequently significant research resources were assigned to develop effective semi-automatic face and face-end operations, eventually accruing in the development of an 'integrated coal face' system. This required basic research on nucleonic sensing devices to steer the machine within the seam approach. Much of this research has found its way into current automated face support and cutting systems. Indeed, it is with some pride that UK and international mining engineers can look back at the significant range of developments which emerged from this period. The list of candidates for mention would be lengthy, but of note is the work on the science and application of in-seam and surface seismic exploration techniques, fundamental work to control dust and noise exposure, together with world-class research on understanding the mechanisms of methane emissions and their control. Allied with much of this science there were significant initiatives to develop and advance remote controlling and monitoring techniques for the production and transport machinery, as well as environmental monitoring systems. The electrochemical gas sensor was a notable development, together with a wide range of intrinsically safe certified fixed and hand-held instruments for mine environmental monitoring. Associated development of computer based monitoring and control systems, the forerunners of current SCADA (supervisory control and data acquisition) systems, were also a major hallmark of UK mining research.

National Coal Board implemented the systems engineering concept in the mid 1970s to design MINOS (Mine Operating System), a highly centralised, hierarchically organised system of remote control and monitoring in mines. Control engineering and computerised information systems were the main technical factors on which MINOS was based upon. Amongst its few other aims, the main objectives of the MINOS system were reduction of repair and maintenance costs; improvement in output quality; improvement in monitoring and information; elimination of human error; and increase in safety levels and improvement of working conditions (Burns et al. 1985; HQTD 1983; HQTD 1986). MINOS consisted of a number of subsystems each of which can be treated as a system in its own right. For example, the coal face system consists of the shearer and armoured face conveyor, roof supports, power supplies, gate end activity, the coal seam, face workers, information about production, delays and machine status. Upon its development, the Barnsley West Side Complex was the first mine complex which was targeted for the full implementation of the MINOS system and the Selby complex was designed to use MINOS system from the start (Burns et al. 1985). The full mechanisation of coal production through the combined use of a power loader, armoured face conveyor and powered roof supports increased the technical complexity of coal extraction and led to the development of the IMPACT (In-built Machine Performance and Condition Testing) subsystem within MINOS. A further development of

note, but which originated in the United States, was the adaptation of rock bolting systems to the deep and variable strata conditions present within the UK. This development, arguably more than any others, introduced a step change in the performance and related cost base of the UK coal mining industry's development phase activities.

Progressive decline in the UK coal industry and the need to respond to the industry's numerous testing, analysis and investigation requirements lead to an increasing diversification of the research and technical support activities undertaken within the National Coal Board and its successor, the British Coal Corporation, which replaced NCB in 1987. The MRDE function witnessed a number of changes to its strategic research and development activities, with operational 'rebranding' to reflect this; firstly Headquarters Technical Department (HQTD) in 1989, then Technical Services and Research Executive (TSRE) in 1990. Operating in conjunction with these functions was an extensive Scientific Control function which provided the principal laboratory services arm and undertook selective research activities including the development and introduction of the tube bundle analysis approach, and the development and evaluation of escape respiratory protective devices. The mechanical and rotary testing, material testing, rope and support testing and rock mechanics laboratory functions at Bretby also had an ongoing and critical role in the research an introduction of several key technologies within the industry. These included measures to improve corrosion resistance, enhancements to lubricants and fire-resistant fluids and materials, design improvements to gearboxes and hydraulic power units, together with the introduction of ferrographic analysis and early condition monitoring techniques. All of these initiatives achieved substantial and sustainable gains in the general robustness, performance and fitness for purpose of the heavy duty mining systems which are now commonplace in the underground mining industry. The close relationship and synergies with the large mining equipment manufacturing sector in the UK should also not be underestimated. This was an essential component of the process of technology development and technology transfer within the industry. Whilst many initial concepts and prototype designs were evaluated internally within NCB/ British Coal, it was generally the mining equipment suppliers who were charged with completing the final development and refinement. As noted below, virtually all active equipment and machine development from the 1990s became the province of the equipment manufacturers—a situation which prevails today. British Coal closed TSRE in April 1994.

At this point, forgiveness is sought from the reader for the unashamedly UK-centric nature of this paper. It is therefore very necessary to point out the excellent work that was also being undertaken at a national level elsewhere. Within the United States, the US Bureau of Mines (USBM) and its successor within the CDC's National Institute for Occupational Safety and Health (NIOSH) are of particular note, particularly in regard to their open, collaborative research approach and initiatives to disseminate their research findings. In mainland Europe, the work of Bergbau-Forschung and other establishments in Germany, Poland, Spain, France and Belgium are noted. In Australia, the work of CSIRO and others is noted. In the Republic of South Africa, the Chamber of Mines and subsequently Mining Technology CSIR and Safety in Mines Research (SIMRAC) have provided a continuum of mine safety related research. Whilst these national initiatives are well-known, there were equally significant programmes conducted in the USSR (principally Russia and Ukraine), Japan, India and China. Indeed, it is anticipated that the focus of mining research and development must shift to look east towards the massive industries of China and India. As a further point of note, it would equally be an omission to understate the extensive research, development and demonstration undertaken within the hard rock mining industries. Here the general freedom from hazardous area electrical system approvals has permitted a range of advanced automation technologies to be evaluated and implemented in advance of the coal industry. Research within Canada, Sweden and Austria is mentioned, but the hard-rock industry deservedly warrants a separate paper on its own specific technological developments.

Post-1994 Industrial Research in the UK

The strategic provision of mining research was materially affected when the British Coal Corporation ceased its research programme on privatisation of the UK coal industry. Whilst it is not intended that the respective merits of public and private ownership models be discussed here, it was evident in the privatisation proceedings that the new owners would not, at least initially, support significant research activities. However, it was recognised that technological stagnation and eventual obsolescence were equally undesirable in the longer term. The strategic response to this was an attempt to relocate the research within academia and possibly within the mining consultancy and specialist mining technical service companies. This was enacted in part with the transfer of much of British Coal's coal preparation (beneficiation) and rock testing and characterisation equipment and laboratory apparatus to the University of

Nottingham. Health and safety related research was taken up in part by International Mining Consultants Ltd. (IMCL). With the restructuring and acquisition of parts of IMCL, there was a further need to maintain a stable but viable level of support for mining research. This was achieved to a significant extent by the transfer of activities to Mines Rescue Service Ltd., described below.

Provision for a rescue service with rescue workers that wear breathing apparatus to rescue/recover mineworkers underground where irrespirable atmospheres exist due to fires and/or explosions can be traced back to the beginning of the last century. Over the years, legislation has ensured that an adequate service and facilities are provided and maintained. The first rescue station was established at Tankersley in the UK in 1902. The UK Mines Rescue Service was maintained as part of the National Coal Board and subsequently the British Coal Corporation. With the privatisation of the UK coal industry, the UK Mines Rescue Service continued as a non-profit limited company known as Mines Rescue Service Limited (MRSL). MRSL provided continuous rescue cover for coal mines under the Statutory Mines Rescue Scheme, where deep mine coal operators pay a levy per tonne of coal mined towards the maintenance of rescue cover and rescue training. The business of the company is managed by a Board of Directors and the day to day running of the company is managed by the Chief Operating Officer. In response to the contraction of the UK mining industry, MRSL has implemented a strategy to maintain its services to the mining industry at an acceptable cost by diversifying into other areas of work whilst maintaining its non-profit status. As mentioned above, MRSL made a strategic decision to support mining research within the UK. In 2001, the Chief Operating Officer, backed by MRSL's Board of Directors and with agreement from the European Commission, acquired two research projects from IMCL which was undergoing a rationalisation process. The two projects entitled 'Fire Fighting Systems' and 'Alarm and Evacuation' were funded under the European Coal and Steel Community (ECSC) Research Programme and were deemed to be within the capabilities of MRSL. Engineers with a research background were employed to lead these projects and the MRSL Research and Consultancy Unit was formed.

From 2001 to-date, the Unit has successfully acquired funding for 16 mining projects. Of these projects, 13 were funded under the ECSC Programme or its successor Research Fund for Coal and Steel (RFCS) Programme. The project research areas have included mine fire fighting involving the development of water mist fire suppression systems, escape and rescue support in mines, resilient wireless network and data transmission systems, radio imaging of coal seams, speech intelligibility improvement in noisy environments, virtual and augmented reality systems, improvements to mine transport systems, mine shaft stability and mine water impacts after closure. The specialist research team, though modest in numbers, has significant prior experience in mining research and disciplines such as general physical sciences and specific disciplines such as computational fluid dynamics, ergonomics, intrinsic safety and communications technology. Two projects were funded by the UK Health and Safety Executive (HSE) for research into thermal physiological impacts of attempting escape whilst wearing escape respiratory protective devices (self-rescuers) in hot and humid conditions, together with the development of suitable wearer expectations training. A further project funded under the EU 5th Framework Research Programme examined the issues and options for supporting personnel evacuating within road, rail and metro tunnels in conditions of smoke. The MRSL Research and Consultancy Unit continues to expand its activities and has arguably offered a small-scale but effective strategic response to the general decline in mining research in the UK.

Mining Research in UK Academic Institutions

Over the past one hundred years there has been a requirement for mining research in terms of seeking improvements in mining systems and productivity. There has also been a requirement for suitable academic institutions and departments to train the engineers, scientists, managers and craftsmen required by the mining industry. In the pre-WW2 years, the output from approximately 1,600 collieries was ~230Mt (Statham 1958). During the period 2008–2009, 18Mt was produced with 10Mt from 10 opencast operations and 8Mt from 6 large deep mines and 6 small collieries. The underground mines have continued to reduce during 2009. In terms of academic establishments which have served the coal mining industry, these were centred in many of the coal mining areas. At the time of nationalisation of the British coal mining industry (1947), eight regional Divisions were formed, each with a number of Areas which in turn had a number of collieries assigned. As the mining industry reduced in size the operational areas merged. Along with this reduction in Areas, there came a reduction in the requirement for mining departments at colleges and universities. During the early 1980s the remaining academic establishments with mining departments included the Royal School of Mines, University of Newcastle, University of Nottingham, University of Leeds, University of Cardiff, University of Strathclyde,

North Staffordshire Polytechnic, Nottingham Trent Polytechnic, with Camborne School of Mines serving the hard rock mining industry. Rationalisation of UK's mining engineering departments in 1988–89 resulted in the closure of the mining departments in Cardiff and Strathclyde and the merger between the two departments in Leeds and Newcastle at Leeds University.

Many of these academic institutions developed well-deserved academic reputations in specific mining fields. The University of Newcastle which served the North East of England area focused its mining research on mine mechanisation, rock mechanics and mine waters. Although the University of Newcastle's mining department no longer exists, their research on mine waters continues to thrive under an environmental engineering departmental banner. North Staffordshire Polytechnic served the Western Area which comprised of the Lancashire, Staffordshire and North Wales coalfields. Its research was focused mainly on pump packing and the use of mine wastes for packing in longwall advancing faces which was the major production method up to the mid 1980s. Towards the late 1980s, when retreat mining became prominent, and the Western Area reduced the number of working collieries, North Staffordshire Polytechnic's mining department closed and mining research ceased. At Nottingham University mining research included mine ventilation and methane control, mineral processing, and rock mechanics. Towards the end of the 1990s, the mining department in Nottingham University was incorporated into a merged School which consisted of chemical, environmental and minerals engineering. Today, the University of Nottingham continues mining-related research in the areas of rock mechanics and mineral processing in two different departments, namely the Civil Engineering and the Chemical and Environmental Engineering. Coal mine subsurface environmental engineering research at the Royal School of Mines has evolved over the years and today mainly deals with clean coal and energy technologies, carrying out research into coalbed methane, enhanced coalbed methane and CO_2 storage in coalbeds, as well as research into CO_2 storage in saline aquifers and depleted natural gas reservoirs.

PART II—THE DEVELOPMENT OF METHANE PREDICTION METHODS IN THE UK

British Coal Firedamp Prediction Model

Starting as early as the 1950s, the European coal mining industry developed a number of methane prediction methods suited to the geological conditions

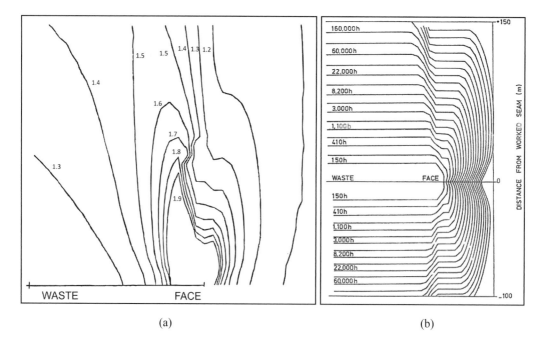

(a)   (b)

Figure 1. (a) Distribution of principal stresses ratio σ1/σ3 (vertical section parallel to and half way between the gate roads, (b) Distribution of time constants t_1 around a 900 m deep working (after MRDE 1980)

of the coalfields in each country (Curl 1978). These empirical approaches to methane emission prediction were all based on the "degree of gas emission" principle, where the percentage of the gas contained in the coal seam at a specific level above or below the workings which flows into the workings is referred to as the degree of gas emission. The gas left in coal seams after mining is considered to be the difference between the *in situ* gas-in-place and the total amount of gas emitted over the entire mining period.

Research carried out by MRDE in the UK during the 1970s and '80s developed a methane prediction method which was unique in Europe that time dependent behaviour of gas emission was introduced to the predictions (MRDE 1980). The method is based on Airey's theory of gas emission from broken coal, which was later extended to include fractured or fragmented coal seams around a working longwall face (Airey 1968;1971). The rate of gas release from the coal blocks was assumed be determined entirely by the emission characteristics of the broken coal, represented by a variable time constant t_1 defined as a function of distance from the face (Airey 1971) as,

$$t_1 = t_o \exp\left(\frac{x}{x_o}\right) \quad x \geq 0 \qquad (1)$$

where x represents the distance ahead of the position of the maximum stress (front abutment), x_o is a distance constant and t_o is the minimum time constant in hours, which occurs at and behind the front abutment position. Therefore, smaller the value of t_1, larger the gas emission rate from a relatively smaller blocks of coal nearer to the working coal face. From a subsequent theoretical work in rock mechanics Airey used the ratio of principal stresses σ_1/σ_3 as a criterion for coal failure around a longwall face and assumed that surfaces of equal σ_1/σ_3 would be coincident with the surfaces of equal t_1 (MRDE 1980), thus providing the theoretical basis to his degree of gas emission surfaces (Figure 1a). Using analytical solutions for the stresses around a coal face given by Berry and Sales (1967), Airey then computed the distribution of time constants around a coal face (Figure 1b), and hence the degree of gas emission from sources seams in the roof and floor, as a function of distance from the face line (Figure 2). As well as accounting for the emission from the roof and floor source seams, Airey proposed gas emission curves for the seam being mined, based on the weekly advance rate, and the coal being transported out of the district on the conveyor in order to calculate the total gas volume entering the longwall district.

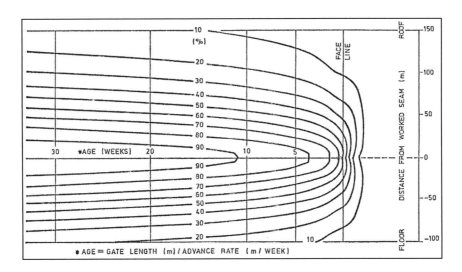

Figure 2. Airey's degree of gas emission (%) curves for the roof and floor of a longwall district at 1,200 m depth (after MRDE 1980)

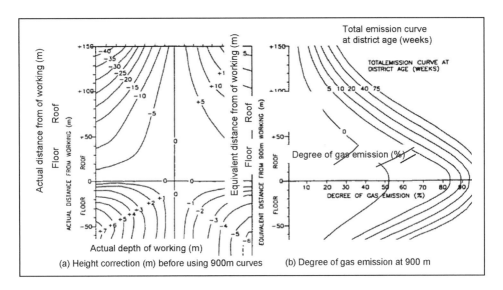

Figure 3. The British Coal methane prediction method—degree of gas emission curves for a 900 m deep longwall coal face and the chart to calculate the correction factors for workings at different depths (after MRDE 1980)

Airey's theory and the accuracy of his degree of gas emission curves were validated through filed measurements of gas quantities and residual gas contents of coal seams at a large number of UK collieries. As Creedy (1981) once wrote, the results of residual gas content measurements carried out in the floor of a mined out area indicated that methane emissions using Airey's prediction method were overestimating the gas quantities released. Based on these validation studies carried out over a period several years the gas emission curves were modified to eliminate the observed over- and underestimates and produced the British Coal firedamp prediction method as it is know today. In order to be comparable with the other European methane prediction methods, Airey's degree of gas emission curves have later

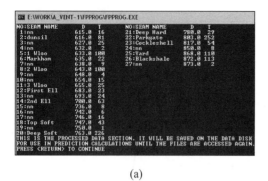
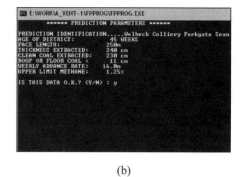

(a)　　　　　　　　　　　　　　　　　(b)

Figure 4. FPPROG user interface illustrating (a) the borehole section showing seam depths and thicknesses, (b) the longwall face design parameters for a case example

been transformed in to a more conventional form where the degree of gas emission is plotted against vertical distance from the working face horizon, with separate lines for different district ages. Figure 3 presents the degree of emission curves for a longwall face at 900 metres depth, and a chart from which corrections could be calculated for other depths (MRDE 1980). As it can be seen from Figures 2 and 3, the final gas emission curves reached represent a significant departure from the original, theoretically based, emission curves Airey proposed. In fact, it can be argued that this final semi-empirical model mimics the unique strata behaviour observed in the UK coalfields more closely than the original model developed by Airey.

The development of the methane prediction method was followed by field efforts towards determining the *in situ* gas contents of all the coal seams in the UK coalfields and a supplementary research programme towards computerising the prediction method for its widespread use by the industry. Research funded by the ECSC Programme at MRDE resulted in the development of FPPROG, the Personal Computer based firedamp prediction model (HQTD 1987). Used mainly as an in-house facility within British Coal, FPPROG came with a borehole/coal seam database and was very easy to use (Figure 4). Selecting the relevant coal seam sequence, and entering longwall face design parameters such as the worked seam thickness, face length and weekly advance rate the user could predict methane emission rates from the roof, floor and worked seams for desired periods (district age) in the life of a longwall district and calculate drainage and air quantity requirements based on a chosen methane drainage capture rate (Figure 5).

The transformation of Airey's methane prediction model from its purely theoretical format (built

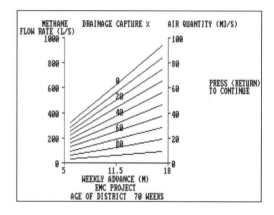

Figure 5. FPPROG output screen illustrating the methane emission rates and air quantity requirements for different drainage capture rates (0 to 100%) at a weekly advance rate of 5 to 18 metres per week in a 70 week old longwall district

upon the theory of gas emission/diffusion in coal and strata mechanics principles) into a semi-empirical format in order to accommodate the unique coalfield characteristics prominent in the UK has confirmed the view that such empirical models cannot be transported and applied at a new coalfield with relatively different geology than that found in the UK.

Academic Research into Methane Prediction and Models Developed

Methane prediction research at UK universities during 1970s and '80s was conducted in parallel with the research of National Coal Board, the emphasis being more on petrophysical properties of coal seams and the application of rock and reservoir

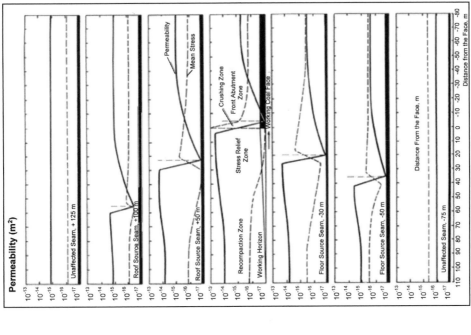

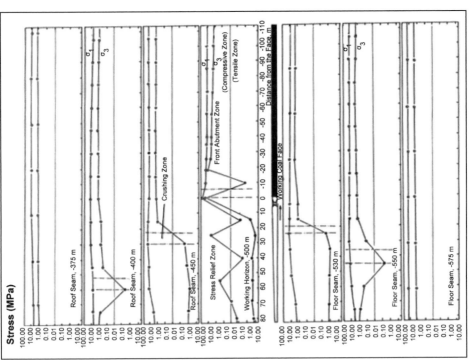

Figure 6.   Maximum and minimum principal stress, mean stress and permeability profiles around a 500m deep working longwall face (after Durucan 1981)

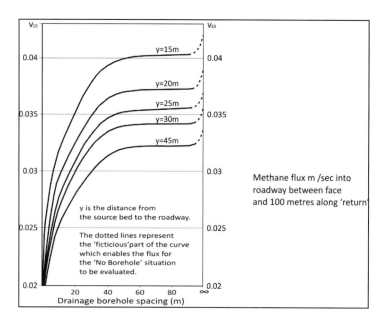

Figure 7. Chart for the evaluation of methane flux into longwall gateroads (after O'Shaugnessy 1980)

engineering principles to modelling the flow of methane around working longwall faces. At the University of Nottingham, Mordecai and Morris (1971), Gawuga (1979) and Durucan (1981) investigated the impact of stress on permeability of coal measure rocks and coal seams as a function of depth and stress abutments around working faces. Durucan (1981) extended his research to study the distribution of principal stresses and the propagation of fractures around longwall faces using finite element analysis and correlated these with his early work on stress-permeability relationship for coal seams to define the position of different stress zones and the maximum permeability areas around working longwall faces (Figure 6). Parallel research carried out by O'Shaugnessy (1980) used finite element analysis to predict methane pressure and flow rates around longwall coal faces and developed a methodology to predict the methane flux in to mine roadways with or without methane drainage (Figure 7).

Research at Imperial College in the '90s extended Durucan's early work and implemented dynamic finite element analysis to study the time dependent behaviour of stresses, the fracture zone and permeabilities around longwall faces and predicted methane emission in to longwall districts (Durucan et al. 1993). A coupled geomechanical reservoir model was developed, which was first used for predicting methane emission and production at operating and abandoned coal mines respectively (Durucan et al. 2004; Martinez-Rubio 2007). In an attempt to demonstrate the advantages of using reservoir models for the prediction of methane emissions in coal mines, rather than the empirical models developed by the European mining industry, Martinez-Rubio (2007) used field data from the UK and the US, each representing a specific coalfield rock and seam characteristics. Figure 8 demonstrates that, when the same UK coalfield reservoir characteristics (mechanical/elastic properties and permeability) are used, both British Coal and the Imperial College models produce the same degree of gas emission curves for a given longwall district. When compared with Figure 8, Figure 9 illustrates the difference between the obtained degree of gas emission curves for cases where the roof is predominantly made of strong, low permeability coal measure rocks (non UK) and those with weaker and thus more easily fractured roof rocks (UK). The coupled geomechanical reservoir model (METSIM) was later extended, first for the assessment of wellbore and far field permeability and flow behaviour of Coalbed Methane (CBM) reservoirs and later (METSIM2) for Enhanced Coalbed Methane (ECBM) and CO_2 storage (Shi and Durucan 2005; Durucan and Shi 2009).

CONCLUSIONS

The coal mining industry and coal production in the UK have shrunk significantly during the post-1994 era since privatisation. This also had a significant impact on the funding available for both the

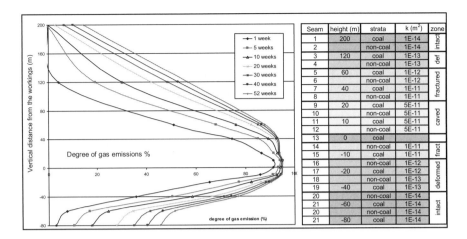

Figure 8. Degree of gas emission curves for coal seams around a longwall panel obtained using the Imperial College reservoir model and the UK coalfield reservoir characteristics (after Martinez-Rubio 2007)

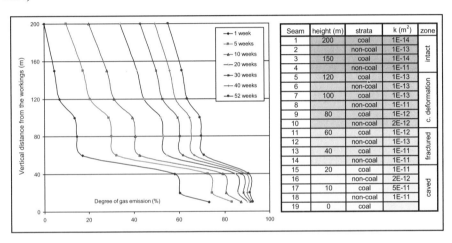

Figure 9. Roof section degree of gas emission curves obtained using permeabilities reported by Esterhuizen and Karacan (2005) (after Martinez-Rubio 2007)

industrial and academic research on mining. Current coal mining research in the UK is mainly carried out at universities and by the MRSL.

REFERENCES

Airey, E.M., 1968. Gas emissions from broken coal: An experimental and theoretical investigation. *Int. J. Rock Mech. Min. Sci.* 5: 475–494.

Airey, E.M., 1971. A theory of gas emission in coalmining operations, *Proc. 14th Int. Conf. Directors of Safety in Mines Research*. Donezk, USSR.

Berry, D.S. and Sales, T.W., 1967. An elastic treatment of ground movement due to mining, Part 2, Transversely isotropic ground movement. *J. Mech. Phys., Solids.* 9.

Creedy, D.P., 1981. Evaluating the gas content of disturbed and virgin coal seams. *XIX Conf. Safety in Mines Research*. Katowice.

Curl, S.J., 1978. *Methane Prediction in Coal Mines*. Rep. ICI'IS/TR 04. London: IEA Coal Research, 80 pp.

Durucan S. and Shi J.Q., 2009. Improving the CO_2 well injectivity and enhanced coalbed methane production performance in coal seams. *Int. J. Coal Geology.* 77(1–2): Elsevier, 214–221.

Durucan, S., 1981. *An investigation into the stress-permeability relationships of coals and flow patterns around working longwall faces*, PhD Thesis, University of Nottingham.

Durucan, S., Daltaban, T.S., Shi, J.Q. and Foley, L., 1993. Permeability characterisation for modelling methane flow in coal seams. *Proc. 1993 Int. Coalbed Methane Symp.*, Birmingham, Alabama. 453–460.

Durucan, S., Shi, J.Q. and Korre, A., 2004. Numerical modelling and prediction of abandoned mine methane recovery: Field application at the Saar coalfield, Germany, *Geologica Belgica—Volume 7*.

Esterhuizen, G. and Karacan, C.Ö., 2005. Development of numerical models to investigate permeability changes and gas emission around longwall mining panel. *Proc. 40th U.S. Symp. Rock Mech.*, ARMA/USRMS paper 05-744, Anchorage, Arkansas.

Gawuga, J.K., 1979. *Flow of gas through stressed carboniferous strata*, PhD Thesis, University of Nottingham.

Headquarters Technical Department, 1983 *Application of computer-based environmental monitoring*. Final report on ECSC research project 7220-AS/807. Burton-on-Trent: British Coal, 166p.

Headquarters Technical Department, 1986. *Computer-based monitoring and remote control of methane drainage systems*. Final report on ECSC research project 7220-AC/815. Burton-on-Trent: British Coal, 15p

Headquarters Technical Department, 1987. *Investigation of firedamp and its emission in coal seams*. Final report on ECSC research project 7220-AC/819. Burton-on-Trent: British Coal, 48p.

M&QA, 1954. *The Mines and Quarries Act*. London: HMSO.

M&QA, 1972. *Mines and Quarries Act and Regulations, the law relating to Safety and Health in Mines and Quarries,* HMSO.

Martinez-Rubio, R., 2007. *Modelling of methane extraction form abandoned coal mines*. PhD Thesis, Imperial College London, 176p.

Mills, M.P., 1999. *The Internet Begins with Coal: A Preliminary Exploration of the Impact of the Internet on Electricity Consumption*. The Greening Earth Society, Arlington, VA

Mordecai, M. and Morris, L.H., 1971. An investigation into the changes of permeability occurring in a sandstone when failed under triaxial stress conditions, *Proc, U.S. Rock Mech. Symp.* 12: 221–239.

MRDE, 1980. *Prediction of firedamp emission*. Final report on ECSC research project 7220-AC/801. Burton-on-Trent: Mining Research and Development Establishment, National Coal Board, 102p.

O'Shaugnessy, S., 1980. *The computer simulation of methane flow through strata adjacent to a longwall face*. PhD Thesis. University of Nottingham.

Saxton, I., 1986. Coalmine ventilation—from Agricola to the 1980s. *Trans. Inst. Min. Eng.*, Part 1: 490–503.

Shi J.Q. and Durucan S., 2005. A model for changes in coalbed permeability during primary and enhanced methane recovery. Paper No: 87230, *Res. Eval. Eng.*, Society of Petroleum Engineers, 291–299.

Statham, I.C.F.,1958. Coal Mining Practice, Volume 1, The Caxton publishing Company Limited, 13–14.

Unwin, I.D., 2007. The measurement of air flow in British coal mines: A historical review. http://www.cmhrc.co.uk/cms/document/air_flow_2007.pdf. Accessed September 2009, 88p.

Hearing Loss in the Mining Industry: The Evolution of NIOSH and Bureau of Mines Hearing Loss Research

R.J. Matetic
NIOSH, Pittsburgh, Pennsylvania, United States

Robert F. Randolph
NIOSH, Pittsburgh, Pennsylvania, United States

Peter G. Kovalchik
NIOSH, Pittsburgh, Pennsylvania, United States

ABSTRACT: Noise is one of the greatest hazards to a miner's health, rivaled only by respirable dust and repetitive trauma. Hazardous noise exposures are more prevalent in mining than in any other major U.S. industrial sector, and, as a consequence, miners report more hearing problems than any other type of worker. NIOSH has made the reduction of noise-induced hearing loss a major strategic goal and is attacking the problem through improved noise controls and interventions for workers. The NIOSH strategy expands on decades of research by its U.S. Bureau of Mines predecessors that identified some of the key issues with underground and surface noise sources. The current NIOSH program expanded significantly in 2000 in response to stakeholder interest and to provide a research complement to the regulatory initiative embodied in the Health Standards for Occupational Noise Exposure published by the Mine Safety and Health Administration in that year. Since then, NIOSH has developed significant new technologies for reducing noise from continuous mining machines and roof bolting machines, which have been the main sources of noise overexposures for underground coal miners. Field studies have shown that NIOSH-developed noise controls reduce workers' noise exposures below the MSHA Permissible Exposure Level (PEL). The facilities and engineering techniques developed for the underground coal efforts are now being applied to underground metal mining, processing plants, and other mining sectors. NIOSH has also addressed the worker's role in prevention by developing the Hearing Loss Simulator, QuickFit earplug tester, and other tools workers and trainers can use to reduce the risk of noise-induced hearing loss. These worker-empowerment technologies motivate and involve miners to implement noise controls, participate in administrative exposure management, and use hearing protection effectively. As it enters its second decade of mining noise and hearing loss prevention research, NIOSH has started to evaluate the impact each of its noise controls and other interventions have had on the hearing loss problem. In addition, NIOSH will direct its resources in partnership with industry stakeholders to address every mining sector within the next ten years.

INTRODUCTION

The U.S. Bureau of Mines (USBM) and National Institute for Occupational Safety and Health (NIOSH) have a long history of identifying and addressing hazardous mining noise. In 1971, the Bureau published findings that 73% of underground coal miners were exposed to hazardous noise (Lamonica, Mundell and Muldoon 1971). A 1976 NIOSH publication showed that miners had "measurably worse hearing than the national average." (NIOSH 1976). Unfortunately, noise and hearing loss are still a problem for today's miners. Mining has the highest prevalence of hazardous noise exposure of any major industry sector (Tak, Davis, and Calvert 2009) and mining is second only to the railroad industry in prevalence of workers reporting hearing difficulty (Tak and Calvert 2008). Noise-induced hearing loss is a special concern for miners because it is permanent and life altering.

Besides hearing loss, noise-exposed workers are more likely to experience speech interference, disturbed sleep interference, excess stress, tinnitus, and decreased work performance (Kryter 1994).

The NIOSH and earlier USBM programs that addressed hazardous noise have consistently followed a doctrine guided by the hierarchy of controls (NIOSH 1996):

1. **Engineering controls** that reduce noise at the source or prevent noise from reaching the worker's location
2. **Administrative controls** that reduce worker exposure through changes in procedures or behavior to reduce their time and proximity to noise sources
3. **Hearing protection devices** that reduce the noise reaching workers' ears

Engineering controls are preferred as they are the most effective defense against noise because they do not rely as much on constant human vigilance as do administrative controls and hearing protection. However, effective and feasible engineering controls can be very challenging to develop. The NIOSH mining hearing loss research program has focused on this challenge from its early years within the U.S. Bureau of Mines to its latest accomplishments.

USBM HEARING LOSS RESEARCH PRIOR TO 1996

Noise control and hearing loss prevention first became a high priority for the U.S. Bureau of Mines in the 1970s, following the enactment of the Federal Coal Mine Health and Safety Act of 1969. This act contained noise standards based on the 1969 amendment to the 1936 Walsh-Healey Public Contracts Act. The Bureau conducted a series of projects during the 1970s and 1980s to characterize the noise problem and develop noise controls and other recommended solutions (Aljoe 1985). During this period, the work was conducted by a small (3–5 person) research team augmented by external research contracts. The team also developed research facilities that included a reverberation chamber and an anechoic chamber at the Bureau's Pittsburgh Research Center. The Bureau research was successful in identifying a number of potential noise control solutions that were summarized in a published handbook (Bartholomae and Parker 1983). Unfortunately, the Bureau-developed solutions were not widely adopted.

NIOSH TRANSITION AND THE MSHA NOISE RULE

In 1996, the health and safety functions of the U.S. Bureau of Mines were transferred to NIOSH and reorganized under the Office of Mine Safety and Health Research. The noise control and hearing loss program remained within what became the NIOSH Pittsburgh Research Laboratory (PRL). The program began to establish closer ties with other hearing loss researchers within NIOSH, primarily those at the Division of Applied Research and Technology (DART) in Cincinnati Ohio. DART produced two important reference documents that helped set the tone for future development of the hearing loss research program in mining. The first was "Preventing Occupational Hearing Loss—A Practical Guide." (NIOSH 1996), which outlined principles and strategies for solving noise problems, notably re-asserting the primacy of engineering noise controls. The second was "Criteria for

a Recommended Standard: Occupational Noise Exposure," which included a Recommended Exposure Limit (REL) of an 8-hour time-weighted-average sound level of 85 dB(A) with a 3-dB exchange rate (NIOSH 1998). The NIOSH REL was based upon population hearing loss risk studies and the underlying physics of sound and physiological responses to noise exposure. The REL remains NIOSH's recommendation for exposure limits and current research is investigating enhancements to address impulse noise and exposure durations significantly longer than 8 hours.

During the late 1990s, NIOSH researchers provided input to the MSHA rulemaking process that would culminate in a new noise rule. For instance, NIOSH provided MSHA with an analysis of audiometric tests for over 60,000 miners which showed that by the age of 60, 70–90% of miners had hearing impairment compared to 12% of the individuals who reach that age without significant noise exposure (Franks 1996, 1997).

MSHA's Health Standards for Occupational Noise Exposure were codified in 30 CFR 62 in 1999 with enforcement beginning in 2000. The new standard, commonly called the "Noise Rule," gave regulatory primacy to engineering and administrative controls. It did this by requiring that mine operators use all technologically achievable engineering and administrative controls to avoid overexposures, and that hearing protectors would no longer be permitted in place of the feasible controls. Feasibility of a control was now determined by an evaluation process that ended with listing the control on a Program Information Bulletin containing "Technologically and Administratively Achievable Noise Controls." (MSHA 2008).

NIOSH HEARING LOSS PREVENTION INITIATIVE

NIOSH identified a need for noise control and hearing loss prevention research for miners that would complement the regulatory effort embodied in the MSHA Noise Rule. The NIOSH Associate Director of Mine Safety and Health Research at the time, R. Larry Grayson, reiterated the primacy of engineering controls in the NIOSH approach (Grayson 1999) and the Institute began building a greatly enhanced research capability. Along the way, NIOSH increased the number and expertise of personnel working on noise and hearing loss issues, expanded the research facilities and resources for them to use, and extended the program's scope with a multi-stage process that would be more likely to generate useful end products for the mining industry.

Operating deflection shape ^
(ODS) analysis at 183 Hz

Beamforming noise source
contour plot at 1.1 kHz

Figure 1. **Vibrating screen assembly inside the NIOSH hemi-anechoic chamber next to microphone phased array used for noise source identification via the beamforming technique. Insets show results of operating deflection shape analysis and beamforming analysis.**

Expanded Human Resources

Prior to 1999, the USBM and NIOSH hearing loss research program employed just 3–5 researchers. In most years, this team worked within a group that primarily addressed dust control and respiratory hazards. After 1999, hearing loss research was elevated to a full branch level (the second level of organization in the NIOSH structure) and began growing though a plan of internal transfers and recruitment. The Hearing Loss Prevention Branch now maintains a staff of 16–24 (with fluctuations due to temporary student personnel and project cycles). Reflecting the emphasis on noise controls, the branch consists of researchers primarily from various engineering disciplines, but also employs audiologists, behavioral scientists, and staff with other health-related credentials, along with several technicians.

Expanded Physical Resources

To deliver research outcomes with greater efficiency and productivity, NIOSH needed to expand and improve their laboratories and instrumentation. The existing reverberation laboratory received a full structural overhaul and upgraded instrumentation (Peterson and Bartholomae 2003). The 1300 cubic meter chamber, now known as the Acoustic Test Chamber (ATC), has been accredited by the National Institute of Standards and Technology's National Voluntary Laboratory Accreditation Program (NVLAP) for precision-grade sound power measurements according to ISO 3741. In 2005, to complement the capabilities of the reverberant ATC, NIOSH constructed a new 1200 cubic meter hemi-anechoic chamber (HAC) that can be used for noise source identification and other studies requiring an environment free of acoustic reflections (Figure 1).

For hearing protection and audibility studies, the research program constructed an Auditory Research Laboratory (ARL) in 1999. This laboratory consists of two test chambers, one that has reverberant surfaces for diffuse sound testing and another with sound-absorbing surfaces for conducting hearing tests and assessing directional audibility. The ARL has received NVLAP accreditation for testing the attenuation of hearing protectors. The program also acquired and equipped a mobile laboratory, the Hearing Loss Prevention Unit, which is a 32-foot trailer containing a 4-person sound-insulated booth that can be used for hearing tests and hearing protector evaluations. The trailer also has a space for training and worker counseling.

Expanded Strategic Plan

Recognizing that a more comprehensive program would be needed to make hearing loss solutions available to miners and the mining industry, the NIOSH mining hearing loss program developed a strategic plan in 2004 that consisted of four major components:

1. Develop and maintain a noise source/mine worker exposure database for prioritizing noise control technologies
2. Develop engineering noise control technologies applicable to surface and underground mining equipment
3. Empower workers to acquire and pursue more effective hearing conservation actions
4. Improve the reliability of communication in noisy workplaces

The plan continued to emphasize the primacy of engineering controls (component 2) while augmenting it with data for identifying research needs (component 1) and providing tools to workers so they could more readily adopt solutions to reduce their exposure (component 3). The communications aspect (component 4) was included to address audibility issues that might involve hearing protector design and use.

The components of the strategic plan have undergone periodic review and revision, and a revised set of goals are under development for a new strategic plan for all of NIOSH's mining research programs. The revised goals will continue to emphasize engineering solutions enhanced by effectiveness evaluations and improved transfer of technologies to the workplace.

NOISE CONTROL DEVELOPMENT AND DISSEMINATION PROCESS

NIOSH develops noise controls through a process designed to yield effective and practical solutions. To date, 14 different noise controls have been successfully developed through the process. Typically, the process involves the following steps:

1. Use risk data and stakeholder input to identify candidate machines or technologies for control development
2. Perform field measurements and task observations to identify which tasks and operator locations are associated with the most noise exposure

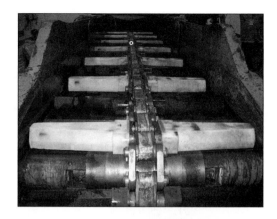

Figure 2. NIOSH-developed coated flight bars on a continuous mining machine conveying system

3. Perform noise source identification to determine which machine components are generating the most noise
4. Develop noise control solutions based on noise control engineering principles
5. Validate whether the designs attain the expected noise reduction through computer modeling and laboratory testing
6. Field test the control in a working mine to assess exposure reduction, practicality, and durability
7. Promote adoption of the control

The first two machines that were addressed through this process are the continuous mining machine (CMM) and the roof bolting machine (RBM). These machines were selected based on MSHA exposure data which shows they are the two underground coal mining machines associated with the largest percentages of noise overexposures.

CMM Controls

In NIOSH field observations of continuous mining machine operators, 86% were exposed to noise exceeding the MSHA PEL (Bauer, Babich, and Vipperman 2006). Noise source identification in the NIOSH laboratories identified the conveying system as the dominant noise source (Kovalchik, Johnson, Burdisso, Duda, and Durr, 2002). One of the noise controls developed for the conveying system was the coated flight bar (Figure 2). Several engineering revisions of this control were developed and tested in the NIOSH laboratories and then field tested in

Figure 3. NIOSH-developed collapsible drill steel enclosure to control noise on roof bolting machines

a working coal mine (Smith, Spencer, Alcorn, and Kovalchik 2006). The coated flight bar alone reduces operator exposures by 7 dB (Smith 2008). Additional reductions based on NIOSH designs and engineering include a dual-sprocket chain, a noise-controlled scrubber fan, vibration-isolated tail roller, and an isolated take-up system.

RBM controls

A similar process was followed for the roof bolting machine. NIOSH researchers conducted field observations to assess overall exposure of roof bolter operators and found that 81 percent had exposures exceeding the MSHA PEL (Bauer, et al. 2006). The bolting task associated with the most exposures was drilling prior to inserting the bolt. The noise control team then performed noise source identification in the PRL Hemi-anechoic Chamber that showed that the drill/roof and drill/chuck interfaces were responsible for most of the radiated noise (Yantek, Peterson, and Smith, 2007). They then developed and field tested a suite of controls that includes a collapsible drill steel enclosure (Peterson 2008) (Figure 3), and isolators for the drill bit and chuck (Kovalchik, Smith, Matetic, and Peterson 2009).

Behavioral Interventions

The NIOSH mining hearing loss research program has also addressed the human side of prevention with a series of behavioral and training interventions. A hearing loss simulator was developed as a way to demonstrate the impact of noise exposure on hearing (Randolph, Reinke, and Unger 2008). The simulator can be used to show that the use of an effective noise control over time can preserve a significant amount of hearing. Although hearing protectors are the last item on the hierarchy of controls, they are still widely needed where feasible and effective noise controls do not yet exist. The NIOSH research program developed a QuickFit device (Figure 4) to help hearing protector users perform a rapid check of earplug attenuation (Randolph 2008).

Future Directions

Additional noise control efforts on vibrating screens used in coal preparation plans are reaching completion following the same process as the work on the CMM and RBM (Yantek, Jurovcik, and Bauer 2006). The research team is now wrapping up work on the underground coal sector with a new project on long-wall mining machines. Another new project will start the process of addressing other mining sectors by developing noise controls for load-haul-dump machines and haul trucks used in underground metal mines. A third new project will evaluate the research program's success at reaching its noise reduction and hearing loss prevention goals, and will develop technology transfer approaches to make noise control advances accessible to every mine where they can be used.

Figure 4. Testing earplug fit with the NIOSH QuickFit device

SUMMARY AND CONCLUSIONS

Hazardous noise and hearing loss are long-standing problems in the mining industry that can most effectively be solved through solutions that reduce noise at the source. The U.S. Bureau of Mines and NIOSH research programs have consistently addressed noise sources through an emphasis on engineering controls. This emphasis was augmented after 1999 by an expanded research program that added field evaluations to prove the effectiveness of controls and a behavioral component to motivate more widespread adoption. To date, the NIOSH program has produced 14 effective mining noise control technologies and several behavioral, training, and management tools. Future developments will expand the program's impact by extending noise control development beyond underground coal to every other mining sector.

REFERENCES

Aljoe, W.W. 1985. *The Bureau of Mines Noise-Control Research Program—A 10 Year Review*, IC 9004 US Bureau of Mines.

Bartholomae, R.C., and Parker R.P. 1983. *Mining Machinery Noise Control Guidelines*.Report No. I 28.16/2:M66/8. Pittsburgh, PA: Bureau of Mines, United States Department of the Interior.

Bauer, E.R., Babich, D.R., and Vipperman, J.S. 2006. *Equipment Noise and Worker Exposure in the Coal Mining Industry.* Pittsburgh, PA: U.S. Department of Health and Human Services, Public Health Service, Centers for Disease Control and Prevention, National Institute for Occupational Safety and Health, DHHS (NIOSH) Publication No. 2007-105, IC 9492.

Franks, J.R. 1996. Analysis of audiograms for a large cohort of noise-exposed miners. Cincinnati, OH, U.S. Department of Health and Human Services, Public Health Service, Centers for Disease Control and Prevention, National Institute for Occupational Safety and Health, Division of Biomedical and Behavioral Science. Unpublished.

Franks, J.R. 1997. Prevalence of hearing loss for noise-exposed metal/nonmetal miners. Cincinnati, OH, U.S. Department of Health and Human Services, Public Health Service, Centers for Disease Control and Prevention, National Institute for Occupational Safety and Health, Division of Biomedical and Behavioral Science. Unpublished.

Grayson, R.L. 1999. Effective prevention of hearing loss in miners. *Holmes Safety Association Bulletin*, January issue, pp. 3–5.

Kovalchik, P.G., Johnson, M., Burdisso, R., Duda, F., and Durr. M. 2002. Noise Control for Continuous Miners, Proceedings from the 10th International Meeting on Low Frequency Noise and Vibration and its Control, York, England.

Kovalchik, P.G., Smith, A.K., Matetic, R.J., and Peterson, J.S. 2009. Noise Controls for Roof Bolting Machines. *Min. Eng.* 61(1):74–78.

Kryter, K.D. 1994. *The Handbook of Hearing and the Effects of Noise.* San Diego, CA: Academic Press.

Lamonica, J.A., Mundell, R.L. and Muldoon T.L. 1971. *Noise in Underground Coal Mines.* BuMines RI 7550.

MSHA. 2008. *Technologically Achievable, Administratively Achievable, and Promising Noise Controls*, Department of Labor, Mine Safety and Health Administration, Program Information Bulletin P08-12 (30 CFR Part 62), http://www.msha.gov/ regs/complian/pib/2008/pib08-12.pdf.

NIOSH. 1976. *Survey of Hearing Loss in the Coal Mining Industry.* Cincinnati, OH: U.S. Department of Health, Education, and Welfare, Public Health Service, Center for Disease Control, National Institute for Occupational Safety and Health, DHEW (NIOSH) Publication No. 76–172.

NIOSH. 1996. *Preventing Occupational Hearing Loss—A Practical Guide.* Cincinnati, OH: U.S. Department of Health and Human Services, Public Health Service, Centers for Disease Control and Prevention, National Institute for Occupational Safety and Health, DHHS (NIOSH) Publication No. 96-110.

NIOSH. 1998. *Criteria for a Recommended Standard: Occupational Noise Exposure. Revised Criteria.* Cincinnati, OH: U.S. Department of Health and Human Services, Public Health Service, Centers for Disease Control and Prevention, National Institute for Occupational Safety and Health, DHHS (NIOSH) Publication No. 98-126.

Peterson, J.S. 2008, *Collapsible Drill Steel Enclosure for Reducing Roof Bolting Machine Drilling Noise.* U.S. Department of Health and Human Services, Public Health Service, Centers for Disease Control and Prevention, National Institute for Occupational Safety and Health, DHHS (NIOSH). Technology News 532.

Peterson, J.S., and Bartholomae, R.C. 2003. Design and instrumentation of a large reverberation chamber. NOISE-CON 2003. Ames, IA: Institute of Noise Control Engineering of the USA.

Randolph, R.F, Reinke, D.C., and Unger, R.L. 2008. *NIOSH Hearing Loss Simulator Instruction and Training Guide.* Pittsburgh, PA: U.S. Department of Health and Human Services, Public Health Service, Centers for Disease Control and Prevention, National Institute for Occupational Safety and Health, DHHS (NIOSH) Publication No. 2008-119.

Randolph, R.F. 2008. *QuickFit Earplug Test Device.* Pittsburgh, PA: U.S. Department of Health and Human Services, Public Health Service, Centers for Disease Control and Prevention, National Institute for Occupational Safety and Health, DHHS (NIOSH) Publication No. 2009-112. Technology News 534.

Smith, A.K. 2008. *Engineering Controls for Reducing Continuous Mining Machine Noise.* U.S. Department of Health and Human Services, Public Health Service, Centers for Disease Control and Prevention, National Institute for Occupational Safety and Health, DHHS (NIOSH). Technology News 531.

Smith, A.K., Spencer, E.R., Alcorn L.A., and Kovalchik, P.G. 2006. Underground Evaluation of Coated Flight Bars for a Continuous Mining Machine. Inter-Noise 2006. The 35th International Congress and Exposition on Noise Control Engineering, 3–6 December 2006, Honolulu, Hawaii. West Lafayette, IN: International Institute of Noise Control Engineering.

Tak, S, Davis, R.D., and Calvert, G.M. 2009. Exposure to hazardous workplace noise and use of hearing protection devices among US workers—NHANES, 1999–2004. *American Journal of Industrial Medicine* 52:358–371.

Tak, S., and Calvert, G.M. 2008. Hearing difficulty attributable to employment by industry and occupation: An analysis of the National Health Interview Survey—United States, 1997 to 2003. *Journal of Occupational and Environmental Medicine* 50:46–56.

Yantek, D.S., Jurovcik, P., and Bauer, E.R. 2006. Noise and Vibration Reduction of a Vibrating Screen. *Trans. Soc. Min. Metal Explor.* 318:201–213.

Yantek, D.S., Peterson, J.S., and Smith. A.K. 2007. Application of a Microphone Phased Array to Identify Noise Sources on a Roof Bolting Machine. NOISE-CON 2007. Reno. NV: Institute of Noise Control Engineering of the USA.

Heat Stress Management in Hot Mines

Petrus Schutte

CSIR, Johannesburg, South Africa

INTRODUCTION

Occupational heat stress is a recognized health and safety hazard in South African mines. The consequences of high occupational heat loads can be expressed in terms of impaired work capacity, errors of judgment with obvious implications for safety, and the occurrence of heat disorders, especially heat stroke which is often fatal.

Engineering strategies to counter occupational heat stress usually involve reducing the environmental heat load (through shielding, refrigeration and increased air flow) to facilitate heat dissipation from the human body, and/or improved mechanization to minimize high metabolic rates. In the South African mining industry, technological and economic constraints often preclude a purely engineering-based approach. Personal protection and administrative controls, such as heat acclimatization and heat stress management, for example, then become the only viable alternatives.

The serious consequences of excessive levels of occupational heat stress were recognized by the South African gold mining industry when the first death from heat stroke occurred in 1924 (1). Steps to combat the heat stress hazard were taken almost immediately and the first form of heat acclimatization was introduced in 1925 (2). This rudimentary method of heat acclimatization was followed by a research program aimed both at obtaining a better understanding of human heat tolerance and at developing improved heat acclimatization procedures.

Research conducted as part of this program since the 1930s has culminated in the development of scientifically based heat acclimatization procedures, heat tolerance test procedures and the Heat Stress Management Programme currently used in the South African mining industry.

In South African mines, work environments having a wet-bulb temperature in excess of 27.4°C are considered to be 'hot' and necessitate the introduction of practices to safeguard miners. This is the heat stress limit for unacclimatized men and is based on an analysis of heat stroke accidents which have occurred in the South African gold mining industry (3).

FORMAL HEAT ACCLIMATIZATION

Formal acclimatization of miners to work in hot underground environments was a widespread practice at South African gold and platinum mines for the greater part of the previous century.

Underground Acclimatization

Prior to 1953 the method used to acclimatize miners was known as 'underground acclimatization' (2). This method consisted of allocating miners to work in a hot production area underground, but under supervision, for 14 days. During the 14-day period the miners' work rate was gradually increased to that undertaken by acclimatized men. This method was not very successful as it was difficult to control the work rate of miners as well as the thermal conditions of the areas in which they worked.

Recognition of the above problems led to the development of the 'two-stage acclimatization' procedure, which took 12 days. The first six days served to condition the miners to physical work in relatively mild heat stress (29.5 to 30.5°C wet-bulb) while the last six days adapted the miners to the most severe levels of the stress that they were likely to encounter subsequently (31.7 to 33.9°C wet-bulb). The acclimatization procedure was closely supervised: body temperatures were recorded and miners developing high levels of heat strain were identified and treated. While 'two-stage acclimatization' was simple in principle, it was difficult to maintain the required thermal conditions always.

Both of the above procedures resulted in the partial acclimatization of miners and a concomitant high heat stroke risk when they were transferred to normal production teams.

Climatic Room Acclimatization

The difficulties encountered in the application of acclimatization procedures underground, particularly in regard to obtaining the right combinations of thermal conditions and work rate, led to the development of climatic room acclimatization procedures.

These procedures were introduced in 1965 and the period of acclimatisation was reduced from 12 to eight days (4). During this period miners worked continuously for four hours a day in a climatic room controlled at 31.7°C wet-bulb, with an air speed of 0.4 m/s. A block-stepping routine was used to obtain the required work rate, which was increased progressively from 35 W (external component) on the first day to 70 W on the eighth day. Hourly measurements of body temperature and daily checks of body mass were made, and any abnormal mass response or high body temperatures were given appropriate treatments. The body temperature profile of a miner was used to determine whether he was acclimatized or not.

The most significant breakthrough in heat acclimatization occurred with the discovery that miners respond more favourably to heat when supplemented with ascorbic acid (Vitamin C). Vitamin C supplementation reduced the acclimatization period from eight to five days, in general, and from eight to four days for previously acclimatized miners (5). A further advantage of the Vitamin C supplementation was the decrease in the number of miners who did not respond to the acclimatization routine from five percent to less than one percent (6).

HEAT TOLERANCE TESTING

Selection procedures based on heat tolerance were first applied in the South African mining industry by Dreosti in 1935 (7). Dreosti's heat tolerance test was based on the body temperature responses of miners shovelling rock for 60 minutes in a saturated environment of 35°C. Miners whose oral temperatures were below 38°C on completion of the test were classified as "heat tolerant." while those with oral temperatures of above 39°C were regarded as "heat intolerant." Miners with oral temperatures of between these two values were classed as "normal." In a sample of 20 000 men 25 percent were judged "heat tolerant." and 15 percent "heat intolerant." Eighty-three percent of the "heat intolerants." could be acclimatized to exhibit heat tolerant responses.

During experiments aimed at the refining of Vitamin C-supplemented acclimatization in the 1970s two important observations were made that paved the way for the development of a test procedure to identify individuals with a high natural tolerance to heat. Firstly, the results showed that, for any given individual, there is an optimum period of acclimatization to heat in which a satisfactory level of adaptation is reached; and, secondly, there are wide individual differences with respect to the optimum period required.

The heat tolerance test that was devised in 1977 (8) involved a block-stepping routine for four hours in a climatic room. Work rate was fixed at 54 W (oxygen consumption of approximately 1.25 l/min)

and the thermal conditions used for the test were 29, 30, 31 and 31.7°C, depending on the temperature of the underground workplaces miners were required to work at on successful completion of the test. Oral temperatures were recorded hourly during the test and miners with temperatures not exceeding 37.5°C were classified as "hyper-heat-tolerant." These "hyper-heat-tolerant." miners had been shown not to benefit any further from formal acclimatization and could be posted immediately to normal underground work. The percentage of miners classified as "hyper-heat-tolerant." depended on the severity of the heat stress applied: at the recommended test temperatures of 29, 30, 31, and 31.7°C the pass rates were 85, 75, 50 and 35 percent, respectively.

In 1991 a new approach to heat tolerance testing was adopted with the introduction of Heat Tolerance Screening (HTS) (9). The primary objective of HTS was to identify gross or inherent heat intolerance (i.e., individuals with an unacceptable risk of developing excessively high levels of hyperthermia during work in heat).

The HTS used in the South African mining industry consists of bench-stepping for 30 minutes in a climatic chamber at an external work rate of approximately 80 W (positive component), in an environment with a dry-bulb temperature of 29.5°C and a wet-bulb temperature of 28.0°C (10). The assessment of relative heat tolerance is based on the body temperature recorded at the end of the 30-minute bench-stepping exercise. Any person whose body temperature does not exceed a given value at the end of the test is classified as "potentially heat tolerant." This implies that that person is fit to undertake physically demanding work in a 'hot' environment (with wet-bulb temperatures of greater than 27.5°C) and that he will be able to acclimatize successfully with regular exposure. Individuals with body temperatures of above the given value on completion of the test are considered to be "heat intolerant." and will not be allocated to work in 'hot' areas.

HEAT STRESS MANAGEMENT

Heat Stress Management (HSM) was introduced to the South African mining industry in 1991. HSM is a multi-faceted approach to promoting health, safety and work performance through minimizing human heat strain and the incidence of heat disorders. It is essentially a risk management approach for reducing risk by removing high-risk individuals from the underground mining population and, through the implementation of countermeasures, striving to minimize the risk of developing heat disorders during work in heat.

HSM consists of two essential elements: the assessment of overall fitness to work in heat, and a natural progression towards heat acclimatization on the basis of safe work practices. These two elements are to a large extent based on the aetiology of heat stroke: the overall fitness for hot environments is measured against an individual employee risk profile and the causal factors in the development of heat stroke in the mining industry are translated into safe work practices.

Overall Fitness for Physical Work in Heat

It is obvious that although most individuals would, in a qualitative sense, exhibit similar responses to heat, the actual response pattern for any given individual is a function of a variety of factors. These determinants can be broadly divided into two categories: *inherent* determinants, i.e., those over which the individual has no control, and *external* determinants, i.e., those over which the individual has some control or which he voluntarily accepts. In the latter category, the most important are nutrition and hydration, whereas the former includes factors such as maximum work capacity, age, gender and body dimensions.

In the present context, overall fitness for work in heat will depend on the outcome of a purpose-designed medical examination, with special emphasis on features that would rule out physical work or exertion in heat (11), and an assessment of heat tolerance by means of HTS (10) in a climatic chamber.

Risk Profile

One of the cornerstones of HSM is the introduction of an individual employee risk profile against which his overall fitness for work in hot environments is measured (12). On the basis of the outcome of the medical examination and HTS, it is quite feasible to develop a 'risk profile' for any employee destined to enter 'hot' working environments in the execution of his duties and responsibilities.

The risk profile consists of the following elements:

- Medical contraindications, i.e., a particular condition, treatment or even a medical history likely to lead to a critical job-related reduction in heat tolerance
- Age (≤50 years), in concert with full-shift exposures to 'strenuous' work in heat
- Obesity (Body Mass Index ≤30)
- Heat intolerance, i.e., inability to successfully complete HTS
- Strenuous work *per se*
- A history of heat disorders

In developing an employee risk profile on the basis of the above elements a threefold approach is recommended, namely:

- A risk profile that features only one of the above elements, especially where it can be controlled or brought under control, should be regarded as 'acceptable'
- The presence of any two elements should be viewed with concern and should not be condoned unless the situation can be ameliorated, for example through specially developed safe work practices
- A profile containing more than two undesirable elements will constitute an unacceptable risk

Safe Work Practices

Within the context of HSM no form of formal heat acclimatization will have preceded the allocation of employees to 'hot' areas of work. Workers will have been screened only for gross heat intolerance and will be expected to commence duties without the advantage of acclimatization. For this reason, special precautions are indicated, the rationale for which is based on the major causes of heat stroke in mining.

A review of the occurrence of heat stroke over a ten-year period indicates that the origin of heat stroke is multi-factorial (13). The main causal factors are interactions between strenuous work, suspect heat tolerance, excessively hot environments, and concurrent dehydration. On the basis of this analysis, a basic framework with the following elements can be derived for work practices in 'hot' environments:

- Monitoring work-place wet- and dry-bulb temperatures on a basis designed to ensure that safe limits are not exceeded and to detect the development of possible trends
- Ensuring that acceptable work rates are maintained in order to avoid the early onset of fatigue (achieved through work-rest cycles: 10 to 15 minutes of rest in every hour) where work is of necessity strenuous and ongoing (e.g., drilling) or by instilling, through constant reminders, a sense of self-pacing
- Ensuring that fluid-replacement beverages (preferably only water or hypotonic fluids) are available at the place of work and that a fluid replacement regimen of at least 2 × 250–300 ml per hour is observed
- Detecting early signs and symptoms of heat disorders and instituting proper remedial action, depending on the precise set of signs and symptoms

- Ensuring that emergency treatment and communication facilities are available and fully functional on a daily basis
- Setting into motion purpose-developed emergency action plans in the event of sudden escalations in environmental temperatures

The above work practices, which should in any event be adopted as standard routine, are especially relevant to employees who return to work after a period of absence, irrespective of duration or reason.

MEASURES OF SUCCESS

Heat has been a constant health hazard in South African gold (and more recently platinum) mines for more than 90 years; the problem cannot be resolved overnight and is likely to remain a potential hazard for some years to come. Measured against its objective, namely, the promotion of the health, well-being and safety of all employees exposed to working conditions potentially conducive to heat disorders, HSM proved to be very successful. If statistics on the incidence of reportable heat disorders (heat stroke and heat exhaustion) are considered, the implementation of HSM has contributed to a decline in heat disorders from 83 cases (21 heat strokes and 62 heat exhaustions) in the three-year period before the general implementation of HSM in 1991 to 33 cases (five heat strokes and 28 heat exhaustions) for the period 2005 to 2007. This reduction is achieved through the elements that form the backbone of HSM: the detection of gross or permanent heat intolerance (approximately three percent of the population assessed) and the control of the progression of 'natural acclimatization on the job' which further reduces the risk of heat disorders.

FUTURE RESEARCH

One of the major objectives of the South African mining industry is to provide the safest work environment possible for miners, especially in view of the changing demographics of the mining population.

In order to achieve the above objective it is important to assess the physiological strain experienced by miners, especially where physical work is performed in hot environments. Very little information is available on the actual physiological strain (the combined strain reflected by the thermoregulatory and cardiovascular systems) associated with mining tasks, especially when these tasks are performed by females or older males. Where information is available, it has been based on the physiological responses of healthy young males.

In the South African context, with its milestones for the employment of female miners, there is a definite need to establish the role of gender with regard to the physiological strain experienced by miners during typical mining activities. In view of their smaller physical work capacity (aerobic capacity) and physical strength, female mineworkers may experience undue physiological strain when performing "prolonged and strenuous." physically demanding tasks as is the case in mining. The current difficulties experienced with the placement of female miners in underground occupations highlight the need for such research.

REFERENCES

1. Cluver, E.H. (1932), "An Analysis of Ninety-two Fatal Heat Stress Cases on Witwatersrand Gold Mines," *South African Medical Journal*, Vol. 6, pp. 15–23.
2. Stewart, J.M. (1982), "Practical Aspects of Human Heat Stress," in *Environmental Engineering in South African Mines,* J. Burrows, R. Hemp, W. Holding and R.M. Stroh, eds. Cape Town, Cape & Transvaal Printers (Pty) Ltd.
3. Strydom, N.B. (1966), "The Need for Acclimatizing Labourers for Underground Work," *Journal of the South African Institute for Mining and Metallurgy*, Vol. 75, pp. 124–126.
4. Wyndham, C.H. and Strydom, N.B. (1969), "Acclimatizing Men to Heat in Climatic Rooms on Mines," *Journal of the South African. Institute for Mining and Metallurgy*, Vol. 70, pp. 60–64.
5. Strydom, N.B., Kotze, H.F., Van Der Walt, W.H. and Rogers, G.G. (1976), "Effects of Ascorbic Acid on Rate of Heat Acclimatization," *Journal of Applied Physiology*, Vol. 41, pp. 202–205.
6. Strydom, N.B. (1982), "Developments in Heat Tolerance Testing and Acclimatization Procedures since 1961," in *Proceedings 12th CMMI Congress,* 3–7 May, Johannesburg.
7. Dreosti, A.O. (1935), "Problems Arising out of Temperature and Humidity in Deep Mines of the Witwatersrand," *Journal of the Chemical Metallurgy and Mineral Society of South Africa*, Vol. 6, pp. 102–209.
8. Schutte, P.C., Rogers, G.G. and Strydom, N.B. (1977), "The Elimination of Heat Acclimatization in Selected Heat Tolerant Individuals," Research report 49/77, Chamber of Mines of South Africa, Johannesburg.

9. Schutte, P.C., Van Der Walt, W.H., Kielblock, A.J., Marx, H.E. and Trethoven, S.J. (1991), "Development of a Short-duration Test for the Detection of Gross Heat Intolerance," Research report 12/91, Chamber of Mines Research Organization, Johannesburg.

10. Kielblock, A.J. and Schutte, P.C. (1998), "A Guide to Heat Stress Management," Safety in Mines Research Advisory Committee (SIMRAC) Research Project Report GAP 505, SA Department of Minerals and Energy, Johannesburg.

11. Donoghue, A.M. and Bates, G.P. (2000), "The Risk of Heat Exhaustion at a Deep Underground Metalliferous Mine in relation to Body-mass Index and Predicted VO^2 Max," *Occupational Medicine*, Vol. 50(40), pp. 259–263.

12. Kielblock, A.J. (2001), "Heat," in *Handbook of Occupational Health Practice in the South African Mining Industry*, R. Guild, R.I. Ehrlich, J.R. Johnston and M.H. Ross, eds. Johannesburg, Safety in Mines Research Advisory Committee (SIMRAC).

13. Kielblock, A.J. (1992), "The Aetiology of Heat Stroke as a Basis for Formulating Protective Strategies," in *Proceedings of the Fifth International Mine Ventilation Congress*, R. Hemp ed. Johannesburg, Mine Ventilation Society of South Africa.

Mine Illumination: A Historical and Technological Perspective

John J. Sammarco
NIOSH, Pittsburgh, Pennsylvania, United States

Jacob L. Carr
NIOSH, Pittsburgh, Pennsylvania, United States

ABSTRACT: Illumination plays a critical role in mining because miners depend on proper illumination to safely perform their work and to see various mine and machinery-related hazards. Open-flame lamps were used in the early days of mining, but these often caused disastrous mine explosions. During 1914, two engineers from the U.S. Bureau of Mines (USBM) formed a new company, Mine Safety Appliances, and initiated work with Thomas Edison for an electric cap lamp. This electric cap lamp was approved in 1915 by the USBM. The seminal works in mine illumination research were dominated by the USBM. They addressed many issues in permissibility, human factors, and researched new lighting technologies. The current mine regulations and test procedures for mine illumination are based on USBM research. Today, illumination technology is drastically changing with light emitting diodes (LEDs) that could revolutionize mine illumination just as was done by the 1915 electric cap lamp. Researchers at the National Institute for Occupational Safety and Health (NIOSH) are leading the way in the human factors and technological aspects of LED research for mine illumination to improve the safety of miners.

INTRODUCTION

Adequate illumination is crucial in underground coal and metal/nonmetal mine safety because miners depend most heavily on visual cues to detect hazards associated with falls of ground, slips/trips/falls (STFs), moving machinery, and other threats. For as long as underground mining has been performed, illumination has been critical to both safety and to the ability of the miners to perform their work. Open flames were used from the earliest days of mining. The Greeks and Romans used oil lamps, and candles were introduced in about the 1st century AD. In about the 16th century, oil wick lamps became common and were used in some mines even until the 1920s (Trotter 1982). Carbide lamps were developed in the 19th century and were used well into the 20th century. The first attempts to make a safety lamp that would not cause explosions in gassy mines began around the turn of the 19th century. Safety lamps provided a significant improvement over open-flame lighting and were the safest method of lighting gassy mines until electric lighting became feasible in the first half of the 20th century. At the turn of the 20th century, the most commonly used lighting technologies were safety lamps for gassy mines and carbide lamps for non-gassy mines (Trotter 1982).

In 1879, Edison developed the first practical incandescent lamp, with an efficiency of about 1.4 lumens per watt (lm/W). During the early 1900s, the new technology of electric lighting began making its way into underground mines in the United States.

Safety was the primary driver of this new technology given the pervasive occurrences of explosions caused by mine gas ignition. These explosions resulted in many mine disasters that claimed the lives of thousands of miners. On May 16, 1910, Congress passed public Law 61-179 that established the USBM. Shortly after, the USBM began research to evaluate the safety of the new technology of electric lighting. In this paper, the seminal works of USBM illumination research are described in conjunction with the development, evolution and spread of electric lighting as well as the technological changes that have taken place since 1910.

ELECTRIC LIGHTING

By the end of the 19th century, electric lighting was widely used in homes and for industrial lighting. Use in mining, however, was delayed several decades. Permanent lighting installations were impractical due to the high cost of wiring the miles of workings and due to the logistics of advancing the lights with the face, and small, efficient batteries did not exist to allow for portable lamps. Additionally, concerns existed about the lamps causing explosions, and efforts to make them explosion proof resulted in heavy cumbersome lamps (Trotter 1982).

There was, however, a pressing demand for safer face lighting as explosions caused by mine gas ignition continued to claim the lives of miners. In 1914, two engineers from the USBM, John Ryan and

Figure 1. The Edison Electric Cap Lamp (circa 1915) from the NIOSH collection of mine illumination. Image courtesy of New South Associates.

George Deike, formed the Mine Safety Appliances Company and began work with Thomas Edison to develop personal electric lamps. By this time, more efficient tungsten filaments had replaced the older carbon filaments, drastically improving light output from the bulbs, and Edison was able to design a small-scale, rechargeable battery that could be carried on a miner's belt (Trotter 1982). The Edison electric cap lamp (Figure 1) was approved by the USBM in 1915 (Clark and Ilsley 1917). Electric lamps gained acceptance and eventually replaced the older technologies.

Permissibility and Ignition of Mine Gases

By the turn of the twentieth century, the safety lamp had drastically improved mine safety. The mining community was hesitant to replace the trusted technology of safety lamps with the new technology of electric lamps until they had been extensively tested and shown to be safe even in gassy environments.

In 1912, a series of tests were conducted by the USBM investigating whether an incandescent lamp bulb with either a carbon or a tungsten filament could ignite mine gases when the glowing filament was exposed (Clark and Ilsley 1913). In these tests miniature and standard bulbs from 8 manufacturers were subjected to four types of tests in which the bulb was smashed, the tip of the bulb was cut off, a hole was punctured in the neck of the bulb and a pre-exposed filament was suddenly subjected to a voltage in a natural gas and air mixture. Ignition was found to occur at mixtures containing as little as 5% natural gas. The probability of ignition for each type of bulb and for each type of test was rated on a scale from "unlikely." to "certain." Based on these results, safety features were built into the cap lamps approved by the USBM. For example, Figure 2 shows a typical method by which breaking the bulb could be made to cause the electrical connection to also be broken using spring-loaded contacts.

Development and Approval of Early Technologies

The next step in USBM research was to establish the test conditions and requirements for the prevention of gas ignition by electric lamps and for determining suitable attributes of electric lamps for mines. The attributes included period of operation, liability to spill electrolyte, liability to ignite gas, weight, light producing capacity, durability and reliability. The permissibility and suitability requirements were critical for electric lamp approvals conducted by the USBM. The testing process for electric lamp approvals along with results of the tests were published by Clark and Ilsley (Clark 1914; Clark and Ilsley 1917). Figure 3 shows the test equipment developed by the USBM. The first requirements were set forth in Schedule 5, issued in 1913, which focused

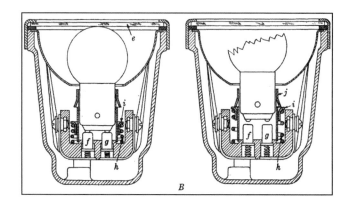

Figure 2. A typical cap lamp head piece design that prevents a burning filament being exposed through the use of spring-loaded contacts (Ilsley and Hooker 1930)

Figure 3. USBM-developed apparatus (circa 1917) for exposed lamp filament ignition testing (Clark and Ilsley 1917)

solely on permissibility. USBM experience necessitated a broader basis of approval so the USBM evaluated several other aspects of electric cap lamps and worked closely with manufacturers during the design and development of early cap lamps. Hence, Schedule 6 was issued in 1914 and the Caeg, Hirsch, and Wico lamp were approved (Clark 1914). Schedule 6 was later revised as Schedule 6A in 1915. Schedule 6A included more requirements for the safety and the practicability of the lamp. The original approvals under Schedule 5 were withdrawn and the lamps were retested in accordance with Schedule 6A. The Edison cap lamp was the first approval under Schedule 6A and by 1917, seven models of lamps had been approved (Clark and Ilsley 1917). By August, 1916, about 70,000 USBM-approved lamps were in use and about 2,000 lamps a week were being installed (Clark and Ilsley 1917). While these cap lamps found great acceptance in the mining industry, their performance (Table 1) pales in comparison to today's modern LED cap lamps that can achieve at least 35 lumens after 10 hours of operation.

The USBM also tested and issued approvals for animal lamps, flood lights, and flashlights (Ilsley, Hooker and Roadstrum 1942). An approval plate was attached to each lamp and a list of approved lamps was periodically published (Howell, Fieldner and Ilsley 1922). The USBM conducted extensive work investigating and reporting on the use of electric lamps at operating mines throughout the country. For example, a survey of the number, location, and type of electric cap and animal lamp installations in the Alabama mining district was conducted in 1935 (Cash 1935). The early work of the USBM provided the industry with confidence in the safety and performance of electric lighting technology thus enabling its acceptance in the mining industry.

Initial Development of Safety Rules and Regulations

In addition to evaluating the new technologies, during the 1910s and afterward, the USBM was charged with developing the safety rules and guidelines that would govern the use of these technologies underground. Safety rules initially developed by USBM covered several aspects of operating electric lamps, but were focused perhaps most heavily on aspects related to permissibility and the prevention of gas and coal dust explosions.

A set of safety rules was published in 1916 for the construction, installation, maintenance and use of several types of electrical equipment including cap lamps underground (Clark and Means 1916), and revised rules were issued in 1926 (USBM 1926) and several times thereafter. The rules for portable lamps were concerned with the enclosure of the bulbs, type and condition of connections, grounding, and type and condition of cords.

In addition to developing the federal standards, the USBM attempted to make state laws known by the industry, and, for example, conducted a review of state mining laws on the use of electricity underground, including the use of stationary, semi-portable and portable electric lamps, in 1920 (Ilsley 1920).

Table 1. Photometric data for early electric cap lamps (circa 1915–1917). Source: Clark and Ilsley

	Average candlepower after burning 1 hour	Average candlepower after burning 12 hours	Total lumens after burning 1 hour	Total lumens after burning 12 hours
Edison cap lamp	1.19	0.83	4.32	3.01
Concordia hand lamp	0.74	0.57	6.92	5.36
GE cap lamp	1.00	0.68	3.63	2.47
WICO cap lamp	0.93	0.69	3.37	2.50
Concordia cap lamp	1.05	0.72	3.80	2.60
Pioneer cap lamp	1.13	0.78	4.10	2.83

In 1935, USBM guidelines were established for the maintenance and inspection of electric cap lamps. An analysis of factors that decrease the efficiency of electric cap lamps was published by the USBM (Hooker and Zellers 1935). Factors decreasing the efficiency of the lamps were attributed to poor lamp house supervision and poor lamp care. The USBM placed responsibility for maintaining proper lamp performance on the mine operators, and provided them with recommendations for inspection and maintenance of batteries, cords, bulbs, lenses and reflectors. The USBM also recommended a procedure for testing lamp performance by using a photometric sphere that the USBM recommended be purchased and used by mine operators. In 1935, guidelines were published on how to maintain permissibility of cap lamps (Zellers and Hooker 1935) which focused on similar types of maintenance and inspection issues.

Through this early work of the USBM in the development and approval of new lamps and in the development of safety rules and guidelines, the industry was provided with a good starting point for the improvement and evolution of underground lighting throughout the 20th and into the 21st century.

FURTHER ADVANCEMENT

Significant advancement was made in the area of mine illumination during the 20th century by both improving existing technologies and developing new technologies. Now in the 21st century, improvements are continuing to be made. In this section, some of the advances in mine illumination since the work of the early 20th century will be discussed.

Advancement of Cap Lamps

The cap lamp steadily became a reliable and trusted piece of equipment used by practically every underground miner. Few major changes in the technology occurred for several years. The major advances for the industry came in the form of new bulb and battery technologies being developed. In the early 1980s, a prototype nickel-cadmium (Ni-Cd) cap lamp battery was developed by the USBM and the Energy Research Corporation. This battery provided about a 48% weight reduction, 15% volume reduction and a useful battery life of 3 to 4 years. A new reflector was designed to accept the new tungsten-halogen type bulbs in the conventional dual-filament incandescent bulbs. The tungsten-halogen lamps were evaluated to replace the conventional incandescent lamps. The tungsten-halogen lamp was shown to be capable of producing the same light output as the conventional lamp while consuming less power. Other improvements to the cap lamp design at this time included

elevation adjustment of the lamp, break-away cords, and segmented cords with coiled ends to allow the cord to better conform to the miner's body (Lewis 1982).

In the 1970s fluorescent cap lamps were developed and tested. The USBM generated performance specifications that included light output, battery capacity, charging time, system weight, and compatibility with existing hardware. A fluorescent cap lamp that met the USBM requirements (Figure 4) was developed by Ocean Energy, Inc. (Ocean Energy Inc. 1975). It was shown to provide superior light output with less degradation over a 10-hour shift as compared to an incandescent lamp. Additionally, bulb life was significantly longer. The fluorescent cap lamp never found widespread use in mining, most likely because of the size, weight, and higher initial costs. However, machine-mounted fluorescent lamps became widespread and are still in use today.

Machine-Mounted and Area Lighting

In the second half of the 20th century, as mining became increasingly mechanized, a clear advantage was recognized in installing lighting systems on mobile equipment such as continuous miners, bolters and LHDs. During the 1970s and 1980s, the USBM conducted or supported several projects investigating the installation of different lighting technologies

Figure 4.   The development of a fluorescent cap lamp enabled the evaluation of this new technology. Source: NIOSH mining photo archive.

Figure 5. Wooden mockups of a mining machine were constructed and used to evaluate machine-mounted lighting. The USBM-developed CAPS software eliminated this costly and time-consuming approach. Source: NIOSH mining photo archive.

on mining machines. In a series of projects conducted or funded by the USBM, lighting hardware was installed and evaluated on conventional, continuous and longwall machinery and for area lighting both in the laboratory and at several low, mid, and high coal and metal/non-metal mining operations throughout the country (Bockosh 1977; Lewis 1982). The USBM evaluated lighting systems for continuous miners and roof bolters (Chironis 1974), longwall faces (Bockosh 1976), conventional mining equipment, and for low coal (Bockosh and Murphy 1976). These evaluations included several types of new lighting technologies such as mercury-vapor and fluorescent systems. In all of these trials, the placement of the lights, the distribution of light, the durability and reliability of the systems, safety concerns, and several other issues were addressed. The USBM also did work with equipment manufacturers investigating the factory integration of lighting systems into equipment as opposed to after-market addition of these lights (Corbitt 1977) and the design and development of permissible DC power supply systems for face lighting systems (Conroy 1976).

Additionally, mining companies, equipment manufacturers and lighting companies independently experimented with the installation of machine mounted lighting systems (Kreutzberger 1977; Patts 1977; Slone 1977). The performance of these systems was scrutinized by the mining companies in terms of durability, light output, ease of installation and maintenance and acceptance by the workers.

Advances Using Computers

Evaluation of new illumination systems was a time-consuming and expensive proposition because it often required construction of full-scale machine mockups (Figure 5) in a simulated mine entry. Thus in 1977, the USBM initiated a cooperative project with the Mathematical Applications Group, Inc. to develop numerical models to simulate machine geometry and luminaire characteristics so that computer-based evaluations could be conducted (Novak and Goldstein 1978). This approach required the use of a goniometer to make experimental measurements of representative samples of luminaires. The numeric model ran on a mainframe computer. Users connected to the mainframe via terminals. While the new approach was validated via comparisons of measurements taken at the Pittsburgh Research Lab using machine mockup installations to the new values obtained by simulation, the mainframe-based numerical model did not find widespread use. By the 1990s, the PC-based software Crewstation Analysis Program (CAP) for the design and evaluation of machine lighting systems for underground

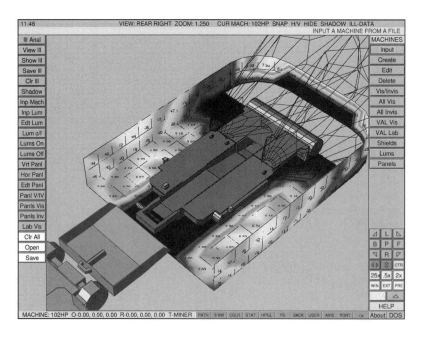

Figure 6. The CAPS software, still in use today, enables a more effective method of evaluating lighting. Source: NIOSH mining photo archive.

coal equipment had been developed by the USBM (Gallagher, Mayton, Unger, Hamrick and Sonier 1996). The CAP software added much more functionality, and was much more user-friendly than the earlier mainframe-based software developed in the late 1970s. Figure 6 depicts what a user of the CAP software would see on their PC. The CAP software drastically changed the way in which underground lighting systems are designed and evaluated, thanks to USBM research on computerized evaluation of mine illumination systems. CAP is still in use today by mine lighting manufacturers and by MSHA. The software is specified in MSHA criteria to acquire lighting survey data for MSHA's Statement of Test and Evaluation applications.

Changes in Safety Rules, Regulations and Guidelines

Safety rules, regulations and guidelines were re-evaluated and changed several times throughout the 20th century. These changes were guided by research efforts investigating the minimum light levels needed by workers, the problems of glare, background illumination and other issues. Some of the guiding research was completed by BCR (du Breuil 1977) and Perceptronics (Crooks and Peay 1982) on USBM contracts.

In 1973, the Mining Enforcement and Safety Administration (MESA) was formed as a new agency in the Department of the Interior. MESA assumed the functions of safety and health enforcement that had previously been carried out by the USBM. In 1977, this agency was moved to the Department of Labor and was renamed the Mine Safety and Health Administration (MSHA). Both MESA and MSHA were responsible for officially setting illumination standards, inspecting lighting installations, issuing equipment approvals and determining violation of illumination standards. The efforts of MSHA during the 1970s to provide guidelines for the design of lighting systems on mobile underground equipment are described in (Wright 1982).

The USBM published an Underground Coal Mine Lighting Handbook in 1982 (Lewis 1986; Lewis 1986). This handbook was prepared to act as a complete reference for underground coal mine operators and equipment designers. The handbook includes sections on the history of underground lighting, technical considerations, light physics, light and vision relationships and the issues of glare.

IMPACTS OF LEGISLATION

Two major pieces of legislation were passed during the second half of the 20th century, which impacted mine illumination. These are the Federal Coal Mine

Health and Safety Act of 1969 and the Federal Mine Safety and Health Act of 1977.

Federal Coal Mine Health and Safety Act of 1969

Regulations for coal mine illumination were established of the *Coal Mine Health and Safety Act of 1969*, Public Law 91-173, also referred to as the *Coal Act*. Section 317 (e) stated that "Within nine months after the operative date of this title, the Secretary (of the Interior) shall propose the standards under which all working places in a mine shall be illuminated by permissible lighting, within eighteen months after the promulgation of such standards, while persons are working in such places. ." Working places were defined as those areas inby the last open crosscut. The USBM was charged with establishing the standards for minimum illumination levels. A notice of proposed rulemaking was published in the Federal Register during December, 1970. This notice prescribed a required minimum illumination of the least 5 foot-candles and no more than 110 foot-candles in all working places. At this time, mine officials and mining equipment manufacturers stated that research and experimentation was needed to scientifically determine if this required minimum illumination was valid, and if the technology existed to achieve it. Numerous public hearings were held in which comments and objections were voiced. Given the concerns and objections of the mining industry, the regulations were temporarily suspended and the USBM and MESA were charged to determine if 5 foot candles was a valid, practical and attainable lighting level. The USBM started two research projects with the National Bureau of Standards (predecessor of the National Institute of Standards and Technology), and the Naval Ammunition Depot to determine minimum levels of illumination (Vines, Klouse, Bockosh, Slone, Evans and Beckett 1977).

In the report by the Naval Ammunition Depot in Crane, Indiana (Hitchcock 1973), 11 jobs (e.g., cutter operator, shuttle car operator, etc.) encompassing the work completed from the working face to the first transfer point were analyzed for both continuous and longwall mining methods. From analysis of these jobs, a total of 114 "visual tasks." were identified and analyzed. Some examples of the visual tasks analyzed include "read dial on methane meter." and "determine position of boom in relation to rib." Data for each task was taken including identification of the item viewed, description of the background, distance to the item viewed, and the time taken to study the item. In addition, the visual properties of several mines were studied to quantify visual parameters such as reflectivity of coal and roof slate, reflectivity of machinery, and light transmission through the

mine atmosphere. Based on these measurements and observations, a simulated mine entry and mining equipment models were constructed with similar visual properties. The visual tasks were then evaluated using the Illuminating Engineering Society (IES) method for prescribing illumination using a contrast reducing visibility meter (Visual Task Evaluator). The threshold visibility luminance for each visual task was set to encompass a 99% probability of detecting the task and to encompass 95% of a 55–65 year-old population with 50% of the measurement variation included. The results of this study showed that there is little difference in the light requirements between jobs. Based on the results of this report, a minimum illumination level of 0.06 foot lamberts was recommended.

In October 1976, the regulations were promulgated which required 0.06 foot lamberts and also specified areas to be illuminated , and the industry was given till July, 1978 to come into compliance with these regulations.

Both prior to and following the announcement of these regulations, the USBM and MESA conducted a number of seminars to educate the industry about the standards and how they apply (Blakely 1976; Chironis 1976; Klouse 1976; Lester 1976; Lester 1976; Crawford 1978). About 44 one-day seminars were given to about 1000 people. The regulations prompted several mining companies to take action to insure their operations were in compliance. For example, Westmoreland Coal constructed an illumination laboratory for testing of equipment on site (Slone 1977).

Federal Mine Safety and Health Act of 1977

With respect to the illumination of underground work areas, the Federal Mine Safety and Health Act of 1977, public Law 95-164 was an extension of the 1969 Coal Act to metal and nonmetal mining. Overall, this act of 1977 established a single piece of comprehensive legislation for coal and metal nonmetal mining. The procedures for determining lighting system acceptance according to these regulations are presented in (Huffman 1982). Described are the methods for determining if systems meet the three requirements of being accepted as permissible, providing adequate illumination, and minimizing discomfort and disability glare.

Shortly after the 1977 act was introduced, the USBM began illumination research for metal/nonmetal mines. The focus of this work was on the establishment of minimal luminance requirements. The research was empirically based on human subject detection of hazards common to metal/nonmetal mines (Merritt 1983). Over 13,000 measurements of

Table 2. Summary of the major milestones and research of the USBM with respect to illumination

Year (s)	Event or Research Area
1910	U.S. Bureau of Mines
1912–1913	Research proves exposed bulb filaments can ignite mine gas
1913–1917	Electric lamp performance and safety research
1913–1973	USBM conducts approvals of electric lamps
1916–1926	Developed safety rules and guidance for electric lights
1935	Guidance for maintenance and inspection of electric cap lamps
1935	Guidance for maintaining permissibility
1969	Development of mine illumination requirements for coal mines
1976–78	Mine lighting conferences and seminars given to over 1000 people
1980	Prototype cap lamp with NiCd battery and tungsten-halogen bulbs
1982	Guidance developed: Underground coal mining lighting handbook
1983	Development of mine luminance requirements for metal/non metal mines
1995	USBM develops the CAP software for PC-based analysis of lighting
1995	End of USBM

visual test performance were used for this research. As a result, various ranges of minimum luminance for safety were recommended for common visual tasks encountered in metal and nonmetal mining. The visual tasks included roof crack detection, rock motion detection, floor and rubble detection, peripheral motion detection, and protrusions detection. Recommended lighting practices were also established for cap lamps, machine mounted luminaires, and portable/fixed luminaires.

MINE ILLUMINATION RESEARCH TODAY

NIOSH carries on the tradition of leading mine illumination research to improve the safety of miners. Today, the new lighting technology is the light emitting diode (LED) and it is poised to revolutionize mine illumination. White LEDs are achieving about 110 lm/W in comparison to, in general, about 15 lm/W for an incandescent bulb. They are robust because they do not have a glass envelope or filament that can break, and they can provide useful light in excess of 50,000 hours of operation as compared to about 1000 to 3000 hours for an incandescent bulb.

NIOSH researchers are investigating the human factors issues and technological aspects of LEDs to improve the safety of miners. Human factor issues of age-related visual performance have been investigated for detection of mine hazards. The results of a comparative study using an incandescent cap lamp, a commercially-available LED cap lamp, and a NIOSH prototype LED cap lamp indicated significant improvements for older subjects when using the NIOSH prototype LED cap lamp. Moving hazard detection improved 15.0%, trip hazard detection

improved 23.7%, and discomfort glare was reduced 45.0% (Sammarco, Reyes and Gallagher 2009). LED cap lamps have also been investigated to better understand the technology to ensure proper application for mining (Sammarco, Freyssinier, Zhang, Bullough and Reyes 2008). NIOSH research is expanding into machine-mounted LED lighting for metal and nonmetal mines and for coal mines.

SUMMARY

USBM mine illumination research addressed many issues including permissibility, human factors aspects of visual performance, electric lamp power sources, and new lighting technologies as summarized by Table 2. Over the years, the USBM research successfully addressed unresolved issues and enabled significant impacts in the mining industry. Based on the USBM research that established that a lamp's glowing filament could cause an ignition, the design and construction of incandescent bulbs was modified and improved to minimize the danger of causing an explosion in the event a bulb was broken. The USBM paved the way for acceptance of electric lighting in the mining industry by establishing that electric lighting was safe and reliable. The USBM led research of new lighting technologies such as tungsten-halogen bulbs and fluorescent lighting, as well as lighting hardware for new lighting technologies for machine-mounted luminaires. Cap lamp improvements included Ni-Cd battery technology that reduced battery weight by 48%. Many of the technologies the USBM researched are still in use today as well the computer-based software CAP that is still the de-facto tool used to evaluate mine illumination. The federal regulations for mine illumination we use today are still based on the luminance requirements established by the USBM.

DISCLAIMERS

Mention of any company or product does not constitute endorsement by NIOSH. The findings and conclusions in this report are those of the authors and do not necessarily represent the views of the National Institute for Occupational Safety and Health.

REFERENCES

Blakely, J.W. 1976. Problems Remain in Underground Lighting; New Regulations Forthcoming. *Coal Mining & Processing*: 58.

Bockosh, G.R. 1976. A Case Study of Longwall Illumination. *Proceedings: Bureau of Mines Technology Transfer Seminars: Coal Mine Illumination. U.S. Bureau of Mines Information Circular 8709.*

Bockosh, G.R. 1977. The Bureau of Mines Research Program on Mine Illumination. *Papers Presented Before The Third Symposium on Underground Mining*. Louisville, KY: National Coal Association: 394–400.

Bockosh, G.R. and J. Murphy, N. 1976. Low Coal Illumination. *Proceedings: Bureau of Mines Technology Transfer Seminars: Coal Mine Illumination. U.S. Bureau of Mines Information Circular 8709.*

Cash, F.E. 1935. Electric Cap Lamps in Alabama Mines. *U.S. Bureau of Mines Information Circular 6865.*

Chironis, N.P. 1974. Underground Mine Lighting. A look at What's New in Concepts and Equipment. *Coal Age*: 66–76.

Chironis, N.P. 1976. New Lighting Regulations to Spark Improved Safety, Production for Coal Mines. *Coal Age*

Clark, H.H. 1914. Permissible Electric Lamps for Miners. *U.S. Bureau of Mines Technical Paper 75.*

Clark, H.H. and L.C. Ilsley 1913. Ignition of Mine Gases by the Filaments of Incandescent Electric Lamps. *U.S. Bureau of Mines Bulletin 52.*

Clark, H.H. and L.C. Ilsley 1917. Approved Electric Lamps for Miners. *U.S. Bureau of Mines Bulletin 131.*

Clark, H.H. and C.M. Means 1916. Suggested Safety Rules for Installing and Using Electrical Equipment in Bituminous Coal Mines. *U.S. Bureau of Mines Technical Paper 138.*

Conroy, G.J. 1976. Operating Lighting Units on Coal Mine DC Power Systems. *Proceedings: Bureau of Mines Technology Transfer Seminars: Coal Mine Illumination. U.S. Bureau of Mines Information Circular 8709.*

Corbitt, S.W. 1977. Factory Integration of an Illumination System into a Continuous Mining Machine. *Papers Presented Before the Third Symposium on Underground Mining*. Louisville, KY: National Coal Association: 401–414.

Crawford, J.W. 1978. Impact of U.S. Legislation on Equipment Manufacturers. *Colliery Guardian International.*

Crooks, W.H. and J.M. Peay 1982. Definition of Illumination Requirements for Underground Metal and Nonmetal Mines. *Proceedings: Second International Mine Lighting Conference of the International Commission on Illumination (CIE). U.S. Bureau of Mines Information Circular 8886.*

duBreuil, F. 1977. BCR Program in Mine Illumination. *Papers Presented Before the Third Symposium on Underground Mining*. Louisville, KY: National Coal Association: 385–388.

Gallagher, S., A.G. Mayton, R.L. Unger, C.A. Hamrick and P. Sonier 1996. Computer Design and Evaluation Tool for Illuminating Underground Coal-Mining Equipment. *J Illum Eng Soc* 25 (1): 3–12.

Hitchcock, L.C. 1973. Development of Minimum Luminance Requirements for Underground Coal Mining Tasks: Research and Development Department, Naval Ammunition Depot.

Hooker, A.B. and D.H. Zellers 1935. Factors that Decrease the Light of Electric Cap Lamps. *U.S. Bureau of Mines Report of Investigations 3292*

Howell, S.P., A.C. Fieldner and L.C. Ilsley 1922. Permissible Explosives, Mining Equipment, and Apparatus, Approved Prior to March 15, 1922. *U.S. Bureau of Mines Technical Paper 307.*

Huffman, F.M. 1982. Underground Lighting System Acceptance Procedures. *Proceedings: Second International Mine Lighting Conference of the International Commission on Illumination (CIE). U.S. Bureau of Mines Information Circular 8886.*

Ilsley, L.C. 1920. State Mining Laws on the Use of Electricity in and about Coal Mines. *U.S. Bureau of Mines Technical Paper 271.*

Ilsley, L.C. and A.B. Hooker 1930. Permissible Electric Mine Lamps. *U.S. Bureau of Mines Bulletin 332.*

Ilsley, L.C., A.B. Hooker and W.H. Roadstrum 1942. Investigations of Permissible Electric Mine Lamps 1930–40. *U.S. Bureau of Mines Bulletin 441.*

Klouse, K. 1976. Measurement Procedures. *Proceedings: Mureau of Mines Technology Transfer Seminars: Coal Mine Illumination. U.S. Bureau of Mines Information Circular 8709.*

Kreutzberger, J. 1977. Illumination Experiences at Pensylvania Mines Corporation. *Papers Presented Before The Third Symposium on Underground Mining.* Louisville, KY: National Coal Association: 368–377.

Lester, C.E. 1976. Mining Enforcement and Safety Administration Review of New Rulemaking in Mine Illumination. *Proceedings: Bureau of Mines Technology Transfer Seminars: Coal Mine Illumination. U.S. Bureau of Mines Information Circular 8709.*

Lester, C.E. 1976. New Underground Lighting Regulations and How they Apply. *Coal Mining & Processing*: 66–69, 100.

Lewis, W.H. 1982. Overview of the U.S. Bureau of Mines Illumination Research Program, 1981. *Proceedings: Second International Mine Lighting Conference of the International Commission on Illumination (CIE). U.S. Bureau of Mines Information Circular 8886.*

Lewis, W.H. 1986. Underground Coal Mine Lighting Handbook (In Two Parts) 1. Background. *U.S. Bureau of Mines Information Circular 9073.*

Lewis, W.H. 1986. Underground Coal Mine Lighting Handbook (In Two Parts) 2. Application. *U.S. Bureau of Mines Information Circular 9074.*

Merritt, J., Perry, T., Crooks, W., Uhlaner, J 1983. Recommendations for minimal luminance requirements for metal and nonmetal mines. U.S. Bureau of Mines Contract Report.

Novak, T. and R. Goldstein 1978. Evaluation of Mine Illumination System using Numerical Modeling. *Proceedings of the Fourth WVU Conference on Coal Mine Electrotechnology*, Morgantown, WV: University of West Virginia

Ocean Energy Inc. 1975. Development of a Fluorescent Cap Lamp System: Ocean Energy, Inc.

Patts, L.D. 1977. Results of Consol's Ongoing Program on Mine Illumination. *Papers Presented Before the Third Symposium on Underground Mining.* Louisville, KY: National Coal Association: 378–383.

Sammarco, J., J. Freyssinier, X. Zhang, J. Bullough and M. Reyes 2008. Technological Aspects of Solid-State and Incandescent Sources for Miner Cap Lamps. *IEEE Industry Applications Society 43rd Annual Meeting.* Edmonton, Canada: IEEE.

Sammarco, J., M. Reyes and S. Gallagher 2009. Do Light-Emitting Diode Cap Lamps Enable Improvements in Miner Safety? *Society for Mining, Metallurgy, and Exploration 111th Annual Meeting.* Denver, CO: Society for Mining, Metallurgy, and Exploration.

Slone, R.E. 1977. Westmoreland Coal Company's Approach to Mine Illumination. *Papers Presented Before The Third Symposium on Underground Mining.* Louisville, KY: National Coal Association: 360–367.

Trotter, D.A. 1982. *The Lighting of Underground Mines.* Trans Tech Publications.

USBM 1926. Safety Rules for Installing and Using Electrical Equipment in Coal Mines. *U.S. Bureau of Mines Technical Paper 402.*

Vines, R., K. Klouse, G.R. Bockosh, R.E. Slone, G. Evans and G. Beckett 1977. Panel Discussion—Illumination. *Proceedings of the 8th Annual Institute on Coal Mining Health, Safety and Research*, Blacksburg, Virginia: Virginia Polytechnic Institute and State University.

Wright, O.J. 1982. Efforts to Design Lighting Systems into Underground Mining Equipment. *Proceedings: Second International Mine Lighting Conference of the International Commission on Illumination (CIE). U.S. Bureau of Mines Information Circular 8886.*

Zellers, D.H. and A.B. Hooker 1935. Maintaining the Permissibility of Electric Cap Lamps. *U.S. Bureau of Mines Information Circular 6832.*

U.S. Bureau of Mines/NIOSH Mining Electrical Safety Research: A Legacy of Protection Against Shock, Fires, and Explosions

Michael R. Yenchek
NIOSH, Pittsburgh, Pennsylvania, United States

Gerald T. Homce
NIOSH, Pittsburgh, Pennsylvania, United States

ABSTRACT: This paper reviews the 100-year history of federal electrical safety research in the U.S mining industry, originally by the US Bureau of Mines, and as carried on today by NIOSH. First, there is a brief historical review of the use of electrical power in mining and an examination of the associated shock, fire, and explosion hazards. These hazards were the impetus for government intervention to improve the safety of U.S mineworkers. The evolution of electrical power systems and equipment in mining then serves as a backdrop for a review of continued electrical safety research. The paper concludes with a recap of the contributions made to mine worker electrical safety by federal research efforts, and thoughts on where the future will lead such research under NIOSH.

INTRODUCTION

This paper reviews the 100-year history of federal electrical safety research in the U.S mining industry, originally by the US Bureau of Mines, and as carried on today by NIOSH. First, there is a brief historical review of the use of electrical power in mining and an examination of the associated shock, fire, and explosion hazards. These hazards were the impetus for government intervention to improve the safety of U.S mineworkers. The evolution of electrical power systems and equipment in mining then serves as a backdrop for a review of continued electrical safety research. The paper concludes with a recap of the contributions made to mine worker electrical safety by federal research efforts, and thoughts on where the future will lead such research under NIOSH.

Early in its history, the lack of federal safety regulations forced the Bureau into an advisory role for the mining industry. Since the Bureau had no regulatory powers, it relied principally on publications, such as technical papers and bulletins, to enlighten the industry about potential hazards. These suggestions were to be the precursors for the mandatory rules and regulations that later governed the use of electricity in mining. Many of the "good practices." were the result of surveys of operators and manufacturers as was the case for the installation of cables in shafts and boreholes (Ilsley and Gleim 1932). At that time, several states had enacted laws governing mining equipment based upon the recommendations of the Bureau of Mines. It was only later, in 1946, that a federal safety code for bituminous coal and lignite mines was enacted. Ultimately, the Federal Coal Mine Safety Act of 1952 provided for annual inspections and gave the Bureau limited enforcement authority. However, it was the passage of the Coal Mine Safety and Health Act of 1969 that greatly influenced and expanded the Bureaus research, with much work being accomplished through grants and contracts. In 1973, the Bureau of Mines electrical safety functions associated with equipment approvals transferred to the Mining Enforcement and Safety Administration (MESA, later MSHA).

The Bureau of Mines was actively involved in electrical safety research since its inception in 1910. An electrical section was organized in 1909 under the technologic branch of the U.S. Geological Survey (Ilsley 1923) and continued under the Bureau. The act establishing the Bureau detailed its scope, which included in part, investigating the methods of mining especially in relation to the use of explosives and electricity (Powell 1922). Technical Paper 4 (Clark 1911), describing the purpose and equipment of the Bureau's electrical section stated that, "There is a distinct branch of electrical engineering in connection with the equipment of mines. The conditions in mines are very different from those on the surface. Not only is it more difficult to install and properly maintain electrical equipment underground, but, unless suitable precautions are observed, the presence of such equipment in mines adds danger to a calling already hazardous." The paper also acknowledged the cooperation of various state inspection departments and equipment manufacturers in safeguarding life and property underground.

By 1910, electricity was already being used underground for haulage, lighting, driving pumps, fans, drills, coal cutting machines, and hoists, for detonating explosives, and for signaling. Both direct and alternating current were used, the former much more extensively. Direct current (DC) was distributed at up to 600 volts, while alternating current (AC) at over 2,000 volts was carried a short distance underground to feed motor-generator sets or rotary converters. Explosives could be detonated by batteries, magneto generators, or power circuits. Signals, which may include lights, bells, and telephones, were operated principally from batteries.

Technical Paper 4 (Clark 1911) recognized the temporary nature of electrical service at many points and the fact that many workers did not appreciate its inherent danger. These dangers from electricity were classified as shock, fire, and explosions. Shock hazards included the trolley wire and other bare conductors along with ungrounded equipment. Fires could result from grounds to coal, flashing of motors, and short circuits. Explosions could originate from arcing and sparking that could ignite methane or dust accumulations.

The electrical section of the Bureau was originally a part of the experimental station in Pittsburgh, PA and was initially staffed by an electrical engineer, an assistant electrical engineer, a junior electrical engineer, and an electrical engineering aid (Powell 1922). The section's purpose (Clark 1911) was to "attempt to discover the causes of accidents from the use of electricity in mines and to suggest means for the prevention of such accidents; to make tests of the safety of electrical equipment under conditions most conducive to disaster; and to make such tests of a general nature as bear upon the safety of electricity in mines, accidents attributed to electrical causes, etc." The laboratory, initially at the Government Arsenal at Fortieth and Butler Streets in Pittsburgh, was primarily for testing mining equipment. It featured direct current capability up to 750 volts and alternating current up to 2000 volts for power, and 30,000 volts for high potential tests. Two galleries were used for testing electrical equipment in the presence of explosive gas. The smaller consisted of a boiler-iron box with connections for gas and air, heavy plate-glass observation windows, and pressure-relief vents. Incandescent lamps and other small items were evaluated in this chamber. The larger was a tube designed to represent a short section of a mine entry, thirty feet long and 10 feet in diameter, and was used for tests of larger apparatus such as motors and switches (Clark 1911). This facility can be contrasted with the present-day Mine Electrical Laboratory, which is located at the NIOSH Pittsburgh Research Laboratory. Established in the early 1980s, it can support full-scale testing of mine electrical power system components, and over the years has been used for projects examining mining safety issues such as fuse performance, electric motor failure, trolley system faults, electric motor circuit protection, electrical cable performance, and overhead power line hazards. It is equipped with electrical power supplies capable of delivering voltage up to 100 kV AC or DC, and current up to 50 kA AC, as well as a secure control room from which tests and experiments can be controlled and observed safely. It has also been used to support Mine Safety and Health Administration investigations. A small rail haulageway at Lake Lynn was used for trolley system research.

This research history is organized by major types of electrical apparatus, i.e., explosion-proof enclosures, motors, trailing cables, trolley wires, etc. Within each category, the work is listed chronologically along with mine technological developments. Space limitations precluded inclusion of some related areas such as communications, lasers, and instrumentation as well as research that was applicable mainly to non-mining industries.

PERMISSIBLE EQUIPMENT

Electrical equipment used in potentially methane-laden atmospheres must be housed in explosion-proof (X/P) enclosures or be of intrinsically-safe design, so as to have insufficient energy, under normal or faulted conditions, to ignite a methane-air mixture. The earliest electrical safety investigations involved the explosion hazards associated with various types of electrical apparatus already in use in gaseous underground coal mines. These included portable electric mine lamps, motors, storage-battery locomotives, mine lamp cords, gas detectors, danger signals, and coal-cutting apparatus. A series of schedules, published by the Bureau of Mines, specified the evaluation fees and requirements for the use of such equipment in gaseous mines and further stated that the Bureau would give its seal of approval to acceptable apparatus as being *permissible*. Schedule 2, approved by the Secretary of the Interior, October 26, 1911 was the first schedule issued and covered testing of motors. The Bureau published revised requirements (Schedule 2A) for test and approval of explosion-proof electric motors in 1915 (Clark 1915). This included tests where the motor casing was filled and surrounded with the "the most explosive mixture of Pittsburgh natural gas and air." Subsequent tests would prove that the results obtained with natural gas and pure methane were essentially the same (Leitch, Hooker, and Yent 1925). The motor would be operated at rated speed and the gas mixture ignited by a spark inside the casing. Gas mixtures would be varied and coal dust added in over fifty tests prescribed to determine the greatest internal enclosure pressure.

Occasionally excessive pressures, due to "pressure heaping." would be recorded especially where two internal compartments were joined via a small interconnecting passage (Gleim 1929). The conditions leading to these excessive pressures were better defined in a subsequent study (Gleim and Marcy 1952). A metallic foam material, investigated for potential use in large volume enclosures as a vent and flame arrestor, performed satisfactorily for methane-air mixtures (Scott and Hudson 1992).

Tests were also conducted on broken incandescent lamps to relate newly introduced tungsten filament temperatures to the "fire damp." (methane-air) gas ignition threshold (Clark 1912) and to show that broken lamps can ignite methane-air mixtures (Clark and Ilsley 1913). These tests also showed that higher wattage carbon filament lamps were necessary to ignite a given methane-air mixture compared with tungsten filaments. They were followed up by tests of electric lamps for miners for permissibility according to Schedule 5, a key criterion of which was that (Clark, 1914) "under no circumstances can the bulb of a completely assembled lamp be broken while the lamp filament is glowing at a temperature sufficient to ignite explosive mixtures of mine gas and air." Electric cap lamps were introduced and approved under Schedule 6. Yet in 1935, more than half of the Nation's miners still used open carbide lights for illumination (Zellers and Hooker 1935).

The first electrically powered coal mining machine, the coal cutter, was seeing widespread use by the 1920s (Morley 1990). Tests to determine the permissibility of cutting machine components were conducted in a new gallery in Pittsburgh (at that time located on Forbes Avenue adjacent to Carnegie Mellon University) which was designed jointly by the Bureau and the Jeffrey Manufacturing Company (Ilsley and Gleim 1920). This chamber shown in Figure 1 (Ilsley, Gleim, and Brunot 1941) consisted of a cylindrical steel tank with an internal diameter of 10 feet and a height of 4½ feet. Four observation windows were spaced equally around the chamber.

The Bureau recognized the potential fire and explosion hazards from batteries used in mine locomotives. Investigations indicated that the safety of the locomotive would depend largely on the proper care of the batteries. Therefore, mechanical endurance tests were devised to prove their adequacy and later were incorporated into Schedule 15 (Ilsley and Brunot 1922). The first storage-battery approval was issued in 1921 (Brunot and Freeman 1923).

Figure 1.   Explosion-proof motor testing gallery, Pittsburgh (Ilsley and Gleim 1920)

Permissible equipment and explosives lists were published annually as Technical Papers. Periodically guides, based upon the latest approval schedules, were issued to facilitate the field inspection of permissible equipment (Gleim 1932; Gleim 1954) and, for machine designers (Gleim, James, and Brunot 1955). By 1923, a number of electric apparatus had been listed as approved including: 5 electric cap lamps, 20 coal-cutting machines, 4 electric drills and 5 battery locomotives (Ilsley 1923). Despite the publication of approval requirements, the widespread use of unapproved equipment in underground mines persisted into the 1920s (Ilsley 1922). From 1910 to 1924, there were 499 fatalities in disasters and fires caused by ignitions from electrical apparatus and circuits (Ilsley 1924). However, by the end of the decade there was a noticeable increase in the use of permissible equipment, even though maintaining them in permissible condition was still a problem (Gleim and Freeman 1932). By the mid 1930s rock-dust distributors, air-compressors, pumps, and room hoists were added to the classes of permissible equipment.

In the early 1980s, a guide was prepared for the design of explosion-proof electrical enclosures. It addressed those aspects of design that affected enclosure strength and ruggedness. Materials such as polycarbonates, as well as adhesives and sealants, which may be degraded by the mine environment, were also analyzed (Staff 1982). Research into the by-products of electrical arcing in X/P enclosures found that hydrogen and acetylene can be formed as certain organic insulators decompose. Research produced acceptance criterion for adhesives and sealants and a design guide for explosion-proof enclosures (Cox and Schick 1985). Performance-based design and test recommendations were published for enclosures containing high-voltage components (Berry and Gillenwater 1986). These recommendations were referenced by MSHA to help formulate new regulations governing the use of high-voltage longwall machines in coal mines (Federal Register 2002).

Intrinsic safety research concentrated on basic studies into ignition mechanisms. Factors influencing the minimum igniting current level were identified. Early work focused on capacitive discharges (Lipman and Guest 1959). Later, spark ignition probabilities were defined for various electrode materials and for resistive, inductive, and capacitive circuits under varying temperature, humidity, and atmospheric conditions (Cawley 1988; Peterson 1992).

TRAILING CABLES

Initially most of the inspection, testing, and approval of electrical equipment concentrated on the explosion-proof integrity of enclosures. However, trailing cables had been employed since shortly after the turn of the century and the Bureau of Mines generally accepted the cables that were specified by the manufacturer for a particular machine. One of their first uses was in connection with cable-reeled, rail-mounted locomotives. These cables, featuring a weatherproof braid, had a short life and were largely superseded by rubber-sheathed cables in the 1920s when three types of cables were in general use: a concentric 2-conductor, a flat or twin duplex, and a round triplex (Ilsley, Hooker and Coggeshall 1932). By 1930, a new emphasis was being placed upon the dangers from wiring and connections outside the enclosures (Ilsley and Brunot 1931). Field inspections had shown that the workmanship of splices in concentric cables was typically poor and that they were seldom vulcanized to exclude moisture (Ilsley and Hooker 1929). Schedule 2C, published in 1930, specified a performance test for two-conductor trailing cables subjected to abuse due to being run over by rail-mounted cars and locomotives (Ilsley 1930). The performance of twin cables was found in tests to be far superior to the concentric design. The overheating of rubber-insulated reeled cables was also investigated by electrical loading and measurement of resultant temperature rise. Revised ampacities were established for various cable sizes (Ilsley and Hooker 1931).

The standard procedure for locating faults in trailing cables into the 1970s was to apply a high level of electric power to blow a hole in the cable at the fault site, a crude and unsafe procedure. The Bureau developed safe, portable devices that locate faults accurately and rapidly based upon infrared, electromagnetic, and time domain reflectometer (TDR) techniques. Three manufacturers marketed electromagnetic devices based upon the original Bureau design (Bureau of Mines 1981).

Industry cable splicing practices were studied in a 1960 survey of 440-V and 4,160-V cables at an Indiana surface mine (Douglas 1960). A follow-up survey of industry practices uncovered deficiencies in cable repair and care (Williams and Devett 1962). In the late 1970s, cable-related research concentrated on improved splicing procedures for trailing cables (Staff 1984). Programs were initiated to redesign splice kits to better protect spliced cables from short circuits and moisture intrusion and to increase mechanical strength. Improved crimp connectors and joining sleeves were designed. This work culminated in splicing workshops conducted for industry personnel and the effort saw 12 major kit manufacturers improve their products based upon Bureau work. The operating characteristics of repaired (spliced) portable power cables were later compared with undamaged cables (Yenchek, Schuster, and Hudson 1995) and deficiencies in the thermal rating of splice kit insulating tapes were noted.

In the 1980s, in cooperation with MSHA, the American Mining Congress and the Insulated Cable Engineers Association, the Bureau initiated an extensive investigation into the operational characteristics of trailing cables powering coal mining machinery. Initially a relationship was established between electric current and resulting temperature rise of the cable conductors (Yenchek and Kovalchik 1989; Yenchek and Kovalchik 1991). These extensive laboratory tests were the basis for the subsequent development of a computerized thermal model for cables operating in open air (Yenchek and Cole 1997). Machine designers could apply this interactive program to determine anticipated cable temperatures given the electrical load requirements of new machines. Accelerated life tests of cable insulation and jacket materials were conducted to determine the long-term effects of elevated temperatures (Yenchek and Kovalchik 1989). Arrhenius models were constructed that related cable thermal life to operating temperature for the cable materials. A correspondence was then established between current load and useful thermal life (Yenchek and Kovalchik 1993). Using this methodology and knowing the electrical characteristics of a particular machine, a designer

Figure 2.   Cable reel thermal test in the Mine Electrical Laboratory

could theoretically predict and compare the available service lives of different cable sizes. Finally, research was conducted to measure the temperature of cables wrapped on shuttle car reels as shown in Figure 2 (Kovalchik, Scott, Dubaniewicz, and Duda 1999). A novel method was devised to measure internal cable temperatures with an imbedded optical fiber. This work allowed for a more accurate calculation of ampacity derating factors for reeled cables, especially under cyclic loads.

ELECTRICAL POWER SYSTEMS

In 1916 in cooperation with professional societies, operators, and equipment manufacturers, the Bureau of Mines published a suggested guide for the installation and use of underground and surface electrical equipment in bituminous coal mines (Clark and Means 1916). In 1923 a suggested guide for electrical safety inspection was published (Ilsley 1923) and in 1926 one for the design of accessories used in permissible equipment (Ilsley and Gleim 1926). A subsequent publication covered surface as well as underground installations and recommended the National Electric Code as a suitable reference (Gleim 1936). Ultimately an electrical safety standard, co-sponsored by the American Mining Congress and the Bureau, was published as a supplement to federal and state regulations (American Standard 1952) and was periodically revised.

Periodically the Bureau would publish warnings to the industry about dangerous electrical system practices. In 1929 a Bureau publication pointed out a hazard associated with equipment use, that of cables and equipment overheating due to low voltage on direct current systems as a result of insufficient copper and poor rail bonding (Ilsley 1929). Incredibly, voltages measured at equipment in some mines were as low as 40 to 80 volts on 250 DC systems! Yet, equipment operating at voltages greater than 250 volts was viewed as potentially more hazardous and difficult to maintain as permissible (Ilsley 1932). Although most mines in the U.S. were still using DC to power machinery in 1960, the use of alternating current was growing due to concerns with fires and voltage regulation on DC systems, concurrent with the lifting of AC distribution voltage limits underground by individual states (Morley 1990). Also current interruption was far easier with AC. The Bureau, recognizing this trend, published recommendations for the use of AC, and endorsed the wye-connected, resistance-grounded neutral transformer secondary configuration due to its inherent safety advantages (Staff 1960). In the 1970s a series of Application Notes (Figure 3) on a variety of aspects of mine power system operation were produced for distribution to the mining community. In 1990, a comprehensive engineering reference was published and was used to instruct university students on the design of mine power systems (Morley 1990). More recently, NIOSH, in an analysis of underground explosions, showed that many were caused by electrical equipment in intake air entries. This led to recommendations to apply the hazardous location provisions of the National Electrical Code in these areas (Dubaniewicz 2007).

GROUNDING

A good grounding system is paramount for the protection of personnel in a mine or mineral processing facility. Prompted by unwillingness on the part of the mining industry to provide adequate grounding of equipment enclosures in the 1920s, a compilation of best practices was published to promote the

Figure 3.   Loose-leaf power system handbook

Figure 4. Ground bed design workshop at Pittsburgh Research Laboratory, 1980

importance of grounding mine electrical systems to minimize shock hazards (Ilsley 1930). Bureau of Mines engineers characterized the effectiveness of ground rods, machines in earth contact, and rail through resistance measurements (Griffith and Gleim 1943). The effectiveness of drill-hole casings as a grounding medium was confirmed through field measurements (Griffith and Gleim 1944). A study, conducted out of the Bureau's Birmingham, Alabama office, showed how the impedance offered by moist earth could increase over time when conducting ground fault currents (McCall and Harrison 1952). To assist in mines in complying with the requirements for a low-resistance grounding medium, the Bureau undertook research to systematize the design of ground beds based upon soil and other environmental conditions. This work led to publication of a guide that greatly simplified the design of driven-rod ground beds (King, Hill, Bafana, and Cooley 1979; Staff 1979) along with workshops (Figure 4) for industry personnel. Other ground bed subjects investigated were ground bed measurement, corrosion of ground rods, and composite bed materials (Staff 1985).

The 1969 Act required a device to monitor ground conductor continuity to prevent accidental electrocution resulting from high frame potentials. Bureau research produced such a device, which served as an industry prototype (Staff 1979). Contractual research in the 1970s developed a method to monitor safety grounding diodes on underground DC circuits, quantified earth contact resistance for various electrodes, and devised a prototype device to measure ground return resistance for open-pit mine machines (Bennett, Sima, and King 1978).

ELECTRICAL MAINTENANCE

Accidents associated with the maintenance of electrical equipment, especially troubleshooting and repair, have, over the years, been a major cause of injuries in mines. Design practices, including dead-front construction, voltage segregation, and interlocks, were identified as means to minimize shock during control box maintenance (Staff 1982). Beginning in the 1970s and continuing well into the 1990s the Bureau of Mines sponsored and conducted research into the early prediction of insulation failure in electrical components, focusing primarily on electric motors. Research efforts concentrated on sophisticated computer-based analysis of motor input power measurements during normal operation, and included contract research (Kohler and Trutt 1987), in-house work at the Pittsburgh Research Laboratory's (PRL) Mine Electrical Laboratory (Homce 1989), cooperative work with the Navy, and extensive field data collection (Homce and Thalimer 1996). The serious problem of electrical burns due to arc flash incidents in mining was addressed by identifying and promoting prevention methods used successfully in other industries (Figure 5) (Homce and Cawley 2007). As part of this work, a safety training package, titled "Arc Flash Awareness." was released on DVD. Thousands of copies were distributed directly, and it had extensive promotion, distribution, and use by other agencies and organizations, including translation into Spanish by the Electrical Safety Foundation International (Kowalski-Trakofler, Barrett, Urban, and Homce 2007).

PROTECTION DEVICES

Devices such as fuses, circuit breakers, vacuum interrupters, and relays are use to protect against fire and shock. A 1951 study out of the Bureau of Mine's Birmingham, Alabama office showed that a fuse in combination with a short-circuiting contactor could provide effectively the same protection against short circuits as a more expensive circuit breaker (Harrison 1951). A resettable current-limiting prototype was built and tested to minimize arcing during interruption of DC currents (Hamilton and Strangas 1978). A portable calibrator was developed for on-line use with DC circuit breakers (Bureau of Mines 1981). In the 1980s, in efforts to improve personnel shock protection, the Bureau investigated the application of sensitive ground fault relays (GFR's) to mine power circuits (Trutt, Kohler, Rotithor, Morley, and Novak 1988). Mine-worthy prototype devices with sensitivities in the milliampere range were designed and

Figure 5. Appropriate arc flash personal protective equipment demonstrated by NIOSH researcher

constructed for both AC and DC circuits (Figure 6) (Yenchek and Hudson 1990). NIOSH investigated how the starting of large AC induction motors might cause nuisance tripping of short-circuit protection on coal mine power systems. A method was devised to provide short-circuit protection without intentional time delays to account for motor starts (Yenchek, Cawley, Brautigam, and Peterson 2002). A recent study concluded that the significant capacitance of shielded cables on mine distribution systems can have a detrimental effect on high voltage ground fault relay selectivity (Sottile, Gnapragasam, Novak, and Kohler 2006).

TROLLEY WIRES

The fire and shock dangers associated with the use of bare trolley wires for rail coal haulage had long been recognized by the Bureau of Mines. A 1932 Bureau study showed that of 71 electrical fatalities in one state during a 5-½ year period, 55 were from contact with trolley wires operating at 250 volts or less. The hazards associated with the use of bare trolley wires were tied to poor installation, lack of guarding, inadequate maintenance, and back-poling (Gleim 1932). During a 19-month period, in 1943–44, 101 mining fatalities resulted from four fires that originated on the trolley system, as the coal industry attempted to meet the demands of war production. A contemporary

Figure 6. Sensitive AC GFR's installed in Safety Research Coal Mine

Figure 7.   Tests on DC haulageway at Lake Lynn Laboratory

Bureau publication suggested the elimination of bare, single-wire trolley by the application of two-wire overhead systems (eliminating the track as a return), circuit breakers that would discriminate between heavy fault and load currents, and diesel or battery-powered locomotives (Griffith, Gleim, Artz, and Harrington 1944). In the early 1980s the Pittsburgh Research Center demonstrated technology that could detect such faults (Staff 1982). Since this approach required additional wiring along the length of the haulage way, it was never adopted by the industry. A final attempt to develop a 'smart' circuit breaker for trolley haulage involved tests at cooperating mines to develop a neural-network detection algorithm based upon the rectifier current signature (Figure 7) (Peterson and Cole 1997). A simple, low-cost means to detect hot trolley insulators was prototyped (Yenchek and Hudson 2000). Although successful, the smart circuit breaker and hot insulator detector were never deployed due to the industry-wide transition from rail to belt for coal haulage.

SURFACE MINING

The electrification of shovels and draglines began before 1920 (Morley 1990). As horsepower requirements increased, so did distribution and utilization voltages. An area of interest to the Bureau of Mines and later, NIOSH, was the electrical shock hazard from accidental contact of overhead electrical power lines, both at surface mines and the surface facilities of underground mines. Such accidents usually involve high reaching mobile equipment like cranes or other equipment such as scaffolds and ladders and are a leading cause of electrical fatalities at mine sites. Early contract work thoroughly analyzed the problem, and looked at available warning-device solutions (Hipp, Henson, Martin, and Phillips 1982) (Morley, Trutt, and Homce 1982). NIOSH developed and patented (Sacks, Cawley, Homce, and Yenchek 1999) a practical low-cost concept to detect line contact and emit a warning (Sacks, Cawley, Homce, and Yenchek 2001). More recent research has evaluated currently available warning devices (Figure 8) (Homce, Cawley, and Yenchek 2008).

ACCIDENT ANALYSES

Periodically, analyses of accidents were conducted which showed a continued need for research aimed at improving electrical safety in mines. A 1936 analysis of electrical accidents in coal mines showed that electricity ranked fourth as a cause of coal mine fatalities, excluding deaths from fires and explosions of electrical origin (Harrington, Owings, and Maize 1936). Ninety percent of electrocutions were caused by contact with the bare trolley wire. Forty-one percent of the explosions for the period 1928–1935, resulting in 476 deaths, were caused by the electrical ignition of gas or coal dust. These were mainly due to the use of open-type equipment or poorly maintained permissible equipment. From 1930–35,

Figure 8. Full scale power line proximity alarm test at Pittsburgh Research Laboratory

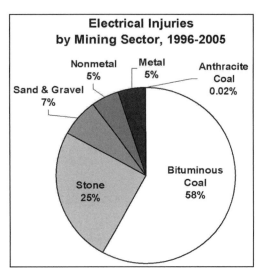

Figure 9. Electrical accident data analysis

fires of electrical origin were the cause in 29% of the fatalities from all mine fires (Harrington, Owings, and Maize 1936). An analysis of electrical accidents in the late 1970s and Bureau research yielded a cost-benefit analysis for electrical safety projects (Tenney, Cooley, and Elrazaz 1982). Recent NIOSH studies of occupational electrical injuries have indicated that, despite dramatic improvements over the years, mining has an electrical injury rate that is still much higher than many other industries. These studies, in addition to highlighting the need for continued improvements in mining electrical safety, were very influential in the broader electrical safety community. The first (Cawley and Homce 2003) was described as a "landmark research study." and "one of the most, if not the most, comprehensive analyses of occupational electrical injuries ever undertaken." in the April 2004 *IEEE Industry Applications* magazine. The second was hailed as "seminal research." in the May/June 2007 edition of the same publication (Cawley and Homce, 2006). Within mining (Figure 9), electricity continued to rank fourth as a cause of death (Cawley 2003). Burns were found to be the leading type of injury, but were rarely fatal. Electrical shock was the overwhelming cause of electrical fatalities.

SUMMARY AND FUTURE DIRECTION

The Bureau of Mines/NIOSH electrical safety research program has focused on a wide variety of topics over its 100-year history. Research has been conducted to address safety issues associated with advancing technology as well as to develop new solutions prompted by regulatory mandates. Future priorities will be driven by surveillance, customer/

stakeholder input, and risk analyses and should include overhead power line, electrical maintenance, and lightning hazards. The technical feasibility of an improved power line proximity warning system that does not depend on electric field detection should be studied. There will be a continuing need to provide effective training for both experienced and younger, inexperienced maintenance personnel to increase awareness of electrical shock and burn hazards and the need for, and type of, protections available. The procedures for electrical maintenance can be systematized by developing job hazard methodologies for critical tasks. Finally, further investigation is needed in the future to achieve a better understanding of lightning and develop practical guidelines and recommendations that can help mine operators reduce the chance of ignitions and protect workers from the resulting hazards.

REFERENCES

American Standard M2.1. 1952. *American Standard Safety Code for Installing and Using Electrical Equipment in and about Coal Mines.* Bureau of Mines Bulletin 514. Washington, DC: U.S. Government Printing Office.

Bennett, J.L., G.R. Sima, and R.L. King. 1978. Electric Shock Prevention. Proceedings of the Fourth WVU Conference on Coal Mine Electrotechnology, Morgantown, WV, Aug 2–4, 1978.

Berry, D.R. and B. Gillenwater. 1986. Development of a High-Voltage Permissible Load Center (contract H0308093, Foster Miller, Inc.) BuMines OFR 58-86, 136 pp.: NTIS PB 86-215803.

Brunot, H.B and H.B. Freeman. 1923. *Permissible Electric Drills.* Bureau of Mines Reports of Investigations 2434. Washington, DC: Government Printing Office.

Bureau of Mines. 1981. *Technical Highlights: Health and Safety Research, 1970–1980.* Supt. of Docs. No. I 28.151.

Cawley, J.C. 1988. *The Probability of Spark Ignition in Intrinsically Safe Circuits.* Proceedings of the Ninth WVU Mining Electrotechnology Conference. Morgantown, WV, July 26–29. p. 128.

Cawley, J.C. 2003. *Electrical Accidents in the Mining Industry, 1990–99.* IEEE Trans Ind Appl Nov/ Dec 39(6): 1,570–1,577.

Cawley, J.C. and G.T. Homce. 2003. Occupational Electrical Injuries in the U.S., 1992–1998, and Recommendations for Safety Research. *Journal of Safety Research*, Vol. 34 pages 241–248.

Cawley, J.C. and G.T. Homce. 2008. Trends in Electrical Injury, 1992–2002. *IEEE Transactions on Industry Applications*, Vol. 44, No. 4.

Clark, H.H. 1911. *The Electrical Section of the Bureau of Mines, Its Purpose and Equipment.* Bureau of Mines Technical Paper 4. Washington, DC: Government Printing Office.

Clark, H.H. 1912. *Ignition of Gas by Miniature Electric Lamps with Tungsten Filaments.* Bureau of Mines Technical Paper 23. Washington, DC: Government Printing Office.

Clark, H.H. 1914. *Permissible Electric Lamps for Miners.* Bureau of Mines Technical Paper 75. Washington, DC: Government Printing Office.

Clark, H.H. 1915. *Permissible Explosion-Proof Electric Motors for Mines; Conditions and Requirements for Test and Approval.* Bureau of Mines Technical Paper 101. Washington, DC: Government Printing Office.

Clark, H.H., and L.C. Ilsley. 1913. *Ignition of Mine Gases by the Filaments of Incandescent Lamps.* Bureau of Mines Bulletin 52. Washington, DC: Government Printing Office.

Clark, H.H., and C.M. Means. 1916. *Suggested Safety Rules for Installing and Using Electrical Equipment in Bituminous Coal Mines.* Bureau of Mines Technical Paper 138. Washington, DC: Government Printing Office.

Cox, P.A. and W.R. Schick. 1985. *A Design Guide for Explosion-Proof Electrical Enclosures. Part 3-Final Technical Report.* (Contract H0387009, SW res. Inst.). BuMines OFR 47C-86, 353 pp.: NTIS PB 86-209525.

Douglas, S.J. 1960. *Testing and Splicing Electric Cables and Frame-Grounding Pit Equipment.* Bureau of Mines Information Circular 7995. Washington, DC: Government Printing Office.

Dubaniewicz, T.H. 2007. The Brookwood Disaster and Electrical Requirements for Hazardous (Classified) Locations. Proceedings of the IEEE Industry Applications Society Annual Meeting, N. Orleans, LA: 1,353–1,359.

Federal Register. 2002. *Electric Motor-Driven Mine Equipment and Accessories and High-Voltage Longwall Equipment Standards for Underground Coal Mines.* Vol. 67, No. 47, pp. 10,972–11,005.

Gleim, E.J. 1929. *Abnormal Pressures in Explosion-Proof Compartments of Electrical Mining Machines.* Reports of Investigations 2974. Washington, DC: Government Printing Office.

Gleim, E.J. 1932. *Guarding Trolley Wires in Mines.* Bureau of Mines Information Circular 6577. Washington, DC: Government Printing Office.

Gleim, E.J. 1932. *Notes Pertaining to Safety Inspections of Permissible Electric Mine Equipment.* Bureau of Mines Information Circular 6584. Washington, DC: Government Printing Office.

Gleim, E.J. 1936. *Electrical Viewpoint in a Complete Safety Survey of a Coal Mine.* Bureau of Mines Information Circular 6894. Washington, DC: Government Printing Office.

Gleim, E.J. 1954. *United States Bureau of Mines Standard for Inspection and Test of Explosion-Proof Enclosures for Mining Equipment.* Bureau of Mines Report of Investigations 5057. Washington, DC: U.S. Government Printing Office.

Gleim, E.J. and H.B Freeman. 1932. *Maintenance of Electrical Mine Equipment from the Viewpoint of the Safety Inspector.* Bureau of Mines Technical Paper 537. Washington, DC: Government Printing Office.

Gleim, E.J., R.S. James, and H.B. Brunot. 1955. *Explosion-Proof Design and Wiring for Permissible Mining Equipment.* Bureau of Mines Bulletin 541. Washington, DC: U.S. Government Printing Office.

Griffith, F.E. and E.J. Gleim. 1943. *Grounding Electrical Equipment in and about Coal Mines.* Bureau of Mines Report of Investigation 3734. Washington, DC: Government Printing Office.

Griffith, F.E. and E.J. Gleim. 1944. *Ground-Resistance Measurements of Drill-Hole Casings.* Bureau of Mines Report of Investigations 3756. Washington, DC: Government Printing Office.

Griffith, F.E., E.J. Gleim, R.T. Artz, and D. Harrington. 1944. *Prevention of Fires Caused by Electric Arcs and Sparks from Trolley Wires.* Bureau of Mines Information Circular 7302. Washington, DC: Government Printing Office.

Hamilton, H.B. and E. Strangas. 1978. *Implementation of a High-Speed, Heavy Current, DC Switching System.* BuMines OFR 56-78. 77 pp. NTIS: PB 283 388/AS.

Harrington, D., C.W. Owings, and E.R. Maize. 1936. *Some Suggestions on the Prevention of Electrical Accidents in Coal Mines.* Bureau of Mines Information Circular 6919. Washington, DC: Government Printing Office.

Harrison, L.H. 1951. *The Short-Circuiting Contactor as an Electrical Protective Device for Coal Mine Service.* Bureau of Mines Report of Investigation 4759. Washington, DC: Government Printing Office.

Hipp, J.E., F.D. Henson, P.E. Martin, and G.N. Phillips, G.N. 1982. *Evaluation of Proximity Warning Devices.* Bureau of Mines Final Report, Contract JO188082. Washington, DC: Superintendent of Documents.

Homce, G.T. 1989. *Predicting the Failure of Electric Motors.* Bureau of Mines Information Circular 9211. Washington, DC: U.S. Government Printing Office.

Homce, G.T. and J.C. Cawley. 2007. *Understanding and Quantifying Arc Flash Hazards in the Mining Industry.* Proceedings for the 2007 Institute of Electrical and Electronic Engineers Industry Applications Society conference, New Orleans, LA.

Homce, G.T. and J.R. Thalimer. 1996. Reducing Unscheduled Plant Maintenance Delays—Field Test of a New Method to Predict Electric Motor Failure. *IEEE Transactions on Industry Applications*, Vol. 32, No. 3, pp. 689–694.

Homce, G.T., J.C. Cawley, and M.R. Yenchek. 2008. *A Performance Evaluation of Overhead Power Line Proximity Warning Devices.* NIOSH Information Circular 9510. Pittsburgh, PA: DHHS (NIOSH) Publication No. 2009-110.

Ilsley, L.C. 1922. *Ten Reasons Why Permissible Equipments Approved under Schedules 2B and 15 are Safer than Unapproved Types.* Bureau of Mines Reports of Investigations 2398. Washington, DC: Government Printing Office.

Ilsley, L.C. 1923. *Electrical Safety Inspection: Suggestions for Mine-Safety Engineers.* Bureau of Mines Reports of Investigations 2541. Washington, DC: Government Printing Office.

Ilsley, L.C. 1924. *Hazards of Electric Sparks and Arcs in Coal Mines.* Bureau of Mines Reports of Investigations 2626. Washington, DC: Government Printing Office.

Ilsley, L.C. 1929. *Hazards from Low or Under Voltage.* Bureau of Mines Information Circular 6201. Washington, DC: Government Printing Office.

Ilsley, L.C. 1930. *Performance Tests for Trailing Cables.* Bureau of Mines Information Circular 6263. Washington, DC: Government Printing Office.

Ilsley, L.C. 1930. *The Grounding of Electric Systems in and around Mines.* Bureau of Mines Information Circular 6318. Washington, DC: Government Printing Office.

Ilsley, L.C. 1932. *250 versus 500 Volts or More for Circuits in Gassy Mines.* Bureau of Mines Information Circular 6556. Washington, DC: Government Printing Office.

Ilsley, L.C. and H.B. Brunot. 1922. *Endurance Tests of Storage Batteries for Use in Permissible Mine Locomotives.* Bureau of Mines Reports of Investigations 2358. Washington, DC: Government Printing Office.

Ilsley, L.C. and H.B. Brunot. 1931. *Suggestions for Wiring Permissible Equipment.* Bureau of Mines Information Circular 6463. Washington, DC: Government Printing Office.

Ilsley, L.C. and E.J. Gleim. 1920. *Approved Explosion-Proof Coal-Cutting Equipment.* Bureau of Mines Bulletin 78. Washington DC: Government Printing Office.

Ilsley L.C. and E.J. Gleim. 1926. *Suggestions for the Design of Electrical Accessories for Permissible Mining Equipment.* Bureau of Mines Bulletin 258. Washington, DC: Government Printing Office.

Ilsley, L.C. and E.J. Gleim. 1932. *Data in Reference to Installation of Cables in Shafts and Boreholes.* Bureau of Mines Information Circular 6595. Washington, DC: Government Printing Office.

Ilsley, L.C., E.J. Gleim, and H.B. Brunot. 1941. *Inspection and Testing of Mine-Type Electrical Equipment for Permissibility.* Bureau of Mines Information Circular 7185. Washington, DC: Government Printing Office.

Ilsley, L.C. and A.B. Hooker. 1929. *A Preliminary Investigation of Rubber Sheathed Concentric-Type Trailing Cables for Mining Machines.* Bureau of Mines Reports of Investigations 2909. Washington, DC: Government Printing Office.

Ilsley, L.C. and A.B. Hooker. 1931. *The Overheating of Rubber-Sheathed Trailing Cables.* Bureau of Mines Report of Investigations 3104. Washington, DC: Government Printing Office.

Ilsley, L.C., A.B. Hooker and E.J. Coggeshall. 1932. *Rubber-Sheathed Trailing Cables.* Bureau of Mines Bulletin 358. Washington, DC: Government Printing Office.

James, R.S. 1950. *Effect of Temperature on Flame-Arresting Properties of Flat Joints in Explosion-Proof Mine Equipment.* Bureau of Mines Report of Investigations 4639. Washington, DC: Government Printing Office.

King, R.L., H.W. Hill, Jr., R.R. Bafana, and W.L. Cooley. 1979. *Guide for the Construction of Driven-Rod Ground Beds.* Bureau of Mines Information Circular 8767, 26 pp.

Kohler, J.L. and F.C. Trutt. 1987. *Prediction of Incipient Electrical-Component Failures.* Bureau of Mines Final Report, Contract JO338028. Washington, DC: Superintendent of Documents.

Kovalchik, P.G, L.W. Scott, T.H. Dubaniewicz, and F.T. Duda. 1999. Dynamic Temperature Measurement of Overheated Shuttle Car Trailing Cables in Underground Coal Mines. *Trans. Soc. Min. Metal Explor*. 306: 1–4.

Kowalski-Trakofler, K., E. Barrett, C. Urban, and G.T., Homce. 2007. *Arc Flash Awareness.* DVD training package. DHHS (NIOSH) Publication No. 2007-116D

Leitch, R.D., A.B. Hooker, and W.P. Yent. *The Ignition of Firedamp by Exposed Filaments of Electric Mine Cap-Lamp Bulbs.* Bureau of Mines Reports of Investigations 2674. Washington, DC: Government Printing Office.

Lipman, K. and P.G. Guest. 1959. *Measurement of Electrical Waveforms of Low-Energy Spark Discharges.* Bureau of Mines Report of Investigations 5460. Washington, DC: U.S. Government Printing Office.

McCall, M.C. and L.H. Harrison. 1952. *Some Characteristics of the Earth as a Conductor of Electric Current.* Bureau of Mines Report of Investigations 4903. Washington, DC: U.S. Government Printing Office.

Morley, L.A. 1990. *Mine Power Systems.* Bureau of Mines Information Circular 9258. Washington, DC: U.S. Government Printing Office.

Morley, L.A., F.C. Trutt, G.T. Homce. 1982. *Electrical Shock Prevention, Volume IV—Overhead-Line Contact Fatalities.* Bureau of Mines Final Report, Contract JO113009. Washington, DC: Superintendent of Documents.

Peterson, J.S. 1992. Influence of Electrode Material on Spark Ignition Probability. Bureau of Mines Report of Investigation 9416. Washington, DC: U.S. Government Printing Office.

Peterson, J.S. and G.P. Cole. 1997. *Detection on Downed Trolley Lines Using Arc Signature Analysis.* NIOSH Report of Investigations 9639. Pittsburgh, PA: DHHS (NIOSH) Publication No. 97-123.

Powell, F.W. 1922. *The Bureau of Mines, Its History, Activities, and Organization.* The Institute for Government Research Monograph No. 3. New York and London: D. Appleton and Co.

Sacks, H.K., J.C. Cawley, G.T. Homce, and M.R. Yenchek. 1999. Alarm System for Detecting Hazards Due to Power Line Transmission Lines. U.S. Patent No. 6,600,426.

Sacks, H.K., J.C. Cawley, G.T. Homce, and M.R. Yenchek. 2001. Feasibility Study to Reduce Injuries and Fatalities Caused by Contact of Cranes, Drill Rigs, and Haul Trucks with High Tension Lines. *IEEE Trans Ind Appl*, 37(3): 914–919.

Scott, L.W. and A.J. Hudson. 1992. Performance of RETIMET Metal Foam Vents on Explosion-Proof Enclosures. Bureau of Mines Report of Investigations 9410. Washington, DC: Superintendent of Documents.

Sottile, J., S.J. Gnapragasam, T. Novak, and J.L. Kohler. 2006. Detrimental Effects of Capacitance on High-Resistance-Grounded Mine Distribution Systems. IEEE Trans Ind Appl Sep-Oct; 42(5): 1,333–1,339.

Staff, Bureau of Mines. 1982. Underground Coal Mine Power Systems, Proceedings: Bureau of Mines Technology Transfer Seminar, Pittsburgh, Pa. Bureau of Mines Information Circular 8893. Sup. of Docs, 88 pp.

Staff, Bureau of Mines. 1984. *Splicing Mine Cables.* Bureau of Mines Handbook 1-84, 92 pp.

Staff, Bureau of Mines. 1985. *Earth Grounding Beds-Design and Evaluation. Proceedings: Bureau of Mines Technology Transfers, Pittsburgh, PA and Reno, NV.* Bureau of Mines Information Circular 9049, 24 pp.

Staff, Health and Safety. 1960. *Recommended Standards for Alternating Current in Coal Mines.* Bureau of Mines Information Circular 7962. Washington, DC: U.S. Government Printing Office.

Staff, Mining Research. 1979. *Mine Power Systems Research (In Four Parts). 2. Grounding Research.* Bureau of Mines Information Circular 8800, 78 pp.

Staff, Mining Research. 1979. *Mine Power Systems Research (In Four Parts). 3. Circuit Protection.* Bureau of Mines Information Circular 8801, 148 pp.

Tenney, B.S., W.L. Cooley, and Z. Elrazaz. 1982. *Coal Mine Electrical Accident Reduction.* Proceedings of the WVU Conf. on Coal Mine Electrotechnology, Morgantown, WV; July 28–30, 1982, p. 137.

Trutt, F.C., J.L. Kohler, H. Rotithor, L.A. Morley, and T. Novak. 1988. *Sensitive Ground Fault Protection for Mines. Phase III: Coordination-Free GFR's for AC Distribution* (contract J0134025, PA State Univ.). 95 pp.

Williams, F.A. and W.R. Devett. 1962. *Prevailing Practices in Splicing Electric Trailing Cables in Coal Mines.* Bureau of Mines Information Circular 8090. Washington, DC: Government Printing Office.

Yenchek, M.R., J.C. Cawley, A.L. Brautigam, and J.S. Peterson. 2002. Distinguishing Motor Starts from Short Circuits through Phase Angle Measurements. *IEEE IAS Transactions*, Vol. 38, No. 1.

Yenchek, M.R. and G.P. Cole. 1997. Thermal modeling of portable power cables. *IEEE Transactions on Industry Applications* 33(1):72–79.

Yenchek, M.R. and A.J. Hudson. 2000. A Passive Means to Detect Hot Trolley Insulators. *Min Eng* 2000 Jan; 52(1): 43–47.

Yenchek, M.R. and A.J. Hudson. 1990. *The Bureau of Mines Ground-Fault Protection Research Program-A Summary.* Bureau of Mines Information Circular 9260, 23 pp.

Yenchek, M.R. and P.G. Kovalchik. 1989. Thermal characteristics of energized coal mine trailing cables. *IEEE Transactions on Industry Applications* 25(5):824–829.

Yenchek, M.R. and P.G. Kovalchik. 1989. Mechanical performance of thermally-aged trailing cable insulation. *IEEE Transactions on Industry Applications* 25(6):1,000–1,005.

Yenchek, M.R. and P.G. Kovalchik. 1991. Thermal characteristics of energized shielded and an unshielded reeled trailing cable. *IEEE Transactions on Industry Applications.* 27(4):791–796.

Yenchek, M.R. and P.G. Kovalchik. 1993. The impact of current load on mine trailing cable thermal life. *IEEE Transactions on Industry Applications* 29(4):762–767.

Yenchek, M.R., K.C. Schuster, and A.J. Hudson. 1995. *Operational characteristics of trailing cable splices.* Bureau of Mines, Report of Investigations 9540. Pittsburgh, PA: U.S. Department of the Interior

Zellers, D.H. and A.B. Hooker. 1935. *Maintaining the Permissibility of Electric Cap Lamps.* Bureau of Mines Information Circular 6832. Washington, DC: Government Printing Office.

Environmental Research by the U.S. Bureau of Mines

L. Michael Kaas

Arlington, Virginia, United States

ABSTRACT: The U.S. Bureau of Mines launched its formal research program in Minerals Environmental Technology in 1979. This was largely in response to the Nation's expanding environmental awareness during the 1960s and 1970s, a host of new environmental and land-use laws and regulations, and the need for improved technologies to deal with the complex pollution control and environmental remediation problems at active and abandoned mining and metallurgical operations. In reality, activities related to the environment were conducted for much of the Bureau's 86 years as the Nation's principal minerals research and information agency.

INTRODUCTION

When the U.S. Bureau of Mines (USBM) was created in 1910, the term "environmental research" was not in our lexicon. In this paper we will see how, from its very beginning, the Bureau was engaged in a variety of activities that would today be considered environmental research. We will describe how the Bureau utilized its expertise gained from decades of research programs to formally respond to the strong National commitment to environmental protection, from the 1960s and 1970s, until it was closed 1996. And finally, we will also look at several of the Bureau's important environmental accomplishments. The references provided give a small sampling of the many Bureau publications on environmental topics.

EARLY ENVIRONMENTAL RESEARCH

The Bureau was created to improve the safety and health of coal miners and to reduce the horrific rate of mining fatalities. However, in 1913, amendments to the Bureau's Organic Act greatly expanded its functions to include mineral processing, utilization, conservation, and economic development. (Holmes 1914) The term "conservation" would evolve over the years to include environmental problems under the Bureau's purview. In the years ahead, other laws would strengthen the Bureau's ability to develop technology for environmental protection and to contribute to the development of environmental and land-use policy.

Early Mine Subsidence Research

One of the earliest USBM publications dealing specifically with an environmental problem was Bulletin 25, "Mining Conditions under the City of Scranton, Pa." (Griffith and Conner 1912) In the Preface, Joseph Holmes noted that by early in 1911, mining in the 11 anthracite beds beneath the city had removed 160.6 Mt (177 million st) of coal and 39.9 Mt (44 million st) of waste rock (Figure 1). This represented only 30% of the coal reserves and it was in the local and national interest that as much of the remaining coal as possible be conserved for recovery without dangerous subsidence or excess cost.

In 1909, mine subsidence had severely damaged a school, which fortunately was unoccupied at the time. This event raised concern for public safety and the future of the anthracite mining industry. An advisory board was formed to recommend solutions to the subsidence problems to local governmental bodies. Members of the board were well-known engineers, John Hayes Hammond, D.W. Brunton, R.A.F. Penrose, Lewis B. Stillwell, and W.A. Lathrop. The board suggested that William Griffith and Eli T. Conner conduct a consulting study. Both were considered well-qualified engineers with knowledge of mining and geology in the anthracite area.

The consultants made an extensive review of mining and geologic maps and other records, and

Source: Griffith and Conner 1912

Figure 1. Anthracite Mining in Scranton, PA, circa 1910

conducted inspections of the underground mines. The report estimated that 10% of the city was threatened by subsidence, including 3 additional schools. Detailed maps and geologic cross-sections showing the mines and coal seams under the city were included in the report.

The consultants recommended that mine voids be filled by flushing with a mixture of water and sand or other solid materials such as culm (coal waste) or crushed rock, beginning with the most endangered areas. They also recommended that a general policy be adopted to support all mined areas under the city. Bureau tests of fill materials were included in the report.

Decades later in the 1970s and 1980s, long after anthracite mining had stopped, Congress would again call upon the Bureau to conduct a series of demonstration projects that would place mine waste in underground mine voids in an effort to reduce the continuing threat of subsidence.

Early Air Pollution Research

Once it was authorized to conduct investigations to increase efficiency, prevent waste, and conserve mineral resources, the Bureau studied the problems of emissions at smelting and ore roasting plants. In several parts of the country, "metallurgical smoke" was causing considerable friction between the metals industry and agriculture. An early Bulletin, designed for a non-technical audience, discussed the gases and particulates in metallurgical smoke, the industrial control practices at the time, and the legal aspects of the problem, but it made no specific recommendations regarding the steps that should be taken to mitigate the problems. (Fulton 1915)

Damage to agriculture in Solano County, California, from the emissions of the Selby Smelting and Lead Company (Figure 2) had resulted in protracted litigation and a court injunction. In an effort to resolve the problems, the Selby Smelter Commission was established. Joseph Holmes, the first Director of the Bureau of Mines, was designated Chairman of the Commission. (Holmes et al. 1915; Manning 1917) Bureau chemists and other staff who had been engaged in research on smelter smoke, along with a lengthy list of scientists from other organizations, were authorized to provide expertise and technical support to the Commission. The final report of the Commission, published by the Bureau as Bulletin 98, included expert papers and the results of investigations. Included were detailed discussions on the effects of smelter smoke on vegetation, livestock, and soil conditions in the Selby smoke zone. The Commission concluded that while the smelter had been in violation of a court injunction, it had

Source: Holmes et al. 1915

Figure 2.   Shelby Smelter before installation of emission controls, circa 1910

subsequently taken steps to install equipment and change operating practices to reduce emissions and was no longer in violation.

In the 1930s and 1940s, the Bureau would again become involved in the issue of smelter emissions. This time it would be in technical support of the international Arbitral Tribunal set up by the Canadian and U.S. Governments to deal with the damage to vegetation in Washington State caused by the sulfur dioxide from the smelters of the Consolidated Mining and Smelting Company of Canada, Ltd., at Trail, British Columbia. (Dean and Swain 1944)

THE ENVIRONMENTAL MOVEMENT OF THE 1960s AND 1970s

For several decades, including the Great Depression, World War II, and the Korean War, the Bureau's programs continued to include projects that would today be considered environmental research. Other than some state regulations, the mining and metallurgical industries were able to operate in a largely self-regulated manner with respect to the environment. That was about to change.

By the 1960s, the Nation's attitude toward the environment was changing. The environmental laws passed in the late 1960s and 1970s, and their continuing amendments, radically changed the ground rules and increased the regulatory burden for mineral exploration and mining. These laws included:

- National Environmental Policy Act of 1969 (NEPA)
- Clean Air Act of 1970
- Federal Water Pollution Control Act of 1972 (i.e., Clean Water Act)
- Endangered Species Act of 1973

- Safe Drinking Water Act of 1974
- Resource Conservation and Recovery Act of 1976 (RCRA)
- Toxic Substances Control Act of 1976 (TSCA)
- Surface Mining Control and Reclamation Act of 1977 (SMCRA)
- Uranium Mill Tailings Radiation Control Act of 1978 (UMTRCA)
- Archaeological Resources Protection Act of 1979
- Comprehensive Environmental Response and Liability Act of 1980 (CERCLA, i.e., Superfund)

Two examples illustrate the scope of the Bureau's ongoing environmental research during this period. Both projects are somewhat unique because the technology was scaled-up from the laboratory bench, to the pilot plant, and to the commercial/demonstration plant scale.

The Citrate Process

The long-term involvement with air pollution from metallurgical facilities led to an extensive research program on flue gas desulfurization. At the time of the research, fossil fuel-burning power plants, and copper, lead, and zinc smelters were grappling with the control of sulfur dioxide emissions to comply with the Clean Air Act. An important accomplishment of the program was the development of the Citrate Process. (Nissen et al. 1985) In the process, sulfur dioxide is absorbed from cleaned and cooled flue gas with a citric acid solution which is in turn treated with hydrogen sulfide to precipitate a marketable sulfur product. The citrate solution is regenerated for reuse.

Extensive laboratory investigations lead to the design of large scale continuous processes. Pilot plants were operated at two smelters in cooperation with the Magma Copper Company in Arizona, and the Bunker Hill Company in Idaho (Figure 3). Based on the success at the Bunker Hill plant, a demonstration plant was constructed and operated at the George F. Weaton Power Station in Monaca, PA, under a cooperative cost-sharing agreement between the USBM, the Environmental Protection Agency (EPA), and St. Joe Minerals Corporation. Shortly after the plant was constructed, St. Joe shut down its nearby Monaca Zinc Smelter which reduced the electrical load on the power plant. The plant operated intermittently for 3 years but the demonstration was eventually discontinued due to lack of funds. Faced with the costly retrofitting of their aging smelters with pollution controls, U.S. producers eventually

Figure 3.   Citrate process pilot plant, Bunker Hill Smelter, Kellogg, ID

chose to close existing plants and construct new facilities with different metallurgical technologies.

Recovering Mineral Values from Urban Ore

The second example shows how Bureau scientists utilized their knowledge of traditional mineral processing methods and equipment to address a more general problem, the recycling of valuable materials from municipal waste. In 1977, Max J. Spendlove of the Bureau's College Park [MD] Research Center received American Institute of Mining, Metallurgical, and Petroleum Engineers' Environmental Conservation Distinguished Service Award for his research on processes and equipment to treat urban waste and industrial refuse for improvement of the environment and recovery of essential metallic materials. He was credited with coining the phrase "urban ore" for municipal refuse. He directed a series laboratory and pilot plant projects which characterized the contents of waste streams and developed the flowsheets used in some of the earliest commercial material recovery facilities (Figure 4). While many of these pioneering commercial plants experienced operational problems and the ups and the downs of the markets for their recycled material, they were the precursors of today's single stream recycling plants. (Stanczyk and De Cesare 1985)

REINVENTION OF THE BUREAU OF MINES

In the late 1970s, the Bureau of Mines was still reeling from the effects of spinning-off long-established programs in mine safety inspection to the Mining Enforcement and Safety Administration (MESA) and eventually to the Mine Safety and Health Administration (MSHA), fuels minerals information to the Energy Information Administration (EIA),

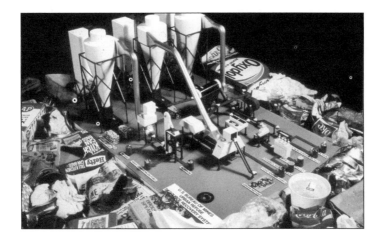

Figure 4. Model of the pilot plant for recovering materials from raw municipal refuse

and research on mining and processing of energy minerals to the Energy Research and Development Administration (ERDA) which became part of the new Department of Energy (DOE) (Figure 5). One component of the Bureau's program that remained was environmental research associated with mining and processing of fuels minerals, metals, and non-metals. (Kaas 2006) The Bureau established a Minerals Environmental Research (MER) program. Because this took place in the middle of the energy crisis, the initial projects were mainly concerned with environmental problems associated with coal mining. The passage of SMCRA had reinforced this approach.

While the funding of the spin-offs increased, the Bureau's funding had remained relatively flat. In 1979, it took a comprehensive look at how its remaining scientists, facilities, and stagnant funding could be redirected to most effectively continue to help meet the mineral needs of the Nation.

Minerals Environmental Technology Research

One of the important parts of the resulting restructuring was the creation of the Bureau's first comprehensive environmental research program, Minerals Environmental Technology (MET). (Wiebmer 1980) The planning for the new program considered the environmental problems associated with present and past mining and processing of all types of mineral commodities. During this same time period, the EPA was gearing-up to regulate non-coal mining and mineral processing wastes under RCRA. Many of the MER projects became part of the resulting MET program which had the following components:

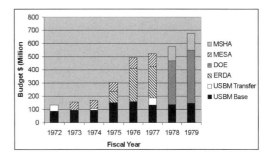

Figure 5. Erosion of USBM functions and funding in the 1970s

Mine Waste Control
Mine waste research focused on control measures to reduce the impacts from mining wastes on surface and groundwater. The effects of coal mine subsidence on aquifers were studied. Research was conducted on control techniques for acid drainage from active and abandoned mines (Figure 6). Tailings dam seepage and control methods were investigated. Cooperative research on underground disposal of coal waste was conducted with the Office of Surface Mining Reclamation and Enforcement (OSMRE) and industry.

Process Waste Control
Processing waste research was directed at finding cost-effective techniques to reduce the environmental impacts of tailings, effluents, and emissions from mineral processing plants. Treatment methods were studied to help curb production of toxic and corrosive wastes. The chemistry of tailings was investigated to improve stabilization and prevent adverse

Figure 6. Neutralizing the acidity of acid mine drainage

effects from weathering. Research was conducted on the removal of fine particles to improve water clarity and permit recycling.

Mined Land Reclamation
Reclamation research included methods to segregate, store, and replace topsoil, and to improve the long-term physical, chemical, and biological stability of reclaimed areas. (Dean et al. 1986a) Revegetation methods were studied for use in western, semi-arid, mined lands and for stabilization of steep slopes such as those found in Appalachia (Figure 7). Research on subsidence prediction techniques used field investigations and case histories from industry coop-erators to provide improved guidance to OSMRE for permitting.

Special Environmental Problems
The variety of problems studied included preven-tion of fugitive dust, noise, and air blast and ground vibrations. Research produced award winning publi-cations on surface mine blasting. (Siskind et al. 1980) Field tests were conducted to find more cost effec-tive methods to control haul road dust. Laboratory research was conducted on the particulate mineral-ogy of asbestos and other minerals.

Elsewhere in the Bureau, metallurgical research continued to study the conservation of mineral resources by improving metal recoveries and foster-ing recycling technologies for industrial and post-consumer waste. (Dean et al. 1986b)

In addition to the research program recom-mended by the Bureau, for several years the Congress provided additional (add-on) funding for demonstra-tion projects that used existing technology to reduce coal mine subsidence by filling the workings with pulverized waste rock. (Walker 1993) The projects included areas in the anthracite region of northeast-ern Pennsylvania where the Bureau had done some

of its earliest environmental research. Other projects studied ways to extinguish or control underground coal fires at abandoned mines. (Chaiken et al. 1983)

Funding for MET research exceeded $20 million in the early 1980s. Congress, the Carter Administration, and the industry were supportive of efforts to become more proactive in address-ing the problems facing mineral producers in the world of greatly expanded environmental regula-tions. Unfortunately, creating a stronger and more visible commitment to environmental research was not without risk. In the mid-1980s, during the first Reagan Administration, environmental research fell out of favor. Perhaps it was a backlash from the aggressive pro-environment policies of the Carter era. Some accused the Bureau of becoming too "green." The results were that the MET Program vanished from Bureau's Fiscal Year (FY) 1984 bud-get, funding environmental projects became very difficult, and many projects were redirected, moth-balled, or eliminated.

Although environmental research was out of favor, Bureau scientists, economists, and commodity specialists continued to work closely with the EPA which was developing new regulations for mining waste under RCRA. Over 43 Gt (48 billion st) of these wastes had accumulated on the landscape and were being generated at the rate of 0.9–2.7 Gt (1–3 billion st) annually. The Bureau wanted to ensure that the final regulations had a sound technical founda-tion. It was the Bureau's position that large volume, low toxicity materials such as overburden, waste rock and tailings should not be considered as hazard-ous waste and should be regulated under Subtitle D rather than under the more stringent Subtitle C.

Environmental Technology Research

Fortunately, the hiatus in environmental research did not last. By the end of the second term of the

Figure 7. Revegetation test plots on tailings

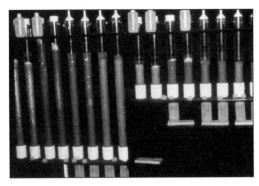

Figure 8. Constructed wetlands to treat acid mine drainage

Figure 9. Column leach tests on mining wastes

Reagan Administration and in the G. H. W. Bush Administration, environmental research was once again a high priority. In the late 1980s, industry was feeling the full economic impact of environmental air, water, and land reclamation regulations. EPA's new mining solid waste regulations were on the horizon. The cost of cleaning-up huge mining and metallurgical Superfund sites posed a substantial liability for some companies. In 1987, the Bureau got back into environmental research by creating the Environmental Technology (ET) program. (Veith and Kaas 1988; Shea-Albin and Fitch 1991) The budget for the revitalized environmental research program again topped $20 million and continued at approximately that level until the Bureau was closed at the end of 1995.

The goals of ET research were to develop technologies and systems to (1) enable cost effective compliance with environmental regulations, (2) clean-up contamination of land and water resulting from past and present mining and mineral processing operations, (3) reduce the quantities of mining and mineral processing waste generated and minimize environmental impacts, and (4) ensure that current and future environmental regulations affecting the minerals industry were based on sound scientific data. The program of ET research is summarized below.

Control of Mine Drainage and Liquid Wastes
Research was designed to minimize or prevent contamination of the Nation's waters by acid mine drainage and toxic constituents from coal, metal, and non-metal mines and waste disposal areas. Abandoned coal mines were generating acid drainage that had polluted over 8,047 km (5,000 mi) of streams in Appalachia. The Bureau perfected the use of constructed wetlands to remove metals from the

Figure 10. Bio-Fix beads for control of hard rock acid mine drainage

effluent. (Hedin et al. 1994) This has become one of the most widely used control methods (Figure 8).

Laboratory studies such as column leach tests were conducted to understand the mechanisms and conditions by which toxic constituents are released from mine waste and tailings (Figure 9).

It was discovered that dead biomass such as sphagnum peat moss could be contained in porous beads (Figure 10). These Bio-Fix beads were demonstrated to be cost-effective for removing cadmium, lead, and other toxic metals from contaminated water at low-flow, hard-rock acid mine drainage sites. (Jeffers et al. 1993).

Solid Waste Management and Subsidence
Research was conducted on the safe disposal of solid wastes from mining and mineral processing operations, minimizing and preventing subsidence, and reclamation of surface mines and waste disposal areas. Extensive field data gathering in eastern and midwestern coal mining areas led to the successful development of subsidence prediction models

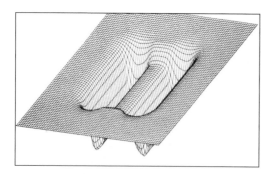

Figure 11. Subsidence models predict surface deformation over longwall panels

Figure 12. Dewatering placer mine discharge after treatment with polyethylene oxide

for longwall and room-and-pillar mining. (Jeran and Adamek 1991) (Figure 11). The software was adapted for personal computers and made available to coal producers (Jeran 1991).

The Bureau worked closely with the Florida phosphate industry to dewater slimes. Slime ponds covered over 405 km^2 (100,000 acres) of land. Technology was developed utilizing environmentally safe polyethylene oxide polymer to flocculate the slimes so that they could be dewatered on screens. (Scheiner et al. 1985) Once shown to be effective in phosphate operations, the technology was taken to Alaska for dewatering placer mine discharge (Sharma et al. 1992) (Figure 12).

Automobile battery casings are a waste product of battery recycling plants. Some of these plants had been designated as Superfund sites. The Bureau developed a process to remove lead from the casings so that they could be safely discarded in landfills. The lead could be removed from solution by electrowinning (Lee et al. 1993).

Elsewhere in the Bureau, pyro-metallurgical research led to the development of techniques by which hazardous materials from municipal waste incinerators could be sealed in glass for safe disposal.(O'Conner et al. 1994) (Figure 13).

Abandoned Mined Land Reclamation Research
A reclamation research program for abandoned mined lands was initiated by OSMRE in FY 1985 and transferred to the USBM in FY 1987. It was designed to primarily benefit state agencies that were attempting to deal with abandoned mined lands and refuse piles, subsidence, poor water quality/acid mine drainage, mine fires, and other environmental problems from past coal mining operations. Projects were proposed by outside organizations and selected by a Review Panel.

Figure 13. Vitrification of hazardous materials from municipal refuse

National Mine Land Reclamation Center
The National Mine Reclamation Center was a consortium of five universities, West Virginia University (the lead school), The Pennsylvania State University, Southern Illinois University, North Dakota State University, and the University of North Dakota. It was created by Congress in FY 1988 and funded under a cooperative agreement with the Bureau. Research conducted through the Center was primarily related to environmental problems in surface and underground coal mining.

The Bureau Becomes a Superfund Potentially Responsible Party

The Bureau itself did not escape the impact of increasing environmental regulations. In 1915, USBM scientists in a small laboratory in Denver had developed the extraction technology to make lower cost radium available to the medical community. Practically overnight, a new industry to recover radium from Colorado carnotite ore was created in the Denver area. (Parsons et al. 1915) The power of the French cartel that had been charging hospitals exorbitant prices for radium was broken. Several plants operated until the 1930s when even lower cost radium became available from the Belgian Congo.

Little did the original Bureau researchers realize that the radioactive tailings from the laboratory would, 70 years later, become part of the EPA's Denver Radium Superfund Site. In the 1980s, the Bureau was named a potentially responsible party for the site of the old laboratory. A brick plant was now on the site and was to be replaced with a shopping center. The Bureau paid for the multi-million dollar clean-up conducted by the EPA.

Land-Use Decisions Concerning the Public Lands

The Nation's increased environmental awareness also signaled a dramatic change in the management of the Public Lands in the lower 48 states and in Alaska. In 1964, the National Wilderness Preservation System was created. Once areas were included in the system they would be off-limits for future exploration and development. A series of laws called for mineral land assessments to support land management decision-making.

* Wilderness Act of 1964
* Forest and Rangeland Renewable Resources Planning Act of 1974
* Federal Land Policy and Management Act of 1976 (FLPMA)
* Alaska Native Claims Settlement Act of 1971 (ANCSA)
* Alaska National Interest Lands Conservation Act of 1980 (ANILCA)

To enable the land managing agencies including the Forest Service and the Bureau of Land Management (BLM) to meet the mandated deadlines for designating lands eligible to be wilderness, the USBM and U.S. Geological Survey (USGS) were directed to conduct joint surveys to assess the mineral values of areas proposed for inclusion in the system. These assessments, although somewhat cursory due to time and funding constraints, identified areas that had the greatest potential for mineral resource development. The land managing agencies would recommend excluding many of them from new wilderness areas, parks, and refuges.

The Bureau was a logical choice for this work. Starting under the Strategic Minerals Act of 1939, it had been actively involved with investigations of mineral deposits for years. In the process of these investigations, comprehensive files had been developed on thousands of known mineral deposits and prospects nationwide. Figure 14 shows the extent of coverage of the files in the Bureau's Field Operations Centers. (USBM Staff 1995) This information helped target areas for the new mineral assessments.

Initially, joint reports on the individual study areas were published by the USBM and the USGS; however, the review process became cumbersome and delayed getting the results to the stakeholders. From the early 1980s, Open File Reports were published by each organization. The Bureau published over 800 reports during that period. (Kaas 1996) Annual funding for Mineral Land Assessments was approximately $7–12 million.

Because of its long history of close cooperation with the Public Land managing agencies, they called upon the Bureau for assistance when they realized the magnitude of the liability for cleaning-up of abandoned mines on the Public Lands. Not only did the Bureau have a computerized inventory of where most of the mines and prospects were located, it also had files on those locations, and technical expertise to help evaluate the potential public safety and/or environmental hazards that might exist. To jump start hazard evaluation efforts, the Bureau distributed a handbook to assist land managers and conducted on-the-job training in the field. (USBM Staff 1994) USBM personnel utilized geographic information systems to assist land mangers and state agencies with studies of watershed and ecosystem areas that had been severely impacted by historic mining. (Coppa et al. 1996) (Figure 15).

Mineral Assessments in Elephant Country

Land management issues were particularly complex in Alaska. At the time of Statehood in 1959, Alaska was primarily Public Land controlled by the Federal Government. Under Statehood, the State of Alaska was permitted to select 421,000 km^2 (104 million acres) of the BLM land. ANCSA allocated 178,000 km^2 (44 million acres) from the unreserved Public Land to the Alaska Native corporations. The Secretary of the Interior was allowed to withdraw 324,000 km^2 (80 million acres) for National Parks, the National Wildlife Refuge System, National Forests, and the Wild and Scenic Rivers System.

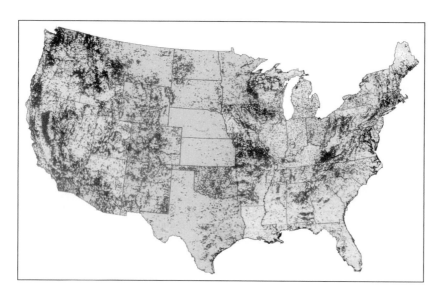

Figure 14. Map of mine, mill, and prospect locations from the Minerals Availability System (MAS)/Mineral Industry Location System (MILS)

Conflicts arose between the Secretary's selections and those of the state and the native corporations. This led to the passage ANILCA in 1980. It added 421,000 km² (104 million acres) to conservation units with 227,000 km² (56 million acres) designated as wilderness and withdrawn from mineral development. Information from USBM mineral land assessments aided the land selection process.

Miners have frequently referred to Alaska as "Elephant Country," where huge mineral deposits were waiting to be discovered. The USBM Alaska engineers and geologists shared that vision. In 1975, while the Bureau was conducting a mineral assessment of lands proposed for the Noatak National Arctic Range, a major mineral discovery was made. (USBM Staff 1975) The prospect was called the Red Dog. When the news release about the discovery hit the street, there was a claim-staking rush to the area. The Canadian mining company, Cominco (now Teck Resources, Ltd.), was able to establish a major land position on the deposit. It took 14 more years for the mine to open.

In 1980, under ANCSA, the Northwest Alaska Native Association (NANA) Regional Corporation selected a 311 km² (120 mi²) block of land that included the Red Dog Deposit. After a lengthy process of negotiations between NANA and Cominco, an agreement was reached to develop the mine. (Figure 16) So far, Cominco/Teck has invested US $1.0 billion of capital in the mine. It has become

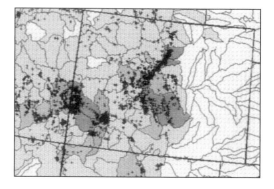

Figure 15. Concentration of mining activity in watersheds on the Public Lands

one of the world's largest producers of zinc concentrate. The workforce numbers 475 of which about half are Inupiaq (Eskimo) Alaska Natives. From the first production in 1989 through the end of 2008, the mine produced approximately 8.7 Mt (9.6 million st) of zinc, 1.7 Mt (1.8 million st) of lead, plus byproduct silver. (Teck Resources 2009) The total value of the lead and zinc produced is estimated to be US$13 billion. In 2009, Teck Alaska, Inc. and NANA announced plans for development of the Aqqaluk Deposit adjacent to the main Red Dog Deposit. This will extend the life of the operation at least until 2031. The total value of mineral production from the

Figure 16.   Red Dog Mine, Alaska (Courtesy of Teck Resources, Ltd.)

Red Dog area will easily surpass the total cumulative funding of the USBM from 1910 to 1996.

The Impact of Land Withdrawals

A USBM 1991 policy analysis of 8 of the western states and Alaska showed that by 1990 of the total land area, 3.5 million km^2 (1.35 million mi^2), 60% (2.1 million km^2 (811,000 mi^2)) was federally controlled. Of the federally controlled land, 20% (439,000 km^2 (170,000 mi^2)) comprised favorable mineral terranes and known mineral deposit areas. Because of land withdrawals and restrictions, only 34% (151,000 km^2 (58,300 mi^2)) of the favorable mineral terranes and known mineral deposit areas on Public Lands were actually available for unrestricted exploration and development. This amounts to only 7% of the total Public Lands. (Hyndman et al. 1991) Since the time of the study, more land has continued to be designated as wilderness each year.

IMPACT OF CLOSURE ON ENVIRONMENTAL RESEARCH

The spin-off of Bureau functions to other government organizations in the 1970s should have signaled that there would be more trouble ahead. The USBM had lost its dominant role in government as the sole provider of minerals research and information. Industry and Congressional support for the Bureau weakened as new agencies became actively engaged in mining matters. Although the Bureau's annual budget in the early 1990s exceeded $100 million, it was dwarfed by the resources available in the DOE, Department of Labor, and EPA.

In 1995, the Congress was seeking to reduce government spending through its Contract with America. The Administration had many unfunded priorities and was looking for ways to redeploy its resources. In the FY 1996 appropriations process, the unthinkable happened. The Secretary of the Interior and the Congress agreed to close the U.S. Bureau of Mines. Traditional USBM supporters did little to stop the process. Closure of the 96 year-old Bureau took place in a mere 90 days. A few programs were transferred to other agencies, but in the end, most of the USBM laboratories and offices were closed, projects, including those in environmental research, were scrapped, and 1,200 Bureau employees were terminated. (Staff 1996)

CONCLUSION

While the U.S. Bureau of Mines and its environmental programs no longer exist, they left behind a legacy of research findings and mineral resource information that will surely continue to pay benefits to the taxpayers for a long time. We have seen how, from its beginning in 1910 and long before the emergence of the environmental movement of the 1960s and 1970s, the Bureau conducted many programs and projects that we would now consider environmental research. We have also seen how, in its last three decades, the USBM expanded its environmental programs to help ensure that the Nation could continue development of its domestic mineral resources. Today, the mineral industries continue to face increasing environmental regulation. Environmental research takes place around the world, albeit in a somewhat fragmented fashion. What is missing since the closure of the Bureau of Mines is an unbiased and authoritative voice for minerals and environmental technology in the Federal Government.

REFERENCES

Chaiken, Robert F., Robert J. Brennan, Bernice S. Heisey, Ann G. Kim, Wilbert T. Malenka, and John T. Schimmel. 1983. *Problems in the Control of Anthracite Mine Fires: A Case Study of the Centralia Mine Fire (August 1980)*. RI 8799. Washington, DC: U.S. Bureau of Mines.

Coppa, Luis V., David A. Ferderer, Indur M. Goklany, and L. Michael Kaas. 1996. Developing Priorities for Verifying and Inventorying Suspected Abandoned Mine Sites on Public Lands. In *Proceedings, 26th International Symposium on the Application of Computers and Operations Research in the Mineral Industry*, September 16–20, 1996, The Pennsylvania State University, University Park, PA. R. V. Ramani, editor. Littleton, CO: SME.

Dean, K. C., L. J. Froisland, and M. B. Shirts. 1986a. *Utilization and Stabilization of Mineral Wastes*. USBM Bulletin 688. Washington, DC: Government Printing Office.

Dean, K. C., J. W. Sterner, M. B. Shirts, and L. J. Froisland. 1986b. *Bureau of Mines Research on Recycling Scrapped Automobiles*. USBM Bulletin 684. Washington, DC: Government Printing Office.

Dean, R. S., and R. K. Swain. 1944. *Report Submitted to the Trail Smelter Arbitral Tribunal*. USBM Bulletin 453. Washington, DC: Government Printing Office.

Fulton, Charles H. 1915. *Metallurgical Smoke*. USBM Bulletin 84. Washington, DC: Government Printing Office.

Griffith, William, and Eli T. Conner. 1912. *Mining Conditions under the City of Scranton, Pa.* USBM Bulletin 25. Washington, DC: Government Printing Office.

Hedin, Robert H., Robert W. Nairn, and Robert L. P. Kleinmann. 1994. *Passive Treatment of Coal Mine Drainage*. IC 9389. Washington, DC: U.S. Bureau of Mines.

Holmes, Joseph A. 1914. The Third Annual Report of the Director of the Bureau of Mines to the Secretary of the Interior for the Fiscal Year Ending June 30, 1913. Washington, DC: Government Printing Office.

Holmes, J. A., Edward C. Franklin, and Ralph A. Gould. 1915. *Report of the Selby Smelter Commission.* USBM Bulletin 98. Washington, DC: Government Printing Office.

Hyndman, Paul C., Roberts, Clark A., Bottge, Robert G., and Barnes, Don J. 1991. The Availability of Federal Mineral Estate in Alaska, Arizona, Colorado, Idaho, Nevada, New Mexico, Oregon, Utah, and Washington. In *Proceedings on Environmental Management for the 1990s*, February 25–28, 1991, Denver, CO. James M. Baker, William M. Greenslade, and Douglas J. Lootens, editors. Littleton, CO: SME. pp. 263–269.

Jeffers, T. H., P. G. Bennett, and R. R. Corwin. 1993. Biosorption of Metal Contaminants Using Immobilized Biomass: Field Studies. RI 9461. Washington, DC: U.S. Bureau of Mines.

Jeran, Paul W. 1991. *SUBPRO (Subsidence Prediction Model Software for Personal Computer).* Pittsburgh, PA: USBM Pittsburgh Research Center.

Jeran, Paul W., and Vladimir Adamek. 1991. *Subsidence Over the End of a Longwall Panel.* RI 9338. Washington, DC: U.S. Bureau of Mines.

Kaas, L. Michael. 1996. *Indices to U.S. Bureau of Mines Mineral Resources Records.* USBM Special Publication 96-2. CD-ROM. Washington, DC: Government Printing Office.

Kaas, L. Michael. 2006. *A History of the U.S. Bureau of Mines, with Some Highlights of its Involvement in Anthracite Mining.* The Mining History Journal. 13:73–92. Golden, CO: Mining History Association.

Lee, A. Y., A. M. Wethington, M. G. Gorman, and V. R. Miller. 1994. Treatment of Lead-Contaminated Soil and Battery Wastes from Superfund Sites. In *Proceedings, Superfund XIV Conference and Exposition*, November 30–December 2, 1993, Washington, DC. Rockville, MD: Hazardous Materials Control Resource Institute. pp. 119–136.

Manning, Van H. 1917. The Yearbook of the Bureau of Mines, 1916. Washington, DC: Government Printing Office.

Nissen, W. I., L. Crocker, L. L. Ogden, and others. 1985. *The Citrate Process for Flue Gas Desulfurization.* USBM Bulletin 686. Washington, DC: Government Printing Office.

O'Conner, William K., Laurence L. Oden, and Paul C. Turner. 1994. Vitrification of Municipal Waste Combustor Residues: Physical and Chemical Properties of Electric Arc Furnace Feed and Products. In *Proceedings of Process Mineralogy XII—Applications to Environment, Precious Metals, Mineral Beneficiation, Pyrometallurgy, Coal, and Refractories.* Warrendale, OH: TMS. Copies were also distributed as USBM OFR 164–94. Albany, OR: USBM Albany Research Center.

Parsons, Charles R., R. B. Moore, S. C. Lind, and O. C. Schaefer. 1915. *Extraction and Recovery of Radium, Uranium, and Vanadium from Carnotite.* USBM Bulletin 104. Washington, DC: Government Printing Office.

Scheiner, B. J. Annie G. Smelley, and D. A. Stanley. 1985. *Dewatering of Mineral Waste Using the Flocculant Polyethylene Oxide.* USBM Bulletin 641. Washington, DC: Government Printing Office.

Sharma, Sandeep K., B. J. Scheiner, and A. G. Smelley. 1992. *Dewatering of Alaska Placer Effluent Using PEO.* RI 9442. Washington, DC: U.S. Bureau of Mines.

Shea-Albin, Valois R., and William N. Fitch. 1991. Mining Waste Research in the U.S. Bureau of Mines. In *Proceedings on Environmental Management for the 1990s*, February 25–28, 1991, Denver, CO. James M. Baker, William M. Greenslade, and Douglas J. Lootens, editors. Littleton, CO: SME. pp. 199–205.

Siskind, D. E., M. S. Stagg, J. W. Kopp, and C. H. Dowding. 1980. *Structure Response and Damage Produced by Ground Vibration from Surface Mine Blasting.* RI 8507. Washington, DC: U.S. Bureau of Mines.

Staff. 1996. *U.S. Department of the Interior Annual Report Fiscal Year 1995.* Washington, DC: Department of the Interior. pp. 49–51.

Stanczyk, Martin H., and Roger DeCesare. 1985. *Resource Recovery from Municipal Waste.* USBM Bulletin 683. Washington, DC: Government Printing Office.

Teck Resources. 2009. *2008 Annual Report* and prior year reports. www.teck.com. Accessed October 2009.

USBM Staff. 1975. *Alaska Mineral Find Reported by Mines Bureau.* News Release, September 3, 1975. Washington, DC: Department of the Interior.

USBM Staff. 1994. *Abandoned Mine Land Inventory and Hazard Evaluation Handbook.* USBM Open File Report 11-95. Spokane, WA: USBM Western Field Operations Center.

USBM Staff. 1995. *Spatial Data Extracted from the Minerals Availability System/Mineral Industry Location System (MAS/MILS).* USBM Special Publication 12-95. CD-ROM. Washington, DC: Government Printing Office.

Veith, David L., and L. Michael Kaas. 1988. Mineral-Related Waste Management in the Bureau of Mines Environmental Technology Program. In *Proceedings—1988 Symposium on Mining, Hydrology, Sedimentology, and Reclamation,* December 5–9, 1988, University of Kentucky, Lexington, KY. Donald H. Graves, editor. Lexington, KY: University of Kentucky. pp. 7–12.

Walker, Jeffrey S. 1993. *State-of-the-Art Techniques for Backfilling Abandoned Mine Voids.* IC 9359. Washington, DC: U.S. Bureau of Mines.

Wiebmer, John D. 1980. The Bureau Rebounds from a Disastrous Decade. *Mining Engineering.* 32(4):391–394. Littleton, CO: SME.

CSIRO Coal Mining Research and Technologies

Hua Guo
CSIRO, Pullenvale, Queensland, Australia

Rao Balusu
CSIRO, Pullenvale, Queensland, Australia

David W. Hainsworth
CSIRO, Pullenvale, Queensland, Australia

Baotang Shen
CSIRO, Pullenvale, Queensland, Australia

Deepak P. Adhikary
CSIRO, Pullenvale, Queensland, Australia

Shi Su
CSIRO, Pullenvale, Queensland, Australia

Jonathan Roberts
CSIRO, Pullenvale, Queensland, Australia

ABSTRACT: Over the last 10 years, CSIRO, through its integrated coal mining research program, has developed a range of advanced mining technologies that have delivered significant benefits to the Australian coal mining industry. This paper provides an overview of these technologies such as longwall automation, highwall mining, dragline automation, coal mine methane drainage, mine fires control, and mine ground water management. This paper also highlights key current coal mining research activities, for example, roadway development, integrated coal and methane extraction, and ventilation air methane capture and utilisation.

INTRODUCTION

The Commonwealth Scientific and Industrial Research Organisation (CSIRO) is Australia's national science agency and one of the largest and most diverse research agencies in the world. CSIRO delivers solutions for agribusiness, energy and transport, environment and natural resources, health, information technology, telecommunications, manufacturing and mineral resources. CSIRO employs a total of 6,500 people in 22 industry sectors and is supported by both government and private funding.

The coal industry is Australia's largest export industry with 2007–2008 exports reaching a record of 252 million tonnes, representing 77% of the country's total black coal production.

One of CSIRO's key objectives is to develop advanced coal production technologies to maximise the value of Australian coal resources in a safe and environmentally responsible way. The CSIRO's coal production research program has an annual budget of about AUD$18 million, involves more than 76 CSIRO researchers, and focuses on four key areas:

- Improving coal mining recovery, efficiency and safety
- Improving coal mining technologies
- Minimising coal mining environmental impacts
- Enhancing coal cleaning

This paper provides examples of CSIRO past and current coal mining technology developments.

MINING TECHNOLOGIES DEVELOPED BY CSIRO

Longwall Automation

Longwall mining is the basis of most of the underground production in Australia and is the focus of several of CSIRO's major research projects.

Automation of the longwall mining process has always held the lure of increased productivity, but more recently has been driven by issues of occupational health and safety. The presence of hazardous gases, respirable dust and the inherent danger of personnel working in close proximity to large mobile mining equipment is increasingly unacceptable.

CSIRO Longwall Automation Project Outcomes
CSIRO's longwall automation research (Reid, 2008), supported by AUD7M of direct industry funding over 7 years through the Australian Coal Industry Research Program (ACARP) has succeeded in:

- Developing automatic face alignment using an inertial navigation-based sensor on the shearer to measure precisely the face geometry and feedback signals used to move roof supports
- Implementing on-line measurement of creep with information incorporated into face alignment correction
- Trial implementation of enhanced horizon control based on the inertial navigation system
- Introducing a broadband communication system to the shearer using wireless Ethernet
- Obtaining new results in locating coal-face features for use in thermal infrared-based horizon control
- Developing a longwall information system (LIS) which integrates information from multiple systems and sensors, and provides high quality visualisation and control interfaces

Visualisation systems incorporating database and graphical user interface software, depicted in Figure 1, allowing multiple users access to tailored information have been developed and an integrated information system to merge automation, geotechnical, mine design and equipment performance has been put in place.

The LASC (Longwall Automation Steering Committee) automation technology has been commercialised, with non-exclusive global licences signed by all of the world's major longwall Original Equipment Manufacturers (OEMs).

CSIRO Longwall Automation Benefits
LASC technology ran in beta form in a number of longwall mines prior to commercialisation. Automatic face alignment was seen as contributing significantly to productivity improvement. At one site, closer maintenance of face profile and the elimination of manual alignment procedures gave an average increase of production of 130tph over a month's operation. Further modelling showed that elimination of manual string line based face alignment alone could enable a longwall to produce a further 435,000 saleable tons per year.

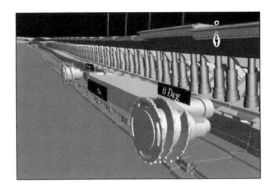

Figure 1.   Longwall visualisation

Highwall Mining

Highwall coal mining is used in open-cut operations, yet involves techniques more akin to underground coal mining. In the Australian context, highwall mining is viable when the cost of overburden removal in strip mining becomes uneconomic as seams dip further from the surface.

Highwall mining systems can be divided into two main categories; those utilising a remotely controlled continuous miner (CM) followed by a coal haulage system (Figure 2) and those based on augering techniques. In both cases the mining operation is controlled remotely and no roof support is installed in the highwall entries. The highwall mining concept is based on a series of parallel, unsupported drives formed with barrier pillars maintained between adjacent holes at some predetermined thickness. Because no active roof support is established, the entries are completed as quickly as possible. Typical figures are 300m penetration in 10–12 hours.

Geomechanics and Layout Design
Pillar/panel stability and span stability are two major geomechanics issues for highwall mining. Comprehensive studies in these areas have been conducted in Australia, led by the CSIRO. Innovative theoretical and numerical developments, together with comprehensive site monitoring and case studies, have been performed by CSIRO (Shen & Duncan Fama 2001). These studies have significantly improved the understanding of how rocks and coal respond to highwall mining. Based on the results of these studies, CSIRO has developed systematic design guidelines that have helped to improve the layout design for enhanced coal recovery with reduced risk (Duncan Fama, Shen & Maconochie 2001).

Figure 2. A continuous miner highwall mining system and typical pillar/entry layout

The Need for Guidance

Optimal recovery depends on tight control of the separation of the highwall drives. Insufficient pillar width can lead to collapse of the drive due to failure of the pillar, while excessive thickness lowers the percentage of the resource that can be extracted. Highly accurate guidance of the remotely controlled process is essential and inertial navigation offers the only practical solution to the problem.

A practical highwall mining guidance method using a ring laser gyro (RLG) based strapdown inertial navigation system (INS) which combines heading deviation and external odometry information to compute three-dimensional position has been developed by CSIRO (Reid, Hainsworth & McPhee 1997)). The INS-based guidance system has been successfully used in two classes of auger highwall mining machines. It has also been successfully used in CM-based highwall mining systems.

Benefits of INS-based Highwall Guidance

CSIRO's techniques have been exported to the USA. One user is the American mining equipment company ICG ADDCAR.

This company has seen benefits in both safety and productivity. The risk of highwall collapse combined with the risk of equipment entrapment has been greatly reduced. The guidance technology has meant the company has been able to confidently increase penetration depths from a maximum of approximately 350m to more than 500m.

Dragline Automation

In open cut coal mining, the rate of overburden removal is almost always the production bottle neck. In most large operations the dragline is used as the primary overburden removal mechanism. Strategies to increase the size and power of draglines appear to have reached their limits and automation is increasingly seen as the next phase in the evolution of dragline efficiency.

Early automation attempts in the late 1980s were based on replaying control signals from an experienced operator. However this is impractical without measurement of the instantaneous bucket swing angle. In 1993, CSIRO researchers funded by ACARP began to investigate bucket swing sensing and the subsequent full automatic control of a dragline's swing motion. Vision systems were found to be ineffective and were replaced by 2D scanning lasers. These sensors were robust to weather, lighting conditions and the mechanically harsh environment experienced at the end of a dragline boom. With this system it became possible to accurately measure the swing angle of the hoist ropes and hence the angle the bucket-rope plane made with the vertical boom plane.

A full scale proof-of-concept automated swing control system was developed on a BE1370 dragline at Tarong coal mine in South East Queensland which performed dig to dump motion. Bucket filling remained the responsibility of the operator. A key component was that the operator could override and regain control at any time.

Three significant conclusions were reached (Corke et al. 2003). Firstly, the system performed more than half its cycles in better time than the average of the operators working over that period. Experienced operators could not tell a manual cycle from an automatic one. Secondly, the performance of the automation system was compromised by the system's lack of accurate terrain information. Thirdly, it was clear that the automation system treated the dragline more gently than human operators.

The lack of real-time awareness was addressed by CSIRO with the development of a Digital Terrain Mapping (DTM) system (Roberts, Winstanley & Corke 2003). Figure 3 shows a DTM produced during trials.

The trial showed that DTMs could be produced routinely in a production environment and that DTMs were also useful to mine planners and surveyors.

In 2005, the CSIRO team demonstrated a fully autonomous tenth-scale dragline that incorporated bucket swing feedback, generation of DTMs, optimal bucket path planning and autonomous digging.

Now that autonomous draglines have been shown to be feasible, the practice of precision dumping has again been raised. The concept of a dragline being capable of accurately dumping its load into a hopper, feeding into a conveyor or into trucks could transform certain mining operations. CSIRO has shown how a tenth-scale model dragline can reliably

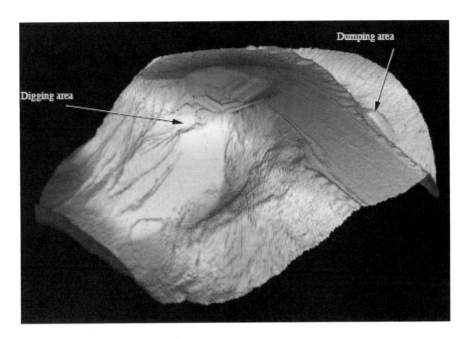

Figure 3. A Digital Terrain Map produced during a four-week production trial

dump material directly into a truck tray (scaled accordingly) (Dunbabin et al. 2006).

Dragline automation is now technically feasible. Economics and the market will determine when it is commercially introduced.

Mine Gas Drainage and Control

High gas emissions from the longwall goaf is one of the main problems at a number of highly gassy coal mines in Australia, resulting in a significant number of face stoppages . To address the issue of goaf gas control, CSIRO's research aimed at developing optimum goaf gas drainage and control strategies for longwall panels. This was successfully carried out by CSIRO with funding from ACARP (Balusu et al. 2004).

Goaf Drainage Research

Initial research showed that longwall gas emissions were increasing substantially over the years despite increasing the number of goaf holes in the panel or decreasing the hole spacing. It was found that the standard practice of draining gas from just one or two goaf holes near the face seemed not to be an efficient strategy to reduce gas delays in the panel. Similarly, bleeder ventilation systems for gas control seemed to be effective only up to 600 to 800m of face retreat from the panel start-up area due to compaction of the goaf.

A series of detailed gas flow characterisation studies, tracer gas investigations and computational fluid dynamics (CFD) modelling investigations were carried out.

The static pressure distribution in the goaf with the traditional practice of two goaf holes gas drainage near the longwall face is presented in Figure 4. Results show that with this type of goaf holes drainage, static pressure in the goaf builds up to 180 Pa. This pressure distribution indicates that all the goaf gas migrates towards the tailgate corner of the goaf and a major proportion of the goaf gas may escape into the tailgate return airway, particularly during low barometric pressure periods. Analysis of the results indicates that goaf gas drainage strategy is also very important in addition to the goaf holes design.

In order to improve the goaf gas drainage system efficiency, deep goaf holes gas drainage strategy was introduced into the modelling simulations. The pressure distribution in the goaf with this new goaf drainage strategy is presented in Figure 5. Results indicated that the goaf gas migrates to a wider area towards the tailgate side and only a minor proportion of goaf gas escapes towards the tailgate return. In addition, the low goaf gas pressure development in the goaf (<80 Pa) helps in reducing the effects of changes in barometric pressure on return gas levels. These results indicate that the optimum gas drainage

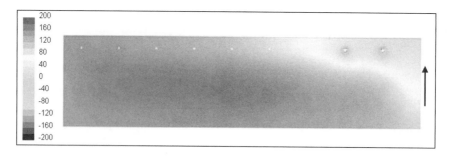

Figure 4. Pressure distribution in the goaf—with traditional goaf gas drainage strategy (gas drainage through two goaf holes near the face)

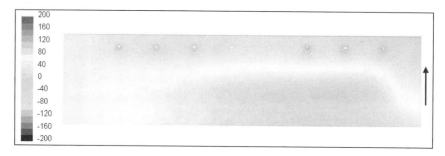

Figure 5. Pressure distribution in the goaf—with new goaf gas drainage strategy (gas drainage through deep goaf holes plus goaf holes near the face)

strategy should incorporate goaf holes near the face as well as deep goaf holes in the panel in order to improve the gas drainage system efficiency.

Benefits of Optimum Gas Control Strategies
Based on the fundamental understanding of goaf gas flow dynamics obtained from the modelling and field investigation, optimum gas control strategies were developed and implemented at the two mine sites.

Total drainage flow rates from the goaf plants increased to 5,500 to 6,000 m³/hr compared with 4,500 to 5,000 m³/hr in the previous panels. In addition, a new goaf fan at the start-up area of the panel was draining gas at the rate of about 2,000 to 3,000 m³/hr. Total goaf gas drainage flow rates reached 1,600 to 1,800 l/s, around 50 per cent improvement over the previous panel's goaf holes performance. It was also observed that gas drainage from three to four holes substantially reduced the problem of frequent blockage of flame arrestors. As a result, goaf gas delays were reduced almost to zero.

Results showed that the top two goaf holes were draining gas at concentrations over 70 per cent, which is well above expectations. Total gas drainage flow rate through the roof goaf holes was around 500 l/s. Analysis showed that the total goaf gas

drainage flow rate increased from around 2,000 l/s in a previous panel to over 3,000 l/s.

Inertisation Strategies for Longwall Sealing
In underground coalmines, the potential for explosion in a goaf immediately after sealing of a longwall panel is a major safety issue. To minimise this risk, some mines started the practice of injecting inert gas into the sealed goafs immediately after sealing the panel. The traditional inertisation methods may take anywhere from two days to a week to achieve complete inertisation in the longwall goafs. There was a need to reduce the explosion potential as quickly as possible during the longwall sealing off periods. A research project to address this issue was carried out by CSIRO and Newlands Colliery with funding from ACARP (Balusu, Wendt & Xue 2002).

In traditional schemes, inert gas was injected into the goaf generally through the MG/TG seals immediately after panel sealing. In three of the six case studies reviewed, these methods were not effective in preventing the formation of explosive gas mixtures. In other cases, oxygen concentrations were above 12 per cent for up to two days after panel sealing. The results showed that injection of inert

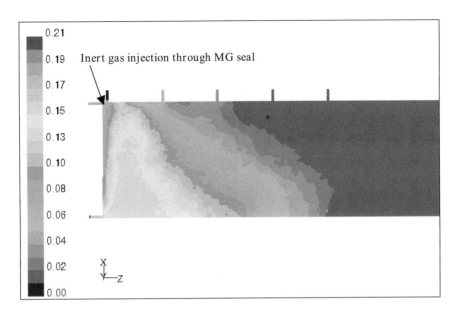

Figure 6.   Oxygen distribution in the goaf one day after sealing with traditional inertisation

gas through MG resulted in migration of face area oxygen towards inbye sections of the goaf.

Computational fluid dynamics (CFD) modelling of the goaf inertisation process was carried out to investigate the effect of various factors on inert gas dispersion patterns. The modelling showed that under normal longwall retreat conditions, ventilation systems, and gas emission flow rates had a major influence on goaf gas flow patterns near the face when compared with the effects of goaf gas buoyancy pressures. There was higher oxygen penetration on the intake side of the goaf. However, when panel airflows were restricted to simulate chock recovery operations, goaf gas composition and buoyancy pressure played a major role on gas distribution in the goaf and resulted in higher oxygen ingress on the lower side of the goaf.

Parametric studies showed that location of inert gas injection is the most important parameter and had the major effect on goaf inertisation. Traditional inertisation schemes involving injection through MG seal resulted in a reduction of oxygen concentration only near the point of injection, near the maingate area. As shown in Figure 6, the oxygen level near the tailgate area was very high at 15 to 17 per cent even after 24 hours of inert gas injection.

Simulations with inert gas injection at 200m behind the face line showed that the efficiency of goaf inertisation improved considerably with this new location. Oxygen levels reduced down to 10 to

12 per cent over a wider area of the goaf within a day of panel sealing.

Further simulations showed that ventilation pressure distribution around the goaf and airflows through the panel during final stages of sealing also would have a significant influence on goaf inertisation after sealing. Results indicate that minimal ventilation through the panel during final stages of sealing and staring of inert gas injection one day before sealing, with doors on return seal still open, results in rapid inertisation.

The optimum strategy consists of starting inert gas injection one day before sealing at 200m behind the face at 0.5 m³/s flow rate. Results of simulations showing oxygen distribution in the goaf one day after sealing are shown in Figure 7. The oxygen level was below 8 per cent at all locations in the longwall goaf and had achieved goaf inertisation within a few hours of panel sealing,

Mine Water Inflow Prediction

Reliable prediction of rock mass deformation, mine stability, mine water inflow and mine gas emission is essential, not only for improving mine safety and reducing production costs, but also for assessment of the environmental impact of mining.

Predictions require an accurate simulation of complex, highly non-linear and irreversible processes including the mechanics of rock deformation and failure due to coal mining and the consequent water flow and gas desorption and flow. CSIRO, in

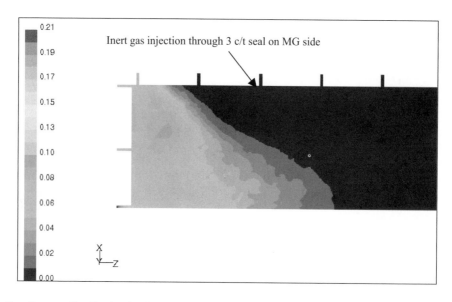

Inert gas injection through 3 c/t seal on MG side

Figure 7. Oxygen distribution in the goaf one day after panel sealing with optimum inertisation

collaboration with the New Energy and Industrial Technology Development Organization (NEDO) and the Japan Coal Energy Center (JCOAL) has developed a three-dimensional numerical model called COSFLOW (Guo et al. 2003; Guo, Adhikary & Craig. 2007). It uses a Cosserat continuum approach for the efficient description of mechanical stress changes and deformation in weak layered rock, typical of coal measures. This mechanical model is coupled with a two-phase dual porosity fluid flow model to describe flow of water and gas through porous rock, desorption/adsorption of gas from the matrix and subsequent flow of water and gas through the fracture network. The coupling includes simulation of permeability and porosity changes with rock deformation.

CSIRO has been actively involved in mine site hydrogeological characterisation, study of subsidence and subsurface rock caving, and predictive simulation of mine subsidence and mine water inflow. COSFLOW was applied to mine water inflow prediction at Springvale Colliery and has been found to have a remarkable capability to simulate mining induced rock deformation, aquifer interference, permeability changes and water inflow into a longwall mine and to produce accurate predictions as discussed below.

In 2003 CSIRO carried out detailed water inflow simulations to predict the mine water inflow at Springvale Colliery. At that time LW407 was being extracted. The COSFLOW predictions made in 2003 are found to agree well with the mine inflow

measurements made during the extraction of future panels LW408, 409, 410 and 411 (see Figure 8).

Figure 9 presents the porewater pressure distribution in the coal seam after mining panel LW408 as obtained from the numerical simulation made in 2003. The location of piezometer borehole SPR31 is marked on the plot. The numerically predicted pore water pressure at that location is about 400 kPa (i.e., water head of about 41m). The piezometer

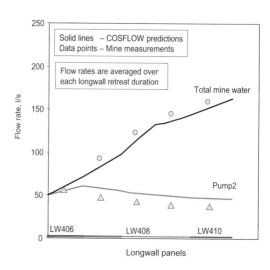

Figure 8. Comparison of numerical prediction with the mine water inflow measurements (After Guo, Adhikary & Gabeva 2008)

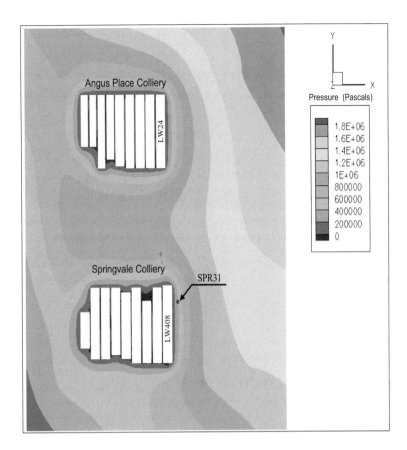

Figure 9. Pore pressure distribution (Pascals) at the mining seam after mining LW408 (The position of the piezometer hole is indicated as SPR31, After Guo, Adhikary & Gabeva 2008)

installed in the coal seam in SPR31 was reading a value of around 43m after the completion of LW408 extraction in 2004.

The consistency of COSFLOW predictions made in 2003 with the current trend of water inflow rates and porewater pressure evolution validates the integrated approach taken in the mine water simulations at Springvale Colliery.

CURRENT KEY RESEARCH ACTIVITIES

Roadway Development

One of the key areas for automation in underground coal mining is the development of the core roadway infrastructure. In many longwall operations the development of longwall gate roads is the production limiting process and the industry is very receptive to ways in which the roadway development process can be improved. CSIRO has been researching ways to improve the speed and efficiency of roadway development for more than ten years.

Automatic Bolting and Drill Monitoring

A major performance bottleneck in roadway development is the time taken in installation of bolts to provide roof and rib support. Moreover the current practice of manually drilling and bolting is one of the most hazardous tasks in underground coal mining. A real need exists for a system to minimise exposure to areas of unsupported roof and improve the production rate of this vital activity.

There has been consistent effort by a number of groups in Australia over the past ten years to implement automatic bolting in coal mining applications, but a viable system has not been produced. One of the fundamental problems is that automation studies have concentrated on automating the existing manual bolting process.

A key to successful development of an autonomous bolting system has been the invention of a self-drilling bolt. The bolt construction incorporates an integral single-use drill bit and a grout delivery channel in the centre of its annular cross section. The

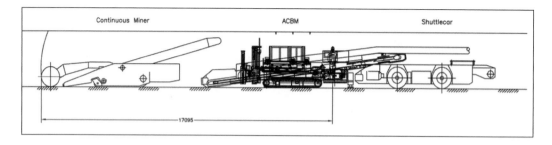

Figure 10. Diagram showing the arrangement of the ACBM with respect to the miner and shuttle car. The ACBM's dimensions are 12m × 3m × 2m and it weighs 50t.

major departure from the manual method is that the bolt is used to drill the hole and then remains fully inserted in the drilled hole while the remainder of the process is carried out. After drilling, grout is pumped into the void around the bolt through the integral delivery channel. After a curing interval, the bolt is tensioned.

The Autonomous Conveyor-Bolting Machine (ACBM)
In a project supported by JCOAL and ACARP, CSIRO created the ACBM, a machine designed to carry out autonomous installation of self-drilling bolts to speed up the bolting process and to remove people from the vicinity of the bolting operation . It is a mobile platform fitted with automatic bolting rigs, coal receiving hopper, automatic bolt storage and delivery system, and a through-conveyor for coal transport.

Figure 10 shows the placement of machinery associated with the rapid roadway development process, with a leading continuous miner, the ACBM and a shuttle car for coal transport.

While the original project goal for the ACBM of underground demonstration of autonomous drilling and bolting has not been achieved, the ACBM has served as a useful test bed for elements of a complete automation system. For example the method for storage and delivery of bolts has been adopted as a model for future studies.

Mining Navigation and Control Systems
ACARP is currently focusing on roadway development through its Roadway Development Task Group (RDTG). Its plan is to develop and direct a roadmap for targeted research and development aimed at roadway development improvement through CM automation, automated installation of roof and rib support, continuous haulage and integrated panel services. The project has a goal to produce a system that is available 20 hours per day and shows average development rates of 10m/hr.

CSIRO is developing the mining navigation and control systems necessary to deliver a remotely supervised, self-steering automated continuous miner. This is based on the technologies developed for the longwall automation work described above. The automation component will also provide a real-time machine position and operational information essential for the successful integration of bolting, mesh and haulage systems. Figure 11 shows the fundamental components associated with the automation system (top) and interaction with the other components (bottom).

This work commenced in January 2009 and has concentrated on requirements analyses for automation systems, characterisation of CM operations over a number of mines and performance testing of navigation units and algorithms.

Integrated Coal and Methane Extraction Technologies

The demand for coal and the capture and utilisation of coal seam methane is anticipated to grow significantly as the world strives to meet a predicted 60 per cent increase in energy requirements over the next 30 years. Conjointly, coal mine methane (CMM) released during mining is a key hazard that is the cause of the most disastrous accidents in coal mines globally today. CMM emissions are a key limiting factor to coal production efficiency and responsible for a significant part of global greenhouse gas (GHG) emissions.

CSIRO is undertaking several major projects funded by ACARP, Australian Government and Huainan Coal Mining Group (China) to optimise coal and methane extraction, improve mine safety, increase productivity and reduce fugitive GHG emissions.

These projects involve comprehensive studies, site monitoring and trials at major mines in Australia and China and will apply and demonstrate a holistic approach to integrated design, planning and opera-

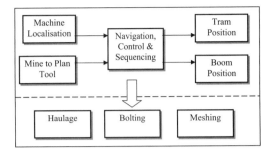

Figure 11. High-level automation system (CSIRO) and information exchange block diagram

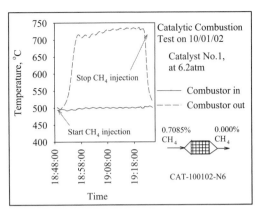

Figure 12. An example of CH_4 catalytic combustion performance

tion of coal production, hazard management, mine methane control and CMM capture for utilisation.

The techniques and strategies developed through these projects can be applied at a large number of gassy underground coal mines to maximise efficiency of captured drainage gas with conventional gas utilisation technologies.

Diluted Mine Methane Mitigation and Utilisation

World wide methane emissions in 2000 from mine ventilation air alone were over 273MMT CO_2-equivalent (US EPA, 2003). Underground coal mining is by far the largest source of fugitive mine methane, with approximately 70 per cent of all coal mining related emissions from underground ventilation air (Moore et al. 1998).

Ventilation air methane (VAM) capture, mitigation and utilisation is a major problem. It represents the largest proportion of methane emissions from coal mines. The air volume flow rate is large while the methane resource is dilute and variable in concentration. For example, a typical gassy mine in Australia produces ventilation air at a rate of approximately 150 to 400 m^3/s, with a methane concentration of 0.3 to 1 per cent.

Ventilation Air Methane Catalytic Turbine (VAMCAT)
Based on a series of methane catalytic combustion experimental trials, a process design for a VAMCAT system has been devised by CSIRO. The new VAMCAT technology has the potential to significantly reduce fugitive methane emissions while also using the methane as a clean energy source. Figure 12 is an example showing methane catalytic combustion performance. It is evident that the methane is fully oxidised into CO_2 and H_2O.

The patented VAMCAT system can be powered by about one per cent methane in air and will generate power while also consuming the mine's fugitive

methane. System analysis showed that a large scale VAMCAT unit can achieve generation efficiency of about 25 per cent.

Pilot-scale Demonstration Unit
A team from CSIRO and the Australian Department of Climate Change, together with China's Shanghai Jiaotong University, Huainan Coal Mining (Group) Co. Ltd. and several contracted manufacturers have developed the first ~25kW prototype pilot-scale demonstration unit which will be trialled at a coal mine in China.

Once this novel gas turbine technology is demonstrated at a coal mine, it will also have application in the mitigation and utilisation of methane from landfill, animal waste and the combustibles in industrial off-gas.

Operational performance data and experience gained from this prototype unit will then be used for the design of a second generation turbine of at least one megawatt output.

CONCLUSIONS

CSIRO has established an enviable track record of success in these areas and has developed a range of technologies that have delivered significant benefits to the coal industry and broader community over the last 10 years.

Challenges facing the coal industry currently are unprecedented. There are strong requirements for the industry to continuously increase mining efficiency and mine safety in a manner that satisfies the ever increasing stringency of environmental performance which is most clearly manifested at present through a focus on fugitive emissions of methane, coal mine subsidence, water management and spoil pile stability and rehabilitation. The nature of these challenges dictates the need for technological and innovative solutions.

CSIRO has the capability and capacity to undertake long-term strategic R&D that will underpin the sustainability of coal production in Australia. CSIRO is continuing its role in the development of new technologies that deliver sustained benefit for Australia.

ACKNOWLEDGMENTS

Acknowledgement is made to ACARP, JCOAL, NEDO and the management and site operational personnel from major mining companies in Australia including Xstrata Coal, Anglo Coal, BHP Billiton and BMA Coal. The major manufacturers of coal mining equipment in Australia including Joy Mining Equipment, Bucyrus and Eickhoff are also acknowledged for their cooperation and support.

The authors wish also to thank their CSIRO research colleagues for their contributions to these projects and Bob Chamberlain for his input to the preparation of this paper.

REFERENCES

Balusu, R., Tuffs, N., Peace, R., Harvey, T., Xue, S. and Ishikawa, H. (2004). Optimisation of goaf gas drainage and control strategies—ACARP project C10017. *CSIRO Exploration and Mining Report 1186*

Balusu, R., Wendt, M. and Xue, S. (2002). Optimisation of Inertisation practice—ACARP project C9006. *CSIRO Exploration and Mining Report 907F, June, 116pp.*

Corke, P., Winstanley, G., Dunbabin, M. and Roberts, J. (2003). "Dragline Automation: Experimental Evaluation through Productivity Trial," in Proceedings of International Conference on Field and Service Robotics, Lake Yamanaka, Japan, July 2003, 249–254.

Dunbabin, M., Corke, P., Winstanley, G. and Roberts, J. 2006, "Off-world Robotic Excavation for Large-Scale Habitat Construction and Resource Extraction," in To Boldly Go Where no Human-Robot Team Has Gone Before: Papers from the 2006 AAAI Spring Symposium series, Tech. Report SS 06-07, T. Fong, Ed.

Duncan Fama, M.E., Shen, B. and Maconochie, P. (2001). Optimal design and monitoring of layout stability for highwall mining. CSIRO Exploration and Mining Report 887F, ACARP project C8033.

Guo H., Adhikary D.P. and Craig M.S., Simulation of mine inflow and gas emission during longwall, Rock Mechanics and Rock Engineering, (2007).

Guo H., Adhikary D.P. and Gabeva D., *Hydrogeological response to longwall mining (ACARP Project C14033)*, CSIRO, Exploration and Mining Report (2008).

Guo H., Adhikary D.P., Xue S., Craig S., Poulsen B., Chen S., Wendt M., Su S., and Mallett C., *Predevelopment studies for mine methane management and utilisation, Exploration and Mining Report*, 1079C, pages 134 (2003).

Moore, S., Freund, P., Riemer, P. and Smith, A. *Abatement of methane emissions.* IEA Greenhouse Gas R&D Programme, June 1998.

Reid, D. C., Hainsworth, D. W., and McPhee, R. J., *Lateral guidance of highwall mining machinery using inertial navigation*, in Proceedings of the 4th International Symposium on Mine Mechanisation and Automation, 1997, pp B6-1-B610 (Brisbane) (1997).

Reid, D. C., Practical automation solutions for improved longwall mining, Australian Coal Summit, Sydney Australia, 2–4 September 2008.

Roberts, J., Winstanley, G. and Corke, P. (2003). 3D Imaging for a Very Large Excavator," International Journal of Robotices Research, July 2003, vol. 22, no. 7/8 pp. 467–478.

Shen, B. and Duncan Fama, M. E. (2001) Geomechanics and highwall mining. World Coal, Vol 10, No.2, pp. 35–38.

US EPA. *Assessment of the worldwide market potential for oxidising coal mine ventilation air methane.* United States Environmental Protection Agency, EPA 430-R-03-002, July 2003.

100 Years of Underground Applied Mining Research in Southeast Missouri (SEMO)

Richard L. Bullock

Department of Mining Engineering & Nuclear Engineering,
University of Missouri Science & Technology, Rolla, Missouri, United States

ABSTRACT: There was a time within the American mining industry that progressive companies were very active in research in mining techniques related to mining applications. Such a company was St. Joe Lead Company, (later Doe Run Company) that performed applied underground mining research in the areas of safety, ground control, drilling, blasting, loading and hauling; all of which lead to tremendous and significant advances in productivity and safety. These research activities took place between the early 1900s and 2009. This paper documents many of the origins of the basic concept which today the underground mining industry takes for granted (and forgets their origin). The resulting mining operations in SE Missouri have been the industry leader in safety, by far the leading recipients of the NIOSH Sentinels of Safety Awards for Underground Mining over the past 50 years. This is no coincidence; it parallels the applied research accomplished by this mining organization.

INTRODUCTION AND HISTORY

St. Joe/Doe Run Company History

St. Joseph Lead Company was formed in 1864 and was the dominant mining group in the SEMO Old Lead Belt (OLB) where they mined for 110 years. St. Joe made the discovery of the major Viburnum Trend in 1955. The Trend was mostly on U.S. Government Forestry land, so St. Joe could was limited to lease about 28 miles of the 42 mile Trend. St. Joe built six mines and four mills. Other companies explored and developed mines: AMAX, Homestake, Cominco/Dresser, ASARCO & Kennecott. Years later, St. Joe was sold, and later evolved into the Doe Run Company. By 1992, all mines in the district were owned and operated by Doe Run.

St. Joe/Doe Run Mining Research Histories

From the very beginning of St. Joe, management demonstrated a progressive attitude towards research and development of mining tools, equipment and mining systems that would improve safety, decrease the work effort, improve cost, improve mined grade and/or reduce cost.

For the celebrating of the 100th anniversary, L.W. Casteel (1964) the SEMO General Manager stated: "It has been a century dedicated to the principle that mining should not be an art but a science; that every process must be improved; that every unknown must be investigated." Through the years, St. Joe's management has pursued these objectives with surprisingly consistent vigor. In 1930, St. Joe created the job of "experimental engineer." For the first time, someone was assigned the job of research, completely unencumbered by other responsibilities. It was his job to test new equipment, make time studies for methods improvement, solve problems, and create new ideas. (Casteel, 1964) There has been an organization or individuals dedicated to these objectives within St. Joe/Doe Run for the past 80 years. They were first called Experimental Department, then Mine Research Department (MRD) and more recently Technical Services Department (TSD). But long before 1930, the mine engineers of St. Joe were investigating and developing new tools and methods for mining. When one looks at the accomplishments that have been made in safety, productivity, cost reduction and resource recovery over the history of the St. Joe/Doe Run companies, one would have to agree that "mining research." has indeed paid off; not only in cost reduction, but they have consistently over the years been one of the leaders in mine safety as this paper will show.

EARLY RESEARCH AND DEVELOPMENT: LOW HEAD ELECTRIC POWERED CABLE SHOVEL (1922)

One of the early efforts of research cooperation between the U.S. Bureau of Mines and the lead-zinc mines of SEMO was that of studying the development of mechanical loading of ore that was just beginning in these mines in 1920. After WWI, hand loading of ore was still the predominant method of mucking. As backbreaking as hand mucking was, labor opposed mechanization of ore loading. But as was pointed out (Van Barneveld, 1924), the net result of this [mechanization] as affecting labor is not, as labor fears, a reduction in the number of men employed, but rather a widening of the definition of ore, the inclusion of material hitherto unprofitable, the expansion of operations, and the consequent hiring of more men at better pay [and safer, easier and more productive jobs]. One such mechanization development which was to have enormous consequences in the mining operation of all of the SEMO from 1922 through to 1967; this was the development of the St. Joe Shovel. The development was the result of cooperative research and development effort between the Thew Shovel Company and St. Joe. St. Joe established the design criteria of what they wanted and Thew designed and fabricated the first machine. It was tested in the mines for one year, when minor changes and improvements were made before putting the machine in mass production. St. Joe and the other mines operating in the area acquired many of these mechanical loaders (see Figure 1). When your author went to work for St. Joe in 1955, there were 22 mines operating, and they had approximately 75 St. Joe Shovels. It is estimated that they loaded at least 300 million tons of ore before they were finally replaced in the new mines at Viburnum by trackless loaders.

GROUND CONTROL RESEARCH

To understand the ground control research, the rock formations of the mineralized Bonneterre (BT) dolomite formation needs to be understood. The BT is broken into 19 zones of different types of dolomite. Lead and zinc mineralization occurred in all of the zones some place in the OLB district. Some zones are very thick bedded and extremely competent rocks (such as the five zone and seven zone formations). The seven zones also contain massive algae reef structures. However, some of the lower zones are thin bedded and the beds are interlaced with shale bands and glauconite. These beds also contain numerous fractures which are filled with water, feeding up from the Lamotte sandstone below the BT. However, in spite of the major differences in rock mass strength, all of the formations were mined with room and pillar (R & P) mining and room widths would vary from 20 to 50 feet wide and the final mining heights varied from about 7 feet to 300 feet. There was very little timber used in the mines. A good description of the ground conditions is found in a paper by Weigel, (1943).

Invention of the Roof Bolt: St. Joe's Major Contribution to the Mining Industry

By the early 1900s, the region's lead mines had penetrated into the lower beds of the BT and found

Figure 1.   St. Joe low-head electric powered cable shovel (Stockett, 1943)

thin, shaly beds that provided poor roof conditions. Pillar spacing had to be greatly reduced and, in some cases, the diameters of the pillars increased. In certain areas, the roof would cave until the natural arch between pillars was reached. If the ore was rich enough to permit this dilution, mining continued. If dilution was not permissible, the area was abandoned. Obviously there existed a real need for some means of ground support. (Casteel, 1964)

In many cases, because of the random distribution of the ore, commercial mineral was found in the back after the initial mining had taken place. In some cases, this upper orebody might extend upwards for a hundred feet or more. Thus to mine these orebodies, the engineered scaffolding system that was known as "trapeze or trap mining." was developed in the early 1900s. The scaffolding was suspended from the back by split wedge anchored eye bolts. It was noticed that on many occasions after mining was completed and the trapeze was removed, but the eye bolts were left, that years later, the back would sometime cave near the area were the traps were run, but where the eye bolts remained the caving would stop and they would be holding up loose rock. Thus without realizing what they had invented, the first roof bolts in the early 1900 were simply eyebolts that held up slabs of loose rock (Figure 2).

Management concluded that long bolts placed close together would support thick layers of loose rock. In 1927, the idea of roof bolting was further perfected. Bolts were made in 8-ft lengths with one end split and the other end threaded. These were anchored in the roof and a square washer placed over the protruding end. This washer was drawn up tight against the roof by means of a threaded nut. It was

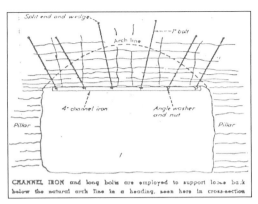

Figure 3. Primitive Roof Truss (Circa 1936) used in St. Joe's mines

Figure 2. St. Joe's engineered scaffolding, known as "trapeze or trap" were used to mine high orebodies left in the back. The eye bolts used to suspend the scaffolding lead to the observations of the affect of the first roof bolts.

now possible to keep ore dilution to a minimum and to safely mine areas that had previously been a disaster. (Casteel, 1964)

After WWII, at least twenty years after St. Joe started using roof bolts throughout their mines, the Chief Engineer from Consolidation Coal visited one of the St. Joe mines and was so impressed he tried them in Consol's No. 7 mine and then the USBM studied SEMO bolting practices and spread the idea to the coal mines in Illinois. (Mark, 2008) Of course there were immediate improvements in the bolts for the coal mines in that they used high strength steel bolts with expansion shell anchors on the end of the bolts. With the elimination of the multitude of props, this allowed for greater mechanization of equipment to be used in the coal mines

Channel Irons: A Primitive Roof Truss

Immediately after the development of the roof bolt, St. Joe developed the system of placing 20 ft (6.1 m)

long four inch (102 mm), channel irons bolted to the back. Each channel iron was held with six, eight to twelve ft (2.4 to 3.7 m) long roof bolts. The two end roof bolts were angled out over the adjacent pillars (Figure 3). In very bad ground, the channel irons would be placed as close as every foot, though this was not normal practice.

Pillar Reinforcement Wire Rope

Because of initial over mining and in some cases additional pillar slabbing of high grade ore, large areas begin failing. Both roof bolts and channel irons were tried and would not stop the pillars from failing. Finally in 1940 St. Joe started experimenting with wrapping the pillars with used hoisting wire rope. (Wycoff, 1950) Then eye bolts would be anchored near the top of the pillar and about every foot down the pillar. Every wrap was pulled tight with a 10 ton (9.1 tn), hand actuated chain hoist and clamped with a series of bolt clamps (Figure 4). Finally when the full pillar was wrapped, gunite (shotcrete) was often applied. There is little doubt that this method of pillar reinforcement, while labor intensive did save large areas from massive failures.

Concrete Pillars

Concrete Pillars Using Screw Jacks for Loading
In approximately 1938, St. Joe started a research project to remove a group of high grade pillars in the Leadwood South Mine. However, it was near two other areas that had been pillar slabbed extensively. The idea of the new project was to build 19, 12 ft (3.7 m) concrete pillars and place 16 one ton jacks on top of the pillars, which were meant to load the pillars. These pillars were all between 30 and 50 feet

Figure 4. Fully wrapped pillar with used mine hoist rope (Wykoff, 1950)

Figure 5. Concrete pillar with 16, 1-ton screw jacks on top for loading (Casteel, 1964)

(9.1 to 15.2 m) tall. There was a method to measure very accurately to the thousandth of an inch to show pillar compression. However, the 16 tons (14.5 tn) of load on top of the pillar was insignificant compared to the weight that was to come from roof convergence and compared to what compression that the pillar had to take before it started to support the back. It was a noble experiment, but the results were too erratic to be of value. Measurements took place from 1939 up through 1960s, with very little change in the pillar height over the years (See Figure 5). In retrospect, these concrete pillars allowed the removal of ore pillars that more than paid for the project, but it was not as profitable as they had hoped.

Concrete Pillars Using Hydraulic Flat Jacks for Loading

In the late 1960s the mine research department set out to build several concrete pillars which used six quadrant flat jacks in these 8 to 12 foot (2.4 to 3.7 m) diameter pillars. Several methods of concrete pillar construction were tried, but the best method was to build the high strength monolithic pillars, with concrete placed on site. The flat jacks were pressurized to pre-stress the pillars. They contained no interior reinforcing, but externally prestressed with 1-inch (25 mm) round mild steel bands on 6-in centers. The

pre-stress developed within the monolithic concrete pillars averaged 1700 psi (117.3 MPa), which was significant. However, it was concluded that the pillars of this type should have been able to been able to have been loaded to 3360 psi (231.8 MPa) and that future pillars of this type should be equipped with a second set of flat jacks that would permit high ultimate loading after the first set eliminates excess structural slack. (Reed and Mann, 1960) (Reed and Mann, 1961).

Roof Bolt Research and Pull Testing to Establish the Best Bolt and Anchor

While St. Joe invented the roof bolt, they by no means optimized the effective use of the concept. The 1-inch, mild steel split wedge bolts were manufactured within the St. Joe shops. However, better bolts were being manufactured throughout industry. In about 1957 the MRD set out to prove the value of using a ¾-inch, high strength bolt, using an expansion shell for the anchor. The group tested approximately 15 different types of expansion shells, in many different rock zones and compared the results with the 1-inch mild steel split wedge bolt, by pull testing and ease of installation and then selected the best combination of bolt and shell for the OLB.

Figure 6.   Tube extensometer (Reed, 1959)

Figure 7.   Tape extensometer (Reed, 1959)

Ground Control Research Related to Pillar Extraction

Ground Convergence Monitoring Measurements
The development of an instrument (that was not yet available to industry in 1955) to very accurately measure the convergence between the back and the floor was absolutely imperative. It was obvious that as pillars failed, the roof sagged. There was no case where the floor heaved to account for the convergence. The accuracy of the measurements that was needed was 0.001 of an inch. The movements toward failure in the BT formation were very slow. Thus it might take months for the room to fail. But characteristic to roof failures, the convergence rate would accelerate towards the end. The research effort determined that movements of 0.001 to 0.002/month where normal and were no cause for alarm. Movements of 0.003 to 0.004 inches/month were cause for alarm and some mitigating action must be taken to stop the convergence, whether it was caused by pillar deterioration or problem with the back span. Movement of as much as 0.007/month meant that the back was in failure mode and all activity in that particular stope area must cease. By that point, you probably could not stop the failure from occurring.

For monitoring of convergence, the Head of the MRD, Dr. John Reed developed and patented the Reed Tube Extensometer and the Tape Extensometer (Figure 6 and Figure 7). These tools were developed and used extensively in many areas where known pillar deterioration was taking place or later in areas

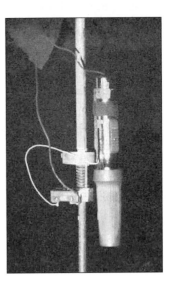

Figure 8.   Stope convergence warning light (Reed, 1959)

where pillars were to be removed. Also developed was a stope convergence warning light, which consisted of two spring-loaded telescoping tubes, a micro-switch, and a flashlight (Figure 8). A movement of a few thousands of an inch would turn on the flashlight and give the miner a warning of back movement. (Reed, 1959) The use of these tools by the Research Department saved several major areas from catastrophic failures.

Investigations of Various Methods of Removing Pillar Ore

Removing Pillars Over 100 Feet Tall

The MRD assisted with methods of pillar removal. A real challenge was those pillars in the Bonne Terre mine that were over 100 feet tall. The first project was a pillar that was about 30 to 40 feet in diameter and approximately 100 feet tall. They ran trapeze out to the pillar and shot the top 15 foot off of the pillar. Then bolted the back and with a wagon drill, drilled long holes down through the pillar and blasted the bottom portion of the pillar at one time. The results were reasonably successful, but the rock was scattered in some parts and there were some very large slabs that had to be redrilled and blasted.

Removing Pillars by Slabbing Only a Portion of Them

In 1959 MRD monitor areas with roof convergence tools where all of the pillars in stopes were slabbed from about 85% extraction down to 95% extraction. This was done without the roof failing and was economically very successful in stopes under about 30 feet.

Removing Whole Pillars By Retreating Method (Reed, 1959)

In 1959, the MRD made a major attempt to get 100% of extraction of an isolated 20-ft high stoping area. For this research project the stope was systematically bolted on 4.5-ft (1.4m) centers . The MRD modified a two boom jumbo to hold a basket for an operator and all of his controls and the other boom to hold a drill for roof bolting and an impact wrench. About 24 pillars were removed and about 85% of the ore was recovered from these pillars. A network of convergence stations was monitored throughout the project, and microseismic counts were monitor on a regular basis. Stope warning lights were used to warn the crews of excessive back convergence. In one area were failure was imminent, convergences of the rate of 0.005 inches per hour were recorded and total convergence of up to 0.849 inches was recorded. That portion of the stope collapsed within 16 hours after all equipment had been removed from the area. The unsupported area averaged about 200 ft. (60m) by 600 ft (180m).The system proved that pillars could be safely removed when the correct bolting system was applied.

Removing Pillars By Using Mine Backfill and Fencing (Lane et al., 2001)

Since the Viburnum Trend mines have been operating for over 50 years, a large portion of the production is

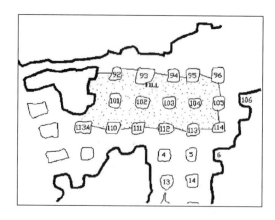

Figure 9.   Test area fill plan at Buick Mine (Lane, 2001)

coming from pillar removal. The procedure has been well document and will not be covered in much detail here. The process entails encapsulating large area of pillars, with cemented rock fill (CRF). However, in this case it also contains fly ash and cycloned mill tailings. Their approach is shown in Figure 9. The area is enclosed with steel fence, reinforced with wire rope and then filled. The perimeter pillars are then removed and then the encapsulated pillars are removed from sublevels below the pillar with long-hole blasting. In some cases the filled area are left as large abutment pillars and the pillars in the general area are removed. But in no case, is the open span more than 150 ft (46 m) exceeded. All of this is modeled in advance using NFOLD model and in the field extensometer reading track the projected convergence. In very narrow areas, pillars can be removed without backfill, but modeling and instrumentation is always closely followed.

Pillar Reinforcement by Fully Cemented Rebar Research and Application

Over extraction again occurred at the Fletcher mine in the Viburnum Trend. The pillars were 20 to 40 feet high. As a result, approximately 400 pillars began to fail as the mine expanded away from the shaft. The first attempt to stop the problem was to bolt the pillars. Meanwhile, the MRD developed and installed a wire extensometer convergence system (Figure 10). This identified the magnitude of the problem. It became obvious, that bolting the pillars was offering little resistance to the failing pillars and the converging roof. The point anchored roof bolts were yielding and letting the pillars fracture. It was realized that what was needed was a very stiff rock bolting system that did not yield except at the point that it was

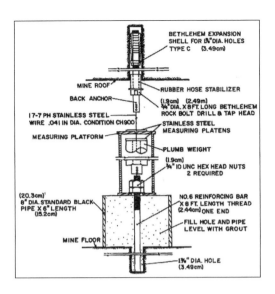

Figure 10. Wire extensometer station that were install in the critical areas of Fletcher Mine (Weakly, 1982)

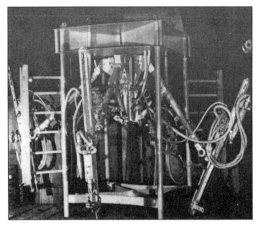

Figure 11. St. Joe's mechanized shaft sinking jumbo (Lutjen, 1954)

stressed. Thus a fully grouted rebar type rock bolt was developed, using a rebar placed into a 1⅜-inch (35 mm) hole, that had been filled to the proper depth with high early strength cement/fly ash grout to fully grout a 10 ft (3 m) bar. The pull test on this system averaged 22.8 tons (20.7 tonnes). The cost in 1972 per pillar with an average of 220 bolts was $500 per pillar. This was approximately 25 percent of the cost to wrap a pillar with heavy wire rope. This solved the problem in stopping the convergence of the back and the pillar deterioration. (Weakly, 1977) (Weakly, 1982)

SHAFT SINKING RESEARCH

In the OLB, St. Joe had many conventionally sunk, timbered shafts. But in 1952, C.K. Baines lead a group of researchers in the design of a new mechanized shaft sinking method, patterned after the more mechanized deep shaft sinking method of South Africa, using a Galloway type stage; except Bain also mechanized the shaft drilling jumbo (see Figure 11). He pioneered the trend to circular concrete lined shafts in this country. However because St. Joe shafts were shallow, the work stage, the jumbo, the forms, everything was pulled out of the shaft between each cycle. The Indian Creek No. 23 shaft was the first sunk by this method, where there were no loss time accidents and at a much less cost per foot than any previous shafts and the shaft was nearly maintenance free. It was an extremely safe and efficient method of

shaft sinking, and St. Joe had their own shaft sinking crew which over the years sank twelve shafts, varying from 723 ft (220 m) to approximately 1,600 ft (490 m). (Bain, 1953) (Lutjen, 1954)

DRILLING AND BLASTING RESEARCH

While the rocks of the BT dolomite drilled very easily compared too many igneous rocks, some of the formations are difficult to break. The rock is tough and in some cases "spongy." and sometimes does not have normal breaking characteristics. In narrow drifts, rounds of only about 6-ft (1.8-m) advance could be broken. In 24-ft (7.3 m) wide stopes, rounds of only about 8 ft (2.4 m) could be broken. When the MRD was organized in 1955, they set out on a two pronged research project to be able to greatly increase the depth of pull in drifting and stoping, as discussed below.

Parallel Hole (Burn) Cut Drill Round Research

The Experimental Department in the very early '50s had tried breaking a long round to two, 8-inch (200 mm) cut holes. These tests results were observed in 1955 and showed that the rock which was in the most difficult rock to break, the algal reef rock, had completely re-compacted and had not broken anything. You could faintly see where the two large holes had been. They concluded from this early experiment, that parallel hole cuts (burn cuts) will not work in the BT. The new MRD set out to find out why; they work everywhere else, why not here?

After an extensive literature research, field work started. One research engineer started drilling with a large auger drill, using both 6 and 8-inch (152 and

200mm) holes. (Yancik, 1956) After many months of testing he was very successful in breaking drill round up to 15 ft (4.6 m) in length in the 11 zone rock he was testing. The problem was not the blasting; it was the cost and time of drilling the large holes that hurt this phase of the project.

The other research engineer started trying many patterns, using hole sizes from 1-⅜ to 4-in (35 to 102 mm). The small hole patterns would break well in some of the rock zones but would not break in the reef rock. Twenty five different patterns were tested. The researcher developed a pattern that would break well in all types of the BT and then tested the successful pattern in full size rounds (Figure 12). The research crew drove a small drift with 25 back to back drill rounds, in the most "spongy," algal reef rock in the OLB. The 25 rounds averaged 11.2-ft (3.4 m). for the 12-ft (3.7 m) rounds. The criteria that are required to successfully break a parallel cut round were published (Bullock, 1957) (Bullock, 1961) many years before the Swedes, who completed excellent research work on parallel hole cuts was published in 1963 (Langefors and Kihlstrom). The St. Joe research on the burn cut pattern revealed that for the BT formations, not only do you need enough open hole relief to receive the blasted material from each charged hole, but to get 95% success you needed to place a large open hole between the charged holes and thus eliminating sympathetic detonation and dead pressing of each hole as it is

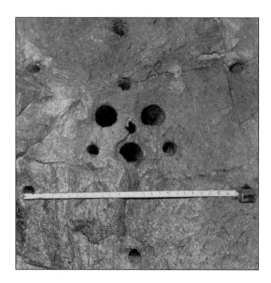

Figure 12.   The successful Bullock Parallel Hole cut and the first reliever holes (Bullock, 1961)

detonated. While the original work was devoted to drifting research, the results were equally applicable to the R & P stoping and later would quadrupled the productivity of the drilling and blasting in stopes. This burn cut pattern has been used to break approximately 20% of the tonnage that has been produced from the Viburnum Trend of New Lead Belt Mines.

Development of the Use of Prilled Ammonia Nitrate Fuel Oil (ANFO) in Small Diameter Holes. (Bullock, 1962)

The MRD group started working on the development of ANFO in small diameter holes down to 1-⅛ inch (28.6 mm) in diameter in 1960. They developed a working agreement with Monsanto (a major AN producer) and started testing ANFO. Monsanto made several densities of AN prills and they had been shooting them in various sizes of pipes. (Yancik, 1960) They discovered that they could shoot some of their prills in pipes as small as 1-inch, but were not sure of the sustained propagation. The joint effort went underground at the Bonne Terre Mine and started successfully shooting holes that were only 1-⅛-inch (28.6 mm) in diameter at the bottom of the hole and breaking the rock easily. Though many different types of prills were tested, their standard porous prill was the most dependable. This was the first time prilled ANFO had been successfully used commercially in holes that small. The Swedes were using crushed ANFO in holes of the same size, but it is more sensitive. This gave St. Joe at least a full year of better safety and low cost advantage of ANFO use over their competition.

About 250 St. Joe drillers were scattered over the district. It was not practical to buy a $5000 commercial placer for all of the miners. Besides, they were much too big for one man to handle in the typical room and pillar stopes. Thus the St. Joe pneumatic pressure pot charger was designed by the MRD and built in the St. Joe shops. (Casteel, 1962) The cost at that time was about $25. They were thoroughly tested and made perfectly safe. A typical miner charging scene is shown in Figure 13.

Percussion Drilling Research

Now that longer rounds could be broken, a faster way was needed to drill the rounds. To efficiently drill the burn cut round, approximately a dozen different rock drills were tested by MRD to determine the most efficient drill to drill the large and small holes in the mid 1950s. The drills that were being used on the current jumbos would drill a 1-⅜-in (35mm) hole at about 22-inches per minute (ipm)(0.55 m/min). The

Figure 13. An operating driller using the St. Joe placer to charge blast holes (Anonymous, 1972)

drill that we finally selected would drill the 1-⅜-inch hole at the rate of 37-ipm (0.94 m/min).

To drill the parallel cut round, a three boom jumbo was needed. There were none on the market at that time that would fit into small drifts. The MRD engineer worked with the Leadwood Mine shop people and converted the chassis of an old two boom jumbo with no remote controls on it, to a full three boom jumbo which allowed one operator to stand behind the console and position and operate the two outside booms, while the center boom and machine was operated from the ground in the center. The jumbo was taken to the Baker Mine and tested for about four month driving drift using the Bullock Burn Cut and performed very well. It was then taken to the new Higdon Mine shaft and drove drift there for another six months driving large drift by the same method.

It was about this time that the Viburnum Trend was discovered and the new No. 27 mine would need new drill jumbos. Based on the concept of the three-boom MRD jumbo that had built and tested, a stope jumbo to drill the same type of drill pattern, four, three boom track mounted Gardner Denver drill were acquired. It took several months for the research department personnel to retrain the experienced drillers how to drill and blast to drill and break the burn cut pattern correctly. With this drilling system,

now all of the drillers were removed from the close proximity of the drilling face and were therefor safer.

St. Joe Imported the First Rotary Percussion Drill (RPD) in the U.S. for Research

In late 1955, the German Hausherr, air rotary-percussion (RP) drill, DK-7 was bought and tested in the Leadwood mines by the MRD. It was getting phenomenal penetration rates in Germany of 4 to 6 feet (1.2 to 1.6 m) per minute, but it used about 600 cu ft. (17.2 cu m) of air per minute. (Bullock, 1962). While this amount of air was totally impractical for use for the replacement of hundreds of drills in the OLB mines, the concept had potential. It was a high thrust machine and it drilled at the rate 3.8 to 6.0 ft. (1.2 to 1.8 m) per min in the BT dolomite.

All of the drill manufacturers in the country came to watch this drill being tested. Only one acknowledged an interest in pursuing research on the concept. So Joy Manufacturing and the MRD group worked together for approximately four years testing various versions of a hydraulic rotary machine with a percussion hammer and the research on this prototype was completed in 1961. This then was the first hydraulic-rotary/air-percussion RPD (Joy AH-4) with a 1-½ inch (38 mm) RP bit, drilled at the average rate of 60 inches (1.52 m) per minute, when using RP type bits, rather than a standard cross bit.

The first, RPD two-boom jumbo was designed and built in the Bonne Terre shop on a Cline truck chassis and both booms and drill feeds could be positioned from the console. Each boom movement and feed movement had electrical potentiometers on each motion so that the operator could dial a number for a particular hole and the drill feed would align on that hole (Figure 14). The first jumbo was sent to the Indian Creek mine where it proved very reliable and performed as expected, except for the potentiometer electronics which did not hold up in the wet environment. Following this test, two boom jumbos were ordered from Joy for the Indian Creek Mine, and three more for the new No. 29 at Viburnum. These jumbo produced 43% more tons per shift than the old three boom jumbos at the No. 27 and No. 28 mines. (Bullock, 1968). When the new Fletcher mine came into production, six Joy RP Serpent, two boom jumbos, were purchased (Figure 15).

All-Hydraulic Drilling Research

St. Joe MRD engineers took a proactive part in the development of the early all-hydraulic RP (AHRP) drills. Information was gathered on 14 different AHRP drills while they were in there early prototype stages of development. While it was not possible to

Figure 14. St. Joe designed and built two boom jumbos using Joy AH-4 RPD drills (Bullock, 1968)

Figure 15. Joy RPD Drill (Bullock, 1968)

test all of these drills, the St. Joe researcher took technical data from these drills and was able to develop Specific Energies calculations which demonstrated the ranking efficiencies of the drills (Figure 16). (Bullock, 1974), (Bullock, 1975) (Bullock, 1976). The data clearly showed that the drill being developed by Ingersoll Rand was one of the more efficient drills, based on Specific Energy, so the MRD developed a cooperative agreement for testing and assisting in the development of their all hydraulic drill. The first HARD II and HARD III drills were tested in St. Joe's mines and showed penetration rates averaging 9-ft (2.7 m) per min with RP bits. As a result of the early research, St. Joe equipped there newer mines with the much safer AHRD, well ahead of the rest of the industry and were at a temporary economic advantage.

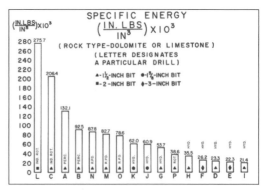

Figure 16. Comparison of specific energies of various types of rock drills and different brands of all-hydraulic drills (Bullock, 1974)

Results of the Drilling and Blasting Research

To illustrate the measure of the success of the drilling and blasting research and the investment in mechanized equipment which was made possible through that research, Figure 17 shows the 440% increase in productivity during the 30 year period when the MRD first started blasting and drilling research and when the results paid off in the new mines of the Viburnum Trend. The safety that was achieved at the same time is discussed later.

RESEARCH AND DEVELOPMENT OF TRACKLESS LOADING AND HAULING IN SEMO

The Introduction of Trackless Equipment Through Research and Development

For the first 75 years of underground mining in the OLB, everything was tied to rail haulage. In the mid-1940s, St. Joe built its first diesel powered, six ton diesel powered truck. It was very similar to the truck shown in Figure 18. While the mines of the OLB were not easily adaptable for truck haulage to rail haulage units, the early attempt was an experimental project for testing diesel powered trucks underground. Water-bath scrubbers were built and tested on this truck and proved successful in reducing CO and carbon particulate (though not to today's standards). The early test using diesel truck haulage were successful. Thus when the Hayden Creek mine was developed, 10 ton (9.1 tonne) LeTourneau diesel trucks were used. When the next OLB step out occurred, the Indian Creek Mine was equipped with St. Joe built 6-ton (5.4-tonne) diesel trucks (Figure 19). These trucks were mostly loaded with St. Joe built scrapper-ramp loaders. Indian Creek

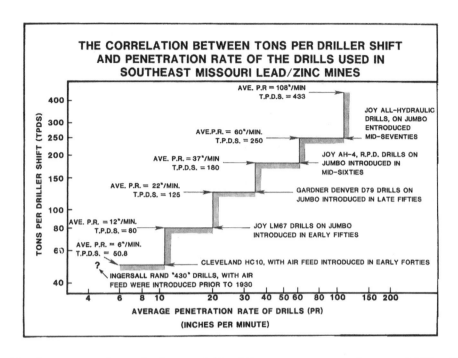

Figure 17. Improvement in tons broken per driller shift vs average penetration rate of drills with the accompanying mechanization (Bullock, 1983)

mine soon went to 10-ton (9.1 tonne) Cline trucks. When the first mine of the Viburnum trend mine was developed, It was first equipped with St. Joe shovels loading 10-ton (9.1 tonne) Cline trucks.

It was about this time (1962) that a new concept was developed by American Zinc Company for the Metaline Falls Mine in NE Washington. This was the first "Load-Haul-Dump." (LHD) vehicle, the "Gismo." As originally developed, it was track mounted tractor power unit, with an articulated loader/hauler unit. The concept was then acquired by Sanford-Day of Pennsylvania, who put the same idea of articulated mobile units on rubber tires and introduced them in the American Zinc mines of Tennessee. St. Joe heard about the concept and immediately went to see the "Transloaders." as they were called. St. Joe bought one of the Sanford Day units and put it into the next Viburnum Trend mine to be developed, No. 28. It was an instant success in that it could load and transport rock at about 150 to 200 tons (136 to 181 tonnes) per shift, but they were very high on maintenance. After a test period, to optimize the Transloaders, St. Joe engineers made 62 design changes in their Viburnum shops, all of which were copied by Joy for all future Transloaders that they manufactured. St. Joe equipped all of the

Figure 18. St. Joe Scrapper-Ramp loading a 6-ton (5.4 tonne) St. Joe diesel truck (Christiansen et al., 1959)

Viburnum mines and Fletcher with Transloaders. These loaders were able to produce 250 tons (227 tonnes) hauling up to 4,000 ft (1220 m) in the mines of St. Joe.

In about 1967, a Caterpillar 980 loader was put underground at Fletcher. This loader proving too small, so the Fletcher mine was then equipped with Cat 988 loaders. However, at that time Cat would not modify their loaders to roll back the bucket to a safe level carrying position for LHD application. So St. Joe had to modify the hydraulics and the linkage to accommodate this requirement. Also, they did not build a bucket that held a full 10-ton (9.1 tonne) load for LHD application; for this, St. Joe adapted the 988 with a Balderson bucket. Now they had a true FEL or LHD whichever was needed. However, at that point, all the stoping was only short hauls, so LHD was all that was needed. This was the first application of 988 to LHD.

Research for Optimization of Trackless Haulage Equipment (Bullock, 1975)

While the mines at Viburnum, Fletcher and Brushy Creek mines were still fairly young, haulage distances and loading/haulage situations were extremely variable; in distance, haul grade and what equipment was available. It was difficult to optimize a loading/ hauling situation since it was changing every day. It was known that trucks would be needed to produce the needed tonnage as the haul distance increased. A research project was set up to help the mine management to determine the optimum combination of when to use:

- LHD
- FEL to trucks (How many trucks and what size)
- FEL-LAF (One loader load one truck and follow it to the dump)
- LHD-HFC (LHD to ore pass, truck haul from chute)

The MRD had the challenge of determining which system to apply in different parts of the mines, determining the correct size of the loaders the trucks. They also had to consider the different brands of equipment and their related cost. A computer simulation package was developed from Caterpillar's materials moving simulation basic program (Morgan, 1972) and adapted to the underground four loading-haulage systems and the St. Joe mine conditions. Time studies were completed on the fixed portion of the cycle. The researcher simulated the four haulage conditions for distances from 100 feet to 36,000 feet (30 m to 11,000 m) and haulage grades from a –9% to a +17%. He also considered three different brands of loaders in the range of 6 to 10 ton (5.4 to 9.1 tonnes) and four different brands of trucks in the range of 20 to 40 tons (18.1 to 36.3 tonnes).

The simulation program was validated in several operating situations and proven to be an accurate predictive tool. It was used for determining which equipment to buy for the mines, how to assign the

Figure 19. The 6-ton (5.4 tonne) Sandford-Day Transloader (Anonymous, 1964)

equipment for different stoping/hauling conditions and even to determine and set production bonuses for the union workers performing loading and hauling.

Research on Tire Management

The first real success at high production in trackless haulage also created the first real tire problem. While St. Joe had been working with the Purcell Tire Company of Desoto, Missouri in recapping and repairing tires since they first introduced trackless haulage into the OLB, Hayden Creek, Indian Creek and then No. 27 mine, these applications were for truck haulage at low speed and the tires were usually on fairly good roads. Now with the Transloader LHD, the front tires were literally in the rock pile and also totally supporting the load of the bucket. Plus the fact, that the tires were then subjected to speeds of 15 to 20 mph (24 to 32 km/h). The MRD started investigating the tire failure problem and developed a computerized tire failure records system, where the recapper recorded the type of failures that were occurring. (Mount, 1976) This effort did greatly improve tire life and the number of recaps that you could get on the tires. As a result, in many cases, St. Joe was able to get four to six recaps, per carcass. It also helped management enforce the tire maintenance program by knowing that most of the tire problems could be corrected by:

- Properly constructed, well graded, well drained, dry roadways
- Maintaining proper inflation pressure on the tires
- Using the properly designed tire for the each application, whether it be FEL, LHD or haulage

Diesel Particulate Research

From 1970, MRD and the safety department conducted the following safety and health related diesel research:

- Epidemiological survey of annuitants exposed to diesels
- Pulmonary function surveys of miners working underground and on the surface
- Safety training and operational surveys
- Diesel engine type, injector selection and preventative maintenance procedures to reduce emissions
- Particulate filters, scrubbers and traps
- Cooperative research with the US Bureau of Mines and Michigan Technological University to measure diesel exhaust NO, NOX, CO, CO_2 and particulate
- Improved ventilation system design based on diesel HP loading and more accurate monitoring techniques
- Fuel mixes and additives
- Cooperating with manufacturers in designing diesel equipment to be used in specific mining tasks

The studies resulted in measurable dilution of diesel exhausts due to improved mine ventilation, reduced emissions of NO, NOX, CO and particulates, and more efficient use of equipment. The surveys concluded that there were no detrimental health effects from working on and around diesel equipment. A lot of the more detailed information on some of the St. Joe diesel contaminant research was covered by Mount (1982).

THE RESULTS OF 100 YEARS OF UNDERGROUND MINE RESEARCH IN MINE IMPROVEMENTS AND MECHANIZATION

This documented 100 years of dedicated applied research and development of mine mechanization has lead to extremely safe, productive and low cost underground mines. From the mid 1950s on, St. Joe made every effort to move from electrical tread mounted and rail mounted equipment to diesel rubber tired equipment. Between 1950 and 1975 the number of diesel units increased from 2 to 170. From about the same time, all of the drilling was done with about 250 column or hand held drills, which were replaced by mechanized drill jumbos. During the period of the rapid expansion of the step out from the OLB, St. Joe was spending (in 2009$), $325,000 for mining equipment to mechanize each underground job. (Bullock, 1973) This mechanization was duplicated in all of the other mines which followed St. Joe in opening mines in the Viburnum Trend. Now that they are all operated by Doe Run, it is fair to compare

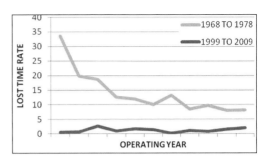

Figure 20.   Comparison of lost time rate for two periods for St. Joe/Doe Run

them collectively as a mining group; basically all the mines benefiting from the research that lead to their current practices. The results are best shown by looking at the comparison of two periods; one when the mechanization was taking place and the other when the mechanization had become standard operating procedure (See Figure 20). While there are of course other factors that must be considered, certainly this mechanization of mining from a labor intensive occupation to one of operator mechanization is one of the most critical factors.

These mines are most proud of their record of mine safety and the records that they have set. They are by far the most awarded of all of the mining groups of the coveted former Holmes Safety Award, now known as the NIOSH Sentential of Safety Awards:

Mine	Number of Winning Years
Fletcher	Six Times
Sweetwater	Six Times
Magmont	Four Times
Buick	Four Times
Brushy Creek	Three Times
Indian Creek	One Time
Casteel-Buick	One Time

Overall, the new lead belt mines have won this coveted award 25 times since 1971. These achievements did not come by way of luck, just good safety management or simply strong mining safety regulation. Of course, all of these factors have had their influence, but there is no better evidence in mining history of the importance of mining research, which lead to mechanizing jobs which put the miner in a position of safety, than has been demonstrated by St. Joe/Doe Run over the past 100 years.

REFERENCES

Anonymous, 1964, St. Joe Prepares for its Second Century of Growth and Progress, EMJ, McGraw-Hill, NY, NY, April, 33 pp.

Anonymous, 1972, "Louses Leading to Trapeze." and "Fertilizer, A Blasting Agent," *St. Joe Headframe, Vol.*7. No. 1, Spring, pp. 1 and 4.

Anonymous, 2000, Looking Back: 1953 Gismo, The Spokesman-Review, Feb. 18, [Newspaper archives]

Bain, C.K., 1953, St. Joseph Lead's Indian Creek Development, Mining Engineering, Transactions 196, pp. 899–904.

Bullock, R.L., 1959, Fundamental Research on Burn Cut Drift Rounds, *Second Annual Symposium on Mining Research, Bulletin N. 95,* University of Missouri-Rolla,.

Bullock, R.L., 1961, Fundamental Research on Burn Cut Drift Rounds, *The Explosive Engineer,* Volume 39, No. 1 & No. 2., Jan–Feb and Mar–Apr, Hercules Company, Wilmington, DL.

Bullock, R.L., 1962, Research on the Underground Application of AN/FO Mixtures in Small Diameter Holes, *International Symposium on Mining Research,* Volume 1, University of Missouri-Rolla.

Bullock, R.L. , 1962, Recent Developments in Drilling Research at St Joe, Fifth Symposium on Rock Mechanics, University of Minnesota, May, pp. 229–240.

Bullock, R.L. , 1968, Rotary-Percussion Drilling, The State of the Art, *Mining Congress Journal,* Volume 54, No. 12, December, pp 42–50.

Bullock, R.L., 1973, Mine-Plant Design Philosophy Evolves From St. Joe's New Lead Belt Operations, Mining Congress Journal, American Mining Congress, Washington, DC, Vol. 59, No. 5, May, pp. 20–29.

Bullock, R.L. , 1974, Industry-Wide Trend Toward All-Hydraulically Powered Rock Drills, *Mining Engineering,* Volume 60, No. 10, pp 54–65.

Bullock, R.L., 1975, Technological Review of the All-Hydraulically Powered Rock Drills, *SME Preprint No. 75-AU-,* SME, Littleton, CO.

Bullock, R.L., 1975, Optimizing Underground, Trackless Loading and Hauling Systems, Doctor of Engineering Dissertation, University of Missouri-Rolla, 132 p.

Bullock, R.L. , 1976, An Update on Hydraulic Drilling Performance, *Proceeding, 1976 Rapid Excavation and Tunneling Conference,* Chapter 33, Society of Mining Engineers of AIME, New York, 1976, pp. 627–648.

Bullock, R.L., 1976, A Survey on Hydraulic Drilling Performance—1976, *Tunnels and Tunneling,* September-October, pp. 38–45.

Bullock, R.L., 1983, Session Summation on Mine Systems Design, *Proceedings, Annual Workshop Generic Mineral Technology Center Mine Systems Design and Ground Control,* (Ed. Karmis, M. & J.R. Lucas), VPI, Blacksburg, VA, pp. 171–177.

Casteel, L.W., 1962, St. Joe's Pneumatic Charger, Mining Engineering, May, 1962, pp. 59–60.

Casteel, L. W., 1964, The First Century of Research by St. Joseph Lead Co., Mining Engineering, July, SME, Littleton, CO, pp. 111 -111D.

Christiansen, C.R., Calhoun, W.A, and Brown, W.F., 1959, Mining and Milling Methods and Cost at the Indian Creek Mine, St. Joseph Lead Co., Washington County, Mo., USBM I.C. 7875, p. 24.

Lane, W.L., et al., 2001, Pillar Extraction and Rock Mechanics at the Doe Run Company in Missouri1991 to 2000, Underground Mining Methods: Engineering fundamentals and International Case Studies, (Ed. Hustrulid and Bullock), SME, Littleton, CO, pp95–101.

Langefors U., and Kihlstrom, B., 1963, *The Modern Technique of Rock Blasting,* Almqvist, & Wiksell, Stockholm, pp. 405.

Lutjen, G.P., (1954) How to Sink a Shaft Safely and Efficiently, *144 New Ideas for Mine, Mill and Smelter Operations,* E & MJ, McGraw-Hill, NY, pp. 27–31.

Mark, C., 2008, The Introduction of Roof Bolting to U.S. Underground Coal Mines (1948–1960): A Cautionary Tale, NIOSH, Pittsburgh, PA, 11 pp.

Morgan, W.C., 1972, Travel Time and Earth Moving Production Computer Program, Caterpillar Tractor Company, Unpublished, Personally Communicated, Peoria, IL.

Mount, W.H., 1976, Tire Management Program is Serious Business at St. Joe Minerals, Mining Engineering, SME, Littleton, CO, pp. 38–40.

Mount, W. H., 1982, Contaminants of the Underground Atmosphere, *Underground Mining Methods Handbook,* Hustrulid, W.A., Ed., SME, Littleton, CO, Sect. 8, Chap. 2, pp. 1629 -1647.

Reed, J.J.; Bullock, R.L. and Yancik, J. J., 1958, "How St. Joe Uses Mining Research," Engineering and Mining Journal, V. 159, No. 5, May, pp. 85–87.

Reed, J.J., 1959, "Case History in Pillar Recovery," Mining Engineering, AIME Transactions, B.11, No. 7, July, pp. 701–705.

Reed, J.J., and Mann, C.D., 1960,"St. Joe Builds Practical Rock Mechanics Tools, Engineering and Mining Journal, V. 162, No. 3, Mar., pp. 100–106.

Reed, J.J., and Mann, C.D., 1960, "Prestressed, Preloaded Concrete Pillars give Better roof Control," Engineering and Mining Journal, V. 161, No. 11, Nov., pp. 88–93.

Reed, J. J., and Mann, C.D.,1961, "Full Scale Testing Develops Efficient Preloaded Concrete Pillars," Journal of the American Concrete Institute, Title No. 58-30, Nov. pp. 625–638

Reed, J.J., 1962, Survey of Developments in the Field of Rock Mechanics, Mining Engineering, SME, Littleton, CO, Apr, pp 60–62.

Reed, J.J., 1979, Case History in Pillar Recovery, Mining Engineering, SME, Littleton, CO, Jul, 701–705.

Stockett, N.A. , 1943, Mining Practices of the St. Joseph Lead Company in Southeast Missouri, Mining Technology, AIME Transactions, New York, N.Y., pp. 59–68.

Van Barneveld, C.E., 1924, Mechanical Underground Loading in Metal Mines, U.S. Bureau of Mines, Mississippi Valley Station, School of Mines and Metallurgy, Rolla, Missouri, 639 pp.

Weakly, L.A., 1977, Room and Pillar Ground Control Utilizing the Grouted Reinforcing Bar System, Monograph on Rock Mechanics Applications in Mining, SME, Littleton, CO, Chap 16, pp. 127–137.

Weakly, L.A., 1982, Room and Pillar Ground Control Utilizing the Grouted Reinforcing Bar System, Underground Mining Methods Handbook, W.A. Hustrulid, Ed. SME, Littleton, CO, pp 1,556–1,560.

Weigel, W.W., 1943, Channel Irons for Roof Control, Engineering and Mining Journal, Vol. 144, No. 5, May, pp. 70–72.

Wykoff, B.T, 1950, Wrapping Pillars With Old Hoist Rope, Mining Engineering, Transactions, Vol. 187, AIME, New York, August , 1950, pp 898–903.

Yancik, J. J., 1956, Development of Large Hole Burn cut Drift rounds, MS Thesis T 1107, University of Missouri School of Mines and Metallurgy, Rolla, Mo. 107 p.

Yancik, J.J., 1960, Some Physical, Chemical & Thermohydrodynamics of Explosive Ammonium Nitrate-Fuel Oil Mixtures, PhD Thesis T 1277, 256 p.

Maintenance Management and Sustainability: A Historical Perspective

José A. Botin

Universidad Politécnica de Madrid (Escuela de Minas), Madrid, Spain

ABSTRACT: Over the last decades, mine and plant maintenance have evolved from considering repair cost as the only management consideration to management strategies focusing on safety, environmental compliance and operational efficiency. This paper will present a historical perspective of the major changes experienced by maintenance management systems and technology in the last decades and how these changes have contributed to safety and sustainability of the minerals industry.

INTRODUCTION

The Association Française de Normalisation (AFNOR) defines maintenance as the actions to maintain or restore an equipment or system to a specific state, or to a state where it is able to provide a determined service.

In general, and very especially in the mining industry, maintenance is a critical factor for operating efficiency and sustainability. Tomlingson (2005)[1] estimated that for an average mining operation, direct maintenance cost exceeds 40 percent of total operating costs. He also states that including "hidden costs." derived from production losses, safety and environmental risks related to equipment failure, the incidence of maintenance may be as high as 60% of total operating costs.

The close relationship between maintenance and safety hazards is well documented in the MSHA accidents statistics[2], which estimated that in 2007, 42% of mine fatalities are directly related to maintenance activities. Furthermore, from my own the analysis of MSHA statistics by accident classification, I have estimated that maintenance is the primary or secondary cause for up to 60% of mine fatalities. A similar conclusion on the importance of maintenance on safety was obtained from a NIOSH study on major aggregate companies, published by Rethi,[3] who stated that 65% of all injuries documented originated from performing construction, maintenance and repair activities.

In recent decades, the steady focus of the mining industry on performance and productivity has fueled a continuous growth in the size and technology levels of mining equipment and systems. As the result, the maintenance function has become a complex production service which must be managed for maximum return on assets, safety and sustainability.

KEY CONCEPTS ON MAINTENANCE MANAGEMENT

Maintenance Management[4] deals with achieving an optimum balance between equipment performance and maintenance cost. In practice, the maintenance manager must deal with three key concepts: *reliability, maintainability and failure*. Furthermore, the optimum *maintenance strategy* for a specific piece of equipment or system must be based in a compromise between availability, reliability and maintenance cost. These concepts are discussed below in this paper.

"Reliability." describes the behavior of a system or equipment during operation. It characterizes its ability to perform its functions under stated conditions for a specified period of time. Similarly, *"maintainability."* describes the behavior of a system under repair and characterizes the relative ease and economy of time and resources with which an item can be brought back to operation.

As a maintenance concept, *"failure."* is and event causing the equipmentt to loose its capacity to satisfy operating standards. It represents the transition from the "under operation." to the "under repair." state. From the management perspective, failure represents the transition from a state controlled by reliability and operating performance to a state controlled by maintainability and maintenance performance.

Overall, maintenance performance is linked to the concept of *"Functionability."* or the proportion of scheduled time during which the equipment or system is operating (e.g., is able to satisfy the operating standards). Functionability may be used to describe the overall behavior of a system both at operation and at repair and may be characterized by the *"Availability."* function.

An equipment or system may fail in many different ways ("*failure modes*" each having different effects on the equipment and different probability of occurrence. In this context, the term *"maintenance policy."* refers to the set of management decisions to cope with a specific failure mode of equipment. For example, a truck tire may fail by "wear." or by "accident." and, while the probability of occurrence of a wear failure increases with tire age (age dependent failure), the probability of an accidental failure remains constant with time (non-age dependent or random failure).

Most authors classify maintenance policies into three groups:

- **Reactive** or Failure Based Maintenance policies (FBM). Where the maintenance actions (repair) are performed after the failure mode has occurred. For example, the accidental failure of a truck tire caused by sharp rocks is random and unpredictable. Hence, the only logical line of action (Maintenance policy) is to replace the tire after failure (Corrective maintenance).
- **Fixed time** or Life Based Maintenance policies (LBM). When maintenance actions are preventive and are planed at fixed time intervals, aiming to act before failure has occurred. For example, wear failure of a pump impeller is age-related. Therefore, a fixed-time maintenance policy aiming to replace the impeller before the pump fails would be adequate.
- **Predictive** or Condition Based Maintenance policies (CBM). Here maintenance actions are also preventive, but their timing is determined by either the failure simulation models prediction or through monitoring of the condition of the equipment relative to the failure mode. Condition assessment may derived from visual inspection or through monitoring of a physical parameter (e.g., vibrations, temperature.) using embedded or portable measuring.

The concept of *"maintenance strategy."* derives from the fact that equipment failure affects the organization in a different way (*failure consequences*). The effect of a failure may be limited to repair costs, but may also imply production losses or may affect product quality, customer service, safety or the environment. In this context, *"maintenance strategy."* is defined as the management focus for the maintenance of a specific equipment or system. The *maintenance strategies* may focus on *economy, availability or "zero failure"* depending on how and how much the functional failure of a equipment affects the over-all business process (e.g., cost, production, safety, environmental risk).

For example, the failure of the ball mill in a concentrator would probably imply important production losses. Therefore, mill maintenance should be planned for maximum *availability*. To the contrary, in case of a flotation cell, where failure consequences would normally be limited to the cost of repair, a strategy focusing on economy (repair cost), would be adequate. Furthermore, for a failure mode that may generate a safety hazard, or unacceptable environmental risks, a "zero failure." strategy (maximum reliability), is required.

Today's leading standard tool for the development of optimal maintenance strategies is *Reliability Centered Maintenance* (RCM),. RCM[5] is a methodology that analyses the equipment functions, identifies their safety, environmental and economic priorities, and directs management focus towards those units which are critical from the reliability, safety, and environment and production point of view. The key conceptual advantage of RCM is that, unlike conventional PM methodologies, RCM focuses in avoiding failure consequences, not necessarily by avoiding failures. Moubray[6] stated that the most efficient failure mode evaluation entails asking seven questions about the system under review, as follows:

1. What are the functions and performance standards of the asset?
2. What are the different failure modes of the equipment or system?
3. What causes each failure mode?
4. What happens when each failure mode occurs?
5. In what way does each failure mode matter?
6. What can be done to predict or prevent each failure?
7. What should be done if a suitable proactive task cannot be found?

RCM standard procedure was standardized in 2000 by SAE-JA1011, "Evaluation Criteria for RCM Processes."

A SUSTAINABLE MAINTENANCE STRATEGY MODEL

In practice, a maintenance strategy for a given equipment or system is represented by the management focus on one of the following three maintenance objectives: (i) minimize repair cost (economy strategy); (ii) minimize downtime (availability strategy) or, (iii) Maximize reliability (zero-failure strategy), where this management focus is materialized through

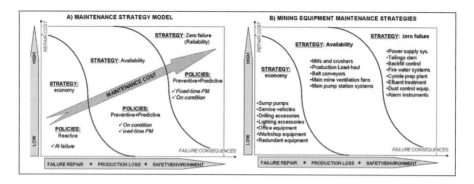

Figure 1. A sustainable maintenance model

the selection of a maintenance policy for each failure mode of the equipment.

The graphical model of Figure 1, represents a simplified graphical approach for the determination of a maintenance strategies an policies for mining equipment using a 2-D decision criteria based on two factors: (i) failure consequences and (ii) failure-related cost.

For example, when failure consequences are limited to the costs (left zone in graph), maintenance focus should be on minimum cost (*economy strategy*), which leads to the use of reactive maintenance policies (FBM). However, when failure can cause significant production losses (central zone in graph), then the maintenance strategy should focus on minimizing downtime (*availability strategy*), which leads to predictive maintenance policies (on condition) for failure modes not related to age and fixed-time PM policies when failure is age dependent (i.e., wear failure).

A management focus on reliability (*zero-failure strategy*) can only be justified when failure represents a safety hazard and/or unacceptable environmental risk (e.g., permissible machinery, tailings dam stability systems, safety instrumentation.). This strategy requires of frequent fixed-time preventive maintenance (LBM), which leads to low availability and high maintenance cost.

In a mining situation (Figure 1B), "availability strategies." are applicable to maintain equipment like mills, crushers, main ventilation fans and other equipment which failure would imply a total or partial shut down of production. "Zero-failure strategies." would be necessary to maintain cyanide preparation plants, tailings dam instrumentation, effluent management systems etc.

The expected cost expected cost C_t of an equipment maintenance strategy may be quantified by the following model[6]:

$$C_t = \sum_i (C_m^i + C_u^i + C_r^i)$$
(1)

For all failure modes *i*

Where:
C_m = Direct maintenance cost
C_u = Profit loss due to equipment unavailability during maintenance
C_r = Financial effects of failure related safety/ environment risk

Therefore, the optimum equipment maintenance strategy is that which minimizes the sum of these three cost elements over all equipment failure modes *i*.

Cost components C_u and C_r are difficult to quantify and often they are not considered for maintenance budgeting and cost control. However, proper consideration of all three cost components when stating management objectives is the decisive driver towards sustainability.

THE MAINTENANCE FUNCTION: A HISTORICAL PERSPECTIVE

Throughout history, the concept of maintenance was restricted to military logistics and to the need to preserve the operational capacity of an armed force both in peacetime and under operational conditions.

In late 18th Century, the Industrial Revolution and the boom of steal manufacturing and mass production assembly lines made evident the need to introduce maintenance as a part of the production cycle. Manufacturing equipment was subject to rapidly increasing failure rates with age where manufacturers would react by overproducing and stockpiling goods so that they could shut down their plant and repair their equipment while maintaining the supply to their customers.

In the period of pre-World War II, industry was not highly mechanized. Downtime did not matter much and maintenance was restricted to fixing the unit when it breaks. Most equipment was simple, over-designed and easy to repair and therefore, there was no need for planned maintenance other than cleaning, servicing and lubrication routines. It was during the post-war period, with the increase in mechanization and complexity of industrial processes and equipment led to the concept of preventive maintenance.

Glein and Freeman[7] (1932), in an early reference to the relation between safety and maintenance in coal mining, stated that *"Careful maintenance and discipline will do much to retard the occurrence of dangerous conditions."* However by 1950, with few exceptions related to safety, mine maintenance was not more than a reactive repair task with a low incidence—less that 20%—on operating costs. Therefore, mine managers considered maintenance as a necessary evil, with little or no consideration to safety, environmental control and risk management.

The modern concepts of *Preventive Maintenance* (PM) were developed in the 1960s with the great expansion of the manufacturing industry and the explosion of domestic consumer expenditure and greater market competition. Plant equipment and machines became complex and downtime became important to production capacity and efficiency, which led to the concept of preventive maintenance, which was first implemented in the USA by Ford Motor Company in 1965. Initially, it consisted of performing maintenance tasks (e.g., inspections, adjustments, cleaning, lubrication, parts replacement.) on fixed-time intervals, without regard to equipment condition.

The implementation of preventive maintenance in the mining industry was initially driven by the federal Coal Mine safety Act (1969)[8] , enforcing health and safety standards which required of preventive inspection and maintenance of permissible equipment to enhance safety. The Act refers to mine maintenance as a critical for the safety and the environment and thus, it may be considered an early reference to sustainability in mine maintenance. Almost 30 years earlier, Karl L. Kroneth[9] (1941) made a reference to the important role of inspection-based preventive maintenance in the reduction of safety hazards in coal mines, stating that *"The proper procedure for maintenance of permissible equipment may be divided into three parts: (i) inspection, (ii) actual maintenance and (iii) changes or improvements."*

Another important reference to the importance of maintenance for safety is made by Taylor in 1973 (Mining Engineering Handbook), stating that *"The*

maintenance department should play a large part in making its plant a safe one in which to work."

Mine maintenance concepts changed sharply in the 1970s and the '80s. In these two decades, mining experienced a sharp increase in production and productivity driven by the shifted in emphasis from underground to surface mining. Mine maintenance managers had to cope with the continuous scale-up of haulage trucks and other production equipment and fast technology developments such as dispatching, blasting, mine planning and mineral processing. In this scenario, the concept of maintenance evolved and preventive maintenance (PM) was introduced as a standard procedure. An early reference to this evolution is addressed by Hawthorne[10] in 1967 when he stated that *"Good and consistent lubrication is an absolute necessity for effective PM. It is also advisable to consolidate various maintenance jobs so that the overall lost production time of a machine is kept to a minimum; downtime on a multimillion-dollar machine is extremely costly from this standpoint."*

In this period, the most significant development was the introduction of *Predictive Maintenance (PdM)* in the 1980s. Predictive Maintenance or "Condition-based." maintenance uses real time monitoring techniques to predict failure when distinguishable symptoms may be detected before failure occurs. This results in less-frequent and less-serious equipment shutdowns, less unscheduled downtime, and better control of safety and environmental risks.

The mining industry followed the same trend and in late 1990s predictive maintenance became standard for maintenance. Haulage truck fleets and other major pieces of mine and plant equipment were equipped with condition monitoring systems to sense for temperature, pressure, levels, RPM, and other "health data." of critical subsystems such as engines, breaks, bearings and tires, which was radio-transmitted to control panels or dispatch centers. Furthermore, off-board diagnostic techniques (e.g., vibration, oils and lubricants, ultrasonic, and thermographic analysis) were implemented for signs of wear, oil contamination, or malfunctioning. The improvements of maintenance procedures and equipment design resulted in a dramatic increase of availability. As an example,[11] in early 1980s the average availability of truck fleets was below 70%. Today, 85% to 90% availability is commonly achievable by well run fleets.

Today, operational efficiency and risk management are the key drivers for maintenance. Therefore, the management approach to maintenance has shifted from "avoiding failure." to "avoiding failure consequences." This philosophy is the base of RCM or "Reliability-centered Maintenance." and, as

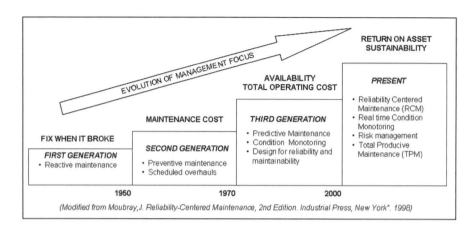

Figure 2.   The evolution of maintenance strategy and management objectives

described above in this paper, has become today's most widely used methodological framework for sustainability in maintenance. Most major mining companies operate sustainable mine maintenance programs designed on the basis of RCM strategies and advanced real time condition monitoring and diagnostic systems.[12]

To summarize, the historical evolution of the strategy and objectives of maintenance can be represented by the graph in Figure 2.

*   *A first generation* (before 1950), when maintenance was basically a handyman production service dealing with the repair of simple, over designed and easy to repair equipment. Only corrective maintenance was considered since proactive maintenance could not be economically justified.
*   *A second generation* (1950–1970); when the concept of "Preventive Maintenance." is introduced as the means of reducing downtime and maintenance costs. Maintenance becomes a well organized, self managed, production support service focusing on maintaining equipment performance. The sharp increase of maintenance cost leads to the implementation of maintenance planning and control systems.
*   *A third generation* (1970–2000); when a significant increase of equipment mechanization and automation, leads to a sharp increase of equipment owning and operating cost. Downtime becomes of very serious concern and management focus shifts to availability as the means to minimize operating cost. New "predictive strategies." are developed, based on the prediction of the condition of equipment through the

use of failure simulation models, inspection and monitoring of equipment condition (CBM).

Today, maintenance has become a complex production support service dealing with equipment and systems of a very high cost to own and operate. Maintenance focus on return on assets and sustainability. Regarding the mining industry, the importance of maintenance is well exemplified in our previous reference to Tomlingson[2] in this paper, who estimates that in an average mine, direct maintenance cost exceeds 40% of total operating costs and could reach 60% if "hidden costs." from production losses and other operating risks related to equipment failure.

CONCLUSION

In recent decades, mine maintenance has evolved from a handyman repair service to become a critical factor for safety, operating efficiency and sustainability. Today, for an average mining operation[2], direct maintenance cost exceeds 40% of total operating costs and this proportion may be as high as 60%, if "hidden costs" derived from production losses, safety and environmental risks related to equipment failure are included. Regarding the importance of maintenance for the reduction of safety hazards, MSHA[3] accidents statistics show that, in 2007, 42% of mine fatalities were related to defective maintenance.

REFERENCES

1.   Tomlingson, P.D. 2005. Maintenance Management: Minimizing Risk to Profitability. SME. Mining Engineering. Jan 2005. Littleton, Co. USA.

2. Mine Safety and Health Administration—MSHA. 2007. Metal and Nonmetal Fatal Accident Review.

3. Rethi, L.L. et al. 1999. Recognition training program for construction, maintenance and repair activities. NIOSH. Pittsburgh, Pennsylvania, USA.

4. Maintenance Management. 1995. Wiley and Sons, New York.

5. Moubray, J. 2001. Reliability-Centered Maintenance. Industrial Press Inc.

6. Botin, J.A. Ed. 2009. Sustainable Management of Mining Operations. pp. 308. SME. Littleton, Co. USA.

7. Gleim, E.J. and Freeman, H.B. 1932. Maintenance of electrical mine equipment from the viewpoint of the safety inspector. US Bureau of Mines.

8. Ramani, R.V. and Mutmansky, J.M. 1999. Mine Health and safety at the turn of the Millennium. Mining Engineering. Jan 1999. SME. Littleton, CO., USA.

9. Kroneth, K.L. 1941. "Maintenance of Permissible Electrical Equipment." Technical publication no 1387. AIME. New York Meeting. Feb. 1941.

10. Hawthorne, A.L. 1968. Surface Mining 1st Edition. Chapter 10.1. Preventive and Breakdown maintenance. Maintenance. SME. Littleton, Co. USA.

11. Golosinski, T.S. and Ataman, I.K. 1999. Total Open Pit Mining Systems: A dream or a Reality?. 16th Mining Congress of Turkey, 1999.

12. Lewis, Michael W. 2000. Remote-control monitoring cuts maintenance costs. Mining Engineering Feb 2000. SME. Litleton, Co. USA.

Innovative Solutions Developed at KOMAG to Improve Operational Safety and Reliability of Mechatronic Systems in Mines

Antoni Kozieł
KOMAG Institute of Mining Technology, Gliwice, Poland

Małgorzata Malec
KOMAG Institute of Mining Technology, Gliwice, Poland

ABSTRACT: The paper describes a role of KOMAG in implementing research results aiming at improving operational safety and reliability of mechatronic systems. Some examples of projects oriented onto investigating a relationship: MAN-MACHINE-ENVIRONMENT are given. A collaboration between the former Bureau of Mines then NIOSH and KOMAG in the domain of dust control is mentioned. The scope of the paper includes machines and equipment for an underground extraction and processing of minerals, control, diagnostic and monitoring systems, work safety, model, laboratory and in-situ testing of machines and equipment. The paper presents regulatory, technical and technological changes introduced in the Polish mining industry.

INTRODUCTION

Research-and-development work, actions aimed at reaching a higher safety level, risk analyses, promotion of positive safety culture as well as continuous control of hazards and resulting risk are some of strategic objectives of the KOMAG Institute of Mining Technology. For many years KOMAG has been involved in scientific, research and implementation projects that result in a systematic improvement of work safety as regards operation of machinery systems in the mining industry.

Research work in the scope of safety, in particular dust hazard control, which was realized together with American partners in 1990ties, can be an example. In 1995 the PIER organization (Partners in Economical Reform) opened the Occupational Safety Office in Katowice, which realized the programme of reducing hazards in the Polish mining industry. That programme enabled, for the very first time, to take advantage of the subjective feelings of employed people to assess the quality of work safety management in a plant (Harrison Daniel 1996).

A determination of basic factors causing a generation of coal dust and a determination of those ranges of activity in the Polish mines that could cause a reduction of dust concentration, were the objectives of the Polish-and-American research project within Maria Sklodowska-Curie II Joint Fund.

The fund offered a financial support to KOMAG and the National Institute for Occupational Safety and Health (NIOSH), Pittsburgh Research Laboratory (Pittsburgh, USA), enabling a realization of scientific-and-technical collaboration.

The results of studies given in conclusions of joint work indicated that "dust level in Polish longwalls could be reduced by an introduction of operational practices and methods of airborne dust control which were used in the USA."

A need to make miners and mining officers aware of a harmful impact of dust was also emphasized and benefits resulting from taking preventive measures in mining processes, used in Polish longwall workings, were presented. The results of specified research work formed the basis for undertaking the next programmes and projects, and their examples are presented in the paper.

NEW DIRECTIONS OF RESEARCH WORK AND NEW CONCEPTS AIMED AT INCREASING EFFICIENCY OF DUST CONTROL IN THE SYSTEMS USED IN SHEARERS AND ROADHEADERS

Warming-up of cut grooves and tips of cutting bits to high temperatures is, apart from dust generation, a side effect of coal cutting process with cutting drum bits. Also sparks can be generated then. In the case of accumulating methane, released from the solid coal during the cutting process, there is a risk of its ignition and a risk of methane-and-air mixture explosion hazard. An analysis of the accidents that had taken place in the Polish mining industry revealed a need of undertaking research work aimed at resolving the above mentioned problem.

Test Stands and Methodology for Testing Fire-fighting Efficiency and Preventive Measures Against Methane Ignition

A development of test stands that would enable to assess the efficiency of spraying installations and

Figure 1. Testing gallery—outside view

systems, used in shearers and roadheaders, was important for realizing the tasks of projects aimed at dust control and at an elimination of methane ignition.

Carrying out of such tests was an innovative solution and it required a development of test stands as well as of the testing methodology. The original concept was developed in the result of collaboration between the Central Mining Institute, "Barbara" Experimental Mine and the KOMAG Institute of Mining Technology.

The stand for testing efficiency of protecting and extinguishing methane ignition was installed in the "Barbara" Experimental Mine, on the surface, in a model of roadway section (see Figure 1). The testing roadway was closed at both ends. On one side an inlet window was installed and on the other side an entry gate was made. Inside the roadway a model of coal wall, that precisely reproduced the shape in natural conditions, was built. The roadway was equipped with indispensable electric equipment and measuring instrumentation to record and analyze the test results.

The methodology included a determination of the following parameters:

* An efficiency of extinguishing of burning gas
* An efficiency of gas ignition prevention

The extinguishing efficiency was measured by a time period required to suppress burning gas since the moment of starting up the spraying installation (starting air compressor and water pump as well as cutting drum).

The efficiency of gas ignition prevention was assessed through an elimination of gas ignition at operating spraying installation and rotating cutting drum.

The first tests were carried out on an experimental air-water installation designed to operate on the KSW-460NE longwall shearer manufactured by the Zabrzańskie Mechanical Works (see Figure 2 and Figure 3).

The obtained results of tests performed on the test stands were positive. The tests showed that keeping the strict relationship between the parameters of compressed air and water is an important factor enabling to achieve the required efficiency of spraying installation.

The positive results were obtained for the following parameters:

* Water consumption $30 \div 34$ l / min, at pressure about 0.4 MPa
* Compressed air consumption $1.3 \div 1.7$ m^3/min, at pressure about 0.4 MPa

Time of extinguishing of burning flames in the distance 0.1 m from the longwall face varied from 5 s to 35 s and when the burning gas was in the half of the cutting drum web depth, extinguishing of flame was instantaneous. For the same parameters, in both positions, there was no gas ignition. The tests, carried out in the rig, using the developed testing methodology, enabled a development of an experimental installation by the manufacturer and getting the certificate, which was indispensable for approving it for underground tests in the "Pniówek." Colliery.

Air-and-Water Spraying System for Longwall Shearers

The essence of the solution of air-and-water spraying system for a longwall shearer consists in using two spraying systems:

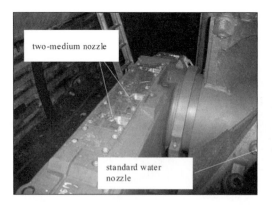

two-medium nozzle

standard water nozzle

Figure 2. View of arm and drum with spraying nozzles

Figure 3. Extinguishing of flames by the spraying system

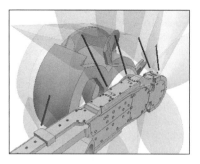

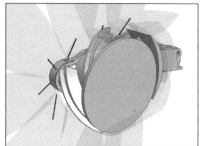

Figure 4. Model of the cutting drum external spraying with air-and-water mist

- External one, installed on the shearer arms, generating the separating water curtain around each cutting drum
- Internal one, in a form of nozzles installed on the plane of the cutting disk with spraying nozzles behind bits

The external spraying system for the KSW-460NW longwall shearer consisted of nine STK-2D two-media nozzles, which were installed on the shearer's arms. Mixing of water and compressed air took place in the nozzles.

The internal spraying system consisted of forty G-243 nozzles, which were installed on each cutting drum.

A model was built and model tests were conducted to verify the range of spraying zone. A required range of spraying streams, depending on a position of the cutting drum with cowls in relation to the coal face and a direction of cutting (see Figure 4), was determined on the basis of simulating spraying zones for seams of height from 1.7 to 2.0 m and cutting drum diameter equal to 1.6 m.

The tests also enabled to determine the locations of two-medium nozzles installation and the angles of nozzles in relation to the axis of cutting drum and the planes of shearer arm. And finally, they enabled to develop a design of the spraying system as well as to carry out the stand tests on a model of a shearer arm and on a real object.

A model of the cutting drum, on which the components of the spraying system were installed, was constructed. The model was placed in a foil tunnel of the following dimensions: 2.8 × 4 × 10 m (see Figure 5).

During the tests the range of spraying and a creation of the tight curtain around the cutting drum were tested at assumed spraying parameters. Water pressure was changed in the range 0.29 ÷ 0.41 MPa, and air pressure was changed in the range 0.29 ÷ 0.41 MPa. The tests enabled to determine water consumption (6 ÷ 11 l/min) and compressed air consumption (650 ÷ 1500 l/min). It was found that pressure as well as water and air consumption should be individually selected for given operational conditions, including

Figure 5. Cutting drum model

the range of spraying streams and a demand for water and air.

Operational tests, which were conducted in W-9 longwall, at seam 357/1 in the "Pniowek" Colliery, were the next stage.

The tests objectives were as follows:

- To verify the correctness of assumed parameters of spraying installation on the KSW-460NE longwall shearer in operational conditions
- To verify the correctness of operation of two-medium spraying nozzles designed at KOMAG
- To compare dust control effectiveness of air-and-water system and conventional water system

The operational tests enabled to determine water and air relationship that guarantees an effective operation of the installation. For the air-and-water system the parameters were as follows:

- Water pressure—0.5 MPa
- Air pressure—0.55 MPa
- Water consumption—70 ÷ 90 l/min
- Air consumption—2.5 ÷ 4.5 m³/min

For conventional water system the parameters were as follows:

- Water pressure—about 0.3 MPa
- Water consumption—about 210 l/min

The measurements of dust level proved similar effectiveness of both systems as regards dust control. However, a significant reduction of water consumption is a big advantage of the innovative spraying system, which has a positive impact on the mining process, transportation of the run-of-mine and process of coal beneficiation.

The tests, previously conducted in the "Barbara" Mine, indicated that an innovative solution of the air-and-water system can effectively prevent methane ignition and effectively extinguish burning methane. The operational tests also confirmed a correct operation of two-medium spraying nozzles.

Air-and-Water Spraying System for Roadheaders (Bałaga, Prostański, Rojek, and Sedlaczek 2009)

Positive results of tests and positive operational effects of air-and-water spraying system for a longwall shearer, confirmed in operational conditions, aroused interest of a manufacturer of roadheaders.

In the result of collaboration between KOMAG and REMAG, JSC a technical documentation of air-and-water installation was developed, its prototype was made and it was installed on the R-200 roadheader (see Figure 6).

External spraying packs were equipped with STK-R two-medium nozzles. The curtain system was included in the roadheader water system. A control divider, which enables cutting off water supply to the air-and-water installation during roadheader stoppage, while maintaining water circulation in the cooling system, was used.

After obtaining positive test results in the "Barbara" Experimental Mine as regards control of gas hazard, the roadheader was implemented in the "Murcki" Colliery in December 2008. So far, the positive experience, resulting from operational tests, made REMAG, JSC install the air-and-water system on the KR-150 roadheader.

Figure 6. Tests of the water curtain generated by the air-and-water system installed on R-200 roadheader

Stability of roof support on inclination	Graphical symbol	Limiting angles of loosing stability for given height H	
		H=2.29 [m]	H=3.9 [m]
In relation to longitudinal axis	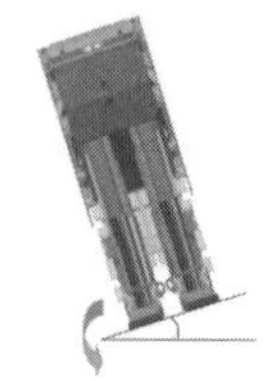	26.8°	17.5°
In relation to transverse axis downwards	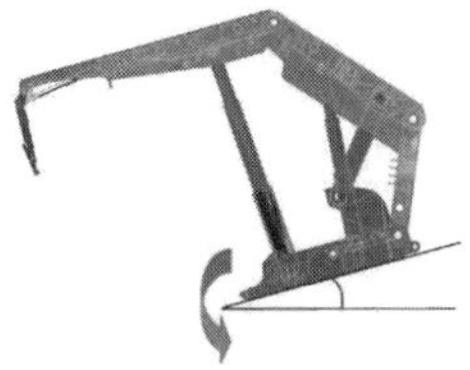	42.3°	27.8°
In relation to transverse axis upwards		44.1°	32.1°

Figure 7.   Values of limiting angles of loosing stability of ZBMD–22/44-POz powered roof support

RESEARCH AND DEVELOPMENT WORK UNDERTAKEN AT THE KOMAG LABORATORY OF MODELLING METHODS AND ERGONOMICS TO INCREASE SAFETY IN THE SCOPE OF MAN-MACHINE-ENVIRONMENT RELATIONSHIPS

Virtual prototyping of mechatronic systems, with special attention paid to man-machine-environment relationships, is one of KOMAG's activities of high priority. Designing of a virtual prototype includes comprehensive tests of computer models of machines and equipment reflecting in-situ conditions, including the operators' work conditions. Due to such an approach it is possible to create safe and human friendly work environment.

Modelling and Visualization of Technical and Health Hazards in the Mechatronic Systems Used in Mines (Jaszczyk, Michalak, Winkler, and Tokarczyk 2007)

Machines and auxiliary equipment, operating underground, are a source of technical hazards. A loss of stability of a machine on uneven floor, breaking of

chain in a flight-bar conveyor are examples of problems caused by a technical factor.

Health hazards are often caused by incorrect technical solutions (e.g., excessive dust generation), by uncomfortable and inconvenient work conditions (e.g., awkward body postures) or are caused by improper human behaviour at the work place (e.g., breaking the rules of wearing anti-dust masks).

Modelling and Simulation of Machines' Operation in Emergency Conditions

Tests of roof support stability on inclined floor are carried out using the MBS (Multibody Systems) method implemented in Visual Nastran 4D software environment. Limiting angle of loosing stability is determined for the maximal and minimal support height (see Figure 7).

A roof support stability during transportation upwards is tested in the case of emergency braking of a transportation platform. When breaking of the catch between the platform and locomotive occurs, rolling down of a free platform takes place. A simulation of emergency braking indicates the importance

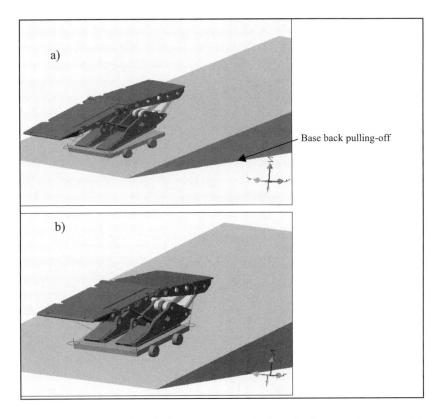

Figure 8. **Effects of transportation platform emergency braking in the case of improper (a) and proper (b) fixation of powered roof support unit**

of the support fixation on a transportation platform and the impact of incorrect fixation (see Figure 8).

Visualization of Dust Penetration into the Respiratory System

Giving information to employees about the effects of not using personal protection devices is one of the important factors in the process of preventing occupational diseases. A visualization of the effects caused by using or not using masks, while staying in the zone of increased dust concentration, has a psychological sound (Figure 9). An employer, expecting increased dust generation, can prepare proper protective measures and should inform employees that use of protective devices is in their own interest and for their own benefit.

Use of Virtual Technologies for Reconstruction of Accidents (Bojara, Jaszczyk, Michalak, and Winkler 2006)

A visualization of risk factors, dangerous situations, failures and accidents plays a significant role in a process of analyses of theoretical and real events. An analysis of dangerous situations is conducted for a reconstruction of accidents, for an analysis of transportation and assembly processes as well as for a proper operation of machines. A visualization of human behaviour and of an operation of machines in given situations can form initial material for a preparation of training materials. A scenario created on the basis of post-accident materials such as:

- Testimonies of witnesses and outsiders
- Sketches of place of accident
- Post-accident reports
- Technological installations
- Technical documentation
- Photographs of place of accident

creates initial material for a visualization.

An analysis of dangerous events requires suitable methods. A theoretical analysis of event with the use of Event Tree Analysis (ETA) method is the first step. A tree of events, which describes the accident

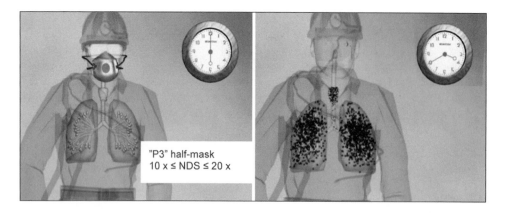

Figure 9. Visualization of results caused by using and not using the mask while staying in the dust zone

Figure 10. Visualization of accident with the use of photographs taken at the place of accident

with division into events significant for the accident, is a result of that analysis. Such a division indicates important elements that should be included in a visualization. Modelling of objects, which take part in the event, is an important stage in the visualization process.

Models of material objects and machines are built on the basis of available CAD documentation or, in the case when the documentation is not available, the PhotoModeler software, which enables to build geometrical models on the basis of photographs of real objects, is used. After a completion of modelling, the accident environment with machines and equipment is transferred to 3DstudioMax software, where human silhouettes are introduced (see Figure 10).

A development of multimedia material in a form of animation or static pictures is the final part of visualization. Precise imagining of all the factors makes a professional visualization. However, it requires moving in a dispersed software environment. KOMAG

has all indispensable software and can realize most complicated visualizations.

Aiding the Risk Assessment Associated with Lack of Adaptation of Machines to Operator's Anthropometrical Features (Jaszczyk, and Michalak 2008)

Planning the space for machine's operator is one of the stages in the designing process of new solutions. Due to the dimensions of machine components (travel system, steering wheel, drive), it is not always possible to change the dimensions of operator's cabin, what impacts a type and method of instrumentation arrangement in the cabin. An analysis of the operator's comfort and range zones is made on the basis of the standards, which include basic definitions, according to the developed algorithm, with the use of PRO-M software. The software modules enable to build a model within the space of the machine under designing. The results of analyses, in a form of graphical files and technical descriptions,

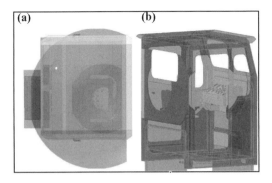

Figure 11. Ergonomic assessment of operator's cabin in a floor-mounted locomotive: (a) model of comfort and range zones, (b) range of comfort zones in operator's cabin

Figure 12. Structure protecting operator of ŁBT-1200 EH/LS-A loader

are entered to the PRO-M software, where the solution correctness is verified or the need to make changes is indicated (see Figure 11). The use of software and additional tools significantly increases the level of designers' knowledge and enables to avoid costly mistakes in the further stages of the machine designing process.

Structures Protecting Operators of Mining Face Machines (Kalita, Prostański, Tokarczyk, and Wyrobek 2006)

Roof fall, which can be dangerous for operators of mining machines, is among others, one of the hazards that occur during coal mining operations. At the stage of designing self-propelled machines, which operate in roadway systems, the structures protecting an operator against falling objects, meeting the requirements for FOPS (Falling Object Protective Structures), have to be included.

A protection against falling objects, whose energy does not exceed 11.6 kJ, is the task of FOPS. An ability of absorbing such energy by a protective structure should be confirmed by tests as regards resistance to impact loads, according to the standard. A protective structure should ensure proper Deflection Limiting Volume (DLV) for the operator, in which there are no plastic and elastic deformations. An example of the structure protecting an operator of ŁBT-1200 EH/LS-A loader, designed at KOMAG, is presented in Figure 12.

The Finite Elements Method (FEM) was used while designing (see Figure 13), and a verification of numerical tests was carried out at the test stand. Positive results of calculations and stand tests enabled to design a cabin for a loader operator according to DLV.

Safety of Underground Mine Transportation Systems (Berkowicz, Dudek, Halama, Michalak, Rajwa, and Winkler 2008)

Safety of mine transportation is controlled by specialists from the field of mining, machinery construction, electrotechnics and electronics, work safety and health protection, who work in a dispersed environment, including coal mines, factories, inspectorates of mines and specialistic institutions. They take part in the process of ensuring safety at different stages of life cycle of transportation systems, starting from designing, through testing, manufacturing, assembly, approval for operation, use, modernization and withdrawal of the machine from operation. They use the knowledge developed in a form of standards, guidelines, regulations, instructions collected on paper or electronic information carriers. Despite of technical possibilities, the knowledge has not been yet collected in internet resources, and due to that it is not available as a whole for all the participants of a safety ensuring process. That is why, it is suggested to create resources, including the knowledge developed in a form of textual files, graphical files and interactive models of workings, machines and equipment integrated with risk factors, which are assigned to them. A development of a given domain and sharing the knowledge resources will be possible due to a participatory mode of ensuring the safety (PTKB), by making available the software platform, which enables a collaboration of specialists from different domains. An example of participatory designing mode realization on the internet platform is presented in Figure 14.

Figure 13. Map of dislocations after a fall of weight of energy equal to 11600 J within the time equal to 0.01152 s

A verification of designs is conducted in the environment of specialistic software such as:

- Finite Elements Method (MSC.Patran)
- Analysis of multi-body systems (MSC. visualNastran)
- Ergonomics analyses (Anthropos- ErgoMAX)
- Biomechanical calculations (3D SSPP 4.3)

Suggested computer tools are embedded in the internet platform and they form resources of specialistic knowledge, which are available outside coal mines. Designing and operational knowledge resources are also available for trainings on mining safety. Their connection with the transportation systems, used in real conditions, is their significant advantage. Trainings can be addressed to precisely specified receivers, e.g., operators of floor-mounted railway systems. Obviously, this model of collaboration has the limitations, which mainly result from utilitarian and commercial understanding of knowledge resources. Thus, it will be necessary to select precisely the key resources important for each user due to their core business. The solutions of intellectual property protection will be important in this case.

USE OF RFID TECHNOLOGY AND DATABASE SYSTEMS FOR EVIDENCE OF MACHINES COMPONENTS (JANKOWSKI, MAKSYMOWICZ, MEDER, SZCZURKOWSKI AND WOREK 2008)

Radio frequency identification (RFID) of people and objects becomes more and more common in the mining industry due to the requirements of work safety and a development of automation of production and logistic processes. The system for recording of time and operational conditions of powered roof supports, which enables processing the necessary data to assess a wear degree of these components, is the first implemented RFID technology in the Polish mining industry (see Figure 15).

The system consists of the following components:

- RFID reader, microcomputer and reading lance
- Passive transponders located in a special casing on the machines components, where each transponder has its own and unique identification number
- Software enabling management of identification system and controlling its data base

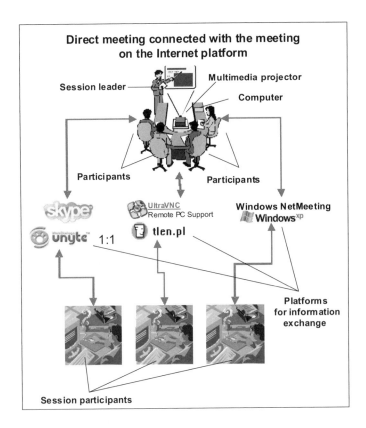

Figure 14. Forms of participatory designing mode realization on the internet platform

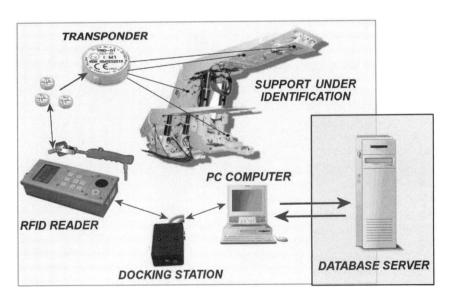

Figure 15. System for identification of powered roof support components on the basis of RFID technology

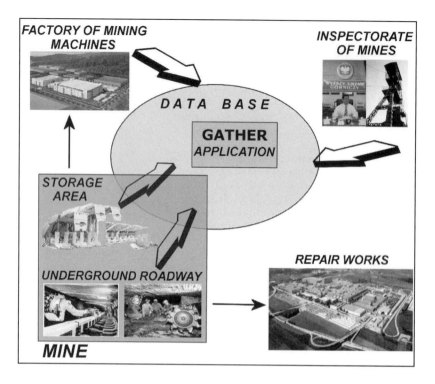

Figure 16. Structure of information system for an identification of mining machinery subsystems

The identification system is adapted for operation in rooms threatened by explosion hazard and is subjected to ATEX Directive requirements (it is intrinsically safe).

The GATHER software, developed by KOMAG, is an inherent part of the system for an identification of powered roof support components. The structure of the system information technology is presented in Figure 16.

RFID system, implemented for operation, is designed for use at a high management level. An effective collaboration with manufacturers of mining machines and mines paves the way for RFID applications in the Polish mining industry.

SUMMARY

New technologies are required to satisfy the Polish demand for hard coal and the need for increasing the effectiveness of the mining industry. "Clean coal technologies," renewable sources of energy and waste management can be examples. A realization of these strategic objectives will cause an increase of power security in Poland, meeting ecological and work safety requirements.

KOMAG will continue the research work aiming at a reduction of energy consumption of mechanical systems, which are used in the mining industry, implementing state-of-the-art mechatronic solutions and low-emission technologies. A broad, interdisciplinary collaboration with foreign institutes, universities and other organizations will help to reach these objectives. The experience gained so far as regards a realization of international projects confirms the need of continuing the joint work.

KOMAG has at its disposal the research personnel of high qualifications in different specializations and very well equipped testing infrastructure. It has implemented quality systems and can compete as regards a complexity of services and consulting activities of interdisciplinary character.

Present needs of the mining industries are challenges for KOMAG and collaborating scientific and industrial partners.

REFERENCES

Bałaga, D., Prostański, D., Rojek, P., and Sedlaczek, J. *Air-and-Water Spraying in Polish Shearers and Roadheaders as a New Method for Control of Dust and Methane Hazards.* Mining Machines 2/2008, Gliwice, June 2008. (in Polish)

Bałaga, D., Prostański, D., Rojek, P., and Sedlaczek, J. *Air-and-Water Spraying System for Shearers and Roadheaders—Solution Still Under Development.* Mining Machines 1/2009, Gliwice, March 2009. (in Polish)

Berkowicz, M., Dudek, M., Halama, A., Michalak, M., Rajwa, S., and Winkler, T. *Securing the Safety of Underground Mining Transportation.* Joint tests of ZINT (Integrated Scientific and Technological Institute) "Fuels—Safety—Environment," a virtual scientific network, in 2006–2007, Central Mining Institute 2008. (in Polish)

Bojara, Sz., Jaszczyk, Ł., Michalak, D., and Winkler, T. *Use of Virtual Technologies in Reconstruction of Accidents in Mining Industry.* Mining Machines 1/2006, Gliwice, March 2006. (in Polish)

Harrison Daniel, J. *Polish-and-American Collaboration in Work Safety Management.* International Scientific-and-Technical Conference, Szczyrk, 17th-19th September 1996.

Jankowski, H., Maksymowicz, L.J., Meder, A., Szczurkowski, M., and Worek, C. *Practical Experience Gained During Implementation of RFID Technology and State-of-the-Art Database Systems for Evidence of Mining Machines Components.* Underground Mining School 2008. Conference Proceedings. AGH University of Science and Technology, Kraków 2008. (in Polish)

Jaszczyk, Ł., and Michalak, D. *Supporting Assessment of Risk Associated with not Proper Adaptation of the Machines to Anthropometrical Features of Operator.* Industrial Safety—Volume 2, Warszawa 2008. (in Polish)

Jaszczyk, Ł., Michalak, D., Winkler, T., and Tokarczyk, J. *Modelling and Visualization of Technical and Health Hazards in the Mining Mechanization Systems.* State Mining Authority 2007. (in Polish)

Kalita, M., Prostański, D., Tokarczyk, J., and Wyrobek, E. *Protective Structures for Operators of Face Machines.* Mining Machines 4/2006, Gliwice, December 2006. (in Polish)

Kozieł, A., and Organiscak, J.A. *Joint Polish and U.S. Study of Longwall Dust Generation Factors and Control Practices.* MINEXPO International 2000—Sessions—Underground Mining and Development (Coal) II—Las Vegas, Nevada, USA—9th–12th October 2000.

Kozieł, A., Collinet, J.F., Organiscak, J.A. and Meder, A. *Joint Polish-and-American Scientific-and-Research Work on Factors Having Impact on Dust Generation in Longwall Panels and Actions for Airborne Dust Control.* Mining Machines, Gliwice, June 2001. (in Polish)

Rozmus, P. and Sobala, J. *Safety Management System and Control of Dust Hazards in Mining Industry.* International Scientific-and-Technical Conference, Szczyrk, 17th–14th September 1996. (in Polish)

Immigrant Labor Forces and Mine Safety

Nancy L. Dorset
West Virginia University, Morgantown, West Virginia, United States

Lawrence A. Hunchuck
The Pennsylvania State University, Uniontown, Pennsylvania, United States

ABSTRACT: The need for bituminous coal miners in southwestern Pennsylvania multiplied rapidly from 1880 through the end of World War I in 1918. For the most part, newly immigrated men from southern and eastern Europe filled these new positions. However, the new immigrants did not speak English and directing them in safe work procedures was difficult if not impossible. This being the peak of the Coal and Coke Industry, the immigrant miners often paid for this lack of training with loss of limb or even their lives. With the new wave of immigration to the U.S. including south of the border, the language barrier is becoming a safety issue once again.

INTRODUCTION

With the onset of the Industrial Revolution, the American Industrialization during the mid 1800s increased the demand for a more economical and efficient fuel to facilitate the needs for business and domestic consumption. This also increased the demand of the metallurgical coal requirements for the growing iron and steel industries. The fuel demand was practically resolved by utilizing the bituminous coal from the Pittsburgh seam. The Pittsburgh coal seam which covers much of southwestern Pennsylvania is nationally known as having some of the finest high-volatile metallurgical coals in the world. Among the bituminous coals, having a high carbon content and low ash content, the initial demand came from the "Connellsville District." This District covers an area of the Pittsburgh seam that encompasses Fayette and Westmoreland counties in Pennsylvania and had the highest-quality coal for making coke. This area became a world-class center for coke within the Pittsburgh Seam at that time. The coal was soft, easily mined and in a seam ranging from 120–130 inches thick and was the first to be exploited.

With this increased fuel demand, massive labor shortages sky-rocketed as the requirements to fulfill the obligations to sustain this need increased. Along with this substantial industrial growth came massive migration of many Americans to the Western Federal Lands being opened to settlement. Simultaneously in Europe, stretching from the Adriatic to the Baltic Sea, there was growing economic, political and social turmoil. The Austro-Hungarian Empire, which was agricultural and emerging from a suppressed social structure as well as being involved in political, cultural and economic conflict, provided the opportunity for meeting the labor demand in the United States. A steady growth in population, a shortage of farmland, and oppressive taxes, created severe economic poverty. Many eastern Europeans were banned from speaking or writing in their native languages. Wars in the Austrian/Hungarian Empire forced military enlistments of all able-bodied men. Legalized persecution threatened ethnic and religious minorities. It was a courageous decision to leave and give up all that was familiar to them and travel across an ocean to a foreign land.

In southwestern Pennsylvania, the rapid expansion of mining operations demanded a rapid increase in the labor force. Thousands of immigrants took advantage of the increasing employment opportunities available in the United States, especially in the industrial area of southwest Pennsylvania. Farmers and craftsmen were to become coalminers, coke yard workers, and steelworkers for southwestern Pennsylvania's growing coal, coke and steel industries. An 1850 census indicated that most of the immigrants in Pennsylvania were from England, Scotland, Germany, Wales, Ireland and France and these people dominated the early mining industry. These immigrants, mainly from the British Isles encompassed almost the entire Pennsylvania mine labor force.

The Southern and Eastern European immigrants started arriving in the coal fields as early as the 1880s, with the timeline comprised as follows: (Coal and Coke Heritage Center–Penn State-Fayette University)

- Slovaks—1880
- Magyars (Hungarians)—1882
- Poles—1890
- Croatians—1890
- Serbians—early 1890s
- Italians—1895

However, the great bulk of immigrants arrived after 1901 including: Russians, Bulgarians, Romanians, Ukrainians, Syrians, Armenians, and Macedonians. Only 14% of these immigrants had any mining experience in their native land (Fay 1919). The immigrants from southern and eastern Europe faced ethnic and racial discrimination from Western Europe and the immigrants which have already settled here. The accident reports and analyses must be interpreted with this in mind. For instance, Albert Fay, a mining engineer at the Bureau of Mines, in his 1919 Coal *Age* article states:

> The Americanism of this great body of labor, and its relation to accidents in the mining industry are two problems of prime importance—the first as affecting citizenship and the growth of this Republic, and the second as an economic problem in mining costs. The former exerts an influence upon the latter, for the former implies education, social welfare, civic pride and a general uplift to the ideals that America stands for. Ignorance, dirty and filthy living conditions, ill-health, disregard for law and order, discontent and lack of civic interest lead to indifference and carelessness, which are perhaps the greatest of all accident cases.

Later, in the same article, Fay accuses the recent immigrants of being ignorant and inexperienced and thus lowering the safety standards and living conditions in the coal patches (which is a German slang word meaning renting to a new tenant). In fact, the coal operators recruited men from southern Italy to come to the United States to serve as strikebreakers as early as 1895. But, on the other hand, Clay F. Lynch, General Superintendent of the Frick Coke Co., sent out a memo to all mine superintendents in 1917 requesting "Have your foreign-born employees understand also that if they have peaceful law abiding and industrious friends elsewhere who are now being subject to persecution or violence on account of their nationality, we will be glad to have them come to the coke region to work for us and to accept the protection we offer." Company agents recruited employees at their ports of entry and encouraged the immigrants to refer job opportunities to friends and relatives in their home lands. The coal companies would pay for their initial transportation cost and the immigrants would reimburse the companies from their regular pays.

The Language Barrier and Mining Accidents

Most of these new immigrants lacked the training and experience needed to perform in the mining industry. Many were farmers or worked as laborers in their native lands. Upon coming to the United States they decided to follow the occupation of mining because the work provided a better means of income than any other mode of employment. Many of them had been here only a short time and knew little or nothing about mines or mining conditions such as: rock formations, mine gases (or damps as they were referred to at that time), the hazards associated with coal dust in association with the handling of black powder—matters about which every coal miner should have been thoroughly informed. To determine whether a piece of slate or roof is or isn't likely to fall often requires a considerable degree of experience, and the majority of the Slovaks, Magyars, and Italians didn't have this experience.

Few, if any, were able to speak English at the time of immigration and there was a high degree of illiteracy in their own language. "The southern and eastern Europeans has had practically no experience in mining," writes Lauck, "He is immediately employed in the mines after reaching this country, and is unable to read printed notices or comprehend oral instructions" (Lauck 1911a). According to an article in the 1914 Connellsville Weekly Courier, "it is because of their ignorance of the language that foremen who bossed these foreigners have to use sign language. "They are routed out in a bunch every morning, taken to the yards, and shown by motions what they are expected to do." (The Weekly Courier 1914). The new immigrants were at first employed to do the rough and most unacceptable work in the mines and on the coke yards; however, they have gradually advanced in the scale of occupations. They generally filled all positions of lesser responsibility in both mining and coke-making, such as pick-miners, machine runners, cutters, loaders, roadmen, brattice-men, etc. The majority of such occupations as drivers, yard foreman, and fire boss, where the responsibility was greater, were filled by Americans, Germans, Irish, or Scotch. Engineers, mine foremen, superintendents, and other positions of larger responsibility were almost exclusively occupied by Americans, Germans, or members of English-speaking races.

In an article for the popular audience, Lauck stated, "So far as evidence exists there seems to be a direct causal relationship between the employment of immigrants and the prevalence of accidents in western Pennsylvania" (Lauck 1911b). The Pennsylvania State Mine Inspector for the Third District states in

Table 1. Fatalities sorted by ethnicity for 11th District

Country	Deaths	Country	Deaths	Country	Deaths
American	26	Polish	48	Russian	13
English	5	Hungarian	26	Lithunian	3
Welsh	2	Italian	45	Belgian	3
Scotch	6	Slavonian	71	Swede	4
Irish	5	Austrian	32	French	4
German	11	Finnish	1		

his official report: "(The immigrants are) ignorant of the business of coal-mining and their term of service as miners was a brief one. They knew little or nothing of the safeguards to be employed …but were ignorant of our language and could not possibly be instructed properly and warned of the danger by the officials in charge of the mines" (Adams 1884).

Lauck notes the immigrant churches offered reading and writing courses in English. Some offered citizenship classes to prepare employees and their family members for the exam necessary for naturalization. But until the immigrants were fluent in English, the different ethnic groups were segregated to different mines or, at the very least, to different places or different shifts at the mine. "In other sections the mixture of races (ethnic groups) in the operating forces produce a Babel of tongues which effectively prevents concerted action" (Lauck 1911b). Because of the language barrier, the mine operator would have to hire several extra knowledgeable men to serve as assistants to the mine foreman.

Roderick included the ethnicity for the fatalities that occurred in the Eleventh Bituminous District, which includes the Connellsville Coal District, from 1900–1906 (Roderick 1906). Unfortunately there is no data to indicate the percentage of each group in the general mining population. However, as seen from Table 1, the southern and eastern European nations account for the majority of fatalities.

By 1914, the first wave of immigrant miners were qualifying as firebosses or as foremen. The Courier Weekly article of 1914 quotes a mine inspector that declares the immigrants swill have "crowded the American miner to the wall" within the next ten years. The article further notes that many miners, both immigrant and American, were taking advantage of correspondence courses in mining. In addition, immigrant miners that had little education in their youth were mastering other subjects available by mail.

Efforts to Increase Mine Safety in the Connellsville Coal District

The H.C. Frick Coke Company was the dominant employer in the Pennsylvania Connellsville coal district. Circulars were sent to all the company's mines on a regular basis to provide information from the main office. In 1891, Thomas Lynch, the President of H.C. Frick Coke Company, introduced the phrase "Safety is the first consideration" at the top of every circular. The motto "Safety First" soon appeared on signs at the mine sites shortly after. This appears to be the origin of this motto that is still in use over 115 years later. As a result of the Mammoth Mine Explosion in 1891, the Frick Coke Company published a set of 25 safety rules to be followed by all of its mines and employees. It was published in all the major languages used throughout the region. As accidents and incidents increased, the number of rules increased. The Coal and Coke Heritage Center has two copies of the booklets in Slovenian and Croatian from the 1920s. *The Coal Miners' Pocketbook* listed most of the rules in its 1916 edition (Foster 1916). A comparison of the Frick rules with the Pennsylvania bituminous coal law shows that the Frick Coke Company safety rules exceeded that required by the law.

The Frick Coke Company also made other efforts to communicate good work practices. Circular 97, dated October 1, 1909 announces that a booklet, Regulations for Systematic Timbering, was available in English, Polish, Slovak, Italian and Magyars. To assist the superintendent in the proper distribution of the booklets, the language and the title of each booklet is identified in the circular. A Frick Coke Company circular dated November 20, 1917 informed the mine superintendents that a motion picture on mine safety will be made available. C.L. Albright dictates that all of the men and their families are to see the film. He encourages the superintendents to "Stir up some enthusiasm" by having the coal patch band to play in conjunction with the showing of the film.

A small article in *Coal Age* explains that the Allegheny Pittsburg Coal Company put up illuminated signs throughout the company towns so the women could read the safety messages and "then lecture their husbands." The article further states that "Often the victim of an accident quarreled with his wife before leaving home that day." (Reynolds 1924).

C.L. Albright, Manager of Accident Relief for the Frick Coke Company, addressed company officials concerning accidents and production at the annual company dinner on January 10th, 1920 (Albright 1920). Although he did not speak specifically to the issue of foreign-born employees, he offers an analysis of mining fatalities in the company's mines. He, related how the fatalities per 1,000 employees declined from 1907, when active safety work began, until 1919. Albright blames the "Great War" (World War I) for taking away its most capable employees. However, he remains concerned about the excuses being made for fatalities at the company's mines. Out of the 49 fatalities that occurred in 1919, "20.4% were chargeable to the hazard of the industry. Conditions were such that there was nothing to indicate impending disaster." Another 33.7% of the fatalities were deemed due the fault of the victim. These fatalities were further broken down into the following categories:

- Bad practice—36.6%
- Carelessness—26.3%
- Bad judgment—26.3%
- Disobedience—10.6%

The rest of the fatalities, 40.9%, were due to company negligence. He listed the reasons as:

- Existence of bad conditions—40%
- Bad judgment—35%
- Bad practice—15%
- Inexperience—10%

Albright noted there was an increasing practice of assigning the cause of fatalities to "carelessness" of the victim. He proposed that "thoughtlessness" was a better term. This implied a "failure to comprehend the consequences that may result from a negligent or overt act." Although safety meetings were already required at all operations, Albright wondered out loud if they had become "operations" meetings where production was stressed and directions for the work at hand were the dominant theme. He suggested that safety meetings be held outside, noting that education was an ongoing process that required attention each working day.

In closing, Albright reported how much the company had spent on safety activities and losses due to injuries, as follows:

- $305,000 used to promote safety
- $300,000 paid out for injuries (an average of $78.21 paid to each injured man)

- $335,000 in lost wages for the injured men
- A loss of 350,000 tons of coal due to injured men off work

This is large expense if one remembers that the values above reflect 1920s dollars. At the time, the highest paid miners were making less than $22 per week if they worked the full six days per week. However, temporary lay-offs and strikes often reduced the number of work days available.

Immigrant Mineworkers in the 21st Century

The number of non-native speakers had greatly increased in the mining industry since the early 1990s. In 2009 the Bureau of Labor Statistics reported that 21% of hourly employees were of Hispanic or Latino ethnicity (BLS 2009). Although the Bureau does not indicate what percentage are non-native speakers, the federal Mine Safety and Health Administration recognized a need to offer a Spanish translation of their web site as early as 2002 (MSHA 2002). The BLS did not break down the statistics far enough to determine the percentage of those from Hispanic or Latino ethnicity that work underground or whether it was coal or metal/non-metal mining.

A review of all mining fatalities from 1995 through January 2009 finds six fatalities where non-fluency in English may have played a part in the accident. All six cases involved mine workers who were native Spanish-speakers although the country of origin is not indicated. It is impossible to determine the number of injuries where language barriers may play a role because MSHA does not require collection of ethnicity or race under 30 CFR Part 50 regulations.

In four cases, citations were issued for lack of adequate training. In all four cases, the victim spoke only Spanish or very little English. In one case, some training had been provided in Spanish, but lack of hazard signs in Spanish was considered a factor in the victim's death. The death of a mineworker at a Utah coal mine in 1999 portrays the difficulties that may face the mining industry as more non-native speakers are employed in the mining industry. Of the eight members on the crew, only the electrician and supervisor were native-born. Four of the remaining crew members, including the victim, spoke only Spanish. Neither the supervisor nor the electrician spoke Spanish so they had to rely on the two Hispanics that knew some English to provide interpretation for the others. The fatal investigation found several violations of the roof control plan that was probably due to the language barrier. A Spanish-speaking mine operator/helper crew had mined beyond the limit of the roof control plan and roof falls had resulted in at least three locations. It was this miner operator that was fatally injured when a rib over-hang fell on him and the mine helper (MSHA, 1999).

The cyclical nature of the mining industry often discourages native-born workers from entering the mining industry. During the last coal boom in the early 2000s, Kentucky coal mine operators petitioned the state legislature to allow them to hire experienced underground miners that did not speak English. In this case, the trained miners were from Ukraine. Five Hondurans and two Mexicans passed the Kentucky surface miner exam in May 2001. The 24-hour training was conducted in Spanish and six of them needed the help of a translator to take the exam (Associated Press, 2001).

How will today's mining industry respond to the continuing increase of employees that speak little or no English? Will mine operators offer English and citizenship classes for the immigrants as it did a hundred years ago? The authors were unable to determine literacy rates for current Hispanic and Latino immigrants, so it can not safely be said whether hazards signs, roof control plans, etc. in Spanish will have a great impact great to insure a safe work environment for all employees.

SUMMARY

Just as America's wealth and quality of life were built by the immigrants that mined the coal and other minerals at the beginning of the 19th Century, immigrant workers will continue to provide the basic commodities that the American economy will depend on in the future. It is our job, as mine operators, mine safety professionals, union officials and enforcement agencies to insure that immigrants receive the proper training and guidance so that they, too, can work safely and return home to their families each and every day. To that end, more research is needed to identify the different immigrant populations that work in each of the mining sectors: underground, surface, coal, metal/non-metal and aggregate.

Another area of interest is how receptive different immigrant groups are to learning English and maintaining permanent residency in the United States. Persons that plan to return back to their homeland may have less personal incentive to become fluent in English—both spoken and written. Many children of immigrants learned English by just watching television. However, in today's media culture, cable and satellite television provide programming in a myriad of languages and may buffer that earlier impact.

In closing, the wise mine operator will explore resources offered by churches, school boards or local community colleges that can provide "English as a Second Language" (ESL) classes to both the miner and his or her family. As the H.C. Frick Coke Company discovered in the early 20th Century, inviting the miners' families to view films about mining safety may continue to be an effective tool to promote safe work habits by the miner himself. The diversity of the mining workforce will continues to expand at or near the same rate as the cultural diversity of the United States. Instead of decrying the immigrant employee as "ignorant" or "careless" as our earlier counterparts did, industry needs to provide the proper training and guidance for all mine employees regardless of their ethnicity.

ACKNOWLEDGMENTS

This paper would not have been possible without the assistance of Pamela Seighman, curator, and Elaine DeFrank, Oral Historian, of the Coal and Coke Heritage Center located at Penn State Fayette, the Eberly Campus in Uniontown, PA. They both spent many hours pulling newspaper and journal articles, copies of coal company in-house communications, artifacts and transcripts of oral histories that provided original sources for our paper. It is our hope that more researchers will make use of the vast wealth of information available at the Center.

REFERENCES

Adams. T. 1884. Third bituminous coal district. In Annual report of the Secretary of Internal Affairs of the Commonwealth of Pennsylvania. Harrisburg, PA: Lance S. Hart, State Printer and Binder.

Albright, C. L. 1920. Address at Seventh Annual Dinner of the Executive Officers, Superintendents and Department Heads of the H.C. Frick Coke Co. and Hostetter Connellsville Coke Co. at the Greensburg Country Club on January 10th, 1920. Unpublished Report.

Associated Press. 2001. Non-union coal companies import foreign miners to break UMW in Kentucky. Urbana-Champaign Independent Media Center:newswire/1132. http://archieve.ucimc.org/newswire/display/1132/index.php. Accessed September 2009

Bureau of Labor Statistics. 2009. Household data annual averages: section 11. Employed persons by detailed occupation, sex, race and Hispanic or Latino ethnicity for 2008, p. 214. www.bls.gov/cps/cpsaat11.pdf. Accessed September 2009

Fay, A. H. 1919. Mine accidents: English-speaking vs. non-English speaking employees. *Coal Age*. Vol.16:777.

Foster, T.J. 1916. *The Coal Miners' Pocketbook*. New York: McGraw Hill Book Company, Inc.

Lauck, W.J. 1911a. Bituminous coal mining. In *Report of the Immigration Commission*, W.P. Dillingham, Commission Chairman. Washington, D.C.: Government Printing Office

Lauck, W.J. 1911b. The bituminous coal miner and coke worker of western Pennsylvania. *The Survey*. Vol. 1, April 1, 1911.

MSHA. 1999. Fatal Investigation: Bear Canyon #1 on July 15, 1999 (District 9). www.msha.gov/FATALS/1999/FTL99C18.HTM. Accessed September 2009.

MSHA. 2002. News Release: MSHA web site now available in Spanish. www.msha.gov/MEDIA/PRESS/2002/NR021022.HTM. Accessed September 2009

Reynolds, C.E. 1924. Untitled. *Coal Age* 26 (8).

Roderick, J. 1906. Report of the Department of Mines of Pennsylvania Part II—Bituminous Coal. Harrisburg, PA: State Printer and Binder.

The Weekly Courier (Connellsville, PA). May 1914. The Connellsville coke regions: their past, present and future. (Special historical and statistical number)

Section 2. Mine Productivity

40 Years of Technical Development at RAG Deutsche Steinkohle

Martin Junker

RAG Deutsche Steinkohle and CEO of RAG Mining Solutions, Herne, Germany

INTRODUCTION

Thank you very much for the invitation to the Symposium in conjunction with the 2010 SME Annual Meeting here in Phoenix. And today, the SME is also celebrating "A Century of Mining Research." (Figure 1).

In the name of my company, RAG Deutsche Steinkohle, I would like to congratulate your Society on this anniversary of original research, which began with the establishment of the U.S. Bureau of Mines in 1910.

We too—the German coal-mining industry—also recently celebrated an important event. It is now 40 years since RAG was established, merging what had formerly been 54 separate mines into a mining company for the whole Ruhr region. This was when the foundation stone was laid for further development of innovative mining technology and safety in Germany.

DEVELOPMENT OF MINING TECHNOLOGY

Underground hard coal mining started in Germany more than 300 years ago. Before the introduction of the steam engine at the start of the 19th century, mining was performed by means of hard physical labor. But technology developed very fast during the industrial revolution as a whole, and naturally also influenced the further development of German mining engineering. Some technical "milestones." in the history of coal mining up to 1969, when our company was established, are shown in this figure. These milestones were also important for the development of other sectors of industry (Figure 2):

- First use of the steam engine in 1785
- First use of the wire rope developed in Germany in 1834/1835
- First use of compressed air, electrical trolley and battery locomotives in 1875, 1882 and 1902
- The first main electrical shaft hoisting system was used in 1894 with sensitive control using a Leonard circuit between 1901 and 1903
- First plow in 1944
- First mine control room in 1956
- Development and first use of a road heading machine for mechanical cut of roadways in 1965
- First rodless borer for opening up vertical blind shafts

In contrast to other coal deposits throughout the world, German hard coal mining has always had to work with unfavorable mining conditions (Figure 3). This is and was the reason why coal mining in Germany, more than in other countries, has been forced to use and develop the most modern

Figure 1.   Two reasons for celebrating

Figure 2. Technical milestones in mining up to the 20th century (examples)

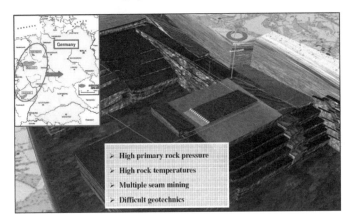

Figure 3. Conditions in the German Hard Coal mining industry

technology which is available for reaching, winning and transporting coal.

While, at the beginning, mining activities took place at depths of between 300 m and 600 m, today mining is carried out at much greater depths of around 1150 m. Driving of roadways happens at depths of more than 1600 m and takes place under difficult conditions including:

- High rock temperatures up to 77°C
- High rock pressures around 45 MPa
- A high number of seams worked

Since its establishment in 1969, RAG has driven forward two important innovation phases (Figure 4):

- Mechanization
- Technization

and is currently pushing forward with the third phase, which is:

- Automation

Mechanization

By mechanization is meant replacement of manual and work-intensive activities in preparation, extraction and infrastructure with the use of machinery and equipment. In this respect the replacement of single prop support systems (in 1970 still used by 55% or all mines) by the hydraulic shield system in the 1979s

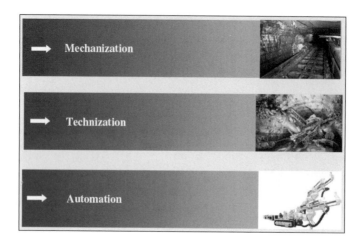

Figure 4. Innovation phases at RAG

Figure 5. Mechanization at the faces

was very important. (Figure 5). This was accompanied by further development of shield supports from the circular arch shield up to the lemniscate shield, which can bear greater support pressures. In the winning itself, there was increasing replacement of the drag-hook plow by the so-called Plow (gleithobel), in the case of the shearers higher performance was achieved with new shearer generations up to EDW 300L and EDW 150/2L.

In the drivage phase, use of road headers (which were first used in 1965) was increased. Because of increase in convergences caused by the greater depths, technical progress for improvement of stability consisted of increased use of concrete materials in back-filling of roadways and roadside packs.

Technization

It was possible to achieve considerable improvements in the 1990s through standardization, innovation and optimization of processes and technology. This created the basis for the third phase in which we now find ourselves—the automation phase. The development of electro-hydraulic control systems for the winning process at the end of the 1980s (1987) meant further progress. With the help of modern computer technology, for the first time it was possible to specify a set cutting depth for the winning machinery and therefore to extract the coal in the most efficient way (Figure 6).

Development of the other winning equipment, drives, conveyors, chains and so on continued in parallel and these were linked by computer for the first time. And an optimized mine layout—the

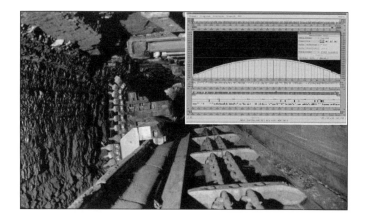

Figure 6.   Visualization at a plow face

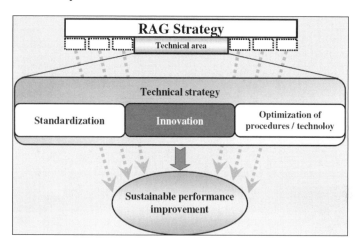

Figure 7.   RAG technical strategy

so-called ideal layout—was also standardized for the first time. In the 1990s, the main focus was on efforts to concentrate mining operations at a few high-performance working faces. To make this high concentration possible, longer longwall faces (today on average around 338 m, with a maximum of up to 470 m) are set up. Because of parallel development of operating equipment with even higher performance—tailormade for these longer faces—it has been possible to achieve a sustainable increase in output and efficiency. Because this new equipment and—very important—the work processes were also consistently standardized with the help of the RAG technology strategy, it was possible to achieve success with further mechanization (Figure 7).

Altogether it can therefore be seen that RAG has a comprehensive engineering and technical management system in different but interlinked fields

of competence, either as a user or as a developer of innovative technology.

Automation

Developments in winning technology are a very good example of the logical transition to the third innovation phase, where automation is achieved through linking of machines and equipment which are already partly-automated.

First Steps

Starting from the first years of the 1980s, the first steps were taken towards making later automation possible. As I have already described, introduction of electro-hydraulic shield support systems were very important during the further development of mechanization (Figure 8). In this phase, for the first time a large variety of process data, such as

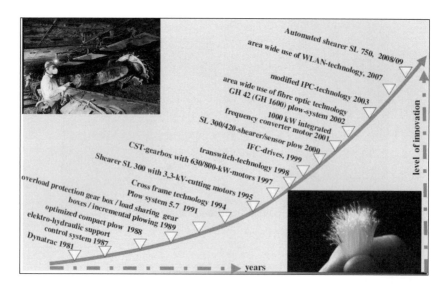

Figure 8.   Milestones of German winning technology since 1980

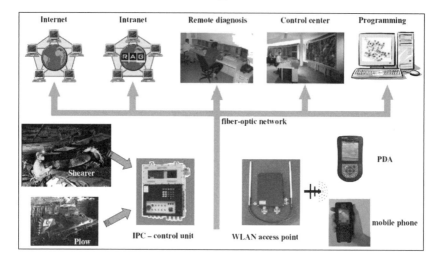

Figure 9.   IT-infrastructure—control technology/automation

information regarding the internal pressures of the hydraulic components, became available. At that time it was not possible to make further use of this data, as the copper cable communications systems soon reached their limit. But with the introduction of overall use of fiber optic technology in 2002, we made sure that high-performance control systems based on IPC technology can be used in all areas of our mining operations. Through the establishment of so-called access points it was also possible to establish exchange of data with mobile pocket PCs using WLAN (Figure 9). Use of this WLAN technology

also made possible wireless transfer of large volumes of data and groundbreaking developments in voice communication, for example in mobile phone communication. In the control centers, the data is not only visualized and archived, but is also used for control purposes. Today, there are PC workstations at all important locations in our mines, where all the data that is required for detection and repair of faults is immediately available.

The development of a centralized electronic information and communication structure was a result of this and it now provides the necessary basis

Figure 10.   Examples of innovative and automatable mining equipment

for rapid exchange of data with the operating equipment, which in turn is being developed constantly. This operating machinery and equipment is not only standardized, but has been made capable of, for example, performing diagnosis and first smaller partial automation functions. However, despite the clear increases in performance, the limitations of this "ever bigger—ever more powerful." mentality became clear, as the actual running time could not be increased to the same extent. First the individual processes used in winning, drivage and logistics were examined in detail and improved (Figure 10). Then it becomes necessary to create the conditions for further partial process automation.

In order to make our range of machinery "more intelligent," it was also necessary to provide it with corresponding sensors and actuators. For it had to be ensured that these "intelligent." machines are able to perform self-diagnosis to the necessary extent with regard to their condition status and their loading states. Therefore "intelligent." machines are constructed in such a way that they are able to sense their current state, environment, position and context within the mining process. Today these partial automation solutions are implemented in all areas of our operations in the form of intelligent machines, such as in actual winning with the partial automation of our shearer-loaders and in drivage machines and in the first use of a driverless overhead diesel monorail locomotive equipped with route sensors and controlled without cables via the WLAN network.

Decentralized Information Processing
The large volume and variety of operating and process data also places greater demands on the control centers above ground. At the present time, the

information that is not yet bundled at a central point is collected on a decentralized basis at individual control stations. Each of the mines has several control stations, for the winning and preparatory work, for conveyor technology and for the infrastructure (Figure 11).

However, decentralized control stations and control rooms means that separate solutions were created for the winning, logistics and drivage areas in the different mines. Island solutions were created and all the relevant items of individual data from the separate machines connected to the system were collected, analyzed and documented. Based on the already very high level of partial automation of the individual items of machinery and equipment, the operating and process data were fed to these decentralized individual control rooms. However, this approach did not permit an integrated approach to all mining processes.

A noteworthy exception to this was the diagnosis station of the so-called shearer-loader online services (Figure 12), which was already commissioned at the beginning of 2000. This station was available for 24 hours a day centrally to all shearer-loader operations of RAG for fault recognition and troubleshooting.

This Shearer Loader Online Services was then developed further by the Online Support Center (Figure 13). Now, as central diagnosis and communication point, it also monitors all the control systems used by RAG in winning operations, in roadway drivage and within the infrastructure in all mines.

Complete new design of the underground data information and control network has created the conditions for linking the various items of our equipment into an integrated network (Figure 14). Through further increases in the level of automation,

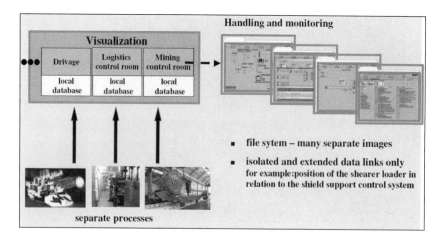

Figure 11.   Current state-of-the-art of control room technology

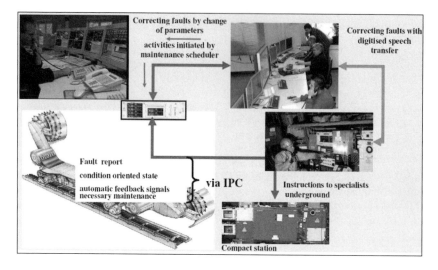

Figure 12.   Shearer Loader Online Services

all our systems can be linked with the environment itself and with other systems. We started from the experiences of mostly decentralized and independent control rooms and their technology, separate information systems and databases. Then it was essential to bring together the data and information for higher-level visualization, diagnosis and control of the individual part processes.

Linking of Innovative Operating Equipment
to an Integrated System

The data from all participating processes, including the associated information from existing databases, are integrated here into a central control system (Figure 15). Monitoring of the coal faces and mining operations by means of plausibility tests, generation of operating data, selection criteria and linking of states are transferred into a transparent and automatic diagnosis and information system. This leads to shorter reaction and action times if defects occur or alarms are registered, and unnecessary work is avoided. Within the framework of this higher-level linking of the available data, it is also important only to use "useful." information. In other words it is vital to use process and operating data which are important for the respective process in an intelligent way.

Through linking with the other areas such as winning, coal transportation, maintenance and infrastructure, all the information can be intelligently

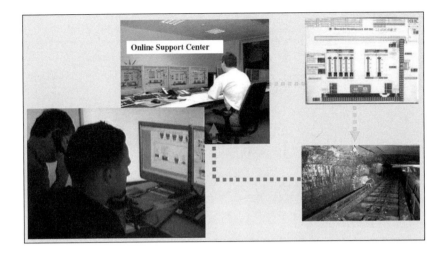

Figure 13. Monitoring and controlling: Online Support Center (OSC)

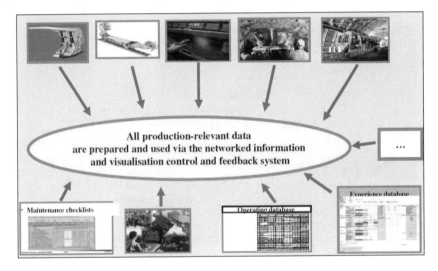

Figure 14. Areas to be linked

networked (Figure 16). This ensures effective monitoring and opens up comprehensive control possibilities. In a project which is unique on the world stage, we have set the scene for bringing together all the partial processes already mentioned into this new control system and visualizing them in a central control room (Figure 17).

Through linking all the information available from databases above and below ground, we can achieve a general overview of the mine as a whole, and also to control individual processes in an optimum way. For example, the speed of the winning machine can possibly be controlled dependent on the constantly monitored methane content of the

atmosphere underground so that switch-offs as a result of methane levels and therefore downtimes can be avoided in future. The aim is to bring together all individual processes within one system and to achieve control of the processes, taking all available operating parameters into consideration.

40 YEARS OF TECHNICAL DEVELOPMENT— 40 YEARS OF MINING COMPETENCE

During the past 40 years we have been able to gather a great deal of experience while working under the difficult conditions which face us in our mines (Figure 18).

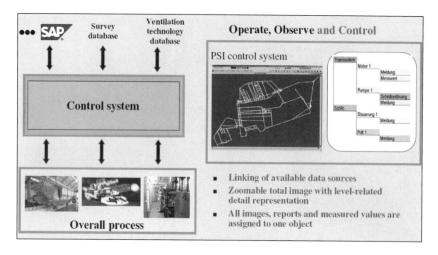

Figure 15.   Future process control center, as from 2010

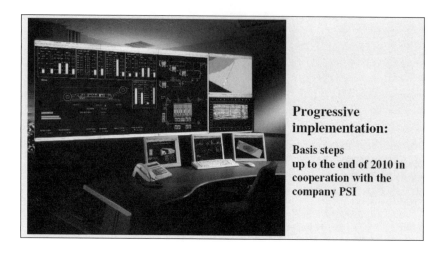

Figure 16.   Process control technology—Future of the mining industry

Based on our experience of handling the most difficult mining situations and the high level of technology used in our mines, RAG established a new subsidiary in April 2009, RAG Mining Solutions (Figure 19).

One of the tasks of RMS is to manage the sale of used German mining equipment to suitable mines throughout the world. On the other hand, RMS also intends to place German mining know-how and services at the disposal of other companies in the form of training and advisory services for complex mining concepts. Through our many activities and the technical innovations we have developed and imple-

mented, we have achieved an outstanding level of competence, where we are world leaders.

We are also extremely experienced and successful in the fields of safety and health and environmental protection (Figure 20). This success is based on a large number of safety programs, safety training programs and management systems, for example in the areas of accident prevention and dust reduction. In fact we have been able to reduce the number of accidents in our mining operations by more than 80% in the last few years. We offer a large number of solution packages for the areas of occupational safety and health and environmental protection. These can

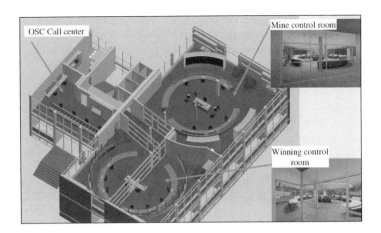

Figure 17.   New Control Center: Process control technology—Mine of the future

Figure 18.   RAG Mining Solutions, March 2009

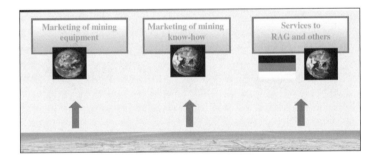

Figure 19.   Business areas of RAG Mining Solutions

help other mining companies to achieve significant improvements in these areas.

Because of the nature of our coal deposits, drivage has to take place at ever greater depths (Figure 21). And we have developed and introduced new support concepts to make this possible. These can also be used as solutions for other mining companies working with different coal deposit and rock pressure situations.

Great depth also means high rock temperature (Figure 22). With our ventilation and climate control experts, climate control problems can be solved on a long-term basis. We offer support from the planning stage right up to commissioning of complex climate control equipment.

We offer detailed customized solutions for particular problem areas within complex winning, drivage and logistics systems (Figure 23). On the

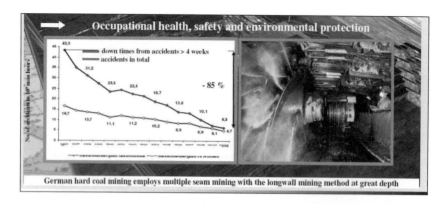

Figure 20. Outstanding competence in occupational safety and health and environmental protection

Figure 21. Outstanding competence in rock control in deep mines

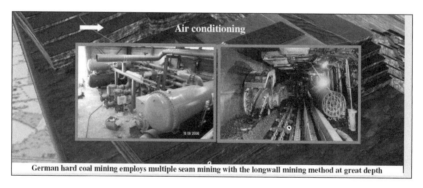

Figure 22. Outstanding competence in air conditioning

one hand, these allow optimum use of machines in the individual sense. But they also make it possible to integrate machinery and equipment into a system network, optimizing complete processes and making them more productive and cost effective.

Our standardized items of machinery are also fitted with partial automation functions to make mining operations yet more efficient (Figure 24). The mining automation experts at RAG are supremely competent and experienced. They can offer their expertise to benefit mining operations throughout the world. Our experts can help to achieve integration of the different functions and processes within a mine. This makes overall opera-

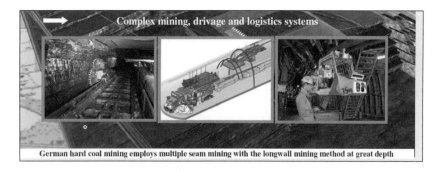

German hard coal mining employs multiple seam mining with the longwall mining method at great depth

Figure 23.   Outstanding competence in complex mining, drivage and logistics systems

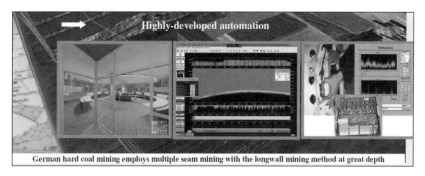

German hard coal mining employs multiple seam mining with the longwall mining method at great depth

Figure 24.   Outstanding competence in highly-developed automation

tions much easier to visualize and control, and greatly improves efficiency.

SUMMARY

Since the establishment of RAG in the year 1969, the German mining industry has provided a large range of innovative technology and has gathered many years of experience in deep mining engineering. Conditions with regard to coal deposits are difficult in Germany. Therefore the German hard coal mining industry always has to make use of the most advanced technologies and processes if it is to win coal efficiently and safely. Following the initial mechanization and technization phase, the German hard coal mining industry is now pushing forward with the automation phase. This will mean that we can operate, monitor and control a coal mine in an integrated way. We are very happy to share the knowledge gained in this process with other international mining companies through our newly-established company, RAG Mining Solutions. This means that the German hard coal mining industry can contribute to and drive forward our common goal—which is to make mining processes safer and more efficient throughout the world.

Improving Dragline Productivity Using a Diggability Index as an Indicator

Mehmet Kizil

School of Engineering, University of Queensland, Brisbane, Australia

ABSTRACT: Draglines are the largest and most expensive machines used in strip coal mines and their productivity has always been a topic of interest. The productivity of a Dragline mainly depends on the cycle time and payload. Cycle time includes fill, swing, dump, return and positioning time. There is no doubt that dig time is one of the most critical factors which is believed to be closely related to fragmentation of the excavated material after blasting. A number of techniques have been developed to quantify the diggability of the material in order to improve productivity through improving blasting. This paper examines the factors that influence diggability of draglines and describes the newly developed dragline diggability index to measure productivity and blasting performance.

INTRODUCTION

At a cost of $50 Million to $100 Million, dragline is the most critical component in the overall system and any improvement in dragline productivity will result in big savings for the mine. Witstanley (2004) claims that 4% improvement in dragline productivity would save a typical Australian coal mine $3.0 Million a year. The largest single mining cost for a strip mining operation is the cost of overburden removal which is affected by:

- Dragline design and performance
- Operator competency and practice
- Mine planning
- Blast design and performance
- Maintenance of equipment

Frimpong, Kabongo and Davies (1996) investigated the effects of blasting design on dragline productivity and cost and concluded that any additional cost to improve blasting operations would improve dragline productivity and result in an automatic cost savings as dragline cost accounts for 60% of the total mining cost whereas drilling and blasting cost account for the remaining 40%.

Historically, there have been a number of ways to assess blast performance and machine impact through the application of diggability. Generally, all assessments can be split into two categories; machine-based performance monitoring and electronic real-time monitoring. The idea of diggability is not a new concept by any means and has been around for almost 30 years. Prior to electronic real-time dragline monitoring, many diggability studies focused on determining a *"diggability index"* for use with equipment selection. Major studies in this field included studies by Scoble and Muftuoglu (1984)

and Karpuz (1990). These studies aimed at determining an index to predict the influence of ground conditions on excavating equipment performance. Scoble and Muftuoglu (1984) aimed to derive a diggability index through the use of geotechnical core logging. From that study, it was determined that rock strength, weathering and block size were the main factors influencing diggability.

Williamson, McKnezie and O'Loughlin (1983) were the first to develop a system to monitor crowd and swing on a DC motor and relay. This system was used on a P&H electric shovel to derive an index of muckpile diggability. The most recent of all diggability studies have focussed on the use of real-time monitoring data to develop what has become known as a diggability index. Since the development of the Tritronics 9000 Dragline Monitor, field and experimental studies in Australia have produced varying results. Australian studies conducted by Torrance and Baldwin (1990), Rowlands and Just (1992), Sengstock (1992), Humphreys and Baldwin (1995), Lumley (2003), Taylor and Kizil (Taylor, 2006), Brown and Kizil (Brown, 2007) and Schumacher and Kizil (Schumacher, 2009) have built the foundation for this investigation.

Most draglines in Australia are equipped with the Tritronics monitoring system which records up to 27 parameters related to the machine and operations. These parameters are stored in a database and transferred to the base for storage and further analysis.

TRITRONICS DRAGLINE MONITORING SYSTEM

Most draglines in Australia use Tritronics monitoring system. Tritronics is a monitoring system designed to capture data from sensors installed on draglines. An onboard computer collects data for each dragline

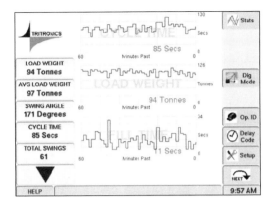

Figure 1. Tritronics Display Screen (Tritronics, 2005)

cycle and records, consolidates, and sends to the office computer via telemetry. This data is stored in the database and can be queried later for different purposes. A comprehensive range of critical parameters (27 in all) are measured and recorded. These include: spot time, fill time, swing time, dump time, return time, cycle time, re-pass count, hoist dep count, swing angle, dump height, start fill height, end fill height, start fill reach, end fill reach, dig index 1, dig index 2, drag energy, drag length, payload weight, suspended load, and operator ID.

The production volume data is calculated from the payload and other information stored in the database. A summary of important data collected such as cycle time, payload and fill time are displayed on a display in the dragline operator's cabin (Figure 1).

The Tritronics has supplied a data file with the details of eight dragline digging cycles with measurements taken every 50ms for an investigation into diggability. The data set contained the following columns of information:

- Hoist Arm Amps
- Drag Arm Amps
- Drag Reference
- Drag Volts
- Drag Length
- Hoist Length

Figure 2 shows a graphical display of the detailed information collected. Drag length parameter can be used to observe the number of cycles and when that cycle started and finished. When a cycle is analysed in more details as shown in Figure 3, the relationships between the loads and various activities can be determined. The figure clearly shows when fill, dump

and return activities of a full cycle start and finish. A similar pattern is repeated for every dragline cycle.

Drag length can be used to identify each digging cycle. At the start and finish of the fill cycle, the drag length parameter curve seems to be levelling. Drag arm amps can also be used to identify the start and finish of a fill as at the start of each fill the drag arm amp seems to rise rapidly followed by drag load peaks.

DIGGABILITY ASSESSMENT

The Tritronics collects an enormous amount of data but the challenge is to turn this data into useful information and subsequently knowledge. There is a great demand for establishing a numerical way of correlating the effectiveness of a blast to overburden removal using a dig index. The dig index will provide a means of correlating drilling and blasting costs to dragline productivity. However, the reliability of the results will depend on the accuracy of the machine and operator related constants.

A number of indices have been developed to measure the performance of draglines using the results of the electronic monitoring systems installed on draglines. In this study, the South African Dig Index, the Tritronics Dig Index and the Kizil Dig Indices are examined.

South African Dig Index

One of those techniques is the recently developed the South African Dig index. The index measures diggability by measuring the rate of energy a dragline uses to dig that material (Edwards, 2004). The dig index rating is generated from the inclination of the ramping current generated by the drag motors on the dragline. The ramping current is believed to be a function of digging conditions and how the material affects the dragline during digging.

The dig index is calculated automatically from the Contel monitoring system installed on draglines and it is based on the inclination of the drag current curve. As it can be seen in Figure 4, shallow inclination suggests easy digging while steeper curve indicates hard digging conditions.

The South African dig index is calculated from the following equation:

$$\text{Dig Index} = \frac{1}{n\,\delta t}\left(\delta I_1 + \delta I_2 + \ldots + \delta I_n\right) \qquad (1)$$

Where δt is time interval and it is fixed at 100ms. The absolute values of δI are used to show the steepness of the falling current.

The above equation gives the average gradient for the drag current plot in the form of a dig index

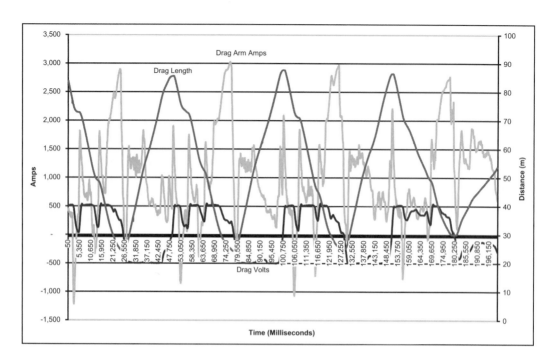

Figure 2. Detail dragline data showing changes in drag length, drag arm amps and drag volts

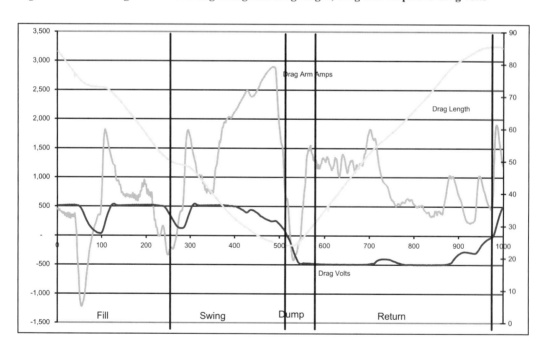

Figure 3. A full dragline cycle starting from digging position

which is a relative number which increases with hard digging conditions. Although Edwards (2004) claims that this index gives reasonable results, it lacks the inclusion of payload in the equation. For example, if two buckets, one full and one half-full, gave the same dig index results, it would be seen as both digging locations having the same digging conditions. In reality, the dragline would have consumed more energy to dig and drag the full bucket than the half full bucket. Therefore, it is recommended that a bucket fill factor is introduced into the South African Dig Index to improve its reliability further. The new index will be:

$$\text{Dig Index} = \frac{1}{\lambda n \delta t}\left(\delta I_1 + \delta I_2 + \ldots + \delta I_n\right) \qquad (2)$$

Where λ is the bucket fill factor and can be calculated from payload and bucket full capacity as follows:

$$\lambda = \frac{\text{Payload}}{\text{Bucket Full Capacity}} \qquad (3)$$

The Tritronics monitors are designed to capture data from sensors for measuring a range of inputs at 5Hz speed. This means that Tritronics is capable of taking one measurement every 200ms. However, when it records the data, parameters are averaged over the cycle period to save on data storage and transfer time. This subsequently results in lose of the detailed data and therefore makes the application of the South African dig index nearly impossible.

The Tritronics Dig Index

The Tritronics monitoring system reports two dig indices; one is just a random number and the second one is based on the drag energy used, payload, digging time and fill distance. The reliability of this index has not been proved yet and no or minimum correlation was determined between this index and productivity.

The Kizil Dig Index

Since 2004, Kizil (2004) has been trying to derive a dig index using the diggability related parameters recorded by the Tritronics. The newly developed Kizil Index is a function of: dig energy used to fill a bucket, fill time, fill distance, payload and vertical bucket lift during the excavation. The actual formula will not be published due to confidentially and intellectual property.

This index calculates a number from 1 to 100; 1 signifying *"easy digging"* and 100 signifying *"very hard digging."* This index essentially has little

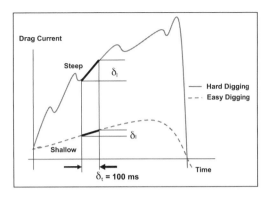

Figure 4.   Measurement of digging condition from dragline current used (Edwards, 2004)

relevance without supporting data and historic performances for a mine site. For example, a dig index of 25 may be perceived as *"good digging,"* but if historical data shows that digging in a similar area yields a dig index of 20, the index of 25 would be considered as harder than normal digging for the dragline.

The index has been applied to a number of cases to prove it's applicability and reliance. Figure 5 shows the relationship between the Kizil Index and the Instantaneous Productivity. Instantaneous productivity is generated from the bucket filling phase of a dragline cycle. Equation 4 outlines the measure of instantaneous productivity.

$$\text{Instantaneous Productivity} = \frac{\text{Net Load Weight} \times 3{,}600}{\text{Fill Duration}} \;(\text{t/h}) \quad (4)$$

Instantaneous productivity is based on the digging phase of a dragline. Since the dig index measures the *"ease of digging,"* it is important that all other parameters of a dragline cycle be ignored to maintain accuracy.

As seen in Figure 5, the Kizil Dig Index provides an exceptional correlation with the Instantaneous Production with an R^2 value of greater than 95%. The results also show that this index is a successful, robust and repeatable index for measuring diggability. An analysis of the Tritronics data shows that the relationship between diggability and productivity is exponential and any improvement in digging conditions would result in a substantial increase in productivity.

The previous version of the Kizil Index has been programmed in a DLL library to be used as an add-on program to the Tritronics monitoring system. This program has been installed on 22 draglines in Queensland to provide the dragline operators with

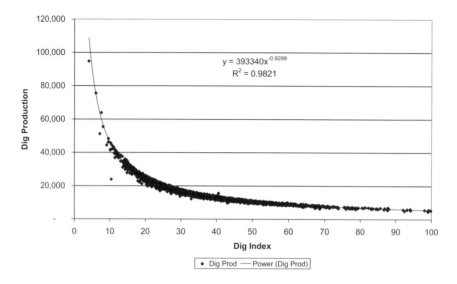

$$y = 393340x^{-0.9298}$$
$$R^2 = 0.9821$$

Figure 5.   The relationship between the Kizil Index and the Instantaneous Productivity

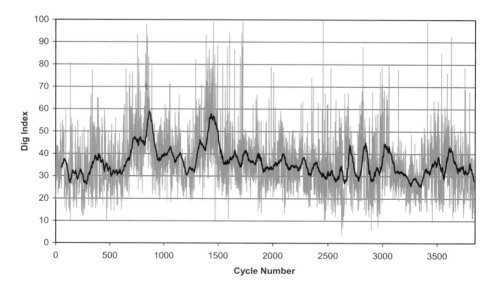

Figure 6.   Real-time measurement of diggability

an indication of diggability in real-time on a screen, similar to the one shown in Figure 6.

CASE STUDY

The drilling and blasting data and dragline monitoring results were collected from an undisclosed Queensland open cut coal mine site. There were four blasts reviewed as part of this study and each of the blasts were contained in the same strip. The configuration of the blasts contained in this study and the dig index calculations are outlined in Figure 7. Dig Index distribution for the block 1 to 4 is also given in Figure 8.

The results showed a good correlation between the drilling rate and the Dig Index. As seen in Figure 9, as the drilling rate is reduced the Dig Index increases which indicates harder digging conditions. A similar trend was also observed between the Instantaneous Production and the Dig Index. As the

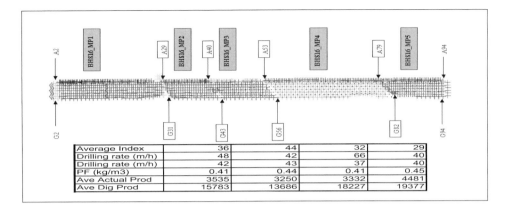

Figure 7.   Case study blast configuration, drilling, and production data, and dig index results

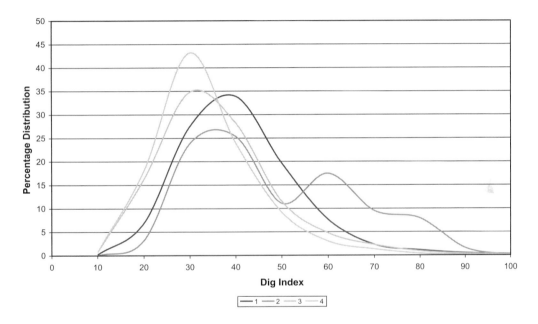

Figure 8.   Dig Index distribution for Block 1 to 4

Dig Index increases, the Instantaneous Production rate is reduced, as seen in Figure 10.

CONCLUSIONS

The initiative of using diggability as a measure of dragline productivity and blast performance has been around for more than 30 year. However, it is only recently people have begun to recognise the potential of quantitative data gained through dragline monitoring systems. A number of indices have been developed mainly based on using the energy used by a dragline to dig a bucket of earth to measure diggability of the material. However, none has been verified to be successful as many factors that influence diggability have not been taken into consideration. This research has highlighted the need for better justification of such parameters used in the calculation of a diggability index. The newly developed Kizil Dig Index has been proved to produce both robust and repeatable results and is being successfully used on many draglines in central Queensland.

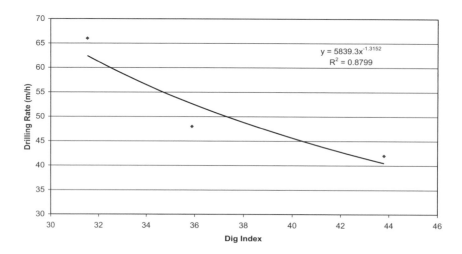

Figure 9. The relationship between the drilling rate and Dig Index

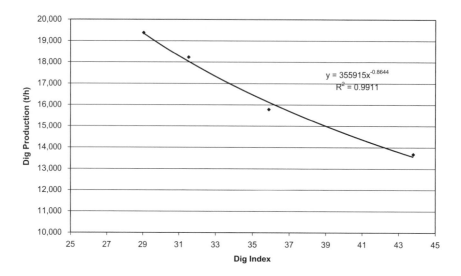

Figure 10. The relationship between the Instantenous Production and Dig Index

It is anticipated that there is now the potential for significant gains in productivity at coal mines which utilise Dragline operation. A follow-on effect should also see improvements in the economic model combining total overburden removal from drilling and blasting stages through to dragline operation. By being able to assess the effectiveness of blasted zones utilising data that is readily available to engineers' onsite, the possibilities for improvements and further research are endless.

REFERENCES

Brown, N. 2007. Post blast evaluation using dragline diggability, Graduate thesis (unpublished), *University of Queensland*, Brisbane.

Edwards, M. 2004. An Analysis on the Effects of Blasting and Bucket Design on Dragline Productivity Through the Use of the DIG Index at Kleinkopje Opencast Coal Mine. *Explo 2004.* Conference organised by AusIMM. 26–28 July. Perth. Western Australia.

Frimpong, M., Kabongo, K.K. and Davies, C. 1996. Diggability, A Measure of Dragline Effectiveness and Productivity. *International Society of Explosives Engineers*. P-95.

Humphreys, M and Baldwin, G, 1995. Blast optimisation for improved dragline productivity, *Journal of Explosives Engineering*, 12(5), pp: 13–32.

Kizil, M.S. 2004. Development of a Dragline Dig Index for the Anglo Coal Strip Mining Operations, Internal report Jun 2004 Anglo Coal Australia, unpublished.

Lumley, G, 2003. Improving Dragline Productivity—Advanced Operating Techniques, in *Course Manual: Anglo Coal South Africa 2003,* 117p (GBI Consulting P/L: Brisbane).

Karpuz, C, 1990. A classification system for excavation of surface coal measures, *Mining Science and Technology*, 11(1) pp: 157–163.

Rowlands, J C and Just, G D, 1992. Performance characteristics of dragline buckets, *in Proceedings Third Large Open Pit Mining Conference*, pp 89–92 (The Australian Institute of Mining and Metallurgy: Melbourne).

Scoble, M J and Muftuoglu, Y V, 1984. Derivation of a diggability index for surface mine equipment selection, *Mining Science and Technology,* 1(4) pp: 305–322.

Sengstock, G W, 1992. Blast profile control for dragline productivity optimisation, *in Proceedings Third Large Open Pit Mining Conference*, pp. 269–274 (The Australian Institute of Mining and Metallurgy: Melbourne).

Schumacher, E. 2009. The assessment of blast effectiveness using dragline diggability, Graduate thesis (unpublished), *University of Queensland*, Brisbane.

Taylor, T. 2006. The effects of cast blast performance on dragline strip diggability, Graduate thesis (unpublished), *University of Queensland*, Brisbane.

Torrance, A C and Baldwin, G, 1990. Blast performance assessment using dragline monitor, *in Proceedings of Third International Symposium on Rock Fragmentation by Blasting*, pp. 1–13 (Australian Mineral Foundation Incorporated).

Tritronics, 2005. Series 3 Dragline Operator Manual, 40p (Tritronics Australia P/L: Sydney).

Williamson, S, McKenzie, C and O'Loughlin, H, 1983. Electric shovel performance as a measure of blasting efficiency, *in Proceedings 1st International Symposium on Rock Fragmentation by Blasting*, pp. 625–635 (Lulea University of Technology: Sweden).

Winstanley, G. 2004. Dragline Swing Automation. *http://www.cat.csiro.au/cmst/CA/Projects/projects.php?dragline.* Accessed on 21/05/2004.

Assessment of Cutoff Grade Impurities: A Case Study of Magnesite Ore Deposit

Suheyla Yerel
Bilecik University, Civil Engineering Department, Bilecik, Turkey

Adnan Konuk
Eskisehir Osmangazi University, Mining Engineering Department, Eskisehir, Turkey

ABSTRACT: One of the most important aspects of magnesite ore deposits is to investigate of the cutoff grade impurities. In this paper, cutoff grade impurities of Beylikova magnesite ore deposit in Eskisehir (Turkey) are assessed by mathematical equations. This paper presents a procedure for using the bivariate statistical distribution to describe the uncorrelated SiO_2% and CaO% impurity values of magnesite ore deposit. Through the results from the statistical distribution, it was determined that there are magnesite tonnage rate according to their cutoff grade impurities. This bivariate statistical distribution is believed to assist the management of magnesite ore deposits.

INTRODUCTION

Determination of optimum cutoff grades is a fundamental issue in mineral extraction as it assigns the boundaries between ore and waste over time (Cetin and Dowd, 2002). Taylor presents one of the best definitions of cutoff grade. He defined cutoff grade as "any grade that, for any specific reason, is used to separate two courses of action, e.g., to mine or to leave, to mill or to dump...." (Taylor, 1972, 1985; Osanloo and Ataei, 2003).

There are many theories for the determination of cutoff grades. But most of the research that has been done in the last three decades shows that determination of cutoff grades with the objective of maximizing net present value (Ataei and Osanloo, 2003). The determination of the optimum cutoff grade of single parameter can be very complex even when price and cost are assumed constant, but it involves the costs and capacities of the several stages of the mining operations, the waste/ore ratios, average grades of different increments of the ore body and so on (Osanloo and Ataei, 2003). In the literature, there are many studies based on cutoff grade theories developed by Lane (1964) and Taylor (1972), which are applicable to mineral deposits. But all of these methods weren't considering the grade distribution of the deposits.

In magnesite deposit, cutoff impurity rates including SiO_2% and CaO% have more importance than the MgO%. In this study, the case in which two independent impurities SiO_2% and CaO% rates in deposit were assessed by using bivariate lognormal independent distribution model and magnesite tonnage rate according to their cutoff grade impurities was calculated.

BIVARIATE LOGNORMAL DISTRIBUTION MODEL WITH TWO INDEPENDENT VARIABLES

A positive random variable x is said to be lognormally distributed with two parameters μ and σ if $y = \log x$ is normally distributed with mean μ and variance σ^2. The distribution of the random variable x is given in equation 1:

$$f(x) = \frac{1}{x\sigma_y \sqrt{2\pi}} \exp\left[-\frac{1}{2}\left(\frac{\log(x)-\mu_y}{\sigma_y}\right)^2\right] \quad x>0 \quad (1)$$

where μ_y and σ_y are the mean and standard deviation of y, respectively. The cumulative distribution function of x can be computed through the normal distribution as follows in equation 2:

$$F(x) = \Phi\left[\frac{\log(x)-\mu_y}{\sigma_y}\right] \quad x>0 \quad (2)$$

in which Φ is the cumulative distribution function of the standard normal distribution. As there is no analytical form of the cumulative distribution function, it can be calculated by directly integrating the corresponding probability density function (Yue, 2002).

If two uncorrelated continuous random variables x_1 and x_2 are lognormally distributed with different parameters (mean and standard deviation) as follows:

$$f(x_1) = \frac{1}{x_1\sigma_{y_1} \sqrt{2\pi}} \exp\left[-\frac{1}{2}\left(\frac{\log(x_1)-\mu_{y_1}}{\sigma_{y_1}}\right)^2\right] \quad x_1>0 \quad (3)$$

$$f(x_2) = \frac{1}{x_2 \sigma_{y_2} \sqrt{2\pi}} \exp\left[-\frac{1}{2}\left(\frac{\log(x_2) - \mu_{y_2}}{\sigma_{y_2}} \right)^2 \right] \quad x_1 > 0 \quad (4)$$

then the joint distribution of these two independent variables can be represented by the bivariate lognormal distribution. The probability density function of the bivariate lognormal independent distribution can be derived using the Jacobian of the transformation and is given by:

$$f(x_1, x_2) = \frac{1}{2\pi x_1 x_2 \sigma_{y_1} \sigma_{y_2}} \left(\exp{-\frac{1}{2}} \times \left[\left(\frac{\log(x_1) - \mu_{y_1}}{\sigma_{y_1}} \right)^2 \right.\right.$$
$$\left.\left. + \left(\frac{\log(x_2) - \mu_{y_2}}{\sigma_{y_2}} \right)^2 \right] \right) \quad (5)$$

where μ_{y_i} and σ_{y_i} are mean and standard deviation of y_i ($i=1,2$) and they can be derived using the following formulae (Yerel, 2008):

$$\mu_{y_i} = \log(\mu_{x_i}) - \left(\frac{\sigma_{y_i}^2}{2} \right) \quad (6)$$

$$\sigma_{y_i} = \left[\log\left(1 + \frac{\sigma_{x_i}^2}{\mu_{x_i}^2} \right) \right]^{\frac{1}{2}} \quad (7)$$

Here μ_{x_i} and σ_{x_i} are the mean and standard deviation of x_i. For the standard values corresponding (Z) to cutoff y_2 and y_1 can be derived as equations 8–9:

$$z_1 = \frac{\log(x_2) - \mu_{y_2}}{\sigma_{y_2}} \quad (8)$$

$$z_2 = \frac{\log(x_1) - \mu_{y_1}}{\sigma_{y_1}} \quad (9)$$

which is the product of dependent probabilities, gives the total joint probability of cutoff impurity rates correspond to the tonnage rate T_c of magnesite ore deposits as equation 10 (Konuk and Yersel, 1998).

$$T_c = T_{x_1} \cdot T_{x_2} \quad (10)$$

CASE STUDY

Description of the Study Area

Eskisehir is an industrialized city located in the western part of Central Anatolia Region which has a population exceeding 600 thousand habitants and covers an area of approximately 13,700 km² (Orhan et al., 2007). The city is located at equal distance from the primary metropolitan city Istanbul and the capital Ankara (Uygucgil, et al., 2007). The study area is Beylikova magnesite ore deposit is located in the southeast part of Eskisehir city, Turkey.

Dataset

In this paper, the bivariate lognormal statistical distribution is applied to a Beylikova magnesite deposit in Eskisehir. In magnesite deposit, $SiO_2\%$ and $CaO\%$ impurities have more importance than the $MgO\%$ (Yerel, 2008). This study, $SiO_2\%$ and $CaO\%$ data were obtained from 40 vertical drill-holes. Descriptive statistics of the $SiO_2\%$ and $CaO\%$ are presented in Table 1.

RESULTS AND DISCUSSION

The one of the most important impurities of a magnesite ore deposits are the $SiO_2\%$ and $CaO\%$. This study show that bivariate lognormal independent distribution model for the magnesite ore deposit can be modeled by the use of independent variables including $SiO_2\%$ and $CaO\%$.

In this study, correlation coefficient was calculated and a value of correlation coefficient indicated that 0.46% (Figure 1). Thus, we assume that $SiO_2\%$ and $CaO\%$ are independent variables. In cutoff grade policy, independent variables may be estimated multivariate or bivariate model. Investigation in the cutoff grade with the Beylikova magnesite ore deposit may be determined using bivariate lognormal independent distribution model. Investigation in the T_c with the Beylikova magnesite deposit may be evaluated by bivariate model. The parameters were calculation by using equations of section 2. These calculations were graphed and showed in Figure 2. The figure shows that, cutoff $SiO_2\%$ and cutoff $CaO\%$ increases as the T_c increases. But the quality of the magnesite decreases.

CONCLUSIONS

The most important impurities of a magnesite deposits are the $SiO_2\%$ and $CaO\%$. The determination of T_c involves the identification of cutoff impurities. In the magnesite deposit, as the cutoff impurity rates increase, T_c of the deposit also increase. However, with T_c increase, the quality of the magnesite is decreased.

This study presented that bivariate lognormal independent distribution model provide useful information for the cutoff impurities in helping them plan their magnesite deposits. This model is believed to assist decision makers assessing magnesite ore deposits.

Table 1. Descriptive statistics of $SiO_2\%$ and CaO%

Parameters	n	Min.	Max.	Mean	Std. deviation	Variance
$SiO_2\%$	135	0.01	1.21	0.224	0.278	0.077
CaO%	135	0.02	4.57	0.547	0.656	0.430

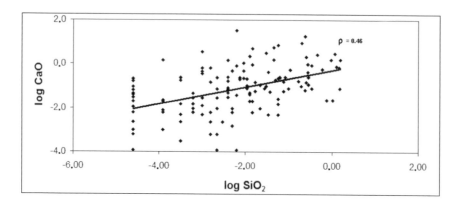

Figure 1. The correlation coefficient between the $logSiO_2\%$ and logCaO%

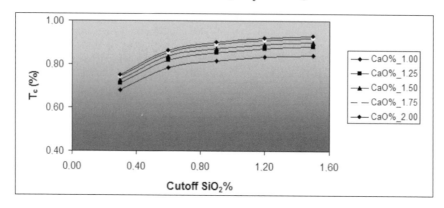

Figure 2. Cutoff $SiO_2\%$ versus T_c

ACKNOWLEDGMENT

This paper constitutes part of the PhD study of Suheyla Yerel.

REFERENCES

Ataei, M. and Osanloo, M. 2003. Determination Of optimum Cutoff Grades Of Multiple Metal Deposits By Using The Golden Section Search Method. *The Journal of the South African Institute of Mining and Metallurgy* 103: 493.

Cetin, E., Dowd, P.A. 2002. The use of genetic algorithms for multiple cutoff grade optimization. *Proceedings of the 32nd International Symposium on Application of Computers & Operations Research in the Mineral Industry,* USA.

Konuk, A. and Yersel, G. 1998. The risks of cutoff impurity decisions in magnesite open pit mining. *5th National Open Pit Mining Conference with International Participation*, Bulgaria.

Lane, K.F. 1964. Choosing the optimum cut-off grade. *Quarterly of the Colorado School of Mines* 59:811.

Orhan, A., Seyrek, E. and Tosun, H. 2007. A probabilistic approach for earthquake hazard assessment of the province of Eskisehir, Turkey. *Nat. Hazards Earth Syst. Sci.* 7: 607.

Osanloo, M. and Ataei, M. 2003. *Using equivalent grade factors to find the optimum cut-off grades of multiple metal deposits.* Mineral Engineering 16:771.

Taylor, H.K. 1972. General background theory of cut-off grades. *Institution of Mining and Metallurgy Transaction* A:160.

Taylor, H.K. 1985. Cut-off grades-some further reflections. *Institution of Mining and Metallurgy Transaction* A:204.

Uygucgil, H., Yerel, S. and Konuk, A. 2007. Estimation of Liquefaction Potential using Geostatistics in Eskisehir. *WSEAS Transactions of Systems* 6(6): 780.

Yerel, S. 2008. *Determination of cutoff grade decision variables in multivariate ore deposits.* Eskisehir Osmangazi University (Ph.D. Thesis), Eskisehir.

Yue, S. 2002. The bivariate lognormal distribution for describing joint statistical properties of a multivariate storm event. *Environmetrics* 17: 811.

An Empirical Relationship to Determine the Performance of Bucketwheel Excavator, Bench Conveyor, and Spreader Systems Verified with Data from the Rhenish Lignite Mining District

Robrecht M. Schmitz
RWE Power, Cologne, Germany

Holger Franken
RWE Power, Grevenbroich, Germany

Stefan Blunck
RWE Power International, Cologne, Germany

ABSTRACT: Continuous mining equipment has been developed, improved and used in the Rhenish Lignite mining district throughout the 20th and 21st century. Mining with the world-wide largest bucketwheel excavators, conveyor belts and spreaders technology, the Rhenish Lignite mining district accumulated a wealth of information with respect to the performance of continuous mining equipment. In this contribution we deduce an empirical relationship developed on the basis of the aforementioned knowledge. This relationship can be used to understand determining factors with respect to long-term realistic performance scenarios of continuous mining equipment validated for Hambach surface mine. Experience confirmed that the relationship is valid for other mines using a bucketwheel excavator, conveyor belts, belt junction and spreader systems as well.

INTRODUCTION

Lignite has been mined in the Rhenish lignite mining district for many centuries. With the development of the briquetting press in the 19th century, lignite could be upgraded to a valuable product. As a consequence of increasing demand, mining continued on an increasing larger scale. The introduction of first the steam engine and later electric motors into excavation equipment, formed the basis of the large scale surface mines. The large number of mines in the Rhenish lignite mining district finally merged into one company in 1959 (today RWE-Power). This merger allowed for the development of highly efficient large mines and accompanying investment in large efficient mining equipment. The mining equipment that perfectly suited the ground conditions in the Rhineland were bucketwheel excavators (BWE), conveyor belts and spreaders (BCS). Large mines (e.g., Fortuna mine) and large machines with a nominal capacity of 100.000 BCM overburden or t lignite/day and later 240.000 BCM overburden or t lignite/day (Hambach mine) were developed.

This short summary shows that mining in the Rhenish lignite mining district is closely associated to BWE-technology. A wealth of information with respect to the performance of this continuous mining technique has been accumulated. It is interesting to analyse if this data can be used to develop a systematic approach to estimate realistic production scenarios for mining operations using or going to use this technology. For this purpose the productivity data of the largest mine (Hambach) in the Rhenish lignite mining district was analysed because Hambach is a typical example of a surface mine especially designed for BCS-systems using a modern BCS-fleet. Therefore Hambach surface mine is introduced in some detail. Some challenges related to the mining operation in Hambach are presented since these are particular for this mine and should be corrected if using the proposed empirical relationship for other mining applications. Then definitions are given to describe the performance of BWE, bench conveyors and spreader chains (BCS). Thereafter the production data of Hambach is analysed and a simple correlation to describe the performance of BCS-systems is verified on the basis of this data. In a subsequent section it is discussed to what extent the correlation derived for Hambach can be used for other BCS applications. It was found that a complex model did not predict the real mine data more accurately than this simple correlation.

Note that an empirical approach is followed to determine the long-term performance of BWE and BCS-systems. The productivity of a singular BWE is not at the focus of this contribution. In contrast to many other mines using the BCS system the mines in the Rhineland use a belt junction system. This belt junction system allows connecting any BWE with any spreader via shuttle heads, thereby partly *decoupling* the BCS-system. The parameters used in the

Figure 1. Aerial photo of Hambach surface mine. Eight large scale bucketwheel excavators are used to excavate the overburden and the coal. The overburden is transported by conveyor belts via the conveyor belt junction to the dump. The coal is transported to the coal stockpile and train loading station. From here the coal is transported to the power stations and coal beneficiation plants. The entire mine turns clockwise to the east with the conveyor belt junction as its approximate slewing centre (pivot).

correlation clearly distinguishes between operational, maintenance and dispositional factors which are mine specific (degree of maintenance quality, material properties, aspects of overburden disposal and mining different coal qualities, weather influences). These parameters have not been intermixed in this paper. The advantage of this notation is that the dedicated reader can fill in values for different factors which he thinks are most appropriate for his application.

THE RHENISH LIGNITE MINING DISTRICT AND HAMBACH SURFACE MINE

The Rhenish lignite mining district is located in between two "branches" of the Rhenish Slate Mountains. During the Tertiary, the Rhenish Slate Mountains were weathered down. The sediments were transported and deposited by rivers traversing the plane area—which was subjected to tectonic subsidence—towards the North Sea. Lush vegetation developed on this plane and along the coastline. During trans- and regressions, processes promoting the development of marshes (Pohl 1992), the dying organic material turned into peat. Due to subsidence of the plane, thick (400 m) peat layers accumulated (Walter 1995). This peat was transformed into 100 m thick lignite deposits. Sand, clay and gravel accumulated during the remainder of the Tertiary and the

following Quaternary, resulting in a several hundred meter thick overburden. Loess was deposited on top of these layers (ENB 2005).

Three active mines are operated in this mining district: Inden, Garzweiler and Hambach surface mines. Hambach surface mine (Figure 1), located west of Cologne, Germany, is one of the largest lignite mines world wide, extending from 300 m below sea level to 300 m above sea level, producing a large amount of coal per year (40 million metric tons) using the largest bucketwheel excavators and a hundred kilometres of conveyor belt. Key data of Hambach surface mine:

- 1978: Start of the mining activity
- 1984: First lignite production
- Total coal content: 2.5E9 t
- Overburden removal in total: 15E9 m^3
- Overall stripping ratio: 6.2 : 1 (m^3 waste : t coal)
- Seam thickness: rd. 60–70 m
- Deepest level (from surface): 370 m (at present), 450 m (final depth)

Coal quality:

- Sulphur content: rd. 0.20–0.35%
- Calorific value: rd. 8,000–11,500 kJ/kg
- Ash content: rd. 2–5%

Equipment:

- 5# BWEs with a capacity of 240.000 BCM overburden or t lignite/day
- 1# BWEs with a capacity of 200.000 BCM overburden or t lignite/day
- 2# BWEs with a capacity of 100.000 BCM overburden or t lignite/day
- 6# spreaders with a capacity of 240.000 BCM overburden/day (corrected for bulking)
- 1# spreader with a capacity of 130.000 BCM overburden/day (corrected for bulking)
- Conveyor belts (2.7 m wide; velocity up to 7.5 m/s; combined length in active mine (without conveyors towards and in the stockpile): 84 km

TECHNICAL CHALLENGES IN HAMBACH SURFACE MINE

Interesting technical challenges and solutions involved in mining in Hambach surface mine are mentioned here below, because these aspects influence the parameters used in the performance correlation presented in the next sections. Some challenges are:

- Stacking of weak soils: With the current stripping ratio of 6 to 1, more than 240 million cubic meters of overburden (and interburden) have to be mined, transported and dumped each year. Because of the size of the excavation equipment, not every sand and clay layer can be mined separately. Therefore mixtures of sand/silt and clay dominate in the daily overburden disposition. Depending on the water content, the relative content of sand to clay in these mixtures and the transportation distance from excavator to the spreader, the mixtures are thoroughly remoulded and, inherently, their consistency will change. The inner dump has a total height of 600 m, measured from the top of the dump to the footwall. The stability of this slope is very important. Therefore remoulded (weak) clay-sand mixtures cannot be stacked straightaway on the dump without taking any preparatory measures. The method used in Hambach to overcome this problem consists in creating large sand basins (length parallel to the conveyor belt: several km; height: up to 15 m; width: 70 to 90 m) on a spreader bench behind which the weak clay and weak clay-sand mixtures are stacked. The clay basin must be covered by sand, or sand-clay mixtures with low clay contents, in order to generate a stable basis for the following bench on the next higher level.

This system (Schmitz et al. 2008) has been in use in Hambach for decades and is continuously optimised: In recent years an innovative scanning and GPS-based system has been developed by the mine to assist mine planners, production and operators (Schmitz et al. 2009).

- Hard concretions in mudstone layers: Since a couple of years cemented concretions are found in the Miocene clays just above the lignite layers. These concretions (siderite nodules) are characterised by a very high unconfined compressive strength (UCS; up to 80 MPa according to DIN 18126 in the lab; classed very to extremely strong according to ISRM standards in the field). Because BWE are traditionally designed for mining soft soils, the excavation of these concretions presents a major challenge. New buckets have been designed to shatter the concretions by introducing high point loads (Schmitz 2007).
- Different coal qualities: The 40 million metric tons of coal Hambach has to produce each year has to be provided in 10 different coal classes. The coal classes are excavated directly from the coal face. This requires excellent mine planning and equipment deployment assisted by IT-technology, processes which have been realised on site.

The above mentioned shows that Hambach presents an excellent example to study BCS-systems and their performance under challenging boundary conditions. Data from Hambach is used in subsequent paragraphs to verify the correlation describing the performance of the BCS-system proposed in the next paragraph.

PERFORMANCE OF BCS-SYSTEMS IN GENERAL: DEFINITIONS

To describe the overall long-term performance of BCS-systems we decided to follow the definitions proposed by Kumar (2006) because of its simplicity and comparability to other mining technologies. Time-availabilities of different components of the BCS are introduced as percentages of calendar or scheduled time (the time when action takes place in the mine i.e., production or maintenance). The system is represented as a series of BCS-chains. Each chain consists of a BWE, several bench conveyors to the belt junction, several bench conveyors from the belt junction to the spreader and the spreader itselft. Since a belt junction system is used, the availability of a spreader is not forcibly dependent on the availability of a BWE upstream in the chain.

The following definitions are introduced:

OEE: Overall equipment effectiveness. The unit is volume or mass mined (excavated, transported and dumped) during a specified period. OEE is dependent on the mining system used.

σ: Time-availability at which overburden or ore is mined at a specified performance. The value is expressed as a ratio.

P: Performance. The unit is production per unit time and depends on the properties of material being excavated and the face geometry (depends on mine planning and dispositional effects and the mechanical availability of the system etc.). The performance P of a BWE indicates the degree of capacity utilisation and depends on several factors related to the mined material (sticky weak material or hard concretions give lower values than dry fine sand), operational conditions (geometry of the face) and other aspects. This paper focuses on the time-availability of a BCS chain, the P value is not the focus of this paper. Values to calculate the P of BWE systems have been provided by Wolski and Golosinski (1987).

Examples for system availabilities:

Dragline: $\sigma_{chain} = \sigma_{dragline}$

If a BCS-system is regarded to be a serial system, this parameter can be expressed as:

BCS: $\sigma_{chain} = \sigma_{bwe} \cdot \sigma_{bench\_conv\_bcs} \cdot \sigma_{spreader}$

σ_{bcs}, the time-availability of a mine in reality is dependent on σ_{chain}, but must be corrected by planned maintenance, planned standstill and dispositional factors. This is done by subtracting the hours attributed to these factors from the scheduled time (explained below).

Finally the proposed OEE for a BCS chain can be expressed as:

$$OEE_{bcs} = P_{bcs} \cdot \sigma_{bcs}$$

The question is if the performance of a BCS-system can be described on the basis of these expressions. This will be verified with actual data from Hambach surface mine. First several other parameters will be introduced.

Basically σ_{mining_system} can be split into:

$\sigma_{scheduled\_time}$: This factor takes the scheduled time into account in which action (maintenance, production etc.) takes place in the mine. The remaining time (no production and no maintenance) consists of holidays (unit: percentage of calendar time).

$\sigma_{planned\_standstill}$: This factor takes mining activities like shifting of conveyor belts and further planned and non-productive activities into account.

This value is highly depended on the geometry and lay-out of a mine.

$\sigma_{planned\_maintenance}$: This factor considers all planned major maintenance activities.

$\sigma_{disposition}$: this factor includes operational and organisational factors concerning selective coal mining as well as restrictions related to the specific disposition of weak overburden etc.

Unplanned maintenance and mechanical breakdowns are included in e.g., σ_{bw} as defined below. Note that all planned maintenance is part of the dataset of the used production database of the mine. Maintenance is considered to be an integral part of the mining process and is scheduled like production and therefore part of the scheduled time. During the remaining time (calendar time minus scheduled time) no activities take place in the mine i.e., corrections as proposed by Wolski and Golosinsky (1987) are not required in this case.

For the sake of simplicity a BWE was split into the following major components (including associated gearboxes and drives) to determine the mechanical availability of the major components which are relevant to the operation:

- Bucketwheel (20 m diameter, 18 buckets)
- Conveyor belts on the bucketwheel excavator, the bridge and hopper car (in total more than 200 m)
- Additional mechanical installations (hoists, crawlers, etc.)
- Other components

The availability of these individual components can be retrieved from the mine production database.

The productivity of BWEs has been the subject of several other publications. We found especially the article by Wolski and Golosinski (1987) of great value and agree with these authors that a standardisation of terminology is very useful for comparison of different mining technologies. In this contribution we focus on the BCS-technology and we:

1. Concentrate not only on the BWE but on the coupled problem of BCS-systems in a state-of-the-art surface mine and supplement additional data to the data compiled by e.g., Wolski and Golosinski (1987).
2. Use real mine data of a state-of-the-art BCS-system.
3. show that the different components of a BCS-system can be subdivided in distinct units with comparable time-availabilities.

VERIFICATION OF THE PERFORMANCE CORRELATION USING DATA FROM HAMBACH SURFACE MINE

Production data from Hambach surface mine is recorded automatically. Data from the period 2003–2007 was extracted from this database (up to 90,000 datasets) and subsequently rearranged according to the definitions mentioned above. Following values were obtained:

1. Scheduled time

The time available for production and maintenance, the basis for all further calculations is given by:

$$T_{scheduled\_time} = \sigma_{scheduled\ time} \cdot calendar\ time$$

$T_{scheduled\_time}$: 8040 h (value used in the calculation)

$\sigma_{scheduled\ time}$ = 92% (value used to compare different BCS-systems)

2. The mining system

$T_{planned\_standstill}$: 768 h (value used in the calculation)

$\sigma_{planned\_standstill}$ = 90% of the scheduled time (value used to compare different BCS-systems)

$T_{planned\_maintenance}$: 1031 h (value used in the calculation)

$\sigma_{planned\_maintenance}$ = 87% of the scheduled time (value used to compare different BCS-systems)

$T_{disposition}$: 925 h h (value used in the calculation)
$\sigma_{disposition}$: 88% of the scheduled time (value used to compare different BCS-systems)

$$T_{mining\_system} = T_{scheduled\ time} - (T_{planned\ standstill} + T_{planned\_maintenance} + T_{disposition})$$

$$\sigma_{mining\_system} = T_{mining\_system} / T_{scheduled\_time} = 66\%\ \text{of the}$$
scheduled time

3. The BWE

The BWE is, as explained above, divided into the following components:

σ_{bwe_total}: 93%
σ_{bw}: 96%
$\sigma_{bwe\_conv\_total}$: 98%
σ_{bwe_inst}: 98%
σ_{bwe_other}: a correction factor

Assuming that: $\sigma_{bwe\ total} = \sigma_{bwe\_conv\_total} \cdot \sigma_{bwe\_inst} \cdot \sigma_{bwe\_other}$, it follows from this data that $\sigma_{bwe\_other} = 100\%$ (the factor is apparently not required) and that the bucketwheel, the conveyor belts and "further installations"

are indeed the major components that influence the overall availability of the BWE.

4. The bench conveyors

The analysis of the above mentioned database shows that the overall σ of the conveyor belts (on average 4 to 5 separate bench conveyors from the BWE to the belt junction, on average 1.4 km long) amounts to:

$\sigma_{bench\_conv\_b}$: 95%

The data-base was used to determine if the length of a separate conveyor or the number of conveyor belts or both play a role. A significant correlation with the length was not found. The number of individual conveyor belts (and transfer points) plays however a major role:

$\sigma_{bench\_conv\_b}$ can be estimated by:

$\sigma_{bench\_conv\_b} = 0.99^n$; n = number of conveyor belts or conveyor flights

If only the number of individual conveyors (equal to the number of transfer points) is taken into account, the value for an individual conveyor amounts to:

$\sigma_{bench\_conv\_b\_i}$: 98.9%

This value can be compared to the conveyor belts on the excavators. Analogue to the above mentioned, the availability of each conveyor belt of the BWE separately can be expressed as:

$$\sigma_{bwe\_conv\_total} = \sigma_{bwe\_conv\_1} \cdot \sigma_{bwe\_conv\_2} \cdot \sigma_{bwe\_conv\_3} \cdot \sigma_{bwe\_conv\_4} \cdot \sigma_{bwe\_conv\_5}$$

$\sigma_{bwe\_conv\_i} = 99.6\%$

and is as such slightly higher than the availability of the individual bench conveyor.

5. The spreaders

Likewise the spreader was analysed by representing the spreader by following determining components:

- Conveyor on the spreader (three belts on the spreader and a belt on the tripper car)
- Additional mechanical installations

Following values were retrieved from the database:

$\sigma_{spreader\_total} = 95\%$

$\sigma_{spreader\_conv\_total} = 99\%$

$\sigma_{spreader\_inst} = 98\%$

$\sigma_{spreader\_other} = 97\%$

Similar to the approach followed above, for each of the four individual conveyors the following value can be determined:

$$\sigma_{spreader\_conv\_i} = 99.7\%$$

Note that a similar value was found for the availability of the individual conveyors on the BWE.

Similar to the BWE analysed above the availability of the spreader can be defined as:

$$\sigma_{spreader\_total} = \sigma_{spreader\_conv\_total} \cdot \sigma_{spreader\_inst} \cdot \sigma_{spreader\_other}$$

The correction factor $\sigma_{spreader\_other} = 97\%$ incorporates other determining components of the spreader system like e.g., the loose connection between the belt wagon and the spreader itself.

The availability of the bench conveyors from the belt junction (again on average 4–5 separate conveyor belts, each 1.4 km long) towards the spreaders have not been analysed and are assumed to be equal to the availability of the bench conveyors from the BWE to the belt junction. Assumed: $\sigma_{bench\_conv\_b} = \sigma_{bench\_conv\_s}$, leading to: $\sigma_{bench\_conv\_bcs} = \sigma_{bench\_conv\_b} \cdot \sigma_{bench\_conv\_s}$

First Verification of the Correlation Describing Performance with Data from Hambach Surface Mine

With the data given above, the average availability of the BWE for production can be calculated:

$$\sigma_{planned\_maintenance} \cdot \sigma_{bwe\_total} = 81\%$$

On a daily basis this value shows that the BWE is available: $81\% \cdot 24h = 19$ h per day. This value was exactly requested by the lignite mining company operating in the Rhenish lignite mining district when the fleet of BWE was ordered from the late 1970th to the early 1990th and this value was guaranteed by the manufacturers that constructed and manufactured the machines.

First Conclusion

The analysed data which span a 5 year period (including tens of thousand individual data sets) show that the availability for production of the BWE and bench conveyors is extremely high in Hambach surface mine irrespective of challenges involved in overburden disposal, hard concretions and coal disposition. Even though the newest BWE has been in operation since over 16 years the specified availability is met.

The BWE and spreader time-availability can be described by analysing the major components: Conveyor belt on the machines and additional mechanical installations and the bucketwheel (only for the BWE).

Second Verification of the Correlation Describing Performance with Data from Hambach Surface Mine

For Hambach surface mine the combination of the values given above presents the OEE of the system from the bucketwheel excavators to the conveyor belt:

$$\sigma_{bcs} = \sigma_{mining\_system} \cdot \sigma_{BWE} \cdot \sigma_{bench\_conv\_bcs} \cdot \sigma_{spreaders}$$
$$= 66\% \cdot 93\% \cdot 90\% \cdot 95\%$$
$$= 53\% \text{ of the scheduled time}$$

with $\sigma_{scheduled\ time} = 92\%$

$\sigma_{bcs} = 48\%$ of the calendar time

The value for the σ_{bcs} calculated in this way underestimates the actual data with a small 1.7% deviation.

This value needs to be multiplied with the P value to obtain the overall equipment effectiveness:

$$OEE_{bcs} = P_{bcs} \cdot \sigma_{bcs}$$

P is strongly dependent on the material to be excavated and can be as high as 100% in sand and 60% or lower in hard clays. P is highest in high cut mode on a large face and decreases when cutting ramps or doing other special tasks (more values are provided by Wolski and Golosinski, 1987).

Second Conclusion

With the simple correlation:

$$\sigma_{bcs} = \sigma_{mining\_system} \cdot \sigma_{BWE} \cdot \sigma_{bench\_conv\_bcs} \cdot \sigma_{spreaders}$$

the performance of a system as complex as Hambach surface mine can be described with a 1.7% failure margin. Obviously the influence of the different conveyor belts is dominant. Grouping all conveyors into one class the formula can be simplified to:

$$OEE_{bcs} = P_{bcs} \cdot \left(\sigma_{mining\_system}\right) \cdot \left(\sigma_{bw} \cdot \sigma_{spreader\_other}\right) \cdot \left(\sigma_{bwe\_inst} \cdot \sigma_{spreader\_inst}\right) \cdot \sigma_{all\_conv\_total}$$

POTENTIAL USE OF THE CORRELATION DESCRIBING PERFORMANCE FOR OTHER BCS APPLICATIONS

The correlation describing performance defined above was developed using data from Hambach surface mine. The correlation has been tested on other data from other mines world-wide using a BCS-system. Besides the interesting comparison of the overall availability of different mines, the system and standardisation developed above was used to compare the individual values for planned standstill, disposition and performance of individual components like the bench conveyors or the conveyors on the machines. If the correlation is applied to other BCS-systems, some values representative of Hambach surface mine need to be changed to suite specific conditions, some values however can be retained.

Example: The analysis above showed that it is apparent that the level of maintenance of e.g., the BWE in Hambach surface mines is excellent. Since being in operation for more than 16 years (most recent BWE) the BWEs still fulfill the specifications. Consequently the values for σ_{bw}, σ_{bwe_inst}, $\sigma_{all\_conv\_total}$ or $\sigma_{spreader\_inst}$ are most likely not higher in other applications or operations in other mining environments.

Major differences will be found in:

- $\sigma_{disposition}$: Depending on the local geology, absence of very weak layers, number of different coal qualities which are required to be mined selectively etc.
- $\sigma_{planned\_standstill}$: Depending on the number of conveyor belt shifts, relocation of the BWEs on the bench and between different benches.
- $\sigma_{scheduled\_time}$: Depending on the number of holidays and working days per week and year.
- P_{bcs}, which amounts to 100% when a nominal block width (as defined by the manufacturer) and specified nominal material is excavated (normally easy to mine materials like loose sands or lignite).

For other applications this expression can be summarised as:

$$OEE_{bcs} = P_{bcs} \cdot \sigma_{scheduled\_time} \cdot (T_{scheduled\_time} - (T_{planned\_standstill} + T_{planned\_maintenance} + T_{disposition}))/T_{scheduled\_time} \cdot \sigma_{bwe\_total} \cdot \sigma_{bench\_conv\_bcs} \cdot \sigma_{spreader\_total}$$

The underscored values are values for Hambach surface mine as given in the text (using $\sigma_{bench\_conv\_bcs}$ = 0.99ⁿ) which are deemed to have applicability in general.

Example: Assuming simple mining conditions with sandy overburden, one coal quality, horizontal extensive seams, no concretions, short conveyor length to spreader, following values can be attained:

P_{bcs} = 100%

$\sigma_{scheduled\_time}$ = 100%

$\sigma_{planned\_standstill}$ = 98%

$\sigma_{disposition}$ = 100%

$\sigma_{bench\_conv\_bcs}$ = 95% (@ n = 5)

Using the other values from Hambach, in this case OEE_{bcs} would be 72%. With BCS-systems that are constructed and achieve over 12,000m³/h, impressive amounts of material can be moved with a OEE_{bcs} equal to 72% for many decades to come.

These values can be used to estimate the long-term productivity of other mine equipment set-ups like the cross-pit spreading method (XPC). Looking at XPC methods in such cases the value increases to:

OEE_{bcs} = 76%, a value higher than in the previous example as a consequence of the absence of bench conveyors.

OVERALL CONCLUSION

Lignite has been mined in the Rhenish lignite mining district for over hundred years. Especially during the 20th century the mines increased dramatically in size embracing modern mining equipment and mining techniques. The most adapted mining technique for the Rhenish Tertiary mining basin that was developed with local mining equipment manufacturers is the BCS-system. The master piece of this process of upscaling is Hambach surface mine. In the late 1970th a new BCS-system was developed for this mine. The production data of Hambach surface mine was analysed and a correlation describing performance was developed. This correlation tested on other BCS applications shows that the time-availability of a complex machine like a BWE can be reduced to its major components: the bucketwheel, the conveyor belt system on the BWE and other technical components like the hoist and the crawlers which were clustered in one group called "additional mechanical installations." Furthermore this simple correlation, which can be used to estimate the long-term productivity of BCS-systems, was introduced and verified with data from Hambach surface mine. If the correlation is used for other applications, following aspects should be considered:

- It was shown that the maintenance of the BCS-system in Hambach was even after many years of service "as specified." If the correlation is used to estimate the performance of other mines and if these mines will be operated with a high maintenance standard the same values are attainable
- The belt junction in Hambach allows connecting any BWE with any spreader thereby

partially decoupling the availability of BWEs and spreaders

- If mining a simple deposit (no weak soils, only a few coal qualities), higher vales for $\sigma_{disposition}$ can be selected
- Depending on the mine lay-out, the number of conveyor belt shifts per year etc., higher values for $\sigma_{planned\_standstill}$ can be chosen
- The time-availability of the bench conveyor system ($\sigma_{all\_conv\_total}$) depends strongly on the number of conveyor belts (the length of the conveyor belts plays a minor role)
- Values for $\sigma_{planned\_standstill}$, $\sigma_{planned\_maintenance}$ and $\sigma_{disposition}$, not used in direct calculations, can be used to compare different mines on the basis of these unitized parameters

Time-availabilities and, depending on the performance, overall equipment effectiveness of over 70% are attainable with the BCS-system over long periods of time.

ABBREVIATIONS AND UNITS

BWE   Bucketwheel excavator
BCS   BWE, bench conveyors and spreaders system
UCS   Unconfined compressive strength
DIN   Deutsche Industrie Norm: German industrial standard
ISRM   Society for Rock Mechanics
OEE   Overall equipment effectiveness
P   Performance. Unit: production per unit time
T   Time (unit: h)
σ Time-availability, ratio of maximum 100% available time, at which a component in a BCS-system allows overburden removal or ore mining at a specified performance. Unit: percentage

Following time-availabilities were used:

σ_{bcs} Time-availability of the bucketwheel excavator, bench conveyor and spreader system

σ_{mining_system} Time-availability reflecting the scheduled time, the planned standstill, the planned maintenance and the disposition into account

$\sigma_{scheduled\_time}$ Time-availability reflecting the scheduled time during which actions (maintenance, production etc.) take place in the mine. The remaining time (no production and no maintenance) consists of holidays

$\sigma_{planned\_standstill}$ Time-availability of mining activities like shifting of conveyor belts and further planned and non-productive activities into account

$\sigma_{planned\_maintenance}$ Time-availability representing all planned major maintenance activities

$\sigma_{disposition}$ Time-availability reflecting operational and organisational factors concerning the coal management as well as restrictions related to the specific disposition of weak overburden etc.

$\sigma_{bwe} = \sigma_{bwe\_total}$ Time-availability of the bucketwheel excavator

σ_{bw} Time-availability of the bucketwheel

σ_{bwe_inst} Time-availability of the hoist and the crawlers and other components of a BWE clustered into the group "additional mechanical installations"

$\sigma_{bwe\_conv\_i}$ Time-availability of one conveyor on a BWE

$\sigma_{bwe\_conv\_total}$ Time-availability of all conveyor belts on the BWE

σ_{bwe_other} A correction factor to correct the time-availability of a BWE calculated on the basis of the time-availability of separate components

$\sigma_{spreaders} = \sigma_{spreader\_total}$ Time-availability of the spreaders

$\sigma_{spreader\_inst}$ Time-availability of the hoist and the crawlers and other components of a spreader clustered into the group "additional mechanical installations"

$\sigma_{spreader\_other}$ A correction factor to correct the time-availability of a spreader calculated on the basis of the time-availability of separate components

$\sigma_{spreader\_conv\_i}$ Time-availability of one conveyor on a spreader

$\sigma_{spreader\_conv\_total}$ Time-availability of all conveyors on a spreader

$\sigma_{bench\_conv\_bcs}$ Time-availability of all bench conveyors system of a whole BCS-chain

$\sigma_{bench\_conv\_bcs\_i}$ Time-availability of one bench conveyor in a system

$\sigma_{bench\_conv\_b}$ Time-availability of all bench conveyors from a BWE to the belt junction

$\sigma_{bench\_conv\_b\_i}$ Time-availability of one bench conveyor from a BWE to the belt junction

$\sigma_{bench\_conv\_s}$ Time-availability of all bench conveyors form the belt junction to a spreader

$\sigma_{bench\_conv\_b\_i}$ Time-availability of one bench conveyor from the belt junction to a spreader

$\sigma_{all\_conv\_total}$ Time-availability of all conveyors (incl. conveyors on the BWE and spreaders) in a BCS-system

n   Number of bench conveyors

LITERATURE

ENB. 2005. Origin of the Lower Rhenish Lignite. Brochure RWE-Power. (in German).

Kumar, U. 2006. Maintenance technology and management. Luleå University of Technology.

Pohl, W. 1992. Lagerstättenlehre. 4th Edition. E. Schweizerbart'sche Verlagsbuchhandlung. Stuttgart.

Schmitz, R.M. 2007. Excavation of rock concretions in Hambach opencast mine: In situ tests. In Proceedings: 11th Congress of the International Society for Rock Mechanics. Edited by: Ribeiro e Sousa, Olalla & Grossmann. 593–596.

Schmitz, R.M., Kübeler, U., Elandaloussi, F., Lau, D., Hempel, R-J. 2008. The introduction of IT into mass mining: the digital mine in Hambach surface mine. In 5th International Conference & Exhibition on Mass Mining. Edited by: Schunnesson, H., and Nordlund, E. 499–508.

Schmitz, R.M., Hempel, R.J., Bertrams, N., Kaydim, M., Eraydin, E., Kübeler, U. 2009. Sixth spreader upgraded with laser scanner- and GPS-technology in Hambach surface mine. Mining Engineering. 61(9): 64–68.

Walter, R. (1995) Geologie von Mitteleuropa. E. Schweizerbart'sche Verlagsbuchhandlung. Stuttgart.

Wolski, J., Golosinski, T.S. (1987) Performance factors of the bucket wheel excavator system. Society of Mining Engineers of AIME, Transactions (280): 1904–1911.

Section 3. Ground Control

Mitigating the Risk of Rockbursts in the Deep Hard Rock Mines of South Africa: 100 Years of Research

Raymond J. Durrheim

CSIR Centre for Mining Innovation and the University of the Witwatersrand, Johannesburg, South Africa

ABSTRACT: South Africa hosts two of the world's richest ore bodies, the gold-bearing conglomerates of the Witwatersrand Basin and the platinum-bearing pyroxenites of the Bushveld Complex. Both ore bodies extend to depths of several kilometers. Gold was discovered near present-day Johannesburg in 1886. Mining-induced seismicity and its hazardous manifestation, rockbursting, were first encountered in the early 1900s when extensive stopes reached depths of several hundred meters. A committee, appointed in 1908 to investigate the cause of the tremors, recommended that seismographs be installed, and scientific observations began in 1910. Research was carried out in an ad hoc way until a mining research laboratory was established in 1964. Since then, research addressing the risks of deep hard rock mining has mostly been conducted under the auspices of the Chamber of Mines Research Organization, the Mine Health and Safety Council, and several collaborative research programs. Research work to mitigate the rockburst risk has focused on three main areas: (i) development of macro-layouts (e.g., sequential grid) and regional support (e.g., backfilling) to control the release of seismic energy through the geometry and sequence of mining; (ii) development of support units and systems (e.g., rapid-yielding hydraulic props, pre-stressed elongates) to limit rockburst damage; and (iii) mine seismology, which seeks to develop techniques to continually assess the seismic hazard and control seismic activity (e.g., through the rate of mining). Implementation of these technologies, together with improvements in training, work organization and regulation, have reduced fatality rates and made it possible to mine successfully at depths exceeding 3.5 kilometers.

RISKS POSED BY ROCKBURSTS

The gold-bearing strata of the Witwatersrand Basin, which crop out near present-day Johannesburg, have produced almost one-third of the gold ever mined (Handley 2004). The Central Rand goldfield was discovered in 1886. Mining-induced seismicity and its hazardous manifestation, rockbursts, were first encountered in the early 1900s when extensive stopes, supported solely by small reef pillars, reached depths of several hundred meters. Gold-bearing conglomerates were found buried beneath younger strata in the East Rand district in 1914, the Far West Rand and Klerksdorp districts in 1937, the Orange Free State in 1946, and the Kinross district in 1955. The dipping conglomerate "reefs" were found to persist to depths of several kilometers. Rockbursts have remained one of the most serious and least understood problems facing deep mining operations, claiming the lives of thousands of mine workers. Despite many technical advances, rockbursts continue to pose a significant risk (Figure 1).

As the South African gold mining industry expanded and the severity of rockbursts increased, the authorities appointed several committees to investigate the problem.

- In 1908, minor damage in a village near Johannesburg led to the appointment of a committee, chaired by the Government Mining Engineer, to "inquire into and report on the origin and effect of the earth tremors experienced in the village of Ophirton." The 1908 Ophirton Earth Tremors Committee found that "under the great weight of the superincumbent mass of rock … the pillars are severely strained; that ultimately they partly give way suddenly, and that this relief of strain produces a vibration in the rock which is transmitted to the surface in the form of a more or less severe tremor or shock" (Anon 1915). The 1908 Committee recommended that the support pillars should be replaced by waste packs.

- The 1915 Witwatersrand Earth Tremors Committee was asked to "investigate and report on: (a) the occurrence and origin of the earth tremors experienced at Johannesburg and elsewhere along the Witwatersrand; (b) the effect of the tremors upon underground workings and on buildings and other structures on the surface; (c) the means of preventing the tremors." The 1915 Committee concluded that "the shocks have their origin in mining operations," and "while it may be expected that severer shocks than any that have yet been felt will occur in Johannesburg, their violence will not be sufficiently great to justify the apprehension of any disastrous effects" (Anon 1915).

Figure 1. Tunnel in a deep South African gold mine damaged by an M_L=3.6 tremor (photograph W.D. Ortlepp)

Figure 2. Apartment block in Welkom destroyed by an M_L=5.2 tremor on 8 December 1976 (photograph 'The Star')

- The 1924 Witwatersrand Rock Burst Committee was appointed "to investigate and report upon the occurrence and control of rock bursts in mines and the safety measures to be adopted to prevent accidents and loss of life resulting therefrom" (Anon 1924). The 1924 Committee made many recommendations concerning general mining policy, protection of travelling ways, and the stoping out of remnants.
- The 1964 Rock Burst Committee was mandated to "study the question of rockbursts and to revise the recommendations of the Witwatersrand Rock Burst Committee (1924)" (Anon 1964). It was considered opportune to conduct a new investigation as "not only had mining depths in excess of 11,000 feet below surface been reached on the Witwatersrand, but the rockburst danger had also revealed itself in the newer mining areas of the Far West Rand, Klerksdorp and the Orange Free State." The recommendations of the 1964 Committee were based on a considerable body of research and practical observations. The necessity for carrying out further research was noted.

South Africa's gold production peaked at 1000 t in 1970, while employment on gold mines peaked at 500,000 in the latter half of the 1980s. Mines reached depths exceeding 3.5 km, the greatest mining depths in the world by far. However, rockbursts continued to claim scores of lives each year. Mining-related seismic events first caused serious damage to surface structures in the Free State district when an M_L=5.2 seismic event damaged a six-storey apartment block in Welkom on 8 December 1976, which was fortunately evacuated before it collapsed (Figure 2).

The largest mining-related seismic event recorded in South Africa occurred in the Klerksdorp district on 9 March 2005. The M_L=5.3 main shock and aftershocks shook the nearby town of Stilfontein, causing serious damage to several buildings (Figure 3) and minor injuries to 58 people. No. 5 Shaft at DRDGold's Northwest Operations suffered severe damage, two mineworkers lost their lives, and 3200 mine workers were evacuated under difficult circumstances. The event prompted the Chief Inspector of Mines to appoint an Expert Panel to investigate some wider concerns regarding the risks posed by gold mining, including:

- Does mining, past and present, trigger or induce large seismic events and will it continue to do so in the future?
- Are the technologies available to manage seismicity adequate in the current situation of remnant mining, deeper mines, and mining within large mined-out areas?
- Are current approaches to planning, design, monitoring, and management appropriate and adequate?

Superficially, the issues were similar to those addressed by the earlier committees. However, much had changed since 1964. The South African gold mining industry was mature and production had been falling for the past three decades (only 255 t of gold was mined in 2007, compared with 1000 t in 1970). New problems were faced as mines approached the end of their lives, ceased operation, and were allowed to flood. Many of the cities and towns in the gold mining districts had grown, and several seismic events with M_L>5 had caused damage to residential, commercial, and civic buildings. In addition, a great

Figure 3.   Buildings in Stilfontein damaged by a M_L=5.3 tremor on 9 March 2005 (photograph R.J. Durrheim)

deal of rock mechanics and seismological research had been conducted.

The Expert Panel (Durrheim et al. 2006) concluded that the M_L=5.3 seismic event was the result of extensive mining that had taken place over several decades. Seismic events will continue to occur in the gold mining districts as long as deep-level mining takes place, and are likely to persist for some time after mine closure, especially while they flood. Regional and in-mine monitoring networks were found to be on a par with those installed in seismically-active mining districts elsewhere in the world, although measures to improve the quality and continuity of seismic monitoring were recommended, particularly when mines change ownership. A range of technologies available to mitigate the risks of large seismic events was identified, although it was noted that particular care should be taken when mining close to geological features that could host large seismic events.

The Bushveld Complex is the world's largest resource of platinum group elements. Seismicity only became a source of concern in the 1990s when mining depths approached 1 km. Knowledge and experience gained in gold mines is being adapted and applied in an endeavour to prevent rockbursting becoming a serious problem.

RESEARCH ORGANIZATIONS

The history of private, government and academic organizations that played a prominent role in rockburst research work is briefly summarized in this section so as not to interrupt the research narrative.

The *Witwatersrand Chamber of Mines* was formed in 1889, only three years after the discovery of gold. It changed its name several times as the mining industry expanded and political structures evolved: to the *Chamber of Mines of the South African Republic* in 1896, the *Transvaal Chamber of Mines* in 1900, the *Transvaal and Orange Free State Chamber of Mines* in 1953, and finally to the *Chamber of Mines of South Africa* in 1967. The Coalbrook Colliery disaster, in which 435 men died, occurred in January 1960. It is the worst accident in South African mining history. The official enquiry found that there was no scientific basis for the design of pillars in coal mines, and highlighted the need for systematic research. Consequently the Transvaal and Orange Free State Chamber of Mines established a *Mining Research Laboratory* (later renamed the *Chamber of Mines Research Organization*, COMRO), to address issues such as pillar design in coal mines and the threats to the gold mining industry of increasing depth and working costs and a stagnant gold price. It was funded on a cooperative basis by the six major mining houses operating in South Africa at that time. By 1986 COMRO employed nearly 700 people, although restructuring in the early 1990s led to the closure of several divisions, a reduction in the staff complement, and finally a merger with the CSIR in 1993.

The *Council for Scientific and Industrial Research* (CSIR) was founded in 1945. Rock mechanics research was carried out in the *National Mechanical Engineering Institute (NMERI)*. As noted above, COMRO merged with the CSIR in 1993, and the *CSIR Division of Mining Technology*

(Miningtek) was formed with a staff complement of about 250. However, support for mining research continued to decline. The *CSIR Centre for Mining Innovation* was constituted in 2009, with a research staff complement of about 40.

The *University of the Witwatersrand*, founded in 1922, has its origins in the *South African School of Mines*, which was established in Kimberley in 1896 and transferred to Johannesburg as the *Transvaal Technical Institute* in 1904, becoming the *Transvaal University College* in 1906, and the *South African School of Mines and Technology* in 1910. The *Bernard Price Institute for Geophysical Research* (BPI) was founded in 1936. Bernard Price was an electrical engineer and founder of the Victoria Falls and Transvaal Power Company. Gold mines were a major customer, but power failures owing to electrical storms were a recurrent problem. The BPI initially carried out research in two main fields: thunderstorm, atmospheric and lightning phenomena; and geophysical investigations for which the Witwatersrand was particularly suitable, such as the seismic waves generated by mining-related earth tremors. Researchers at the BPI played a leading role in establishing the discipline of mine seismology. The *Department of Geophysics*, founded in 1955, complemented the research activities of the BPI and became the base for mine seismology research work after the closure of the BPI in 2003.

The *South African National Seismological Network* (SANSN), operated by the *Council for Geoscience* (formerly the South African Geological Survey), was established in 1971. The SANSN monitors the entire country, and presently consists of 23 stations. The earthquake catalogue is complete above local magnitude 2. More than 90 per cent of the located events occur within the gold and platinum mining districts. Eight of the SANSN stations are deployed near these districts, yielding a location error of 1 to 10 km in these regions. Studies of seismic activity have shown that the southern African sub-continent is a tectonically stable intra-plate region, characterized by a relatively low level of natural activity (Brandt et al. 2005).

ISS International Ltd. (ISSI), a company specializing in technologies to monitor and model the rock mass response to mining, was founded in 1990. Directed by Aleksander Mendecki, it has become a world leader in mine seismology technology, with more than 100 systems installed worldwide. ISSI has conducted many research projects for mining companies, SIMRAC and collaborative research programs.

RESEARCH ENDEAVOURS

In this section we chronicle the research work that has been carried out over the last century to determine the cause and mitigate the effects of rockbursts in the deep hard rock mines of South Africa.

First Surface Seismographs (1910)

The first investigation into the causes of mine tremors was conducted in 1908 by the Ophirton Earth Tremors Committee. Based on the recommendations of the 1908 Committee, two seismographs were installed in 1910, one at the Union Observatory in Johannesburg and the other in the village of Ophirton. Seismograms recorded by the 200 kg Wiechert horizontal seismograph at the Union Observatory were analyzed by Wood (1913, 1914), who concluded that the source of the tremors was close to Johannesburg and probably on the Witwatersrand itself. The Ophirton seismograph was moved to Boksburg in 1913, where tremors were beginning to be felt. Philip Gane of the BPI analyzed the diurnal distribution of nearly 15,000 events, and found that their incidence peaked at blasting times (Gane 1939). An analysis of more that 29,000 mine tremors recorded in the period 1938–1949 produced similar results (Finsen 1950, cited by Cook 1976).

First Surface Seismograph Network (1939)

Fatalities due to rockbursts in gold mines remained a serious concern, and in 1938 the Chemical, Mining and Metallurgical Society of South Africa convened a Scientific Discussion Meeting where the need for rigorous scientific research into the fundamental mechanics of rockbursts was recognized. An array of seismographs covering the northern rim of the Witwatersrand Basin was deployed in 1939 by researchers at the newly established BPI. Data were transmitted by radio to a central point, where continuous 24-hour registration coupled with an ingenious device that triggered distant seismographs allowed all the larger mining-related events to be located accurately in space and time (Gane et al. 1946). It was shown that the epicenters of tremors were confined to areas that had recently been mined.

Initiation of Coordinated Research (1953) and the First Underground Seismograph Network (1961)

By 1948 it had become apparent that isolated and purely practical attempts to solve the rockburst problem were inadequate. The mining methods advocated to minimize rockbursts were subject to compromise and contradiction. For example, the recommended support methods ranged from filling the stopes with waste as solidly as possible, to complete caving of worked-out areas.

In 1953 the Transvaal and Orange Free State Chamber of Mines sought the aid of scientists at the CSIR and the University of the Witwatersrand (particularly the BPI), and assumed sponsorship of all rockburst investigations. The achievements of the program are summarized in a landmark paper by Cook et al. (1966). In retrospect, three overlapping phases can be discerned:

1. *Observations of a largely empirical nature,* e.g., observations of the pattern of fracturing in laboratory tests and underground, in situ measurements of stress, statistical relationship between mining variables and rockbursts, and networks for seismic monitoring and development of seismic location techniques. The first underground network was installed at East Rand Proprietary Mines (ERPM, see Cook 1963, 1964 and 1976) in 1961 to monitor a kilometer of working face at a depth of 2.5 km (Figures 4, 5 and 6a and 6b). A similar nine-seismometer network was established at Harmony mine in the Free State goldfields in 1964 (Joughin, 1966). A portable high-resolution seismic network was also used to study de-stressing blasts at ERPM.

2. *Attempts to attribute rational significance to the documented experience,* e.g., analytical studies based on elastic theory, development of analogue techniques for solving the elastic response of complicated mine outlines.

3. *Formulation of a rockburst mechanism.* It was postulated that the rock mass was divided into two domains: a region of continuous rock remote from the excaviovn where behaviour is elastic and predictable, and a region close to the excavation where the behaviour is non-elastic. The transition from the elastic to the non-elastic region involves fracture and the release of energy.

A final phase of *controlled underground experiments* was proposed to test the hypothesized rockburst mechanism and to vary the mining parameters to minimize the effects of rockbursts.

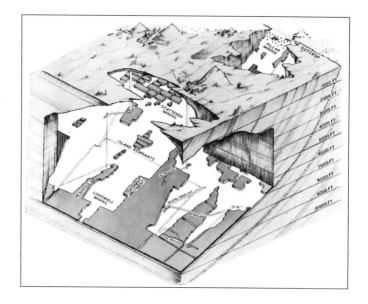

Figure 4. **Anatomy of a typical deep Witwatersrand Basin gold mine (a pictorial view of ERPM from Cook et al., 1966)**

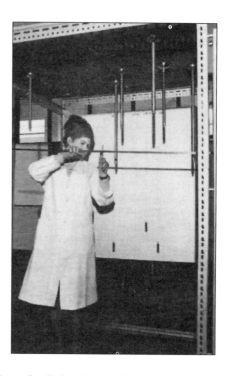

Figure 5. **Delay-time analogue computer used to find focus of seismic events (from Cook et al., 1966)**

Figure 6a.   Plan (1000 ft grid) of F longwall East and G longwall West, ERPM, and foci of 445 seismic events (Cook et al., 1966)

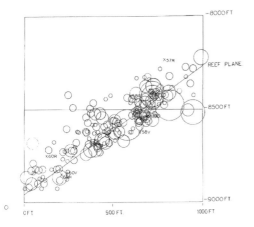

Figure 6b.   Dip section of F longwall East, ERPM, and foci of located seismic events (Cook et al., 1966)

The Mining Research Laboratory of the Transvaal and Orange Free State Chamber of Mines (founded in 1964 and renamed the Chamber of Mines Research Organization, COMRO) assumed responsibility for rockburst research, which focused on three main areas:

1. *Mine layout*, aimed at minimizing the effect of rock pressure at the design stage. The MINSIM computer program was one of the outstanding research products. Upgraded versions of the boundary element elastic code are still widely used.
2. *Support units and systems*, aimed at reducing falls of ground and the extent of rockburst damage. Rapid-yielding hydraulic props, developed by 1970, were a breakthrough in the support of stopes exposed to seismic activity. Backfilling was another major theme.

3. *Rockburst control*, which was concerned with developing instruments to monitor seismicity and engineering techniques to control the rockburst risk.

COMRO's contribution to various projects over the next three decades is described below.

Regional and Research Seismic Networks in the 1970s

Klerksdorp Regional Seismic Network: The increasing level of damage and injuries due to rockbursts in the Klerksdorp mining district led to the establishment of a permanent seismic network in 1971 as a joint venture between the Chamber of Mines and the four mining companies in the area (Van der Heever 1984). Only four seismometers were initially installed, which proved to be too few for accurate locations as the area monitored was about 200 km². By 1982 the network consisted of 32 stations. In addition, an 18-station micro-network was installed to monitor a seismically hyperactive area of 0.1 km³. Underground communication was by means of electrical cables up to 10 km long, while surface communication was by radio rather than by wire (Scheepers 1984).

Western Deep Levels and the Rockburst Research Project: The first attempt to monitor seismicity at Western Deep Levels mine was in 1965, when a network of surface and underground seismometers was established. The network operated consistently for two periods, March 1966 through February 1967, and January 1969 through May 1969 (Seaton and Hallbauer 1971). A new seismic monitoring system was developed at Western Deep Levels in 1974, and in 1977 the Rockburst Research Project, jointly sponsored by the Chamber of Mines and Anglo American Corporation, was initiated. This system utilized four tri-axial accelerometers to monitor micro-seismic events. Such monitoring could only occur between 20h00 to 6h00 owing to noise from rock drills and blasting. Data was recorded on magnetic tape, which was brought to the surface for playback. In 1979, cables were installed to enable data to be transmitted to the surface where digitizing, processing and storage on digital magnetic tape took place. Brink and Mountfort (1984) reported that four events (M_L=0.3, 0.4, 1.5 and 2.5) were predictable in hindsight, and that men could have been withdrawn prior to the events without losing more than one shift, and expressed the opinion that it was possible to "predict rock bursts with confidence." This prompted a major expansion of the project, with the objective of developing a "real time monitoring system" capable of timely predictions. A pilot project was initiated in 1980. A micro-seismic network consisting of five tri-axial accelerometers was installed to monitor events in the magnitude range $-4 < M_L < 0$ in a 1 km longwall, and a mine-wide 24 tri-axial geophone network was installed to monitor all events with $M_L > 0$.

Source Mechanism Studies: From 1969 to 1979 the BPI team of Art McGarr, Steve Spottiswoode, Rod Green and Nick Gay contributed significantly to the emerging discipline of mine seismology (e.g., McGarr 1971; Spottiswoode and McGarr 1975; McGarr et al. 1975; Gay and Ortlepp 1979 and McGarr et al. 1981). The magnitudes of the stresses driving violent failure and the dimensions of the ruptures in the rock were determined for the first time. It was found that the source mechanism of many mining-induced tremors is similar to the mechanism of shallow natural earthquakes.

Doornfontein Research Networks: During the late 1970s, COMRO installed two research networks at Doornfontein mine (Pattrick, 1984): a 200 m array consisting of 19 geophones to monitor trials of a mechanical non-explosive mining machine; and a 2 km array consisting of 13 geophones to monitor seismic events in a larger area of about 2 km². Data were recorded on two 24-hour magnetic tapes that were played back and digitized in a surface laboratory. In 1981, a 12-channel 20 m array was installed to monitor the non-violent sub-audible fracturing ahead of advancing stopes. All of these networks were temporary, and operated for periods of a few months to a few years.

Routine In-Mine Monitoring (1978 Onwards)

The first fully mine-owned and -staffed seismic network was installed at Blyvooruitzicht Mine in 1978 (Spottiswoode 1984). In 1982 the Gold Fields group established a network on its mines in the Far West Rand region. The collapse of an apartment block in Welkom following an $M_L = 5.2$ event on 8 December 1976 prompted Anglo American to install a permanent regional seismic network on its mines in the Free State district. By April 1980, 24 geophones were installed covering an area of about 300 km², yielding a location accuracy of 300 m in plan and 500 m in depth (Lawrence, 1984).

Quantitative Seismology (1990 Onwards)

The success achieved during the 1970s and 1980s of using seismology to better understand the source mechanisms of mining-related seismic events led to improvements in mine layouts and support design. Mine seismology moved from the realm of pure research to become a practical and indispensable tool for production purposes. By the early 1990s, real-time seismic monitoring using digital networks had become the standard within deep gold mines. The primary objectives were: (i) rapid response to rockbursts to limit the loss of life, (ii) assessment of the seismic hazard, (iii) back analysis of large and/or damaging seismic events, and (iv) research to improve knowledge of the rockbursting phenomena and support experimental development of technologies to mitigate the risk.

SIMRAC (1991 Onwards)

The *Safety in Mines Research Advisory Committee* (SIMRAC) was established in terms of the Minerals Act (Act 50 of 1991) with the principal objective of advising the *Mine Health and Safety Council* (MHSC) on the determination of the safety risk on mines and the need for research. SIMRAC has the responsibility to identify research projects, impose a levy on mines to fund such research, conclude agreements for carrying out such projects with research organizations, monitor project progress, and communicate the results of research to all parties concerned. SIMRAC identified rockbursts and rockfalls as serious safety hazards, particularly in gold mines. From 1991 to 2004, more than R250 million was spent on rock-related research, representing some 500 man-years of effort. This large body of work, mostly carried out by CSIR Miningtek and ISSI, is briefly summarized below. For a more comprehensive reviews see Adams and Van der Heever (2001) and Durrheim et al. (2005).

1. *Basic scientific research*
    This was conducted to determine the properties and behavior of the rock mass and the engineering materials that are used to stabilize and support it.
    *   *Rock properties and rock mass behavior:* Research work involved field and laboratory investigations and numerical simulations. Advanced computer codes, such as WAVE (simulation of wave propagation and the elastodynamic interactions between faults and stopes) and DIGS (simulation of large-scale assemblies of interacting cracks), were developed and used to improve the fundamental understanding of rock failure mechanisms (Napier et al. 1995, 1998 and 2002).
    *   *Mine seismology:* Techniques were developed to analyze seismograms and seismicity patterns. Early work concentrated on the quantitative description of the seismic source and seismicity (e.g., Mendecki et al. 1996), with attention subsequently being given to the rockburst damage mechanism (e.g., Durrheim et al. 1998). Attempts were made

to predict seismic events, but the success rate was not high enough to be used to withdraw crews on a routine basis (De Beer 2000). The focus then turned to the management of risk through the continuous assessment of the seismic hazard, the creation of rockburst-resistant excavations through optimum layouts and support systems, and the integration of seismic observations with numerical modeling to achieve better simulations of rock mass behavior (e.g., Mendecki et al. 2001).

2. *Engineering research*
This was conducted to produce practical methodologies and technologies to improve safety.
- *Layouts and design criteria:* There was a major change in layout philosophy in deep gold mines in the early 1990s—the favored orientation of stabilizing pillars changed from strike to dip, as the pre-development of tunnels allowed hazardous structures to be detected and bracket pillars to be planned prior to stoping (Vieira et al. 2001). These new layouts were evaluated, as well as other regional support systems such as backfill and bracket pillars. Considerable work went into the development of reliable mine design criteria. The validity of concepts such as Energy Release Rate (ERR), Excess Shear Stress (ESS) and Volumetric ESS were tested, and the concepts expanded or new concepts introduced (e.g., Generalized ERR, Local Energy Release Density).
- *Support components and systems:* Many new components and systems were evaluated. Pre-stressed yielding elongates and packs made of synthetic materials were widely introduced, while thin spray-on liners were evaluated as a possible replacement for shotcrete. A new methodology to design stope support systems was developed (Roberts et al. 1995) and widely implemented.
- *Mining methods:* Preconditioning was field-tested. It was demonstrated to reduce the hazard of face-bursting and was widely implemented.

3. *Transfer of knowledge and technology*
- Reports are available on CD and or may be downloaded from the website (*http://www.mhsc.org.za/*)
- A small library of Textbooks and Guidelines has been published
- Numerous product launches, seminars and workshops have been held

4. *Implementation*
Several projects sought to support the industry implement research findings and new technologies. For example, problems associated with the use of rapid yielding hydraulic props were investigated (Glisson and Kullmann, 1998), and methods for training underground workers in strata control were developed (Johnson et al., 2000 and 2002).

A study was commissioned to assess the scope, quality, and impact of South African rock-related research in general, and the SIMRAC research program in particular (Durrheim et al. 2005). The review team concluded that SIMRAC had succeeded in identifying the major research needs and had coordinated a comprehensive program of research. In particular, it was found that SIMRAC-sponsored research had contributed to the development of several important new technologies, such as systems for seismic monitoring and analysis, dip-pillar mining layouts, pre-conditioning, and pre-stressed elongates. Furthermore, SIMRAC-sponsored research had also contributed to the work of the collaborative research programs and the formulation of the mandatory "codes of practice to combat rockfall and rockburst accidents." The relatively small impact of the research effort on safety statistics was attributed to the increasing depth of mining, the increasing proportion of remnant and pillar mining, the failure to achieve short-term prediction, the long lead-time for new knowledge to be implemented, and shortcomings in the knowledge and technology transfer process. It was noted that some of the work was of a very high quality, attested by the award of Rocha medals to four CSIR researchers between 1999 and 2005, *viz.* Arno Daehnke, Francois Malan, Lindsay Linzer and Mark Hildyard. (Rocha medals have been awarded annually since by the International Society of Rock Mechanics for an outstanding PhD thesis. Richard Brummer of COMRO was awarded a Rocha medal in 1990.)

DeepMine Collaborative Research Program (1998–2002)

The *DeepMine Collaborative Research Program*, which sought to create the technological and human resources platform to mine gold safely and profitably at depths of 3 to 5 km, was initiated by Güner Gürtunca, Director of CSIR Miningtek, in 1998. The 5-year R66 million program of research was sponsored by AngloGold, Durban Roodepoort Deep, Gold Fields, the Chamber of Mines of South Africa,

163

CSIR, and the Department of Trade and Industry. Research work that directly addressed the rockburst risk, mostly carried out by CSIR Miningtek and ISSI, is briefly reviewed below using the DeepMine research methodology as a framework (for a more comprehensive review see Durrheim 2007).

1. *Define the ultra-deep environment*
    *   No new rock types are expected at ultra-depth, although the thicknesses of the various strata are expected to change, and the ore body to become increasingly quartzitic. Although the intensity of fracturing and convergence is expected to increase significantly with depth, the size of the fracture envelope will remain much the same.
    *   The relationship between the depth of mining and seismic activity was investigated using data acquired in the Carletonville district, where the same reef has been mined from 150 m to 3400 m below surface. The relationship was not robust enough to make firm predictions about the level of seismicity at ultra-depth.
2. *Establish systems criteria*
    *   Criteria providing indicators of seismic hazard were evaluated. It was found that the critical values of parameters such as Energy Release Rate (ERR) and peak particle velocity (PPV) were poorly defined.
3. *Assess if currently available technologies meet the systems criteria*
    *   An assessment of the effect of face advance rate on seismicity indicated that the maximum safe mining rate is dependent on the geotechnical area. Slightly higher mining rates may be sustained at ultra-depth without an increase in the hazard of face bursting because time-dependent fracture processes will take place more rapidly.
    *   An assessment of the effect of the mining method (e.g., drill and blast, continuous non-explosive rock breaking) on seismicity indicated that the overall seismic hazard, which is dominated by the larger seismic events, is independent of the rock-breaking process.
    *   An assessment of seismic prediction capability indicated that while predictions were better than random, this was not sufficiently reliable for implementation as a management tool.
    *   Commonly used rockburst-resistant support systems (e.g., yielding timber elongates, or timber packs with rapid yielding hydraulic props, or backfill and rapid yielding

hydraulic props) were shown to meet the systems criteria for the support of narrow stopes at ultra-depth, although areal coverage would become increasingly important as depth increases.

4. *Conduct research*
    Fill the knowledge or technology gaps if the ultra-deep environment or systems criteria were not known or the currently available technology did not meet the criteria.
    *   Stress was successfully measured at a depth of 3352 m, 700 m deeper than any previous stress measurement in a South African gold mines.
    *   A new methodology was developed to estimate the seismic hazard at ultra-depth, based on the concept of Volumetric Energy Release and Potential Damage Area.
    *   Prototype software developed to integrate seismic monitoring and numerical modeling as an aid to seismic management (Spottiswoode 2001, Hoffman et al. 2001, Lachenicht et al. 2001 and Wiles et al. 2001).
    *   Borehole radar technology was developed to detect hazardous structures in environments, such as the Venterdorp Contact Reef, where favourable electrical property contrasts exist between the footwall rocks and the overlying lavas (Du Pisani and Vogt 2004). Small faults and other changes in reef topography could be accurately located, giving prior warning of changes in mining conditions.
    *   A holistic strategy to manage seismicity was formulated. The proposed strategy had three main components: (i) reduction of seismicity by using geophysical methods to identify seismogenic structures (dykes, faults) ahead of mining, and design the layouts accordingly; (ii) the use of appropriate support systems to create rockburst-resistant excavations; and (iii) the continuous monitoring of seismic hazard. It was concluded that it would be possible to adapt rock engineering technologies (mine layouts, stope and tunnel support systems, etc.) to manage the levels of seismicity expected at ultra-depth, provided there was foreknowledge of potentially seismogenic structures (Durrheim, 2001).

Semi-controlled Earthquake-Generation Experiments (See SA, 1995–present)

In 1991 Louis Nicolaysen, Director of the BPI, submitted a proposal "Semi-controlled experiment on seismic events" to the International Association of Seismology and Physics of the Earth's Interior

An Industry Guide to Methods
of Ameliorating the Hazards of
ROCKFALLS AND ROCKBURSTS

1988 EDITION

Chamber of Mines Research Organization

Figure 7. Cover of the "Rockfall and Rockburst Guide" (Anon, 1988) showing some of the technologies used to ameliorate the seismic hazards, viz. rapid yielding hydraulic props, timber packs, computer simulations of mine layouts, and backfill.

(IASPEI), which was taken up by the Japanese seismological community. Since 1995 Japanese-South African cooperative research projects have been monitoring the earthquake generation process in close proximity to hypocenters (e.g., Ogasawara et al. 2002 and 2009; Yamada et al. 2005 and 2007). Amongst the significant observations were large, sudden changes in strain associated with large events (Ogasawara et al. 2005), and strain forerunners of seismic events (Yasutake et al. 2008).

KNOWLEDGE DISSEMINATION

Several landmark publications have documented advances in understanding of the causes of rockbursts and the development of measures to mitigate the risk. The Association of Mine Managers published two volumes (1933 and 1975) containing papers and discussions addressing the rockburst issue. *An Industry Guide to Methods of Ameliorating the Hazards of Rockfall and Rockbursts* (Anon, first edition 1977; second edition 1988) was prepared by COMRO at the request of the Association of Mine Managers.

These guides summarized the state of the art of rock engineering in South African tabular hard rock mining conditions (Figure 7). They were superseded by *A Handbook on Rock Engineering Practice for Tabular Hard Rock Mines* (edited by A.J. Jager and J.A. Ryder, 1999) and *A Textbook on Rock Mechanics for Tabular Hard Rock Mines* (edited by J.A. Ryder and A.J. Jager, 2002), published by SIMRAC. Mine seismology and rockbursting were specifically addressed in two books published in 1997: *Rock Fracture and Rockbursts: an illustrative study* by W.D. Ortlepp (1997), and *Seismic Monitoring in Mines* (edited by A.J. Mendecki, 1997).

Several noteworthy conferences addressing the rockburst issue have been held in South Africa. The Association of Mine Managers of South Africa held the *Symposium on Strata Control and Rockburst Problems of the South African Goldfields* in 1972. *The First International Symposium on Rockburst and Seismicity in Mines* was held in 1982 and the *Fifth Symposium* in 2001, the quadrennial symposium becoming the prime international venue for the sharing of rockburst research (see review by Ortlepp, 2005). The *Tenth Congress of the International Society of Rock Mechanics* and the *Second International Seminar on Deep and High Stress Mining* were held in Johannesburg in 2003 and 2004, respectively.

CONCLUSION

Gold was discovered in 1886 near present-day Johannesburg, and mine tremors soon posed a risk. The first scientific measurements of mining-related seismic events were made with a seismograph in 1910, exactly a century ago. Efforts to understand the causes of mining-related seismicity and to mitigate the effects of rockbursts were first coordinated in the 1950s, when the Chamber of Mines mobilized experts at CSIR and the University of the Witwatersrand to research the phenomena. Over the next half century, COMRO, CSIR, the BPI, ISSI and other research organizations devised new mine layouts, improved support elements and systems, and developed real-time digital seismic networks to monitor the response of the rockmass to mining (Figure 8). Mining at depth would have been impossible without these advances, and a significant reduction in fatalities and injuries has been achieved.

Not all research efforts have been successful. An obvious means of reducing the rockburst risk would be to reduce the exposure of workers to hazardous conditions in the face area. Numerous rock-breaking technologies have been tested in the past two decades under the auspices of COMRO, CSIR and various collaborative research programs. These range from incremental improvements to the conventional drill-and-blast method (rigs, jigs and remote controls) and long-hole drilling, to fully-mechanized narrow reef mining systems (impact rippers, activated and mini-disc cutters) and low-energy explosives and propellants. While some technical successes were achieved, none of these methods have been implemented on a large scale.

South Africa's gold production peaked at 1000 tonnes in 1970. Inevitably, ore bodies have been depleted and production has declined to about 250 tonnes, levels that are comparable with the output in the 1920s. Public and private for rockburst research has also reduced, so it is not surprising that the research capacity has declined drastically. COMRO and the BPI have closed, as have laboratories for the testing of rock properties, support elements, and backfill. One positive result is that many researchers have joined the ranks of practitioners and collaborators, aiding the transfer of knowledge.

Nevertheless, there are several very good reasons why the capacity to do research into mining at

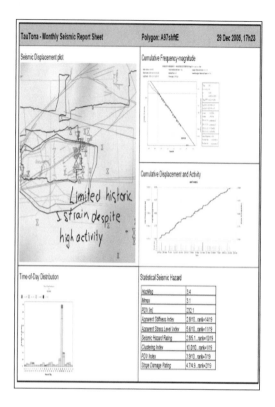

Figure 8.   Example of a modern seismic report sheet used for monthly planning (Courtesy of ISSI and TauTona gold mine). Short-term hazard assessments are also issued daily.

deep and high-stress conditions should not be lost. Gold continues to make a significant contribution to the South African economy through wages, tax and foreign exchange earnings. Furthermore, it is estimated that South Africa hosts 40 per cent of the world's gold resource (Chamber of Mines, 2007), much of it in reefs that were below pay limits at the time of mining or that are at ultra-depth. The gold price has climbed to record levels in recent years, which could make the mining of these resources attractive. The Bushveld Complex hosts almost 90 per cent of the world's platinum group metal resources (Chamber of Mines, 2007), and output has expanded tremendously in recent decades. Mining is already reaching the depths where seismicity poses a risk. The latest published statistics (Chamber of Mines, 2007) report that in 2007 the South African gold and platinum mines employed 169 057 and 186 411 people, respectively, while fatality and injury rates remain higher than international safety benchmarks.

ACKNOWLEDGMENTS

The prime risk of writing a history of rockburst research in South Africa is failure to do justice to the contributions that have been made by so many people in the course of a century. The author takes sole responsibility for any omissions. The following colleagues, all characters in the rockburst research story, are thanked for their insights: Steve Spottiswoode, Van Zyl Brink, Mike Roberts, John Napier, Olaf Goldbach, Terry Hagan, Matthew Handley, Aleksander Mendecki, Gerrie van Aswegen, Kevin Riemer and Paul van der Heever. I would also like to salute the following colleagues who infected me with an enthusiasm for rockbursts and mining seismology: Güner Gürtunca, Tony Jager, Rod Green, Art McGarr, Artur Cichowicz, Andrzej Kijko, Lindsay Linzer, Francois Malan and the late Nick Gay and Dave Ortlepp.

The South African Institute of Mining and Metallurgy and the Chamber of Mines of South Africa are thanked for permission to publish figures from the paper by Cook at el. (1966) and the cover of the "Rockfall Guide" (Anon, 1988), respectively.

REFERENCES

Adams, D.J. and Van der Heever, P.K. 2001. An overview of seismic research co-ordinated by SIMRAC since its inception. In *Proc. Fifth Int. Symp. on Rockbursts and Seismicity in Mines.* Edited by G. van Aswegen, R.J. Durrheim and W.D. Ortlepp. Symposium Series S27, p. 205–212. Johannesburg: The South African Institute of Mining and Metallurgy.

Anon. 1915. *Report of the Witwatersrand Rock Burst Committee 1915.* Pretoria: The Government Printing and Stationery Office, Union of South Africa.

Anon. 1924. *Report of the Witwatersrand Earth Tremors Committee 1924.* Cape Town: The Government Printer, Union of South Africa.

Anon. 1964. *Recommendations of the Rock Burst Committee 1964.* Cape Town: Department of Mines, Republic of South Africa.

Anon. 1988. *An Industry Guide to Methods of Ameliorating the Hazards of Rockfall and Rockbursts.* 2nd edition. Johannesburg: Chamber of Mines Research Organization.

Association of Mine Managers of the Transvaal. 1933. *Some aspects of deep level mining on the Witwaersrand Gold Mines with Special Reference to Rockbursts.* Johannesburg: Transvaal Chamber of Mines.

Association of Mine Managers of South Africa. 1975. *Symposium on strata control and rockburst problems of the South African Goldfields.* Johannesburg: Chamber of Mines of South Africa.

Brandt, M.B.C., Bejaichund, M., Kgaswane, E.M., Hattingh, E. and Robin, D.L. 2005. *Seismic History of Southern Africa.* Seismological Series No. 37. Pretoria: Council for Geoscience.

Brink, A.v.Z. and Mountford, P.I. 1984. Feasibility studies on the prediction of rockbursts at Western Deep Levels. *Proc. First Int. Sym. on Rockbursts and Seismicity in Mine.* Edited by N.C. Gay and E.H. Wainwright, SAIMM Symposium Series No. 6, pp. 317–325. Johannesburg: South African Institute of Mining Metallurgy.

Chamber of Mines. 2007. *Facts and Figures 2007.* Johannesburg: Chamber of Mines of South Africa.

Cook, N.G.W. 1963. The seismic location of rockbursts. In *Proc. Fifth Symp. Rock. Mech.*, pp. 493–516. Pergamon Press.

Cook, N.G.W. 1964. The application of seismic techniques to problems in rock mechanics. *J. Int. Rock. Mech. Min. Sci.* 1: 169–179.

Cook, N.G.W. 1976. Seismicity associated with mining. *Eng. Geol.* 10:99–122.

Cook, N.G.W., Hoek, E., Pretorius, J.P.G., Ortlepp, W.D. and Salamon, M.D.G. 1966. Rock mechanics applied to the study of rockbursts. *J. S. Afr. Inst. Min. Metall.* 66: 435–528.

De Beer, W. 2000. *Seismology for rockburst prediction.* Final Report, project GAP 409. Johannesburg: Mine Health and Safety Council.

Du Pisani, P. and Vogt, D. 2004. Borehole radar delineation of the Ventersdorp Contact Reef in three dimensions, *Exploration Geophysics*, 35: 278–282.

Durrheim, R.J. 2001. Management of mining-induced seismicity in ultra-deep South African gold mines, *Proc. Fifth Int. Symposium on Rockbursts and Seismicity in Mines.* Edited by G. van Aswegen, R.J. Durrheim and W.D. Ortlepp, SAIMM Symposium Series S27, pp. 213–219. Johannesburg: South African Institute of Mining Metallurgy.

Durrheim, R.J. 2007. The DeepMine and FutureMine Research Programmes—Knowledge and Technology for Deep Gold Mining in South Africa, in *Challenges in Deep and High Stress Mining*, pp. 131–141. Edited by Y. Potvin, J. Hadjigeorgiou and D. Stacey. Perth: Australian Centre for Geomechanics.

Durrheim, R.J., Milev, A., Spottiswoode, S.M. and Vakalisa, B. 1998. *Improvement of worker safety through the investigation of the site response to rockbursts.* Final Report, project GAP 201. Johannesburg: Mine Health and Safety Council.

Durrheim, R.J., Brown, E.T., Kaiser, P.K. & Wagner, H. 2005. *A holistic assessment of SIMRAC rock-related research to date.* Final Report, project SIM 04-02-06. Johannesburg: Mine Health and Safety Council.

Durrheim, R.J., Anderson, R.L., Cichowicz, A., Ebrahim-Trollope, R., Hubert, G., Kijko, A., McGarr, A., Ortlepp, W.D. and Van der Merwe, N. 2006. *Investigation into the risks to miners, mines, and the public associated with large seismic events in gold mining districts.* Pretoria: Department of Minerals and Energy.

Finsen, A. 1950. Union Observ. Circ. 110 (July 1950).

Gane, P.G. 1939. A statistical study of the Witwatersrand earth tremors. *J. Chem. Metall. Min. Soc. S. Afr.* 40: 155.

Gane, P.G., Hales, A.L. and Oliver, H.A. 1946. A seismic investigation of Witwatersrand earth tremors. *Bull. Seism. Soc. Am.* 36: 49–80.

Gay, N.C. and Ortlepp, W.D. 1979. Anatomy of a mining induced fault zone. *Bull. Geol. Soc. Am., Part 1.* 90:47–58.

Glisson, F.J. and Kullmann, D.H. 1998. *Problems associated with the use of rapid yielding hydraulic props,* Final Report, project GAP 442. Johannesburg: Mine Health and Safety Council.

Handley, J.R.F. 2004. *Historic Overview of the Witwatersrand Goldfields.* Howick: Handley.

Hofmann, G., Sewjee, R. and Van Aswegen, G. 2001. First steps in the integration of numerical modeling and seismic monitoring, *Proc. Fifth Int. Symp. on Rockbursts and Seismicity in Mines.* Edited by G. van Aswegen, R.J. Durrheim and W.D. Ortlepp, SAIMM Symposium Series S27, pp. 397–404. Johannesburg: South African Institute of Mining Metallurgy.

Jager, A.J. and Ryder J.A. (editors). 1999. *A Handbook on Rock Engineering Practice for Tabular Hard Rock Mines.* Johannesburg: Safety in Mines Research Advisory Committee.

Johnson, R.A., Hagan, T.O., Squelch, A.P. & Jaku, E. 2000. *Effective training methods in strata control for underground workers.* Final Report, project GAP 442. Johannesburg: Mine Health and Safety Council.

Johnson, R.A., Jaku, E. Squelch, A.P., Barnett, H. & Hagan, T.O. 2002. *Trial training in strata control for underground workers.* Final Report, project GAP 851. Johannesburg: Mine Health and Safety Council.

Joughin, N.C. 1966. *The measurement and analysis of earth motion resulting from underground rock failure.* Ph.D. Thesis, University of the Witwatersrand, Johannesburg, South Africa.

Lachenicht, R., Wiles, T. and Van Aswegen, G. 2001. Integration of deterministic modeling with seismic monitoring for the assessment of the rockmass response to mining: Part II Applications, *Proc. Fifth Int. Symp. on Rockbursts and Seismicity in Mines.* Edited by G. van Aswegen, R.J. Durrheim and W.D. Ortlepp. SAIMM Symposium Series S27, pp. 389–395. Johannesburg: South African Institute of Mining Metallurgy.

Lawrence, D. 1984. Seismicity in the Orange Free State gold-ming district. *Proc. First Int. Symposium on Rockbursts and Seismicity in Mines.* Edited by N.C. Gay and E.H. Wainwright, SAIMM Symposium Series No. 6, pp. 121–130. Johannesburg: South African Institute of Mining Metallurgy.

McGarr, A. 1971. Violent deformation of rock near deep level, tabular excavations—seismic events. *Bull. Seism. Soc. Am.* 61:1,453–1,466.

McGarr, A., Green, R.W.E. and Spottiswoode, S.M. 1981. Strong ground motion of mine tremors: some implications for near-source ground motion parameters. *Bull. Seism. Soc. Am.* 71:295–319.

McGarr, A., Spottiswoode, S.M. and Gay, 1975. Relationship of mine tremors to induced stresses and to rock properties in the focal region. *Bull. Seism. Soc. Am.* 65:981–993.

Mendecki, A.J. (editor), 1997. *Seismic Monitoring in Mines.* London: Chapman and Hall.

Mendecki, A.J., Ilchev, A., Napier, J.A.L. and Sellers, E.J. 2001. *The integration of seismic monitoring with numerical modelling,* Final Report, project GAP 603. Johannesburg: Mine Health and Safety Council.

Mendecki, A.J., Mountford, P., Dzhafarov, A.H., Sciocatti, M., Niewiadomski, J., Radu, S., Van Aswegen, G., Funk, C, Maxwell, S.C. and Young, P. 1996. *Seismology for rockburst prevention, control and prediction.* Final Report, project GAP 017. Johannesburg: South African Mine Health and Safety Council.

Napier, J.A.L., Hildyard, M.W., Kuijpers, J.S., Daehnke, A, Sellers, E.J., Malan, D.F., Siebrits, E., Ozbay, M.U., Dede, T. and Turner, A. 1995. *Develop a quantitative understanding of rockmass behaviour near excavations in deep mines.* Final Report, project GAP 029. Johannesburg: Mine Health and Safety Council.

Napier, J.A.L., Malan, D.F., Sellers, E.J., Daehnke, A, Hildyard, M.W., Dede, T. and Shou, K-J. 1998. *Deep gold mine fracture zone behaviour.* Final Report, project GAP 332. Johannesburg: Mine Health and Safety Council.

Napier, J.A.L., Drescher, K., Hildyard, M.W., Kataka, M.O., Malan, D.F. and Sellers, E.J., 2002. *Experimental investigation of fundamental processes in mining induced fracturing and rock.* Final Report, project GAP 601b. Johannesburg: Mine Health and Safety Council.

Ogasawara, H., Yanagidani, T. and Ando, M. (editors) 2002. *Seismogenic Process Monitoring,* Rotterdam: Balkema.

Ogasawara, H., Takeuchi, J., Shimoda, N., Ishii, H., Nakao, S., van Aswegen, G., Mendecki, A.J., Cichowicz, A. Ebrahim-Trollope, R., Kawakata, H., Iio Y., Ohkura, T., Ando, M. and the Research Group for Semi-controlled Earthquake-generation Experiments in South African deep gold mines, 2005. High-resolution strain monitoring during M~2 events in a South African deep gold mine in close proximity to hypocentres. In *Proc. 6th Int. Symp. on Rockbursts and Seismicity in Mines.* pp. 385–391. Edited by Y. Potvin, and M. Hudyma, M. Perth: Australian Centre for Geomechanics.

Ogasawara H., Kawakata, H., Ishii, M., Nakatani, M., Yabe, Y., Iio, Y. and the Research Group for the Semi-controlled Earthquake-generation Experiments at deep gold mines, South Africa (SeeSA). 2009. The semi-controlled earthquake-generation experiments at deep gold mines, South Africa—Monitoring at the proximity to elucidate seismogenic process. *Journal of Japanese Seismological Society, Ser. 2,* in press (in Japanese with English abstract).

Ortlepp, W.D. 2005. RaSiM comes of age: a review of the contribution to the understanding and control of mine rockbursts, In *Proc. 6th Int. Symp. on Rockbursts and Seismicity in Mines.* pp. 3–20. Edited by Y. Potvin, and M. Hudyma, Perth: Australian Centre for Geomechanics,

Ortlepp, W.D. 1997. *Rock fracture and rockbursts: an illustrative study.* Monograph Series M9, Johannesburg: South African Institute of Mining and Metallurgy.

Pattrick, K.W. 1984. The instrumentation of seismic networks at Doornfontein gold mine. *Proc. First Int. Symposium on Rockbursts and Seismicity in Mines.* Edited by N.C. Gay and E.H. Wainwright, SAIMM Symposium Series No. 6, pp. 337–340. Johannesburg: South African Institute of Mining Metallurgy.

Roberts, M.K.C., Eve, R.A., Squelch, A.P. and Taggart, P.N. 1995. *Efficient stope and gully support system design.* Final Report, project GAP 032. Johannesburg: Mine Health and Safety Council.

Ryder J.A. and Jager A.J. (editors), 2002. *A Textbook on Rock Mechanics for Tabular Hard Rock Mines.* Johannesburg: Safety in Mines Research Advisory Committee.

Seaton, J. and Hallbauer, D.K. 1971. *An analysis of seismic data recorded at Western Deep Levels Limited.* Unpublished report. Johannesburg: COMRO.

Scheepers, J.B. 1984. The Klerksdorp seismic network—monitoring of seismic events and systems layout. *Proc. First Int. Symposium on Rockbursts and Seismicity in Mines.* Edited by N.C. Gay and E.H. Wainwright, SAIMM Symposium Series No. 6, pp. 341–345. Johannesburg: South African Institute of Mining Metallurgy.

Spottiswoode, S.M. 1984. Source mechanisms of mine tremors at Blyvooruitzicht gold mine. *Proc. First Int. Symposium on Rockbursts and Seismicity in Mines.* Edited by N.C. Gay and E.H. Wainwright, SAIMM Symposium Series No. 6, pp. 29–38. Johannesburg: South African Institute of Mining Metallurgy.

Spottiswoode, S.M. 2001. Keynote address: Synthetic seismicity mimics observed seismicity in deep tabular mines, *Proc. Fifth Int. Symposium on Rockbursts and Seismicity in Mines.* Edited by G. van Aswegen, R.J. Durrheim and W.D. Ortlepp, SAIMM Symposium Series S27, pp. 371–377, Johannesburg: South African Institute of Mining Metallurgy.

Spottiswoode, S.M. and McGarr, A. 1975. Source parameters of tremors in a deep-level gold mine. *Bull. Seism. Soc. Am.* 65:93–112.

Van der Heever, P.K. 1984. Some technical and research aspects of the Klerksdorp seismic network. *Proc. First Int. Symposium on Rockbursts and Seismicity in Mine.* Edited by N.C. Gay and E.H. Wainwright, SAIMM Symposium Series No. 6, pp. 349–350. Johannesburg: South African Institute of Mining Metallurgy.

Vieira, FM.C.C., Diering, D.H. and Durrheim, R.J. 2001. Methods to mine the ultra-deep tabular gold-bearing reefs of the Witwatersrand Basin, South Africa. In *Underground Mining Methods: Engineering Fundamentals and International Case Studies*, pp. 691–704. Edited by W.A. Hustrulid and R.L. Bullock, Society for Mining, Metallurgy, and Exploration, Inc.

Wiles, T. Lachenicht, R. and Van Aswegen, G. 2001. Integration of deterministic modeling with seismic monitoring for the assessment of the rock-mass response to mining: Part I Theory. *Proc. Fifth Int. Symp. on Rockbursts and Seismicity in Mines*. Edited by G. van Aswegen, R.J. Durrheim and W.D. Ortlepp, SAIMM Symposium Series S27, pp. 379–387. Johannesburg: South African Institute of Mining Metallurgy.

Wood, H.E. 1913. On the occurrence of earthquakes in South Africa. *Bull. Seism. Soc. Am.* 3:113–120.

Wood, H.E. 1914. Witwatersrand earth tremors, *J. Chem. Metall. Min. Soc S. Afr.* 14: 423.

Yamada, T., Mori, J.J., Ide, S., Kawakata, H., Iio, Y. and Ogasawara, H. 2005. Radiation efficiency and apparent stress of small earthquake in a South African gold mine. *J. Geophys. Res,* 101(B01305), doi:10.1029/2004JB003221.

Yamada, T., Mori, J. J., Ide, S., Abercrombie, R.E., Kawakata, H., Nakatani, M., Iio, Y. and Ogasawara H. 2007. Stress drops and radiated seismic energies of microearthquakes in a South African gold mine. *J. Geophys. Res.* 112(B03305), doi:10.1029/2006JB004553.

Yasutake, G., Ogasawara, H., Kawakata, H., Naoi, M. and the Research Group for SeeSA. 2008. Slow strain-steps observed in a potential M<3 source area at a ~3.1 km depth, Mponeng gold mine, South Africa (2). *Abstracts, 7th General Assembly of Asian Seismological Commission and Seismological Society of Japan, 26 Nov 2008*, Tsukuba: X3-017.

BIBLIOGRAPHY

Chronological list of MSc and PhD theses that address the cause and mitigation of rockbursts in South African mines.

Logie, H.J. 1948. *The recording and study of Witwatersrand earth tremors.* Ph.D. thesis, University of the Witwatersrand, Johannesburg, South Africa.

Henry, S.J. 1950. *The application of electronics to tremor seismology.* PhD Thesis, University of the Witwatersrand, Johannesburg, South Africa.

Cook, N.G.W. 1962. *A study of failure in the rock surrounding underground excavations.* PhD Thesis, University of the Witwatersrand, Johannesburg, South Africa.

Deist, F.H. 1966. *The development of a nonlinear continuum approach to the problem of fracture zones and rockbursts and feasibility study by computer.* Ph.D. Thesis, University of the Witwatersrand, Johannesburg, South Africa.

Joughin, N.C. 1966. *The measurement and analysis of earth motion resulting from underground rock failure.* Ph.D. Thesis, University of the Witwatersrand, Johannesburg, South Africa.

Heunis, R. 1976. *Improvements in the design of location systems and the development of a seismic method for the delineation of geological dykes in mines.* M.Sc Thesis, University of the Pretoria, Johannesburg, South Africa.

Van Proctor, R.J. 1978. *An investigation of the nature and mechanism of rock fracture around longwall faces in a deep gold mine.* Ph.D. Thesis, University of the Witwatersrand, Johannesburg, South Africa.

Green, R.W.E. 1979. *A data acquisition and processing system, with applications to seismological problems in southern Africa.* Ph.D. Thesis, University of the Witwatersrand, Johannesburg, South Africa.

Hagan, T.O. 1980. *A photogrammetric study of mining-induced fracture phenomena and instability on a deep-level longwall stope face with variable lag lengths.* M.Sc. Thesis, Rand Afrikaans University, Johannesburg, South Africa.

Spottiswoode, S.M. 1980. *Source mechanism studies on Witwatersrand seismic events.* Ph.D. Thesis, University of the Witwatersrand, Johannesburg, South Africa.

Arnott, F.W. 1981. *Seismicity in the Welkom area, O.F.S. (with special reference to the origin of the 1976-12-8 event).* M.Sc. thesis, University of the Witwatersrand, Johannesburg, South Africa.

McDonald, A.J. 1982. *Seismicity of the Witwatersrand basin.* M.Sc. Thesis, University of the Witwatersrand, Johannesburg, South Africa.

Van der Heever, P.K. 1982. *The influence of geological structure on seismicity and rockbursts in the Klerksdorp goldfield.* M.Sc. Thesis, Rand Afrikaans University, Johannesburg, South Africa.

Pattrick, K.W. 1983. *The development of a data acquisition and pre-processing system for microseismic research.* M.Sc. Thesis, University of the Witwatersrand, Johannesburg, South Africa.

Legge, N.B. 1985. *Rock deformation in the vicinity of deep gold mine longwall stopes and its relation to fracture.* PhD thesis. University of Wales.

Brummer, R.K. 1987. *Fracturing and deformation at the edges of tabular gold mining excavations and the development of a numerical model describing such phenomena.* Ph.D. Thesis, Rand Afrikaans University, Johannesburg, South Africa.

Hagan, T.O. 1987. *An evaluation of systematic stabilizing pillars as a method of reducing the seismic hazard in deep and untra-deep mines.* Ph.D. Thesis, University of the Witwatersrand, Johannesburg, South Africa.

Handley, M.F. 1987. *A study of the effect of mining induced stresses on a fault ahead of an advancing longwall face in a deep level gold mine.* M.Sc. Thesis, University of the Witwatersrand, Johannesburg, South Africa.

Brawn, D.R. 1989. *A maximum entropy approach to underconstraint and inconsistency in the seismic source inverse problem, finding and interpreting seismic source moments.* Ph.D. Thesis, University of the Witwatersrand, Johannesburg, South Africa.

Goldbach, O.D. 1990. *The use of seismogram waveforms to characterize the fracture zone around a mine excavation.* M.Sc. Thesis, University of the Witwatersrand, Johannesburg, South Africa.

Webber, S.J. 1990. *Quantitative modeling of mining induced seismicity.* M.Sc. Thesis, University of the Witwatersrand, Johannesburg, South Africa.

Cole, R.S.J. 1991. *Design of a seismic data acquisition system and automatic triggering software.* M.Sc. Thesis, University of the Witwatersrand, Johannesburg, South Africa.

Piper, P.S. 1991. *An assessment of backfill as a means of alleviating the rockburst hazard in deep mines.* M.Sc. Thesis, University of the Witwatersrand, Johannesburg, South Africa.

Finnie, G.J. 1993. *Time-dependant seismic hazard in mining.* M.Sc. Thesis, University of the Witwatersrand, Johannesburg, South Africa.

Squelch, A.P. 1994. *The determination of the influence of backfill on rockfalls in South African gold mines.* MSc. thesis, University of the Witwatersrand, Johannesburg, South Africa.

Daehnke, A. 1997. *Stress wave and fracture propagation in rock.* Ph.D Thesis. Vienna University of Technology.

Ferreira, R.I.L. 1997. *Quantitative aspects of mining induced seismicity in a part of the Welkom Goldfield.* M.Sc. Thesis, University of the Witwatersrand, Johannesburg, South Africa.

Simser, B.P. 1997. *Numerical modeling and seismic monitoring on a large normal fault in the Welkom goldfields, South Africa.* M.Sc. Thesis, University of the Witwatersrand, Johannesburg, South Africa.

Hemp, D.A. 1998. *An investigation of the application of seismic tomography to the study of deep level South African gold mines.* M.Sc. Thesis, University of the Witwatersrand, Johannesburg, South Africa.

Malan, D. F. 1998. *An investigation into the identification and modelling of time-dependent behaviour of deep level excavations in hard rock.* Ph.D thesis, University of the Witwatersrand, Johannesburg, South Africa.

Roberts, M.K.C. 1999. *The design of stope support in South African gold and platinum mines.* Ph.D Thesis, University of the Witwatersrand, Johannesburg, South Africa.

Güler, G. 1999 *Analysis of the rock mass behaviour as associated with Ventersdorp Contact Reef stopes, South Africa.* M.Sc. Thesis, University of the Witwatersrand, Johannesburg, South Africa.

Haile, A.T. 1999. *A mechanistic evaluation and design of tunnel support systems for deep level South African mines* Ph.D thesis, University of Natal.

Andersen, L.M. 2001. *A relative moment tensor inversion technique applied to seismicity induced by mining.* Ph.D. Thesis, University of the Witwatersrand, Johannesburg, South Africa.

Hildyard, M.W. 2001. *Wave interaction with underground openings in fractures rock.* Ph.D. Thesis. University of Liverpool.

Kataka, M.O. 2002. *Simulation of ground motions of a large event using small events as empirical Green's functions.* Ph.D. Thesis, University of the Witwatersrand, Johannesburg, South Africa.

Hanekom, J.W.L. 2002. *The effect of two different mining sequences on a seismically active structure.* M.Sc. Thesis, University of the Witwatersrand, Johannesburg, South Africa.

Naicker, N. 2003. *The relationship between mine seismicity and depth of mining.* M.Sc. thesis, University of the Witwatersrand, Johannesburg, South Africa.

Vieira, F.M.C.C. 2004. *Rock engineering-based evaluation of mining layouts applicable to ultra-deep, gold bearing, tabular deposits.* Ph.D. Thesis, University of the Witwatersrand, Johannesburg, South Africa.

Toper, A.Z. 2005. *Destressing/preconditioning to control rockbursts in South African deep-level gold mines.* PhD thesis, Université Laval, Quebec.

Coal Mining Research in Australia—A Historical Review of Developments

Bruce K. Hebblewhite

The University of New South Wales, Sydney, Australia

ABSTRACT: The Australian coal mining industry has a rich history, with the earliest known underground coal mines commencing soon after white settlement in 1788 on the east coast of Australia. Today, the industry is a dominant part of the Australian economy, providing the single largest source of Australian export income, with the industry being the world's largest black coal exporter. Technology and skills have developed to a highly sophisticated level –both for underground mines and open-cuts; such that Australia is also a world leader in mining technology and services, as well as mining-related specialist skills.

   This paper provides an overview of the development of the industry from early times through to the present. It includes a review and case studies of a number of significant research developments and their application in the industry. Such developments include not only equipment and "hard technology" outcomes, but also some of the very important research outcomes with respect to developing a high level of understanding of the different mine environments, and being able to apply such understanding to improved mine design and safer and more efficient mining operations. The paper also tracks some of the more significant trends over more recent decades with respect to the different forms of research providers and specialist research organisations that have played a lead role in these Australian research developments.

INTRODUCTION

This paper aims to provide an historical overview of developments in coal mining research in Australia, over recent decades and beyond. The paper is restricted to black coal only, excluding the large scale mining of brown coal (lignite) reserves in Victoria, together with the supporting research and services. Some of the earlier development details are quite sketchy and do not bear scrutiny with respect to exact dates or formal referencing. Clearly the major coal mining research effort in Australia has occurred within the last 20 to 30 years, and so this era will be addressed in more detail. As an overview, the paper does not attempt to document every significant research achievement, or even every active research group, person, program or organisation. It provides a series of highlights of selected research activities and achievements in particular fields of mining research, together with a summary of research funding programs and organisations. For those who carried out undoubtedly significant work which is not referenced in this paper, the author extends his apologies on the basis of the limited scope outlined above.

AUSTRALIAN COAL INDUSTRY

Historical Developments

Coal was discovered near Newcastle, New South Wales, on the east coast of Australia in 1791 by escaped convicts, and over the next decade coal deposits were reported at many other sites in the Illawarra and the Hunter Valley. Newcastle coal was first exported in 1799, but coal mining did not become a major industry until the 1850s. Coal production continued to expand, primarily in New South Wales, into the 20th century, however during the early years of the Great Depression, coal production fell by a third from a mid-1920 high of 14 million tonnes. Exports of black coal, which had exceeded one million tonnes a year until the mid-1920s, fell to about 50,000 tonnes by the late 1940s as petroleum products increasingly replaced coal in industry and railways (Hampton & Allen, 2009). War requirements led to an urgent need to increase coal production but the industry was, at best, inefficient. Any move to increasing mechanisation to enable much-needed changes to mining in Australia, however, demanded technologically skilled mining engineers, who were in short supply.

   In 1929 a joint Commonwealth and New South Wales Royal Commission into the coal industry was established (Rutledge, 1981). "The Davidson Commission reported two basic problems: an overcapacity and over-manning of the mines and a spirit of hostility between capital and labour. It recommended the closure of some mines, the transfer of surplus labour to other occupations and the establishment of a joint Commonwealth and New South Wales Board to control the industry by way of determining wage rates, labour conditions and a maximum selling price for coal. In 1938–39 Justice Davidson also chaired the New South Wales Royal Commission into the safety and health of workers in coal mines, and many of its recommendations were incorporated in the

Coal Mines Regulation (Further Amendment) Act of 1941." "In 1946 Justice Davidson issued the report of his third Royal Commission into the national coal industry, which finally resulted in the establishment of the Joint Coal Board to control the industry and to represent the interests of the Commonwealth and New South Wales governments," according to Hampton & Allen (2009).

While Australia may have been doing well in metalliferous mining, post-war, the coal industry was seen to be 'fundamentally inefficient and technologically out of date, producing only some 11 million tonnes from 138 mines with a yearly output per employee of about 621 tonnes' (AGSO, 1998).

Current Status

In 2008 the Australian coal industry is now one of the most technologically advanced in the world, as well as being a world leader in standards of mine safety performance and management. Total production of black coal in 2008 reached over 420 million tonnes (run of mine), 56% coming from Queensland and 41% from NSW. Open-cut mining accounted for 76% of this total production. Australia exported over 70% of total production, making it the largest exporter of coal in the world. In 2008, longwall mining accounted for over 99MT of annual underground production, coming from 29 longwall mines, with some of the highest labour productivities in the world, and production levels up to 7 million tonnes per annum.

RESEARCH FUNDING SCHEMES

Australia has seen a number of broadly based research funding programs operating over the past 40–50 years. These have included the following major offerings:

- NCRAC
- ACA(Research)
- NERDDP
- AMIRA
- JCB Health & Safety Trust
- ACARP
- ARC

These schemes are briefly outlined below.

NCRAC—Full details of this scheme are unclear, but the National Coal Research Advisory Council provided a low level of competitive research funding in the 1970s, and possibly earlier, to support the primary safety initiatives of underground coal mining.

ACA (Research)—From the early- to mid-1970s, the coal industry peak body, the Australian Coal Association, funded their major industry research organisation, Australian Coal Industry Research Laboratories Ltd (ACIRL), to carry out research into coal mining—focussed on both coal mine safety and mining systems and performance. The ACA research funds were raised through a levy on annual production tonnage across the industry (in both NSW and Queensland). This amount was initially approximately 0.5c/annual ROM tonne produced in the mid-1970s, but by the early 1980s, it had increased to approximately 1.5c/tonne. However, with the increase in funding came a broadening of the research focus—firstly to include coal preparation, and then subsequently to broaden further into coal processing (coal carbonisation and combustion research).

NERDDP—The National Energy Research, Development and Demonstration Program covered the full spectrum of energy research, however coal-related research was a major component. NERDDP was a compulsory scheme initiated by the Federal Government in the early 1980s to bolster Australia's energy industry efficiency and technologies. In the case of coal, the NERDDP scheme was funded from the coal industry by a compulsory 5c/tonne levy on annual ROM coal production. NERDDP funded research in at least four major fields related to coal, being coal mining, coal preparation, utilisation and marketing. As the NERDDP scheme came into operation, the ACA progressively withdrew their dedicated support for research activities at ACIRL.

Over the life of NERDDP, in spite of criticisms from the industry, due to being forced to pay a compulsory levy, the distribution of which was not entirely within their control, the NERDDP scheme delivered some major research outcomes which took the industry forward in terms of technology levels and operational standards. Some of these are discussed later in the paper.

AMIRA—In parallel with NERDDP, an independent body, Australian Mineral Industry Research Association, representing a collection of industry funding partners, offered brokered research projects for which individual companies signed up to support particular projects of interest. AMIRA operated in all the major disciplines of interest to the industry, and covered both coal and metalliferous mining sectors. AMIRA continues to operate to this day, however its role in coal mining research has virtually disappeared since the 1990s.

JCB Health & Safety Trust—The Joint Coal Board H&S Trust has operated a nationally competitive research funding program since the late 1980s, with regular calls for research proposals specifically oriented towards mine safety. This program is funded through investment income earned by the JCB resulting from their Coal Mines Insurance investment portfolio.

ACARP—The Australian Coal Association Research Program was established by the coal industry in the 1980s, as a replacement for the NERDDP scheme. It is funded on exactly the same basis as NERDDP—a compulsory levy of 5c/tonne on annual black coal production—but is managed directly by the industry through an administration unit and contracted project management teams. The industry was required to make certain commitments to the government regarding the nature and ongoing future of the ACARP program, as a condition off the government disbanding the NERDDP scheme and handing over the levy funds raised from the industry. ACARP operates through a series of different technical committees, made up of volunteer industry representatives who set annual research priorities. Following an annual call for project proposals, the committees review applications through a two stage process, before awarding successful projects with funding for up to three years. In recent years, ACARP has also introduced a program of targeted "landmark projects" in priority research areas, where major projects are funded through invited research providers. An example of this type of project is the longwall automation project funded through CSIRO, and their collaboration partners.

ARC—The Australian Research Council is the principal funding body of the Federal government which provides research funding to university researchers across all disciplines. As such, coal mining research opportunities are a very small part of the ARC program, but nevertheless ARC represents an ongoing funding opportunity for mining groups within the university sector.

MAJOR RESEARCH PROVIDERS

The number and form of major coal mining research providers has changed dramatically over the last fifty years in Australia, in a manner not dissimilar with trends observed in other countries around the world, such as the demise of the US Bureau of Mines; the British Bretby Research Group; Bergbau-Forschung in Germany and the Chamber of Mines Research Organisation in South Africa. Australia has moved from having one or two dedicated research organisations to the current situation where research is carried out across numerous institutions of all shapes and sizes.

Australian Coal Industry Research Laboratories Ltd.—ACIRL was established in 1965 by the Australian Coal Association, together with the Federal and state governments, as a company limited by guarantee and shares, to provide research and technical/consulting services for the coal industry. In the mining sector, ACIRL first developed its mining research group in the late 1960s, based at Bellambi, NSW with a strata control research group focused on geotechnical site characterisation research, and longwall feasibility analysis using a large physical modelling facility. In the 1970s this group began to expand, supported by the ACA Research funding, and by early 1980s had a team of over 60 professional research staff, led by Dr Bruce Hebblewhite, operating out of five sites in NSW and Queensland researching a broad spectrum of mining topics ranging through mine design and planning, strata control, gas drainage and long hole drilling, mining methods, spontaneous combustion, surface mining simulation, dragline optimisation and many other fields. In fact during the 1980s, ACIRL was the primary research resource for the Australian coal industry, and was also engaged internationally on a number of occasions (including New Zealand, China, South Africa, Korea, UK)—both for research and consulting/advisory projects. However, as with other parts of the world, the model of a centrally supported major national research organisation began to lose favour by the early 1990s and the ACIRL mining research group effectively ceased to operate by the mid-1990s.

CSIRO—CSIRO has been the other major national research provider for coal mining research in Australia, although its fortunes and levels of coal mining research activity have varied over the years. Initially through the Division of Applied Geomechanics, and more recently through the Division of Exploration and Mining, the federal government funded CSIRO has been able to attract industry research funding, leveraged by its high level of government funded infrastructure and basic research capacity support. Early work was in the field of geomechanics, but this spread into mine gas related work, and is now predominantly in the field of mining systems and automation and similar technologies.

Beyond the above two major organisations, since the 1980s, the research sector has spread to a much broader group since the establishment of ACARP funding, to the extent that individual mining companies, plus industry consulting organisations, as well as universities and some state government research groups are active providers of research today. Some examples of these groups are as follows:

Industry Groups—Historically, the two major companies that operated large in-house technical services teams were BHP and CRA—both operating on the south coast of NSW servicing the gassy and deep coal mines of the Illawarra District. In each case, these two operations were led at various times by two industry research luminaries—Dr Allan Hargraves (BHP), and Dr Ripu Lama (CRA). Both were particularly influential in both strata control and gas-related research and left their mark on many of the mining practices and understandings that the industry benefits from today.

State Government Groups—Three different groups come to mind from within the two major state governments of NSW and Queensland. In NSW, the subsidence research activity carried out from within the NSW Department of Mineral Resources, led by Dr Lax Holla, pioneered some of the early work on characterising coal mine subsidence in NSW (following on from the work of Mr Bill Kapp from the BHP research group). Also in NSW the research activity into explosibility and intrinsic safety and flameproof testing carried out at the Londonderry Research Centre, west of Sydney, in the 1980s and 1990s, was of world class. Similar mine safety and safety technology developments have been more recently carried out by the Queensland government Safety In Mines Testing And Research Station, SIMTARS.

Consultants—There are numerous consultants who have been active in coal mining research over the years—ranging from one man operations, to international corporations. Just a couple of examples of successful consulting research contractors include: Strata Control Technology (SCT), Coffey Partners International, Strata Engineering—all in strata control; GeoGas and Sigra in mine gas and gas drainage systems/drilling technologies; Mine Subsidence Engineering Consultants (MSEC)—in subsidence assessment and prediction, and many more.

Universities—Three universities have dominated the coal mining research field, but with others also involved. The University of New South Wales (UNSW) has for many years worked on strata control and geotechnical matters; rock cutting/excavation engineering; then extending into dust, gas monitoring and windblasts; and more recently carried out significant research into topics of satellite-based radar technologies for mine subsidence monitoring and development of virtual reality technologies for innovative mine safety training. The University of Queensland has worked on spontaneous combustion assessment techniques, as well as a range of innovative equipment and operations research developments, through the CRC Mining group. The University of Wollongong has maintained an active research program into strata control and ground reinforcement in particular, as well as ongoing longwall mining research.

RESEARCH DEVELOPMENTS

There have been many significant highlights in coal mining research in Australia. A small selection of these is outlined below.

Introduction of longwall mining (1960s–1980s)—During and since the period when British longwall equipment was first introduced to Australia in the 1960s, and the early equipment was proven to be grossly under-rated in terms of support capacity, there was significant research carried out to understand what was different about Australian mining conditions. In 1969 ACIRL duplicated the German physical modelling facilities operating in Essen, Germany, with a facility at Bellambi, north of Wollongong. This enabled a 1:10 scale model of a vertical section through the centreline of a longwall panel to be constructed using various mixtures of sand, plaster and cement. These models were constructed and comprehensively analysed for most of the early Australian longwalls and provided great visual and semi-quantitative insights into strata caving behaviour and support performance. The early understanding of the problem of periodic weighting under massive roof strata was first identified using these models.

Long-Hole Directional Drilling—ACIRL was responsible for introducing the first long hole drilling capability into the local coal industry in the early 1980s, with the purchase and trialling of an Acker "Big John" drill rig from the USA, funded by NERDDP. This work was highly successful and was soon complemented with the use of down-hole motors and borehole survey tools to enable directional drilling to be achieved at hole lengths of in excess of 1,000m, entirely in-seam. This technology, introduced from the oil and gas industry, provided a major leap forward in both in-seam gas drainage and in-seam exploration. In more recent years, work by groups such as Sigra and CRC Mining (supported by ACARP and other industry funding) has seen the successful introduction of both surface to seam deviated drilling, and tight radius drilling to further enhance seam gas drainage and exploration capabilities.

Pillar Design—After a series of multiple fatalities in NSW bord and pillar operations in the early 1990s, the JCB provided a major $1.6M research funding to UNSW to develop an improved understanding of the mechanics of coal pillar mechanics and related strata control issues. By 1998 UNSW had identified major findings in relation to safe pillar mine operations, and had developed an empirically-based UNSW Pillar Design Procedure, which was later to become the main standard for pillar design across the Australian industry. Through the findings of this research, and the related industry training workshops, the UNSW work has been directly attributed for a major drop in the number of pillar-related fatalities in Australian coal mines since the mid-1990s.

Mine Subsidence—In the early 1990s it was realised that surface subsidence behaviour in areas where the surface terrain was not flat, but impacted by significant topographic variations, was not well understood. This led to a number of research projects to identify and characterise what was to become known as "non-conventional" subsidence effects, and hopefully to develop prediction capabilities. The subsidence consulting group, MSEC, together with the major mining companies operating in the Southern Coalfield of NSW (notably BHP-Billiton Illawarra Coal) have subsequently collected tens of thousands of subsidence monitoring data values, and analysed these to produce what is known as an "incremental profile" subsidence prediction technique based around the non-conventional behaviour of these regions where mechanisms such as valley closure, valley floor uplift, or upsidence, and far-field horizontal movements have been consistently identified.

Longwall Automation—Through one of the Landmark Projects, ACARP has funded a multi-million dollar program of research, co-ordinated through CSIRO, to develop the relevant technologies required for an automated longwall face operation. This work has involved significant industry input through participating collieries such as the Beltana Mine of Xstrata Coal, in the Hunter Valley, where the longwall face is now capable of operating in a fully automated mode as one of the most advanced longwall operations of its type in the world. In conjunction with ACARP, CSIRO is now commercialising these technologies through licence agreements with each of the major longwall equipment manufacturers.

Virtual Reality (VR)—A dedicated team of software engineers, fine arts graduates and project managers at UNSW has been working since 2001 on identifying technologies and application areas for the use of virtual reality systems in the coal mining industry. Seed funding has been provided internally by UNSW, followed by development funding from ARC, JCB Health and Safety Trust and ACARP. As a result of nearly fifty man-years of development effort, UNSW has pioneered the development and implementation of world-class three-dimensional,

immersive, stereo, virtual and mixed reality training and educational modules for the mining sector. Together with hardware developed by the UNSW iCinema group which designed and built the world's first 360 degree immersive reality stereo projection theatre—AVIE (Advanced Visualisation and Interactive Environment)—the UNSW Mining Engineering team are providing state of the art VR technologies across a broad spectrum of industry applications, and many more opportunities and potential innovations are still being identified.

CONCLUSIONS

Australia has seen a large number of changes in the forms of research funding; the nature of the research providers; and the types of research application for coal mining—especially over the last 30–40 years. Each of the funding schemes, and each of the approaches to research activity has had both merits and problems from time to time. However, in spite of these, the Australian coal industry has advanced to a position now where it must be close to number one in the world in terms of technology, innovation and research support. Furthermore, the industry performances—both in terms of productivity and safety, are also at or close to the top, on a world scale, which must reflect positively on the many research achievements made over recent decades. Some of these achievements and industry performances have translated as immediate gains as a result of research activity. Others are more indirect, and longer term, however the evidence and inevitable links between strong, practical research, and a high performing industry are clear for all to see. Acknowledgment is clearly due both to the research teams, their organisations and the research funding providers who have enabled this work to take place.

REFERENCES

Hampton B & Allen B (2009), History of the School of Mining Engineering, The University of New South Wales (*in press*).

Rutledge M (1981), 'Davidson, Sir Colin George Watt (1878–1954)', *Australian Dictionary of Biography*, Vol 8, Melbourne University Press, Melbourne, 1981, pp223–224.

Australian Geological Survey Organisation (1998), *A Century of Mining in Australia*, Australian Mining Industry Report, 1998–99, republished by the Australian Bureau of Statistics (ABS cat no 8414.0).

The Impact of Ground Control Research on the Safety of Underground Coal Miners: 1910–2010

Christopher Mark
NIOSH, Pittsburgh, Pennsylvania, United States

Thomas M. Barczak
NIOSH, Pittsburgh, Pennsylvania, United States

ABSTRACT: Ground falls claimed the lives of 50,000 us coal miners during the 20th century, more than all other types of underground accidents put together. While seldom garnering headlines, ground control research has been an important focus area for the U.S. Bureau of Mines (USBM) (and now the National Institute for Occupational Safety and Health (NIOSH)). These organizations have played a key role many central developments, including:

- The transition from wood posts to roof bolts
- The requirement that every mine employ a Roof Control Plan
- The development of shield supports for longwalls
- The application of empirical pillar design methods
- The development of improved standing support systems

These and other successful interventions required that the technology, economics, and mining culture all intersect. The paper concludes with a discussion of some current safety technologies that represent the next steps forward for ground control.

INTRODUCTION

The 20th century witnessed two revolutionary transformations in underground coal mining technology. At the beginning of the century, hand loading was nearly universal. Mechanical loading began the first great change in about 1930 (Figure 1), and this process was completed when continuous miners supplanted drilling and blasting during the '50s and '60s (Anon, no date). The rapid growth of longwall mining during the last two decades of the century was the second revolutionary development.

Neither of these new mining methods could have succeeded without equally dramatic developments in ground control technology. If roof bolting had not replaced timbers, mechanized room-and-pillar mining could not have reached its potential. Modern longwall mining is similarly inconceivable without heavy duty, self-advancing shield supports.

Clearly, these innovations in support technology have had an impact on mine safety as well. While roof falls once claimed hundreds of lives each year, today annual fatalities are usually numbered in the single digits (Pappas et al., 2000). The fatality rate has also improved, measured on either a per-ton or a per-hour basis (Figures 2 and 3). Yet a careful analysis of the history shows that improvements in safety did not occur in lock-step with technological developments. The *safety culture*, which encompasses regulatory mandates, company-specific safety policies, and other social and legal aspects, has also been central because it determines the level of risk that miners take (or are exposed to) while underground.

GROUND CONTROL DURING THE HAND-LOADING ERA

At the turn of the century, roof support was considered the responsibility of each individual miner. It was his duty to "examine his working place before beginning mining work, to take down all dangerous slate, and make it safe by properly timbering the area before commencing to mine coal." The mine operator was responsible for delivering timbering materials to each working place, and the foreman checked that it was installed properly on his daily visit (Paul and Geyer, 1928).

The miners often had considerable discretion about the amount of support they installed. In weak shale roof, posts were set 2.5 to 5 ft apart (Paul et al., 1930), but where the roof rock was strong, no posts might be set at all. An important element in early roof support systems were "safety posts," which were set at the end of the track to protect miners while they

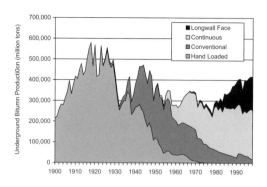

Figure 1. Underground coal mining methods of the 20th Century

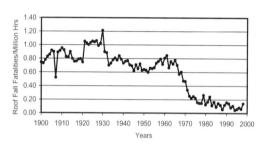

Figure 2. Roof fall fatalities in underground coal mines

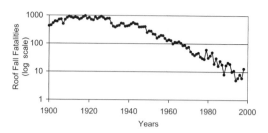

Figure 3. U.S. roof fall fatality rate for underground coal mines (fatalities per million employee hours)

Figure 4. Hand loading with a safety post

loaded the coal or prepared the face for the next shot (Figure 4) (Paul et al., 1930). These temporary supports required extra time and effort, and their use was often at the discretion of the miner.

Under these circumstances, it is understandable that much of the blame for roof fall accidents was placed on the inexperience and carelessness of the miners themselves (WVU Press, no date). In 1912, the USBM asked:

> How can you, the miner, escape harm from roof falls?" The answer was: "Be careful...do not take the risk of loading a car before putting up a prop...set extra posts, even though they are in your way (Rice, 1912).

Over time, however, safety professionals began to recognize that "a condition responsible for many fatalities from falls of roof is the absence of any policy on the part of management with respect to systematic methods of roof inspection and support" (Paul, 1927). Encouraging mine managers to prepare, promulgate, and enforce a *systematic timbering*

plan became a key element in Bureau's roof control efforts.

The USBM also exhorted miners to "comply with systematic methods of timbering, where such systems have been adopted, and exercise judgment in placing additional posts for your own protection." But as long as the typical mine foreman was responsible for about 80 miners, and seldom spent more than 5 minutes with each one during a shift, enforcing timber plans would present a challenge.

THE IMPACT OF MECHANIZED MINING ON GROUND CONTROL

Between 1930 and 1948, the portion of underground coal that was loaded by machine rose from less than 10 pct to nearly ⅔ (Dix, 1988). In many ways, machine mining actually made roof support more difficult. Loading machines required a prop-free front in which to work. The machine operator was usually protected by posts and crossbars, but the helper had to venture into the unsupported face zone (Figure 5) (Coal Age, 1945). A detailed and widely reported Bureau of Mines study conducted in 1951 concluded that "mechanical operations are,

Figure 5.   Machine loading with timber support

to a considerable degree, more dangerous from the standpoint of roof falls" then hand loading, "notwithstanding that much closer supervision is maintained in such operations." The study also found that 74% of roof fall fatalities occurred wit)hin 25 ft of the working face, and that 3 out of 4 of these took place by the last permanent support (between the last support and the face) (Forbes et al.,1951). The same study also concluded that:

> Regardless of roof conditions, minimum standards of roof support suited to the conditions and mining system of each mine should be adopted and followed…. The judgment of the person should never be substituted for the minimum support required in the systematic roof support plan (Forbes et al.,1951).

The difficulties posed by traditional timber supports increased as the early track-mounted loading machines were replaced by crawler-mounted ones. When rubber-tired shuttle cars that carried coal away from the face were introduced in the late 1930s, timbering became a critical bottleneck in the mechanical mining process. Little wonder, then, the Bureau's Edward Thomas wrote in 1948 that:

> The more progressive mining companies are continually searching for improved types of roof support that will give maximum protection and at the same time offer minimum interference with the preparation and loading of coal (Thomas et al., 1948).

THE INTRODUCTION OF ROOF BOLTING

Even as Thomas wrote those words, the roof bolt was emerging as the leading candidate "temporary legless support" in machine mining (Thomas et al., 1948). As a support, roof bolts are superior to timbers because

"timbers offer support after the strata they are supporting have failed; whereas roof bolts reinforce the roof rock, which contributes to its own support."

The St. Joe Lead Company was the first major mining company to make extensive use of roof bolts, beginning in the 1920s. Early in 1947, C. C. Conway, Chief Engineer for the Consolidation Coal Company in St. Louis, visited one of the St. Joe mines near Bonne Terre, Missouri, and was impressed with the roof bolts he saw there (Jamison, 2001). He determined to try them at Consol's Mine No. 7 near Staunton IL. The first roof bolts were installed in Mine No. 7 using hand-held stopper drills (Figure 6). The anchors were expansion shells "similar to those used to support trolley wire," though slot-and-wedge type anchors like the ones "ordinarily used in the metal mines" were also employed.

The USBM was apparently involved in the roof bolt trials at Mine No. 7 almost from the beginning. Subsequently, the Bureau initiated a major effort to promote the use of roof bolts nationwide. They touted both the safety and efficiency advantages of the new supports (Thomas, 1951). Safety features included:

- A systematic support "within inches" of the working face
- Can't be dislodged by blasting or equipment;
- Improved ventilation (because of less air resistance)
- Reduced accumulation of explosive coal dust (because places can be cleaned more thoroughly) (Thomas et al., 1949)

Economic advantages included:

- A reduction in the time required to load a place by 15–50%
- Potential for widening rooms

Figure 6.   Installing roof bolts with a hand-held stopper drill

Figure 7. An early roof bolt machine

- Faster haulage
- Reduced labor cost for roof support and material handling

Since the Bureau was without regulatory powers, it had since 1910 "mastered the art of prodding operators into implementing new technologies that resulted from its scientific investigations (Aldrich, 1995). The Bureau's roof bolting campaign had to overcome numerous barriers to the new technology. The first was the cost and availability of the required equipment. The development of carbide alloy insert bits was essential because it made it possible to drill holes cheaply in hard rock (Thomas, 1956). Hand-held stopper drills were already available at most mines, but they now required a mobile source of compressed air. Many of the first mines to install roof bolts built their own cars to carry the drills, compressor, bolts, and other supplies from face to face (Figure 7). The bolts themselves were not readily available, and sometimes had to be fabricated in the mine's own shop. Finally, miners were used to the reassuring presence of heavy timbers, and roof bolting seemed to be "reverse in principal to the old methods," because it "appears at first glance to approximate holding oneself up by one's bootstraps."

Between 1949 and 1955, numerous case histories of the successful application of roof bolting from all over the coalfields were reported in the mining press. For example, a major Bureau study in five southern West Virginia mines found that they "produced over two million tons of coal without a fatal accident and only 4 lost-time accidents, as compared with two fatal and 71 lost time accidents over a similar period when conventional timbering methods were used. Production increases ranged from 0.86 to 10.7 tons per man shift." That these results were achieved during the dangerous process of pillar recovery made them even more impressive. In northern WV, the success of roof bolting in pillar work

allowed 7–10% more coal to be recovered (Flowers, 1953).

Such figures led many safety professionals to concur with Joe Bierer of the West Virginia Department of Mines, that:

> Herein lies a wonderful opportunity for the coal industry to bring about an epochal advance in safety for the mineworker, a humanitarian accomplishment to compare with the great social advances of recent years. No such immediately effective and readily confirmed benefit has derived from any other measure ever conceived, or devised, for safety in coal mines (Bierer, 1952).

Unfortunately, even as roof bolting continued its dramatic expansion, overall fatality rates stubbornly refused to go down. The total number of roof fall fatalities declined, but three out five mining jobs also disappeared between 1948 and 1960. Each of the remaining miners actually had a greater chance of being killed in a roof fall than his counterpart in 1948.

The Bureau advanced two explanations for the frustrating lack of progress. The first was that when miners go beyond the last support, they are unprotected, regardless of what type of support is used. A 1954 Bureau study had found that more than 50% of roof fall fatalities occurred in the unsupported space between the last row of supports and the face (Sall, 1955).

The second explanation for the lack of progress was that "many mines are now bolting where the method is marginal in the sense that perfect anchorage cannot be obtained." Leon Kelly, a Bureau engineer in Vincennes IN, described this process in remarks he made to the 1950 Annual Meeting of the IL Mining Institute. He cited examples from three mines in his own experience in which the level of support had been reduced, and concluded:

> In each case, when bolts were first used at the mine, everyone was more or less afraid of them and the pattern that was adopted was followed religiously. As time went on, and none of the bolts fell out, they were taken for granted and it was assumed that bolts would hold up the roof as long as there were bolts in the roof. Some operators are beginning to tell me that we are all overbolting, and naturally when they feel that way they will reduce the either the number of bolts or the length of the

bolts they use…. If failures are accepted as a calculated risk, it is only a matter of time until a serious accident occurs (Kelly, 1951).

Looking back, it seems clear that simply replacing timbers with bolts would not be sufficient to substantially reduce roof fall rates. The success of any support system in a particular application depends not just on the *type* of support, but also on the density of the pattern, the capacity of each unit, when they are installed, the quality of the installation, and many other factors including the span, the rock quality, and the ground stress.

Simply stated, roof bolts can only prevent roof falls if enough of them are installed. That costs time and money. The mine operator's natural tendency was to adjust the expenditure to achieve an *acceptable* level of roof fall risk. Unless the mining culture was changed to reduce the acceptable level of risk, competitive pressures would mean that the new technology would be adapted to obtain the same results as before.

Moreover, it seems clear that roof bolting actually introduced a vicious new hazard, silica dust. The Bureau publicized research that showed dry drilling, with either pneumatic stoper or electric rotary drills, could result in silica dust concentrations up to *200 times* the recommended level of 5 million particles per cubic ft of air. Such concentrations were a serious menace to not just the drill operators, but also anyone working downwind in the return air. But in 1957, the Bureau reported that of the 424 mines using roof bolts, just 8% employed water to allay dust, 35% employed dry dust collectors, and nearly half employed no means of dust control other than respirators (Figure 8). There is little doubt that the prevalence of occupational lung disease among the generation of miners who worked in the dusty,

Figure 8. Dry drilling using respirators for dust protection

mechanizing mines of the postwar period was higher than among most others in the past. Unfortunately, it seems that silica dust from roof bolting must have contributed to this terrible human toll.

THE 1969 ACT

On Nov. 20, 1968, a massive gas and dust explosion destroyed the Farmington Mine and killed 78 miners. The disaster led directly to the Federal Coal Mine Health and Safety Act of 1969, which was signed into law by Richard Nixon. Under the Act, Federal inspectors were given much expanded enforcement powers, and a detailed set of health and safety standards was made mandatory for all mines (Lewis-Beck and Alford, 1980). The key ground control provision, one that Bureau ground control specialists had been advocating for two generations, was that each mine was required to develop a Roof Control Plan which included systematic support throughout the mine. Strict guidelines regarding bolt spacing, bolt length, entry width, and other ground control parameters were also included. Working under unsupported roof without safety posts was banned.

The results were quick and dramatic. In the six years following 1968, roof fall fatality rates plummeted by two-thirds, and they stayed approximately constant at that new level for the next 15 years. The improvement might have been due to regulatory enforcement, or to changes in safety standards implemented by the operators themselves, or to mandated safety technologies like protective canopies and Automated Temporary Roof Support systems. But there can be little doubt that the reduction was not due primarily to new technology, but rather to widespread application of technologies that were already available.

ROOF SUPPORT AND LONGWALL MINING

Modern longwall mining began in 1952 with a Bureau-sponsored trial at the Stotesbury mine of Eastern Gas and Fuel Associates near Beckley, WV (Haley and Quenon, 1954; Mason, 1976). This marked the beginning of mechanized extraction of a relatively large coal panel using a German coal planar. Roof control for the retreating longwall face was provided by wood cribs and mechanical 40-ton-capacity friction props with I-beam caps. Three panels were successfully mined between 1952 and 1958 producing an average of 530 tons per shift (Haley and Quenon, 1954).

The primary constraint to increased production from longwall operations during this era was roof support advance. Mechanical friction props and

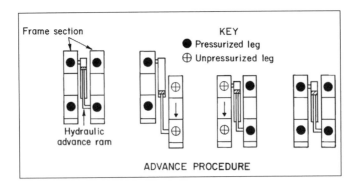

Figure 9. Self-advancing frame-type roof support

wood cribs were the predominant support system, and they had to be manually removed from roof contact, and in the case of the wood cribs, reconstructed as the face advanced. Support capacity during this era typically ranged from 25 to 75 tons per jack supplemented by 100 ton capacity wood cribs. The jacks were often damaged during the panel extraction and tended to be unstable in heavy strata conditions, which further reduced their effectiveness.

The next innovation was the introduction of powered roof supports, which were hydraulically pressurized and advanced without manual labor. Eastern Associated Coal Company installed the first hydraulic self-setting and self-advancing roof support system in 1960 in a 52-in Pocahontas coal seam at their Keystone mine in southern West Virginia. These initial self-advancing roof supports were frame-type constructions where two hydraulic jacks were connected by a beam to form a frame construction as shown in Figure 9 (Chironis, 1977a). The units were operated in sets of two with two frames linked together by a central shifting ram. As one frame remained set between the roof and floor, the other frame was lowered and advanced, thus creating a self advancing support system. Longwall installations using self-advancing frame supports were successful in strata conditions that provided easy caving of the roof behind the supports. However, there were also very serious failures when mines attempted to install these supports under more competent roof strata, such as massive sandstone and limestone geologies.

The first high-capacity support system was installed in Island Creek's Beatrice mine in 1966. This system featured 560-ton units that provided excellent control of the competent sandstone roof, and operated for a period of ten years without failure while producing a respectable 1,000 tons per shift of raw coal (Kuti, 1972). Nationwide, support capacity

increased by a factor of 2 to 4 during the next few years, with four-leg supports providing resistances as high as 700 tons.

An improvement in support design was also made with development of the chock support. The chock support can be thought of as a mobile crib. It essentially tied two frame supports together with a rigid canopy and semi-rigid base. This design improved structural stability and increased roof contact area over the previous frame-type systems. The increased roof cover helped reduce accidents caused by material falling from the immediate roof, especially during support advance. However, several failures were still reported due to the inadequacy of the chock support systems to operate in competent strata that caved in large pieces and exerted rotational moments or horizontal displacements to the support structure.

It was not until the introduction of the shield support in 1975, with its improved lateral stability and more roof cover (Figure 10), that the last major hurdle to longwall utilization was overcome. The first shield-supported face installed in the United States was by Consolidation Coal Company (Consol) in 1975 at their Shoemaker mine near Moundsville, West Virginia. From the very beginning, encouraging results were obtained. The face produced 750,000 tons in its first year of operation (Chironis, 1977b). The outstanding successes of these initial shield installations created a feverish quest for utilization of this technology. Shield utilization rose dramatically from their introduction in 1975 to 1982, when 83 pct of the longwall faces were shield supported. By 1985, just 10 years after introduction, nearly 90 pct of the operating longwall faces were supported by shields.

The basic shield design itself underwent several developmental changes during this era (Figure 11). The most significant design improvement was the

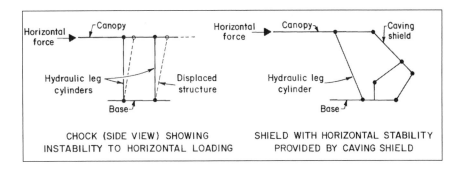

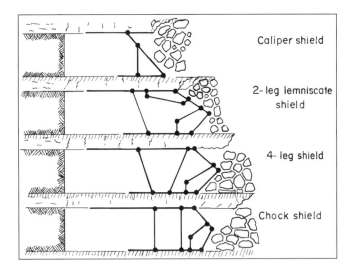

Figure 10. Comparison of chock and shield support constructions and stability

Figure 11. Modifications of basic shield design

incorporation of the lemniscate linkage into the caving shield assembly. Previous designs had the caving shield connected directly to the base, which caused the tip of the canopy to transcribe an arc as the shield was lowered and raised. The lemniscate system provided vertical travel of the canopy throughout its operating range, which minimized the unsupported span in front of the shield.

The USBM was extensively involved in promoting the utilization of longwall mining during the 1970s. Bureau sponsored research totaled approximately 67 million dollars, which would be the equivalent of about $254 million today (BLS inflation calculator, http://data.bls.gov/cgi-bin/cpicalc.pl). Included in these efforts were: early demonstrations of the shield support system at York Canyon and Old Ben; demonstrations of thin seam longwall mining, single entry gateroad design, multi-lift extraction of a

thick seam, steeply pitching seam extraction, and an advancing longwall trial.

In 1979, the Bureau under contract to MTS Corporation built a sophisticated load frame at a cost of $6.8 million ($20.2 million in 2009) for conducting research on longwall support systems. The Mine Roof Simulator, as it is called (Figure 12), has the capability to provide controlled loading of up to 1,500 tons vertically and 800 tons horizontally over a 20-ft by 20-ft platen area to full scale support structures. This facility was instrumental in helping to identify design deficiencies in the early generation shield supports and to develop testing protocols taking advantage of the active loading capabilities of the simulator. Much of this effort centered around achieving proper load transfer through the caving shield lemnicsate assembly, particularly for horizontal shield loading (Barczak and Schwemmer, 1988a).

Figure 12. Mine Roof Simulator unique active loading frame used for full-scale support testing

As the shield continued to grow in capacity and the leg cylinders became increasing larger to accommodate the increase in capacity, two-leg shield designs became the standard in the industry. The advantages of the two-leg shield over the four-leg design is more efficient loading, less maintenance, and active horizontal loading whereby the inclination of the leg cylinders can impart through the canopy a confining force toward the longwall face (Barczak and Garson, 1986). This acts to restore the loss of confinement of the roof strata in the face area created by the formation of the gob as the immediate roof caves. The disadvantage of the two leg design is increased toe pressure that can create problems in soft floor conditions. This problem is mitigated by increased contact area and specialized base lifting devices when necessary.

In addition to the increase in capacity, the shields have also increased in size. They have increased in length to accommodate one-web-back operations and larger face conveyors and deeper shearer webs, making these units over 20 feet in length by the end of the 20th century. The standard canopy width by the year 2000 was 1.75 m, replacing the 1.5 m designs that were prevalent during the 1990s. The first 2.0-meter wide shield was installed in Foundation Coal's Cumberland mine in 2003. This trend of increasing canopy width to 2.0 m is likely to continue. The 2.0-m width may represent an upper limit with current shield construction materials since wider shields weigh more, thereby requiring more effort to transport during face moves.

The question of is bigger better or how much capacity is needed remains a debated issue as the largest shield capacity is now 1,750 tons. A consequence of the increase in capacity by increasing the leg cylinder diameter has been a proportional increase in shield stiffness (Barczak and Schwemmer,1988b).

This results in increased loading per unit of convergence, so convergence which is beyond the direct control of the shield causes an increase in shield loading in proportion to the increase in shield stiffness.

Less debatable has been the desire to maintain consistent and high setting forces. Modern electrohydraulic control technologies are capable of maintaining a set-to-yield ratio in the 0.6 to 0.8 range, provided there is sufficient flow capacity from the pumping system and the support hydraulics are well maintained. Unfortunately, the increase in shearer speed, automated shield advancement with multiple shields moving and simultaneous conveyor advancement, and wider longwall faces have placed increasing demands on the hydraulic power distribution system and consistent setting pressure remain a concern at many current longwall operations.

DESIGN OF LONGWALL GATE ENTRY SYSTEMS

On December 19, 1984, 28 miners entered the longwall section at the Wilberg Mine near Huntington, UT. They hoped to break the world's record for one-day production from a longwall. Instead, a fire broke out at an air compressor located near the mouth of the section, and deadly smoke filled the headgate entries. Only one miner escaped.

During the subsequent investigation many safety experts insisted that miners died, not because there was a fire, but because they had no escape route. The miners had been forced to try to escape through the headgate because the tailgate from the Fifth Right longwall had been blocked by roof falls (Cocke, no date). As a result, new regulations were introduced that required longwall mines to maintain safe travel ways on the tailgate side of longwalls (30CFR75, 2000).

At the time, there was very little theory available to assist mine planners in sizing longwall pillars. Traditional pillar design methods were not appropriate because the long and narrow geometry of typical gate entry layouts makes a classic squeeze highly unlikely. In addition, the abutment loads applied to longwall pillars are significantly greater and more complex than the tributary area loading assumed by traditional pillar design methods. Effective longwall pillar design requires some way to estimate the abutment loads during all phases of longwall mining, from development all the way into the tailgate.

Most importantly, however, the stability of the tailgate is not purely a pillar design problem. Studies conducted as early as the 1960s had concluded that "whether or not the stress [from an extracted longwall panel] will influence a roadway depends more

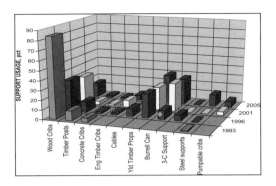

Figure 13. Distribution of tailgate support since 1993 showing change in support practices from conventional wood cribs and posts into a variety of alternative support technologies

on the strength of the rocks which surround the roadway itself than on the width of the intervening pillar" (Carr and Wilson, 1982).

The USBM developed the Analysis of Longwall Pillar Stability (ALPS) method to answer these needs. First, design estimates of abutment loads were obtained through an extensive series of underground stress measurements. The greater challenge was how to integrate pillar design into a broader design method that focused directly on tailgate performance.

The solution was to employ a sophisticated empirical technique. The Bureau conducted studies at more than half of all the longwalls in operation in the U.S., and documented historical gate entry performance at all of them. A new rock mass classification system, the Coal Mine Roof Rating (CMRR), provided a quantitative measure of the structural competence of the roof (Molinda and Mark, 1994). Multivariate statistical analysis provided simple, quantitative guidelines for sizing longwall pillars, and showed that when the roof is strong, smaller pillars can safely be used.

The Bureau conducted an extensive program to disseminate the ALPS technology to the industry. ALPS was implemented as a user-friendly computer software package, which was widely distributed free of charge. A number of hands-on computer training sessions were held throughout the coalfields. Since the mid-1990s, ALPS has been considered the industry standard for longwall gate entry design.

In addition to this pillar design effort, an extensive development in standing support to provide increased stability in the tailgate entries occurred in the early 1990s. Timber supports in the form of wood cribs and posts were all that was used prior to mid-1980, when trials of concrete supports were

common. Then, the 1990s saw a revolution in new support technologies, providing a wide range of support capabilities. Figure 13 (figure shows a distribution of several types of tailgate supports since this time.

The soft response of the wood cribs that allowed too much roof movement to occur under abutment loading led a search to develop stiffer and higher capacity supports for longwall tailgate applications. However, a valuable lesson was learned from concrete supports that had a compressive strength and material modulus an order of magnitude higher than the conventional wood cribs. Any standing support cannot control the behavior of the overburden. Plain and simple, you cannot prevent all of the closure of a mine opening with any practical and economical support system, and if the support cannot survive this uncontrollable convergence, it will fail prematurely and not provide adequate roof support. These super high capacity and stiff concrete cribs were crushed in several longwall tailgates (Figure 14) (Barczak, 2006).

This lesson has transpired into a goal to utilize the ground reaction concept for support design (Figure 15). The basic concept is to match the performance characteristics to the ground response, whereby the convergence developed in the longwall tailgate is proportional to the amount or capacity of support used. Numerical models are used to develop

Figure 14. Failure of high capacity concrete crib in longwall tailgate

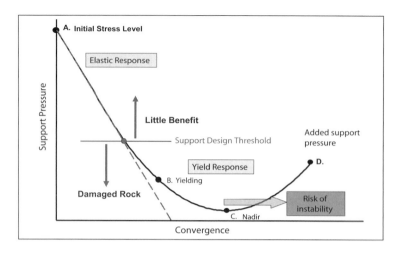

Figure 15.   Ground reaction curve for support design

the ground response curves which are calibrated with in mine measurements of support loading and convergence. The onset of onset of strain-soften rock response, indicating that the rock has been loaded beyond its elastic range and is now yielding causing damage to the rock mass, defines the support threshold. Providing more support than this provides little benefit since the support cannot impact the elastic ground response and will produce a negligible reduction in tailgate convergence. Conversely, too little support will result in greater convergence and increased risk of instability leading to roof fall.

The performance characteristics of the various roof support products are determined from full-scale testing in the NIOSH Mine Roof Simulator through a rigorous testing protocol that determines the limitations of the support system. This data is incorporated in the NIOSH Support Technology Optimization Program (STOP) which facilitates selecting the most effective support for a particular ground condition (Barczak, 2006 and 2008).

NIOSH subsequently employed the same basic empirical approach to develop solutions for several other complex ground control problems. The software packages, particularly the Analysis of Retreat Mining Pillar Stability (ARMPS) and Analysis of Multiple Seam Stability (AMSS), have been widely adopted by the mining community. Each is based on extensive data bases of real world case histories, gathered from in-mine studies. Because they are so firmly linked to reality, they have met the mining community's needs for reliable design techniques that can be used and understood by the non-specialist.

Unfortunately, the 2007 Crandall Canyon mine disaster provided an unwelcome reminder of

the importance of pillar design. The MSHA report on the disaster pointed to the "flawed" mine design as the root cause, because "the stress level encountered exceeded the strength of a pillar or group of pillars near the pillar line." The report documented how the two pillar design software packages used to develop the mine plan, ARMPS and LaModel, had been employed improperly. Following the disaster, MSHA distributed a series of memorandums and other documents that strongly encouraged mine planners to use ARMPS in the pillar design process. The memorandums state that any design that does not meet the NIOSH criteria should be considered "complex and/or non-typical," and be subject to more extensive MSHA review.

RECENT ACCOMPLISHMENTS AND FUTURE CHALLENGES

Fatalities inby supports: Fifty years ago, miners working inby supports were the victim in half of all underground roof fall fatalities. The practice of going inby supports was banned by the 1969 Act, but through the 1990s such accidents still accounted for nearly half of all roof fall fatalities. During the past six years, however, there have been just been two inby incidents. The progress is associated with the industry-wide educational campaign under the slogan "Inby is Out!" Another factor contributing to the reduction in inby fatalities is the development of the concept of the "Red Zone." In many roof control plans, miners must now stay outby the second row of bolts, particularly when making extended cuts or turning a crosscut.

Pillar recovery: Progress has also been made in pillar recovery. Historically, retreat mining has been less safe than other underground mining techniques. Although the intentional caving that occurs is an unavoidable part of the retreat mining process, premature caving can cause roof falls that put miners at risk. A NIOSH study published in 2003 found that retreat mining elevated a miner's risk of being killed in a roof fall by a factor of three.

The same study also found that many of the fatalities had occurred where the mine was actually following its Roof Control Plan. As a result, NIOSH, together with MSHA, advocated that retreat mines increase the roof support they employ during retreat mining by leaving final stumps, using Mobile Roof Supports, and installing longer and stronger roof bolts. The widespread adoption of such safer retreat mining techniques seems to have made a difference. During the past four years there has been just one fatal roof fall, compared to an average of two per year during the previous decade.

Rock and rib falls: More than 400 miners continue to be injured each year by rock falling from between supports, and 100 more are injured by rib falls. Together, these two categories also account for a large percentage of recent ground fall fatalities. Available technologies such as roof screen, rib bolting, and inside control roof bolters could dramatically reduce injury and fatality rates if they were used more widely. Further advances in these areas will likely be the next big advance in ground control safety.

Disclaimer

The findings and conclusions in this paper have not been formally disseminated by the National Institute for Occupational Safety and Health and should not be construed to represent any agency determination or policy.

REFERENCES

Aldrich, M. 1995. Preventing 'the Needless Peril of the Coal Mine': The Bureau of Mines and the Campaign Against Coal Mine Explosions, 1910–1940. *Technology and Culture, Society for the History of Technology*, pp. 483–518.

Anon. Information taken from 1900–1922 from U.S. Geological Survey; 1923–1976 from USBM Mineral Handbooks; 1977–1998 from DOE/EIA Coal Industry Annuals.

Barczak, T.M. 2000. Optimizing secondary roof support with the NIOSH Support Technology Optimization Program (STOP). In *Proceedings of the 19th International Conference on Ground Control in Mining*. Morgantown, WV, pp. 74–83.

Barczak, T.M. 2006. A Retrospective Assessment of Longwall Roof Support With a Focus on Challenging Accepted Roof Support Concepts and Design Premises. In *Proceedings of the 25th International Conference on Ground Control in Mining*. Morgantown, WV, pp. 232–244.

Barczak, T.M. and Schwemmer, D.E. 1988a. Horizontal and Vertical Load Transferring Mechanisms in Longwall Supports. *U.S. Bureau of Mines RI 9188*.

Barczak, T.M. and Schwemmer, D.E. 1988b. Stiffness Characteristics of Longwall Shields. *Bureau of Mines RI 9154*.

Barczak, T.M., Esterhuizen, G.S., Ellenberger, J.L., and Zhang, P. 2008. A First Step in Developing Standing Roof Support Design Criteria Based on Ground Reaction Data For Pittsburgh Seam Longwall Tailgate Support. In *Proceedings of the 27th International Conference on Ground Control in Mining*. Morgantown, WV, pp. 349–359.

Bierer, J. 1952. Roof Bolting for Safety. WV Department of Mines.

Carr, F. and Wilson, A.H. 1982. A New Approach TO THE Design of Multi-Entry Developments for Retreat Longwall Mining. In *Proceedings of the Second Conference on Ground Control in Mining,* Morgantown, WV, pp. 1–21.

Chironis, N.P. (ed.). 1977a. New System at Old Ben Coal Opens Longwalling to Midwest. *Coal Age Operating Handbook of Underground Mining*. McGraw-Hill, p pp. 46–47.

Chironis, N.P. (ed.). 1977b. Consol's Record Production Runs Give Impetus for Use of Shield Supports for U.S. Longwalls. *Coal Age Operating Handbook of Underground Mining*. McGraw-Hill, pp. 68–73.

Coal Age. 1957. Roof Bolting in 1956. *Coal Age,* 4:88.

Cocke, E. No Date. The Wilberg Mine Fire. Utah History Encyclopedia, http://www.media.utah.edu/ UHE/w/WILBERGMINE.html).

Code of Federal Regulations. 2000. *Mandatory Safety Standards—Underground Coal Mine*s. 30CFR75.

Dix, K. 1988. What's a Coal Miner to Do? The Mechanization of Coal Mining. *University of Pittsburgh. Press.*

Esterhuizen, G.S. and Barczak, T.M. 2006. Development of Ground Response Curves for Longwall Tailgate Support Design. In *Proceedings of the 41st U.S. Rock Mechanics Symposium* (Golden Rocks 2006), June 17–21, Golden, CO, Paper 06-935.

Forbes, J.J., Back, T.L., and Weaver, H.F. 1951. Falls of Roof: The No. 1 Killer in Bituminous Coal Mines. *U.S. Bureau of Mines IC 7605.*

Flowers, A.E. 1953. Successful Roof Bolting at Idamay. *Coal Age* 10:76–80.

Haley, W.A. and Quenon, H.A. 1954. Modified longwall Mining With a German Coal Planar. Progress Report 2: Completion of Mining, in Three Adjacent Panels in the Pocahontas No. 4 Bed, Helen, WV.

Jamison, W. 2001. Telephone interview with the author.

Kelly, L.W. 1951. Discussion on "Progress in Roof Bolting." In *Proceedings of the 1950 Annual Meeting of the Illinois Mining Institute,* pp. 27–30.

Kuti, J. 1972. Outlook of Longwall Mining Systems. *Coal Age,* 77(8):64–72.

Lewis-Beck, M.L. and Alford, J.R. 1980. Can Government Regulate Safety? The Coal Mine Example. *89Amer. Pol. Sci. Rev.,* 74:751–752.

Mason, R.H. 1976. The First Longwall Systems Were Primitive Compared to The Ones Now in Use. *Coal Min. & Procelili.,* 13(12):59–64.

Molinda, G.M. and Mark, C. 1994. Coal Mine Roof Rating (CMRR): A Practical Rock Mass Classification for Coal Mines. *U.S. Bureau of Mines IC 9387.*

Pappas, D.M., Bauer, E.R., and Mark, C. 2000. Roof and Rib Fall Incidents and Statistics: A Recent Profile. In *Proceedings of the New Technology for Coal Mine Roof Support,* NIOSH IC 9453, pp. 3–22.

Paul, J.W. 1927. Stop, Look, and Listen! The Roof is Going to Fall. *U.S. Bureau of Mines IC 6032.*

Paul, J.W. and Geyer, J.N. [1928]. State Laws Related to Coal Mine Timbering. *USBM Technical Paper 421.*

Paul, J.W., Calverly. J.G., and Sibray, D.L. 1930. Timbering Regulations in Certain Coal Mines of Pennsylvania, West Virginia, and Ohio. *U.S. Bureau of Mines Technical Paper 485.*

Rice, G.S. 1912. Accidents from Falls of Roof and Coal. *U.S. Bureau of Mines Miner's Circular 9.*

Thomas E., Seeling, C.H., Perz, F., and Hansen, M.V. 1948. Control of Roof and Prevention of Accidents from Falls of Rock and Coal: Suggested Roof Supports for use at Faces in Conjunction with Mechanical Loading. *U.S. Bureau of Mines IC 7471.*

Thomas, E., Barry, A.J., and Metcalfe, A. 1949. Suspension Roof Support: Progress Report 1. *U.S. Bureau of Mine IC 7533.*

Thomas, E. 1951. Suggestions for Inspection of Roof Bolt Installations. *U.S. Bureau of Mines IC 7621.*

Thomas, E. 1956. Roof Bolting Finds Wide Application. *Min. Eng.,* 11:1080.

Sall, G.W. 1955. Four Years of Roof Fall Accidents. *Min. Cong. J.* 9:79

WVU Press. Work Relations in Hand-Loading Era, 1880–1930.

The Rise of Geophysics for In-Mine Imaging in South African Gold and Platinum Mines

Declan Vogt
CSIR, Johannesburg, South Africa

Michael van Schoor
CSIR, Johannesburg, South Africa

ABSTRACT: Geophysics has been used for at least a century to discover new orebodies, including significant deposits such as the West Wits line discovered by Krahmann in the 1930s. It is only since the mid 1980s, though, that geophysics has become an important tool for mining itself.

The paper discusses the rise of various geophysical techniques for in-mine use in South African tabular gold and platinum orebodies, from 3D seismic reflection, used for feasibility studies; to techniques that are applied in-mine, particularly radio and electrical methods, including borehole radar and electrical resistance tomography. Barriers to adoption are discussed, including the lack of skilled clients, the need for reliability and the need for good integration with mining operations.

INTRODUCTION

The Early Days

Geophysical techniques have been used in South Africa for discovering new orebodies almost since the discovery of the techniques themselves. The Witwatersrand Basin, the world's largest known gold deposit, outcrops on surface for 58 km, east and west of Johannesburg. However, further west, it is covered by younger rocks. Boreholes had shown the existence of gold "reefs," as they are known locally, in the West Wits but it was not until 1930 that Dr Rudolph Krahmann was able to show that the position and extent of the reefs could be determined using a magnetometer, and the West Wits Line was born (Krahman 1936, Walker 1950). It has a strike distance of 64 km, and is home to what was once dubbed the world's richest gold mine, Driefontein, which poured its 100-millionth ounce of gold in 2005 (Anon. 2005).

Later that decade, gravity measurements were used to identify a potential portion of the Witwatersrand Basin about 200 km to the south of Johannesburg. Drilling confirmed the find, which later became the Free State goldfield. Almost all of the other remaining gold fields in the Witwatersrand were eventually discovered by geophysics—magnetics or gravity, alone or in combination (Roux 1967).

While geophysics is well known as a technique for discovering new orebodies, its use is less well known in the later stages of mining.

The Mining Life Cycle

For our purposes, the life cycle of a mine, as defined by Hartman and Lacy (1992) and others, can be simplified to:

- Exploration
- Feasibility
- Construction
- Operations
- Closure

In this paper, we discuss the application of geophysics at each stage in the life cycle, in order to make mining safer and more productive. Our discussion is largely limited to the tabular hard-rock deposits of South Africa, particularly its gold and platinum mines. The first step of exploration is discussed above, and does not come into the realm of in-mine imaging.

FEASIBILITY

As the mine gets closer to production, the risk in the project drops, so information that lowers risk is most valuable early in the life cycle. For this reason, it is at the feasibility stage that the greatest use of geophysics takes place—to reduce the risk of a poor mining decision.

In the feasibility stage, the mineral resource manager needs to determine that there are sufficient resources in the ground to make mining a financially successful proposition. The two factors to be considered are the size of the orebody and the difficulty of mining. In South Africa, exploration techniques are being successfully applied at a higher resolution for feasibility studies, particularly to map dislocations to reef. In particular, magnetics and electromagnetics have been used to map and characterise dolerite dykes and faults in platinum and coal prospects (Campbell 2006, du Plessis and Saunderson 2001). More recently, testing of a Full Tensor Magnetic Gradiometer has shown that it is particularly sensitive to dykes, and can also characterise non-magnetic dykes for Bushveld platinum prospects (Rompel 2009). Knowing the true thickness of dykes is an important part of calculating geological losses.

It is in the use of seismic reflection techniques for hard rock mining that South Africa has been a pioneer, with "what is probably the most extensive use of reflection seismics in the hard rock mining environment globally." (Pretorius 2009). 2D seismic lines proved that the major economic horizons in the Witwatersrand Basin and the Bushveld Complex are seismic reflectors or have suitable proxies. With the introduction of 3D imaging came the resolution that would enable mine planning. Early surveys such as those at Oryx (de Wet and Hall 1994) and Vaal Reefs (Pretorius et al. 2000) showed the quality that could be achieved, and later surveys prevented at least one mine planning disaster.

Pretorius (2009) states that "today, at least one phase of 3D seismic imaging would be considered mandatory on Anglo-managed mine developments in the Witwatersrand Basin and Bushveld Complex."

CONSTRUCTION

During mine construction, the most important need is for high resolution geotechnical information. The construction risk is highest at the shaft, and lowers as the development moves towards the edges of the mine. For shaft sinking, a combination of seismic surveys done during feasibility, geological logging from the shaft borehole, and geophysical wireline tools applied in the shaft borehole, provide near total coverage of the volume to be excavated for a shaft (Kaldine 2007). Borehole radar and acoustic or optical televiewer are both good tools for mapping structure near the shaft borehole: borehole radar can map out to tens of metres away from the borehole (Mabedla and Trofimczyk 2009) while a televiewer maps structure in the immediate vicinity of the borehole (Mahlatji and Krynie 2009). In one case, the shaft was relocated on the basis of geotechnical evidence, including geophysical logging, that it would run within 15 m of a low strength lamprophyre dyke (Trofimczyk and du Pisani 2009).

The other mining risk is in the positioning of surface mine infrastructure. Here, typical geophysical tools for geotechnical applications can be applied, such as gravity, electrical resistance tomography (ERT) or ground penetrating radar (GPR) to look for voids below preferred building locations. This would only be done if there were good geotechnical reasons to do so. Techniques such as gravity or ERT are also useful for determining if voids such as sinkholes are developing near or under existing buildings (Van Schoor 2002), and could prevent disasters such as the sinkhole that swallowed the reduction works at West Driefontein gold mine and killed 29 people in 1962 (de Bruyn and Bell 2001).

OPERATIONS

The cost and inconvenience of doing in-mine geophysics are barriers to its use, despite the obvious value of the information that can be obtained. Once mining is underway, the typical information requirements are for orebody grade and geometry, and for geotechnical information about the host rock. In many environments, the location of the orebody is not well known, and in-mine geophysics provides excellent guidance for mining operations, for example finding sulphide occurrences in typical distributed massive sulphide deposits. For these problems higher resolution versions of standard exploration tools can be used such as electromagnetics, magnetics or even gravity.

In South African gold and platinum mines, the overall geometry of the tabular orebody is well known, and the orebody is continuous on a large scale for kilometres in both strike and dip. Mine operators are therefore sometimes overconfident in their understanding of the geology. Mine operators, particularly on platinum mines, also have a very long life-of-mine, and are less concerned by geological losses than operators close to the end of the life of mine.

However, operators are now beginning to appreciate the value of in-mine geophysics, and are starting to introduce some techniques on a routine basis. The change has occurred as particular tools have become more reliable and easier to apply, and as mining pressures have changed. For example, GPR is particularly suitable as a tool for determining hangingwall conditions and the likelihood of hangingwall failure, and hence in determining the need for the appropriate support. In an industry increasingly concerned with safety, GPR is becoming a routine tool, and is likely to be written into the codes of practice for some mines in the near future.

A good review of high resolution techniques for in-mine use is contained in Van Schoor et al. (2006).

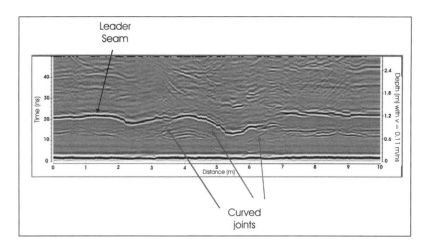

Figure 1.   A GPR image of a chromitite stringer in the hangingwall of the UG2, showing evidence of curved joints (from Van Schoor et al. 2006)

Ground Penetrating Radar

GPR is a high resolution technique that reflects short radio pulses off target horizons in rock that have different electrical properties to the host rock. The transmit and receive antennas are usually co-located in a single box. Data is collected by moving the antennas along a survey line, and presented in an image such as that in Figure 1, where position along the survey line is along the x-axis, and time of the reflection into the rock is on the y-axis. If the velocity is known, the depth of the reflection can be calculated. The colour scale in the image represents the strength of the reflectors. The image in Figure 1 is inverted, as it represents a radargram looking up into the roof of an excavation. Usually, the time and depth scales start at 0 on the top of the image, increasing downwards.

GPR works best in resistive host rocks that contain reflective target horizons. The quartzites of the Witwatersrand Basin and anorthosites of the Bushveld Complex are both highly resistive and very suitable for radar use (du Pisani and Vogt 2003). Radar resolution and range are linked to the operating frequency. At a frequency of 500 MHz, the resolution is better than 10 cm, with a range of 5 m to 8 m. At higher frequencies, better resolution comes at the cost of decreased range, and vice versa.

A good application of GPR is in the hangingwall of the Bushveld Complex (Vogt et al. 2006). There are a number of thin chromitite layers that occur in the hangingwall of the UG2 reef that are subject to weakness and that can fail if unsupported. The chromitite is an excellent reflector, making GPR a very effective tool for mapping the potential parting planes. In Figure 1, the distance to the horizon

is clearly visible. If the distance to the horizon is known accurately, it can be used to confirm that roof-bolts used for support are long enough to reach the solid rock above the unstable ground in the immediate hangingwall.

In Bushveld Complex mines, additional hazards to hangingwall stability are caused by flexural slip structures, often known colloquially as domes. GPR can detect flexural slip structures if the fault contains infill that is different to the host rock. Figure 1 shows some evidence of flexural slip structures, labelled as curved joints.

GPR is also useful for determining the location of voids in rock. In mining, one of its most practical ad-hoc uses is to find lost boreholes. When drilling from one level to another, for a drain hole, or raise bore pilot hole, or for any other purpose, if the borehole does not end up where it is expected, GPR can be used to find it easily in the sidewall, footwall or hangingwall. The conventional alternative is to drill many short holes in an attempt to reach the lost borehole. For this application, GPR can greatly reduce the search time and in an application like a raise borehole, it can save significant costs by bringing the raise into operation more quickly.

In Figure 2, a GPR survey has been used to find a lost raise bore hole. The modelled response is illustrated, together with the measured response, and the excellent correlation can be seen. The measured radargram is not of particularly good quality, with many other reflectors present due to the fractured nature of the sidewall. However, the coherent hyperbola that is characteristic of a cylindrical target is clearly visible against the noise. The hyperbola is not

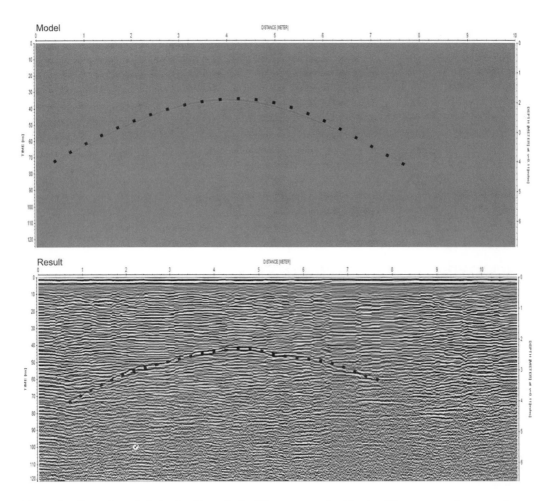

Figure 2. Detecting a hole in the sidewall of a Witwatersrand gold mine. The upper radargram shows the modelled response of a raise bore hole, while the lower radargram shows the measured response.

symmetric about its axis because the raise bore hole that it is imaging is not parallel to the survey line.

GPR is also sensitive to fractures. Often the fractures are not imaged individually, particularly if they are small and closely spaced, but the character of the GPR image changes when fracturing has occurred. In one study (Grodner 2001), GPR was used to determine how much fracturing was caused by preconditioning. Preconditioning is the practice of setting off small explosive charges ahead of the mining face, to transfer stresses away from the immediate face area. It is normally undertaken where the mining face is under high stress, as it is in deep level gold mines. In Figure 3, the character of the GPR survey before and after preconditioning shows how the fractured the rock is clearly visible. If GPR is used for fracture density determination, it must be calibrated at sites with known density, but it is particularly effective at mapping changes in fracturing, for example as in this case after blasting.

Borehole Radar

While GPR works well for geotechnical hazards, it has two problems in determining the geometry of reef geology in tabular orebodies:

- To achieve significant range of tens of metres, the frequency has to be dropped below 100 MHz. At this frequency, antennas become impractically long for use underground.
- Radar, in common with seismic reflection, works best when applied parallel or sub parallel to the horizons of interest. For perpendicular targets, only the top of the target returns a radar

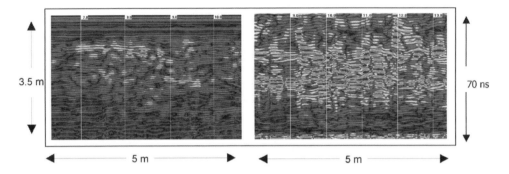

3.5 m   70 ns

◄──────── 5 m ────────►   ◄──────── 5 m ────────►

Figure 3.   GPR scans of unpreconditioned (left) and preconditioned (right) faces (Grodner 2001)

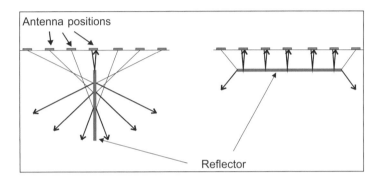

Antenna positions

Reflector

Figure 4.   Radar works best when applied near parallel to targets of interest

response, as illustrated in Figure 4. In South African tabular orebodies, there is no convenient access-way where GPR can be applied to easily map the topography of the reef.

The solution is borehole radar—GPR applied in a borehole. Once the antenna is in a borehole, its physical size ceases to be important, and the borehole can be drilled exactly where it is required to map a reef horizon. Figure 5 shows how a borehole drilled under the reef and approximately perpendicular to it can map features that are not visible using GPR from the face.

Borehole radar was first applied in South Africa in 1995, when trials were run with the Swedish Malå system. These trials showed that there are strong reflectors associated with the Ventersdorp Contact Reef (VCR) in the Witwatersrand Basin, and with both the Merensky and UG2 reefs in the Bushveld Complex. Other significant gold reefs are often associated with reflective markers, for example the Carbon Leader often has a shale layer a metre or two above in its hangingwall. The Malå system was not capable, at the time, of operating at the high

temperatures found deep underground in gold mines, and also had some logistical problems, leading to the development of two competing borehole radar systems, the South African CSIR Aardwolf (Vogt 2006), and the Australian-South African Geomole (Bray et al. 2007). Both are in regular use on South African platinum and gold mines, and Geomole is active in-mine elsewhere (e.g., Kemp et al. 2009).

In both gold and platinum, the most immediate use for borehole radar is to identify dislocations to the planar reef: faults, dykes, and potholes. These dislocations are of a size that is too small to be easily identified in 3D seismic images, but that can significantly hinder day-to-day mining operations. In particular, it is very useful to know whether the throw of a fault is less than 3 m, where mining can continue through the fault, or greater than 3 m, when redevelopment will be required. If this information can be supplied at least a month before the working face reaches the fault, planning can take into the account the stoppage that will occur when the working face reaches the fault.

In gold, the topographic information provided by borehole radar can also be used to refine the gold

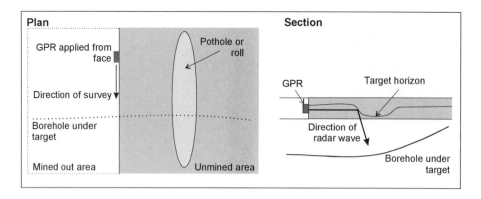

Figure 5.   How borehole radar applied under the reef and sub-parallel to it can map undulations in the target, where GPR signals from the mining face are deflected away (Vogt et al. 2006)

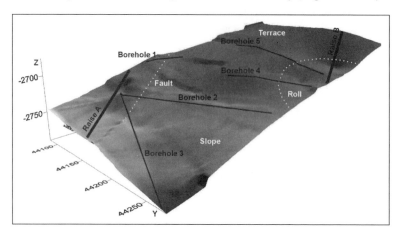

Figure 6.   A three dimensional reconstruction of the VCR, delineated by borehole radar results from five boreholes, and from survey results along two raise lines. The scale is in metres (du Pisani and Vogt 2004).

grade model (Figure 6). For example, in the VCR, the gold is associated with terraces and with low lying areas, and there is little mineralization in the slopes or "rolls." as they are known. The rolls are also associated with bad ground conditions, so borehole radar can enable planning that allows for the rolls not to be mined, making mining safer, and lowering dilution (du Pisani and Vogt 2004).

In the Bushveld Complex, potholes cause significant loss of ground, and are also associated with poor ground conditions. The potholes are slumps in the reef horizon that cut across the normal order of the stratigraphy. Borehole radar can be used to delineate the edges of known potholes. It is not straightforward to discover potholes using borehole radar, because a single radar survey gives just a single line of elevation on the reef, meaning that several

boreholes have to be drilled underneath a block of ground to ensure that no potholes escape detection. However, if a pothole is known or suspected, borehole radar can easily determine its depth and horizontal edges. Just the saving caused by not mining into a pothole unnecessarily, with related costs and dilution, is often enough to justify the use of borehole radar. A thorough examination of the financial benefit of using borehole radar in the Bushveld was undertaken by du Pisani (2007).

Electrical Resistance Tomography

ERT is a modern adaptation of the well known electrical resistivity method that uses inversion theory to produce images of conductivity versus depth for lines on surface. On surface it is relatively cheap to acquire, and usually works very well. CSIR has been experimenting with the application of ERT underground and between boreholes. When using ERT on surface, electrodes are placed on one side of the body to be imaged. In mine, electrodes can be place on two or more sides, improving resolution.

In the Bushveld Complex, if all the electrodes can be placed on the reef horizon, the apparent resistivity will increase if the reef slumps into a pothole, because the paths taken by current between the electrodes will be longer as they follow the reef horizon (Van Schoor 2005b).

A case study from Van Schoor (2005a) is illustrated in Figure 7. A block of unmined ground about 85 m long and 32 m wide was thought to be reachable on all four sides. In fact, the bottom tunnel, marked SPD on the figure was not accessible. The mine plan on the left of Figure 7 shows inferred targets based on available geological information. 24 electrodes were placed around the block, and an ERT survey was conducted. The resistivity image is on the right of Figure 7. The resistive anomaly marked A correlates well to the observations from which the pothole in the mine plan was inferred, and appears to correlate with the pothole occurrence in the mined out area at the northeast corner of the survey. The extremely resistive anomaly B is interpreted as the combined effect of the small pothole near electrodes 16 and 17, the large pothole in the southeast, and the Iron-Rich Ultramafic Pegmatoid (IRUP) in the southwest. The fact that the IRUP extends further

north than indicated in the mine plan was confirmed by a post survey underground visit.

ERT delivers images that are not of a high resolution, but for targets like potholes, it can scan a large area on reef, so it can be used for discovery. It promises to be an excellent way of quickly determining whether a mining block contains a pothole. If it does, follow-up work can be done using borehole radar to determine the details of the pothole.

Seismic Methods and Seismology

In the 1990s in-mine seismic methods were tested for use underground in South African mines. A number of techniques, including seismic refraction (Wright et al. 2000), seismic reflection, seismic tomography and an in-mine implementation of vertical seismic profiling, called mine seismic profiling or MSP were used. Unfortunately, many of the experiments do not appear to have been reported in the literature. Today, none of these techniques are in regular use, because of the logistical problems that were encountered. MSP in particular (Goldbach 2007) showed promise as a method of providing information about the reef from existing development, but was very slow and difficult to conduct. There appears to be no theoretical reason not to use these techniques, and if small and easily deployed equipment is developed, the techniques may be revisited.

Seismic monitoring is undertaken routinely in deep gold and platinum mines to determine

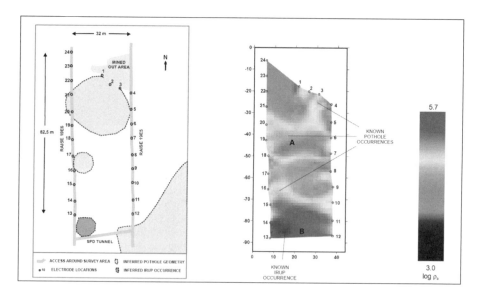

Figure 7. A case study of ERT use. The prior information and electrode locations are plotted on the geological map on the left, with the ERT image of conductivity plotted on the right (van Schoor 2005).

where seismic events occur, and to estimate risk. Seismology is not considered further here, as it is a subject in its own right and has evolved somewhat separately from the in-mine geophysics discussed here. It is reviewed extensively in a paper by Durrheim in this volume, or see Linzer (2007).

MINE CLOSURE

At the end of the life of a mine is the closure phase. In the case of good record keeping, required on mines currently operating, there is little need for geophysics, except to monitor waste dumps or slimes dams. For this purpose ERT (Rucker et al. 2009) and airborne radiometrics (Coetzee and Larkin. 2009) are particularly useful. Where older mines stopped operating without filing closure documents, there is often uncertainty as to their location. If they are shallow, ERT is an excellent tool for mapping the workings (Chirenje and Dop. 2009; Chambers et al. 2007).

The geophysics used in mine closure is not unique to mining, but uses the tools and techniques developed in other fields such as geohydrology and engineering geophysics to detect targets of interest after a mine has closed.

CHALLENGES

Pretorius (2009) raises the lack of skills in mining companies. Even at head office level, most mining companies have lost their entire geophysical skills base. This makes it very difficult to sell geophysical services to a knowledgeable client, which is a requirement when introducing a new technique. CSIR's experience is that the initial stages of developing a new technique for in-mine use require a high level of commitment from the mining company, but also some level of technical expertise to appreciate the use, limitations and problems of the technique.

As techniques become more mature, for example GPR, they become part of routine operations on the mine. At this point, the reliability of the instruments becomes important. Many geophysical instruments are not developed with the harsh in-mine environment in mind, and quickly become unreliable. In addition, the in-mine market is not sufficient to set up local support for instruments, so time is lost shipping instruments back to their overseas manufacturers for repairs. For in-mine geophysics to become viable, it is essential that manufacturers set up support infrastructure in South Africa particularly as there are local operations that do repair geophysical equipment.

Related to the challenge of reliability is the challenge of logistics. For example, during the developmental phase, a survey might take several days. While that is acceptable as part of an experiment, it is not possible in a production environment. As tools come close to production use, it is essential that the logistics behind them become as simple as possible, and preferably that they match the mines' current mode of operation.

CONCLUSION

In-mine geophysics is building on the platform created by 3D-seismics, and is starting to provide high-resolution mapping of in-mine features such as reef geometry and potentially grade, and host rock mechanical properties. The challenge for further uptake of in-mine techniques is to provide the information cost effectively and quickly enough for mining decisions to be made on the data. As service providers refine their operations, geophysics will become part of the standard arsenal of the modern mining company, at all scales from fracture mapping over metres to mapping geological structures for shaft positioning over kilometres.

REFERENCES

Anon., 2005. "Driefontein shines on—100 million ounces," in *Mining Review Africa*, No. 5, pp. 48–49.

Bray, A., Sindle, T., Mason, I.M.M., Palmer, K., Cloete, J.H., Steenkamp, J., du Pisani, P. and Trofimczyk, K., 2007. "Using slimline borehole radars from cover holes," in *Proceedings of the 10th SAGA Biennial Technical Meeting and Exhibition*, Wild Coast, 22–26 October, pp. 184–189.

Campbell, G., 2006. "High resolution aeromagnetic mapping of "loss-of-ground." features at platinum and coal mines in South Africa," *South African Journal of Geology*, December, 109, no. 4, pp. 439–458.

Chambers, J.E., Wilkinson, P.B., Weller, A.L., Meldrum, P.I., Ogilvy, R.D. and Caunt, S., 2007. "Mineshaft imaging using surface and crosshole 3D electrical resistivity tomography: A case history from the East Pennine Coalfield, UK," *Journal of Applied Geophysics*, 62: 4, pp. 324–337.

Chirenje, E. and Diop, S., 2009. "The Use of Multi-electrode Resistivity Tomography to Map Shallow Old Gold Mine Workings in the Primrose Area, Boksburg." in *Proceedings of the 11th SAGA Biennial Technical Meeting and Exhibition*, Swaziland, 16–18 September, pp. 264.

Coetzee, H., and Larkin, J., 2009. "Application of the airborne radiometric method in radiation protection applications." in *Proceedings of the 11th SAGA Biennial Technical Meeting and Exhibition*, Swaziland, 16–18 September, pp. 453–455.

De Bruyn, I.A. and Bell, F. G., 2001. "The occurrence of sinkholes and subsidence depressions in the Far West Rand and Gauteng Province, South Africa, and their engineering implications," *Environmental and Engineering Geoscience*, 7:3, pp. 281–295.

De Wet, J.A.J., and Hall, D. A., 1994. "Interpretation of the Oryx 3-D seismic survey," *South Africa Institute of Mining and Metallurgy*, 3, pp. 259–270.

Du Pisani, P., 2007. *The financial benefit of using borehole radar to delineate mining blocks in underground platinum mines*, MSc dissertation, University of Pretoria, http://upetd.up.ac.za/thesis/available/etd-02092009-141519/, retrieved 6 September 2009.

Du Pisani, P. and Vogt, D., 2003, "The radar frequency electrical properties of a selection of Bushveld and Witwatersrand Basin rocks." in *Proceedings of the Eighth SAGA Biennial Technical Meeting and Exhibition*. Kwa Maritane, Pilanesberg, South Africa.

Du Pisani, P., and Vogt D., 2004. "Borehole radar delineation of the Ventersdorp Contact Reef in three dimensions," *Exploration Geophysics 35*, pp. 319–324.

Du Plessis, S.J. and Saunderson, R. D., 2001. "The successful prediction of non-magnetic dykes using a Dighem survey." in *Biennial South African Geophysical Association Conference*, October, Alpine Heath Resort, Drakensberg, Natal.

Durrheim, R.J., 2009. "Mitigating the risk of rockbursts in the deep hardrock mines of South Africa," this volume.

Goldbach, O., 2007. "In-mine seismics," in *Proceedings of the 10th SAGA Biennial Technical Meeting and Exhibition*, Wild Coast, 22–26 October, pp. 197–203.

Grodner, M., 2001. "Delineation of rockburst fractures with ground penetrating radar in the Witwatersrand Basin, South Africa," *International Journal of Rock Mechanics & Mining Sciences*, 38, pp. 885–891.

Hartman, H.L. and Lacey, W. C., 1992. "Mineral Prospecting and Exploration," in *SME Mining Engineering Handbook, 2nd Edition*, pp. 205–209.

Kaldine, F., 2007. "An investigation into the use of integrated geophysical methods to reduce risk at a new platinum mine shaft site in the Bushveld Complex, South Africa." in *Proceedings of the 10th SAGA Biennial Technical Meeting and Exhibition*, Wild Coast, 22–26 October, pp. 60–65.

Kemp, C., Smith, K., Bray, A., Mason, I. and Sindle, T., 2009. "Imaging an Orebody Ahead of Mining Using Borehole Radar at the Snap Lake Diamond Mine, Northwest Territories, Canada." in *Seventh International Mining Geology Conference*, Perth, WA, 17–19 August.

Krahmann, R., 1936. "The geophysical magnetometric investigations on the West Witwatersrand areas between Randfontein and Potchefstroom, Transvaal." *Trans. Geol. Soc. S.A.*, 39, pp. 1–44.

Linzer L.M., Bejaichundb, M., Cichowicz, A., Durrheim R. J., Goldbach O. D., Kataka, M. O., Kijko A., Milev A. M., Saunders I., Spottiswoode, S. M. and Webb, S. J., 2007. "Recent research in seismology in South Africa," South African Journal of Science 103, September/October, pp. 419–426.

Mabedla, B., and Trofimczyk, K., 2009. "The Integration of Borehole Geophysical logs for Geotechnical Risk Assessment at the Paardekraal 13-Level Ventilation Shaft Project." in *Proceedings of the 11th SAGA Biennial Technical Meeting and Exhibition*, Swaziland, 16–18 September, pp. 179–187.

Mahlatji, M.N. and Krynie, W., 2009. "Techno-advantages of acoustic televiewer logging on shaft boreholes." in *Proceedings of the 11th SAGA Biennial Technical Meeting and Exhibition*, Swaziland, 16–18 September, pp. 172–178.

Rompel, A.K.K., 2009. "Geological Applications for FTMG," in *Proceedings of the 11th SAGA Biennial Technical Meeting and Exhibition*, Swaziland, 16–18 September, pp. 39–42.

Pretorius, C.C., Trewick, W. F., Fourie, A., and Irons, C., 2000. "Application of 3-D seismics to mine planning at Vaal Reefs gold mine, number 10 shaft, Republic of South Africa," *Geophysics*, 65, pp. 1,862–1,870.

Pretorius, C.C., 2009. "Exploration and Mining Geophysics and Remote Sensing in 2009—Where have we come from and where are we going to?." in *Proceedings of the 11th SAGA Biennial Technical Meeting and Exhibition*, Swaziland, 16–18 September, pp. 330–340.

Roux, A.T., 1967. "The application of geophysics to gold exploration in South Africa," in *Mining and Groundwater Geophysics*, reprinted in *Economic Geology report number 26*, Geological Survey of Canada, pp. 425–428.

Rucker, D.F., Glaser, D.R., Osborne, T. and Maehl, W. C., 2009. "Electrical Resistivity Characterization of a Reclaimed Gold Mine to Delineate Acid Rock Drainage Pathways," *Mine Water Environ.*, 28, pp. 146–157.

Trofimczyk, K.K. and Du Pisani, P., 2009. "Integration of Borehole Radar and Acoustic Televiewer Data in Geotechnical Boreholes—A Case Study of the Use of Downhole Geophysical Data in Mitigating Risk to a new Mining Shaft Development," in *Proceedings of the 11th SAGA Biennial Technical Meeting and Exhibition*, Swaziland, 16–18 September, pp. 149–154.

Van Schoor, M., 2002. "Detection of sinkholes using 2D electrical resistivity imaging," *Journal of Applied Geophysics*, 50(4), pp 393–399.

Van Schoor, M., 2005a. "In-mine electrical resistance tomography for imaging the continuity of tabular orebodies," *EAGE Near Surface Conference*, Palermo, Italy, 5–8 September, unpaginated.

Van Schoor, M., 2005b. "The application of in-mine electrical resistance tomography (ERT) for mapping potholes and other disruptive features ahead of mining," *Jour. South African Instt. Mining and Metall.*, v.105(6), pp. 447–452.

Van Schoor, M., Du Pisani, P. and Vogt, D., 2006. "High-resolution, short-range, in-mine geophysical techniques for the delineation of South African orebodies," *South African Journal of Science*, 102, pp. 355–360.

Vogt, D., 2006. "Borehole radar system for South African gold and platinum mines." *South African Journal of Geology*, 109, pp 521–528.

Vogt, D., van Schoor, M. and du Pisani, P., 2006. "The application of radar techniques for in-mine feature mapping in the Bushveld Complex of South Africa," in *The Journal of The South African Institute of Mining and Metallurgy*, 105, 391–399.

Walker, W.M., 1950. "The West Wits Line," in *South African Journal of Economics*, 18, pp. 16–35.

Wright, C., Walls, E.J. and Carneiro, D. de J., 2000. "The seismic velocity distribution in the vicinity of a mine tunnel at Thabazimbi, South Africa," *Journal of Applied Geophysics* 44, pp. 369–382.

Continuous Assessment of Rockfall Risk in Deep Tabular Mines

Van Zyl Brink

CSIR, Johannesburg, South Africa

ABSTRACT: Rockfalls in the deep tabular mines in South Africa are a large contributor to underground accidents. The CSIR, South Africa, is developing technology to provide a continuous assessment of the rockfall risk at a 0.5 meter spatial resolution. A risk contour of the hangingwall/roof may potentially be determined prior to, and updated during, each underground shift.

The technology comprises a neural network-based sounding device (for roof sounding prior to barring); thermal imaging for the detection of potentially loose rock; underground localization/positioning; and general geotechnical instrumentation. The in-stope sensors are deployed in a Zigbee wireless sensor network. The implementation of a mobile sensor system on an autonomous or remote control platform is at an early stage.

This paper describes the embedded concept; the development process; the subsequent implementation and evaluation of the technologies providing continuous risk assessment of hangingwall/roof stability.

INTRODUCTION

Rockfalls are defined as an *uncontrolled fall (detachment or ejection) of ground of any size that causes (or potentially causes) injury or damage* (Minerals Council of Australia, 2003). Rockfalls pose unacceptable risks to the South African mining industry. Using a methodology described by Terbrugge et al. (2006) and data published by the Department of Minerals and Energy (2009), a fatal injury rate for South African miners exposed to rockfalls of worse than 10^{-3} is obtained. This is at least ten times higher

Figure 1.   Preparation for blasting after hand operated drilling—an image taken in a conventional deep level stope (from Chamber of Mines of South Africa Annual Report 2005–2006)

than the acceptable fatal injury rate for voluntary risk as stated by Wong (2005).

The nature of the orebody, particular in gold and platinum, is that of a thin tabular seam in a hard rock environment. The mining process is mainly based on conventional drilling and blasting (as depicted in Figure 1), resulting in a significant number of mining personnel being exposed to fall of ground hazards.

The 2004–2005 Mine Health and Safety Council annual report provided the information used in the following interpretation (MHSC, 2005). Information from this annual report was supplemented with information from an unpublished internal (CSIR) database of all rockfall and rockburst accident investigations for the period 1990 to 2000.

Figure 2 and Figure 3 show that gold mining contributes the greatest number of fatal rockfall accidents in terms of both risk and absolute numbers.

A significant component of managing rockfall risk is embedded in current mining 'best practices'. It is empirical knowledge based on many decades of mining experience. This paper is not arguing against the validity or the merits of this approach. It is, however, suggesting a sensing and monitoring strategy to assess and quantify rockfall risk. This should greatly enhance the miner's ability to manage rockfall risk whilst being exposed to a fractured and non-homogeneous hanging/roof subject to gravity and stress driven deformation.

MECHANISM OF ROCKFALLS

For the purpose of this paper and specifically for making a distinction in early warning risk parameters and

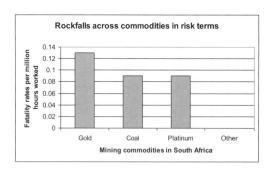

Figure 2.   Rockfall fatal incidents expressed in terms of fatality rate per million hours worked

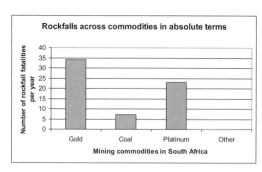

Figure 3.   Rockfall fatal incidents expressed in terms of number of fatalities

the monitoring thereof, some general differentiation between the mechanisms of falls of ground is made:

- Small falls—These rockfalls are approximately only 25kg and larger, but still have the potential to result in a serious injury or fatality (Stacey and Gumede, 2006). These falls typically occur between support units or in the unsupported area of a newly exposed hangingwall within meters from the face (Brink and Roberts, 2007). In most cases the disturbing force or trigger, is only gravity, but could also include ejection due to local face bursting; or even shakedown from nearby seismic events.
- Large falls—Rockfalls with a mass exceeding hundreds of tons, occasionally happen in platinum mines (see an example of such a fall in Figure 4). These falls are exclusively gravity driven and occur in a particular geological environment with a dome-shaped thrust plane intersecting vertical mine induced fracturing.
- Strong ground motion driven falls—These falls of ground are directly associated with mining induced seismicity. The number of fatal incidents caused by this group is in the same order as gravity driven rockfalls. An example of such a strong ground motion driven fall of ground is given in Figure 5.

FAILURE OF DESIGN AND/OR PROCESS

Stacey (1989) states that one or more of three situations must be present for a fall of ground:

- Rocks are falling where there is no support
- Rocks are falling between supports
- Rockfalls involve failure of the installed support system

Figure 4.   Example of a large 'dome' failure in a platinum mine

Stacey (1989) further states that in all three cases the design of the support system must be inadequate and therefore an improved or different design approach is necessary to correct the situation and ensure safe working conditions. This may be summarised by attributing all falls of ground to a failure in the support design, or to a failure in the process of supporting. The first mentioned situation may be attributed to inappropriate and inadequate support design whilst the latter two may be caused by non-adherence to the standards and/or non-conformance with the procedures of the mine.

Rockfall accident investigations are often blessed with perfect hindsight with respect to whether the reason for the fall is a failure of design and/or a failure in adhering to the practices and procedures of the mine. The ability to continuously assess the risk of failure of either the support system or any non-compliance will greatly empower the decision maker to act pro-actively.

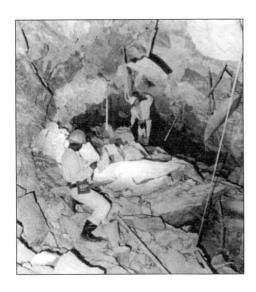

Figure 5.   A haulage devastated by strong ground motion (picture from David Ortlepp)

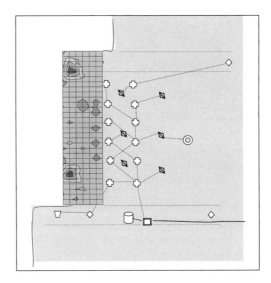

Figure 6.   The suggested visualization of in-stope rockfall risk assessment

CONTINUOUS RISK ASSESSMENT OF ROCKFALL RISK

Standards Australia and Standards New Zealand (2004) defines risk management as '*the systematic application of management policies, procedures and practices to the tasks of identifying, analyzing, treating and monitoring risk.*"

CSIR's model to determine in-stope rockfall risk is depicted conceptually in Figure 6. The figure shows a stope on the left, with a number of sensors in a network, indicated by the symbols. The network is connected to a local computer at middle bottom that further communicates with a computer on surface. The product of all the data from the sensors and other sensors elsewhere in the mine is the colour map superimposed on the working place, indicating the rockfall hazard as a function of position.

To achieve in-stope management of rockfall risk the following basic system characteristics are required:

- The ability to recognise that part of the hangingwall that has the potential to cause harm, i.e., the recognition of rockfall hazard
- The ability to recognise the likelihood, or probability, that some specific harm arising from an incident will occur, i.e., the risk of a rockfall accident
- The ability to, continuously and in semi real-time, do both of the above

The project team came to the conclusion that wireless sensor networking is the required technology to monitor a range of parameters that will allow the recognition of in-stope hazard and the assessment of risk because of the extreme difficulties of maintaining a wired sensor network. A multidisciplinary team from the CSIR, South Africa has developed a scalable risk/hazard monitoring system, based on the wireless sensor network concept, to be applied in deep and high-stress mining environments. This system, called AziSA, is a non-propriety and open standard specification that includes a mining data-acquisition, communication and decision-support system, with the internationally accepted concept of a wireless sensor network embodied in the system (Stewart et al. 2008).

Existing knowledge limits the observer to the following options for detecting possible precursors:

- Direct measuring of movement or deformation (and specifically differential movement)
- Measuring of the rate of change of such movement
- Sensing of the energy released in fracturing, friction on loss of cohesion
- Sensing of the rate of the change in the fracturing, friction on loss of cohesion process
- Observing of a differential change in temperature relative to the surrounding rock mass
- Sensing of differential support loading
- Quantification of rock mass rating and recognition of changes

Figure 7. An analogous situation to early warning

Figure 8. An analogous situation to probabilistic risk assessment

Early Warning Versus Probabilistic Risk Assessment

This project makes a clear distinction between a possible early warning alarm of an impending rockfall and a probabilistic rockfall risk assessment. A basic analogy for describing the distinction between early warning and risk assessment can be found in the 'domino effect' as illustrated in Figure 7 and Figure 8. The first example (in Figure 7) shows the first domino being pushed over. Should an observer be able to sense the force applied, the rotational movement, etc. he/she would be able to provide an early warning of an impending instability of the series of dominoes. The second example (in Figure 8) shows the same domino sequence, but from a different perspective—the observer does not see the first domino in the sequence. In this case the observer cannot provide an early warning of the instability as, within his or her view, no precursor exits.

The observer, however, may observe the orientation of the dominoes; the critical distance apart; and the metastability of each domino. On this basis, the observer may assess the probability that a small disturbance may cause a large instability.

The early warning of rockfalls, specifically of a single block or slab that may come down, is difficult. Early warning may require the direct monitoring of, say, the keyblock in a hangingwall failure. Risk assessment allows for remote sensing, the inference of observations to failure symptoms.

Strategy/Methodology for Monitoring of Rockfall Risk

Wireless Sensor Network Concept

A sensor network in the AziSA system is a coordinated monitoring infrastructure comprising a distributed combination of resources, e.g., sensors, sensor platforms, communication infrastructure and data management and decision support. In order to add meaning and value to a simple sensor measurement, certain attributes have to be associated with the specific measurement. Every measurement has to have information associated with the sensor such as identification; type of sensor; its sensing limits; time and date; and geographic position. Data without this associated information is of no value.

All the in-stope sensors in the AziSA system form a wireless sensor network (WSN), which is a computer network consisting of spatially distributed autonomous devices using sensors to cooperatively monitor physical conditions. The sensors are also low power and battery driven (Römer and Mattern, 2004).

The AziSA specification makes it easier to integrate underground sensors and actuators into monitoring and data networks. It allows for four classes of devices, ranging from small wireless sensors to intelligent controllers. A Class 4 or 3 device is typically a relatively simple sensor which sends measurements to an aggregator, called a Class 2 device. Normally these sensors are part of a low-power wireless network using the Zigbee communication protocol. The difference between Class 4 and Class 3 is that a Class 3 device can make decisions on its own data, and is capable of operating in standalone mode and of raising alerts. Both Class 3 and 4 are sensor platforms that may contain different detectors. A good example of a Class 4 device would be a basic sensor which measures closure or support loading. A Class 3 device would be something like a microseismic monitor which measures some conditions and can recognise a local alarm state.

The Class 2 device is an aggregator that controls a sub-network of sensors, possibly making decisions based on their data, and sending the data to a

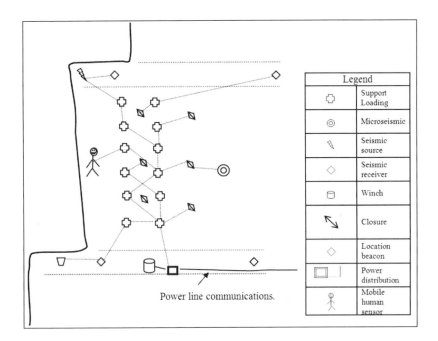

Figure 9. **A schematic indication of the wireless sensor installation in a stope, showing the Zigbee in-stope communication and the power line communication to the shaft**

Class 1 device (the controller) on surface. Typically an active panel may have a number of Class 3 and 4 monitoring devices and one Class 2 aggregator. A Class 1 device is the system controller, responsible for managing several Class 2 devices, storing data over a long time and presenting information for decision making to the outside world. Typically it would manage a database system and be situated on surface.

Positioning/Localisation

As mentioned above, in order to add meaning and value to a simple sensor measurement that exists in AziSA, certain attributes have to be associated with the specific measurement.

One of these attributes is the location at which the measurement was taken. A battery-operated location device has been developed that will operate in a noisy and enclosed space and will determine two-dimensional coordinates relative to a number of beacons with known locations. A study was conducted to gain insight into which existing methods would be most applicable and practical for this particular application and, it was decided to use a trilateration system to solve for the location of a measurement. The location device relies on at least three beacons that are placed around the mining stope. Each beacon simultaneously transmits an electromagnetic pulse and an acoustic pulse. This allows the sensor

to resolve the acoustic travel time and with similar information from two other beacons the sensor's position can be resolved in two dimensions (Ferreira 2008).

Communication

In-Stope Communication. Initial investigation showed that in-stope communication would have to be wireless. The environment is simply too hostile to consider wired sensors. Investigation into short-range low power wireless networks revealed two major players: the TinyOS system from UC Berkeley and Zigbee, which is an open standard. It was decided that Zigbee would be the protocol of choice for in-stope communication as it was poised for a major uptake which would make it commercially attractive. The AziSA message scheme was implemented over Zigbee on the small wireless sensors (Class 3 and 4) so that they could interface with the rest of the AziSA network via the Class 2.

Figure 9 shows a schematic representation of in-stope communication. Each sensor was converted to a Zigbee sensor, which allows for wireless sensor networking. Each sensor may act as a network router and network paths are dynamically determined and are automatically re-arranged should any sensor fail. At the nearest electrical power, an aggregator (Class 2 device) is placed that functions as a bridging device between the in-stope Zigbee and the power line communication (or any other Internet Protocol communication) to the shaft area. Figure 9 also shows a mobile sensor, in this case a sensor carrier by a miner.

Stope to Shaft Communication. Extensive mining in a tabular orebody leads to long distances from stope to the mining shaft. In many cases no fast communication cabling exists and it is envisaged that Power Line Communication (PLC) will be used to take data back to the shaft. PLC essentially transforms the power grid into a network carrying broadband communication in the form of an IP (Internet Protocol) signal. Commercial equipment is to be used and an evaluation of such equipment in an underground environment is being undertaken. Where fast communication infrastructure exists, PLC communications will switch over to the mine network as close to the stope as possible.

Data Management and Decision Support. The Class 2 aggregator will be mounted at the closest power point to the stope, typically the power supply to the winch. It is a relatively basic computing device, and will support no long-term data storage. The configuration of the sensors in the network for which it is responsible will be stored locally. Sufficient data reported by the sensors will be stored locally to enable the Class 2 to make the necessary decisions with which it is tasked, and to buffer data for transmission to the Class 1 controller. If the communication network fails, the Class 2 can continue to take measurements for a period of days to weeks until its own storage is filled.

The main data management is undertaken on the Class 1 controller, which is being implemented on standard server class computers and thus has more resources available to it. The Class 1 receives data from Class 2 aggregators and transfers it to the data management system. The Class 1 also responds to user queries by retrieving data from the data management system and presenting this to the user in an appropriately meaningful and useful form.

The Class 1 will typically be networked with a number of application servers that will register themselves with the Class 1 to be notified about incoming data of particular types. The application servers will each apply the decision support algorithms required for their particular application, and will also communicate to the people operating the system, or to the miners, as required. Communications in the AziSA system is fully bidirectional, so application servers can use the network to pass alarms all the way back to the stope.

When a sensor joins an AziSA network, it registers with the system controller, and is assigned a unique identifier for future reference. This registration will include the identification and position of the sensor and its component detectors. Each detector will be associated with a physical phenomenon concerning which measurements are made, and with information relating to its calibration and for the processing of raw data values to derive information about the particular phenomenon. The detector will also be registered as reporting data in a particular form: a physically stationary device might send discrete measurements (e.g., single temperature or humidity values) or bursts of measurements (e.g., images or seismic waveforms), while a mobile device might report data from a different physical location on each occasion, and provision is made for the storage of all such data in the Class 1. The Class 1 design is such that it can store data that is not defined when the system is installed, as long as it complies to the AziSA standard. There is no requirement to change the Class 1 to install new types of sensor.

Decision support is primarily based on a probabilistic risk assessment methodology. A similar methodology was used at Moonee Colliery for goaf warning (Brink and Newland, 2001) and in assessing rockburst and rockfall risk for the MHSC (Brink et al., 2002a).

Development of Probabilistic Risk Assessment Using an Expert System Philosophy

Bayes' Theorem was adopted as an expert system philosophy for a rock engineering risk assessment project (Brink, 2005). The basic concept of this philosophy is that a hypothesis should be developed that is tested against some evidence. In this case, the hypothesis is that a rock-related incident (rockfall or rockburst damage) will occur within the next eight hours. A prior probability, $P(H)$, is set to be the probability that such a rock-related incident may occur during the next eight hours (the probability of the hypothesis being true) without any associated evidence.

Evidence associated with the hypothesis being true is, for example, the observation that a high seismic event rate is prevailing. The probability of observing such evidence while the hypothesis is also true, $p(E|H)$, is determined from previously observed information. Similarly, the probability of observing the evidence, but with a hypothesis that is not true, $p(E|\sim H)$, is set.

On the basis of Bayes' Theorem, the probability of observing the hypothesis H if the evidence E has been observed is the probability $p(H|E)$.

$$p(H \mid E) = \frac{p(E \mid H)p(H)}{p(E \mid H)p(H) + p(E \mid \sim H)p(\sim H)} \tag{1}$$

where *p(H)* is the prior probability of the hypothesis being true and *p(~H)* is the prior probability of the hypothesis not being true.

Similarly, the probability of observing the hypothesis H if the evidence E has *not* been observed is *p(H|~E)*:

$$p(H \mid \sim E) = \frac{p(\sim E \mid H)p(H)}{p(\sim E \mid H)p(H) + p(\sim E \mid \sim H)p(\sim H)} \tag{2}$$

The initial importance (or Rule Value) of each item of evidence is given as:

$$RV = \mid p(H \mid E) - p(H \mid \sim E) \mid \tag{3}$$

A typical expert system will have a number of possible hypotheses and even more items of evidence that can be associated with these hypotheses. For the purpose of testing the hypothesis of a large rockfall and/or burst within the next eight hours, the expert system only considers this single hypothesis, tested against the available items of evidence.

The main advantage of employing an expert system approach is that it can cater for uncertainties. An expert system can also allow for the actual response to whether specific evidence is observed to between a 'yes' or a 'no'. (Naylor, 1998).

Initially, the hypothesis starts off with the given prior probability. This prior probability is updated by the application of one item of evidence, which may or may not be observed with a given degree of uncertainty, to produce a new posterior probability. This new posterior probability is then used as the new prior probability for a re-application of the same equation to the next item of evidence. The process continues until all the items of evidence have been accounted for.

Monitoring Strategy

The planned monitoring strategy to provide early warning and/or risk assessment of rockfalls makes a clear distinction between the early warning and risk approaches. This section describes the measuring/monitoring options that are to be implemented.

Early Warning Sensing

Parameters that may provide short-term indications of an impending rockfall are the sensing of the energy released or the rate of change in fracturing and friction on loss of cohesion. A practical way to monitor the energy released in fracturing, etc. is the monitoring of the acoustic emissions or microseismicity. Differential movement of part of the hangingwall (or roof) and specifically the rate of change of such movement has the potential to be used as an early warning input.

Microseismic Monitoring. An established and proven technology in coal mining is the GoafWarn device, a stand-alone microseismic unit with internal alarm condition recognition (Brink et al., 2002b). Application of this device is to provide a local alarm of imminent goafing. Figure 10 shows an example

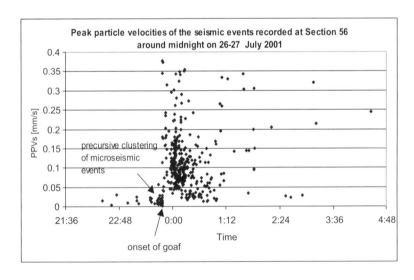

Figure 10. **An example of the precursive clustering of microseismicity immediately prior to the onset of a large goaf**

of precursive clustering of microseismicity immediately prior to the onset of a large goaf. The recognition of the precursive cluster is the basis of operation of this device. The warning period ranges from approximately two hours to even less than a minute. This technology was recently tested in hard rock mining with promising early results.

Differential Movement. An ability to monitor differential movement in the hangingwall of a tabular stope may greatly assist in the recognition of precursory movement or deformation prior to a rockfall. In open stopes or in surface slope stability, laser or radar scanning is an ideal monitoring technique. The closeness of the scanner to the rock surface to be scanned, limits the effective scan surface underground and as such face scanning is probably not practical in a tabular orebody.

Electronic Sounding. The electronic sounding device is a new development that allows for the cognitive decision-making process in sounding of the rock by an operator to be duplicated in an electronic device. The device is illustrated in Figure 11. The sounding device may be part of the mobile human sensor as shown in Figure 9. It gives an immediate indication of the degree of detachment of the specific part of the rock mass to the miner and allows for immediate action in terms of 'making safe.' It is also an AziSA device in that frequency response (Figure 12.) as well as the two-dimensional position is transmitted to the central database. In principle, a supervisor of the dayshift (production shift) is in the position to have a full risk assessment contour, based on the sounding process, at his disposal prior to the start of the shift

Thermal Image Sensing. Thermal imaging through infrared sensing has the proven ability to sense portions of the rock mass that might have cooled down faster from the virgin rock temperature to the temperature of the air ventilation, schematically shown in Figure 13. Faster cooling is an indication of the degree of detachment of a specific part of the rock mass. shows a greyscale example of such a thermal image.

Stationary infrared sensing does not make much sense due to its limited coverage of the hangingwall so a current development is the integration of infrared sensing with the sounding device. Each tap on the hangingwall will be associated with an infrared image of that particular part of the rock mass. Both sets of data, plus the geographic positioning, are to be transmitted to the central database and will act as input parameters into the risk assessment process.

Closure Measurement. Continuous closure monitoring is included in the monitoring strategy. A number of wireless closure sensors are to be installed in each panel in order to quantify the overall closure profile. Malan (2003) showed the value of continuous closure measurements for improved support design and hazard assessment. Figure 9 schematically showed the incorporation of closure measurement in the stope risk assessment. A relatively inexpensive closure meter with a resolution of 0.01 mm has been designed and implemented.

Support Loading. Shield loading on the longwall shearer in coal mining is routinely monitored. By analogy, in this application, the loading on individual support elongates will be monitored. It is foreseen that differential loading of support will provide input into the rockfall risk quantification. The

Figure 11.   The electronic sounding device for hangingwall stability

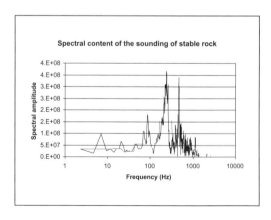

Figure 12.   Output of the recorded response to sounding in the frequency domain and used as input in the neural network

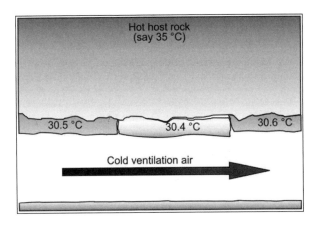

Figure 13.   An illustration of the concept employed in the thermal imaging recognition of a potential rockfall hazard

support loading sensing is an AziSA function and again will be implemented with wireless continuous monitoring.

Correlation Between Seismic Velocity and Rock Mass Rating. Sjøgren et al. (1979) state that their studies have confirmed earlier assumptions that there is a strong correlation between longitudinal velocity and fracturing and that the velocities can be used to give rather accurate predictions of the quality of rock masses. As shown in Figure 9, a seismic source and receiver are placed across the panel as close as possible to the face. Regular seismic velocity measurements will be made to infer the quality of the rock mass.

CONCLUSIONS

This paper describes a current research project to identify early warning and risk assessment of imminent rockfalls in South African mines. The mechanisms of falls of ground and risks were characterised in broad terms; possible precursive parameters and the appropriate sensing technology were described. Known technologies have been adopted and integrated in a system based on the wireless sensor network concept. The validity of individual monitoring tools has been established but the system integration; the real-time and wireless sensing implementation of the system; and overall system effectiveness in providing early/risk assessment have yet to be established.

REFERENCES

Brink, A.v.Z. and Newland, A.R. (2001). Automatic real time assessment of windblast risk. ACARP project C8026, Australian Coal Association Research Programme.

Brink, A.v.Z. (2005). Technology transfer of the risk assessment tool for rock-related risk issues. SIM 03 02 01, Department of Mineral and Energy Affairs, South Africa.

Brink, A.v.Z., Roberts, M.K.C. and Spottiswoode, S.M. (2002a). Further assessment of seismic hazard/risk in the Bushveld Complex platinum mines and the implication for regional and local support design. SIMRAC Report GAP 821, Department of Mineral and Energy Affairs, South Africa.

Brink, A.v.Z., Canbulat, I. and Spottiswoode, S.M. (2002b). Development and evaluation of a warning device for pending roof instability in coal mines. Proceedings of the SANIRE 2002 Symposium—'Re-Defining the Boundaries'.

Brink A.v.Z. and Roberts M.K.C. (2007). Early warning and/or continuous risk assessment of rockfalls in deep South African mines. 4th International Seminar on Deep and High Stress Mining, Perth, Australia, 7–9 November 2007.

Department: Minerals and Energy, Republic of South Africa (2009), Annual Report.

Ferreira, G. (2008). An implementation of ultrasonic time-of-flight based localization, 2nd International Conference on Wireless Communication in Under-ground and Confined Areas, Val-d'Or, Quebec, 25–27 August.

Malan, D.F. (2003). Experimental validation of a mine-wide continuous closure monitoring system as a decision making tool for gold mines. SIMRAC Report GAP 852, Department of Mineral and Energy Affairs, South Africa.

Naylor, C. (1998) XMaster®: An Easy-to–use expert system shell for Windows. Expert update: The Bulletin of the British Computer Society's Specialist Group on Knowledge-Based Systems and Applied Artificial Intelligence, Autumn 1998 Volume 1 Number 2.

MCA (Mineral Council of Australia) (2003). Management of Rockfalls Risk in Underground Metalliferrous Mines, A Reference Manual developed by the Australian Centre for Geomechanics, Minerals Council of Australia, Canberra.

MHSC (2005). Mine Health and Safety Council, South Africa, Annual Report, April 2004–March 2005.

Römer, K. and Mattern, F. (2004). The Design Space of Wireless Sensor Networks. IEEE Wireless Communications, vol. 11 (6), pp. 54–61.

Sjøgren, B., Øfsthus, A. and Sandberg, J. (1979). Seismic classification of rock mass qualities, Geophysical Prospecting, vol. 27, pp. 409–442.

Stacey, T.R. (1989). Potential rockfalls in conventionally supported stopes—a simple probabilistic approach, J S. Afr. Inst. Min. Metall., vol. 89, no. 4, April 1989, pp. 111–115.

Stacey, T R and Gumede, H. (2006). Evaluation of risk of rock fall accidents in gold mine stopes based on measured joint data, Jl S. Afr. Inst. Min. Metall., vol. 107, no. 5, pp. 345–350.

Standards Australia and Standards New Zealand (2004) AS/NZS 43360:2004. Sydney. NSW. ISBN 0 7337 5904 1.

Stewart, R, Donovan, SJ, Haarhoff, J and Vogt, DR. (2008). AziSA: an architecture for underground measurement and control networks, 2nd International Conference on Wireless Communications in Underground and Confined Areas, Val-d'Or, Quebec, Canada, 25–27 August 2008, pp 4.

Terbrugge, P.J., Wesseloo, J., Venter, J. and Steffen, O.K.H. (2006). A risk consequence approach to open pit slope design, J S. Afr. Inst. Min. Metall., vol. 106, no. 7, 2006. pp. 503–511.

Wong, W. How did that happen?, Professional Engineering Publishing Limited, London and Bury St Edmonds, UK. 2005.

Pillar and Roof Span Design in Stone Mines

Gabriel S. Esterhuizen
NIOSH, Pittsburgh, Pennsylvania, United States

Dennis R. Dolinar
NIOSH, Pittsburgh, Pennsylvania, United States

John L. Ellenberger
NIOSH, Pittsburgh, Pennsylvania, United States

ABSTRACT: Underground stone mines in the United States use the room and pillar method of mining in relatively flat laying deposits. Roof and rib falls account for 15% of all lost time incidents. This paper presents the result of a study to improve pillar and roof span design methods so that unexpected ground instabilities will be reduced. A pillar design equation and related design guidelines are presented based on the observation of pillar performance at 34 different underground operations. A roof span design procedure is also proposed that systematically addresses the main issues related to roof instability. The guidelines are based on the observation of actual pillar and roof span performance in stone mines in the Eastern and Midwestern United States and therefore only apply to stone mine design within these regions.

INTRODUCTION

Underground limestone mines in the United States use the room-and-pillar method to extract limestone formations that are generally flat lying. Pillar stability is one of the prerequisites for safe working conditions in a room-and-pillar mine. Unstable pillars can result in rock sloughing from the pillar ribs and can lead to the collapse of the roof if one or more pillars should fail. In addition, the roof span between pillars should be stable to provide safe access to the working face. Falls of ground from unstable roof and pillar ribs account for about 15% of all lost working days in underground limestone mines (Mine Safety and Health Administration 2009). In the past, pillar and roof span dimensions were largely based on experience at nearby mines, developed through trial and error or designed by rock engineering specialists. This paper presents the results of research carried out by the National Institute for Occupational Safety and Health (NIOSH) in cooperation with participating underground stone mines to develop guidelines for designing stable pillars and roof spans in stone mines. The research is based on the observation of failed and stable pillars and roof spans at 34 operating stone mines supplemented by laboratory testing and monitoring of rock mass response.

STONE PILLAR DESIGN

In a room-and-pillar mine, the pillars are required to provide global stability which can be defined as providing support to the overlying strata up to the ground surface. In addition, local stability, which is defined as stable ribs and roof spans between the pillars, is required to provide safe working conditions. Pillar design typically estimates the pillar strength and the pillar stress, and then sizes the pillars so that an adequate margin exists between the expected pillar strength and stress. The factor of safety (FOS) relates the average pillar strength (S) to the average pillar stress (σp), as follows:

$$FOS = \frac{S}{\sigma_p} \qquad (1)$$

When designing a system of pillars, the FOS is critical to stability, because it must compensate for the variability and uncertainty related to pillar strength and stress and varying dimensions of rooms and pillars. The selection of an appropriate safety factor can be based on a subjective assessment of pillar performance or statistical analysis of failed and stable cases (Salamon and Munro, 1967; Mark, 1999; Salamon et al., 2006). As the FOS decreases, the probability of failure of the pillars can be expected to increase. In practical terms, if one or more pillars are observed to be failed in a layout, it is an indication that the pillar stress is approaching the average pillar strength. The relationship between FOS and failure probability, however, depends on the uncertainty and variability of the system under consideration (Harr, 1987).

Table 1. Uniaxial compressive strength of limestone rocks collected at mine sites

Group	Average MPa (psi)	Range MPa (psi)	Samples tested	Representative limestone formations
Lower Strength	88 (12,800)	44–144 (6,400–20,800)	50	Burlington, Salem, Galena-Plattesville
Medium Strength	135 (19,600)	82–207 (11,900–30,000)	100	Camp Nelson, Monteagle, Plattin, Vanport, Upper Newman, Chickamauga
High Strength	220 (31,800)	152–301 (22,000–43,700)	32	Loyalhanna, Tyrone

Table 2. Summary of mining dimensions and cover depth of mines included in study

Dimension	Average	Minimum	Maximum
Pillar width, m (ft)	13.1 (43.0)	4.6 (15.0)	21.5 (70.5)
Pillar height, m (ft)	11.1 (36.5)	4.8 (15.8)	40.0 (124.6)
Width-to-height ratio	1.41	0.29	3.52
Cover depth, m (ft)	117 (385)	22.9 (75)	670 (2,200)

PILLAR STABILITY ISSUES IN STONE MINES

A survey of stable and unstable pillars in underground stone mines within the Eastern and Midwestern United States identified the causes of pillar instability to provide data for estimating pillar strength (Esterhuizen et al., 2006). Mines that were likely to have unstable pillars owing to their depth of working or size of pillars were identified as targets for the survey. Data were collected that included both the intended design dimensions and the actual pillar and room dimensions in the underground workings. In older areas of mines, where the original design dimensions were unknown, the measured dimensions were assumed to adequately represent the design. The approximate number of pillars in each layout was recorded to the nearest order of magnitude and the depth of cover determined from surface topography and mine maps. The Lamodel software package (Heasley and Agioutantis, 2001) was used to estimate of the average pillar stress in cases where the tributary area method was considered inappropriate. Data from one abandoned limestone mine, which was not observed as part of this study, was added to the records owing to its great depth and reported stable conditions (Bauer and Lee, 2004).

At each mine rock samples were collected to determine the uniaxial compressive strength (UCS) and the rock mass was classified using the Rock Mass Rating (RMR) system (Bieniawski, 1989; Hoek et al., 1995). The UCS results were grouped into three categories based on the average strength obtained at the individual mine sites and are shown in Table 1. The data shows that there is a considerable variation in the intact rock strength of the formations mined. The RMR varied between 60 and 85 out of a possible 100, which indicates the relatively strong rock mass conditions found in these mines.

All the pillar layouts surveyed could be considered to be successful in providing global stability by supporting the overburden weight up to the ground surface, (Esterhuizen et al., 2008). However, not all the pillar layouts were fully successful in providing local stability in the form of stable roof spans and pillar ribs. A total of eighteen cases of single unstable pillars among otherwise stable pillars were observed. These failed pillars are a small percentage of the more than one thousand pillars that were directly evaluated. Roof instability was observed at almost every mine, but was not necessarily widespread at the individual mines. Table 2 summarizes the dimensions and cover depth of the pillar layouts that were investigated.

The survey further showed that the following factors can contribute to pillar failure or instability:

- Large angular discontinuities that typically extend from roof-to-floor in a pillar. Sliding can occur along these discontinuities which can significantly weaken these slender pillars (Esterhuizen, 2006). Of the 18 unstable pillars observed, seven were affected by these large angular discontinuities. Figure 1 shows a pillar that is weakened by two angular discontinuities that contributed to failure of the pillar at a relatively low pillar stress.

- Weak bands within pillars that can extrude resulting in progressive spalling of the pillar ribs (Esterhuizen and Ellenberger, 2007). Figure 2 shows a pillar that has been severely compromised by this mechanism of failure. It

Figure 1. Partially benched pillar that failed along two angular discontinuities. Width-to-height ratio is 0.58 based on full benching height and average pillar stress is about 4% of the UCS.

Figure 2. Pillar that had an original width to height ratio of 1.7 failed by progressive spalling. Thin weak beds are thought to have contributed to the failure. Average pillar stress is about 11% of the UCS.

appears that moisture on the weak beds was a contributing factor in this failure. Other pillars in the immediate surrounding area were unaffected, while a number of other pillars at this mine appeared to have been affected in the same manner.

High pillar stress caused by deep cover or high extraction ratios can cause spalling of the pillar ribs (Lane et al., 1999; Krauland and Soder, 1987; Lunder, 1994; Pritchard and Hedley, 1993). It was found that spalling can initiate when the average pillar stress exceeds about 10% of the uniaxial compressive strength of the pillar material. Pillars tend to take on an hourglass shape when spalling initiates. Figure 3 shows a pillar that has failed and taken an hourglass shape due to rib spalling. Figure 4 is a series of pillars at one of the deeper stone mines showing the effects of minor rib spalling.

The observed unstable pillars were typically surrounded by pillars that appeared to be stable, showing minimal signs of disturbance. Therefore, the failed pillars represented a very small percentage (typically less than 1%) of the total number of pillars at any particular mine. The observations lead to the conclusion that the failed pillars represent the low end of the distribution of possible pillar strengths, and not the average pillar strength. As a result, the average safety factor of the layouts containing the failed pillars can be expected to be substantially higher than that of the failed pillars.

PILLAR STRESS AND STRENGTH ESTIMATION

The stability of a pillar can be evaluated by comparing the average stress in the pillar to its strength. The average stress in pillars that are of similar size and are located in a regular pattern can be estimated with relative ease using the tributary area method. The overburden weight is simply assumed to be evenly distributed among all the pillars. The average pillar stress (σp) is calculated as follows:

$$\sigma_p = \gamma \times h \times \frac{(C_1 \times C_2)}{(w \times l)} \qquad (2)$$

Where γ is the specific weight of the overlying rocks, h is the depth of cover, w is the pillar width, l is the pillar length and C_1 and C_2 are the heading and cross-cut center distances respectively. This provides an upper limit of the pillar stress and does not consider the presence of barrier pillars or solid abutments that can reduce the average pillar stress. In conditions where the tributary area method is not valid such as irregular pillars, limited extent of mining or variable depth of cover, numerical models such as Lamodel (Heasley and Agioutantis 2001) can be used to estimate the average pillar stress.

Figure 3. Partially benched pillar failing under elevated stresses at the edge of bench mining. Typical hourglass formation indicating overloaded pillar.

Figure 4.   Stable pillars in a limestone mine at a depth of cover of 275m (900 ft). Slightly concave pillar ribs formed as a result of minor spalling of the hard, brittle rock.

Estimating pillar strength is more difficult and has been the subject of much research in the mining industry (Hustrulid, 1976). Owing to the complexity of pillar mechanics, empirically based pillar strength equations, that are based on the observation of failed and stable pillar systems, have found wide acceptance (Mark, 1999). A similar approach was followed to develop a pillar strength equation for underground stone mines. Very few pillar failures have occurred in stone mines, therefore the observations alone were inadequate to develop a stone pillar strength equation from the field data. A pillar strength relationship that was originally developed by Roberts et al. (2007) was used as a starting point. Their pillar strength relationship was based on the observation of a large number of collapsed and stable pillars in Missouri Lead Belt mines where the lead mineralization is hosted in dolomitic limestone rocks (Lane et al., 1999). The rock conditions and mining dimensions are similar to those found in stone mines. This relationship was expressed as a power equation incorporating the UCS of the rock, the pillar width and pillar height and was modified to account for the potential impact of large angular discontinuities (Esterhuizen et al., 2008). The final equation is expressed in the following form:

$$S = 0.92 \times UCS \times LDF \times \frac{w^{0.30}}{h^{0.59}} \qquad (3)$$

where UCS is the uniaxial compressive strength of the intact rock, LDF is a factor to account for large angular discontinuities and w and h are the pillar width and height in feet (when using dimensions in meters the 0.92 factor becomes 0.65). If no large discontinuities are present the LDF will equal 1.0, in other cases, the value of the LDF can be calculated using the following equation:

$$LDF = 1 - DDF \times FF \qquad (4)$$

where DDF is the discontinuity dip factor shown in Table 3, and FF is a frequency factor related to the frequency of large discontinuities per pillar as shown in Table 4.

RECTANGULAR PILLARS

Rectangular pillars are used in stone mines to provide ventilation control and to assist with roof control. Rectangular pillars can be expected to be stronger than square pillars of the same width. Strength adjustments to account for the increased strength of rectangular pillars have been suggested by several researchers (Galvin et al., 1999; Wagner, 1992; Mark and Chase, 1997). A numerical model study that simulated brittle rock failure in limestone pillars (Dolinar and Esterhuizen 2007), indicated that slender pillars are not expected to benefit as much from a length increase as wider pillars because of a lack of confinement. The numerical model results

Table 3. Discontinuity dip factor (DDF) representing the strength reduction caused by a single discontinuity intersecting a pillar at or near its center, used in equation 4

Discontinuity dip (deg)	Pillar width-to-height ratio								
	≤0.5	0.6	0.7	0.8	0.9	1.0	1.1	1.2	>1.2
30	0.15	0.15	0.15	0.15	0.16	0.16	0.16	0.16	0.16
40	0.23	0.26	0.27	0.27	0.25	0.24	0.23	0.23	0.22
50	0.61	0.65	0.61	0.53	0.44	0.37	0.33	0.30	0.28
60	0.94	0.86	0.72	0.56	0.43	0.34	0.29	0.26	0.24
70	0.83	0.68	0.52	0.39	0.30	0.24	0.21	0.20	0.18
80	0.53	0.41	0.31	0.25	0.20	0.18	0.17	0.16	0.16
90	0.31	0.25	0.21	0.18	0.17	0.16	0.16	0.15	0.15

Table 4. Frequency factor (FF) used in equation 4 to account for large discontinuities

Average frequency of large discontinuities per pillar	0.0	0.1	0.2	0.3	0.5	1.0	2.0	3.0	>3.0
Frequency factor (FF)	0.00	0.10	0.18	0.26	0.39	0.63	0.86	0.95	1.00

Table 5. Values of the length benefit ratio (LBR) for rectangular pillars with various width-to-height ratios

Width-to-height ratio	0.5	0.6	0.7	0.8	0.9	1.0	1.1	1.2	1.3	1.4
Length benefit ratio (LBR)	0.00	0.06	0.22	0.50	0.76	0.89	0.96	0.98	0.99	1.00

indicate that the benefit of an increased length is likely to be zero when a pillar has a width-to-height ratio of 0.5 and it gradually increases to a maximum as the width-to-height ratio approaches 1.4.

The so called "equivalent-width method," proposed by Wagner (1992) was selected as a basis for calculating the length benefit of rectangular pillars in limestone mines. According to this method, the length benefit is expressed as an equivalent increase in pillar width, which then replaces the true pillar width in the pillar strength equation. A modification is made, called the length benefit ratio (LBR), which is a factor that increases from zero to 1.0 as the width-to-height ratio increases from 0.5 to 1.4. The modified form of Wagner's equivalent-width equation is proposed as follows:

$$w_e = w + \left(\frac{4A}{C} - w \right) \times LBR \qquad (5)$$

where w is the minimum-width of the pillar, A is the pillar plan area, C is the circumference of the pillar and LBR is the length benefit ratio. Table 5 shows the suggested relationship between width-to-height ratio and the value of LBR. The calculated value of we is used in the pillar strength equation instead of the true width. When pillars are square, we will equal the pillar width w.

PILLAR FACTOR OF SAFETY CONSIDERATIONS

Equations 1 to 5 were used to calculate the strength and FOS of all the pillars that were recorded in the field studies. The results are presented in Figure 5, which displays the FOS against the width-to-height ratio. Various symbols were used to indicate currently operating and abandoned layouts, failed pillars and the approximate number of pillars in the various layouts. Abandoned layouts may have been because of stability concerns or changes in operating procedures. The axis displaying the FOS was cut-off at 10.0 causing thirteen cases with FOS values greater than 10.0 not to be displayed.

The results show that the calculated average FOS of all the failed pillars is 2.0, which includes the cases that were intersected by large angular discontinuities. The average FOS of those pillars that are intersected by large discontinuities is 1.5. The minimum FOS for the stable layouts is 1.27 which is one of the disused layouts. It can also be seen that only one of the current pillar layouts has a FOS of between 1.0 and 1.8.

Based on current experience it would appear that almost all the stable pillar layouts have FOS values of greater than 1.8. Layouts that approach the FOS=1.8 line, in Figure 5 are typically deep layouts (> 180m (600 ft)) and are subject to relatively high

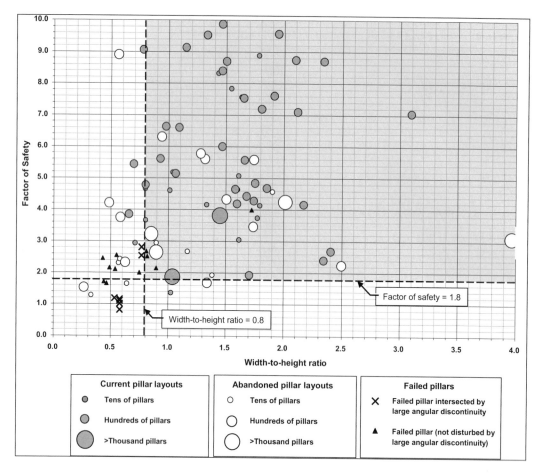

Figure 5. Chart showing the factor of safety against width-to-height ratio using Equation 3. Current abandoned pillar layouts are shown as well as single failed pillars. The recommended area for pillar design is shaded.

average pillar stresses where rib spalling can become an issue. Therefore, it would be prudent to design pillars that have safety factors of at least 1.8 to remain within the range of known successful layouts.

PILLAR WIDTH-TO-HEIGHT RATIO CONSIDERATIONS

Figure 5 shows that there has been a natural tendency for mines to avoid slender pillars. Nine of the layouts that had width-to-height ratios of less than 0.8 are no longer in use for various reasons, while only four mines are currently operating with these slender pillars. In addition, this study and other investigations have shown that slender pillars are more severely affected by the presence of discontinuities than wider pillars (Esterhuizen, 2000 and 2006). Studies have also shown that as the width-to-height ratio

decreases below 0.8, the confining stresses within a pillar approach zero and brittle fracturing can occur throughout the unconfined pillar core (Lunder, 1994; Martin and Maybee, 2000; Esterhuizen, 2006). The confining stress can further be reduced if low friction contact surfaces exist between the pillar and the surrounding rock. The fracturing and spalling failure mechanism is poorly understood and it seems prudent to avoid designing pillars that might fail in this manner.

Inspection of Figure 5 reveals that a number of stable layouts exist that have large safety factors (>3.0) and the width-to-height ratios are less than 0.8. These layouts are mostly at very shallow depths of cover, typically less than 60 m (200 ft). These pillars were found to be either very narrow, as little as 4.5 m (15 ft) wide, or very tall, up to 38 m (125 ft) high. The strength and loading of narrow pillars are both sensitive to small variations in the over-break, blast damage and pillar spacing. Large tall pillars, on the other hand, have high ribs which can represent a safety hazard and the roof becomes inaccessible and poorly visible with increased severity of potential rock fall impacts. The strength of these slender pillars is also more adversely impacted by the presence of unfavorable discontinuities than wider pillars. Therefore, it is not advisable to design layouts with such slender pillars, even if the calculated factors of safety are high.

PILLAR DESIGN GUIDELINES

The results shown in Figure 5 formed the basis for developing the design guidelines that follow. The guidelines are empirically based; therefore, their validity is restricted to rock conditions, mining dimensions and pillar stresses that are similar to those included in this study. Therefore, they should be applicable to the greater majority of limestone mines in the Eastern and Midwestern United States. The guidelines for designing stable pillar layouts are as follows:

1. Confirm that the rock conditions and the rock strength are similar to those observed in the stone mines that were part of this study. A geotechnical investigation including rock mass classification, joint set analysis and rock strength testing is recommended.
2. The pillar strength, loading and safety factor can be estimated using equations 1 to 5. The recommended factor of safety against pillar failure is 1.8 and can be seen in Figure 5 to represent the lower bound of current experience.
3. Pillars having a width-to-height ratio of less than 0.8 should be avoided. Slender pillars are susceptible to the weakening effect of angular discontinuities and are inherently weaker than wider pillars because the pillar core is unconfined.
4. The effect of large discontinuities is accounted for in the pillar strength equation. Further investigation will be required if more than about 40% of the pillars in a layout are likely to be intersected by one or more large discontinuities that dip between dip 30 and 70 deg.
5. Pillars should be designed so that the average pillar stress does not exceed 25% of the UCS to remain within the limits of past experience. The design approach is entirely based on past experience; therefore, no comment can be made

about pillar stability when pillars are loaded beyond 25% of the UCS.
6. Pillar design cannot be carried out without considering roof stability. Roof spans directly impact the pillar stresses, because wider roof spans imply higher stresses in the pillars. As part of the pillar design, an evaluation of roof span stability and likely maximum stable spans should be conducted, as described in the next section of this paper.

The shaded zone in Figure 5 indicates the area in which Equation 3 is likely to produce stable pillar layouts and coincides with the current experience in limestone mines and observations of failed pillars. The design recommendations are based entirely on the observed performance of stone pillars. Therefore, pillars that plot to the left or below the shaded area in Figure 5 are beyond the validity of these design guidelines and specialist rock engineering advice should be sought.

ROOF SPAN DESIGN

In room-and-pillar mines, the roof between the pillars is required to remain stable during mining operations for haulage as well as access to the working areas. In underground stone mines, the size of the rooms is largely dictated by the size of the mining equipment. Underground stone mines use large mining equipment to operate economically and require openings that are on average 13.5-m (44-ft) wide by approximately 7.5 m (25 ft) high to operate effectively. The dimensions of the desired roof spans are largely pre-determined and design is focused on optimizing stability under the prevailing rock conditions. If the rock mass conditions are such that the desired stable spans cannot be achieved cost effectively, it is unlikely that underground mining will proceed. NIOSH research into stone mine roof stability has focused on identifying the causes of instability and techniques to optimize stability through design.

ROOF STABILITY ISSUES IN STONE MINES

Observations were carried out at 34 operating stone mines to identify the factors that contribute to roof instability. Data were collected on rock strength, jointing and other geological structures, room and pillar dimensions, roof stability and pillar performance. Two to five data sets were collected at various locations at each mine site. A data set describes the stability of the roof and pillars in an area of approximately 100×100 m (300×300 ft). The range

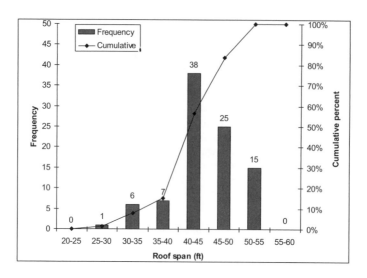

Figure 6.   Distribution of roof span dimensions measured at 34 different underground stone mines

of roof spans observed is shown in Figure 6, which shows that the 92% of the mine openings are more than 10 m (35 ft) wide. Of the mines visited, 41% were installing support in a regular pattern while the remaining 59% did not install any roof support, or only rarely installed spot bolting if required.

All but four of the 34 mines visited had experienced some form of small-scale rock falls or larger roof falls. The observed areal extent of smaller scale rock falls, less than 1m in length, was as follows:

- Thin slabs/slivers: 11% of the total roof area
- Joint bounded blocks: 6% of the total roof area
- Bedding defined beams/slabs: 11% of the total roof area

In the remaining 72% of roof area observed, the roof was stable with no sign of current or past instability. Figure 7 shows an example of a 13-m (43-ft) wide, naturally stable excavation with excellent roof conditions. Most of the smaller scale instabilities are addressed by scaling, rockbolting or screen installation as part of the normal support and rehabilitation activities.

In addition to small scale rock falls, large *roof falls*, which typically extend over the full width of the opening, were observed at 19 of the 30 mines that experienced small scale roof instability. Roof falls were categorized by identifying the most significant factor contributing to each fall. A summary of these factors and the relative frequency of occurrence of each are presented below:

- Stress: Horizontal stress was assessed to be the main contributing factor in 36% of all roof falls observed. These falls are equally likely to occur in shallow or deep cover. A roof fall related to stress-induced damage was observed at a depth of as little as 50 m (150 ft) in one case. The characteristics of stress-related roof falls are described in Iannacchione et al. (2003) and Esterhuizen et al. (2007). Figure 8 shows a typical ellipsoidal stress related fall and Figure 9 shows an example of how such a fall progressed through the mine workings in a direction perpendicular to the major horizontal stress.

- Beams: The beam of limestone between the roof line and some overlying weak band or parting plane failed in 28% of all observed large roof falls.

- Blocks: Large discontinuities extending across the full width of a room contributed to 21% of the roof falls.

- Caving: The remaining 15% of the roof falls was attributed to the collapse of weak shale inadvertently exposed in the roof or progressive failure of weak roof rocks.

Although the large roof falls only make up a small percentage of the total roof exposure, their potential impact on safety and mine operations can be very significant. Most cases of large roof falls required barricading-off or abandonment of the affected entry. When large roof falls occur in critical excavation areas, the repair can be very costly.

Figure 7.   Naturally stable 13-m (44-ft) wide roof span in a stone mine

Figure 8.   Example of large elliptical shaped roof fall that was related to high horizontal stress in the roof

MAXIMUM STABLE ROOF SPAN

The majority of roof spans in operating mines falls within a narrow range of 10 m to 17 m (35 ft to 55 ft), and is related to the space needed to effectively operate large loaders and haul trucks. Very few of the mines used roof spans wider than 15 m (50 ft), so it is not clear whether the stability limit is approached at 17 m (55 ft) or whether it simply satisfies the practical requirements for equipment operation. Given that a large proportion of the mines are able to mine without installed support, it seems to indicate that wider spans can be achieved if additional supports are used. Whether these larger spans would be cost effective will of course depend on the support costs.

One way of assessing the potential maximum span is to compare the stone mine data to experience in other mine openings around the world. The Stability Chart originally developed by Matthews et al. (1980), modified after Potvin (1988), Nickson (1992) and Hutchinson and Diederichs (1996) was used as a basis for comparison. The Stability Chart plots a modified Stability Number N' which represents the rock mass quality normalized by a stress factor, an orientation factor and a gravity adjustment. Stability zones have been indicated on the chart based on 176 case histories from hard rock mines around the world. The following stability zones are indicated:

- Stable—support generally not required
- Stable with support—support required for stability, the support type is cable bolting

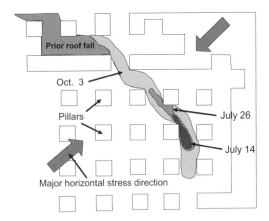

Figure 9.   Plan view showing the development of a stress related roof fall in the direction perpendicular to the direction of the major horizontal stress, after Iannacchione et al. (2003)

- Transition—stability not guaranteed, even with cable bolt support
- Unsupportable—caving occurs, cannot be supported with cable bolts

Figure 10 shows the stability chart with the stone mine case histories and stability categories. In this chart the actual heading width is shown instead of the "hydraulic radius" which is customarily used. The conversion from hydraulic radius to heading

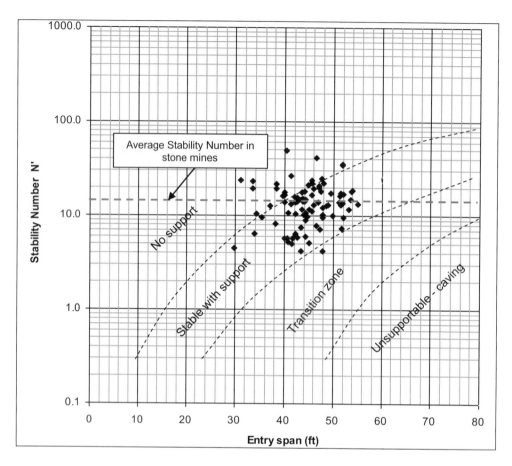

Figure 10. Stability chart showing stone mine case histories and stability zones, modified after Matthews et al. (1980), Potvin (1988), Nickson (1992) and Hutchinson and Diederichs (1996)

width assumes the heading is a parallel-sided excavation. The chart also indicates the average Stability Number for stone mines as a horizontal dashed line.

It can be seen that the majority of stone mine case histories plot in the region of "stable" to "stable with support" and only one is located in the transition zone. This agrees reasonably well with the observed stability and support used in stone mines, although stone mines have been able to achieve stability with light support compared to cable bolting used in the hard rock mine case histories. Based on the average Stability Number for stone mines, it would appear that stable supported excavations can reliably be achieved with spans of up to about 20 m (65 ft) using cable bolt supports. Cable bolt lengths in the hard rock mines are typically greater than half the excavation span, which would imply 10-m (33-ft)-long or longer cable bolts to achieve stability in a 20-m (65-ft) wide stone mine entry. Unsupportable

and caving conditions are indicated when the span increases to about 27 m (90 ft). These results are in line with current experience. It appears that stone mines are working near the span limit that can reliably be achieved using rock bolts as the support system. Increasing the spans beyond the 15–17 m range (50–55 ft) is likely to incur considerable cost and productivity implications as cable bolting would become necessary.

STABILITY OF THE IMMEDIATE ROOF BEAM

The stability of excavations in bedded deposits is closely tied to the composition of the first beam of rock in the roof. An assessment of the data collected showed that 25 of 34 mines were attempting to maintain a specific thickness of limestone beam in the immediate roof. In some cases the upper surface of the beam was a pronounced parting plane while

218

in others it was a change in lithology, typically when the limestone beam is overlain by weaker rocks. A constant thickness of roof beam is achieved either by probe drilling to determine the thickness of the roof beam or by following a known parting plane or marker horizon.

The average roof beam thickness in mines that were able to mine without regular support was 2.25 m (7.4 ft), while the average beam thickness in the mines that were using regular roof support was 1.3 m (4.3 ft). Several of the mines that used regular support do so to alleviate the effects of horizontal stress, which is not related to beam thickness. If these mines are removed from the data, the average beam thickness in mines that use regular support drops to 0.8 m (2.6 ft). These results seem to indicate that mines with a relatively thin beam of limestone in the immediate roof are more likely to encounter unstable roof and regular roof bolting becomes necessary. There was no correlation between roof beam thickness and excavation span.

The beam thickness is obviously not the only factor to consider when deciding on roof reinforcement. Other aspects such as roof jointing, bedding breaks, blast damage, groundwater and horizontal stress can contribute to roof instability resulting in the need for rock bolt support. However, experience seems to indicate that a roof beam thickness of less than about 1.2 m (4 ft) is highly likely to be unstable and a regular pattern of rock bolt supports will be required to maintain the roof stability.

HORIZONTAL STRESS CONSIDERATIONS

This study showed that horizontal-stress-related roof instability can occur at any depth of cover. This is not unexpected, given that the horizontal stresses are caused by tectonic compression of the limestone layers, which is not related to the depth of typical limestone mines (Dolinar, 2003; Iannacchione et al., 2003). Observations show that the tectonic stresses in limestone formations that outcrop may have been released over geologic time by relaxation towards the outcrop (Iannacchione and Coyle, 2002). Consequently, outcropping mines can have highly variable horizontal stress magnitudes which depends on the amount of relaxation that occurred over geologic time and the distance from the outcrop.

A review of horizontal stress measurements in limestone and dolomite formations in the Eastern and Midwestern US and Eastern Canada, (Dolinar 2003; Iannacchione et al. 2002) has shown that the maximum horizontal stress can be expected to vary between 7.6 MPa (1,100 psi) and 26 MPa (3,800 psi) up to depths of 300 m (1,000 ft). Limited information

is available at greater depths. The orientation of the maximum horizontal stress is between N60°E and N90E in 80% of the sites. This agrees with the regional tectonic stress orientation as indicated by the World Stress Map Project (2009). The minimum horizontal stress is approximately equal to the overburden stress.

An analysis of the impact of horizontal stress on beams of rock that may exist in the roof of stone mine workings showed that horizontal stress can be expected to cause buckling of thinly bedded roof strata (Iannacchione et al., 1998). Further analyses using numerical models showed that brittle spalling (Kaiser et al., 2000) of the roof rocks under near-uniaxial loading conditions can explain roof failure at the stress levels encountered in stone mines (Esterhuizen et al., 2008).

Once a stress-induced roof fall has occurred, it can be costly and difficult to arrest the extension of the fall into adjacent areas. Avoidance of these falls through layout modifications has proved to be very successful in several operating mines (Iannacchione et al., 2003). It is first necessary to establish the direction of the major horizontal stress, which can be determined by various stress measurement techniques or can be inferred from stress-related roof failures (Mark and Mucho, 1994). The layout is then modified so that the main development direction is parallel to the maximum horizontal stress and the amount of unfavorably oriented cross-cut development is minimized (Parker, 1973). A further modification that has proved to be successful is offsetting the cross-cuts and increasing the length of the pillars, so that a continuous path does not exist along which a roof fall can progress across the layout. Modifying a layout in this manner will not necessarily eradicate all stress-related problems, but has been shown to considerably reduce these problems (Kuhnhein and Ramer, 2004).

Figure 11 shows a mine layout that has been optimized for horizontal stress. The main heading direction is parallel to the maximum horizontal stress, pillars are elongated so that unfavorably oriented cross-cuts are minimized, the cross-cuts are narrower than the headings and the cross-cuts are off-set so that potential stress related roof falls will abut against solid pillar ribs, rather than snake through the layout.

ROOF REINFORCEMENT

The survey of roof support practices showed that grouted rock bolts are the most widely used form of support. Bolt lengths are typically between 1.8 m (6 ft) and 2.4 m (8 ft), and where the bolts

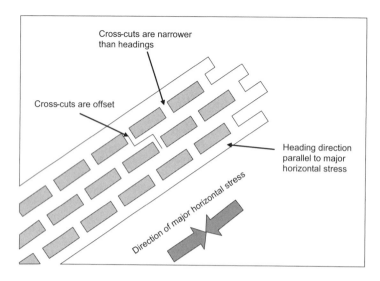

Figure 11. **Diagram showing room and pillar layout modified to minimize the potential impact of horizontal stress related damage**

are installed in a regular pattern, the most commonly used bolt spacings are either 1.7 m (5 ft) or 1.8 m (6 ft). As with most other roof bolting designs in strong rocks, high strength and stiff bolts are more likely to provide the desired rock reinforcement than low strength and low stiffness systems (Iannachione et al., 1998). Roof screen or other supplemental supports such as cable bolts are rarely used.

Roof reinforcement in the relatively strong bedded rock encountered in stone mine can have one or more objectives. Depending on the geological conditions the support system can be expected to:

- Provide suspension support for a potentially unstable roof beam
- Provide local support to potentially unstable blocks in the roof
- Combine thinly laminated roof into a thicker, stronger unit
- Provide surface control when progressive spalling and small rock falls occur

The above support functions can usually be achieved by the 1.8 m (6 ft) and 2.4 m (8 ft) bolts used in the stone mines. When poor ground is encountered locally or when horizontal stress related roof failures occur, intense bolting, steel straps and cable bolts have been used with mixed success to halt the lateral extension of these large roof falls.

From a design point of view, a stone mine is unlikely to be economically feasible if heavy support such as cable bolts and screen would be required on

a daily basis. Such rock conditions would probably require reduced excavation spans and the support costs would be prohibitive. Therefore, the first objective in designing an underground stone mine should be to confirm that the rock mass quality is adequate for creating the typical 10-m to 17-m (35-ft to 55-ft) roof spans without resorting to elaborate support systems.

ROOF SPAN DESIGN GUIDELINES

Designing stable roof spans for underground stone mines should be conducted using basic engineering principles. Good geotechnical information, combined with a pragmatic assessment of the likely modes of instability and providing the required support is likely to produce a stable initial design. Once an initial design has been developed, monitoring of its performance can be carried out to optimize the design.

Useful information can be obtained from neighboring mines that are operating under similar conditions. A particularly useful piece of information would be to identify whether horizontal-stress-related roof problems exist and the orientation of the stress related damage. This information can go a long way in selecting the orientation of the main headings in the proposed mine. The following steps should be followed to carry out a mine layout and roof span design:

1. Geotechnical characterization: Designing stable roof spans for stone mines can be successfully carried out if adequate geotechnical

investigations are conducted ahead of the design. Such investigations are best conducted by experienced ground control specialists and are likely to include rock strength testing, core logging, bedding layering assessment, joint orientation assessment and rock mass classification. If horizontal-stress-related issues are expected, stress measurements can assist in providing an indication of the orientation and magnitude of the maximum horizontal stress.

2. Confirm rock mass quality: Using the results of a rock mass classification or direct inspection of workings, confirm that the rock strength and rock mass quality is similar to that found in Eastern and Midwestern stone mines. The RMR should exceed a value of 60.0 and the rock strength should exceed 45 MPa (6,400 psi).

3. Selection of mining direction: The direction of the headings in the production areas should be favorably oriented to any expected horizontal stress and the prevalent jointing. As with any underground excavation layout, it is preferable to intersect the main joint strike direction by at least 45 degrees. Since room and pillar mines have two orthogonal directions of mining, the heading direction should be favored over the cross-cut direction when selecting the orientation of the layout. If the orientation of the maximum horizontal field stress is known, and stress related problems are anticipated, the heading direction should be oriented parallel to the direction of major horizontal stress, with due consideration of joint orientations and cross-cut stability. Often it will be a compromise to select the final heading orientation. Other modifications to the pillar layout should be considered, as discussed below.

4. Modification of pillar layout: A simple square pillar layout with headings and cross-cuts of equal width is sufficient in most cases. However, if horizontal-stress-related instability is expected, the pillar layout can be modified in improve the likelihood of success. Possible layout modifications were shown in Figure 11.

5. Selection of mining horizon: The location of the roof-line relative to pronounced bedding planes or lithology changes should be identified next. Experience has shown that if the immediate roof beam is less than 1.2-m (4-ft) thick, it is very likely to be unstable. Thicker roof beams may be required if excessive horizontal stresses are encountered. Persistent parting planes can be selected to form the roof-line if they are present at a convenient location in the formation being mined. Using a pre-existing parting plane as the roof line helps to act as a marker and usually provides a clean breaking surface for blasting operations. Many of the mines that do not use roof supports have a natural parting as the roof line.

6. Selection of roof span: Past experience has shown that stable roof spans in the range of 10 m to 15 m (35 ft to 50 ft) have been regularly achieved in underground stone mines. NIOSH studies have shown little correlation between mining roof spans and rock quality, mainly because there is such a small range of rock qualities in operating mines. For an initial design, it might be prudent to design for no more than 12 m (40 ft) spans. The spans can be increased incrementally, if warranted by monitoring of actual roof performance. There is little experience with spans that are greater than 15 m (50 ft).

7. Support considerations: Depending on the characteristics of the immediate roof, basic support in the form of patterned rockbolts may be required. The importance of the thickness of the first beam in the roof, the orientation of excavations relative to the maximum horizontal stress and characteristics of rock joints will determine whether and how much support is required. Mines that do not use bolting are located in formations with a favorable combination of geological conditions, and they conduct blasting practices that maintain an unbroken roof horizon.

8. Monitoring and confirmation: Once a roof design has been finalized and mining is underway, monitoring should be implemented to verify the stability of the roof. Monitoring results can be used to identify potential stability problems before they occur and may indicate that a change in the design is required. Monitoring technologies that are available include borehole-video logging (Ellenberger, 2009), roof deflection monitoring (Marshall et al., 2000), roof stability mapping using the Roof Fall Risk Index (RFRI), (Iannacchione, et al. 2006) and micoseismic monitoring of rock fracture, (Iannacchione and Marshall, 2004; Ellenberger and Bajpayee, 2007).

DISCUSSION AND CONCLUSIONS

A study of pillar and roof span performance in stone mines that are located in the Eastern and Midwestern United States showed that various stability issues can be addressed by appropriate design. Pillars can be impacted by rock joints, large angular discontinuities and can exhibit rib spalling at elevated stresses. Thin weak beds in the pillars, although rare, can have a significant impact by reducing pillar strength. If the roof strata are bedded, beam deflection and buckling can result in roof failure. The roof can also be impacted by large discontinuities and the effects of horizontal stress.

A pillar design procedure is proposed that takes into consideration the rock strength, pillar dimensions and the potential impact of large angular discontinuities. Based on current performance of pillars in stone mines, a safety factor of 1.8 is suggested for pillar design with a lower limit, width-to-height ratio of 0.8.

A roof span design procedure is also proposed that systematically addresses each of the main stability issues. The procedure focuses on selecting an appropriate mining horizon and mining direction. The importance of the thickness of the first bed in the roof and the likelihood for added rock bolting is described. Layout modifications are described that can be made to reduce the incidence of horizontal-stress-related instability.

Both the pillar design and roof span guidelines require that a good understanding be obtained of the geotechnical characteristics of the formation being mined. The essential data are the uniaxial compressive strength of the rock, characteristics of the discontinuities and the rock mass classification. Knowledge of the magnitude and orientation of the stress field can assist in orienting the layout appropriately.

The design procedures are based on observation of the actual performance of pillars and roof spans in stone mines within the Eastern and Midwestern United States. The guidelines should only be used for design under similar geotechnical conditions.

Disclaimer

The findings and conclusions in this paper have not been formally disseminated by the National Institute for Occupational Safety and Health and should not be construed to represent any agency determination or policy.

REFERENCES

Bauer, S.J. and Lee, M. 2004. *CAES Monitoring to Support RMRCT*. Department of Energy, Report DE-FC26-01NT40868.

Bieniawski, Z.T. 1989. *Engineering Rock Mass Classifications*. Wiley, New York.

Brady, B.H.G., and Brown, E.T. 1985. *Rock Mechanics for Underground Mining*. George Allen and Unwin, London.

Dolinar, D. 2003. Variation of Horizontal Stresses and Strains in Mines in Bedded Deposits in the Eastern and Midwestern United States. In *Proceedings of the 22nd International Conference on Ground Control in Mining*, Morgantown, WV, pp. 178–185.

Dolinar, D.R. and Esterhuizen, G.S. 2007. Evaluation of the Effect of Length on the Strength of Slender Pillars in Limestone Mines Using Numerical Modeling. In *Proceedings of the 26th International Conference on Ground Control in Mining*, Morgantown, WV, pp. 304–313.

Ellenberger J.L. 2009. A Roof Quality Index for Stone Mines using Borescope Logging. In *Proceedings of the 28th International Conference on Ground Control in Mining*, Morgantown, WV, pp. 143–148.

Ellenberger, J.L. and Bajpayee, T.S. 2007. An Evaluation of Microseismic Activity Associated with Major Roof Falls in a Limestone Mine: A Case Study. Preprint No. 07-103. Littleton, CO: SME.

Esterhuizen, G.S. 2000. Jointing effects on pillar strength. In *Proceedings of the 19th International Conference on Ground Control in Mining*, Morgantown, WV, pp. 286–290.

Esterhuizen, G.S. 2006. Evaluation of the Strength of Slender Pillars. *Trans. SME* 320:69–76.

Esterhuizen, G.S., and Ellenberger, J.L. 2007. Effects of Weak Bands on Pillar Stability in Stone Mines: Field Observations and Numerical Model Assessment. In *Proceedings of the 26th International Conference on Ground Control in Mining*, Morgantown, WV, pp. 336–342.

Esterhuizen, G.S., Dolinar, D.R., Ellenberger, J.L., Prosser, L.J. and Iannacchione, A.T. 2007. Roof Stability Issues in Underground Limestone Mines in the United States. In *Proceedings of the 26th International Conference on Ground Control in Mining*, Morgantown, WV, pp. 320–327.

Esterhuizen, G.S., Dolinar, D. R. and Ellenberger, J. L. (2008). Pillar Strength and Design Methodology for Stone Mines. In *Proceedings of the 27th International Conference on Ground Control in Mining*, Morgantown, WV, pp. 241–253.

Galvin, J.M., Hebblewhite, B.K. and Salamon, M.D.G. 1999. University Of New South Wales Coal Pillar Strength Determinations For Australian And South African Mining Conditions. In *Proceeding of the Second International Workshop on Coal Pillar Mechanics and Design*. National Institute for Occupational Safety and Health, IC 9448, pp. 63–71.

Harr, M.E. 1987. *Reliability Based Design in Civil Engineering.* McGraw-Hill, New York.

Heasley, K.A. and Agioutantis, Z. 2001. LAMODEL—A Boundary Element Program for Coal Mine Design. In *Proceedings of the 10th International Conference on Computer Methods and Advances in Geomechanics,* Arizona, pp. 9–12.

Hoek, E., Kaiser, P.K. and Bawden, W.F. (1995). *Support of Underground Excavations in Hard Rock.* A.A.Balkema, Rotterdam.

Hustrulid, W.A. 1976. A Review of Coal Pillar Strength Formulas. *Rock Mechanics and Rock Eng.* 8(2):115–145.

Hutchinson D.J. and Diederichs M.S. 1996. *Cablebolting in Underground Mines.* Bitech Publishers Ltd., Canada.

Iannacchione, A.T. and Coyle, P.R. 2002. An Examination of the Loyalhanna Limestone's Structural Features and their Impact on Mining and Ground Control Practices. In *Proceedings of the 21st International Conference on Ground Control in Mining,* Morgantown WV, pp. 218–227.

Iannacchione A.T., Dolinar D.R., Prosser L.J., Marshall T.E., Oyler D.C. and Compton C.S. 1998. Controlling Roof Beam Failures from High Horizontal Stresses in Underground Stone Mines. In *Proceedings of the 17th International Conference on Ground Control in Mining,* Morgantown, WV, pp. 102–112

Iannacchione, A.T., Dolinar D.R., and Mucho, T.P. 2002. High Stress Mining Under Shallow Overburden in Underground U.S. Stone Mines. In *Proceedings of the International Seminar of Deep and High Stress Mining,* Brisbane, Australia: Australian Centre for Geomechanics, pp. 1–11.

Iannacchione, A.T., Marshall T.E., Burke L., Melville R. and Litsenberger, J. 2003. Safer Mine Layouts for Underground Stone Mines Subjected to Excessive Levels of Horizontal Stress. *Min. Eng.*4:25–31.

Iannacchione, A.T., Batchler, T.J. and Marshall, T.E. 2004. Mapping Hazards with Microseismic Technology to Anticipate Roof Falls—A Case Study. In *Proceedings of the 23rd International Conference on Ground Control in Mining,* Morgantown, WV, pp. 327–333.

Iannacchione, A.T., G.S. Esterhuizen, G.S., Schilling, S. and Goodwin, T. 2006. Field Verification of the Roof Fall Risk Index: A Method to Assess Strata Conditions. In *Proceedings of the 25th International Conference on Ground Control in Mining,* Morgantown, WV, pp. 128–137.

Kaiser, P.K., Diederichs, M.S., Martin, D.C., and Steiner, W. 2000. Underground Works In Hard Rock Tunneling And Mining. Keynote Lecture, *Geoeng2000,* Melbourne, Australia, Technomic Publishing Co., pp. 841–926.

Krauland, N. and Soder, P.E. 1987. Determining Pillar Strength from Pillar Failure Observations. *Eng. Min. J.* 8:34–40.

Kuhnhein, G. and Ramer, R. (2004. The Influence of Horizontal Stress on Pillar Design and Mine Layout at Two Underground Limestone Mines. In *Proceedings of the 23rd International Conference on Ground Control in Mining,* Morgantown, WV, pp. 311–319.

Lane, W.L., Yanske, T.R. and Roberts, D.P. 1999. Pillar Extraction and Rock Mechanics at the Doe Run Company in Missouri 1991 to 1999. In *Proceedings of the 37th Rock Mechanics Symposium.* Amadei, Kranz, Scott & Smeallie (eds), Balkema Rotterdan, pp. 285–292.

Lunder, P.J. 1994. Hard rock pillar strength estimation an applied approach. M.S. thesis, University of British Columbia, Vancouver, B.C..

Mark, C. 1999. Empirical Methods for Coal Pillar Design. In *Proceeding of the Second International Workshop on Coal Pillar Mechanics and Design.* National Institute for Occupational Safety and Health, IC 9448, pp. 145–154.

Mark, C. and Mucho, T.P. 1994. Longwall Mine Design for Control of Horizontal Stress. *U.S. Bureau of Mines Special Publication 01-94,* New Technology for Longwall Ground Control, pp. 53–76.

Mark, C. and Chase, F.E. 1997. Analysis of Retreat Mining Stability (ARMPS). In *Proceedings of the New Technology for Ground Control in Retreat Mining.* National Inst. for Occupational Safety and Health, Information Circular 9446, pp 17–34.

Marshall, T.E., Prosser, L.J., Iannacchione, A.T. and Dunn, M. 2000. Roof Monitoring in Limestone—Experience with the Roof Monitoring Safety System (RMSS). In *Proceedings of the 19th International Conference on Ground Control in Mining,* Morgantown, WV, pp. 185–191.

Martin, C.D. and Maybee, W.G. 2000. The Strength of Hard Rock Pillars. *Int. J. Rock Mechanics and Min. Sci.* 37:1,239–1,246.

Mathews, K.E. Hoek, D.C. Wyllie, D.C. Stewart, S.B.V. 1980. *Prediction of stable excavation spans for mining at depths below 1000 metres in hard rock.* Report to Canada Centre for Mining and Energy Technology (CANMET), Department of Energy and Resources; DSS File No. 17SQ.23440-0-90210.Ottawa.

Mine Safety & Health Administration. 2009. www.msha.gov/stats.

Nickson, S.D. 1992. Cable support guidelines for underground hard rock mine operations. M.S. thesis, University of British Columbia, Vancouver, B.C.

Parker, J. 1973. How to Design Better Mine Openings: Practical Rock Mechanics for Miners. *Eng. Min.* 174(12):76–80.

Potvin, Y. 1988. Empirical open stope design in Canada. Ph.D. dissertation, University of British Columbia, Vancouver, B.C.

Pritchard, C.J. and Hedley, D.G.F. 1993. *Progressive Pillar Failure and Rockbursting at Denison Mine*. Rockburst and Seismicity in Mines, Young (ed.). Balkema, Rotterdam, pp. 111–116.

Roberts, D. Tolfree D. and McIntyre H. 2007. Using Confinement as a Means to Estimate Pillar Strength in a Room and Pillar Mine. In *Proceedings of the 1st Canada-US Rock Mechanics Symposium*, Vancouver, eds. Eberhardt, Stead & Morrison, V 2, pp. 1,455–1,461.

Salamon, M.D.G. and Munro, A.H. 1967. A Study of the Strength of Coal Pillars. *J South African Inst. Min. Metall.* 68:55–67.

Salamon, M.D.G., Canbulat, I. and Ryder, J.A. 2006. Seam-Specific Pillar Strength Formulae for South African Collieries. In *Proceedings of the 50th US Rock Mechanics Symposium*, Golden CO, Paper 06-1154.

Wagner, H. 1992. Pillar Design in South African Collieries. In *Proceedings of the Workshop on Coal Pillar Mechanics and Design,* U.S. Bureau of Mines IC 9315, pp. 283–301.

World Stress Map Project. 2009. http://dc-app3-14.gfz-potsdam.de/pub/introduction/introduction_frame.html.

Laboratory Investigation of the Potential Use of Thin Spray-on Liners in Underground Coal Mines

Simon Gilbert
School of Mining Engineering, The University of New South Wales, Sydney, Australia

Serkan Saydam
School of Mining Engineering, The University of New South Wales, Sydney, Australia

Rudrajit Mitra
School of Mining Engineering, The University of New South Wales, Sydney, Australia

ABSTRACT: TSLs can be used for surface/ground support offering remote, rapid and easy spraying technique. They are polymer based liners that can be generally applied onto the rock with a thickness of 3 to 5 mm. They can be applied rapidly with high areal coverage preventing early reaction against ground movement. TSLs may prohibit the initiation and propagation of fractures and key blocks. This improves the rock strength and hence the excavation stability. The extent of improvement would be even more remarkable for jointed rock mass.

This study aims to examine the use of TSLs in underground coal mines as a ground support tool through a laboratory approach using coal samples. Two laboratory tests were performed: Coated core test and Direct pull adhesion test.

INTRODUCTION

Mining is one of the key drivers of the Australian economy. Australia is a world leader in underground coal mining and is the world's largest exporter of coal worth nearly $A22.5 billion in export earnings in 2006–2007. While longwall mining accounts for approximately 18% of total coal production, this proportion is likely to increase as open cut operations reach their economically viable limits (Cram, 2008). Over the years various means have been employed to ensure the stability of the underground roadways and other excavations as the surrounding rock mass is subjected to high levels of stress (Craig, et al., 2009).

In addition to vertical stresses, the rock mass in most of the coalfields in eastern Australia is subjected to high horizontal stresses induced by tectonic forces within the earth's crust. These stresses are redistributed around the longwall extraction. They also cause the bedded sedimentary rock layers to bend and buckle due to stress re-distribution around the roadway (Craig et al., 2009). Ground support can be achieved using a number of tools in underground coal mining such as rockbolts, mesh (or screen) and shotcrete. However, straps and mesh are not chemically bonded to the rock mass. They are attached to other reinforcement elements. Thus, supporting systems such as straps and mesh are only activated once significant ground movement has been occurred (Potvin, et al., 2004). Applications of mesh in the underground environment have been extensive. The installation of mesh is labour intensive, time consuming and expensive (Tannant, 2001). Shotcrete can also slow down production due to curing time required to reach an effective supporting strength. Logistical problems are also encountered with the application of shotcrete. Laurence (2004) emphasises the incapability of current rib support systems that interfere with productivity. He mentions about the requirement for operators to be located close to unsupported strata, which is not effective in many circumstances and are relatively expensive.

Over the past 20 years, the technology of Thin Spray-on Liners (TSL's) has emerged as a promising alternative to the use of shotcrete and mesh as a ground support tool in underground mines. Extensive studies have been performed to understand the supporting capabilities and mechanisms of TSL's. These include both laboratory investigations and full-scale trials. However, these studies have primarily been concerned with metalliferous applications, with limited research for underground coal applications. Standard testing procedures for evaluating the effectiveness of TSL's as a ground support tool have also not yet been established.

TSLs are a novel technology in the area of ground support. A TSL is defined as a thin chemical based coating or layer that is applied to mining excavations at a thickness of 3 to 5 mm (Saydam and Docrat, 2007). They are generally applied by mixing and spraying a combination of liquid/liquid or liquid/

powder components onto the rock face as quickly as possible, where a TSL sets quickly and develops a strong bond with the rock.

This paper aims to increase the understanding of TSL behaviour, identify applicable areas for TSL's in underground coal mines, and to develop standard testing procedures for evaluating TSL performance.

POTENTIAL USE OF TSLs FOR COAL MINES

Coates (1970) suggested that if the applied surface support is air tight, entry of air will be prevented or limited and dilation will be restricted. Stacey (2001) mentioned that for a rock mass to fail, dilation must take place, with opening occurring on joints and fractures. If such dilation can be prevented, failure will be inhibited. Stacey (2001) also indicates that *"although this is unlikely in a static loading environment, in dynamic loading situations, in which rapid entry of air into the rockmass will be restricted, it is possible that an air tight TSL might promote stability."*

Stacey (2001) stated that the effect of block interlock mechanism is related to rock support membranes. He reports that the penetration of a TSL into joints and cracks will inhibit movement of blocks. Stacey (2001) and, Borejzo and Bartlett (2002) also mentioned the penetration behaviour of TSLs in fractured and cracked zones.

Goaf seals may be considered as airtight and watertight. However, leaks in goaf seals can create potentially dangerous environments leading to significant safety and economic problems. Leakage can occur through cracks and fractures created in the seal after it has been installed. This may result in displacement of ribs, roof and floor due to the increased stresses in underground coal mines (Gerard, 2007).

Considering TSLs as an active support, they can penetrate the coal rib, preventing the movement of blocky ground or spalling ribs. TSLs' flexible properties allow them to adjust to reasonable rib and roof movement; accordingly a significant support may be provided even after the longwall has passed by.

Gerard (2007) identified TSLs' further usage as applications in the maingate area to improve safety and productivity considering these areas is one of the most highly trafficked areas in the longwall and is also located in a region which is undergoing the fastest rate of rib deterioration.

Material handling and application advantages that TSLs have over shotcrete are listed below:

- Cost per square metre is approximately the same as the applied cost of 50mm of shotcrete

- Mass of material to handle per square metre of support is 25 times less
- Equipment cost is 5 to 10 times less
- TSL can be available underground immediately when needed
- Area covered per hour using hand spraying is 10 to 20 times higher

A disadvantage of TSLs compared to shotcrete is that they are unable to perform as a rigid, structural arch. Shotcrete can provide this support mechanism due to its thickness, high compressive strength, and its rigid nature. Not only are TSL's too thin, but also too flexible to provide a stiff, compressive form of support. If shotcrete is required to stabilise goaf seals, TSL's could be used in conjunction with the shotcrete as additional support and a sealant.

Gerard (2007) also mentions several possible applications using TSLs as a ground support tool in underground coal mines:

- Preventing rib degradation caused by goaf abutment pressures
- Stabilising goaf seals and cut-throughs
- Maintaining tailgate integrity
- Prevention of weathering in mains with a long design life

Most of the TSLs have proven their good sealant properties (Northcroft, 2006; Saydam and Docrat, 2007, and Morkel and Saydam, 2008). A major threat to the integrity of a heading with an extended design life is deterioration from weathering and spalling. Some TSLs may prevent weathering process, however further testing and trials on the long term performance of TSLs would have to be carried out to ensure their success in this area of application. TSLs perform better as a sealant than shotcrete. CSIRO conducted some tests on a TSL and found that this liner not only has an extremely low permeability to gas, but the water permeability of the liner is significantly less than that of water-tight rated concrete (Hawker, 2001). For this reason, TSLs have been used in tunnelling operations with significant ground water issues to decrease the water permeability of shotcrete. This was done by applying a TSL between two layers of shotcrete (Hawker, 2001). Degville (2003) explains the reason as water cannot travel laterally along the shotcrete—TSL interface because of the strong adhesive properties of TSLs. TSLs have also been used to protect the kimberlite areas in diamond mines from water and humidity (Saydam and Docrat, 2007).

Implementing TSL's into the goaf seal construction cycle would not prove to be a problem as its application is relatively quick, as only a small area is required to be sprayed at a time. Spalling of the ribs in front or behind the seal may cause problems at times as this may continue back to the seal. To prevent this, shotcrete is sometimes sprayed onto the ribs adjacent to the seal. This aims to prevent spalling as well as provide extra protection against leakage. Mesh may also be used, and though it does prevent some spalling, the problem with using mesh is that it is a passive form of support and offers no protection against leakage of gas or water. While shotcrete does act as an active support, it lacks the flexible properties of a TSL to cope well with continuing movement of ground within the cut-through after application.

Ground conditions in the tailgate tend to deteriorate fairly quickly due to the increased pressures applied to the pillars and longwall block by the neighbouring goaf. The tailgate serves as a passage for return air carrying dust and gases produced at the longwall face, therefore maintaining the integrity of the tailgate during the life of the longwall block is very important. Some tailgates may be required to function as bleeder headings once the longwall block has been mined past, and applying a TSL may help prolong the life and performance of such a heading (Gerard, 2007).

LABORATORY TESTS

The use of TSLs has been common in the civil engineering field for many years. Support is growing in the mining industry but widespread application is still lacking. Therefore it is important to develop standard testing procedures (Yilmaz et al., 2003). According to Naismith and Steward (2002), Yilmaz (2007) and Lau et al. (2008) the following requirements should be satisfied from a well-designed TSL testing procedure:

- Simple to prepare sample
- Cost effective
- Repeatable
- Practical
- Adaptable
- Reprocessing (recycling)
- Representative of relevant properties and behaviour
- Relate to in-situ performance
- Statistically valid data should be generated.

The following tests recommended by Yilmaz (2007) are considered being most relevant in defining TSL mechanical properties and research effort so far has mainly concentrated on some of these tests:

- Tensile strength and elongation
- Compressive strength

- Shear strength
- Tear strength
- Bond (adhesion) strength
- Bend (buckling) strength
- Impact (abrasion) strength

In this study two testing methods were used on coal samples for ground support application—Coated Core Test and Adhesion Test. These tests were deemed to be appropriate due to the fact that they cover all the requirements for standard testing procedures. Three different liners, from two different companies were each tested. Due to confidentiality agreement, product names are not disclosed. The laboratory tests were undertaken at the UNSW School of Mining Engineering Geomechanics Laboratory. The liners for the laboratory tests were manually applied as opposed to being sprayed on. Coal was sourced from a coal mine in NSW in Australia and the suitability of three TSL's for application in the coal mining environment was assessed.

Adhesion Strength Test

Adhesion strength is a key property of TSL behaviour. It is relevant to all liner applications in the underground coal mining environment. These include:

- Minimising rib and roof degradation
- Reinforcing pillars
- Stabilising goaf seals

The direct pull adhesion test is designed to assess the adhesion strength properties of TSLs following their placement onto rock or similar surfaces (Archibald, 2001). In the past, many adhesion test procedures were proposed by researchers to assess the characteristic adhesion/bond strength of TSLs. Mercer (1992), Tannant et al. (1999), Espley-Boudreau (1999), Archibald (2001), Kuijpers et al. (2004), Öztürk & Tannant (2004) and Saydam & Docrat (2007) conducted various pull tests to measure adhesion between TSLs and various substrate materials on in-situ and laboratory conditions. The testing procedure for the adhesion test is derived from the Safety in Mines Research Advisory Committee (SIMRAC) project guidelines (Kuijpers et al., 2004) in South Africa. This procedure was adapted from the ASTM D4541 standard for the pull strength of thin coatings.

The PAT GM01-Elcometer testing machine was used in this test as had been used by Saydam & Docrat in 2007. The machine has a hydraulic adhesion/tensile test apparatus that works on a distributed force pull off system. The apparatus can apply a maximum force of 6.3 kN. The steel test dollies are 28.2 mm (diameter) which is the standard size supplied with the equipment. For the size of dollies used, the maximum test pressure is 0–10 MPa. The test elements were glued to the surface with Araldite epoxy. A dolly is then glued with an epoxy resin to the surface of the liner, and the area surrounding the dolly is cored/drilled. Following the application of the liner and the dollies, a tensile force normal to the rock surface is applied. This is provided until de-bonding occurs. The test area was over-cored using a coring bit, so the pull force would be contained over the area under the test element, thereby providing more accurate results. Figure 1 provides a schematic of the test to illustrate the manner in which the test element is fixed onto the coating.

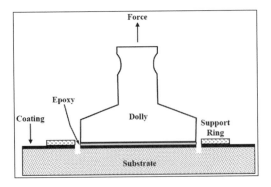

Figure 1.   Schematic of a test element glued onto a substrate (Saydam & Docrat, 2007)

Due to the brittle nature of coal it was difficult to follow the SIMRAC project guidelines for sample preparation. This was because the coal strength was significantly compromised by drilling around the dolly and into the coal. Consequently, an adapted testing procedure was undertaken for this research which included applying the liner to the coal blocks with the aid of plastic rings. These rings were made from plastic sheets to the exact diameter of the 28.2 mm dollies. The benefits of undertaking this procedure were that the thickness of the liner can be accurately controlled, and the TSL can be applied to the exact area of the dollies. This allowed the adhesive strength to be determined solely from the area of the liner directly beneath the dollies. This concept is the primary aim of coring in the SIMRAC project guidelines as adhesion strength is directly related to the area that the liner occupies. Figures 2a–f illustrate the modified sample preparation procedure used in this research.

All three TSLs were utilised during the adhesion testing on different coal blocks. The liner thickness was consistently kept at 4 mm and each block was scaled and the dusted before applying the liner. The effect of applying the liner was investigated as normal and parallel to the bedding planes. This represents the insitu environment of coal which is a sedimentary rock where seams are formed in bedding planes. Due to this characteristic of the coal deposits, TSLs can be applied to the coal surface in the underground coal environment in two orientations. These include normal to the bedding planes and parallel to

Figure 2.   Adhesion test sample preparation

Figure 3. Orientation of bedding planes

Figure 4. Execution of the adhesion Test

the bedding planes. For this project the orientation of the bedding planes is defined in the same convention utilised for tensile strength tests. Figure 3a depicts the normal to bedding plane orientation and Figure 3b depicts the parallel to bedding plane orientation. In regards to this, it was hypothesised that the adhesive strength would be greatest for the normal to bedding application. A total of 21 tests were undertaken with a curing time of 7 days. Figure 4 shows the test execution.

Table 1 indicates the measured resistance (adhesion strength), and the nature of failure is quantified by describing the structure of the actual failure surface. Failure does not necessarily have to happen between the liner and the host rock (which is coal in this case), but can also occur along other surfaces. Internal failure of the host rock, debonding between the liner and the epoxy, internal failure of the epoxy or the liner and debonding between the steel disc and the epoxy are the alternative failure modes. The epoxy and the contact between epoxy and liner as well as the contact between epoxy and steel are not

supposed to fail. Such failure is normally associated with an error in preparation.

Figure 5 shows the average test results comparing the bedding direction. As expected the results from the normal to bedding planes are lower than that from the parallel to bedding planes. The results also show that TSL3 has the highest adhesion strength of the products tested on the coal samples.

Coated Core Tests

Coated Core Test was first designed by Espley et al. (1999) to simulate the confinement that TSLs can have on pillars. It was also developed to quantify the hypothesised increases in strength and structural stability achieved by applying the liner around pillars. Later on Archibald (1992), Tarr et al. (2006) and Lau et al. (2008) conducted this test with different liners on different substrates and liner thicknesses.

The main objectives of the test were summarised by Potvin et al (2003):

*   TSL's ability to contribute towards structural reinforcement of pillars by reducing the damage resulting from potential pillar-bursts by absorbing some of the stored strain energy
*   Post failure behaviours of pillars change from violent and brittle manner to smooth and ductile manner
*   To demonstrate ability of the liners to accommodate large strain ranges and generate continuous tensile support resistance

The cores were coated with an average thickness of TSL of 5mm and left to cure for 7 days. The cores were placed in the testing machine and load was applied at a constant displacement rate. The test execution can be seen in Figure 6. For uncoated specimens, catastrophic failures were observed and there were no residual strength behaviour as expected. Post-yield failure was observed on the TSL-coated

Table 1.   Overall results from adhesion strength tests

Samples	Failure Mode	Adhesion Strength (MPa)	Bedding Direction
TSL1-4	99.6% Coal, 0.04% Liner	0.45	
TSL1-5	98.38% Coal, 1.62% Liner	0.30	
TSL1-6	100% Coal	0.35	
TSL1-7	99.76% Coal, 0.24% Liner	0.50	
TSL1 Average		*0.40*	
TSL2-10	100% Coal	0.70	
TSL2-6	100% Coal	0.30	
TSL2-7	97.49% Coal, 2.51% Liner	0.45	Normal to Bedding
TSL2-8	100% Coal	0.50	
TSL2-9	99.92% Coal, 0.08% Liner	0.70	
TSL2 Average		*0.53*	
TSL3-11	97.88% Coal, 2.12% Liner	0.80	
TSL3-6	95.71% Coal, 4.29% Liner	0.70	
TSL3-8	76.76% Coal, 23.24% Liner	0.80	
TSL3 Average		*0.77*	
TSL1-10	98.69% Coal, 1.31% Liner	0.75	
TSL1-11	79.62% Coal, 20.38% Liner	0.55	
TSL1-9	93.11% Coal, 6.89% Liner	1.00	
TSL1 Average		*0.77*	
TSL2-11	91.40% Coal, 8.60% Liner	0.95	
TSL2-12	99.15% Coal, 0.85% Liner	0.90	Parallel to Bedding
TSL2-13	89.71% Coal, 1.29% Liner	0.70	
TSL2 Average		*0.85*	
TSL3-10	99.89% Coal, 0.11% Liner	1.10	
TSL3-7	44.32% Coal, 55.68% Liner	1.50	
TSL3-9	92.12% Coal, 7.88% Liner	1.05	
TSL3 Average		*1.22*	

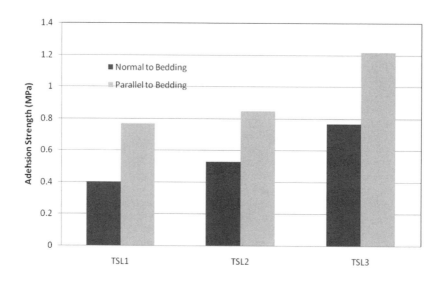

Figure 5.   Adhesion test average results for three different TSLs

samples in all the tests. Although the substrate was broken, it continued to deform further. The load and the displacement of the broken rock that supported by TSL is measured. Figure 7 shows a sample after the test.

Coal was cored from blocks of approximately 500 mm × 500 mm × 500 mm. This process proved to be difficult as approximately half of the cores significant macro cracks with the remainder half being relatively intact. Arising from these observations was a new hypothesis for the coated core test unique to this research project. This hypothesis was to use the cracked samples to represent fractured pillars and examine the impact of liner confinement in this situation. Comparisons between the peak strength and residual strength of the intact and fractured sample could then be drawn. Fully unconfined (uncoated) and passively confined (TSL-coated) samples conducted to axial strain were sufficient to propagate violent (representing the outburst situation) failure of the core specimens. The peak strengths of the coated cores were variable due to heterogeneity of the coal samples. Some examples of samples with prominent micro-fractures can be observed in Figure 8.

Measuring the load and deformation of the liners altering after the rock failed was important to make comments on the support ability of the liners. This can be clearly seen in Figures 9 and 10. The displacements in the graph show the liners' support effect.

As seen from the results in Table 2, there is no significant difference in the peak strength between uncoated and TSL-coated coal samples. Lau et al. (2008) also commented on the similar findings with sandstone samples. It was observed that the coated samples exhibited significantly less severe damage response and controlled post-yield failure

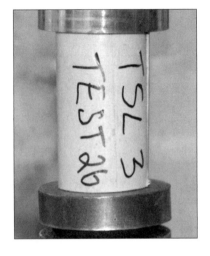

Figure 6.   Coated core test execution

Figure 7.   Coated cores after the compression test

Figure 8.   Samples with prominent micro-fractures

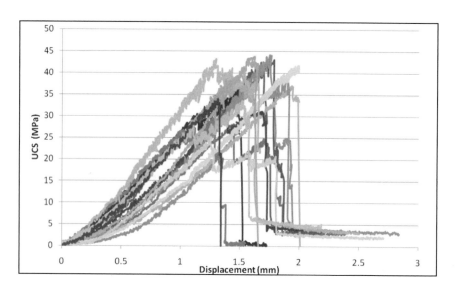

Figure 9.   Load vs displacement curve for Coated Core test with intact (no cracked) coal samples

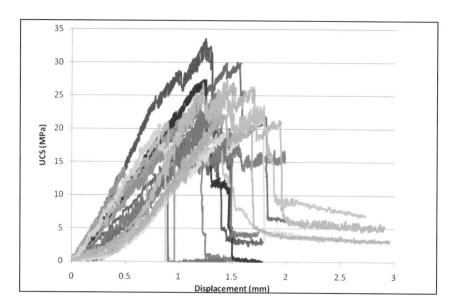

Figure 10.   Load vs displacement curve for coated core test with cracked samples

progression in comparison to the non-coated samples. Analysis of the non-coated cores confirmed this where each sample suddenly failed during the testing procedure. In most of the non-coated tests the samples actually exploded illustrating violent and brittle behaviour.

The controlled post-yield failure progression of the coated samples in this test also induced increases in the residual strength. This was observed for both the intact and the cracked samples. On average the residual strength was greater in the cracked core tests. Additionally, it was discovered that TSL's 1 and 2 provided a higher residual strength on average over all. Both of these TSL's provided an average residual strength of 3.8 MPa which was higher than TSL 3's average residual strength of 3.4 MPa. From these results, it can be seen that the liner has a positive effect on reinforcing coal pillars by increasing the residual strength.

Table 2. Comparison of liners' average peak strengths

Samples	Peak Strength (MPa) Intact Samples	Standard Deviation Peak Strength (MPa) Intact Samples	Peak Strength (MPa) Cracked Samples	Standard Deviation Peak Strength (MPa) Intact Samples
Uncoated	39.01	4.22	18.22	3.67
TSL1	35.49	7.65	20.49	3.85
TSL2	35.23	3.79	26.16	4.66
TSL3	36.55	6.91	23.72	3.35

CONCLUSIONS

In conclusion, the two laboratory testing procedures undertaken were discovered to be highly relevant in understanding the mechanisms of the TSLs. In each investigation the key mechanisms of block interlock, slab enhancement and air tightness were prevalent. Thus, the primary objective of investigating the support mechanisms of TSLs was achieved through these tests.

The adhesion test provided the most practical results. This is because the adhesion testing procedure is highly representative of TSL application in the insitu environment. The adhesion results from this research indicated that all three TSLs tested could be implemented in the underground coal mining environment. This is because the primary mode of failure was through the internal failure of the substrate. Many benefits to the industry were observed in the adhesion testing and these were primarily in regards to the development of a new testing procedure for assessing the adhesion strength in coal. Furthermore, the effect of bedding plane orientation has not been previously studied for adhesion testing. This proved to be a key concept, as it was discovered that if the liner is applied normal to the bedding planes in the underground environment then the adhesion strength could potentially be half what is achieved in the parallel orientation.

The coated core test provided a good visual representation of the support action provided by TSLs. However, the real world applications of this test appear to be limited. Nevertheless, the results of the coated core test indicated that the residual strength of pillars can be increased by provided by coating them with TSLs. This is because the laboratory tests indicated that the residual strength was increased. Additionally, through the implementation of using cracked cores an interesting observation was made that the residual strength of the samples is greater in the cracked coal situation. This has been hypothesised to be a result of a gradual failure of the cracked coal samples in comparison to the intact samples. However, it is strongly recommended that further research is done into the nature of this observation.

Previously none of the studies have conducted laboratory tests on coal samples. Further laboratory tests need to be conducted. Both tests have complex failure mechanisms and further confirmation can be achieved by numerical simulations. Success of these studies will lead to field trials before using them in the underground coal environment.

ACKNOWLEDGEMENTS

This research was supported by 2009—University of New South Wales, Engineering Faculty Research Grant.

REFERENCES

Archibald J.F. (1992). Assessment of wall coating materials for localized wall support in underground mines. Report of Phase 4 Investigations for the Mining Industry Research Organization of Canada (MIROC).

Archibald, J.F. 2001. Assessing acceptance criteria for and capabilities of liners for mitigating ground falls. MASHA, Health and Safety Conference. Sudbury, Canada.

Borejszo, R. and Bartlett, P. (2002) Developments and the future of thin reactive liners since the previous conference in Australia, in 2nd Int. Seminar on Surface Support Liners: Thin Sprayed Liners, Shotcrete, Mesh. Sandton, South Africa, 2002. SAIMM, Sect. 13, pp. 1–10.

Coates, D.F. Rock Mechanics Principles, Mines Branch Monograph 874. (Revised 1970), Department of Energy, Mines and Resources, Canada. 1970.

Cram, K. 2006. Australian Black Coal Mining Operations. In Proceedings 5th Longwall Conference, 30–31 October 2006, Pokolbin, Australia.

Craig, P, Saydam, S. and Hagan, P. "Stress corrosion or rockbolts in Australian coal mines." In press, Proc. 43rd US Rock Mechanics Symposium, 28 June -1 July 2009, Ashville, USA.

Degville, D, 2003. Technical and operational improvements expand the scope for Tekflex, in 3rd International Seminar on Surface Liners (Section 25). Hadjigeorgio, J. (ed.). Quebec City, Canada.

Espley-Boudreau, S.J. 1999. Thin spray-on liner support and implementation in the hardrock mining industry. M.Sc Thesis, Laurentian University, School of Engineering, Sudbury, Ontario.

Gerard, C. 2007. Applications for thin spray-on liners in Australian underground coal mines, Research Thesis, Bachelor of Engineering (Mining), The University of New South Wales, Sydney, NSW, Australia.

Hawker, R, 2001. Tekflex—new areas of application, in International Seminar on Surface Support Liners 2001 (Section 17), Australian Centre for Geomechanics: Perth, Australia.

Kuijpers, J.S., Sellers, E.J., Toper, A.Z., Rangasany T., Ward T., van Rensburg A.J., Yilmaz, H. & Stacey, T. R. 2004. Required technical specifications and standard testing methodology for Thin Sprayed Linings. Safety in Mines Research Advisory Committee (SIMRAC). Final Report. November 2004. Johannesburg, South Africa.

Lau, V, Saydam, S, Cai, Y, Mitra, R, 2008. Laboratory investigation of support mechanism for thin spray-on liners. In Proceedings 12th International conference of the International Association for Computer Methods and Advances in Geomechanics (IACMAG), Goa, India, 1–6 October.

Laurence, D.C. TSL trials at an Australian underground coal mine. In Y. Potvin, D. Stacey & J. Hadjigeorgiou (eds). Surface Support in Mining. Australian Centre for Geomechanics. Western Australia, Australia. p. 135–140.

Mercer, R.A. 1992. The investigation thin polyurethane linings as an alternative method of ground control. M.Sc. Thesis, Department of Mining Engineering Queen's University Kingston, Ontario, Canada.

Morkel, J. and Saydam, S. The influence of potassium on the weathering properties of kimberlite and the information provided by different testing methods. International Journal of Rock Mechanics and Mining Sciences, March 2008.

Naismith, A. and Stewart, N.R. (2002) Comparative testing of ultra-thin structural liners, in 2nd Int. Seminar on Surface Support Liners: Thin Sprayed Liners, Shotcrete, Mesh. Sandton, South Africa, 2002. SAIMM, Section 14.

Northcroft, I W, 2006. Innovative materials and methods for ground support, consolidation and water sealing for the mining industry, The Journal of The South African Institute of Mining and Metallurgy, 106(12) p. 835–844.

Öztürk, H. & Tannant, D.D. 2004. Influence of rock properties and environmental conditions on adhesive bond to a thin spray-on liner coated cores. In Y. Potvin, D. Stacey & J. Hadjigeorgiou (eds). Surface Support in Mining. Australian Centre for Geomechanics. Western Australia, Australia. p. 135–140.

Potvin Y, Stacey D, and Hadjigeorgiou J (2004) Thin spray-on liners (TSLs). In: Potvin, Y, Stacey, TR, Hadjigeorgiou, J (eds) Surface support in mining. Australian Centre for Geomechanics.

Saydam S. and Docrat, Y.S. "Evaluating The Adhesion Strength of Different Sealants on Kimberlite," International Society of Rock Mechanics, July 2007, Lisbon, Portugal.

Stacey, T.R. (2001) Review of membrane support mechanisms, loading mechanisms, desired membrane performance, and appropriate test methods. Journal of The South African Institute of Mining and Metallurgy. October 1. Johannesburg, RSA.

Tannant, D.D., Swan, G., Espley, S., & Graham, C. 1999. Laboratory test procedures for validating the use of thin spray-on liners for mesh replacement. Canadian Institute of Mining, Metallurgy and Petroleum, 101st Annual General Meeting, Calgary, Alberta, published on CD-ROM.

Tarr, K, Annor A and Flynn, D, 2006. Laboratory testing procedures for evaluating thin spray on membrane liners. Mining and Mineral Sciences Laboratories, National resources Canada, Sudbury, Ontario.

Yilmaz, H., Saydam, S. & Toper, A.Z. 2003. Emerging support concept: thin spray-on liners. Chamber of Mining Engineers of Turkey 18th International Mining Congress and Exhibition of Turkey. June 10–13, 2003 Antalya, Turkey.

Yilmaz, H. 2007. Shear-bond strength testing of thin spray-on liners, The Journal of the South African Institute of Mining and Metallurgy, 107 (8) p. 519–530.

Geotechnical Instrumentation Research Leads to Development of Improved Mine Designs

Hamid Maleki

Maleki Technologies, Inc., Spokane, Washington, United States

ABSTRACT: This paper recognizes the 100th year of research accomplishments by the mining industry resulting from close cooperation among mining companies, geotechnical consultants and researchers from the former United States Bureau of Mines in stability assessments and development of improved and productive mine designs. The paper provides an overview of dramatic developments in geotechnical instrumentation, and its applications in the U.S. and around the world for enhancing the understanding of ground response to mining-induced changes in stress, resulting in improvements in ground control technologies.

INTRODUCTION

Development of improved mine designs has become an achievable goal in the 21st Century thanks to significant improvements in geotechnical measurement techniques, great advances in numerical modeling and decades of experience and evaluations in mining projects all across the world. This success is attributed to close cooperation between mining companies, academia and research organizations. The researchers from the USBM made significant contributions by implementing comprehensive underground measurement programs, and making the results available to the mining community to enhance our understanding of strata mechanics in complex mining methods involving caving, load transfer, seismicity and placement of backfill (Maleki 2006, Seymour et al. 1988).

The focus of this paper is on pioneering research on development and utilization of geotechnical instrumentation and measurement techniques that have enhanced our understanding of strata mechanics and ground response to mining-induced changes in the stress field. The focus is on four fundamental areas: (1) Development of stress measurement hardware and procedures highlighting the impact of far-field stress on mine stability and its control through prudent mine orientation, among other options (2) Application of geophysical instrumentation for monitoring mining-induced seismicity and changes in material properties through adoption of microseismic and tomographic techniques (3) Development and utilization of ground support instruments for enhancing the understanding of reinforcement mechanisms and (4) Mine-wide utilization of other instruments for monitoring ground deformations and stability.

APPLICATION OF STRESS MEASUREMENT TECHNIQUES

Significant mining took place during the first half of the 20th Century using common engineering principles and the experience in neighboring operations, the trial and error approach and observation techniques. Historically, the mining industry depended on quick observations of strata deformation and failure, bolt failure, and lithologic changes (Hilbert, 1978) to evaluate stability; these preliminary evaluations sometimes were used to make significant changes in mine layout and support systems, which affected the economics of a mine. Because of the subjectivity involved in these visual observations, the mining process was slowed, creating inefficiencies and safety concerns during the time a decision is being reached.

It was not; however, until the 1970s that the impact of in-situ stress field on stability of the mine openings was fully recognized. Central to the understanding of stress-induced failure mechanism (Aggson 1978, 1979, Maleki et al. 1991) was the development of stress measurement techniques including both hydraulic fracturing and overcoring stress measurements (Bickle 1993, Doliner 2004).

The USBM's borehole deformation gauge and the over-coring technique used in many US mines have been shown to be accurate and reliable in determining in situ stress when adequate numbers of measurements are taken with complete relief. The method basically consists of (1) Drilling a 38-mm-diam pilot hole at the bottom of a 153-mm-diam borehole. (2) Setting the borehole deformation gauge in the pilot hole. And (3) Overcoring the gauge with a 153-mm-diam bit. Instead of the reusable USBM deformation gauges, the Hi-Cell has found applications particularly in Australian Collieries and around the world. Instead of completing the measurements through the underground openings, during the 1990s, Gray (2003) developed the SIGRA tool so that the overcoring measurements can be completed from the surface using a downhole overcoring technique.

The gravity component of the in situ stress field can be generally estimated from the depth of cover (vertical stress) and Poisson's ratio. Horizontal stresses are influenced by gravity through Poisson's effect and by the tectonic regime. For a typical Poisson's ratio of 0.15, gravitational horizontal stress is 18 pct of vertical stress. Therefore, for a 600-m-deep U.S. coal mine, the stress is approximately 15 MPa vertically and only 2.6 MPa horizontally. Tectonic processes and global plate movements also contribute to horizontal stress, and in some underground mines, horizontal stress can be three to five times higher than vertical stresses. In this situation, the horizontal stresses induce stability problems. A discussion of stability problems caused by depositional setting of mines is excluded from this paper but the interested reader is referred to Maleki 1988b.

In mines operating under biaxial horizontal stress field, mine orientation is the best approach to control the stress-induced stability problems including compressive type of failure commonly called "cutters" in the coal mines (Maleki et al. 1991, 1993). If horizontal stresses are high in comparison to rock strength, mines can experience ground control problems. In general, the experience is that the longwall panels should be oriented parallel to maximum horizontal stress to minimize roof stability problems along the gateroads. To optimize headgate stability during the retreat, Su and Hasenfus (1995) suggest that the best panel orientation occurs when maximum horizontal stress is at small deviation (20 to 25 degrees) from the gateroad alignment and the headgate is in the stress shadow of the gob. To increase awareness of horizontal stress problems, NIOSH (Mark 2003) has expanded on the three-dimensional, finite-element modeling completed by Su and Hasenfus. Maleki Technologies Inc. (MTI, 2003) under contract from NIOSH implemented these guidelines on Windows platform (Figure 1) while addressing the rock mechanics aspect of stress concentration and relief using field measurements and three-dimensional stress analyses (Maleki et al. 2003).

To illustrate the use of stress measurements in mine designs and stability assessments, results from a case study (Maleki et al. 2003) are summarized here addressing variations in measured secondary principal horizontal stress across a western US coal mine; distance between farthest profiles is approximately 1.5 km. In addition to three sets of overcoring measurements at the mine level, Haimson (1972) completed a set of hydraulic fractures within a 1,830-m-deep vertical borehole to calculate the stress field and its relationship to earthquakes in the area. These measurements were integrated for this paper to address variations in both magnitude and orientation of premining horizontal stresses. Stress profiles

were also used to identify arching mechanisms in the mine roof.

Maximum (P) and minimum (Q) measured stress profiles are presented in Figures 2 through 5 in conjunction with biaxial measurements of Young's modulus (top). In situ horizontal stress can be calculated from these measurements by a statistical treatment of those measurements far enough from the entry to exclude stress concentrations, in conjunction with Young's modulus obtained from measurements using the biaxial chamber.

The stress measurements from site 1 (Figure 2) were obtained from a thin, massive limestone; because of the very high stiffness of the limestone, which is equal to 58 Mpa, there is significantly greater stress concentration in this unit than in any other unit in the mine roof. The in situ stress field agrees with the stress trend measured at sites 2 and site 3 (Figures 3 and 4 and Table 1), confirming a stress field oriented approximately N76°E with a maximum horizontal stress of 14.7 MPa and a minimum stress of 9.9 MPa. Maximum stress is oriented parallel to the cleat and increases slightly from site 2 toward site 3.

Stress measurements were also obtained at site 4 (located at 30 m distance to site 3) (Figure 5), suggesting a shallow pressure arch in the mine roof at this location. The roof consisted of a massive, stiff siltstone that concentrated stresses as high as 28 MPa at a distance of 1.5 m above the roof line. This stress magnitude and distance relate to the stress concentration (arch) in the mine roof and agree with the moderate stress field of 13.6 to 14.7 MPa measured at site 2 and site 3 when differences in roof rock properties are taken into account using numerical modeling techniques. Stress concentration and roof arch appear to have shifted approximately 3 m into the roof at site 3 (Figure 4); this shifting was probably influenced by inelastic behavior of carbonaceous rocks in the immediate roof and formation of crack (cutter type) failure by the ribs.

The relationship between the horizontal stress field and cover depth is illustrated in Figure 6. A slight stress increase occurs over a distance of approximately 1.5 km, as one moves from site 2 toward site 3, which is both at greater depth and closer to a syncline. (The syncline is located about 1½ km to the north of site 3.) The stress increase can reasonably be accounted for by the increase in cover and/or measurement accuracy. Thus, the in situ stress field is not influenced by the syncline at this location. However, hydraulic fracture measurements (Table 1) suggest significant stress increase with depth and distance to the N50°E regional strike-slip fault. This increase is more than would be expected from overburden weight and so is likely to be the

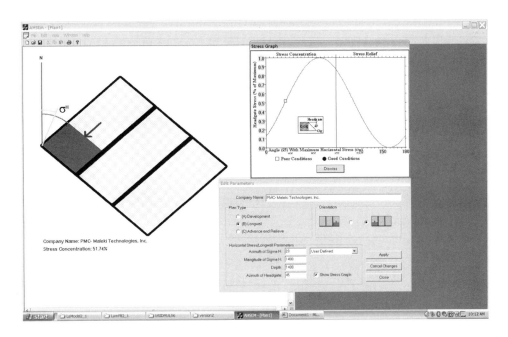

Figure 1.   Horizontal stress concentration results, softer developed by MTI

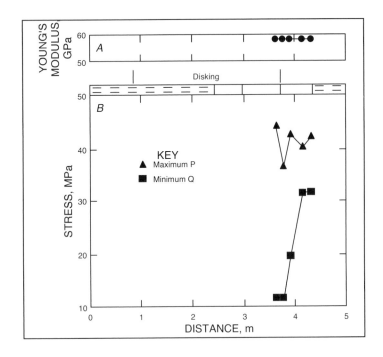

Figure 2.   Horizontal stress and deformation modulus profile in mine roof at site 1

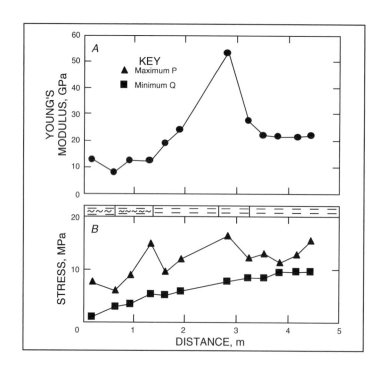

Figure 3. Horizontal stress and deformation modulus profile in mine roof at site 2

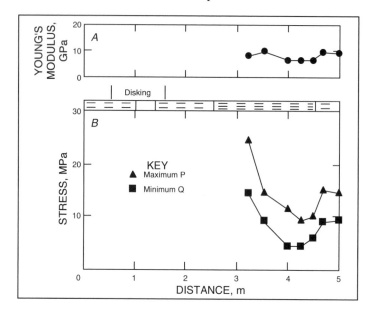

Figure 4. Horizontal stress and deformation modulus profile in mine roof at site 3

Table 1. Maximum and minimum horizontal stress field and depth of cover

Location	Maximum, MPa	Minimum, MPa	Direction of maximum	Depth of cover, m
A—Overcoring site:	42	21	N88W	210
Site 1	13.6	9.6	N79E	184
Site 2	14.7	9.9	N73E	350
Site 3				
B—Hydraulic fracture, Rangely oilfield	59	31	N70°E	1,830

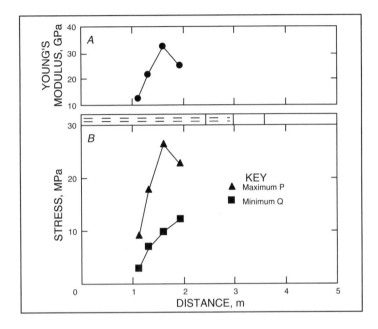

Figure 5. Horizontal stress and deformation modulus profile in mine roof at site 4

result of stress concentrations near the fault or the method of measurement overestimating the stress. These hydraulic fracture measurements were taken 7 km from the mine site and are presented in Figure 6 to compare with mine site measurements.

APPLICATION OF GEOPHYSICAL MEASUREMENT TECHNIQUES

Geophysical monitoring techniques have also evolved considerably during the last century, benefiting from great improvements in the area of data acquisition and processing. These measurements have mostly been used to monitor changes in stability of rock masses by measuring rock deformation and fracturing. Stability assessment is generally carried out by comparing measurements between initial (or expected) conditions and predetermined critical levels for the material under investigation.

For instance, if changes in seismic velocities reach a critical level of 17 percent, the roof is expected to behave inelastically, requiring supplementary support (Maleki, 1993b).

On the large scale, geophysical techniques have become increasing popular in monitoring mining-induced seismicity, in view of the Crandall Canyon Mine disaster (Arabasz and others 2005), more recently. These techniques listen and record rock "noise: under stress. For a good discussion of these techniques, the interested reader is referred to publications by Swanson 1995, Estey 1995 and Maleki 1995. The focus here is on the application of tomographic techniques to stability evaluations and improvements in designs.

Fracture initiation has been monitored using microseismic emissions (Repsher and Steblay, 1985), and fracture density has been recently studied using a controlled source and tomographic

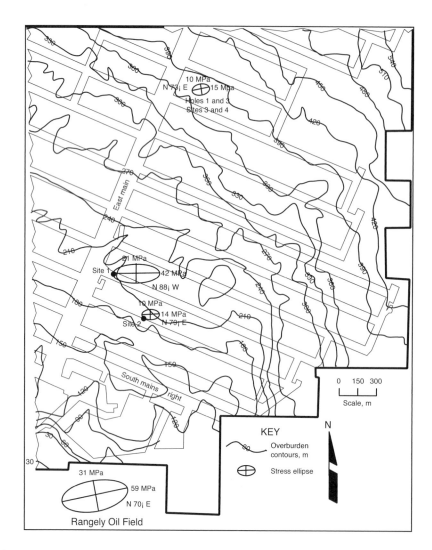

Figure 6.   Measurement location and stress ellipsoid

techniques (Maleki, et al., 1993). The latter approach has allowed changes in roof stability to be compared to roof deformation and fracturing, which can significantly influence dynamic properties such as velocity, attenuation, Young's modulus, and Poisson's ratio.

Controlled source techniques have also been used to identify fracture zones and zones of high stress in pillars (Jung, Ibrahim, and Born, 1991; Friedel, Jackson, Tweeton, and Olson, 1993; Westman, 1993, Maleki and Hollberg 1995); the latter is based on experimental evidence showing an increase in seismic velocity with stress increase.

Figure 7 illustrates typical geometries used in a controlled-source tomographic survey to evaluate roof and floor stability. Pillar surveys are similar

except that there is usually access to all sides of a pillar, while contact is limited to three sides (two source boreholes and the roof) for roof or floor surveys. The impact source is generally some sort of hammer that transmits energy to the rock within source boreholes or directly to the pillar. Accelerometers are attached to the mine roof, floor, or pillar to receive the signal generated by the impact. These techniques quantify wave velocity within a volume of rock (area of interest, Figure 7) in contrast to obtaining point measurements when using sagmeters. Using the crosshole seismic tomography today it is possible to study the structure under investigation by a high density of waves traveling between sources and receivers. This is analogous to the use of a computerized axial

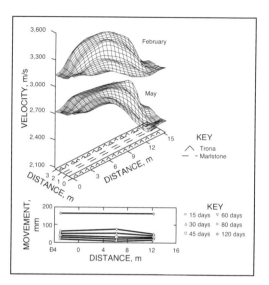

Figure 8. History of roof velocity and deformation along B-B' during 4-month monitoring period. Top, velocity; bottom, deformation.

Table 2. Calculated changes in dynamic properties

Factor	Percentage of change
Seismic velocity	21
Maximum amplitude	22
Dominant frequency	48
Deformation modulus	51
Poisson's ratio	125

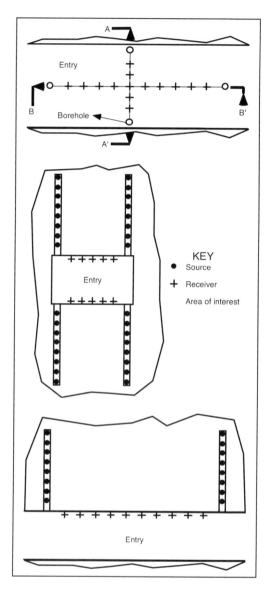

Figure 7. Typical three-sided seismic surveys of roofs and floors. Top, plan view; middle, cross section along A-A'; bottom, cross section along B-B'.

tomography (CAT) scan of the human body, but sound waves are generated instead of X-rays.

The components of a crosshole seismic survey developed for routine underground applications are described by Maleki and others 1992. A PC-based, 13-channel system with programmable gains capable of acquiring up to 150,000 samples a second was used for data acquisition. This system utilized an analog digital board, amplifiers, filters, and a trigger unit (Maleki, Ibrahim, Jung, and Edminster, 1992).

Selected results from a comprehensive case study is presented here to show the application of the tomographic techniques for assessing changes in structural stability in a trona mine where tomographic surveys were complimented with deformation measurements and fracture mapping. Significant changes in seismic wave velocities, amount of roof separation, and number of roof fractures occurred during the monitoring period. Supplementary support, consisting of 2.4-m-long fully grouted resin bolts and straps to control roof deformation, was installed on day 80. Roof and floor movements and pillar dilation were all large, indicating rock failure and complex strata interactions around the mining excavation.

Wave velocities were reduced by 22 percent at the intersection and by 15 percent at mid-pillar during the 120-day monitoring period. Figure 8 and Table 2 demonstrate the calculated changes in dynamic properties.

A preliminary relationship between velocity changes and supplementary support requirements was established for the case study using these measurements. On day 80, wave velocities were reduced by approximately 17 percent at the intersection because of increased fracturing and possible roof destressing (Maleki, 1993b). The roof had deformed 60 mm since development. Changes in wave velocity greater or equal to 15 percent may be used as a criterion for allocating supplementary support.

APPLICATION OF DEFORMATION MEASUREMENT TECHNIQUES

Deformation measurements have found wide applications for assessment of roof stability in coal, hard rock and evaporates within the last three decades (Maleki, 1988a, 1993a) and yield pillar performance; these measurements are inexpensive, simple, and reliable but require some protection of the instruments and cables from fly rock and mining equipment. Development of failure zones in mine pillars is similarly associated with rib fracturing and dilation, which is measured using dilation points and tape extensometers (Maleki, et al. 2003).

Deformation measurements are obtained with extensometers and roof-to-floor convergence meters. These measurements have become very popular for assessment of roof or pillar stability in underground mines, nuclear waste repositories (Maleki 2009), and strategic storage facilities for oil and liquid gas, among others.

Measurements of amount and rate of roof deformation have been extensively used in both longwall and room-and-pillar retreat mining systems (Maleki, 1993a, 1988a). Because there is a large database involving a variety of geologic and stress conditions, there are preliminary criteria for the routine use of this technique in underground coal mines.

Measurements in seven U.S. mines have indicated that most coal-measure rocks deform 2 to 7 cm prior to roof collapse and then accelerate to a critical velocity called the "critical rate." Between the initiation of the critical rate and the actual roof fall, a period of time elapses that is called time to caving.

The rate of roof movement is a good indicator of roof stability, because it indicates a change in the stability of the whole mining system. A change from a steady rate of movement to a high (critical) rate can indicate roof block detachment, loss of support effectiveness, roof shearing by pillars, or impending pillar failure.

A history of relative roof deformation is presented in Figure 9 for the middle entry of a longwall headgate, to illustrate the application of these simple, inexpensive measurements. The span was 6 m; the immediate roof consisted of mudstone overlain by a fluvial sandstone, and the primary support was 1.2-m-long mechanical bolts and straps.

Roof deformation was monitored by two wire sagmeters (Maleki 1981), with anchors 1 (at 2.5 m) and 2 (at 1 m) positioned above and below the bolt anchor horizon, respectively. Relative roof movements were calculated (Figure 9). The difference between the two curves reflects bed separation in the mine roof.

After development, roof movements reached 0.05 cm/day, which resulted in roof flaking. Installation of posts reduced the rate of roof movement, but did not stop bed separation at the sandstone contact. As the longwall face approached within 100 m of the instrument, bed separation and rate of roof movement significantly increased. Installation of additional support slowed movement, but another accelerated roof movement and bed separation cycle occurred as the longwall face came within 30 m of the instrument. This was followed by a roof fall that occurred approximately 4 days after the last accelerated roof movement. The critical rate of roof movement and the time-to-caving are best established by actually monitoring unstable roofs.

A critical rate of roof sag of 0.05 cm/day was determined on the basis of 142 measurements in three mines with 6-m spans and stable pillars (Table 3). This critical rate was based on roof sag measurements along the gate roads of five panels where no secondary support had been installed. Installation of secondary supports generally stabilized roof movements; therefore, measurements taken near secondary supports were excluded from the study. Maleki (1993a) has addressed these same caving parameters for another four pillar-pulling operations where spans are gradually increased to 15 m and pillars crush to residual strength.

REINFORCEMENT MEASUREMENT TECHNIQUES

The aforementioned methods are most suitable for monitoring changes in rock mass deformation and stress conditions. There are other useful measurements developed within the last three decades that indicate strata movement, but that do not directly lend themselves to evaluations of structural stability. Measurements of bolt tension (Maleki, 1992b), for instance, provide indirect indications of roof movements, but also valuable insight into reinforcement

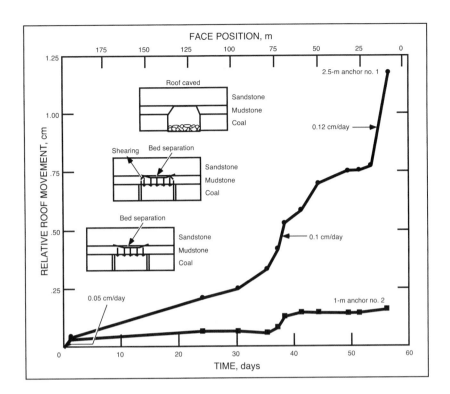

Figure 9. History of roof deformation for a longwall gate road

integrity and load transfer from the rock to bolts and loss of anchorage leading into roof stability problems.

To illustrate the use of instrumented fully grouted rebars, results from a case study is presented (Maleki et al. 1994). Six to eight instrumented bolts were installed in the two test sections, conventional bolting and an alternative pattern using higher density of roof bolts. The mine used 2.5-m-long No.7 Dwyidag bolts with 1.5 m of resin grout in a 1.2 × 1.2-m bolting pattern. These bolts had a yield and ultimate capacity of 182 and 260 kN, respectively. The yield strength of six slotted bolts was 977 kN, which was 7.1 pct higher than the 912-kN strength of five regular bolts; therefore, the measured loads on the instrumented bolts in the conventional section would be lower than the loads on the regular bolts. These bolts contained six pairs of strain gauges mounted at 0.6, 1, 1.4, 1.7, 2, and 2.2 m along the entire bolt length; special procedures were developed to protect the strain gauge wires, and the bolts were carefully oriented with respect to the entry to allow an evaluation of bending moments.

The instrumented bolts were milled with a 6-mm-wide by 3-mm-deep slot along each side and then strain gauges were installed and calibrated to

Table 3. Critical rate of sag for different mines using 6-m spans

Mine	Critical rate, cm/day	Time to caving, days	Number of measurements
1	0.06	2 to 8	115
2	0.08*	NA	24
3	0.05	9 to 11	3

*No failure was recorded but area was unstable.

correlate voltage change to load change (Signer and Jones, 1990). Three bolts were tested to failure to determine the strength and yield point after the slots had been milled. The slotting process reduced the yield strength to 163 kN and the ductility to between 9 and 12 pct. When data from the instrumented roof bolts were reduced, the correlation coefficients from the axial calibrations were used to convert voltage changes to load changes. This process was accurate to ±0.4 kN. The strain gauges used are functional to approximately 50,000 microstrain, and the slotted bolt will yield at approximately 2,300 microstrain.

The axial and bending loads at the locations with strain gauges were compared to assess the support-rock

243

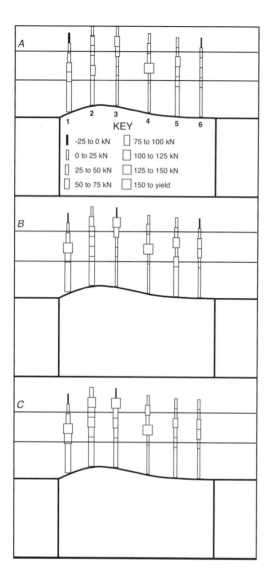

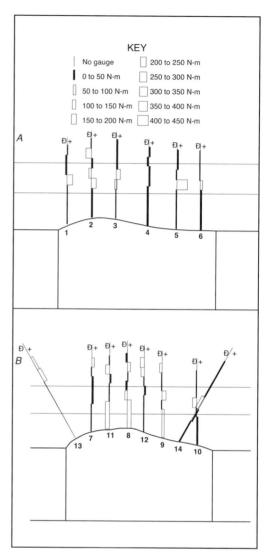

Figure 10. Roof bolt load profiles after bolt installation for conventional section A: 1 day, B: 1 week, C: 2 months

Figure 11. Bolt bending moment profile one week after installation A: Conventional section, B: Alternative section

interaction. The axial bolt loads measured immediately after installation were generally less than 20 kN and increased rapidly in response to mining and strata deformations. Figure 10 compares mean axial bolt load history for the conventional test section.

The bending moments generated by the bolts at 1 week after installation (Figure 11) confirmed that there was significant lateral movement toward the entry within both the immediate and intermediate roof. In the first test section, the moments were maximum within and near the rider-seam-coal interface. In the alternative section with increased reinforcement density, the moments were more uniformly distributed along the entire length of the bolts. The first, conventional section allowed more rock movement. This could produce failure planes or surfaces that would create higher localized bending stresses in the bolts. If the rock remained intact, then the bending stresses would be more evenly distributed (Figure 11B).

CONCLUSIONS

Development of improved mine designs has become an achievable goal in the 21st Century thanks to significant improvements in geotechnical measurement techniques, great advances in numerical modeling and decades of experience and evaluations in mining projects all across the world. This success is attributed to close cooperation between mining companies, academia and research organizations.

Using in-mine measurements of stress, deformation and seismic wave velocity in coal and trona mines, the importance of pioneering research on development and utilization of geotechnical instrumentation and enhancing our understanding of strata mechanics and ground response is demonstrated. Case studies demonstrate the successful, practical use of the following techniques in stability evaluation and mine designs: (1) stress measurements influencing roof stability and bolt loading depending on the magnitude of stress, rock strength and orientation of stress in respect to the workings (2) geophysical instrumentation showing significant changes in roof stability during development particularly at an intersection in one study mine in evaporates (3) mechanical sag meters for predicting potential for roof falls and (4) instrumented fully grouted bolts enhancing the understanding of reinforcement mechanisms depending on support density.

REFERENCES

Maleki, H., S. Signer and M. Jones, Evaluation of Support Performance in a Highly Stressed Mine. Proc. 13th Intern. Conf. on Ground Control in Mining, Morgantown, WV, Aug. 2–4, 1994.

J.R. Aggson. In-Situ Stress Fields and Associated Mine Roof Stability. Mini-Symposium Series No. 78-1, SME_AIME Annual Meeting, Dallas, TX, 1978, 7 pp.

J.R. Aggson. Stress-Induced Failures in Mine Roof. U.S. Bureau of Mines, RI 8338, 1979, 15pp.

D.R. Dolinar. Analysis of the Variation of Horizontal Stresses and Strain in Bedded Deposits in the Eastern and Midwestern United States. MS Thesis, West Virginia University. 2004, 120 pp.

Maleki, H. 2009, Prediction of Deformations and Stability in a Nuclear Waste Repository, 3rd AMIREG International Conference, Assessing the Footprint of Resource Utilization and Hazardous Waste Management, Greece.

Estey, L. 1995. Overview of USBM microseismic Instrumentation Research for Rock Burst Mitigation at the Galena Mine, Mechanics and Mitigation of Violent Failure in Coal Mines and Hard Rock Mines. Edit Maleki, USBM Special Publication 01-95, 1995

Su and Hasenfus 1995. Regional Horizontal Stress and Its Effects on Longwall Mining in the Northern Appalachian Coal Field, Proceeding of International Ground Control in Mining, WV University.

Maleki, H., M. Stevenson, D. Spillman, R. Olson, Development of Geotechnical Procedures for the Analysis of Mine Seismicity and Pillar Designs, 22nd International Conference in Ground Control in Mining, Morgantown, WV, 2003.

Maleki, H. Coal Mine Ground Control, Ph.D. Thesis, Colorado School of Mines, 1981.

MTI 2003. Window based computer program developed for NIOSH related to Analysis of Horizontal Stress in Mining, Internal report by Maleki Technologies, Inc.

W.J. Arabasz, S.J. Nava, M.K. McCarter, K.L. Pankow, J.C. Pechmann, J. Ake, and A. McGarr. Coal-Mining Seismicity and Ground-Shaking Hazard: A Case Study in the Trail Mountain Area, emery County, Utah. Bulletin of the Seismological Society of America. 95(2), 18–30.

Swanson P., 1995, Influence of Mining-induced Seismicity on Potential for Rock Bursts. Mechanics and Mitigation of Violent Failure in Coal Mines and Hard Rock Mines. Edit Maleki, USBM Special Publication 01-95, 1995

Maleki H., J Dubbert and D Dolinar, 2003, Rock Mechanics Study of Lateral Destressing for the Advance-and-Relieve Mining Method. 22nd International Conference in Ground Control in Mining, Morgantown, WV, 2003.

Maleki H. and K. Hollberg 1995. Structural Stability Assessment Through Measurements. Proc. ISRM Workshop on Rock Foundations, Tokyo, Japan, 1995.

Maleki, H. An Analysis of Violent Failure in U.S. Coal Mines, Case Studies. Mechanics and Mitigation of Violent Failure in Coal Mines and Hard Rock Mines. Editor, USBM Special Publication 01-95, 1995

Maleki, H. 2006, Caving, Load Transfer and Mine Designs in Western U.S. Mines, ARMA/ USRMS 06-693, The 41st U.S. Symposium on Rock Mechanics, 50 Years of Rock Mechanics— Landmarks and Future Challenges.

Gray, I, 2004, Report to SIMRAC Demonstration of Overcore Stress Measurement From Surface Using the SIGRA 1ST Tool, Internal report by SIGRA Pty. LTD.

Bickel, D. L. Rock Stress Determinations from Overcoring an Overview. U.S. Bureau of Mines Bull. 694, 1993, 146 pp.

Friedel, M. J., M. J. Jackson, D. R. Tweeton, and M. S. Olson. Application of Seismic Tomography for Assessing Yield Pillar Stress Conditions. Paper in Proceedings of 12th Conference on Ground Control in Mining, ed. by S. S. Peng (Morgantown, WV., Aug. 3–5, 1993). Dept. of Min. Eng., WV Univ., 1993, pp. 292–296.

Hilbert, D. The Classification, Evaluation, and Projection of Coal Mine Roof Rocks in Advance of Mining. Pres. at ann. meet. AIME, 1978, 27 pp.; available from H. Maleki, Spokane Research Center, Spokane, WA.

Jung, Y., A. Ibrahim, and D. Born. Mapping Fracture Zones in Salt. Geophysics: The Leading Edge of Exploration, Apr. 1991, pp. 37–39.

Maleki, H. The Application of Deformation Measurements for Roof Stability Evaluation. Soc. Min. Eng. AIME preprint 93-197, 1993a, 7 pp.

Case Study of the Inelastic Behavior of Strata Around Mining Excavations." Rock Mechanics in the 1990s. Paper in 34th U.S. Symposium on Rock Mechanics, ed. by B. Haimson (Madison, WI, June 27–30, 1993). Geol. Eng. Prog., Univ. of WI-Madison, v. 2, 1993b, pp. 717–720.

Detection of Stability Problems by Monitoring Rate of Roof Movements. Coal Age, Dec. 1988a, pp. 34–38.

Ground Response to Longwall Mining. CO School Mines Q., v. 83, No. 3, 1988b, 14 pp.

In Situ Pillar Strength and Failure Mechanisms for U.S. Coal Seams. In Proceedings of the Workshop on Coal Pillar Mechanics and Design, compiled by A. T. Iannacchione, C. Mark, R. C. Repsher, R. J. Tuchman, and C. C. Jones (Santa Fe, NM, June 7, 1992). U.S. Bureau of Mines IC 9315, 1992a, pp. 73–77.

Significance of Bolt Tension in Ground Control. Paper in Rock Support in Mining and Underground Construction: Proceedings of the International Symposium on Rock Support, ed. by P. K.

Kaiser and D. R. McCreath (Sudbury, ON, June 16–19, 1992). A. A. Balkema, 1992b, pp. 439–449.

Maleki, H., W. Ibrahim, Y. Jung, and P. A. Edminster. Development of an Integrated Monitoring System for Evaluating Roof Stability. Paper in Proceedings, Fourth Conference on Ground Control for Midwestern U.S. Coal Mines, ed. by Y. P. Chugh and G. A. Beasley (Mt. Vernon, IL, Nov. 2–4, 1992). Dept. of Min. Eng., S. IL Univ., 1992, pp. 255–271.

Maleki, H., Y. Jung, P. A. Edminster, and W. Ibrahim. Application of a Static and Geophysical Monitoring System for Detection of Stability Problems. Paper in Proceedings of 12th Conference on Ground Control in Mining, ed. by S. S. Peng (Morgantown, WV, Aug. 3–5, 1993). Dept. of Min. Eng., WV Univ., 1993, pp. 125-132.

Repsher, R. C., and B. Steblay. Structure Stability Monitoring Using High Frequency Microseismics. Paper in Research and Engineering Applications in Rock Masses. Proceedings of the 26th U.S. Symposium on Rock Mechanics, ed. by Eileen Ashworth (SD School of Mines and Tech., Rapid City, SD, June 26–28, 1985). A. A. Balkema, v. 2, 1985, pp. 707–713.

Wawersik, W. R., and C. Fairhurst. A Study of Brittle Rock Fracture in Laboratory Compression Experiments. Int. J. Rock Mech. and Min. Sci., v. 7, No. 5, 1970, pp. 561–575.

Westman, E. C. Characterization of Structural Integrity and Stress State Via Seismic Methods: A Case Study. Paper in Proceedings of 12 Conference on Ground Control in Mining, ed. by S. S. Peng (Morgantown, WV., Aug. 3–5, 1993). Dept. of Min. Eng., WV Univ., 1993, pp. 322–327.

Maleki, H., R.W. McKibbin, and F.M. Jones. Stress Variations and Stability in a Western U.S. Coal Mine. Pres. at SME ann. mtg., Albuquerque, NM, Feb. 14–17, 1994, 7 pp.

Maleki, H., A. Weaver, and R. Acre. Stress-Induced Stability Problems—A Coal Mine Case Study. Paper in Rock Mechanics as a Multidisciplinary Science. Proceedings, 32nd U.S. Symposium (OK Univ., Norman, OK, July 10–12, 1991). A.A. Balkema, Rotterdam, Netherlands, 1991, pp. 1,057–1,064.

Raleigh, C.B. Earthquake Control at Rangely, Colorado. Transactions, Am. Geophy. Un., v. 52, No. 4, 1971.

Signer, S.P., and S.D. Jones. Case Study on the Use of Grouted Support in a Two-Entry Gate Road. Paper in Rock Mechanics Contributions and Challenges: Proceedings of the 31st U.S. Symposium, ed. by W.A. Hustrulid and G.A. Johnson (CO School of Mines, Golden, CO, June 18–20, 1990). A.A. Balkema, Rotterdam, Netherlands, pp. 145–152.

Seymore, B, D, Tesarik and M Larson, Stability of Backfilled Cross-panel Entries During Longwall Mining. 17th Conference in Ground Control, 1988, 10 pp.

Review of Rock Slope Displacement-Time Curve and Failure Prediction Models

Hossein Masoumi

School of Civil and Environmental Engineering, University of New South Wales, Sydney, Australia

Kurt J. Douglas

School of Civil and Environmental Engineering, University of New South Wales, Sydney, Australia

ABSTRACT: Rock slope failures usually develop over a considerable period of time. The ability to ascertain the length of this period and indeed the likelihood of a slope collapsing is of considerable importance to those managing rock slopes. This paper provides a review of the existing methods for determining the pre-failure deformation of rock slopes. It includes a discussion of the different phases of displacement-time curve and failure prediction models. These models are classified into three categories, namely, empirical, semi-empirical and theoretical. A number of empirical and semi empirical models have been proposed by various authors but nothing specific to rock slopes has been provided at a theoretical level. Many of the models have been developed based on soil mechanics and as such have limitations when applied to rock slopes. The authors consider developing a theoretical model to improve the accuracy of predicting the time to collapse. To accomplish this, a better understanding of failure mechanisms, stress redistribution and geological and environmental factors is considered crucial.

INTRODUCTION

Monitoring displacements of large slopes is often performed. The interpretation of this data and prediction of possible collapse presents a considerable challenge to practitioners. This paper presents an overview of published failure prediction models. It highlights current uncertainties and proposes areas for future consideration. The authors believe that an understanding of the failure mechanisms of rock slopes is crucial. For this work, the general classifications of Wyllie and Munn (1978) have been adopted and are shown in the first section.

Slope failure usually develops over a long time period. The slope behaviour leading up to failure can be illustrated by using a displacement-time curve. This curve consists of three phases as primary, secondary and tertiary stages as presented in the second section.

Various authors have proposed empirical and semi-empirical models to predict the time to failure of slopes. These are presented and discussed. A theoretical approach is also briefly discussed where a constitutive model for the creep behaviour of a material can be developed and then applied to a slope using finite element methods or similar.

SLOPE FAILURE MECHANISMS

The term failure may refer to the first movement of a slope, a displacement which may or may not result in a catastrophic event, or it may refer to the final collapse of the slope. The term collapse implies the sudden movement, disaggregation and associated large displacement of the slope. In many papers, the terms failure and collapse are given the same meaning and this is applied here.

To have a chance of predicting the time to collapse, the initial anticipation of the slope failure mechanism is crucial. Failure mechanisms mainly depend on the type of slope (Soil or rock) and its geotechnical structure. Most slope failures can be defined by a simple division into circular, planar and toppling mechanisms (Figure 1).

A circular failure (Wyllie and Munn 1978) or rotational failure (Hutchinson 1988) is the most common type of failure mechanism in homogenous materials (Hoek and Bray 1977). This may occur in soil, weathered and soft rock, waste dumps and much less frequently in highly fractured rock.

A planar failure (Wyllie and Munn 1978) or translational failure (Hutchinson 1988) generally occurs where there is a definite structure to the soil or rock mass. This may be bedding or fissuring in soils or defects in rock masses.

A toppling failure can occur in rock masses where relatively closely spaced defects dip into the slope, although, in some cases they may dip out of the slope (Wyllie and Munn 1978).

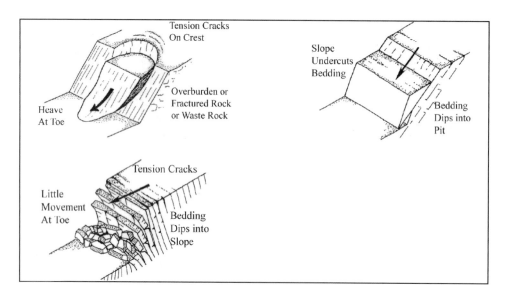

Figure 1.   Circular, planar and toppling mechanisms (Wyllie and Munn 1978)

DISPLACEMENT-TIME CURVE (CREEP)

The pre-failure behaviour of a slope must be understood for a prediction of the time to collapse to be made. This behaviour is typically shown by a displacement-time curve to illustrate the development of the failure. In material and mechanical engineering this behaviour is defined as creep. Creep refers to the deformation of solid materials under constant stress over a long time period. Three stages of creep may be identified in the displacement-time curve, namely, primary, secondary and tertiary creep (Figure 2).

Primary Phase

This stage is recognisable as primary creep, the decelerating stage or the elastic response/rebound phase. It occurs as a response to the removal of material and hence stress. The elastic parameters of the material are the most important for quantification of this phase. During this stage, slopes may stabilise or they may progress to the secondary stage. Current information regarding this phase is generally limited to empirical information from excavated slopes.

Secondary Phase

The secondary phase is also identified as the steady stage, secondary creep or constant phase. This stage often represents the longest period of the overall displacement-time curve. Prediction of the transition point between the secondary and tertiary phases is of significant interest however, none of the current predictive models are able to forecast this point.

Difficulties exist in predicting the length of this phase due to the limited amount of empirical information available. Typically slopes are only monitored after a stability issue has been identified, at which time the secondary phase has usually already commenced (Glastonbury 2002). Some researchers (Saito and Uezawa 1961) use this phase to predict the final time to collapse.

Tertiary Phase

The tertiary stage is also known as tertiary creep, the accelerating stage or the tertiary stage. The information from this phase is broadly used to predict the time to collapse and the inputs of most prediction models are logged in this stage.

The stage begins with a sudden acceleration of the slope and ends with the slopes collapse. Environmental factors such as rainfall, temperature and climate change may lead to the quicker onset of this stage whilst geomechanical and geological parameters may impact on the duration of this phase.

FAILURE PREDICTION MODELS

The prediction of the time to collapse of a slope is a major concern in slope engineering and to date there is no universally accepted model that can be expected to adequately predict the time to failure for all slopes. This is due to the number of variables and uncertainties in slope engineering and the limited amount of data available to address them.

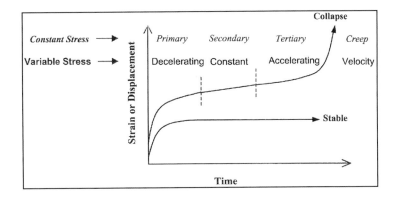

Figure 2.   Representation of the displacement-time curve for slope deformation (Glastonbury 2002)

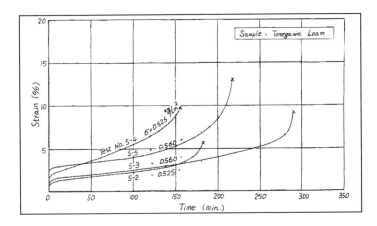

Figure 3.   Results of creep-rupture tests, relating percentage strain with time (Saito and Uezawa 1961)

This section provides a summary review of some of the existing models. These models have been classified into three categories namely, empirical, semi-empirical and theoretical.

Empirical Models

Several authors have attempted to fit existing displacement-time data to mathematical functions. These can typically be divided into two groups or forms, namely, exponential and power models. The major issue regarding these models is their predictive ability. Often, the proposed experimental models may not properly predict the behaviour of new slopes. This might be due to many reasons, such as differences in environmental conditions, geology and failure mechanisms.

Saito and Uezawa's Model
Saito and Uezawa (1961), the first researchers in this field, investigated the pre-failure behaviour of

soil over a long period in the laboratory and then extended the results to field scale. The experiments were limited to compressive and triaxial tests at a constant load of 70% of the Unconfined Compressive Strength (UCS).

The samples were from the same geographical region but soil types were different. Figure 3 shows the creep curves of four samples with three major phases, consistent with Figure 2. The nature of the curves are similar, including the gradients in the different stages, however the durations of the secondary stages vary by up to a factor of two.

Saito and Uezawa (1961) subsequently plotted the results on a logarithmic graph (Figure 4) and proposed the first empirical failure prediction model (Equation 1). This graph includes both laboratorial and practical experiments.

$$\log t_r = C - m \log \dot{\varepsilon} \tag{1}$$

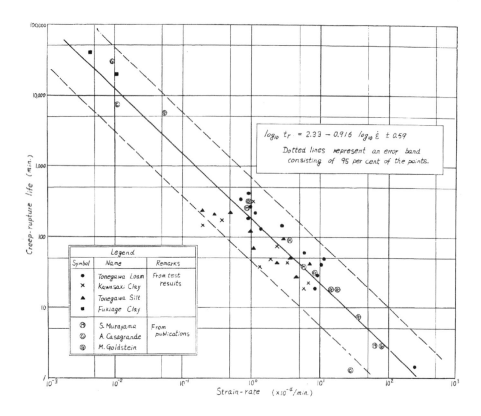

Figure 4. Relationship between creep-rupture life (time to collapse) and strain rate (Saito and Uezawa 1961)

Where, t_r is the time to collapse, $\dot{\varepsilon}$ is the strain rate, m and C are constants. The equation of constants (Equation 2) was chosen based on the data in Figure 4 with error bands containing 95% of the failure points.

$$\log t_r = 2.33 - 0.916 \log \dot{\varepsilon} \pm 0.59 \qquad (2)$$

In the model, $\dot{\varepsilon}$ is the strain rate in the secondary stage and is usually assumed constant. Glastonbury (2002) chose the strain rate at the onset of tertiary stage (The boundary between secondary and tertiary phases) in this model. Careful monitoring would be required to accurately note this boundary prior to collapse.

Although, the model provided an acceptable fit to the experiments at that time, Glastonbury (2002) showed it was not appropriate in some rock slopes. Using his own database of rock slopes, Glastonbury found that a large percentage of his data fell outside the error bands nominated by Saito and Uezawa (1961) (Figure 5).

The initial constants chosen by Saito and Uezawa (1961) were predominately based on

testing of soils. Glastonbury (2002) data shows that although the form of Equation 1 may have some merit, Equation 2 can not be extrapolated to rock slopes. This is not surprising due to the different nature of the materials.

Saito (1965) noted the existence of water in his slopes however, in his laboratorial creep tests all the tests were conducted dry. The application of results from testing dry samples to the non-dry condition may not lead to acceptable outcomes, unless some modifications are considered prior to the application.

Saito's Model

Saito (1969) proposed a new model for slopes in the tertiary stage. Saito (1969) modified Saito and Uezawa's (1961) relationship to Equation 3.

$$\Delta l = l_0 a \log \frac{t_r - t_0}{t_r - t} \qquad (3)$$

Where, Δl is the relative displacement between two measured points and l_0 is the initial distance between two measured points. By selecting three optional

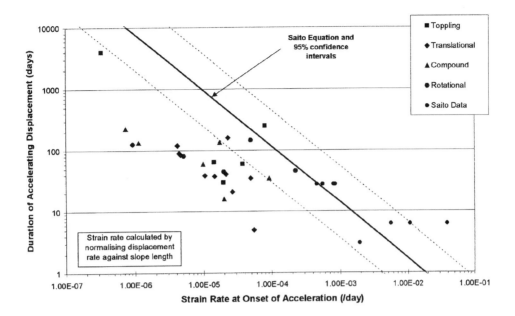

Figure 5. Relationship between strain rate at onset of acceleration and duration of acceleration phase (Glastonbury 2002)

points during the tertiary stage as shown in Figure 6 under the condition:

$$\Delta l_2 - \Delta l_1 = \Delta l_3 - \Delta l_2 \qquad (4)$$

The model could be simplified to Equation 5.

$$\frac{\Delta l_2 - \Delta l_1}{\log \dfrac{t_r - t_1}{t_r - t_2}} = \frac{\Delta l_3 - \Delta l_2}{\log \dfrac{t_r - t_2}{t_r - t_3}} \qquad (5)$$

Saito suggested an iterative procedure where the equation is updated with new data should be adopted to provide the most accurate estimate for the time to collapse.

Care must be taken in selecting the data used for input into the model by Saito. It has been noted in some more recent studies that authors have incorrectly used tertiary stage data for application to the Saito and Uezawa's (1961) model when it is only appropriate for the Saito's (1969) model.

Saito (1969) concluded that predictions should be estimated using the Saito and Uezawa's (1961) model during the secondary stage and then verified continuously by using Saito's (1969) model, once the tertiary stage had commenced.

Fukunzo's Model

Fukunzo (1985) proposed a model based on the inverse of the surface displacement velocity. The

model was only applicable during the tertiary creep stage.

Fukunzo (1985) modified Saito's model to Equation 6.

$$\log\left(t_f - t\right) = A' - B' \cdot \log V \qquad (6)$$

Where, $A' = -\log\{A(\alpha - 1)\}$, $B' = \alpha - 1$ and:

$$\dot{V} = A \cdot (V)^{\alpha} \qquad (7)$$

Where A and α are constants, $\dot{V}$ is the acceleration and V is the velocity. The A controls the extension of the inverse number of the velocity-time curve and α controls the shape of the curve (concave, convex and linear).

He then proposed Equation 8 to be applicable during the tertiary stage.

$$\frac{1}{V} = \left\{A\left(\alpha - 1\right)\right\}^{\frac{1}{\alpha - 1}} \cdot \left(t_f - t\right)^{\frac{1}{\alpha - 1}} \qquad (8)$$

The variation of the velocity-time curves is illustrated diagrammatically in Figure 7 for different α values

Although Fukunzo (1985) considered the effect of rainfall, it was not explicitly considered in the final relationship.

Fukunzo (1985) based his equations on the results of three laboratory scale soil slope models.

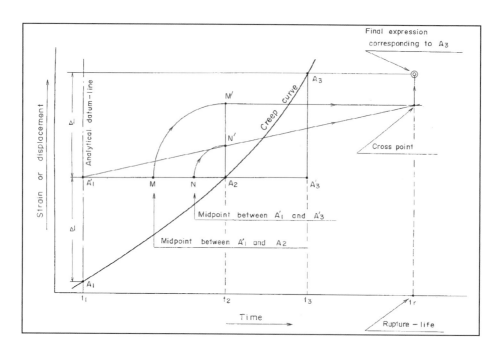

Figure 6. Graphical analysis for rupture-life (time to collapse) in the tertiary creep range (Saito 1969)

Further validation of this model would be required before it was applied to field-scale rock slopes.

Zavodni and Broadbent's Model
In a different approach, Zavodni and Broadbent (1980) proposed a model for rock slopes, based on the results from an open pit copper mine.

The main distinction of this model is the prediction of velocity or displacement rate at the failure point rather than the time to collapse. However, the failure time could subsequently be identified from the graph in Figure 8.

The preliminary proposed model was given as Equation 9.

$$\frac{V_{mp}}{V_o} = K \qquad (9)$$

Where, K is a constant (Average 7.21, ranges between 4.6–10.4), V_{mp} is the velocity or the displacement rate at the midpoint of the accelerating stage (Progressive failure) and V_o is the velocity or the displacement rate at the onset of the accelerating stage.

The main model is given as the Equation 10.

$$V_{col} = K^2 V_o \qquad (10)$$

Where, V_{col} is the velocity or the displacement rate at the collapse point.

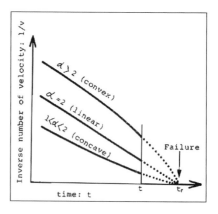

Figure 7. Variation of the Fukunzo model for different a (Fukunzo 1985)

As this model applies only in the tertiary stage, there is no possibility to predict the time to collapse before the onset of tertiary creep. To predict an accurate K, recognition of the midpoint in tertiary creep is required. When failure has not occurred this midpoint is not identifiable. Recognition of the exact onset point of the accelerating stage is difficult. Since the model is sensitive to this point, a poor estimation can lead to a poor prediction. The displacement-log time curve (Creep curve) was assumed linear so as to

252

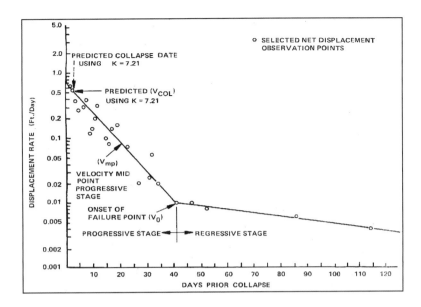

Figure 8. Typical displacement rate versus time record of a large-scale rock failure proceeding to collapse (Zavodni and Broadbent 1980)

simplify the calculations. This may increase the error in any predictions made. Note also that this method has only been validated for one particular mine. Care should be taken when applying this method to other conditions.

Glawe and Lotter's Model
Glawe and Lotter (1996) proposed Equation 11 for predicting the time to failure of slopes failing due to toppling. The equation was developed by curve fitting to data from a failed slope (Figure 9).

$$D = At^3 + Bt^2 + Ct + E \qquad (11)$$

Where A, B, C and E are constants, D is the displacement rate and t is the time of the onset of the tertiary creep.

Hayashi and Kawabe's Model
Hayashi and Kawabe (1996) modified Saito's (1969) model based on new data. The main investigation was focused on landslide and equilibrium analysis.

$$l - l_0 = \frac{l}{b}\ln\frac{\left(t_r - t_0\right)}{\left(t_r - t\right)} \qquad (12)$$

Where, l_0 is the initial displacement, l is the displacement at the time t, t_0 is the initial time, t_r is the time to collapse and b is the constant.

Since Hayashi and Kawabe (1996) carried out tests on soil experiments alone, all uncertainties,

with regard to rock slopes, in Saito's model also exist here.

Glastonbury's Method
Glastonbury (2002) used a comprehensive dataset of pre-failure deformation behaviour of rock slopes to produce bounds on the range of expected slope behaviour (Figures 10 to 13). Glastonbury divided his dataset of rock slopes, based on failure mechanisms (Hutchinson 1988) into four groups namely, translational, rotational, toppling and compound modes of failure.

The graphs are only applicable during the tertiary stage. Glastonbury (2002) provides valuable information on a wide range of slopes in a range of geotechnical conditions.

Semi Empirical Models

These models typically consist of an empirical statistical approach with some modifications based on a theoretical analysis.

Cruden and Masoumzadeh's Model
Cruden and Masoumzadeh (1987) modified Saito and Uezawa's (1961) model by using the DW statistical distribution method (Durbin and Watson 1951) to better estimate the equation coefficients (Equation 13).

253

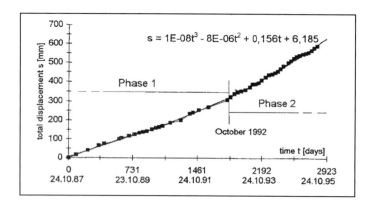

Figure 9. Displacement record for a site, a good fit is given by the function provided (Glawe and Lotter 1996)

$$\dot{\varepsilon} = \frac{C}{\left(t_f - t\right)^n}$$

(13)

Where C and n are constants and other parameters are as defined as for Saito and Uezawa's (1961) model.

The model was validated using results from an open pit coal mine. It was not verified for other geological environments.

There is some uncertainty as to which stage should be used for obtaining the strain rate data. It appears that the strain rate during secondary creep should be used however; this is not definitively discussed in the paper.

Voight's Model
Voight (1989) based on several slopes and laboratory testing, extended Fukunzo's (1985) model comprehensively to other slopes and solid media for broader applications. This resulted in Equation 14.

$$\varepsilon = \frac{1}{A(\alpha-2)}\left\{\left[A(\alpha-1)t_f + \dot{\varepsilon}_f^{1-\alpha}\right]^{\left(\frac{2-\alpha}{1-\alpha}\right)}\right.$$

$$\left. - \left[A(\alpha-1)(t_f-t) + \dot{\varepsilon}_f^{1-\alpha}\right]^{\left(\frac{2-\alpha}{1-\alpha}\right)}\right\}$$

(14)

Where, A and α ($\alpha > 1$, $\alpha \neq 2$) are constants, ε is the strain, $\dot{\varepsilon}$ is the strain rate, t is the optional time and t_f is the time to collapse.

A normalised and simplified relationship is provided as Equation 15.

$$t_f - t = \frac{\dot{\varepsilon}^{1-\alpha}}{A(\alpha-1)}$$

(15)

Where, A and α ($\alpha > 1$) are constants, t and $\dot{\varepsilon}$ are the optional time and the strain rate at that time, respectively.

As discussed for Fukunzo's model α controls the extension of the displacement curve and A controls the range of concavity, linearity or convexity of the graph (Figure 14). Other characteristics are similar to Fukunzo's (1985) model.

The model was initially proposed for slopes however, Voight (1989) worked on developing a relationship that was general in nature with potential applications for a wide range of materials. With an appropriate estimation of the constants, it can be applied to isotropic and anisotropic solid media including those in slopes.

Voight (1989) based on displacement data from several failed soil and rock slopes, calculated the appropriate range of α ($2.0 < \alpha < 2.2$) but he did not define a boundary between rock and soil slopes.

Like earlier models, Voight's model suffers from insufficient consideration of geological factors and failure mechanisms.

In subsequent research Crosta and Agliardi (2002) proposed better values for A and α by considering the influence of environmental factors such as geological and seasonal parameters. They recommended using the 'Rosenbrock pattern search' method (Rosenbrock 1960) in the regression analysis to estimate A and α. Table 1 presents the calculated results by Crosta and Agliardi (2002).

Gunzburger's Model
Gunzburger et al. (2005) investigated the influence of environmental factors on the instability of rock slopes. The main focus was on the impact of temperature change on the instability of slope over the long term. Gunzburger et al. (2005) proposed Equation 18

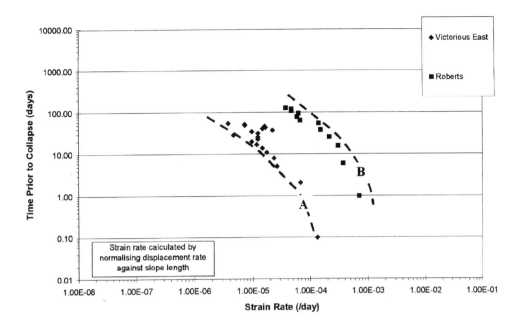

Figure 10. Strain rate-time behaviour for rotational rock slope failures (Glastonbury 2002)

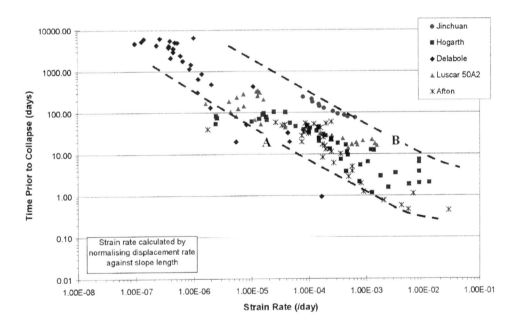

Figure 11. Strain rate-time behaviour for toppling failures in rock slopes (Glastonbury 2002)

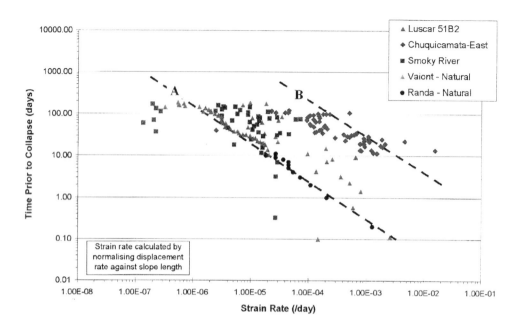

Figure 12.   Strain rate-time behaviour for internally sheared compound slides in rock (Glastonbury 2002)

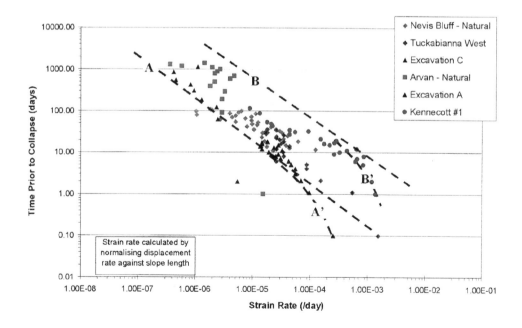

Figure 13.   Strain rate-time behaviour for translational rock slides (Glastonbury 2002)

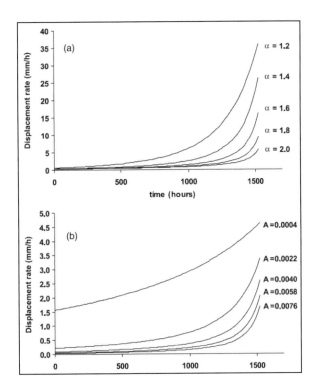

Figure 14. Sensitivity analysis of the Voight's model with respect to parameters A and α (Crosta and Agliardi 2002)

which illustrates the dependency of the displacement rate on temperature.

$$U(t) = \frac{(1+v)\alpha A\delta}{(1-v)\sqrt{2}} \sin\left(\omega t - \frac{\pi}{4}\right) \tag{18}$$

Where, $U(t)$ is the displacement function, v is the Poisson's ratio, t is the seasonal time. The other parameters were taken from the Carslaw and Jeager (1959) relationship (Equation 19).

$$T(x,t) = T_0 + Ae^{-x/\delta} \sin\left(\omega t - \frac{x}{\delta}\right) \tag{19}$$

Where, $\delta = \sqrt{2\alpha/\omega} \cong 0.19$ cm is the typical penetration depth of the temperature along the slope body, T_0 is the initial temperature and $T(x,t)$ is the temperature function.

The model shows the displacement rate as a function of temperature and the temperature, itself depends on climate and seasonal changes. Temperature also causes erosion; therefore, it could be pointed that the model includes erosive effects indirectly.

Table 1. Results of the regression analysis performed by "Rosenbrock pattern search" method of the entire time series (Crosta and Agliardi 2002)

	A	α	t_f	R
Upper scarp				
D5	0.0005	2.008	1650	0.993
D7	0.0013	2.092	1480	0.994
D20	0.0053	2.047	1440	0.995
D21	0.0042	2.036	1440	0.992
Lower scarp				
D13	0.0049	2.028	1440	0.985
D26	0.0043	1.627	1450	0.989
D28	0.0154	1.776	1450	0.986
D29	0.0020	1.660	1390	0.942

Theoretical Method

At present, there does not appear to be a definitive theoretical method for rock slope failure prediction. The published theoretical models typically focus on the stress-strain behaviour of solid media with some models proposed for understanding the behaviour of rock media over extended periods of loading. Shao et al. (2003) is discussed here as an example.

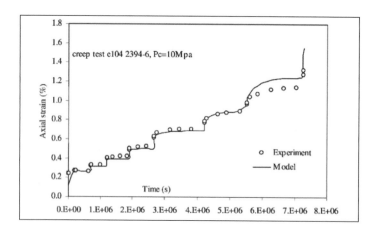

Figure 15. Simulation of a creep test with various phases of deviatoric loading until failure (Shao et al. 2003)

Shao et al. (2003) firstly developed a constitutive model (Equation 20) for understanding the creep behaviour of solid rock media and then used this within a finite element framework to predict the long term behaviour of a slope.

$$\delta\sigma_{ij} = C_{ijkl}^{epd} \delta\varepsilon_{kl} + \Psi_{ij}^{\varepsilon}\zeta\delta t \qquad (20)$$

Where, C_{ijkl}^{epd} is the tangent elastoplastic stiffness tensor of the damaged material, Ψ_{ij}^{ε} represents the second order tensor defining the variation of stresses during the relaxation process and ζ is the differentiation of the Equation 21 that defines the role of time in the model. For detailed information the authors recommended a review of the original paper (Shao et al. 2003).

$$\zeta(t) = \zeta\left(\varepsilon_{ij}^{p}(t), t\right) \qquad (21)$$

The success of this model depends on a good understanding of the plasticity behaviour of solid media. Research in this respect is ongoing by the authors.

Shao et al. (2003) used their constitutive model in a finite element program to assess the behaviour of a model slope. Figure 15 shows that a good fit was obtained for the experiment.

The constitutive model proposed is for a continuum. Therefore, further work is required before this could be applied in practice to a rock mass where the effect of defects is critical.

CONCLUSIONS

A general overview of published slope failure prediction models has been presented. These models are typically empirical or semi-empirical and as such rely heavily on their supporting data. Users should carefully consider which is the appropriate model for their particular geological conditions.

Most of the models discussed use strain rate as the basic parameter of interest. This is possible to monitor and measure in the field. However, Szwedzicki (2003) suggests that understanding changes in field stresses and stress localizations within slopes is important when assessing the future behaviour of slopes. Further research is required to understand how stresses change within slope rock masses over time before the empirical and theoretical models discussed above can be improved.

The methods discussed have mostly been developed from experience with soil slopes or soil laboratory testing. Their application to rock masses where defects (rock structure) dominate the behaviour is questioned. Further work is required to calibrate the models for various types of rock mass. Methods that predict the behaviour of rock slopes should also explicitly consider different failure mechanisms as per Glastonbury (2002).

REFERENCES

Carslaw, H. and Jeager, J. 1959. *Conduction of heat in solids*. Oxford. Clarendon Press.

Crosta, G. B. and Agliardi, F. 2002. How to obtain alert velocity thresholds for large rockslides. *Physics and Chemistry of the Earth*. 27: 1,557–1,565.

Cruden, D. M. and Masoumzadeh, S. 1987. Accelerating creep of the slopes of coal mine. *Rock Mechanics and Rock Engineering*. 20: 122–135.

Durbin, J. and Watson, G. S. 1951. Testing for serial correlation in least squares regression. *Biometrika.* 38: 159–176.

Fukunzo, T. 1985. A new method for predicting the failure time of a slope. In *Fourth International Conference and Field Workshop on Landslides.* Tokyo.

Glastonbury, J. P. 2002. *The pre and post-failure deformation behaviour of rock slope.* Civil and Environmental Engineering. Sydney. The University of New South Wales.

Glawe, U. and Lotter, M. 1996. Time prediction of rock slope failures based on continuous monitoring. In *Mechanics of Jointed and Faulted Rock.* Balkema.

Gunzburger, Y., Merrien-Soukatchoffa, V. and Guglielmi, Y. 2005. Influence of daily surface temperature fluctuations on rock slope stability: case study of the Rochers de Valabres slope (France). *International Journal of Rock Mechanics and Mining Sciences.* 42: 331–349.

Hayashi, S. and Kawabe, H. 1996. A model for acceleration of slide to slope failure In *Seventh International Symposium on Landslides.* Trondheim. Norway Balkema.

Hoek, E. and Bray, J. 1977. *Rock Slope Engineering.* London. IMM.

Hutchinson, J. N. 1988. General Report: Morphological and geotechnical parameters of landslides in relation to geology and hydrology. In *Fifth International Symposium on Landslides.* Lausanne. Switzerland. Balkema.

Rosenbrock 1960. An automatic method for finding the greatest or least value of a function. *The Computer Journal.* 3: 175–184.

Saito, M. 1965. Forecasting the time of occurrence of a slope failure. In *Sixth International Conference on Soil Mechanics and Foundation Engineering.*

Saito, M. 1969. Forecasting time of slope failure by tertiary creep. In *Seventh International Conference on Soil Mechanics and Foundation Engineering.* Mexico City.

Saito, M. and Uezawa, H. 1961. Failure of soil due to creep. In *Fifth International Conference on Soil Mechanics and Foundation Engineering.*

Shao, J. F., Zhu, Q. Z. and Su, K. 2003. Modelling of creep in rock materials in terms of material degradation. *Computers and Geotechnics.* 30

Szwedzicki, T. 2003. Rock mass behaviour prior to failure. *International Journal of Rock Mechanics & Mining Sciences.* 40: 573–584.

Voight, B. 1989. A relation to describe rate-dependent material failure. *Science.* 243: 200–203.

Wyllie, D. C. and Munn, J. F. 1978. The use of movement monitoring to minimize production losses due to pit slope failures. In *Proceeding of the First International Symposium on stability in Coal Mining* Vancouver. Canada. Miller Freeman.

Zavodni, Z. M. and Broadbent, C. D. 1980. Slope failure kinematics. *CIM Bulletin.* (April, 1980): 69–74.

Investigation into the Stress Shell Evolving Characteristics of Rock Surrounding Fully Mechanized Top-coal Caving Face

Ke Yang

Anhui University of Science and Technology, Huainan, Anhui, China

Guangxiang Xie

Anhui University of Science and Technology, Huainan, Anhui, China

Jucai Chang

Anhui University of Science and Technology, Huainan, Anhui, China

ABSTRACT: A key issue in underground mining is to understand and master the evolving patterns of mining-induced stress, and to control and utilize the action of rock pressure. Numerical and physical modeling tests have been carried out to investigate the distribution patterns of stress in the rock surrounding a fully mechanized top-coal caving (FMTC) face. There is a macro-stress shell (MSS) composed of high stress exists in the rock surrounding, which bears and transfers the loads of overlying strata, acts as the primary supporting system of forces, and is the corpus of characterizing three-dimensional and macro-rock pressure distribution. Its evolution is a mined-induced high stress developing and dynamically equilibrating process in spatial rocks during mining and its external and internal shape changes with the variations in the working face structure as the face advances. Within the low-stress zone inside the stress shell, which only bears parts of the load from the lower-lying strata, will produce periodic pressures on the face instead of great dynamic pressure even if the beam ruptures and loses stability. The results show that the FMTC face is protected by the stress shell of the overlying surrounding rock. We reveal the mechanical nature of the top coal of an FMTC face acting as a "cushion." The strata behaviors s are under control of the stress shell. Drastic rock pressure in mine may occur when the balance of the stress shell is destruction or the forces system of the stress shell transfers.

INTRODUCTION

To sum up, some important and significant fruits have been obtained regarding rock pressure on FMTC faces. They are of benefit to the development of FMTC technology.[1-11] However, there are still questions that are frequently asked and so need further interpretation, as to what mechanism is responsible for the easing of mine pressure and for the formation of bearing stress. Increase in FMTC mining thickness is bound to produce changes in the height of the caving zone and the fractured zone. This is a dynamic process which results in variations in the patterns governing the stress distribution in the upper sagging zone. All these changes and variations are caused by something that affects the distribution of abutment pressure and strata behaviors, that is, the macro stress field, which is of much more significance.

For a further investigating into the patterns of macro stress field in surrounding rock of FMTC face during mining from the large-scale and three-dimensional space, numerical and physical modeling tests are carried out with the following details: (i) physical modeling of stress distribution patterns in the vicinity of face during mining; (ii) numerical modeling of stress distribution patterns in the vicinity of face during mining adopting FLAC3D soft.

GEOLOGICAL CONDITIONS

Xieqiao FMTC Face 1151(3) works Seam C13-1 which averages 5.4m thick, dipping 12°~15°, averaging 13°. The mining height is 2.6m, and top-coal is 2.8m thick. The strike length is 1674m and the dip length is 231.8 meters. The structure of coal seam is stable in the face, and develops generally two-layer carbonaceous mudstone or tonstein. The roof and floor of coal seam distributes as follows: The main roof is powdery packsand; the immediate roof is comprised of mudstone or and coal; the immediate floor is mudstone; the main floor is siltstone (Figure 1). The face is at an elevation of -588~662m.The Face is separated with the goaf of Face 1141(3) by an air return and a protective coal pillar as wide as 5m.

PHYSICAL MODELING

Experimental Design

Physical modeling test is based on geological and technological conditions of FMTC Face 1151(3).

Based on similarity criterion with practical situation of in situ and experimental models [12], considering the size of the model and measurement convenience, geometric proportion is 1:100, the bulk density proportion and the stress proportion are 1:1.67.

The length, width, height of the model is respective 420cm, 25cm and 200cm with plane stress model. The two boundary coal pillar, mining strike-ward length, mining depth and simulated strata height of experimental model is respective 110cm, 200cm, 648cm and 170cm. Total 36 BX120–50AA resistance strain gages are used for observing the stress changes in surrounding rock with installing six stress measurement lines in intervals of 40cm (I, II, III, IV, V, VI) on the model. 7v14 data acquisition system is used for collecting data and seven hydraulic jacks are adopted for applying loading for the rest of overlying strata according to gravity compensation loading. The mining procedure is simulated by cutting the seam from left to right in intervals of 50mm. The results of the test are obtained as follows.

Results of Physical Modeling Test

Stress Distribution Patterns of the Roof and the Floor

The stress variation curves of roof and floor on the strike are as shown in Figure 3. As can be seen, the stress increases drastically at observation points 100 meters ahead of the face, more so 60 meters ahead, and reaching their peaks 8~24m ahead, except at points 137.8m from the seam floor. Afterwards, the stress in the rock mass begins to drop at points -2.5m, 8.7m, 23.2m and 48.1m from the seam floor, except at two points 67.7m and 85.4m from the floor, where the stress remains basically unchanged. After the mining, the stress will keep dropping, except at the point of 137.8 m, where the stress remains

basically unchanged, so much so that the rock mass is rendered a zone of stress-dropping.

Stress Distribution Patterns of Strata as the Face Advances

By analyzing Figure 4, patterns governing stress variations in the surrounding rock when the face has advanced different distances are obtained as follows:

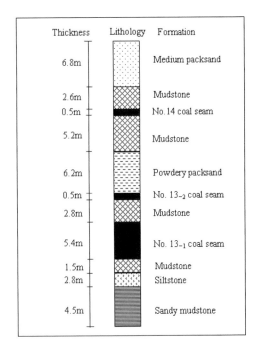

Figure 1. The generalized stratigraphic column of 1151(3) face

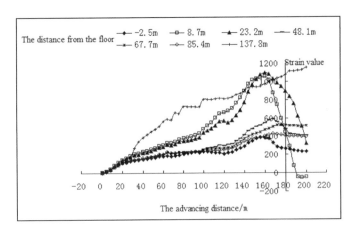

Figure 2.   The stress variation curves of overlying strata and floor on the strike

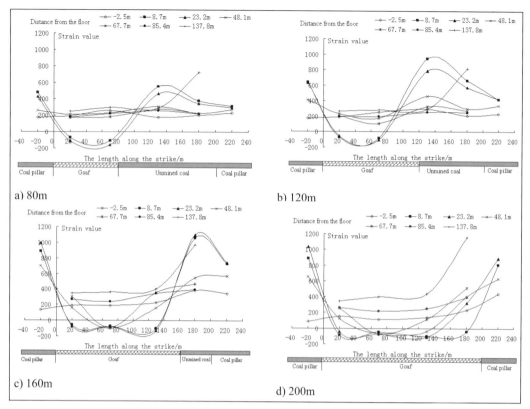

a) 80m

b) 120m

c) 160m

d) 200m

Figure 3. The stress variation curves at the different advancing distances

1. In the process of coal mining, there is a stress-increase zone in virgin rock and coal mass within a certain area behind the open-off cut and ahead of the working face, and the area changes constantly. When the face has advanced a certain distance, this area remains basically unchanged.

2. In the process of coal mining, roughly speaking, the overlying strata near the working face and above the goaf, at a certain level, stay highly stressed. The stress continues ahead and upwards with the advancing of the working face. But at a certain distance behind the face, the stress keeps stable at a certain height of the strata (137.8 meters above the seam floor in our experiments). However, the stress of the strata below the goaf keeps low throughout the process.

3. High stress occurs in the virgin rock and coal mass, both behind and ahead of the working face, and in the strata overlying at a certain height. This fully proves that there is a dynamic stress arch in the surrounding rock. The arch bears and transfers pressures and loads from the overlying strata.

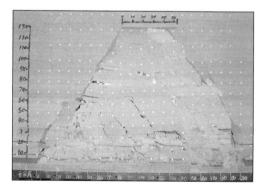

Figure 4. The deformation of overlying strata

The inbreak form and the height of the two zones (caving zone and fractured zone) are shown in Figure 4. The height of the two zones in our physical modeling test is taken as 130m. The results show that the voussoir beam formed by the main roof lies in a stress-decrease zone below the stress arch. The profile formed by the caving zone and fractured zone of the surrounding rock is thus regarded as the negative camber of the stress arch, which lies in the virgin rock and coal mass at both ends of the working face and in its overlying sagging zone.

NUMERICAL MODELING OF STRESS DISTRIBUTION IN SURROUNDING ROCK OF FMTC FACE

Model Construction

In this investigation, fast Lagrangian analysis of continua (FLAC^{3D})[13] code is used. In this study, the 3D model is 314.17m (height) ×500m (strike length) ×600m (dip width). The model simulates Seam 13-2 and its roof and floor, 13° dipping, 5.4m thick., 231.8m long of the face dip. The model is divided into 95332 3-D elements and 112739 nodes. Displacement of the model are restricted on boundaries, horizontal ones by the four sides and the vertical ones by the bottom, with vertical loadings exerted from above, in simulation of the weight of overlying strata.

Results of Numerical Simulation

Stress Field Characteristics of Surrounding Rock on the Strike

It is can be seen from the principal stress vector field (Figure 5) that there is a stress arch composed of high stress undles in the surrounding rock of FMTC face on the strike. The primary mechanical characteristic is that the principal stress value of the stress shell is larger than that inside and outside. The skewback of the stress arch lies in coal seams ahead of the face, forming front abutment pressure. The voussoir beam formed by the main roof lies in the stress-decrease

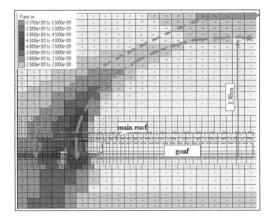

Figure 5. The principal stress vector field of surrounding rocks in the center of face on the strike (MPa)

zone below the stress arch. The principal stress in the main roof is larger than in other strata, indicating that breakage and instability of the main roof will cause stress re-distribution which exerts indirect effects, though not serious, on the top-coal and the face. Nestled within the low-stress area below the macro stress arch of the surrounding rock, the FMTC face is subjected to a tendency of mine pressure easing, different from the case of an fully-mechanized face.

The height of the stress arch on the strike is 130 meters, 24.1 times the mining thickness, in the center of FMTC Face 1151(3). The height of arc along the strike by numerical modeling is the same as the height by physical modeling.

Stress Field Characteristics of Surrounding Rock Along the Dip

Figure 6 show the maximum principal stress distribution at different places of the surrounding rock ahead of and behind the face along the dip. It can be seen that there is a stress arch composed of high stress bundles in the surrounding rock along the dip. The form of the stress arch changes with the face advancing, as follows:

1. Out of the mining influence range, the skewback of the stress arch formed by the mining of the upper section lies on the edge of the virgin coal seams below (Figure 6a). Inside the stress arch, there is a low stress area formed by the goaf of the upper section; outside, there is an initial stress zone of the virgin rock and coal mass along the dip.
2. At the peak of the abutment pressure ahead of the face, there is a large-scale stress concentration zone on the edge of coal seams where strike-ward and dip-ward abutment pressure confer. So the skewbacks are formed in large coal seams and their roof strata (Figure 6b).
3. Near the face, as a result of the unloading of the rock mass on the edge of coal wall and above the face and of abutment pressure transfer to the downside virgin coal seams, a larger-scale stress arch is developed, with a new skewback formed on the edge of the downside virgin coal seams and the original one shrinking upward (Figure 6b,c). The stress arch above the face is 62m high, 11.5 times the mining thickness.
4. Behind the face, farther from the face, the former skewback shrinks further and the stress arch expands over the whole mining area (Figure 6d,e).

There are stress arches in the surrounding rock of FMTC faces both on the strike and on the dip.

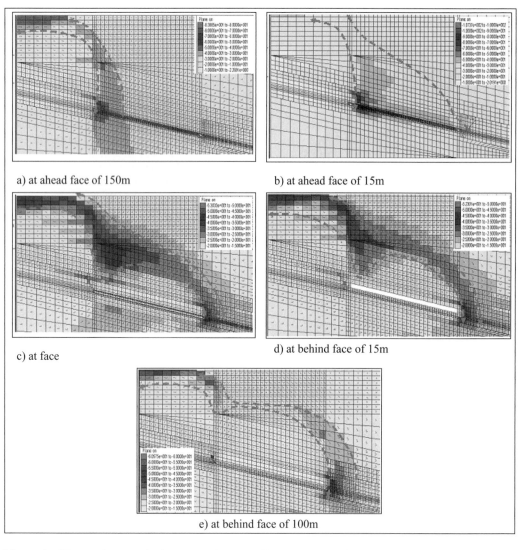

a) at ahead face of 150m   b) at ahead face of 15m

c) at face

d) at behind face of 15m

e) at behind face of 100m

Figure 6. The maximum principal stress field at different distances from face to the dip (MPa)

The two arcs make up of a ellipsoidal shell in three-dimensional space of the face. It can be seen that there is also a macro stress shell composed of high stress. The schematic plot of the stress shell in surrounding rock is as shown in Figure 7.

MECHANICAL CHARACTERISTICS OF STRESS SHELL AND FMTC FACE

1. There is a macro stress shell composed of high stress bundles in the surrounding rock of an FMTC face. The stress shell is mechanically characteristic of the following:

   a. The stress shell, made up of high stress bundles, is not an objective entity.
   b. The principal stress of the stress shell is larger than that of the rock mass inside and outside.
   c. The stress shell lies in the virgin rock and coal mass in the vicinity of the working face and in the sagging zone of the overlying strata. The shell skewback forms abutment pressure behind and ahead of the face and on the edge of the virgin seam on the sides. The stress of the shell skewback is the bearing stress of the face and its vicinity.

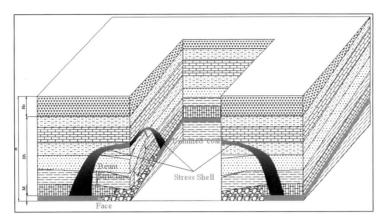

Figure 7.   The schematic plot of the stress shell in surrounding rock

d. The shape of the stress shell changes with the advancement of the working face. Throughout the process of mining, the stress shell changes all along. When a new face is prepared, the skewback of the former stress shell formed in the neighboring face above becomes the inner skewback, which goes higher and higher as the working face approaches and gradually disappears behind the face.

e. The stress shell is geometrically asymmetrical and non-uniform as far as stress distribution is concerned. Its span does not agree with its thickness. Generally speaking, stress is greatest at the skewback, second at the vault and still less at the shoulders. The stress shell is not fully enclosed. The voussoir beam lies in the stress-decrease zone below the macro stress shell. The inner profile formed by the caving zone and the fractured zone of the surrounding rock is the negative camber of the stress shell.

2. Main mechanical characteristics of FMTC face

An FMTC face is mechanically characteristic of its location in the low-stress zone protected by the stress shell of the overlying surrounding rock. The overlying rock mass passes its loads and pressures through the stress shell onto the working face and its vicinity. Stress peak value of coal seam transfer to the downside coal mass, and coal seam of the face is unloading. Hence the stress shell is the primary supporting body and the voussoir, which only bears partial loads of the strata under the stress shell. Breakage and instability of the voussoir, being situated in the low-stress zone under the shell, can give rise to periodic pressures, instead of

great dynamic pressures, onto the face. Only off-balance of the stress shell can result in violent strata behaviors, such as shock bump or rock burst. Since an FMTC face is always situated in the low stress zone below the stress shell, strata behaviors tends to be eased.

The strata behaviors of the face and its neighboring gates are under control of the stress shell. Reasonably adjusting structure parameters of working face can improve dynamic balance of surrounding rock stress shell. It plays a positive role in protecting working face and reducing rock pressure influence. The discovery and analysis of the stress shell has revealed the mechanical nature of the top coal of an FMTC face acting as a "cushion."

CONCLUSIONS

1. There is a macro stress shell composed of high stress in the surrounding rock of an FMTC face. The stress of the shell is higher than its internal and external stress. The stress shell lies in the virgin rock and coal mass in the vicinity of the working face and in the sagging zone of the overlying strata. The shell skewback forms abutment pressure behind and ahead of the face and on the edge of the virgin seam on the sides. The stress of the shell skewback is the abutment pressure of the face and its vicinity. The shape of the stress shell is related to the structure of the working face.

2. An FMTC face is mechanically characteristic of its location in the low-stress zone protected by the stress shell of the overlying surrounding rock. Consequently, strata behaviors tends to be eased. The stress shell, which bears and transfers the loads of overlying strata, acts as

the primary supporting structure. The voussoir beam, situated in the stress-decrease zone below the stress shell, only bears partial loads of the strata below. Breakage and instability of the voussoir can give rise to periodic pressures, instead of great dynamic pressures, onto the face. The strata behaviors of the face and its neighboring gates are under control of the stress shell. Only off-balance of the stress shell can result in violent strata behavior, such a shock bump or rock burst. The discovery and analysis of the stress shell has revealed the mechanical nature of the top coal of an FMTC face acting as a "cushion."

REFERENCES

1. Qian Minggao, Shi Pingwu, et al. Rock pressure and strata control [M]. Xuzhou: Press of China University of Mining and Technology, Nov. 2003 (in Chinese)
2. Jin Zhongming. Theory and technology of top-coal caving method [M]. Beijing: Coal Industry Publishing House, Apr. 2001 (in Chinese)
3. Heping Xie, Zhonghui Chen, Jiachen Wang. Three-dimensional numerical analysis of deformation and failure during top coal caving [J]. Int. J. of Rock Mech. & Min. Sci. 1999, 36:651–658
4. Rajendra Singh and T.N. Singh. Investigation into the behaviour of a support system and roof strata during sublevel caving of a thick coal seam [J]. Geotechnical and Geological Engineering 17: 21–35, 1999.
5. Qian Minggao. Analysis of key block in the structure of voussoir beam in longwall mining [J]. Journal of China coal society, 1994, 19(6):557–563 (in Chinese)
6. Miao Xiexing and Qian Minggao. Analysis of integral mechanical model of support and surrounding rock in top-coal caving face [J]. Journal of coal, 1998, 7(6):1~5(in Chinese)
7. Wu Jian, Meng Xianrui, Jiang Yaodong. Development of Longwall Top-coal Caving Technology in China [J]. Proceeding of 99 International Workshop on Underground Thick-seam Mining. 1999, 6:101–112
8. Lu Ming-xin, Hao Hai-jin, Wu Jian. The balance structure of main roof and its action on top coal in longwall top coal caving work-face [J]. Journal of China Coal Society, 2002, 27(6):591–595 (in Chinese)
9. Jian Wu, Shichang Zhao. System behavior analysis of the ground movement around a longwall, Rock Mechanics as a Guide for Efficient Utilization of Nature Resources [J]: Proceedings of the 30th U.S.Symposium, 1989
10. Qian Minggao, Miao Xiexing, XUJialin, et al. Theory of key stratum in ground control [M]. Xuzhou: China University of Mining and Technology Press, 2003. (in Chinese)
11. Yan Shaohong. Study on the movement of top-coal and roof in top-coal caving face. [Ph.D. Thesis] [D]. Beijing: China University of Mining and Technology, 1995.(in Chinese)
12. Gu DZ. Physical modeling methods [M]. Xuzhou, People's Republic of China: Press of China University of Mining and Technology, 1996.
13. Manual of FLAC3D, Itasca Consulting Group, Inc, 1997.

Analysis of a M_L 4.0 Mining Induced Seismic Event

Michael Alber
Ruhr-University Bochum, Bochum, Germany

Ralf Fritschen
DMT GmbH & Co. KG, Essen, Germany

INTRODUCTION

Like in many other regions worldwide, coal mining in Germany is accompanied by mining induced seismic events. Today three areas hold active coal mines, one of them is located in the Saarland, in the far west of Germany. Although industrial mining in the Saar region started in the middle of the 18th century, only a few cases of mining induced events are reported before the 1990th. In 1997, however, several strong seismic events were felt by the population of villages in the area of the Saar mine. As a result, the Saar mine decided to install a local seismic network in order to determine the peak particle velocities (PPVs) on site and to investigate possible relations between mining and seismicity. Quickly it became clear that the observed seismicity was directly related to mining activities in the seam Schwalbach. Almost all seismic events were located in the vicinity of active longwall faces. In 2006, mining of the seam Schwalbach started in the new field Primsmulde, planned as the mine's new main production field. Mining in field Primsmulde was executed by the longwall method. Here a double longwall was used with an overall span of 700 m and an expected length of 2,500 m. Panel I was typically 100 m ahead of panel II.

In 1997, the seismic network of the mine consisted of just four surface stations. The network was not optimized for locating seismic events, but was primarily installed to determine the intensities of the mining induced events. The intensities were measured as PPV values at buildings basements. The recording was done with 3C geophones having flat velocity amplitude responses in the range between 1 Hz and 80 Hz. Over the course of the measurements the network was continuously extended. When the network was completed with underground stations, event depths and epicenter positions could be determined with an accuracy of 50 m.

From the experience of mining induced events in other fields (Fritschen 2010) a research program was started to investigate the fracture behaviour of rocks under longwall conditions in the vicinity of the seam. However, in the course of mining the double panel it turned out that the numerous mining induced seismic events originate in strata some 300 m above the seam as well as at seam level.

GEOLOGICAL SETTING OF THE MINE

The mine is located on the western part of the Saar-Nahe-basin. The SW-NE striking Saar-Nahe-Basin in SW-Germany is one of the largest Permo-Carboniferous basins in the internal zone of the Variscides. Its north extension is denoted by the Hunsrück Boundary Fault. The structural style within the basin is characterized by normal faults parallel to the basin axis and orthogonal transfer fault zones. The carboniferous rocks located in the N' part of the Saar-Nahe-Basin and show typically synclines and anticlines. The maximum sedimentary accumulation in the basin is today some 7.5 km. Coal bearing measures start in the youngest Westfal D up to Stefan C and are extracted in deep mines at depths around 1,400–1,500 m below surface.

At the mine scale the strata are dipping gently to N/NE with some 15°. The longwalls are in virgin rock mass with no previous mining operations. Stresses at the site were measured by the hydraulic fracture method and reported by MeSy (2008):

$$\sigma_v = 0.0265 \times z \quad (\text{MPa})$$

$$\sigma_h = 0.553 \times \sigma_v \quad (\text{MPa}), \text{ direction } 57°$$

$$\sigma_H = 1.155 \times \sigma_v \quad (\text{MPa}), \text{ direction } 147°$$

For the average depth z = 1411 m of the longwall under investigation the following in situ-stresses are assumed: σ_v = 37.4 MPa, σ_h = 20.7 MPa and σ_H = 43.2 MPa. Those stresses are aligned in accordance to the general stress orientation in central Europe (Reinecker et al. 2005). The ratio σ_H/σ_h is about 2, which is valid for many intraplate crustal regions for a depth of approximately 1 km (Rummel, 2002). In the local situation, the major horizontal stress is oriented perpendicular to the long axis of the basin and in this case, perpendicular to the long axis of the longwall panels. The stress values obtained were not reduced by pore pressure estimates since all drill

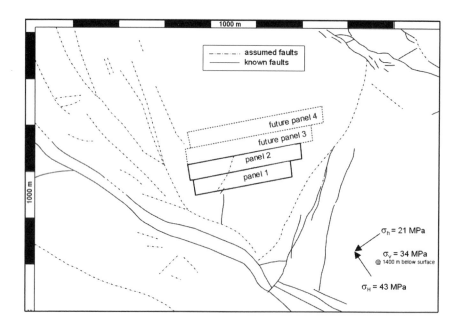

Figure 1. Plan view of the coal field with faults, panels and in-situ stresses at depth

holes were reported to be dry and the mine personnel reported "bone-dry conditions."

The coal bearing strata in the Saar-Nahe-Basin belong to the Upper Carboniferous (Pennsylvanium, 318 to 299 mya). Sediments in the basin consist of 7500 m thick molasse-type coal bearing strata with conglomerate, sandstone, siltstone and claystone as the dominant rock types. The thrusts are all oriented NE-SW and are mainly steep dipping. The thrusts are dissected by many steep dipping NW-SE striking normal faults. The setup of the field along with local faults as well as the in-situ stresses at depth is given in Figure 1. During development of the gates numerous drilholes were executed in the hangingwall as well as in the footwall and the jointing pattern was observed. The orientations of those regional tectonic structures on different scales are summarized in the discontinuity rosette plots in Figure 2.

GEOMECHANICAL SETTING

Near Seam

The lithology around the longwall is known through numerous exploration boreholes and geological mapping while developing the gateroads. Additionally two high quality double barrel core drillings were executed: one borehole 50 m below the seam and one drilling 120 m upwards. The typical coal measure rocks sandstone (fine to coarse grained) and

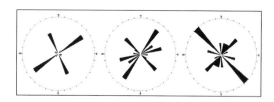

Figure 2. Orientation of tectonic elements: Major faults at 50 km scale (left), faults at 10 km scale (middle) and joints at 2 km scale (right)

siltstone in varying mixtures were encountered. The strata were all dry, typically flat lying and gently dipping with rather constant thickness. Of major concern were a few fluvial channels deposits known within the strata. Those sandstone channels were thought to be very strong and it was believed that their failure might have been the cause for mining induced seismic events (Fritschen et al. 2004). Thus, the focus was first on the geomechanical characterization of the immediate 100 m hanging wall. Towards this, an extensive laboratory program with the focus on uniaxial and triaxial strength tests was carried out (Kahlen and Alber, 2008). The results are summarized in Figure 3 and it may be seen that the coarse-grained channel sandstones do not exhibit significantly higher strength as compared to the surrounding rock types. In contrast to the common belief are the fine-grained sandstones and siltstones

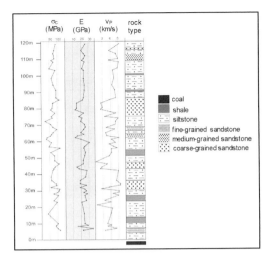

Figure 3. Schematic view of rock types and some petrophysical parameters (s_C = uniaxial compressive rock strength, E = static Young's modulus, v_P = compressional wave velocity) from double core drilling into the hanging wall strata

which form strong beds and exhibit higher stiffness than the channel rocks.

The cores have been evaluated using the Geomechanics Classification after Bieniawski (1989) without using the rating adjustment for discontinuity orientation. The lowest RMR was found in the coals seam (RMR = 25) and the coal measure rocks are of good quality (50< RMR <75) throughout with only a few exceptions. The Rock Mass Ratings were used in conjunction with the Hoek-Brown failure criterion to estimate the uniaxial compressive rock mass strength σ_{CM} as well as the rock mass deformation modulus E_M (Hoek and Brown, 1997). With the exemption of a strong layer of fine-grained sandstone 10 m above the hanging wall strata have more or less similar rock mass strength as well as rock mass stiffness.

300 m Above Seam

As mentioned in the introduction, numerous mining induced seismic events originate in strata some 300 m above the seam. Accordingly, a double barrel drill was executed from the tailgate of longwall I up to 320 m height to obtain high quality rock cores for further investigation. Special focus was placed on the mechanical characterization of last 30 m of cores as the possible source of seismicity. The laboratory program included tests on Brazilian Tensile Strength (no. 17), Uniaxial Compressive Strength (no. 23), Triaxial Compressive Strength (no. 95) and residual Triaxial Compressive Strength (no. 87). In addition, dry densities, compressional wave velocities and shear wave velocities were determined (no. 66). For triaxial testing the confining pressures were up to 25 MPa to place the rocks in virgin as well secondary stress conditions that may be encountered at a depth of 1200 m below surface, i.e., 300 m above the double panel.

The rock types recovered from drilling at 300 m above the panels include interbedded siltstones, fine- to coarse-grained sandstones as well as conglomerates. It was first attempted to characterize each rock type separately but the interbedding was so close that only few specimen with only one individual rock type could be prepared. The Residual Triaxial Compressive Strength Tests were used to estimate the rock mass strength. The firstly conducted Triaxial Compressive Strength Tests left the specimen in a fractured state. The specimens were again triaxially stressed up to their residual strength. This strength of a specimen fractured in the most unfavourable inclination with respect to the stresses is assumed to be the lower boundary of the rock mass strength. The strength data and elastic parameters of sandstone and siltstone are summarized in Table 1.

Seismic Events and Their Analyses

Events Near the Seam

In 2004, the mine started to develop the coal seam Schwalbach in the new coalfield at a depth of approximately 1400 m. The development of the field started two years before coal extraction with the driving of roadways, which later formed the longwall panels. As with the older fields the extraction of the seam Schwalbach was planned as a double panel system. It was not known if this system in the new part of the mine would induce seismic events of considerable intensity because at that time both double panel systems with strong seismic events (old field I) and double panel systems without significant seismicity (old field II) had been observed (cf. Fritschen 2009). A reliable explanation of the contrary seismicity found in these two fields was not available at the time. However, it was soon clear that the extraction of coal in the new field could lead to strong seismic events. In May 2005, after driving the first roadways a seismic event of M_l = 3.3 was induced. This was the first time that an event, which was considerably felt at surface, was induced in a German coal mine just by developing a new field. This event was localized in the middle between two roadways approximately at seam level. No other seismicity was observed after that event in the field, until mining started in October 2006. But even then seismicity was moderate with events that were hardly detected and not felt at surface. The situation changed dramatically in June

Table 1. Average geomechanical data from laboratory tests of two rock types from 300 m above seam A (s_C = uniaxial compressive strength; r= bulk density; v_P =compressional wave velocity; E = Young's Modulus; m_i = constant in the Hoek-Brown failure criterion as well as peak (intact) and residual strength Mohr-Coulomb strength parameters

Rock Type	Sandstone, coarse	Siltstone/ Claystone
σ_C (MPa)	64–102	30–40
v_P (km/s)	3.7–4.2	3.5–4.5
E (GPa)	19–36	8–19
m_i	10.3	8.6
intact c, ϕ	19 MPa, 50°	1.2 MPa, 45°
residual c, ϕ	10.7 MPa, 44°	5.7 MPa, 37°

2007, approx. 9 month after the start of longwall operations, when a Magnitude 3.6 event with a PPV of 29 mm/s was induced in the seam level, more than 300 m in front of the double seam system.

In order to evaluate the possible failure mechanism of both events that took place at the level of the seam (± 25 m) the 3D boundary element package EXAMINE 3D (Rocscience, 2007) was employed to estimate stresses between the development gates as well as ahead of the double panel. The gates of dimension 5 × 5 m did not greatly influence the state of stress, a slight reduction (<<1 MPa) of the major principal stress was numerically evaluated. By the same token, the numerical model shows stresses 300 m ahead of the face similar to the in-situ stresses. A slight reduction (<<1 MPa) of the minor principal stress oriented towards the double panel may however be assumed. Figure 4 shows plots of induced stresses vs. rock mass strength and it may be concluded that even for the worst rock mass condition, that is the low quality shale immediately above seam Schwalbach, the rock mass is stronger than the stresses. Even if the failure may be initiated, there will be not enough energy released from these poor quality strata to produce events of magnitudes 3.3 and 3.6, respectively. It was concluded from those numerical models that there must exist some unknown zone(s) of weakness in the rock mass. Application of the Mohr-Coulomb-criterion suggests a mobilized angle of friction $\phi_{mob} \approx 20°$ which would resist shearing of conjugate planes as shown in the insert of Figure 5. Particularly the SE-NW striking plane coincides with discontinuity/fault orientation as shown in Figure 2 and slickensided faults in shale with very low friction angles are well known in coal measure rocks (Alber 2006, Barton, 2007).

Events Above the Seam

The events described above marked the beginning of a series of strong seismic events, culminating in a M_l = 4.0 event with a PPV of 94 mm/s on Feb. 23, 2008.

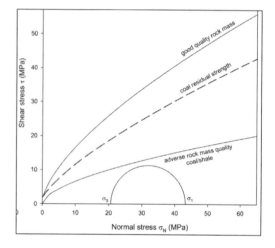

Figure 4. Induced stresses for the seismic event caused by gate development. The strength of most unfavourable rock mass (interbedded coal and shale) is higher than the applied stresses.

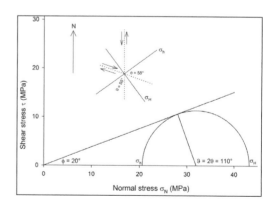

Figure 5. State of stress for the first seismic event while development of the gateroads. The dotted lines in the insert show the most probable orientations of planes of failure.

270

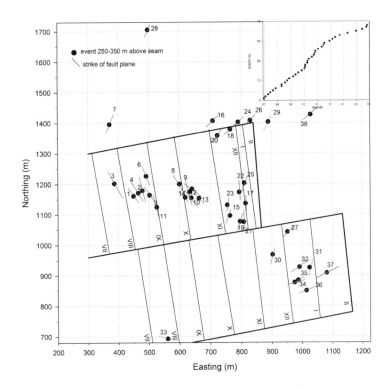

Figure 6. Plan view of the longwall panels and the location of the seismic events numbered by the occurrence. The roman numbers indicate the position of the face by month and the insert shows the events in time.

Mining was immediately stopped, notwithstanding, a last single event of PPV 1.2 mm/s occurred the next day. The M_1 4.0 event caused considerable damage to buildings on the surface. Figure 6 comprehensively shows 38 strong seismic events with their locations with respect to the panel faces as well as their position above and below the panels. The roman numbers indicate the position of the faces by month. Furthermore, the strike of the typically very steep dipping fault planes is shown. The fault plane solutions, i.e., the determination of focal plane orientation and direction of relative movement on that plane, were calculated using the polarities of the recorded P-waves and the polarisation angles of the S-waves.

Figure 6 shows that the events typically took place 100–150 m in front of the excavation faces. Subsequent events are of similar strike of the fault plane and organized in clusters. The events locations are typically high above the panels and few are at panel level. The latter are due to their minor PPV omitted from further analyses.

In a first step the stresses at the respective depth were estimated with the 3D boundary element package EXAMINE 3D (Rocscience 2008) and compared with the intact and residual rock strength (cf. section geomechanical setting). Figure 7 shows the numerical setup and the planes in which the stresses were evaluated. Panel 1 is 100 m ahead of panel 2 and the planes cover the an area (750 m long × 750 m wide) starting from 250 m outby panel 2 to 400 m inby panel 1. The principal stresses in those planes are plotted in Figure 8 versus the lowest boundaries of intact rock strength and residual rock strength, respectively. Clearly the stresses are well below the strength and no failure of rock or rock masses might be expected. It can be concluded from these numerical models that not rock or rock mass failure but slip on existing planes/zones of weakness might be the source of seismic events. The planes/zones of weakness are presumably in similar directions as shown in Figure 2. The tectonic elements originated during formation of the basin in the Variscan orogenesis and are still stressed in the same direction by the Alpine orogenesis.

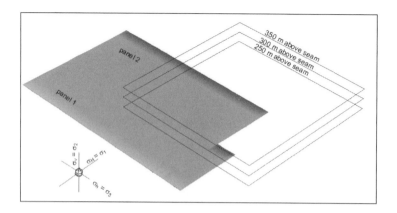

Figure 7. Setup of longwall panels, stresses, and planes above the seam for stress data acquisition

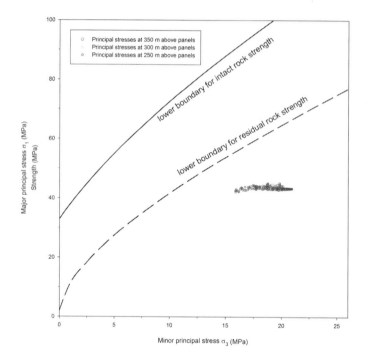

Figure 8. Principal stresses at different positions above the panels and strength estimates of the rock mass

In a next step two planes of weakness with orientation (dipdirection/dip) 30°/90° and 150°/90°, respectively, were selected to represent tectonic elements on all scales and subjected to numerical analyses for their behaviour around the longwall panels. Towards this end the ubiquitous joint approach (Kazakidis and Diederichs 1993) was used. This approach allows interpreting relevant stress data such as normal and shear stresses with respect to an omnipresent joint of given orientation. The planes of weakness were assumed to be cohesionless, as it is typically the case with joints or faults.

Figures 9 to 11 show the mobilized friction angle ϕ_{mob} on the selected planes at various positions around the longwall panels. The mobilized friction angle indicates the demands on the strength of

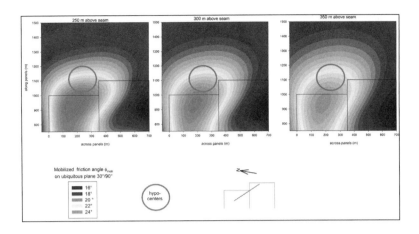

Figure 9. Mobilized friction on NW-SE striking planes of weakness

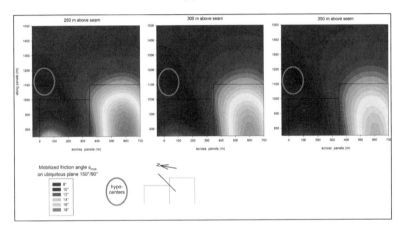

Figure 10. Mobilized friction on NE-SW striking planes of weakness

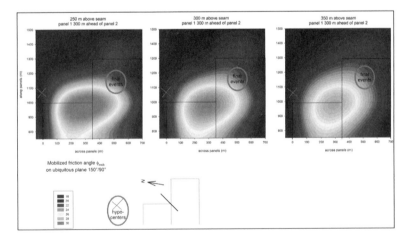

Figure 11. Mobilized friction on NE-SW striking plane of weakness during the final seismic events

the discontinuity in terms of friction; if it exceeds the critical friction angle, the plane slips. The first events (Figure 9) on NW-SE striking planes may be grouped into 2 clusters high above the seam, occurring some 150 m ahead of the face of only panel 2. The mobilized friction angle at those locations was calculated to be 21°–23°, i.e., any plane with a lower friction angle would fail. The next cluster (Figure 10) of events was localized at the NE part of panel 2, above the seam and well ahead of the face an involved NE-SW striking planes. Here the mobilized friction angle was estimated to be as low as 8°. The next cluster occurred high above seam, some 100 m ahead of the longwall face 2 close to panel 1. The strike of the fault planes is NEN-SWS and is close to tectonic features. The secondary stresses on those fault planes mobilized a friction angle of 30 degrees. The final events, which took place when panel 1 was 300 m ahead of panel 2, were mainly located 150 m behind panel 1 and high above the seam (Figure 11). Here, the mobilized angle of friction was calculated to be in the range of 22°–26°.

DISCUSSION AND CONCLUSIONS

Due to an extensive seismic network with both surface and subsurface stations, numerous mining induced seismic events could be located and analyzed for their fault plane solution. Sophisticated geological and rock engineering data from underground mapping and high quality core drilling allowed numerous laboratory tests that led to rock strength estimates at depth. Rock stress measurements yielded results consistent with stress data from central Europe and enabled numerical modelling with the goal of analysing the failure mechanisms of the seismic events. Mining of such a huge underground excavation with dimension 700 m wide and more than 1,000 m long lead to large scale stress redistribution up to 350 m above the seam. This means that stresses at a distance of up to 350 m above or below the longwall face are different from the in-situ stresses. By the same token, stresses at the seam level diminish soon to the in-situ levels. It has been shown that it is very unlikely that the events occurred as a result of stress driven failure of the rock mass as the triaxial rock mass strength of even the most adverse rock mass is higher than the induced stresses. Consequently, all events may be contributed to the category "activation of faults" as discussed by Hasegawa, Wetmiller and Gendzwell (1989), Boler, Billington and Zipf (1999), Gale et al. (2001). Moreover, all events appear to be the consequence of static failures i.e., the (shear) stresses exceeded the (frictional) strength and no seismic

event at another location induced failure processes elsewhere.

The analyses of an event induced by driving solely the roadways suggest that an SE-NW striking plane may slip when the friction angle of 20° was exceeded. This low frictional strength is common in slickensided faults in shale and the plane in consideration is in accordance with the existing ones as shown in the discontinuity rosette plots (cf. Figure 2). The analyses of the events which occurred during the longwall excavations revealed some astonishing insights in the rock mass around the two panels:

- The events appeared in clusters either high above the seam or at seam level
- Aside of the final events, hypocenters were located well ahead of the faces
- The events did NOT occur at locations where the highest friction angle would be mobilized

The existence of several clusters of events suggests that in the rock mass exist distinct zones with planes of weakness, which are activated by the advancing longwalls. The mobilized friction angles were very different as shown in Figure 12 ranging from only 8° up to 30°. Those friction angles are well below Byerlee's law (Scholz 2002) in the low stress range where $\phi \approx 40°$ is the critical friction angle when faults start to slip. However, slickensides in fine-grained sediments such as shale and siltstone may indeed show very low friction angles. The notion that there exist distinct zones with planes of weakness is further supported by the fact that slip was not induced in locations where the highest fraction angle was required, i.e., in locations where the Factor of safety was low. Moreover, the orientation of the planes of weakness coincides nicely with the orientation of the tectonic elements (insert Figure 12).

In a recently published paper (Alber et al. 2009) it has been shown that multiple seam mining with remnant pillars and gob-solid boundaries may lead to stress concentrations which exceed the strength of the rock mass leading to mining induced seismic events of magnitudes up to $M_L = 3.0$. Here, where no old mine workings are involved, it is evident that not the rock mass strength but the strength of existing planes of weakness within the rock mass is exceeded, leading to the activation of faults which were probably generated during the Variscan orogenesis.

This research has shown that even when mining in a virgin rock mass, barely elevated stresses levels may lead to mining induced seismic events. Existing faults planes at distances of several 100 m above and in front of the excavation face may be activated. The strength of the fault planes may be expressed by

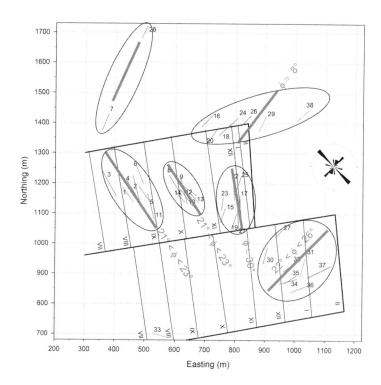

Figure 12. Cluster of seismic events with similar strike of the fault planes and associated mobilized friction angles. The thick lines show the average strike of the hypocenter fault planes. The insert shows the orientation of tectonic elements at panel scale.

extremely low friction angles. It appears in the presented case that even without mining activities the earth's crust is close to failure. Nevertheless, general understanding of the mine's geology, a strong rock engineering program along with a high quality seismological network are necessary to evaluate the earth's response to the disturbances in the stress field by mining activities. Only with such integrative efforts it is possible that seismological observations can be explained by rock engineering computations.

REFERENCES

1. Alber M., Fritschen, R., Bischoff, M. and T. Meier. 2009. Rock mechanical investigations of seismic events in a deep longwall coal mine. *Int. J. Rock Mech. & Mining Sci.* 46: 408–420

2. Alber, M. 2006. Deriving Input Parameters for Numerical Modeling of Fully Grouted Rock Bolts. Proc. 25th Conf. Ground Control in Mining, S. Peng, C. Mark, W. Khair and K. Heasley (eds.) Morgantown, WV, USA: 34–39.

3. Barton, N. 2007. Rock quality, seismic velocity, attenuation, and anisotropy. Taylor & Francis/ Balkema, Leiden.

4. Bieniawski, Z.T.1989. Engineering Rock Mass Classifications. John Wiley and Sons, New York.

5. Boler, F.M.; Billington, S. and Zipf, R.K. 1997. Seismological and Energy Balance Constraints on the Mechanism of a Catasrophic Bump in the Book Cliffs Coal Mining District, Utah, U.S.A. *Int. J. Rock Mech. & Mining Sci. 34* (1): 27–43.

6. Fritschen R., Loske B., Uhl O., Polysos N. (2004). Seismological and geological investigations of seismic events at the Ensdorf mine (in German). Das Markscheidewesen (21).

7. Fritschen R. 2010. Mining induced seismicity in the Saarland, German. Accepted for publication in PAGEOPH special publication.

8. Gale W.J., Heasley K.A., Iannacchione, A.T., Swanson P.L., Hatherly P., King A. 2001, "Rock Damage Characterization from Microseismic Monitoring," Proceedings of the 38th U.S. Symposium of Rock Mechanics, Washington DC: 1,313–1,320.

275

9. Hasegawa H.S.; R.J. Wetmiller and D.J. Gendzwell. 1989. Induced Seismicity in Mines In Canada—An Overview. Pure & Applied Geophysics 129 (3–4): 423–453.

10. Hoek E. and Brown E.T. 1997. Practical estimates of rock mass strength. *Int. J. Rock Mech. & Mining Sci. & Geomechanics Abstracts* 34 (8): 1,165–1,186.

12. Kahlen E. and Alber M. 2008. Failure behavior of coal measure rocks under longwall loading conditions. Unpublished report to DMT, Essen.

13. Kazakidis V.N. and M.S. Diederichs. 1993. Understanding jointed rock mass behaviour using a ubiquitous joint approach. *Int. J. Rock Mech. & Mining Sci. & Geomechanics Abstracts* 30 (2): 163–172.

15. MeSy 2008. Report on hydro-fracture experiments in the footwall of seam A at the Saar mine. Unpublished report in German.

16. Reinecker J., Heidbach O., Tingay M., Sperner B. & Müller B. 2005. The release 2005 of the World Stress Map (available online at www.world-stress-map.org).

17. Rummel F. 2002. Crustal Stress Derived from Fluid Injection Tests in Boreholes. in: In-Situ Characterisation of Rocks (eds. V.M. Sharma and K.R. Saxena), Balkema, Lisse: 205–244.

18. Rocscience (2008) EXAMINE^{3D}, 3D Stress Analysis for Underground Excavations, Version 4.0998, Toronto, CA.

19. Scholz C.H. 2002. The mechanics of earthquakes and faulting (2nd edition). Cambridge University Press.

Microseismic Monitoring in Mines

Maochen Ge

Missouri University of Science and Technology, Rolla, Missouri, United States

ABSTRACT: In the late 1930s, two scientists from the U.S. Bureau of Mines (USBM) discovered to their surprise that a stressed rock pillar appeared to emit micro-level sounds. This early discovery led to the development of the microseismic technique, one of the most important mine safety and ground control techniques. The paper provides a brief review of wide applications of the microseismic technique in mines, its significant impact on mine safety, critical issues for efficient use of the microseismic technique, and the challenges it faces. Also discussed in the paper are the technique's early developmental history and its impact on modern science and engineering.

INTRODUCTION

Discovery and Early History of the Microseismic Technique

In 1938, two young scientists from the U.S. Bureau of Mines (USBM), L. Obert and W. I. Duvall, were conducting an experiment to determine the seismic velocity in mine pillars in the lead-zinc mines located in Northeastern Oklahoma. The geophones used for the survey, however, were unexpectedly and frequently triggered by unknown sources (Obert 1977). Following this initial discovery, Obert and Duval carried out a series of laboratory and field studies and verified that those unknown sources were micro-level signals generated by stressed rocks. This early pioneer work by Obert and Duval led to the development of a unique mine safety and ground control technique, the microseismic technique.

The major development work on the microseismic technique began in the early 1960s. During this time period, South African researchers developed a 16 channel system with the source location capability for studying the rockburst problem at deep gold mines (Cook 1963). This early study convincingly demonstrated the feasibility of the rockburst location by the microseismic technique, the central element of mine microseismic monitoring.

In the middle 1960s, the USBM initiated a major research program in order to make the microseismic technique an efficient tool for mine safety monitoring. The system developed by Blake and Leighton (1970) had a flat frequency response in the range of 20–10,000 Hz, which was suitable for a wide range of field conditions. The USBM source location algorithm was then developed (Leighton and Blake 1970; Leighton and Duvall 1972). Because the algorithm is non-iterative, it avoids the problem of divergence and is suitable for daily operations where the source

location must be carried out automatically. The hardware and software developed from this program, as well as the research and field tests carried out during this period, laid the foundation for the industrial use of the microseismic technique

From the middle 1980s to early 1990s, severe rockburst problems occurred spontaneously in Canadian mines. In order to contend with this problem, over twenty rockburst-prone mines installed microseismic systems for the daily monitoring purpose. Meanwhile, a large scale rockburst research program, sponsored by the Canadian federal government, several provincial governments, and major mining companies, was carried out. This joint effort by the mining industry, the government, and the university researchers, greatly improved the microseismic technique used for daily operations, and fundamentally changed the role of this technique in the Canadian mining industry. The microseismic technique was no longer merely a research tool, it became a standard monitoring technique for rockburst prone mines.

Characteristics of the Microseismic Technique

In comparison with other monitoring techniques, the microseismic technique has three distinctive advantages. The first benefit is source location. With the microseismic technique one can pinpoint the origin of recorded events. Locating the origin of the event is the focus of many monitoring problems. The second important feature is the global monitoring capability, that is, one can monitor a large area with a limited number of sensors. In comparison, most measuring techniques used in rock mechanics are point techniques in that they cannot provide any information beyond the point where the sensor is located. The point techniques are often constrained when the location to be measured is not accessible or the sensor location cannot be predetermined. The microseismic technique, however, is not constrained by these problems as long as the sensor array is properly designed and deployed. The third important feature of the microseismic technique is that it contains real time monitoring capability.

APPLICATIONS IN MINING AND OTHER ENGINEERING FIELDS

Since the 1980s, the microseismic technique has been used to solve a wide range of engineering problems. This section provides a brief review of some of these applications. The focus is the benefit of this technique to the mining industry.

Benefit of the Microseismic Technique to the Mining Industry

As it has been briefly discussed earlier, the microseismic technique has three distinctive advantages: real time monitoring, global coverage, and event location. These advantages make the microseismic technique an ideal tool for studying a wide range of ground control and mine safety problems.

Ground control: The microseismic monitoring technique has been used to study a variety of ground control problems, including pillar stability, roof conditions, the effect of various support techniques, and the mechanism of backfill. At many rockburst prone mines, this technique has become a primary tool for assessing overall ground conditions. Maps of microseismic activity are used to evaluate potential ground control problems as well as the further monitoring needs by other methods.

Rockburst study: Rockbursts are a complex phenomenon and the microseismic technique is the basic means utilized to study this very difficult problem. The function of the microseismic technique in this study is multiple, including identifying the cause of the rockburst problem, delineating high risk regions, mapping seismic activities, and evaluating the effect of design measures for alleviating rockburst potential. Precise timing of a rockburst is inherently difficult. In the author's opinion the degree of success for a timing predication will largely depend upon local conditions. A recent study performed at a limestone mine showed a consistent trend of the microseismic activity proceeding convergence in the roof fall areas, a strong indication of the feasibility of using the microseismic technique as an early warning system at this mine site (Iannacchione et al. 2004).

Verification of mine design and ground control measures: When the locations of rockbursts and associated activities are known, the cause of rockbursts may be objectively assessed in terms of geology, mine layout, mining method and stress field. The mine may be re-planned and the mining method may be altered to alleviate the rockburst potential. The effect of remedial measures can be assessed immediately by the intensity of the associated microseismic activity.

Mine rescue operations: When a rockburst occurs, the first question asked by mine management is "where is the burst and how big is it?." If the mine has a microseismic monitoring system, management can determine almost on a real-time basis whether the burst is an immediate threat to underground workers. With this information, a rescue plan can be quickly formulated. A microseismic system that was designed for locating trapped miners for mine rescue operations has been used by the US Mine Safety and Health Administration (MSHA) since the 1970s.

Risk management: One of the most important functions of the microseismic monitoring program is risk management. At a mine without such a system, a rockburst or anomalous seismic activity could easily cause panic and result in a significant production delay. When a rockburst or anomalous seismic activity can be located and its hazard potential can be evaluated, the problem often becomes more manageable. As a result, production disruptions are often eliminated, avoiding costly loss of revenue.

Daily Microseismic Monitoring Program in Canadian Mines

In the 1980s, many rockburst prone mines in Canada established daily microseismic monitoring programs in response to a sudden surge of rockburst problems. A daily monitoring program in general has three components: a sensor array covering the mine, typically consisting of 32–64 channels, a real time data acquisition and analysis system, and technical staff for the system maintenance (mainly sensors and cables) and preliminary data analysis and reporting.

Although the monitoring efficiency was low in the 1980s, the effort to improve the monitoring efficiency was steady. From the early 1990s, daily microseismic monitoring gradually emerged as a main-stream mine safety and ground control tool and became the primary monitoring means at rockburst-prone mines. According to a survey conducted at those mines, *"they would have serious reservations about their ability to mine in such ground without a microseismic monitoring system"* (Bharti Engineering Associates Inc. 1993).

In addition to the benefit of the technical advance, the success of the daily microseismic monitoring program in Canadian mines is also due to the strong leadership and support provided by the mine management. This can be easily understood by glancing at a typical daily operation.

Each morning, the system operator prepares a brief report on the microseismic activity during the past 24 hours. This report, along with other information, is forwarded to the engineering department for the detection of any new concerns as well as for the general assessment of the ground condition. The findings are then passed to the underground supervisor or ground control staff so that the identified areas can be examined at the beginning of each shift. On the desks of the mine manager and chief engineer, there are designated computer terminals for displaying the seismic activity of the mine. In case of any major or emergency events, they can access the needed information without any delay.

The most encouraging development of microseismic monitoring programs used in Canadian mines is the culture change. Microseismic data is no longer the "property" of the management and engineers. It is encouraged to be used by everyone. Various communication channels are used for this purpose. At the Kidd Creek mine, an online service was developed for the real-time access of microseismic data. In 1993, the mine changed its safety regulation after high-level location accuracy was achieved through a major system update (Ge, et al. 1998). Under this new safety regulation, it is mandatory that all shift supervisors check maps of microseismic activity for the previous 8- and 24-hour periods before they go underground. A computer terminal was specifically placed at the underground entrance for this purpose.

Innovative Applications in Other Engineering Fields

Because of its unique technical features, the microseismic technique is also used extensively in many other engineering fields, including highway slopes, tunnels, dams, aging civil structures, radioactive waste disposal, underground reservoir storage, and geothermal reservoirs. The most important applications in the recent years are tight gas reservoir engineering and carbon dioxide sequestration.

Literatures on the Microseismic Technique

Publications on the theory and application of the microseismic technique can be found from science and engineering journals and conference proceedings in geophysics, acoustic emission, mining, rock mechanics, petroleum engineering, and civil engineering.

Two series of proceedings necessitate mention here. One is for the conferences entitled *Acoustic Emission/ Microseismic Activity in Geological Structures and*

Materials, which is dedicated entirely to the subject of acoustic emission and microseismic techniques. The conferences were organized and hosted by Hardy and Leighton and held in 1975, 1978, 198, 1985, 1991 and 1996 (Hardy 2003). The second series is *Rockbursts and Seismicity in Mines*. The microseismic technique was one of the main subjects of the symposium, which began in 1984 and is held every four years.

There are two books dedicated to the subject of the microseismic technique and monitoring: *Acoustic Emission/Microseismic Activity* by Hardy (2003) and *Rockburst: Case Studies from North American Hardrock Mines* (Blake and Hedley 2003).

EFFICIENT USE OF THE MICROSEISMIC TECHNIQUE

An efficient use of the microseismic technique depends upon many factors. The discussion provided here is limited to those which have played an important role for the development of the microseismic technique during the last 30 years and are important for any major field applications in the future. These factors are size of monitoring systems, sensor array geometry, data processing, seismic event location, and optimization method.

Size of Monitoring Systems

The development of the microseismic technique can be divided into three stages: discovery and early studies prior to 1960, initial technical development in the 1960s and 1970s, and large scale development and application commencing in the 1980s.

The main factor that led to the large scale development and application of the microseismic technique in the 1980s is the rapid growth of electronic and computer technologies that made large microseismic systems available at affordable prices. For instance, the size of the systems deployed by Canadian mines for rockburst monitoring in the middle to later 1980s were typically 32 to 64 channels. At Creighton Mine, two 64-channel systems were used. In contrast, the systems used in the 1960s and 1970s were very small, mostly four or five channels.

Large systems are important in many ways. The most important benefit is the significantly improved capability of signals detection. Microseismic signals are, in general, very small and the most challenging problem faced by the technique is how to capture these small signals. Because of this, the efficiency of a microseismic system is first measured by its capability of signal detection. If a system has difficulties detecting the expected signals, the value of the system diminishes. The difficulties encountered

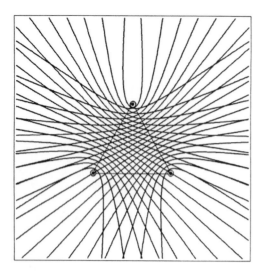

Figure 1.   The hyperbolic field associated with a triangular sensor array, where circulars denote sensor locations (after Ge, 1988)

in many of early applications were largely due to this problem.

The signal strength is affected by many factors. The most important one is the travel distance since the signal strength decays exponentially with the distance it travels. Because of this, the most efficient means to improve the signal strength, and therefore the detectability of small signals, is to reduce the signal travel distance. This is possible only when large systems are available as more sensors can be deployed in monitoring areas.

The improvement of the ability to detect signals by large systems is also due to the fact that large systems are much more robust in handling the direction effect of signals. The signal strength is strongly tied to its propagation direction, which varies periodically from the maximum(s) to minimum(s). A sensor may not be able to detect a signal even it is close to the source of that signal if the sensor happens to be at a location in the minimum energy direction of that signal. The possibility of this situation is significantly reduced when large systems are used.

The other important benefit of large systems is their positive impact on the sensor array geometry. As it will be discussed in the text section, the sensor array geometry is the most important factor affecting the source location accuracy. In order to optimize the configuration of a sensor array, the number of available sensors plays a very important role. It would be very difficult or even impossible to design a good

array if only few sensors are available. This is not a concern with the utilization of large systems.

The impact of large systems on the progress of the microseismic technique since the 1980s is fundamental. Without these large systems, many modern applications would be impossible. If we consider the progress of the microseismic technique during the past 30 years, the most important one is to use a system that is sufficiently large for the project. One should always keep in mind the critical advantages offered by large systems when planning a microseismic monitoring project.

Sensor Array Optimization

Sensor array geometry refers to the configuration of the sensors deployed in the monitoring area. The fundamental importance of the sensor array geometry is that it determines resolvability of the data recorded. In terms of the event location, it has a dominant effect on the location accuracy and reliability.

The importance of the sensor array geometry on the event location lies in the fact that the input errors, such as those associated with arrival times, velocity models and sensor coordinates, are inevitable and the accuracy of the source location largely depends on the ability to limit the impact of these initial errors. From a mathematical point of view, this ability is measured by the stability of the associated mathematical system. For the microseismic source location, the stability of the source location equations is governed by the sensor array geometry (Ge. 1988). A good array can effectively minimize the impact of initial errors while a poor one may significantly aggravate this impact. This effect is illustrated in Figure 1.

In Figure 1 there are three sensors shown as the vertices of the triangle. A source can be uniquely determined based on the P-wave arrival times at these three sensors. The difference between the arrival times between any two sensors can be mathematically represented by a hyperbola and the origin of an event is the intersection of two hyperbolas. If there is an input error occurred, there will be a change in the arrival time difference. Graphically, it means the event will be located on an adjacent intersection and, therefore, the potential location error can be evaluated by the density of the hyperbolic field at the location.

It is clear from the figure that the location accuracy is a function of the array configuration. The accuracy is best at the center of the array and rapidly decreases away from the array. The least accurate areas are those behind the sensors, where the accurate location becomes very difficult. The pattern of the source location accuracy as demonstrated in

Table 1. Effect of outliers on source location accuracy

Event No.	Arrival status[†]	Outlier excluded (m)			Outliers included (m)[*]		
		x	y	z	x	y	z
39	OPPPPPPPPPPP	5720	5651	2080	—	—	—
40	OPPPPPPPPPPP	5717	5669	2084	—	—	—
41	PPPPPPPPPPPO	5727	5663	2092	—	—	—
43	PPPPPPPPPPPP	5736	5675	2088	5736	5675	2088

* "—" denotes the coordinates which are out of the mine boundary.
† P represents P- wave arrivals, and O outliers.

Figure 1 clearly shows that microseismic monitoring is not an arbitrary process of picking up signals. In order to determine the event location accurately, one has to pay close attention to the sensor array geometry.

A common problem for many practitioners is an unawareness or lack of understanding of the importance of sensor arrays. In the author's opinion, this is one of the primary reasons responsible for the poor source location accuracy in many cases. If one needs to improve the source location accuracy, the first step is to examine whether the sensor array is optimized.

Data Processing

The microseismic data recorded at a mine site can be extremely complicated. Noises generated by mining machines, transportation vehicles, ventilation fans, and blasting activities may severely contaminate the recorded data. Even microseismic activities may cause serious problems when the signals interfere with each other under burst conditions. Furthermore, a good portion of recorded signals may be caused by S-wave arrivals instead of P-wave arrivals as would be normally assumed. If these signals are used without discrimination, it will induce a systematic error in the data base. Because of these problems, the raw data must be carefully processed before it can be meaningfully used for any other purposes. The two issues that are particularly important are detection of outliers and identification of P- and s-waves.

Outlier Detection
Outliers are those arrivals which are not due to the physical source that triggers the majority of the stations during an event time window. While outliers have complex causes, the most common ones are interference of seismic activities and culture noises. Human errors, such as recording errors on sensor coordinates, can also be a factor.

Outliers are extremely harmful for the determination of a meaningful source location. The presence of even just one outlier in an event, regardless of the size of the event, is likely to ruin the solution. To

demonstrate the frequent occurrence of outliers and their severe consequence, four real events recorded at a mine site are presented in Table 1.

The four events presented in the table were caused by a nearby blast and recorded within six seconds. They are considered strong events at the mine as at least 12 sensors were triggered. The information presented in the table include the sequence number of these events, the arrival status for the first 12 triggered channels, the event locations calculated without and with the outliers. The arrival status is denoted by letters P and O, representing P- wave arrivals and outliers, respectively. Among these four events, three contain outliers. There are manifested as the first arrival for events 39 and 40, and the last arrival for event 41 (it can be shown that all of these three outliers were due to the interference by other local events).

The devastating effect of these outliers can be seen by comparing the two sets of location results. When the outliers are included in the calculation, the locations are all out of the mine boundary, and the errors are in the order of more than 10 km. However, when the outliers are excluded, all four events have very similar locations. If the blast site (5725, 5660, 2088) is used as an approximation of the origin of these events the errors for these three events are only 13m, 13m, and 5m, respectively.

Identification of P- and S-wave Arrivals
In addition to outlier detection, another important task in data processing is the identification of P-and S-wave arrivals. The first arrival detected by a sensor is not necessarily due to a P-wave as it has been assumed for most microseismic studies. In fact, many of the first arrivals are due to S-waves. If the P-wave arrival assumption were used for these S-wave arrivals, the input data for event location would be severely contaminated.

In order to solve this problem, Ge and Kaiser (1990) developed a phase association theory for the interpretation of the physical status of arrival picks without waveforms. The theory consists of two parts,

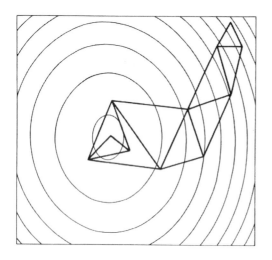

Figure 2. An example of the Simplex moving on the error space contour plot (after Caceci and Cacheris, 1984)

arrival time difference analysis and residual analysis. The data analysis method based on this theory provides a unique means of identifying arrival types and has been adopted by many mines, resulting in substantial improvement of the monitoring accuracy and efficiency (Ge 2005).

Source Location Algorithm
The source location algorithm is the search engine for the origin of recorded events and is the center of microseismic monitoring. The efficiency of this search engine is measured by a number of criteria, including the convergence character, the ability to accommodate different velocity models, the ability to incorporate different optimization schemes, and the calculation speed. A detailed discussion on the subject was given by Ge (2003a, 2003b).

Currently, there are three major algorithms used for microseismic monitoring: the Geiger's method, the USBM method, and the Simplex method. The Geiger's method, developed at the beginning of the last century (Geiger, 1910, 1912), is the best known and most widely used source location method. With this algorithm, the solution is approached iteratively based on the derivative information. The Geiger's method is efficient and flexible, and is used almost universally for local earthquake locations. The major problem of the method is that it is prone to divergence, which may lead to a breakdown of the computing process. Because of this, the method is not suitable for application to daily monitoring.

The USBM method is a widely used mine-oriented source location method, developed by the

U.S. Bureau of Mines' researchers in the early 1970s (Leighton and Blake, 1970; Leighton and Duvall, 1972). The method is simple and easy to use, and, because of its non-iterative algorithm, this method is not prone to divergence. The major problem for method is that it cannot simultaneously handle P- and S-wave arrivals, thus severely limiting its data utilization capability.

The Simplex method is a generic search algorithm which searches the minimums of mathematical functions through function comparison. The algorithm was originally developed by Nelder and Mead (1965), and introduced for the source location purpose in the late 1980s by Prugger and Gendzwill (1988). The mathematical procedures and related concepts in error estimation for this method were further discussed by Ge (1995).

The Simplex method starts from the idea that, for a given set of arrival times associated with a set of sensors, an error can be calculated for any point by comparing the observed and the calculated arrival times. An error space is thus the one in which every point is defined by an associated error, and the point with the minimal error is the event location.

The process of searching for the minimal point with the Simplex method is unique. It begins by setting a Simplex figure, and then moves the figure in the error space based on a set of rules related to the errors on its vertices. The solution is said to be found when the Simplex figure falls into the depression of the error space (Figure 2). The Simplex is a geometric figure which contains one more vertex than the dimension of the space in which it is used. A simplex for a two dimensional space is a triangle, a simplex for a three dimensional space is a tetrahedron, etc.

In comparison with other source location algorithms, the Simplex method has two distinctive advantages. First, it has a very robust convergence character. Divergence is essentially not an issue for this method. Second, the method is very flexible and is capable of accommodating a wide range of conditions and requirements. For instance, one can use P- and S-wave velocities simultaneously, incorporate different velocity models, and make use of either the least squares or absolute value optimization method. In the author's opinion, the Simplex method should be considered as a primary source location method because of its rare combination of the many important features required for a reliable, accurate and robust source location.

Optimization Methods

Source location in a great majority of cases is an overdetermined problem and its solution must be statistically defined. As such, the location accuracy may be significantly affected by the regression method that is utilized. Each regression method is, in fact, an error definition and locating an event is a process to search for the point with the minimum error.

For the microseismic source location, there are two commonly used methods: the least-squares method and the absolute-value method. The least-squares method is by far the best known method. The method assumes the Gaussian distribution for errors. For the microseismic source location, this assumption is difficult to fulfill since most microseismic events are small. The consequence is that the method may be overly sensitive to large errors and lose its function.

Unlike the least-squares method which defines the total error as the sum of *squares* of mismatches, the absolute-value method defines the total error as the sum of the absolute values of mismatches. This method is not overly sensitive to those relatively large errors. Anderson (1982) demonstrated that the absolute value method is a much more robust method for the local earthquake location problem. The author has compared the effect of the least squares method and the absolute value method for a large number of the calibrated microseismic events and found that the absolute value method often yields the better solutions when there are relatively large and isolated errors (Ge 1997).

CONCLUSIONS

The discovery of microseismic activity in stressed rocks by two USBM's researchers in 1938 and their pioneer study on this phenomenon led to the development of the microseismic technique, one of the most important geotechnical monitoring techniques. The mining industry has benefited tremendously from this technique due to its important role in ground control and mine safety. In addition to the mining industry, the microseismic technique is also used to solve many critical problems in other fields of science and engineering. Two high profile problems where this technique has been applied are carbon dioxide sequestration and tight gas reservoir engineering.

Although the principle of the microseismic technique is relatively simple, the use of this technique is not merely installing sensors and collecting data. There are basic rules one has to follow in order to have a reliable and efficient monitoring program. These rules are (1) use a sufficiently large system, (2) optimize the sensor array, (3) carefully process raw data before it is used for any further analysis, (4) select an efficient and robust location algorithm (Simplex method is highly recommended), and (5) select a compatible regression method.

Despite the major progress during the past thirty years, the microseismic technique still faces many challenges. For instance, the Crandall Canyon incident in 2007, which killed six miners and three rescue workers, raised two important issues. One is how to monitor and assess the dynamic loading for those deep coal mines, and the other is that the US mining industry needs a reliable rescue technique for detecting and locating trapped miners. In order to solve these problems, the government, the mining industry and the university researchers are obliged to work together.

REFERENCES

Anderson, K. R., (1982). Robust earthquake location using M-estimates, *Phys. Earth and Planet Int.*, 30, 119–130.

Bharti Engineering Associates Inc. 1993. *A technological review of the seismic and microseismic monitoring systems in Ontario mines*.

Blake, W., and Hedley, D.G.F. 2003. *Rockbursts: Case studies from North American hard-rock mines*. Society for Mining, Metallurgy, and Exploration, Inc., Littleton, Colorado.

Blake, W., and F. Leighton. 1970. Recent developments and applications of the microseismic method in deep mines. *Proc. 11th Symposium on Rock Mechanics*, pp. 429–433, AIME, New York.

Caceci, M.S. and Cacheris, W.P. 1984. Fitting curves to data (the Simplex algorithms is the answer), *Byte* 9, 340–362.

Cook, N.G.W. 1963. The seismic location of rockbursts. *Proc. 5th Rock Mechanics Symposium*, Pergamon Press, Oxford, 493 -518.

Ge, M. 2005. Efficient mine microseismic monitoring. *International Journal of Coal Geology*, 64, 44 -56.

Ge, M. 2003a. Analysis of source location algorithms Part I: Overview and non-iterative methods. *J. of Acoustic Emission* 21, 14–28.

Ge, M. 2003b. Analysis of source location algorithms Part II: Iterative methods. *J. of Acoustic Emission* 21, 29 -51.

Ge, M., 1997. Comparison of least squares and absolute value methods in AE/MS source location: a case study. *Int. J. Rock Mech. Min. Sci.*, 34, No. 3/4, pp. 637.s

Ge, M. 1995. Comment on "Microearthquake location: a non-linear approach that makes use of a Simplex stepping procedure" by A. Prugger and D. Gendzwell. *Bull. Seism. Soc. Am.* 85, 375–377.

Ge, M. 1988. *Optimization of Transducer array geometry for acoustic emission/microseismic source location*. Ph.D. Thesis, The Pennsylvania State University, Department of Mineral Engineering, 237P.

Ge, M., and Kaiser, P.K. 1990. Interpretation of physical status of arrival picks for microseismic source location. *Bull. Seism. Soc. Am.* 80, 1,643–1,660.

Ge, M., Seldon, M., Guerin, G., and Disley, N. Microseismic monitoring at Kidd Mine. 1998. Proc. 6th Conference of Acoustic Emission/Microseismic Activity in Geological Structures and Materials. Trans Tech. Publications, Clausthal-Zellerfeld, Germany, 299–308.

Geiger, L., 1912. Probability method for the determination of earthquake epicentres from the arrival time only. *Bull. St. Louis. Univ.* 8, 60–71.

Geiger, L., 1910. Herbsetimmung bei Erdbeben aus den Ankunfzeiten. *K. Gessell. Will. Goett.* 4, 331–349.

Hardy, H.R. 2003. *Acoustic emission/microseismic activity*. A.A. Balkema, Lisse, Netherlands.

Iannacchione, A.T., Coyle, P.R., Prosser, L.J., Marshall, T.E., and Litsenberger, J. 2004. The relationship of roof movement and strata-induced microseismic emissions to roof falls. SME Annual Meeting, February 23–25, 2004, Denver, Colorado, Preprint 04-58, 8 pages.

Leighton, F., and Blake, W. 1970. *Rock noise source location techniques*. USBM RI 7432.

Leighton, F., and Duvall, W.I. 1972. *A least squares method for improving rock noise source location techniques*. USBM RI 7626.

Nelder, J.A., and Mead, R. 1965. A simplex method for function minimization. *Computer J.* 7, 308–313.

Obert, L. 1975. The microseismic method: discovery and early history. *Proc. 1st Conference of Acoustic Emission/Microseismic Activity in Geological Structures and Materials*. Trans Tech. Publications, Clausthal-Zellerfeld, Germany, 11–12.

Prugger, A., and Gendzwill, D. 1989. Microearthquake location: a non-linear approach that makes use of a Simplex stepping procedure. *Bull. Seism. Soc. Am.* 78, 799–815.

Section 4. Mine Ventilation

How Mining Disasters Changed the Face of Mining in Australia

Paul Harrison
Simtars, Ipswich, Queensland, Australia

Stewart Bell
Queensland Mines and Energy, Brisbane, Queensland, Australia

Darren Brady
Simtars, Ipswich, Queensland, Australia

Jan Oberholzer
Simtars, Ipswich, Queensland, Australia

Ray Davis
Simtars, Ipswich, Queensland, Australia

ABSTRACT: The establishment of the Safety in Mines Testing and Research Station (Simtars) in Queensland was borne out of two underground coal mine disasters, the Box Flat disaster of 1972 and the Kianga No. 1 disaster of 1975, where a total of 30 lives were lost. The fledgling research centre had barely opened its doors, when in 1986 an explosion occurred at the Moura No. 4 Mine with the loss of 12 lives. A further explosion at the Moura No. 2 Mine in August 1994, in which 11 miners lost their lives, dramatically re-emphasised the hazardous nature of underground coal mining. Subsequently, Simtars played a major role in investigating the causes and implementing the research outcomes associated with these disasters.

This paper discusses the impact these disasters had on changing the way safety is managed in mines in Queensland from the investigation of the flame safety lamp after Moura No. 4, the development of gas monitoring technology and management practices, inertisation techniques, rock on rock frictional ignition, implementation of the outputs of the five task groups from the Moura No. 2 Inquiry, the conducting of emergency exercises, major legislative changes in both coal and metal/non metal mining, safety management systems and risk management, through to the development of a mine self escape vehicle. In so doing, it raises the question "Why does it take a mining disaster and, particularly one in our own neighbourhood, to incite us to implement and uphold the systems and institutions necessary to banish them to a faint memory?"

INTRODUCTION

History is important.

> We study the past to understand the present; we understand the present to guide the future.
>
> —William Lund, Historian

The mining industry in Australia today, and Queensland in particular, enjoys one of the most enviable safety records of anywhere in the world. This was not always the case. Prior to 1994, the record was nothing to be proud of. In the very early 90s one large hard rock mine in Queensland recorded 13 separate fatalities in an 18 month period. An underground explosion at Moura No. 2 coal mine in 1994 changed all that (Figure 1 to Figure 3 illustrate the improving trend in safety performance since that time).

HISTORY

Mine Disasters

The Queensland mining industry has endured a number of disasters. Listed here is a sample of them:

On *February 4, 1893* at the Eclipse Colliery at Ipswich in south-east Queensland seven mine workers drowned whilst trying to retrieve mine materials when water from heavy rains collapsed the roof.

On *March 21, 1900* at the Torbanlea Colliery at Maryborough in south-east Queensland five mine workers were killed in a methane explosion.

On *September 5, 1908* at the Mt Morgan gold mine in central Queensland seven mine workers were killed in a fall of ground.

On *November 4, 1908* at the Mt Morgan gold mine in central Queensland five mine workers were killed in a fall of ground.

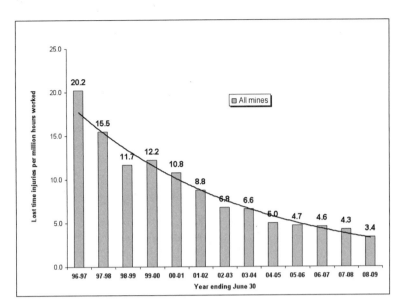

Figure 1. Lost time injury rate for all Queensland mines post Moura No. 2 disaster

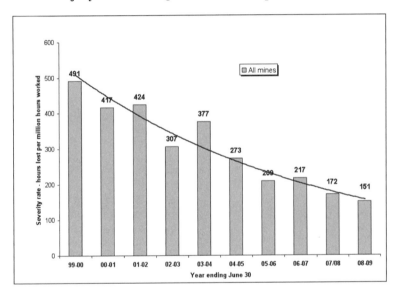

Figure 2. Severity rate for all Queensland mines post Moura No. 4 disaster

On *February 6, 1919* at Cardiff Colliery at Ipswich in south-east Queensland three mine workers were killed when they made an inspection of the workings, which had been closed for a few days, and a gas explosion was caused by a faulty safety lamp.

On *September 19, 1921* a coal dust explosion at the Mt Mulligan Mine in far north Queensland killed 75 mine workers. This is still Queensland's worst mine disaster and has recently been commemorated

by the establishment of a miners memorial service which is held annually on the anniversary of this disaster.

On *April 23, 1928* at the Redbank Colliery at Ipswich in south-east Queensland three mine workers were killed in a methane explosion.

On *May 16, 1936* at Hart's Aberdare Colliery at Ipswich in south-east Queensland four mine workers were killed in a methane explosion.

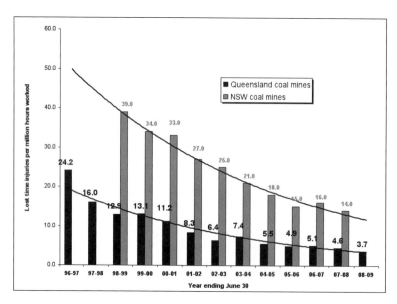

Figure 3. Lost time injury rate comparison for Queensland and NSW coal mines post Moura No. 2 disaster

On *February 1, 1945* at the New Ebbw Vale No. 3 Colliery at Ipswich in south-east Queensland four mine workers were killed in a methane and coal dust explosion.

On *October 13, 1954* at the Collinsville Mine in central Queensland, a carbon dioxide outburst killed 7 mine workers.

On *July 31, 1972* the Box Flat Mine in Ipswich attempted to seal a spontaneous combustion fire from workings. A methane accumulation behind the sealed area exploded killing 14 mine workers underground and 3 at the mine entrance. An 18th worker died some months later as a result of injuries sustained.

On *September 20, 1975* at the Kianga No. 1 Mine in central Queensland, again sealing off a spontaneous combustion fire resulted in methane accumulation behind the sealed area. The explosion killed 13 mine workers.

On *March 12, 1982* at the Laleham No. 1 Mine in central Queensland three mine workers were killed in roof fall.

On *July 16, 1986* at the Moura No. 4 Mine in central Queensland, a roof fall in an unventilated area pushed methane accumulations into workings where a flame safety lamp initiated an explosion that killed 12 mine workers.

On *August 7, 1994* Moura No. 2 in central Queensland, again sealing off of spontaneous combustion fire resulted in methane accumulation. The mine exploded with 24 mine workers underground; 13 escaped, 11 were trapped. A second explosion 12 hours later sealed the fate of the trapped miners. The mine was sealed never to be reopened.

There were other mine fires caused by spontaneous combustion but these were controlled before developing into the disastrous events that occurred at Mt Mulligan et alia.

The Safety in Mines Testing and Research Station (Simtars)

The need for a safety in mines testing and research organisation was recognised in the official inquiries into the Box Flat and Kianga mine disasters and featured as key recommendations (Warden's Inquiry 1972; Warden's Inquiry 1975).

In 1982, the Queensland Government approved the establishment of the Safety in Mines Testing and Research Station (Simtars) as a separate autonomous entity within the Queensland Department of Mines, now the Department Employment, Economic Development and Innovation.

Before that time, research and technical matters relating to mine safety and health were co-ordinated by the Chief Research Engineer within the Department of Mines' Mines Inspectorate. However, the research and technical support itself was provided by the Department of Health.

Simtars' mission was to prevent disasters, fatalities, lost-time injuries and occupational disease in

Figure 4. The Safety in Mines Testing Station at Redbank, 35 km west of Brisbane

mining and allied industries through the provision of research, technical services and training.

The official opening of the world-class Simtars facility in Ipswich, just west of Brisbane, on *September 13, 1988* marked 6 years of careful planning, construction and development (see Figure 4).

In 1986, while the Simtars facility was in the final stages of construction, the Moura No. 4 Mine disaster occurred. In 1994, when commitment to continued support of Simtars may have been wavering, the explosion at Moura No. 2 Mine occurred, dramatically re-emphasising the hazardous nature of mining and the need for continued vigilance.

Following both of the Moura explosions, Simtars travelled to the mines and provided on-site gas analysis and played an important role in the subsequent investigations, the official inquiries and the implementation of their outcomes.

Today, Simtars employs 90 staff and has a portfolio of expertise and services that is unique in Australia and recognised by the international mining community.

History No Longer Repeats

A common theme runs through the recommendations of the official inquiry reports from the disasters of the last 40 years. The reports lamenting the need to keep repeating the same recommendations. This came to an end after the 1994 Moura No. 2 disaster.

Following the Moura No. 4 disaster, attention was focussed on the cause of the explosion. Following the Moura No. 2 disaster, the whole approach to mining in Queensland was turned on its head. This was a watershed moment in the history of

mining in Australia and the drive towards zero harm gained momentum.

AFTER MOURA NO. 4

The Davy Flame Safety Lamp

The cause of the Moura No. 4 Mine explosion perplexed investigators. There was no evidence of a spontaneous combustion initiated fire as in previous disasters. It was hypothesised that a roof fall in an unventilated area of the mine pushed methane accumulations into workings where a flame safety lamp carried by a Mine Deputy ignited the explosion.

Simtars conducted the research that supported this possibility and resulted in flame safety lamps which had been used in mining in Queensland for over 150 years being banned and replaced with modern electronic gas detectors (Golledge 1991). This was an unpopular finding as generations of miners had developed a strong bond with the iconic safety symbol. Some modern folklore was borne out the situation with claims that Simtars was only able to prove the ignition of methane by a flame safety lamp under unrealistic laboratory conditions with the lamp partially disassembled. On the contrary, in over 200 separate tests, Simtars was able to achieve ignition of methane by a flame safety lamp in its normal operating mode more than 10 percent of the time (see Figure 5). Clearly an unacceptable risk when better alternatives were now available.

Mine Gas Monitoring

Australian underground coal mines set the benchmark for some of the world's best practices in gas monitoring. In particular, the Queensland underground coal mining industry has arguably the most comprehensive gas monitoring systems in the world. Each Queensland mine utilises a combination of three systems, real time telemetric sensors, tube bundle analysis, and gas chromatography (GC) in order to monitor and manage the state of the underground atmosphere (Brady 2008).

Differing opinions were expressed at the Moura No. 4 Inquiry (Warden's Inquiry. 1987) as to the ability required to maintain and operate a gas chromatograph at a mine. At the time, a fledgling Simtars investigated the feasibility of implementing the recommendation and found that too many technical problems existed to ensure a successful outcome. Instead, Simtars' then Senior Chemist and mine gas analysis expert, Stewart Bell (now Queensland's inaugural Mine Safety and Health Commissioner), proposed the concept of a virtual gas chemist (Brady, Harrison and Bell 2009). As history shows the Moura No. 4 report's recommendation was not

Figure 5.   Flame safety lamp ignites methane

implemented into the Queensland mining industry exactly as written, but instead Mr Bell's virtual
chemist solution which complied with the Warden's
intent was introduced. This more effective approach
involved the installation of GCs at each underground
coal mine rather than at rescue stations linked back
to a "master" GC at a central location.

One of the reasons it was possible to go further
than the Inquiry recommended was that the timing
coincided with the emergence of onboard communication capabilities in gas chromatographs. With
this communication capability available, Simtars
developed a scheme called Camgas (the Computer
Assisted Mine Gas Analysis System). This allowed
expert gas chemists based at the Simtars facility at
Redbank to oversee GC operation at each mine site
remotely, thereby reducing the expertise required at
the mine site and assisting in the ongoing operability at the mine site of this complex and often finicky
piece of analytical equipment. Effectively Camgas
provided each mine site with an expert gas chemist
24/7 without the need to have one on the payroll—a
virtual chemist.

The first Camgas system was installed at Cook
Colliery in 1989. In the twenty years since, there
have been quantum leaps in the technology used for
GC analysis in both hardware and software. This
has seen major improvements in the efficacy of the
GC as tool at mine sites that can be used proactively
rather than just reactively after a major incident. A
mistaken assumption that was noted in the inquiry
report from the Moura No. 2 disaster (Warden's
Inquiry 1987). Camgas systems are now in use in
Australia, China and India (see Figure 6).

In addition to Camgas, automated real time
and tube bundle monitoring systems are in use at all

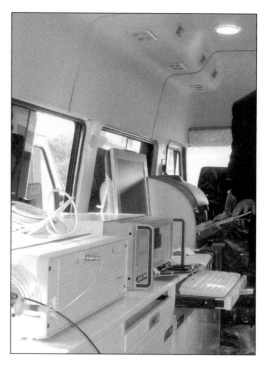

**Figure 6.   Simtars Camgas system installed in a
mobile laboratory in China**

Queensland underground coal mines (required by
legislation following Moura No. 2). In 1990 Simtars
established Safegas, the first automated mine gas
analysis system that incorporated a data interpretation and decision support system tailored specifically
for coal mining.

These systems are used to monitor underground
conditions and initiate predetermined actions to control the situation should the conditions warrant it.
Integral to the efficaciousness of these systems is the
decision support element which helps process and
interpret the large volumes of data acquired. Safegas
systems are now in use in Australia, New Zealand
and the United States (see Figure 7).

AFTER MOURA NO. 2

As Simtars provides the technical support for the
Queensland Mines Inspectorate there were various
issues affecting and involving Simtars that came
from the various task groups set up to action the
Inquiry recommendations following the investigation into this disaster (Warden's Inquiry. 1995). The
following describes the task groups and the actions
taken by them.

Figure 7. Experimental Safegas system at NIOSH's Lake Lynn experimental mine

Task Group 1

Task Group 1 was charged with the development of safety management plans that incorporated the following:

- Ventilation management
- Gas management
- Methane drainage
- Emergency evacuation
- Spontaneous combustion
- Strata management

Simtars has been actively involved in the following areas building on work from its original charter and the recommendations of the earlier Moura No. 4 Inquiry report:

- Gas management
- Spontaneous combustion

Task Group 2

Task Group 2 developed guidelines covering the following:

- Protocols governing withdrawal of persons
  - Sealing of goaf areas
  - Active goaf areas
  - Emergency communications
- Re-entry of a mine or part of a mine
  - Information collection prior to re-entry
  - Monitoring of mine atmospheres during re-entry

- Standards with respect to mine atmospheres during recovery operations
  - Battery powered special purpose vehicle
- The conducting of emergency procedure exercises
  - Training
  - Mutual assistance.

Simtars has been actively involved in the following areas:

Sealing of Goaf Areas

A number of questions associated with establishment of trigger levels and issues with people remaining underground in relation to explosion proof seals were asked. Simtars has assisted in development of trigger levels and is currently addressing issues relating to explosion proof seals through research. These levels are being reassessed in light of the findings after the Sago disaster.

Active Goaf Areas

A number of areas were identified for research that have been addressed or are being addressed by Simtars:

- Inertisation of active goafs (see Figure 8)
- Preventing goaf gases from contaminating the working face
- Sealed area atmospheres
- Seam gas occurring in a goaf
- Seal construction standards
- Sealed area and goaf monitoring
- Determination of ignition sources

Figure 8.   Diesel boiler used routinely to inert active goafs

Information Collection Prior to Re-Entry and Monitoring of Mine Atmospheres During Re-Entry
Before deploying anyone underground following an incident that has resulted in the withdrawal of personnel, significant assessment of the status of the underground environment is required. Just as important as the interpretation of the results is the assessment of the validity of the results being interpreted. Things to consider include:

- Are the samples still coming from the indicated location
- Are the tubes intact (particularly important following an explosion and it's potential impact on a tube bundle system)
- Have tube bundle sample tubes undergone recent integrity testing
- How long does it take for sample to reach the surface
- Are the available sample locations representative of the incident

During a mine emergency it is essential that samples are representative of the area they are assumed to come from and the delivery system is capable of providing ongoing samples. Mines will often turn to collecting gas samples from boreholes in emergency situations. This requires running sampling tube down the borehole to the point of interest or at the very least taking samples from lined boreholes.

There is the possibility that following a major explosion the mine's own tube bundle system may be compromised underground due to tube damage or an insufficient number of sample points exist in the area of interest. To overcome this Simtars commissioned the construction of a mobile mine gas analysis laboratory equipped with a 10 point tube bundle system. This mobile laboratory was commissioned in 1997 and is currently being replaced by a state

of the art 20 point tube bundle system mounted in a fifth wheel trailer.

Battery Powered Special Purpose Vehicle
The value of vehicular transport in assisting persons to escape from a mine in an emergency was highlighted in the Moura No. 2 disaster. Simtars has been heavily committed to research in this area but using diesel rather than battery power which was tested but proved to be unsuitable with the technology currently available.

Training to Facilitate the Conducting of Emergency Exercises
Simtars has been at the forefront of developing the recommendations here:

- Maximising the objectivity and realism of the exercise
- Maximising the security of the exercise
- Providing a consistent approach
- Providing timely feedback to the industry
- Designing the exercise
- Identifying training needs

Task Group 3

Task Group 3 was established to consider actions pertaining to the recommendations which applied to training. Recommendation 4 has been the one focused on by Simtars to date. It was recommended that all employees be effectively trained to:

- Recognise indicators of specific mine hazards, such as spontaneous combustion, and their control
- Become sufficiently familiar with mine gases, and associated risks

The identification and prevention of these risks must be part of a compulsory approved training scheme, as well as part of the mine induction process.

Task Group 4

Task Group 4 was established to consider actions in respect of Inquiry recommendations 9 and 10 which applied to mines rescue strategy development. The major conclusions and recommendations are summarised here:

- Provision of new escape and rescue systems will be of limited value unless people in danger or participating in rescue can make the appropriate decisions when confronted with an emergency situation. Planning, preparation and training for such emergencies is essential to improving their chances of survival.

- It was required that every underground mine should develop a self escape management plan as part of a safety management plan to provide all persons underground with the capability to reach a place of safety, recognising the difficult environmental conditions following an accident. Training initiatives in the form of new generic resources and training guidelines should be developed to support the use of these plans.
- Escape routes, alternative routes and facilities are to be planned, developed and equipped as part of self escape management plans. An oxygen based self escape system is required.

A total of 21 recommendations were made by this task group. The 3 addressed by Simtars to date are:

- Recommendation 1 which states that every underground mine should develop a self escape management plan and an aided rescue management plan. The interrelationship between the escape and rescue plans must be examined and incorporated into the mine safety management plan.
- Recommendation 8 which states that fixed tube bundles and gas chromatographs should be made available at all mines as the primary method of measuring post incident mine atmosphere conditions.
- Recommendation 18 which states that a steering committee should be established to encourage and oversee the development of emergency rescue vehicles.

Task Group 5

Task Group 5 was established to report on 2 areas of the Inquiry recommendations:

- The requirement for the purchase and use of mine inertisation equipment
- The development of appropriate standards of construction of mine seals

Simtars has been actively involved in research and development in both of these areas.

OTHER DISASTERS

Both the Moura No. 2 and No. 4 disaster inquiries recommended closely monitoring and taking heed of overseas developments in mine safety. It is the authors' view that this area has been a major flaw in the approach of many mine safety research organisations around the world (including Simtars). Why is it that issues and solutions that arise in other jurisdictions are generally ignored despite the laws of nature being oblivious to any such jurisdictional boundaries?

Bucking this trend, following the Sago disaster in the United States in 2006, the National Institute for Occupational Safety and Health looked beyond the United States borders for solutions to the problems that disaster created.

Likewise in Queensland, although overseas occurrences do not necessarily directly affect issues locally, the Sago disaster was deemed of great significance. The main reason for this was the re-look and changed attitudes with regard to seals. As the strength criteria of the Ventilation Control Devices used in Queensland have been based on work done in the USA such a change raised significant issues locally.

Simtars personnel spend significant time in India and China working with governments and mining companies in those countries and this has provided them with exposure to a wide range of mine fires and other events not normally seen in developed mining nations.

KEY RESEARCH CONDUCTED

Since its inception, Simtars has successfully completed nearly 50 safety and health focussed research projects and collaborated with other research institutions on numerous others.

Following Moura No. 4 research was undertaken by Simtars to test the propensity of rock/rock and rock/metal strikes to ignite methane gas mixtures to test the probable cause of the Moura No. 4 explosion. As previously mentioned the cause was ultimately attributed to the flame safety lamp leading to the lamp being taken out of service and replaced with electronic methanometers. However, the frictional ignition work has proven invaluable in developing risk controls for longwall shearers in some mines where frictional ignition has since proven to be an issue (Oberholzer 2000).

There has been ongoing study into improving the available information with respect to spontaneous combustion (Cliff, Rowlands and Sleeman 1996; Cliff and Bofinger 1998; Cliff et al 2000) (see Figure 9). This has been a common theme of the recommendations from a number of disaster enquiries, alongside gas monitoring and training, even before the establishment of Simtars. This work has positioned Simtars very well with regard to the most up to date spontaneous combustion knowledge and how

Figure 9. Large scale (16 tonne) test rig for spontaneous combustion testing

Figure 10. Explosion propagation tube

to apply it in the field. It has resulted in the development of early indicators of spontaneous combustion to allow mines to become aware of a problem when there is still time to take preventative measures. It has fed into the design and continuous improvement of gas monitoring systems (previously mentioned in this paper), gas monitoring procedures, gas management protocols and training. Work continues to ascertain if evidence of spontaneous combustion in a mine can be detected at lower temperatures using emerging technologies and gases not previously examined (Clarkson, 2007).

Studies have been conducted by Simtars into the nature of gas and coal dust explosions in a purpose built explosion propagation tube (see Figure 10). The tube was originally developed to study the required limits of stonedust but, due to its design strength and ease of operation, it is now providing valuable information about deflagrations and the possibility of a deflagration transitioning into a detonation. This is recent (yet to be published) work following the Sago disaster.

A significant advance in emergency response was the introduction of the Polish GAG jet inertiser. Simtars brought one to Australia and ran a demonstration project at the Collinsville No 2 underground coal mine which convinced the industry of its benefits (Bell et al 1997).

Simtars was also involved in a project to demonstrate the application of waste gases from diesel boilers to mine inertisation (Brady 1997).

For the last two decades Simtars has contributed significantly to the development of Australian and international standards for, and improvements in, explosion protection techniques through various research projects and participation in the International Electrotechnical Commission's IECEx Scheme as a certification body for explosion protected electrical equipment (Byers et al 1996; De Kock and Birch

2002; De Kock, Birch and Tang 2002; Davis et al 2008) (see Figure 11).

Currently, Simtars is working on behalf of the Queensland Mines Rescue Service on the design and development of a mine self escape vehicle for exiting a mine safely and rapidly after a fire or explosion (Bell et al 2004; Binnie, Davis and Watkinson 2006; Davis, Ellis and Humphreys 2007).

Other problems currently being addressed include matters as diverse as:

- seal strengths
- safer self rescue changeover
- a self rescuer simulator
- ageing affects of circuits on intrinsic safety
- reliability engineering
- robotics for remote search and rescue

CHANGES RESULTING FROM THE INQUIRY INTO THE MOURA NO. 2 MINE DISASTER

As mentioned earlier in this paper, the effects of the Moura No. 2 Mine disaster on the mining industry in Australia was profound. This was perhaps a result of the cumulative effect of previous disasters and the common issues that arose from each of them coupled with a determination by all of the stakeholders that things had to change.

Following this disaster, the management of mining safety and health in Queensland was completely overhauled. The Mines Inspectorate underwent a major review and modernisation. A former director of Simtars, Peter Dent, was tasked with the implementation of the Inspectorate Review recommendations. The Mines Inspectorate has since been subjected to a comprehensive review approximately every 5 years to ensure it remains relevant, effective and impartial.

Following the most recent evaluation, a new position of Mine Safety and Health Commissioner was established. The inaugural commissioner is another former director of Simtars, Stewart Bell.

Completely new legislation was introduced using a risk based approach to cope with the hazards on a mine and required the establishment of safety management plans to cope with the principal hazards; ultimately leading to the development of company wide safety management systems (Coal Mine Safety and Health Act 1999; Coal Mine Safety and Health Regulation 2001; Mining and Quarrying Safety and Health Act 1999; Mining and Quarrying Safety and Health Regulation 2001). At the time it was introduced, there were no published standards available for safety management systems so the Mines Inspectorate developed their own called Safeguard. Today there are Australian and International Standards available. The use of the risk management approach has since been extended to the whole mining industry.

Training of mining staff at all levels to be aware and cope with various levels of hazards was mandated. Statutory positions, such as Ventilation Officer, were assigned required competency based training.

A rigorous gas monitoring and management regime was implemented. Inertisation using diesel boilers now forms part of the daily routine of mine operations in Queensland preventing the potential for occurrence of spontaneous combustion. Rules relating to explosive mixtures and methods to control high risk of explosions behind seals were introduced in legislation.

The GAG inertiser is now an integral part of the Queensland Mines Rescue Service's mine emergency response. It has been used to control mine fires at the Blair Athol Mine in Queensland and the Southland Mine in New South Wales. It's most serious recent outing was to extinguish a fire at Loveridge in West Virginia.

The strength of ventilation control devices and seals was mandated, although the Sago disaster findings put these under question (Simtars is currently conducting research into the risk this might pose to Australian mines).

Every mine in Queensland has an emergency response plan which must be tested annually (Watkinson and Brady 2008). One mine each year is subjected to a statewide Level 1 emergency exercise which tests the mine's response plan as well as the response of all external parties such as the Mines Inspectorate, government emergency services,

Figure 11.   Flameproof enclosure fails explosion protection testing

Simtars, etc. The particular mine knows that it will be subject to an emergency exercise but not when.

CONCLUSION

Mine disasters have significantly influenced the shape of today's mining industry in Australia; a mining industry that is now one of the safest in the world. Borne from such disasters, Simtars, has played an important role in that shaping, but why did it take 3 more disasters and the loss of 36 more lives after the establishment of a safety in mines organisation was recommended for that shape to evolve?

With plans afoot to harmonise safety and health legislation throughout Australia, great care needs to be taken to ensure that we use our understanding of the present to guide the future and that we are acutely aware that the present is grounded firmly in the past.

History is important.

REFERENCES

Bell, S., Davis, R., Watkinson, M., Humphreys, D. and O'Beirne, T. 2004. *Mines Rescue Vehicle Testing and Development Extension*. Australian Coal Association Research Program No: C12004, Safety in Mines Testing and Research Station (Simtars), Redbank.

Bell, S., Humphreys, D., Harrison, P., Hester, C. and Torlach, J. 1997. *A report on the Demonstration of the GAG-3A Jet Inertisation Device at the Collinsville No 2 Coal Mine on 11, 16, 17 and 18 April 1997*. Australian Coal Association Research Program No: C6019, Safety in Mines Testing and Research Station (Simtars), Redbank.

Binnie, P., Davis, R. and Watkinson, M. 2006. *Mines Rescue Vehicle—Demonstration Phase.* Australian Coal Association Research Program No: C14024, Safety in Mines Testing and Research Station (Simtars), Redbank.

Brady, D. M. 2008. The role of gas monitoring in the prevention and treatment of mine fires. *Proceedings from 12th U.S./North American Mine Ventilation Symposium*, Reno.

Brady, D., Harrison, P. and Bell, S. 2009. 20 Years of Onsite Gas Chromatographs At Queensland Underground Coalmines. *Proceedings from the Ninth International Ventilation Symposium,*, Delhi.

Byers, K., Birthwhistle, D., Crisp, J. and Weber G.1996. *Safety Aspects of IP55 Enclosures used in High Fault Level Underground Coal Mines.* Australian Coal Association Research Program No: C4023, Safety in Mines Testing and Research Station (Simtars), Redbank.

Clarkson, F. 2007. *Detection of Heating of Coal at Low Temperature: Stages 1 and 2.* Australian Coal Association Research Program No: C10015, Safety in Mines Testing and Research Station (Simtars), Redbank.

Cliff, D. and Bofinger, C. 1998. *Dissemination of Information on Spontaneous Combustion.* Australian Coal Association Research Program No: C6001, Safety in Mines Testing and Research Station (Simtars), Redbank.

Cliff, D., Clarkson, F., Davis, R., Bennett, T. and Smalley, M. 2000. *Better Indicators of Spontaneous Combustion.* Australian Coal Association Research Program No: C5031, Safety in Mines Testing and Research Station (Simtars), Redbank.

Cliff, D., Rowlands, D. and Sleeman, J. 1996 *Spontaneous Combustion in Australian Underground Coal Mines.* Safety in Mines Testing and Research Station, Reprinted 2004, Redbank.

Coal Mining Safety and Health Act 1999. Office of the Queensland Parliamentary Counsel, Reprinted as in force on 4 September 2009, Reprint No. 3B, Brisbane.

Coal Mining Safety and Health Regulation 2001. Office of the Queensland Parliamentary Counsel, Reprinted as in force on 1 July 2009, Reprint No. 3C, Brisbane.

Davis, R., Birch, J., Chowdhury, A., Parmar, B. and Soady, D. 2008. *Effect of Equipment Placement in a Flameproof Enclosure.* Australian Coal Association Research Program No: C13024, Safety in Mines Testing and Research Station (Simtars), Redbank.

Davis, R., Ellis, C. and Humphreys, D. 2007. *Performance Testing of Diesel Engines in High Methane Atmospheres.* Australian Coal Association Research Program No: C15029, Safety in Mines Testing and Research Station (Simtars), Redbank.

De Kock, A. and Birch, J. 2002. *Verification of Spark Ignition Test Apparatus for High Current IS Circuits.* Australian Coal Association Research Program No: C8004, Safety in Mines Testing and Research Station (Simtars), Redbank.

De Kock, A. Birch, J. and Tang, T. 2002. *Intrinsically Safe (IS) Active Power Supplies.* Australian Coal Association Research Program No: C9001, Safety in Mines Testing and Research Station (Simtars), Redbank.

Golledge, P. 1991. *Research on the Intrinsic Safety of the Flame Safety Lamp in Atmospheres of Methane and Coal Dust.* National Energy Research Development and Demonstration Program (NERDDP) End of Grant Report QSR9004. Safety in Mines Testing and Research Station (Simtars), Redbank.

Mining and Quarrying Safety and Health Act 1999. Office of the Queensland Parliamentary Counsel, Reprinted as in force on 4 September 2009, Reprint No. 3B, Brisbane.

Mining and Quarrying Safety and Health Regulation 2001. Office of the Queensland Parliamentary Counsel, Reprinted as in force on 1 July 2009, Reprint No. 3A, Brisbane.

Oberholzer, J. 2000. *Commissioned Study on Frictional Ignition.* Australian Coal Association Research Program No: C7029, Safety in Mines Testing and Research Station (Simtars), Redbank.

Warden's Inquiry. 1972. *Report on an Accident at Box Flat Colliery on Monday, 31 July 1972.* Brisbane.

Warden's Inquiry. 1975. *Report on an Accident at Kianga No. 1 Underground Mine on Saturday, 20 September 1975.* Rockhampton.

Warden's Inquiry. 1987. *Report on an Accident at Moura No. 4 Underground Mine on Wednesday, 16th July, 1986.* Rockhampton.

Warden's Inquiry. 1995. *Report on an Accident at Moura No. 2 Underground Mine on Sunday, 7 August 1994.* Brisbane.

Watkinson, M. and Brady, D. 2008. Ten years of conducting Level 1 simulated emergency exercises in Queensland's underground coal mines, *Proceedings from 12th US/North American Mine Ventilation Symposium,* Reno.

History of the Research in Polish Coal Mining Industry and Its Present Trends

Józef Dubiński
Central Mining Institute, Katowice, Poland

ABSTRACT: Poland belongs to the developed mining countries of the world and in the past the mining of various mineral resources was considered as its basis for economical and civilization development. Many conditions connected with the specifications of Polish coal basins had substantial impact on the development of the research and mining science in Poland. The paper presents history of the research in Polish hard coal mining having focused on well documented period of the XXth century, starting from its '20s. This history was divided into few characteristic development periods, which were dominated by different research priorities. Present period, which started in 1989 with the changes of political and economical system in Poland can be characterized by the intensive restructuring processes of both: hard coal mining and the mining science. When selecting research priorities more and more attention is being paid on the strategy of sustainable development, which must be deployed by present mining in order to be socially accepted. It is another time, when the work safety problematic gained special importance. There is certain progress in the development of mining specializations focused on technological and environmental aspects. All above aspects are considered the new challenges for the Polish mining science, which contributed into establishing at least few important scientific schools. The effects of their activities are visible in the mining itself and its surrounding. Based on the example of cooperation between Central Mining Institute of Katowice and US Bureau of Mines the paper emphasized role of many years cooperation with American research bodies. It also presents the most important directions of presently implemented and future research works. Among the others they incorporate works focused on the technologies enabling better utilization of the resources, i.e., extraction of the thin coal seams and coal parcels with irregular shapes. Also development of underground coal gasification technology acquires the importance. In the domain of safety, the works are focused on the increase of the efficiency of the mining hazards' monitoring systems and using the achievements of the automatics and robotics for the safe realization of the mining works in the areas with the high level of mining hazards occurrence.

INTRODUCTION

Mining and processing industry of mineral resources commonly defined as a mining has centuries-old tradition in Polish territory. Mining history origins are very old and are well documented. Usually they are perceived with salt mining, which started in Bochnia and Wieliczka regions already in XIII century as well as with silver and lead ores mining in the district of Bytom and then Olkusz. Getting closer towards present-day it must be mentioned that development of hard coal mining in Polish territory started already in XVI century—in the area of Lower Silesia. However in XVIII century hard coal mining started also in Upper Silesia region and it can be said, that with various intensiveness it lasts until now. Its most dynamic development can be recognized after 1945, when mining industry became unquestionable basis not only for Polish energy sector but also for all national economy badly destroyed during Second World War. Similar situation was also observed in the most European countries.

Having analyzed the importance of hard coal mining for Poland it must be emphasized, that this industry was always extremely important for economical and civilization development of the country. Also today, despite of significant decrease in hard coal production—especially after 1989—Polish energy safety still relies on coal fuel; i.e., on hard and brown coals.

Above mentioned many centuries' mining activity should be recognized as a big effort of many generations of people connected with this industry. It must be mentioned, that very significant input into this activity belongs to the Polish mining science. The evidence of above can be found in many historical sources and list of meritorious Polish scientists—miners- starting from Stanisław Staszic until contemporaries is very long. The role of mining science in the development of Polish mining industry of mineral resources was always well-thought-of, both by the mining industry and also by the state authorities. Characteristic feature of this industry, which is endless struggle against the underground forces of nature makes it continual striving for theirs' better understanding and control. Thus, there is common understanding among the miners, that without

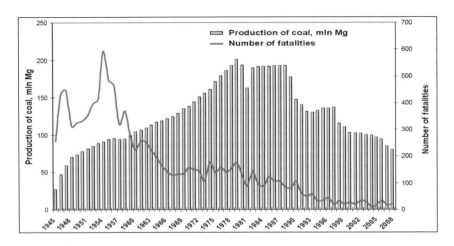

Figure 1. Hard coal production and number of fatalities during the years 1945–2008 (source: State Mining Authority)

appropriate research, measurements and experiments i.e., without science—the exploitation would be difficult or simply impossible. So, the domain of mining activity is connected not only with the development of new and effective exploitation and processing technologies of gained mineral resources but also with pretty complex work safety and environmental issues in the mining areas.

The large scope of mining problematic makes it highly interdisciplinary as solving more and more difficult mining problems always required participation and cooperation of many specialists from various branches and also usage of knowledge and achievements of other scientific disciplines. The mining process itself is covering several components, starting from opening and development of the deposit, its exploitation, transport of mined material, its processing, which enriches final product. All those domains of mining activities in the history of Polish hard coal mining went through its development-path, shaped up by mutual relations and close cooperation with the scientific community.

Considering already mentioned large scope of the leading subject of this paper concerning history of the development of research in Polish mining, despite of focusing just on hard coal mining—some issues will be mentioned just briefly, however the primary attention will be drawn on the work safety issues. High accident rate in this sector of mining industry, including fatal accidents—what is shown on Figure 1—posed in front of mining science big challenge aiming at counteracting and limitation the factors causing this unfavorable work safety state. Visible trend of decreasing the fatal accidents

number confirms effectiveness of actions in the domain of safety improvement. It seems, that in this specific domain Polish experience and solutions have their own important input and significance in the development of the world mining science.

CHARACTERISTICS OF POLISH HARD COAL MINING

The hard coal mining for many decades already have been playing extremely important role in Polish economy and especially in the energy and fuels sectors. Presently, share of hard coal in the electricity production is 55%, and Polish energy sector is consuming half of its annual production, which is little more than 40 mln Mg. Dominant importance of hard coal as a domestic source of primary energy started after 1945 what resulted in its systematic increase of production in the following years (see Figure 1).

Presently, hard and brown coals are recognized as a solid basis for the energy safety of Poland, which has the lowest ratio of economy dependence on the import of energy sources among all 27 European Union countries.

Polish hard coal resources, which are being presently exploited, are located in two basins i.e., in the Upper Silesian Coal Basin (USCB) and in the Lublin Coal Basin (LCB). In the year 2000 many centuries' mining activity was totally finished in the Lower Silesian Coal Basin (LSCB), mainly due to the high non-profitability. Figure 2 shows location of above basins.

It must be emphasized, that Poland has rich hard coal reserves. Geological resources are estimated for 60 billion Mg, however balance reserves

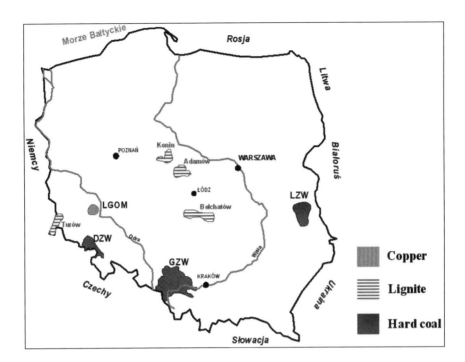

Figure 2. Location of major Polish mining basins

for 43 billion Mg including 16,12 billion Mg in the areas already developed, and 26,3 billion Mg tones is located in the areas not developed so far. Steam coal resources are estimated for ⅔ of total resources, and coking coals are little less than ⅓. Commercial resources' base located only in the developed areas is 4,2 billion Mg (PIG. 2008). It is estimated, that above resources' base can secure hard coal production at the level of 80 mln Mg annually for approximately another 50 years; in case of new mining investments in the not developed areas this figure can be significantly increased. In USCB there is full range of technological types of the hard coals—starting from the steam coals through the coking-ones up to the anthracites. The quality of coal is good, the average ash content varies from 11 up to 17%, and total sulfur content from 0.5 till 2.3%. In LCB however there are coals less carbonified where the ash content in average is 14.5% and total sulfur content is between 1.2 and 1.45%.

It must mentioned, that hard coal exploitation is always accompanied by significant amount of mining waste materials. For example in 2008 it was 32,48 mln tones, including 31,90 mln tones which was utilized for many purposes. Apart from above, there is certain volume of water which is pumped out in order to drain the coal mines. In 2008 it was 243,9 mln m³, including

47,61 mln m³ which was utilized, and 196,29 mln m³ was disposed into the streams and rivers.

Mining and geological conditions in majority of Polish coal mines are generally determined as difficult-ones due to the factors as below:

- Deep exploitation level, in some of the coal mines even exceeding 1000 meters depth
- Multi seams' characteristics of the deposits resulting in arising abandoned coal pillars and exploitation edges in the adjacent seams, having negative impact on the state of stresses in the rock mass
- Many years exploitation resulting in mining out substantial part of the deposit
- Tendency of the coal seams to the bursts, self-combustion and high methane content
- Concentration of the exploitation on the relatively small acreage, in most cases in the built-up area

Only very few coal mines have the conditions, which can be considered as favorable-ones. Performed analysis of the relations between the hard coal resources' base and the level of natural mining hazards shows, that only 35% of the resources can be classified as the resources with safety exploitation conditions, however even 20% are the resources located in the

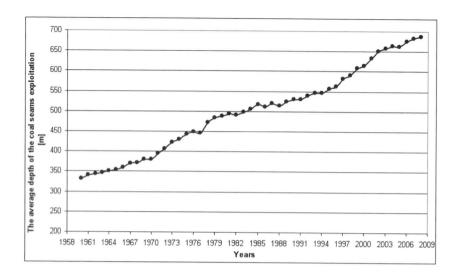

Figure 3. Change of average depth of the coal seams exploitation in Poland in the years 1958–2009

conditions with the highest level of natural hazards (Dubiński, J. et al. 1999).

The consequence of difficult geological and mining conditions is the occurrence of practically all known mining hazards like: seismivity and bursts, endogenous fires, injections and explosions connected with methane emission, dust, water, climatic, radiation, roof falls and others. As an example, Figure 3 shows increase of the average depth of the coal seams exploitation in the USCB coal mines, which has strong impact on the level of many natural mining hazards (Konopko, W., et al. 2008).

Besides, during last years it can be observed more often occurrence of so called associated hazards, which can be characterized by their concurrent appearance i.e., at the same place and at the same time a few different hazards: e.g., methane hazard, bursts and fire hazards (Bradecki, W., Dubiński, J. 2005) (see Figure 4).

Presently, the organizational structure of Polish hard coal mines covers 32 coal mines (see Figure 5), which are incorporated in the following units (mining companies):

- Jastrzębska Spółka Węglowa—6 coal mines
- Katowicka Grupa Kapitałowa—6 coal mines
- Kompania Węglowa—16 coal mines
- Południowy Koncern Węglowy—2 coal mines
- Independent coal mines: Bogdanka and Siltech—2 coal mines

Significantly decreased number of the coal mines in the last twenty years is the result of restructuring

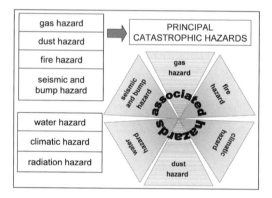

Figure 4. Principal natural mining hazards in Polish hard coal mines

processes driving among the others to the joining and liquidation of the coal mines. The works concerning liquidation are conducted by the specialist firm called: "Spółka Restrukturyzacji Kopalń."

Keeping the important position of the hard coal in Poland is justified by the following reasons:

- Lowest production cost of electricity from the coal fuel
- Access and amount of domestic coal resources providing its sufficiency in the next decades
- Technical and organizational readiness to continue the exploitation of already developed deposits

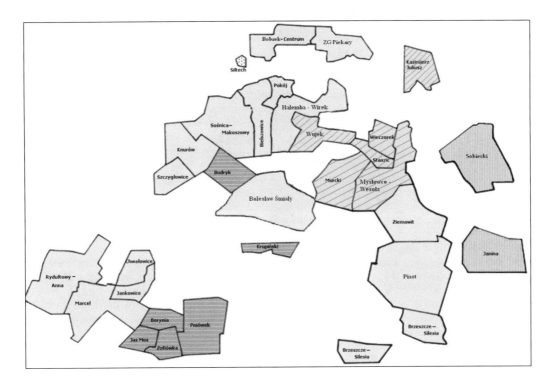

Figure 5. Location of the hard coal mines in USCB

- Realistic possibility of developing new prospective resources
- Potential chances to improve energy production efficiency from the coal fuel
- Implementation of "clean coal" technology, making the coal environmental friendly fuel

DEVELOPMENT OF POLISH MINING SCIENCE IN THE XXth CENTURY

Considering importance of the problem as well as access to many source materials necessary to analyze growth of Polish mining science, the subject characteristics will be focused on the XXth century and on the last years of present century. Above assumption is also affected by the fact that Poland regained its independence only in 1918 after 123 years. Despite of above it must be mentioned, that in the time period before 1918 Polish mining engineers and scientists, who got their education in the universities of Austria, Germany or Russia had significant input into the development of the world mining.

The important dates from the first decades of the XXth century were: the founding in 1919 University of Mining in Kraków and establishing in 1925 Experimental Mine "Barbara," which is presently part of the Central Mining Institute. It must be emphasized, that before 2nd World War it was the only scientific and research unit working for mining industry in Poland. Its establishing and many years of activities for the development of mining science was connected with Professor Wacław Cybulski. Above mentioned institutions were established to support from the education and scientific point of view development of mineral resources' mining industry in Poland. Special research tasks were given to Experimental Mine "Barbara," which establishing was inspired by increasing number of serious problems in the domain of work safety, especially visible in the underground hard coal mining. Methane, coal dust and fires were at these days the reasons of many tragic mining catastrophes in many countries of the world and also in European and Polish mining. The issue, which was also very important at this time was safety connected with the usage of the explosive materials for the purposes of mining works but also technical and organizational development of the mining rescue.

During the post-war period i.e., after 1945 in the development of Polish mining science certain changeability of priorities and directions of conducted research was observed. They were adjusted to the changing in time production tasks and demand of hard coal mining. Despite of above several characteristic periods of this development stage can be distinguished.

The first period—the post war period these were the years 1945–1970. Then, especially in the '50s and '60s of the XX century coal and steel became symbols of reconstructing Europe. Also at this time in Poland an extremely dynamic development of hard coal mining industry could be observed—Figure 1—supported by newly established scientific bodies working for this branch of industry. These were for example: Central Mining Institute and mining faculties of Silesian and Wroclaw Technical Universities. At this time systematic increase of research potential of Polish mining science and scope of the problems it had to face could be observed. In the first years of this period, the conducted researches were mainly focused on the reconstruction and development of hard coal mining and especially on increasing its efficiency and work safety.

It must be emphasized, that in these years mining science played extremely important functions, both: advisory and control over the exploitation processes, work safety, production of machines and equipment but also over proper utilization of the deposits. It played also important role, when creating basis for the Polish mining legislation, especially in the domain of work safety. The researches were focused on testing the phenomenon of coal dust explosions and fires, behavior of rock mass surrounding the mining excavations but also on ground surface deformations together with its protection as well as on the coal quality issues. In the later part of this period the priority was given to the research focused on implementing in the coal mines the latest techniques and mining technologies, including research on coals' and surrounding rocks breaking characteristics, establishing basis for moving from mining by blasting techniques into coal cutting processes by usage of the very first longwall shearers—which were just invented. Also at this time the very first research and implementation works concerning mechanization of the mining process and organization of work in the exploitation excavations could be observed. All above mentioned activities have positive impact on the changes in the work safety domain.

Despite of above however, in the '60s the increase of bumps hazard can be observed, what results in the very first research works focused on coal tendency to bumping and development of the low diameter drillings aiming at evaluating the state of this hazard in practice. It was also the beginning of the far flung observations and measurements of the rock mass pressure symptoms aiming at elaboration

the methods of optimal selection the roof supports both for the coal faces and roadways. The research focused on the surface subsidence and especially on forecasting its deformation were still continued. The scientific achievement in this domain was the theory of Budryk-Knothe, which with very few modifications is being used until today to forecast the impact on the ground surface of coal exploitation with using longwall system.

There is certain growth in importance of the water and gas, coal and rocks' outbursts hazards, which clearly intensify in the coal mines of the LSCB. In the domain of mining techniques of the coal seams there was certain development of the concepts and projects of mining by slices of the thick seams.

Special attention must be drawn on the fact that in this period of time there were very first interesting trials of underground coal gasification research, both in the laboratories and underground in specially adjusted for this purpose incline. It must be emphasized, that majority of the research results were implemented in the coal mines and other plants working for the mining sector. One of the examples can be the research on the mining explosives, which allowed for elaboration and implementation of the new improved formulas of the explosives—safe when using in the atmosphere of coal mine methane.

The new coal mines, which were designed and opened at this time would not be able neither to be constructed nor operate properly without assistance of the scientific bodies, which elaborated appropriate methane sensors and special methane hazard protection systems.

Having assessed this difficult period of Polish hard coal mining development history, one can easily notice significant contribution of mining science with its systematically increasing research potential into practically each of mining domains. It was the time of establishing basis of so called "Polish mining school" and many specialist schools with such mining specializations like: aerology, rock mechanics, work safety in mining, examination of mining exploitation on the ground surface and its protection, coal washing as well as its physical and chemical properties and others. They were connected with the names of their originators, distinguished scientists. Having limited the list of the names of the people from this period we must mention following professors: Witold Budryk, Wacław Cybulski, Bolesław Krupiński, Stanisław Knothe, Tadeusz Laskowski, Jerzy Litwiniszyn, Błażej Roga, Antoni Sałustowicz.

Another, distinctive period in the history of Polish hard coal mining and mining science development these are the '70s and '80s of the XXth century, and practically this period ends up in 1989, which is

Table 1. Mining catastrophes in the years 1970–2009

Date	Coal Mine	Number of Accidents		Causes of the Tragedy
		Fatal	Other	
23.03.1971	Rokitnica	10	—	Rock burst
28.12.1971	Siersza	4	1	Rock fall
28.06.1974	Silesia	34	—	Coal—dust and methane explosion
07.09.1976	Nowa Ruda	17	—	Outburst of gases and rocks
10.10.1979	Dymitrow	35	5	Coal—dust explosion
30.10.1979	Silesia	22	15	Fire
28.11.1982	Dymitrow	20	9	Fire
22.12.1985	Wałbrzych	18	1	Methane explosion
04.02.1987	Mysłowice	19	27	Coal—dust and methane explosion
10.01.1990	Halemba	19	20	Methane ignition
17.09.1993	Miechowice	6	2	Rock burst
16.12.1996	Bielszowice	5	7	Rock burst and methane explosion
06.02.2002	Jas-Mos	10	2	Coal—dust explosion
23.03.2002	Rydułtowy	3	7	Rock burst and methane explosion, fire
07.11.2003	Sośnica	3	7	Methane explosion
22.11.2005	Zofiówka	3	5	Outbursts of methane
27.07.2006	Pokój	4	6	Rock burst
21.11.2006	Halemba	23	—	Methane and coal—dust explosion
04.06.2008	Borynia	6	12	Methane ignition and explosion
18.09.2009	Wujek–Śląsk	19	34	Methane ignition and explosion

characteristic date in Polish history. It was the date of changing the post-war political system and starting the new period connected with the transformation of Polish economy towards market economy.

From the mining and mining science point of view it has two quite different parts. Generally speaking the decade of '70s was significantly different than '80s. The first decade was characterized by extremely dynamic development of Polish hard coal mining, which never took place in the history of this branch of industry so far. At the end of above decade Polish hard coal mining reached its annual output at the level of 200 mln Mg. It was the time when in majority of the coal mines the longwall coal exploitation systems together with advanced coal face mechanized systems were introduced. They enabled to increase the work efficiency and concentration of the output.

Unfortunately, together with the increase of the exploitation depth there is significant increase of natural mining hazards, and especially rock bumping and gas hazards. In order to recognize and fight with them the support from the mining science was essential. The geophysical and specialist monitoring systems of different natural mining hazards were widely introduced in this period. They were accompanied by the development of the long and short term prophylactic technologies. Above solutions became standard in the domain of recognition and fighting rock

bumping, methane and fire hazards. Further works in the domains of coal mines mechanization and automation were progressing. The urgent demand for the construction of new coal mines arisen in above period. In many cases they were constructed in the very difficult hydrogeological conditions and in the heavy methane environment hazard.

The mining science community was heavily involved into above activities having offered them many innovative solutions. It must be emphasized, that hard coal mining in above period was pretty opened for innovative technical solutions and technologies improving production effectiveness, work safety and quality of the coal. It can be indisputably concluded, that the '70s of the XXth century were the most dynamic period considering both: thematic scope as well as number of conducted and implemented research for the mining sector.

Despite of significant progress in the domain of work safety, their level was still far from the satisfactory. The accident rate was still high and the cases of tragic mining catastrophes were pretty often what can be seen in the Table 1 (upon State Mining Authority Statistics).

Unfortunately the second decade of this period i.e., the '80s of XX century, are characterized by significant slowing down the development rate of both: the coal mining and the conducted mining research. The most important research in this period, which

Table 2. Technical and social effects of restructuring processes in the Polish hard coal mining sector in 1989–2008

Index	Years					
	1989	1995	2000	2005	2006	2008
Coal output (mln tonnes)	177.4	135.4	102.2	97.1	94.4	83.6
Employment (×1000)	407	272	155	123	119	113
Numer of coal mines	70	63	42	33	32	32
Number of open longwalls	861	415	183	130	128	123
Efficiency(tonnes/w.days)	3,957	4,869	6,635	8,011	8,071	7,338
Average exploitation depth	524	557	614	662	675	687
Number of accidents	13,365	11,159	2,756	2,117	2,321	2,551
Number of fatal accidents	87	36	28	15	45	24

were continued these were the methane and coal dust explosion hazards research, what was the result of several catastrophes caused by above factors, which arose in this period (Table 1). The coal dust problematic was in this time extended by the evaluation of the dusts, which were recognized as dangerous for the health.

The research concerning the remaining mining hazards i.e., rock bumping, fire and outbursts resulting from their continuously increasing rate were also conducted. One of the most important-ones, which must be mentioned these were the works on the new solutions in the domain of roof supports, and especially the arch supports. More attention was being paid on the problematic concerning environmental aspects of the mining activity. The research concerning the improvement of the quality parameters of the coal and its chemical processing (liquefying of coal) are also being conducted.

The high level of output in more and more difficult geological and mining conditions makes the hard coal mining sector to use the assistance of mining science in the emergency cases, which concern many domains of this sector.

The last decade of previous century and the very first years of the XXIst century this is the period of functioning both: hard coal mining and mining science in the new political and economical conditions. The breakthrough in 1989 initiated in Poland processes of difficult political and economical changes. After already mentioned crisis of the '80s there was a stage of market economy building and in consequence cumulation of such phenomena like: restructuring, liquidation of numerous factories or decreasing their production, commercialization and privatization of up till now state owned companies, but also resignation from artificially keeping up level of employment, driving to the increase of unemployment.

In the above mentioned period Polish hard coal mining went under especially drastic restructuring processes, which aim was to bring the sector to economical profitability and competitiveness of Polish hard coal as a primary energy fuel in the primary fuels' open market. Table 2 shows technical and social effects of restructuring processes, which took place in the hard coal mining sector during 20-years' period of 1989–2008.

Also Polish science, including scientific and research units dealing with mining went under adequate restructuring processes concerning adjusting its research potential and thematic of conducted research to the actual needs of the mining sector. It must be mentioned that at this time in the field of scientific research new formulas of financing the research appeared—these were so called: own grants, intentional, developing and ordered projects—received as a result of the competition procedure. An important moment of this period was elaboration and passing in 1994 the new legislation called Geological and mining law, which together with appropriate executive resolutions and instructions took advantage of many years experience and results of the mining research. In 2001 this act was amended and adjusted to the requirements of European directives of new attitudes, when the mining science was used again.

Present circumstances of functioning the mining sector confronted the mining science with the new tasks, which major goal was first of all conducting by the industry the activities in accordance with the principle of sustainable development (Dubiński, J., Drzęźla, B. 2005). The necessary means to achieve above goal should be first of all the innovative technologies and technical solutions. Such politics of acting resulted in increased importance of the domains connected with mining activities and especially with work safety and environmental protection, considered as the key elements of sustainable development—and their realization became challenge for the mining science.

Having concentrated just on the most important research problems concerning hard coal mining, realized in above period of time, the following of them must be mentioned:

- Safety liquidation of the coal mines or their parts
- Concentration of the production in the conditions of occurrence high risk of individual natural mining hazards and especially rock bumping, methane and fire as well as associated hazards
- Fighting with climatic and radiation hazards
- Coal beds' drainage ahead of mining and economical utilization of CMM
- Elaboration of technical solutions enabling to decrease production costs (new machines and equipment, materials, technologies)
- Revitalization of post-mining territories
- Techniques of protecting building estates located within mining territory
- Technologies of purification the coal mine waters (saline waters and radium contaminated)

As a result of conducted research the new solutions were founded, which truly and significantly contributed into development of the work safety. The most important are (Dubiński J. 2002):

- Methodology of determining maximum safety daily output from the individual longwall in the conditions of high methane hazard
- New method of estimating current state of fire hazard
- Method of seismic geo-tomography for monitoring rock dumping hazard
- Rules of safety exploitation in the conditions of associated hazards occurrence
- Technology of utilization drainage methane for the purposes of central air-conditioning (Pniówek coal mine)
- Technology of removal the radium from the coal mine waters
- New construction solutions in the domain of arch and roof supports

Parallel with above there were conducted research connected with the development of mechanization and automation of individual links of production cycle. In this domain the new designs of longwall shearers and transportation means as well as many others were elaborated. There were elaborated a few new measurement systems in the domain of monitoring principal natural mining hazards occurring in the hard coal (e.g., mining seismology, endogenous fires). Their implementation in the coal mines significantly improved standard of the work conditions at different underground posts and have positive impact on general work safety level. Many of above listed solutions elaborated in Poland have been also implemented in the hard coal mines abroad.

COOPERATION BETWEEN GIG AND U.S. BUREAU OF MINES IN THE DOMAIN OF MINING SAFETY

History of mutual contacts and cooperation between Polish and American mining science has long traditions and started before 1945. Its special element and good example is cooperation between Central Mining Institute and United States Bureau of Mines.

An important factor, which formalized it and drove into significant growth of cooperation was Joined Polish American Maria Skłodowska-Curie Fund. It must be emphasized, that essential part of this cooperation was focused on the issues of work safety in the mining and consisted in joined realization of the specific research tasks, joined seminars and consulting when the obtained results were discussed. They took place both in Poland and in the USA. The most important tasks realized in the '90s of the XXth century there were the following-ones:

- Coal bed methane drainage in order to improve work safety and preserve the environment
- System of automatic dams in order to slow down the gases and coal dust explosions in the underground excavations
- Elaboration of physical precursors of strong seismic phenomena caused by mining operations
- Implementation of P.C. system called "Bumping" to elaborate protection in the roadways exposed to the bumps
- Increase of natural radiation caused by mining activities

Above activities proved, that international, partnership cooperation enables quick and effective transfer of the know-how and jointly elaborated solutions can be characterized by more universal appliance. It also created chance for long term mutual professional contacts.

FUTURE OF POLISH MINING SCIENCE
Present Research Potential

In the interwar period, when in Poland there were only two institutions conducting the research for developing mining sector of mineral resources, including hard coal mines—i.e., University of Mining and Metallurgy in Krakow and experimental mine "Barbara" in Mikołow—the research potential of Polish mining science could be estimated as rather modest-one. After 1945 certain actions were undertaken, which improved the situation both in the research and education domains. Presently there is several different scientific and research units, which are part of basic organizational structures of Polish science. Some of them totally and some only in the selected mining specializations, conduct the research covered by the mining science. Some of them are listed below:

- Research and Development Institutes (under Ministry of Economy)
  - Central Mining Institute (Polish acronym: GIG)
  - Institute of Innovative Technologies
  - Institute of Mining Technology
  - Institute of Open Cast Mining (Polish acronym: POLTEGOR)
  - Institute of Oil and Gas (Polish acronym: INiG)
- Higher Technical Education Institutions
  - AGH–University of Science and Technology with its leading Faculty of Mining and Geo-Engineering together with five other faculties, where mining science is being lectured in the limited extent
  - Silesian Politechnic with the Faculty of Mining and Geology
  - Wrocław Politechnic with its Faculty of Geo-Engineering, Mining and Geology
- Institutes of Polish Academy of Sciences
  - Institute of Rock Mass Mechanics
  - Institute of Mineral Resources and Energy Management
  - Institute Geophysics

Organization, which integrates community of specialists connected with mining science is the Faculty of Earth Sciences and Mining Sciences and Committee of Mining under auspices of Polish Academy of Sciences.

Staff potential of above mentioned units with reference to the employees conducting the research for mining must be evaluated as significant, both from the quantitative and qualitative point of view. It fully covers present needs of Polish mining sector of mineral resources. According to the survey from 2008 it covers:

- 108 professors
- 85 holders of postdoctoral degrees
- 400 doctors
- 800 technical-engineering employees

The available research equipment fulfils world standards in the domains of measurement, laboratory base and specialist experimental fields including experimental mine. Also PC base together with the latest specialist software is fully capable to conduct all necessary computational works in the domains concerning problems of present mining.

The above listed scientific and research units gain the financial means for the research mostly from the industrial contracts (about 70%) and from the domestic budget by the means of different research projects and statutory activities. After 2004 the new source of financing the research became the budget of the European Union with its research projects gained under the consecutive framework programs and the target programs (e.g., from Research Fund for Coal and Steel).

Research Priorities

Since 2004 Polish hard coal mining has been acting under the EU structures and all its rules of functioning are subject to binding in the EU directives, norms and other legal acts. Nevertheless, dominant in all above actions became already mentioned conformity with the strategy of sustainable development. This means care of not only economical aspects of conducted production activities but also care of natural environment exposed for mining activities and of social aspects of the production.

It must be emphasized, that in the circumstances of progressing loss of resources, one of the key elements of environmental protection is becoming reasonable utilization of the coal resources. Social aspect of the production process however, including in its essence care of the human being manifest itself both by the care of working conditions and including its safety and hygiene but also of holding down the jobs. Polish experience from the last years clearly shows, that presently it is impossible to conduct mining operations without social acceptance i.e., without acceptance of local government and the residents themselves. The important factor in the communication between mining and the society, which has significant impact on the social perception of this industry is genuine presenting of mining image by various mass media (press, radio, TV).

It must be emphasized with satisfaction, that awareness of properly operating mining necessity is becoming more and more common and properly understood by the mining sector management staff and its supervision bodies. It all has its end in the research priorities being formulated by the scientific units cooperating and working for the benefit of the hard coal mining. The most important of above were formulated below:

- Improvement of existing solutions in the domain of safety exploitation of the coal beds in the conditions of natural hazards occurrence, especially those with catastrophic character

- Air conditioning of the coal mines
- Development of ahead of mining techniques recognizing structures of the coal beds
- Technologies of effective exploitation thin and irregular coal beds as well as geologically disturbed parts of the deposit
- Mechanization and automation of the basic production centers together with the common usage of the latest tools of informatics, mechatronics and robotics when controlling technological processes
- Explosive-proof safety with reference to the new machines, equipment and materials utilized in mining
- Technologies of mining terrains' ground surface protection and natural environment against the consequences of mining activities
- Development of clean coal technologies

CONCLUSIONS

1. Analysis of hard coal mining history clearly shows that it played significant role in the economic and civilisation development of Poland. Also today Polish energy safety is still based on the own hard and brown coal resources and mining industry of mineral resources is important sector of Polish economy.

2. Polish mining science was always strongly supporting development of domestic mining industry of mineral resources, having contributed into continuous modernisation of the mining technologies, mainly in the aspect of improvement their efficiency and work safety. The tasks and research priorities were changing in time depending on the needs of mining and existing circumstances.

3. Principal achievements of Polish mining science pertaining to the content, which have international extent, are first of all those connected with the domain of work safety and especially with the recognition and fighting with such catastrophic mining hazards like methane, coal dust, bumping and fires.

4. It must be emphasized the fact of mutual cooperation between Central Mining Institute and research centers of US Bureau of Mines, which was realized under Polish-American Maria Skłodowska-Curie Fund, contributing to the international dissemination of obtained results.

5. The important moment in the development of Polish mining science was the year 1989, which changed both political and economic systems in Poland. It had significant impact on the hard coal mining and science sectors, including mining science. Restructuring processes, which were undertaken formed the image of all mining sector and mutual relations between hard coal mining and mining science.

6. Present hard coal mining in Poland is conducting its activities in more and more difficult mining and geological conditions as well as in the complex environmental conditions, which limit possibility of free coal exploitation. It creates the new research priorities, which generally fit into the domain of sustainable development of mining. Problems of work safety, rational utilization of the reserves, environmental protection in the mining terrains are still considered the most important subjects among the research tasks to be undertaken.

7. Existing geo-political reality and especially Polish membership in the European Union make favorable conditions for conducting the research for mining industry under international research consortia. The experience in this domain proves that it can have positive impact on universal applicability of obtained results as well as bigger chances for their commercialization and implementation.

BIBLIOGRAPHY

Bradecki, W., Dubiński, J. 2005, *Effect of restructuring of Polish coal-mining industry on the level of natural hazards.* Archives of Mining Sciences, Kraków, 50, 1, p. 49–67.

Dubiński, J. et al. 1999. *Output concentration and mining hazards.* Ed. Central Mining Institute, Katowice (in Polish).

Dubiński J. 2002. *Scientific challenges of hard coal mining on the threshold XXI century.* Proc. of Underground School of Mining. Szczyrk. T.I. pp. 9–23, (in Polish).

Dubiński, J., Drzęźla, B. 2005, *Problems of sustainable development in the Polish hard coal-mining industry.* Proc. of Mining Workshop "Natural mining hazards." Ed. PAS, Kraków, pp.105–117, (in Polish).

Konopko, W., et al. 2008, *Annual Report on state of basic natural and technical hazards in the hard coal-mining industry.* Ed. Central Mining Institute, Katowice (in Polish).

PIG. 2008. *Balance of Raw Material and Groundwater Resources in Poland, as at 31.XII.2007 r.* PIG Warszawa.

Disaster Prevention in Deep Hard Coal Mining: A German Review

Per Nicolai Martens
RWTH Aachen University, Aachen, Germany

Walter Hermülheim
RAG Aktiengesellschaft, Herne, Germany

ABSTRACT: Disaster prevention in German hard coal mines has since the beginning of the industrial extraction in the 19th century focused on measures to fight against methane. Over the years, different techniques were developed and applied to avoid accidents with high numbers of fatal casualties. Three of these are discussed in detail: the prevention of explosions, the establishing and training of Mine Rescue Brigades and the prevention of mine fires.

INTRODUCTION

The introduction of the steam engine not only opened huge coal reserves to mankind but also pitted the Western European miners of the 18th century against a considerable new threat—methane gas. The history of disaster prevention in deep hard coal mining consequently is a history of the struggle against this enemy.

Milestones on this way are:

- *Begin of 19th century*—Flame safety lamps
- *End of 19th century*—Establishment of expert committees and safety in mines research organizations
- *Begin of 20th century*—Establishment of mine rescue brigades and central mine rescue stations; invention of suitable breathing apparatus
- *Twenties*—stone dusting and stone dust barriers
- *Thirties*—electric safety lamps and first safety explosives
- *Fifties*—filter self rescuers and improved fire safety measures, e. g. non-flammable hydraulic fluids
- *Sixties*—coal dust binding and water explosion barriers
- *Seventies*—explosion proof seals and nitrogen-inertization of mine fires
- *Eighties*—self-extinguishing conveyor belts

In less spectacular steps, ventilation and gas-monitoring systems as key elements against the methane threat as well as flame safe electrical installations were continuously improved over the whole time.

Over the last two decades, the main emphasis has been placed on safely controlling goaf fires caused by spontaneous combustion, on supplementary measures against gas and conveyor belt fires and on the introduction of a new generation of filter self rescuers.

The disaster control measures described are part of an integrated occupational health, safety and environmental protection program of the German coal mining industry.

MILESTONES

As it is not too easy to sum up the developments regarding mine safety, which took place over the entire last century in German hard coal mining, in a condensed paper, the spotlight is on the key milestones in the fields of explosion prevention, mine rescue and fire prevention.

In the 19th century, when hard coal mining moved from gallery mining rather close to the surface to deep shaft mines and the threat from methane which seeps from the coal became more and more dominant.

Explosion Prevention

The horrific mine disasters of the late 19th and early 20th century made a scientific approach to the problems connected to methane and coal dust necessary. During that time the first mine safety research centers were established in Germany.

In 1908 smaller facilities in what was then the state of Prussia evolved into the Bergbau-Versuchsstrecke (BVS), essentially a central mine safety research facility, at Dortmund-Derne (Figure 1). The main focus of this was to examine the devices used in mining with regard to explosion prevention. Today the BVS still exists within Dekra-EXAM, one of Germany's larger testing and certification agencies. Since 1927 the Versuchsgrubengesellschaft research facility researched fending measures against underground explosions, e.g., dust binding procedures and explosion barriers. This research facility ended its operations on the research mine Tremonia in Dortmund

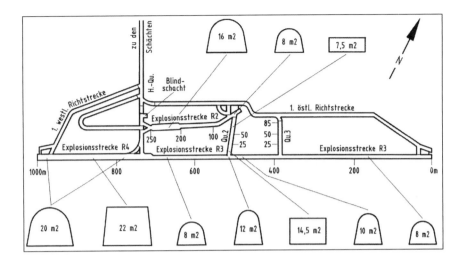

Figure 1.   Layout of BVS at Dortmund

in 1996 after the current water explosion barriers, which have just lately been subject to an EU regulation, had reached full operating maturity.

The invention of the rubber hose by the British veterinarian John Dunlop in 1888 was not only of great benefit to cyclists, but it also for the first time made it possible to extensively sprinkle underground mine workings with water. Unfortunately, over the decades this proved to be an unsuitable approach because the coal dust is hydrophobic and has the tendency to become airborne even when sprinkled.

It was only the stone dusting of the 1920s with which it was possible to sufficiently render the deposited coal dust inert even if it was raised by turbulences. 20 weight per cent of coal dust require at least 80 weight per cent of pulverized limestone or dolomite.

In addition to that, stone dust barriers were installed in the underground roadways. These were essentially wooden platforms which held stone dust. A so-called main barrier holds 400 kg of pulverized rock for each square meter of the associated roadway. This dust is kicked up by the shockwave moving ahead of the flame front of a coal dust explosion and cools the flame front sufficiently to extinguish it.

In the mid-thirties the explosive compounds of the time were modified by adding inert substances, e.g., common salt, to lower the detonation temperature. In the 1950s flame-safe blasting agents with reverse pairs of salts became available. These agents detonate selectively, which means that the explosive will only fully ignite when fully enclosed. This way it is well protected against accidental ignition. Such blasting agents are safe for the angled mortar test,

with which the protection against such accidents can be determined.

After the last severe explosion in Germany at Luisenthal in 1962 water explosion barriers were introduced instead of stone dust barriers. A water barrier is not only effective against coal dust explosions, but also against methane explosions and is triggered by pressures as low as 200 mbar. A fully fledged so-called concentrated barrier holds 200 l of water for each square meter of the associated roadway. These barriers are set up in intervals of 400 m all over the underground roadway system. Even more effective are the so-called segmented water barriers, which consist of smaller groups of open water containers arranged in even intervals of about 20 to 30 m along the entire length of roadway. Such barriers hold 1 l of water for each square meter of associated roadway and extend to cover intersections and forks. These barriers cover the entire area of the associated roadway, and so an explosion is extinguished very close to its point of origin. Dangers associated with the shockwave of the explosion and with the afterdamp are considerably reduced.

The high levels of coal dust generated by modern production and heading methods can no longer be controlled with stone dusting. For about 50 years dust binding is used to render the deposited coal dust inert. For headings and gate roads its use is mandatory. The coal dust is sprayed with a solution of water, magnesium chloride and a wetting agent to dampen it and keep it down so it can not be kicked up by the shockwave of an explosion.

The first measure against explosions with regards to ventilation was the construction of the so called ventilating sections 100 years ago. The ventilating sections were parts of the mine working which were independent from each other in matters of ventilation and in which only up to 100 men were allowed to work simultaneously. From the very beginning until today a powerful exhaust ventilation is at the very base of effective explosion prevention. The airflow which is achieved in deep shaft mines where high levels of gas are encountered can reach 2 to 3 m^3/s for each ton of production per day. All relevant airflows are surveyed for CH_4, CO and air velocity by fixed stationary sensors. The measured values are transmitted to the control room above ground and evaluated by the online computer system there. This way critical and alarm values can be set, tendencies can be observed (e.g., the generation of CO by an unnoticed fire) and freak values caused by blasting can be filtered out.

In the case of multiple seam mining in gaseous seams, which is common in Germany, the methane seeping from the mined seam and from the neighboring seams can only be reliably controlled with an advance ventilation layout and with active gas extraction. Two thirds of the extracted methane are used as fuel for block heating stations. Here mining does its part of avoiding its own harmful effect on the climate.

From the 1970s on abandoned mine workings or areas affected by fire are sealed off in an explosion-proof manner. Research on full-scale samples conducted by the Versuchsgrubengesellschaft research association has shown that the involved seals must, besides being tight and durable, also be able to withstand static pressures of up to 5 bars. All seals are equipped with explosion-proof seal tubes so ventilation can be kept up during construction of the seal and needs only to be cut off after the seal has cured. Depending on the quality of the involved materials the seal is between two and four meters thick in an average roadway.

Mine Rescue

Today's mine rescue is closely connected to the development and introduction of a breathing apparatus suitable for work and rescue and also self-rescue. The first suitable design for a self-contained breathing apparatus for rescue and self-rescue was the *Pneumatophor by Walcher-Gärtner* presented in Austria in 1895. As early as in the first decade of the 20th century the *Dräger 1904/09* rebreather displayed many of the design features which are even today typical for a self-contained device operated with compressed oxygen. The Dräger BG 160 A and Auer MR II/32 from the thirties had quite similar characteristics as modern systems, given the two

hour duration required at the time opposed to the four hour duration of today's systems.

The first volunteer mine rescue teams were formed as early as in the late 19th century in the more progressive mining companies in Germany. A. Meyer, at the time the director of the Shamrock mine in Herne, is therefore seen as the founder of the German mine rescue. In 1905 and in 1908 the mining companies Rheinelbe and Rheinpreußen founded the first professional mine rescue teams. At the same time central offices for mine rescue were founded in Beuthen (today Bytom, PL), Aachen and Leipzig 1907, in Essen 1910 and in Clausthal 1911. According to the Bundesberggesetz all companies which were active in underground mining were obliged to associate with these offices. The central offices were responsible for organizing the mutual help of the mine rescue teams within the area of responsibility, for the proper training of the mine rescuers and their leaders, for furthering the development of the (breathing) equipment and for advising the mining companies in matters of disaster prevention. They also controlled and organized operations in the case of a rescue operation.

Apart from the few professional mine rescue teams, of which the *Technische Sonderdienst* of the RAG Deutsche Steinkohle, a direct successor to the Rheinelbe-Wehr founded in 1905, was the last, mine rescue is a pure volunteer organization. A major hard coal mine has a larger mine rescue force with 10 to 14 teams (in each team there is one team leader and 4 mine rescuers). At any given time four teams can be called to duty within half an hour with a text message based alarm procedure (Figure 2).

The development of new self-contained breathing apparatus for mine rescue led to a new generation of greatly improved rebreathers, e.g., Dräger BG 4 or MSA Auer Air Elite. What remains from the point of view of the rescue men is that even the latest

Figure 2. German mine rescue team

devices weigh at least 15 kg. The weight could not be reduced for more than five decades and some older devices, such as the legendary Dräger BG 174, were even lighter than today's systems.

Mine rescue teams in Germany are also equipped with fireproof clothing which protects against methane deflagration and with a light wire-bound, independent and interference-proof telephone. They are also equipped with hand-held devices for measuring CH_4, CO, CO_2, O_2 as well as temperature and humidity. The necessary fire fighting equipment is stored below ground at known points of danger, e.g., belt conveyors or road heading machines.

Work is especially demanding for the mine rescue teams under hot and humid conditions. In the 80s extensive research was undertaken in climate chambers at the central office at Essen (today Herne) to develop climate tables from which the maximum operating time with regards to temperature and humidity at the site can be read. This takes into account the clothing worn by the rescue men (from flameproof to light clothing). When the core temperature of the body exceeds $38,5°C$, the relevant rescue man will be pulled back from duty. There was no case of an accident caused by heat exhaustion since the tables have been introduced.

Regular fitness tests on an ergometer and at least five exercises per year with full breathing apparatus which are carried on to full exhaustion make sure the members of the mine rescue team are sufficiently, means outstandingly, fit. The standard exercise stretches over two hours at a temperature of 30° C and at a physical strain corresponding to a respiratory volume of approximately 35 l per minute. Because the average age of the mine rescue men is increasing in Germany as well as everywhere else, a test for physical strength was developed together with sports and ergonomics researchers. Every rescue man is required to undergo this test regularly. The test is conducted with equipment which is also used in power training.

Filter self rescuers have been used in Germany since the fifties as respiratory devices for self-rescue. This requires that effective ventilation remains present even after an explosion to provide sufficient levels of oxygen. This can only be achieved by having explosion proof ventilating systems. It is also necessary to keep the fire loads, especially those resulting from plastics, low. Otherwise fires will generate high amounts of particulate matter, which will clog the filter systems of the respirators used for self-rescue. These measures will also improve the range of vision of those escaping from affected mine workings. If all these conditions are met, the filter self rescuer has the indispensable advantage that it can be worn on the belt at any time. The physiological properties of the filter self rescuer could be dramatically improved over the last years. Examples for the systems used in German hard coal mining are the Dräger 990 and the MSA Auer W 90 which have a breathing resistance of well under 5 mbar and a duration which far exceeds the legally required 90 minutes.

Precise planning of the escape routes and careful calculation of the escape times make sure that each person in the mine works can reach airflows which are unaffected by any event in under 90 minutes by walking from his workplace. The movement speed calculated takes the geometry and dip of the mine workings and the climatic conditions into account, but will be under 50 m/min in any case.

To wander from the subject briefly, the self-rescue devices used in non-coal mining in Germany are mainly oxygen self rescuers on KO_2-Basis. There lower oxygen levels will be encountered in the case of a fire because the ventilation systems are not as powerful as in the coal mines. The German potash mines are affected by carbon dioxide eruptions. In hard coal mining oxygen self rescuers are used only in exceptional cases e.g., when there is a danger of gas eruptions or as emergency systems for the mine rescue teams.

Immediately after an accident in the mine an—often preliminary—operation control must immediately take action to avert danger from the involved personnel and to implement the emergency procedures. Every member of the general mine management, e.g., supervisors from different fields of operations or management staff members may find themselves in the position of "preliminary director of rescue operations." Until the permanent rescue operation control (management, central office for mine rescue, ventilation engineer) has been assembled it is of paramount importance to take all necessary measures to keep the personnel safe according to the preset rescue plans and to avoid making premature wrong decisions, which may hamper later rescue operations. It has proved to be extremely efficient to regularly train the necessary skills with all executives which may be called to such duty. Such an exercise takes only two to three hours and encompasses a scenario which is then dealt with in real time.

Fire Protection

In German hard coal mining protection against fire is traditionally a branch of the mine rescue services, which is dealt with by the respective central offices for mine rescue.

The Marcinelles mine disaster of 1956 in Belgium brought the subject of preventive fire protection to attention all over Europe (Figure 3). Since

Figure 3. Mine disaster in Marcinelles 1956

then hydraulic systems which may leak pressurized hydraulic fluid underground in the case of a malfunction may only be operated with flame-retardant fluids. These are either hydrous polymer solutions or anhydrous synthetic solutions based on chlorinated hydrocarbons.

The "forests" of wooden roof supports which were usual in earlier hard coal mining are gone now for several decades because they pose a constant fire hazard. A fire spreading through them is nearly impossible to control. The same applies for gate side packs, which these days are constructed with inert mortar matter. All necessary fire-fighting devices are stored on site underground ready for use not only by the mine rescue, but also for the miners themselves. There is also a regular practical firefighting training for as many of the employees working underground.

Since the 1980s only *self-extinguishing* conveyor belts are used underground. Such a conveyor will burn if there is a sufficient fire present, but it must extinguish at least 10 m downwind from such a fire. This property is achieved by adding flame-retarding additives to the material and is tested in an extensive experiment when the material is approved.

To be able to detect smoldering fires on the ground under conveyors earlier, it is planned to add a heat detection system based on a temperature sensitive light conductor under the conveyor to the present CO-based fire detection systems. With this especially the conveyors in roadways with high airspeeds can be monitored. These conveyors are always problematic from a fire prevention point of view, because recent events in Germany have shown

that airspeeds exceeding 3 m/s tend to overwhelm the self-extinguishing properties of the material.

All production and road-heading machines are equipped with devices for fighting gas fires. The personnel operating these machines are sufficiently trained in fighting such fires. A recently developed high-pressure nozzle, which is connected to the sprinklers of the shield support, makes it possible to put out gas fires at the longwall with water alone. Road heading machines are equipped with hard lines for firefighting and also carry significantly more than 50 kg of dry powder as a backup. Since the crews have been extensively trained in fighting gas fires with these devices no road-heading machine has been lost to a gas fire.

Sprinkling cutting production and road-heading machines is not only directed against respirable particulate matter, but is also geared towards suppressing the ignition of explosive air-methane mixtures. The corresponding parameters, e.g., pressure and volume of the sprinklers, are monitored from the surface control room.

All diesel vehicles are equipped with automated dry powder fire extinguishers. The underground maintenance and refueling facilities for diesel vehicles are equipped with water fog systems, which are able to reliably control a diesel fire. These measures meet the dangers connected to particulate matter resulting from burning diesel fuel, which may be critical for an escape with the filter self rescuer.

For the safe control of fires caused by self-ignition in the goaf a system of targeted goaf inertization with nitrogen was developed in the seventies and brought to such a degree of maturity in the nineties that under certain conditions production can safely be maintained while a fire-fighting operation is on the way. Preconditions for this are an advance ventilation layout, a low threshold for the concentration of ignitable gasses in the goaf, for instance under 50% of the lower explosion limit, and a careful maintenance of these concentrations in by adding nitrogen to the stray ventilation airstream in the goaf according to the values gathered from continuous surveillance of the gas concentrationUnder the protection of such a goaf inertization self-ignited fires are then finally brought under control by means of mortar matter injection into the goaf through drilled holes. The required mortar matter, often several 1000 tons, is hydraulically fed from above ground via high pressure hose. The fringes of the goaf, for instance behind the gate side pack or behind the shields, are increasingly sealed off with phenol resin foam to exclude oxygen in the case of a fire.

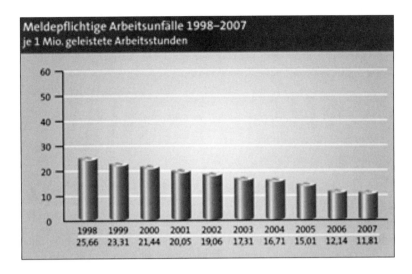

Figure 4. Accidents per million working h, Bergbau-BG 2009

CONCLUSION

While the technical solutions may be fascinated by all, which have been devised even in "unproductive" areas such as mine safety and disaster protection, it must never forget that above all, motivation and training of the involved personnel is the key to success.

Therefore, the measures taken in the fields of occupational health and safety and protection of the environment are embedded into an extensive program of training and motivation, which basically involves all employees. The demands placed on the individual employees within this program increase with the degree to which these take responsibility within the company. So for instance the promotion to chief supervisor underground requires the successful execution of an extensive task in occupational safety.

It is also advisable to school supervisors working underground systematically in regards to fire and explosion prevention as well as in ventilation technology. This personnel can then take on responsibilities for monitoring these areas. In Germany this is required by law.

It is not intend to go as far as the philosopher and engineer Ludwig Wittgenstein in his *Tractatus Logico-Philosophicus* nearly 100 years ago and boast *that in general the problems are now solved for good*. At least the threat which the hard coal miner faces from methane, and which has been mentioned in the beginning, has now after more than 100 years of strenuous research and development successfully been eliminated. The current accident statistics of less than 20 accidents which require report for one million of working hours underline that massive disasters are now a thing of the past (Figure 4).

NIOSH and its predecessor USBM have played a crucial part in achieving a similar success in the USA. In the name of the German *Mine Safety Community* I want to congratulate wholeheartedly to this success and to the anniversary.

In a modern and developed society the demands for a safe and healthy workplace are increasing more and more. So, regardless of our success so far, we have to strive for our goal of *no accidents and occupational diseases*. On this way I wish us all good luck.

Glückauf!

How Will We Respond to a Mine Disaster in 2010?

Paul Harrison
Simtars, Ipswich, Queensland, Australia

Stewart Bell
Queensland Mines and Energy, Brisbane, Queensland, Australia

Darren Brady
Simtars, Ipswich, Queensland, Australia

ABSTRACT: During the Laleham Colliery fire in the early 1980s the response team (then from the Queensland Department of Health) was transported in the mine owner's aircraft. During the Moura No. 4 disaster in 1986 a response team from the newly formed Safety in Mines Testing and Research Station (Simtars) was transported in a Royal Australian Air Force Chinook helicopter and during the Moura No. 2 disaster in 1994 the Queensland Government Ministerial jet was used on two separate occasions to transport Simtars personnel and large volumes of equipment to the Moura mine under emergency conditions.

This paper discusses how the experiences with the Moura disasters have been used to dramatically improve the emergency response capability of government agencies and mine owners through technology improvements, the implementation of MEMS (Mine Emergency Management System) and MEMS training, emergency exercises, restructuring of the rescue service, and mine self escape recovery capability.

INTRODUCTION

Following the Moura No. 2 mine disaster in 1994, a major overhaul of safety management and regulation in Queensland mines has dramatically improved the industry's safety record and reduced the likelihood of a catastrophic event. This improvement has been brought about by the introduction of a risk management approach and the implementation of safety management systems.

However, every system can fail. It is imperative to have in place plans to deal with that failure. When that failure is a potentially catastrophic event, a tested emergency response plan is required.

Mine disasters at Moura No. 4 coal mine in 1986 and Moura No. 2 coal mine in 1994 highlighted the need for a more rigorous and sophisticated approach to emergency response than the traditional mines rescue team model of the past.

While a mines rescue team is still a critical element in any emergency response, the risk management approach quantifies the risk to which these teams may be exposed. This dramatically illustrates the dangers that these rescuers face. In some circumstances, the risk to the team will be so unacceptably high that a rescue attempt will not be allowed by the regulator.

Every tool available to increase the chance of any trapped miners' survival and reduce, or even eliminate, the risk to the rescue team needs to be utilised.

PREVENTION IS BETTER THAN CURE

Australian underground coal mines set the benchmark for some of the world's best practices in gas monitoring. In particular, the Queensland underground coal mining industry has arguably the most comprehensive gas monitoring systems in the world. Each Queensland mine utilises a combination of three systems, real time telemetric sensors, tube bundle analysis, and gas chromatography in order to monitor and manage the state of the underground atmosphere.

These systems are in place to prevent mine disasters, at least ones related to mine gases. To date, they have proven to be very successful in this endeavour. However, if the unthinkable happens, when the mine emergency response team from Simtars arrives at the mine the gas monitoring systems will yield invaluable historical data as well as being vital tools for the assessment of the current mine status. The key to modern mine gas monitoring is trend interpretation and the realisation that something in the mine atmosphere has changed initiating immediate action to understand the cause of the change and to put in place appropriate control measures. That is the inherent strength of routine gas monitoring (Brady 2008).

EMERGENCY PROCEDURES

Mines

No emergency procedure existed at the Moura No. 4 coal mine which would automatically alert and call up experts and equipment. This and the absence of other procedures for use in emergencies lead to the identification of the need for mine managers to prepare detailed emergency procedures to be used to assist in identification of the occurrence, evacuation of employees, notification of officials and relevant statutory requirements. These systems must be in place before an emergency occurs. Trying to cobble something together after the event has started never works. Also the need for duty sheets for persons assigned specific tasks, and lines of delegation in the case of absence of statutory officials. The procedures should also include involvement of outside services such as police, state emergency services, and hospitals, and rapid transportation of persons and equipment, medical officers and specialist assistance (including gas analysis) to the minesite.

The Moura No. 4 Inquiry raised serious concerns that no consideration was given to any alternatives to a wind blast as the cause of the emergency for a considerable time following the event. A methane explosion had actually occurred. Further serious concerns were raised about actions taken by the first two rescue teams who, although equipped with self-contained breathing apparatus, took no steps to use that equipment nor to determine the nature of the carbon monoxide contaminated atmosphere which they were entering. It appears that the first rescue team failed to take any gas monitoring equipment with them and in fact were transported into the mine by a miner who was not equipped with or trained in the use of self-contained breathing apparatus.

The following extracts from the Inquiry report highlight the deficiencies and the risks in the response (Warden's Inquiry 1995):

> During the hearing the members of the Inquiry formed the view that basic procedures should exist throughout the Queensland underground coal mining industry to cover disaster emergencies.

> Evidence indicated that the first qualitative analysis of mine air was not available for some 11 hours after the detection of the incident, and that no emergency procedure existed which would automatically alert and call up experts and equipment.

At Moura this delay did not contribute to events after the explosion but it is obvious that a complete understanding of qualitative and quantitative ventilation conditions is essential if correct decisions are to be made following incidents of this nature. The need for such a service has been demonstrated as has the need for mine managers to have expert advice and access to all services. The development of more detailed disaster procedures is essential.

The Inquiry was given details of the method of work employed at the mine and the events immediately before and after the explosion took place. Members of the Inquiry are aware of the extreme circumstances which existed at the mine immediately after the incident and the understandable desire on the part of the other employees to render assistance to the miners trapped in the mine.

Nevertheless serious concern must be expressed that those responsible failed for some time to consider alternative causes for the disaster other than a wind blast from the goaf area of the Main Dips Section of the No. 4 Underground Mine. Further serious concern relates to the instructions given to and action taken by the first two rescue teams who, although equipped with self-contained breathing apparatus, took no steps to use that equipment nor to determine the nature of the obviously hostile atmosphere which they were entering.

Indeed the evidence shows that the first rescue team failed to take any gas monitoring equipment with them and in fact were transported by a miner who was not equipped with or trained in the use of self-contained breathing apparatus.

Incidents like these highlight the need for mines to have systems in place for dealing with emergencies. We are also very much aware now of known unknowns and the need to take these into account whenever taking action in an emergency. Prior to any re-entry into a mine following a disaster now there would be a significant investigation team assessing the available information and a decision made using risk management techniques. Mines rescue services in Australia are jointly researching along with other organisations including Simtars the information required to facilitate a comprehensive understanding of the situation and the best way to ensure an informed decision on re-entry is made.

There is now a requirement under the Queensland mine safety legislation (Coal Mining Safety and Health Regulation 2001, s149) for a mine to have a Principal Hazard Management Plan for Emergency Response.

These emergency response plans have been evolving since the recommendation was made that they be implemented. The Level 1 exercises conducted in Queensland that test a mine's emergency response plan have highlighted that some emergency procedures work better than others. Queensland's mine safety legislation also requires mines to conduct their own emergency exercises, to test the effectiveness of their emergency management procedures and the readiness and fitness of equipment for use in an emergency (Coal Mining Safety and Health Regulation 2001, s171).

Mine Emergency Management System (MEMS)

There has been a move in Queensland to use the Mine Emergency Management System (MEMS), actively promoted by the Queensland Mines Rescue Service.

MEMS is based on the Incident Command (Control) System (ICS) commonly used by emergency service providers (particularly fire services) throughout the world. MEMS is designed to give a structured approach to the management of a mine disaster. The system works on management by objectives, where objectives are identified and communicated to those involved in the incident.

MEMS has a set structure headed by the Incident Controller who assumes responsibility for the incident. This includes the development and implementation of strategic decisions. Under the Incident Controller are three separate groups all headed by its own controller who reports to the Incident Controller. The three groups are Planning, Operations and Logistics.

The Planning group collects, evaluates and distributes information related to the incident both current and forecasted. It develops control strategies and alternative control strategies (risk assessed). The Operations group manages the operations directly applicable to achieving the objectives. Past emergencies, real and simulated, have identified the need and advantage of collecting information from those involved in the incident. This important task is formally assigned to the Operations group. The third group is Logistics. Logistics is responsible for the procurement and maintenance of human and physical resources.

MEMS focuses on limiting the number of persons reporting to a particular person to enable him or her to supervise successfully (span of control). MEMS also includes provisions for fatigue related changeover of key personnel. A significant incident is likely to extend over several shifts and it is crucial that personnel not be required to participate in response teams for extended periods. Success is also dependent on not relying on a single person to fulfil a particular role as their absence may compromise a response.

To be effective any system deployed by a mine for use during an emergency needs to be trialled and those with responsibilities and duties under the system need to be trained and have sufficient understanding of their roles.

Mines Inspectorate

The Mines Inspectorate is the first external point of contact for a mine during an emergency assuming any urgently required medical treatment has been rendered. A duty inspector is on call around the clock and follows a fully documented flow charted procedure depending on the type of incident. The duty inspector will initiate further inspectorate response as well as any response required of third parties such as police, ambulance, mines rescue or Simtars. The procedure also details experts available to provide advice on a range of specific topics such as gas outburst, spontaneous combustion, strata conditions, and the like, as the need arises. These additional experts may not be immediately available as there is no requirement for them to be on call.

The Mines Inspectorate will have input into the incident management at the mine site, however, the mine owner is responsible for dealing with the emergency. The Mines Inspectorate does not take over management of the emergency as is the case in some other jurisdictions but they have the power to stop mines rescue teams entering a mine if they are not satisfied that the risks have been appropriately mitigated.

Safety in Mines Testing and Research Station (Simtars)

Simtars has a team on call around the clock to respond to mine emergencies. The Mines Inspectorate will mobilise Simtars in the event of a mine fire, explosion or incident involving mine gases. Upon arriving on site, Simtars will provide advice to mine management, however, their formal line of reporting is to the Chief Inspector.

During the Moura No. 4 disaster in 1986, the Simtars facility wasn't yet complete. Despite this a

Simtars team was assembled and transported to the mine on a Royal Australian Air Force Chinook helicopter. More formal arrangements were in place for the Moura No. 2 disaster in 1994. The Queensland Government Airwing provided access to the ministerial jet to transport the response team and their equipment to the mine.

Simtars still maintains an agreement today with the Government Airwing's Chief Pilot to use the ministerial jet for emergency response. A special agreement with the Civil Aviation Safety Authority provides for the carriage of compressed gases and dangerous goods via the Government Airwing during a mine emergency.

Due to the availability of gas monitoring systems at all underground coal mines, the equipment carriage requirements have reduced substantially since 1994. The most significant equipment now carried is a backup micro gas chromatograph.

Simtars can also deploy a mobile mine gas laboratory which is a road going tube bundle monitoring system. Simtars emergency preparedness and emergency procedures are testing during trial callouts run at random throughout the year and also at the Level 1 Emergency Exercise.

Queensland Mines Rescue Service

The Queensland Mines Rescue Service (QMRS) is an accredited mines rescue organisation. They have there own guidelines relating to mine emergency response (Guidelines for Mines Rescue Operations 2007). These guidelines are not used in isolation but do set some already established operational procedures and guidelines that can be utilised in an emergency. It is likely that prior to any deployment of a mines rescue team the incident management team controlling the incident would, in conjunction with QMRS, make the decision based on risk management practices.

The skills and knowledge of the QMRS would also be utilised by incident management teams during any disaster to assist in determining options available for taking control of the situation or the best avenues for rescuing trapped or endangered personnel.

It is a requirement under Queensland mining legislation for coal operators to have an agreement with an accredited mines rescue corporation (Coal Mining Safety and Health Regulation 2001, s172). The agreement must include amongst other things the procedure for coal mine operators to help each other in an emergency, how inertisation equipment controlled by the accredited corporation is to be used at the mine, and the operational procedures developed by the accredited corporation to be followed by the corporation in carrying out the mines rescue services at the mine.

In the event of an emergency the affected mine mobilises its own employees who are QMRS Team Members. The mine also notifies the QMRS using its dedicated emergency call out system. QMRS then organises the transport of personnel and equipment to the affected site and requests assistance for additional mines rescue members from neighbouring mines.

The QMRS is also responsible for the provision of mine inertisation services using the GAG jet inertisation device. If high flow inertisation is required (or likely to be required) QMRS will activate a call-out of GAG trained operators and the deployment of one of the GAG units to site.

LEVEL 1 EMERGENCY EXERCISES

The Moura No. 2 Mine Inquiry identified the requirement for emergency exercises to be conducted at underground coal mines. A mine's emergency response procedures are only going to be effective if they are tested and practised. In December 1996 the Approved Standard for the Conduct of Emergency Procedures Exercises was released. This document provides guidelines for conducting mine site emergency exercises as well as the requirement to test the state-wide emergency response in what is termed a Level 1 Mine Emergency Exercise. That Approved Standard which was issued under the old Act (Coal Mining Act 1925) has been replaced by a Recognised Standard under the new Act (Coal Mining Safety and Health Act 1999). The new standard is Recognised Standard 08 Conduct of Emergency Exercises. Level 1 Mine Emergency Exercises have been conducted according to these Standards at a different mine on an annual basis since 1998.

Figure 1. Mines rescue team assembles at a Level 1 Emergency Exercise

Figure 2. GAG jet inertisation device is warmed up on the surface at Collinsville Mine

The Level 1 Exercise is a learning opportunity for mines and associated state emergency services to test their response times, communication systems and interactions with the aim of continual improvement of the whole system. The exercise scenarios are prepared to meet pre-determined objectives of the exercise such as testing self contained self rescuer change over, first response, incident control, call out of mines inspectorate, mines rescue and Simtars (see Figure 1). The information realised from an exercise not only improves the way that the tested mine responds to future incidents but benefits the entire industry as the report, with all recommendations, is freely available.

According to Watkinson and Brady (2008), the running of emergency exercises in Queensland has led to many improvements and engendered several research projects aimed at providing a better response in the event of an emergency.

INERTISATION

The benefit of inertisation during emergency situations has been identified in Queensland's past two disaster inquiries (Warden's Inquiry 1987; Warden's Inquiry 1995) and is a technique utilised around the world in various types of emergency response. Due to the problems associated with the distance from major cities and resources available to Queensland's underground coal industry, inertisation techniques suitable for Queensland needed to be found. During the Moura No.4 disaster liquid nitrogen was used but there were significant problems in providing and transporting sufficient nitrogen to enable the mine to be inerted quickly.

Figure 3. During trials one cubic metre of coal is set alight underground for the GAG to extinguish from the surface

The Moura No.2 Inquiry pointed out that even if the mine had elected to inert the 512 Panel there was no provision readily available for them to do so. Since then considerable improvements have been made in the availability of inertisation in Queensland. Following successful testing by Simtars of a Polish GAG-3A jet engine inertisation system (Bell et al 1997), the Queensland Government procured two such units which are maintained and operated by the Queensland Mines Rescue Service (see Figure 2 and Figure 3). These units are capable of delivering approximately 20 m^3/s of inert gas, and each underground mine is required to have means of safely coupling to the GAG and the ability to rapidly seal off air intake into the mine to inert the mine if required.

Figure 4. Early attempt at a portable chromatograph not very reliable

Mines routinely use low flow inertisation techniques (approximately 0.5 m³/s) as part of their gas management techniques and most have their own units on site or at least have arrangements for their immediate deployment and use. This capability for timely inertisation assists the mine in combating mine fires or explosive atmospheres during emergency situations.

CONTINUOUS GAS MONITORING

Australia and Queensland in particular is in as good a position if not better than anywhere else in the world with respect to continuous gas monitoring. All underground coal mines utilise realtime and tube bundle gas monitoring systems. Providing these remain intact following, or during, any incident, they provide the incident management team with valuable information on the status of the underground environment without the need to send a rescue team into an unknown environment to make the same assessment.

The use of tube bundle systems also allows the collection of samples from underground for further analysis such as gas chromatography, without the need to actually put any person at risk. The ability to use these in situ sample lines provides information significantly sooner than if boreholes are required.

ONSITE ULTRAFAST GAS CHROMATOGRAPHS

Gas chromatography is the only technique currently available which provides for the complete analysis of all components of a gas mixture during an emergency. Queensland underground coal mines are well equipped for chromatographic sample analysis as the result of recommendations from the Moura No. 4 Inquiry. This followed problems and delays experienced with the portable gas chromatograph (Figure 4) used in the response to that disaster and resulted in the development by Simtars of the Computer Assisted Mine Gas Analysis System (Camgas) whereby all underground coal mines in Queensland have their own gas chromatograph and trained operators supported around the clock telemetric from Simtars home base at Redbank.

Access to chromatographic results enables incident control teams to make reliable informed decisions on the status of the underground environment.

Moura No. 2 mine had an onsite gas chromatograph that was used to assess the status of the underground atmosphere following the 1994 explosion. Although the information provided by gas chromatographic analysis using systems similar to the one at Moura was invaluable in the interpretation of the underground environment, the long analysis time of 20 to 30 minutes per sample that typified these earlier systems (Figure 5) did not lend itself to rapid throughput of large numbers of samples. This is a distinct disadvantage during emergency situations.

With improvements in technology analysis time has reduced to several minutes and detection limits have improved greatly (Figure 6). This much improved run time greatly enhances the ability to gather adequate information and make a more

Figure 5. Earlier gas chromatographs suffered from long analysis times

Figure 6. Four channel ultrafast gas chromatograph

informed assessment of the underground atmosphere in the event of a disaster. The improved instrumentation and Simtars encouraging a proactive approach to gas analysis by gas chromatography has created a realisation in the mining industry that chromatographs can be used for routine assessment of the underground environment as well as for emergencies (Brady, Harrison and Bell 2009). This has increased the skill set of operators and ensures ongoing operational readiness of the equipment. As a consequence, mines are well placed should chromatographic analysis be required during an emergency.

This change in focus of the technique from emergency application to routine daily use actually led to some sections of the Queensland coal industry raising concerns that the chromatograph and its operators would no longer be able to cope with a mine fire. To dispel these concerns Simtars implemented a program to evaluate the mines' ability to analyse and interpret samples typical of an intense mine fire. No issues were found (Brady and Usher 2008).

In addition to a mine's own chromatograph, Simtars emergency response kit is equipped with a portable unit that, unlike previous models, is suitable for transportation to a mine site in the event of a disaster and can be used in addition to, and for validation of, the onsite gas chromatograph.

SIMTARS MOBILE LABORATORY

As far back as the report on the 1972 Box Flat disaster it was identified that "qualified persons and equipment for rapid analysis of gas samples be available in times of emergency" (Warden's Inquiry 1972). Further to this the Inquiry into the 1975 Kianga Mine explosion recommended that "All mines have available at short notice the means of analysing the air

samples obtained while dealing with an outbreak of fire below ground. This end may be accomplished by either mobile laboratories or laboratories established in each mining locality" (Warden's Inquiry 1975).

During a mine emergency Simtars is responsible for mine gas monitoring. The ability to provide this service has changed dramatically from pre-Simtars days when the service was provided by Queensland's Department of Health Laboratories. Gas analysis techniques and instrumentation available for use during an emergency have improved dramatically, and continue to improve. This has meant that the team of chemists responding to an incident are able to deploy a fully equipped automated mobile gas analysis laboratory, a far cry from the classical gas measurement techniques, such as Haldane and Orsat apparatus, used several decades ago.

When responding to the Moura No. 4 disaster in 1986, the team from the fledgeling Simtars was equipped with the already discussed primitive portable gas chromatograph. The response equipment also included a series of portable Maihak Sifor II infrared analysers and a Servomex paramagnetic oxygen analyser. Collected samples were introduced into the analysers and results recorded.

The team that responded to the Moura No. 2 disaster made use of the Camgas gas chromatograph onsite and again used the portable infrared and oxygen analysers. Following response to this disaster Simtars reviewed the options available for providing emergency gas analysis and identified that a mobile laboratory equipped with its own ten point, tube bundle system and fitted with an ultrafast gas chromatograph would be the best outcome with respect to the provision of professional analytical support. Although Queensland mines all have their own tube bundle systems there is the possibility that following a major explosion the system may be compromised

Figure 7. Simtars new mobile gas laboratory under construction

320

Figure 8.   Concept drawings of Simtars new mobile gas laboratory

Figure 9.   Mine vehicle fitted out for self escape on trial at an underground mine

Figure 10.   Prototype zero visibility navigation system on self escape vehicle

due to tube damage or an insufficient number of sample points may exist in the area of interest.

This mobile gas laboratory was commissioned in 1997 and has undergone continuous improvement since. This original laboratory has been replaced by a state of the art 20 point tube bundle version commissioned in early 2010 (Figure 7 and Figure 8). The new mobile laboratory is totally self sufficient, even having its own diesel powered generator onboard. This generator allows instrumentation to be warmed up during transit to the site of the emergency. This means the equipment is ready to commence analysis immediately upon arrival.

The laboratory can be operated remotely via a cellular broadband connection or over a wired Ethernet. The laboratory's server can be patched into a mine's gas monitoring system to expand the capacity of the system while still allowing it to be supervised and controlled from the one location (typically the mine control room) Satellite communications are also included to cater for those situations where a cellular connection is not available.

Because of the limited availability of accommodation in Queensland's mining towns, which would certainly be severely strained during a disaster, the laboratory includes its own separate living quarters.

SELF ESCAPE PHILOSOPHY

Self escape is an important element of a response to an emergency situation. Much emphasis in the annual Level 1 Emergency Exercise is given to testing the efficacy of the particular mine's safety and health management system for self-escape as required under the regulations (Coal Mining Safety and Health Regulation 2001, s168).

It has been recognised that the chances of success of a self escape strategy will be greatly enhanced by the provision of motorised transport. The distance

from the working face to the mine portal can be up to 10 km. This is a significant distance to traverse on foot even under normal conditions. Under emergency conditions in low or, perhaps, zero visibility wearing a self contained self rescuer this becomes an arduous task indeed.

Simtars has been conducting research into modifying the existing fleet of diesel personnel vehicles (Figure 9) to enable them to be operated under conditions that might prevail during a major fire or after an explosion (Bell et al 2004; Binnie, Davis and Watkinson 2006; Davis, Ellis and Humphreys 2007). These vehicles will often be parked nearby to the work area and hence would be immediately available for self escape should an incident occur.

Simtars has demonstrated that the diesel engines in these vehicles can be safely operated in methane concentrations over 10% in air (Davis, Ellis and Humphreys 2007). The final stage of the project requires the development of a viable intrinsically safe or flameproof navigations system to cope with the potential for zero visibility conditions (Figure 10).

CONCLUSION

The Queensland mining industry has come a long way over the past 23 years. We now have a situation where coal mines in remote locations can operate safe in the knowledge that they understand their underground atmospheric environment and that expert technical assistance is available to them at all times. We have systems and technology in place to prevent every coal miner's worst nightmare—an underground fire or explosion. However, should the unthinkable happen again we also have systems in place that provide a far greater prospect of survival than at any time in the past.

REFERENCES

Bell, S., Davis, R., Watkinson, M., Humphreys, D. and O'Beirne, T. 2004. *Mines Rescue Vehicle Testing and Development Extension*. Australian Coal Association Research Program No: C12004, Safety in Mines Testing and Research Station (Simtars), Redbank.

Bell, S., Humphreys, D., Harrison, P., Hester, C. and Torlach, J. 1997. *A report on the Demonstration of the GAG-3A Jet Inertisation Device at the Collinsville No 2 Coal Mine on 11, 16, 17 and 18 April 1997*. Australian Coal Association Research Program No: C6019, Safety in Mines Testing and Research Station (Simtars), Redbank.

Binnie, P., Davis, R. and Watkinson, M. 2006. *Mines Rescue Vehicle—Demonstration Phase*. Australian Coal Association Research Program No: C14024, Safety in Mines Testing and Research Station (Simtars), Redbank.

Brady, D. M. 2008. The role of gas monitoring in the prevention and treatment of mine fires. *Proceedings from 12th U.S./North American Mine Ventilation Symposium*, Reno.

Brady, D., Harrison, P. and Bell, S. 2009. 20 Years of Onsite Gas Chromatographs At Queensland Underground Coalmines. *Proceedings from the Ninth International Ventilation Symposium*,, Delhi.

Brady, D. and Usher, D. 2008. How Well Can We Cope with Gas Analysis During a Mine Fire. *Proceedings from the 2008 Queensland Mining Industry Health and Safety Conference*, Townsville.

Coal Mining Safety and Health Act 1999. Office of the Queensland Parliamentary Counsel, Reprinted as in force on 4 September 2009, Reprint No. 3B, Brisbane.

Coal Mining Safety and Health Regulation 2001. Office of the Queensland Parliamentary Counsel, Reprinted as in force on 1 July 2009, Reprint No. 3C, Brisbane.

Davis, R., Ellis, C. and Humphreys, D. 2007. *Performance Testing of Diesel Engines in High Methane Atmospheres*. Australian Coal Association Research Program No: C15029, Safety in Mines Testing and Research Station (Simtars), Redbank.

Queensland Mines Rescue Service. 2006. *Guidelines for Mines Rescue Operations*. Revision Date:- 30.09.2007. Document Number:- MRS 001GUIDS Revision Number: 0-9 UPDATED:- September 2006, Dysart.

Recognised Standard 08. 2009. *Conduct of Emergency Procedures Exercises*. Queensland Government, Brisbane.

Warden's Inquiry. 1972. *Report on an Accident at Box Flat Colliery on Monday, 31 July 1972*. Brisbane.

Warden's Inquiry. 1975. *Report on an Accident at Kianga No. 1 Underground Mine on Saturday, 20 September 1975*. Rockhampton.

Warden's Inquiry. 1987. *Report on an Accident at Moura No. 4 Underground Mine on Wednesday, 16th July, 1986*. Rockhampton.

Warden's Inquiry. 1995. *Report on an Accident at Moura No. 2 Underground Mine on Sunday, 7 August 1994*. Brisbane.

Watkinson, M. and Brady, D. 2008. Ten years of conducting Level 1 simulated emergency exercises in Queensland's underground coal mines, *Proceedings from 12th US/North American Mine Ventilation Symposium*, Reno.

Prevention of Methane and Coal Dust Explosions in the Polish Hard Coal Mines

Krzysztof Cybulski

Central Mining Insitute, Katowice, Poland

Józef Dubiński

Central Mining Insitute, Katowice, Poland

INTRODUCTION

Methane in hard coal deposits and the associated explosion hazard is one of the most dangerous phenomena that accompany the process of hard bituminous coal mining. In spite of a considerable progress in identifying and control of methane hazard, its increase can be observed in many mining areas.

Mainly, it results from mining at greater depths in coal seams highly saturated with methane. Moreover, in consequence of increased concentration of mining production, methane emission occurs in a smaller number of extracted longwalls, causing an increase of their absolute methanity.

Methane causes negative ecological consequences, increasing greenhouse effect when released directly to the atmosphere. On the other hand, the gas is high calorific and ecological energy fuel.

A correct identification of the methane hazard and its effective control is of fundamental importance for carrying out safe mining process. The statistics of mining catastrophes caused by methane or methane and coal dust explosions is the most tragic in the history of underground mining industry, both in terms of fatalities, serious accidents and material losses. Investigations of methane explosions gave rise to the development of methods of methane hazard identification, as well as methods and means of its elimination, either by adequate ventilation or methane drainage.

In the new economic reality of the Polish mining industry it also becomes necessary to make risk analysis of methane and coal dust explosion in longwalls with a high concentration of production.

Regardless of that, detailed analyses concerning the scientific progress and mining practice in the area of methane hazard prediction and elimination, including technologies of methane drainage, assessments of the technical development of monitoring equipment and application of explosion-protection devices are made. Among 31 hard coal mines in Poland (the state at the end of 2008), as many as 27 are methane coal mines. In 15 of them the category IV, the highest category of methane hazard, with methane content above 8 m^3/Mg, has been found.

From the coal output in 2008, amounted to about 83.6 million Mg, about 66.6 million Mg is the output from methane seams; that makes nearly 80% of the total.

The year-average increase of mining depth in Polish coal mines is 8 m/year, and methane content of extracted seams in the period of last 10 years increased by 2,56 m^3/Mg, resulting in increase of methane hazard within the mining areas.

In 2008, from the rock mass influenced by mining 880.90 million m^3 of methane was emitted, that means an average emission of 1679.10 m^3 CH_4 per minute.

Methane has been collected from the following sources:

- 158.6 million m^3—from longwalls environment, (57.8%)
- 107.1 million m^3—from behind the stoppings, (39.1%)
- 8.5 million m^3—from roadways, (3.1%)

The effectiveness of methane drainage was 31.1%. From the total of 274.2 million m^3 of methane acquired in methane drainage, 156.5 million m^3 of this gas have been utilized. The specification of development of absolute methanity, the amount of collected and managed methane, as well as the output hard coal mines over the years 1998–2008 is presented in Table 1.

The gradual increase of methane released to mining areas within the last dozen years forced the necessity of drawing up safety criteria applied at the stage of design and exploitation. The scientific-research work performed in last years has been directed at solving the issues on methane hazard formation in the conditions of increasing concentration

Table 1.   The development of total methanity, volume of collected and managed methane

Specification	Year										
	1998	1999	2000	2001	2002	2003	2004	2005	2006	2007	2008
Total methanity (million m³/year)	763,3	744,5	746,9	743,7	752,6	798,1	825,9	851,1	870,3	878,9	880,9
Amount of collected methane (million m³/year)	203,6	216,1	216,1	214,3	207,3	227,1	217,2	255,3	289,5	268,8	274,2
Amount of managed methane (million m³/year)	152,7	136,9	124,0	131,5	122,4	127,8	144,2	144,8	158,3	165,7	156,5
Number of hard coal mines	57	47	42	42	42	41	39	33	33	31	31
Output of hard coal (million tonnes)	115,9	110,4	102,5	102,6	102,1	100,4	99,5	97,1	94,3	87,4	83,6

of coal production. An objective assessment of the hazard should be based on the results of research work, experience acquired during previous mining work, as well as prognostic computations.

METHANE HAZARD AND STRATEGY OF METHANE PREVENTION IN HARD COAL MINES

In the light of legally binding mining regulations in Poland, four categories of methane hazard in the underground mining plants extracting hard coal are specified. The developed deposits or their parts are rated among:

- First category of methane hazard, if the presence of methane of natural origin has been found in the quantity from 0,1 to 2,5 m³/Mg per clean carbon substance
- Second category of methane hazard, if the presence of methane of natural origin has been found in the quantity above 2,5 m³/Mg, but not higher than 4,5 m³/Mg, per clean carbon substance
- Third category of methane hazard, if the presence of methane of natural origin has been found in the quantity above 4,5 m³/Mg, but not higher than 8m³/Mg, per clean carbon substance
- Fourth category of methane hazard, if the presence of methane of natural origin has been found in the quantity above 8m³/Mg, per clean carbon substance, or when sudden outflow of methane or methane and rocks outburst occurred

The workings in methane panels in underground mining plants that mine hard coal are classified the following way:

- Not at risk from methane explosion, with the degree "a" of methane explosion hazard, if the concentration of methane in the atmosphere above 0,5% is excluded
- With the degree "b" of methane explosion hazard, if in normal conditions of ventilation, the concentration of methane in the atmosphere above 1% is excluded
- With the degree "c" of methane explosion hazard, if even in normal conditions of ventilation the concentration of methane in the atmosphere may exceed 1%

First of all the classification imposes the restrictions for application of energy-mechanical equipment.

The legally binding mining regulations in Poland limit the permissible methane content in mine atmosphere. As an example, in the longwall areas they are the following:

- Methane content in a longwall heading with the fresh air current (including the methane content in the working with the current "refreshing" the current of used air flowing out of the longwall) must not exceed 1%
- Methane content in the longwall on the section of at least 10 m from the fresh air intake must not exceed 1%. The methane content in the longwall along its remaining length must not exceed 2%
- Methane content in that working in the distance from the longwall outlet to a crossing with the nearest ventilation-active working must not exceed 2%

If the longwall heading with the current of used air is directly connected with a group current of used air, then the methane content in this heading must not exceed 1.5%.

In the Polish hard coal mining, the quantity of methane being released to longwall environment is predicted basing on dynamic forecast of absolute methanity, included in the No. 14 Instruction of the Central Mining Institute. This method is based on determining the desorbable methane resources, both in the mined out coal seam, and in undermined and overmined strata, and on computing the quantity of methane release to workings and longwall goafs, depending on the operational longwall advance.

As the longwall environment (Figure 1) it is recognized:

- Mined-out coal seam in the plot of planned longwall
- Longwall working
- Headings taking air in and out in the longwall reach
- Spatial area of the rock mass limited by the zone of desorption
- Workings that are in contact with the zone of desorption, capable of taking out methane being released

- Post-mining goaf being in contact with the zone of desorption
- Goaf adjacent to the longwall or the longwall headings

The accepted method, by determining the methane release from individual strata, also allows to making precise design of potential methane drainage, both for the roof and floor strata. In the view of absolute methanity forecast, at the stage of longwall planning in methane-containing seams, it is necessary to determine in advance the possible acceptable quantity of mining production that may ensure the safety of the crew after the longwall has been set in operation.

The scope of prevention for methane hazard control includes:

- Selection of appropriate longwall ventilation system
- Required delivery of air for methane hazard elimination
- Technology of methane drainage of the longwall environment
- A method of methane hazard monitoring
- Adequate control of methane accumulations in places of possible ignition or explosion initiation

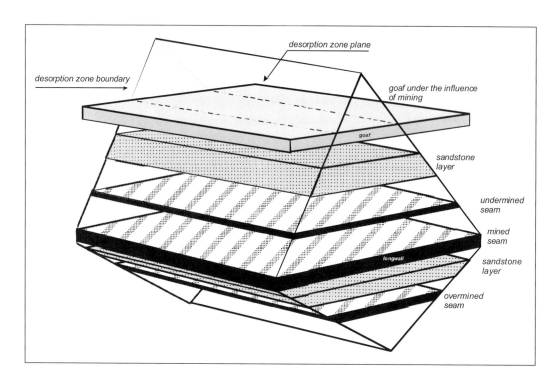

Figure 1.   Schematic diagram of a longwall environment

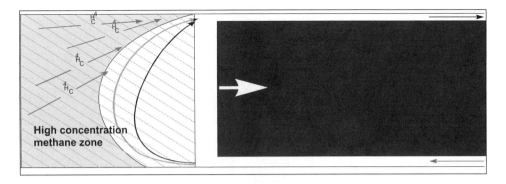

Figure 2. Distribution of methane concentration in goaf of longwall ventilated in the "U" form on the body of coal

The criteria that ensure safe mining are fulfilled by following the procedure included in the Instruction No. 17 of the Central Mining Institute entitled: "The rules of longwall drivage in the methane hazard conditions," both for planned and operated longwalls. "The rules ..." introduce the term of criterial methanity, the value of which should be calculated for longwalls situated in coal seams classified as II, III and IV categories of methane hazard. Under the term of criterial methanity, such a value of absolute methanity is understood, at which in given conditions of ventilation and methane drainage and with the nonuniform methane release to longwall environment, the admissible methane content in the current of used air will not be exceeded.

The method of computing the criterial methanity depends on the system of longwall ventilation applied (form "U" on the coal body, or "Y" with the "refreshing" air flowing out of the longwall). The system of longwall ventilation in "U" form on the body of coal (Figure 2) creates potentially high methane hazard. However, it is a very advantageous in conditions where a hazard of endogenous fire exists in the goaf because of limitation of deep migration of oxygen-rich air in the goaf, where coal has been left in the zone of direct fall of the roof.

In the light of legally binding regulations in Poland, the method of a longwall ventilation with taking the used air out along the body of coal can be applied when:

- The sum of streams of methane volume flowing out of the longwall and its goaf to the longwall heading does not exceed 25 m³/min, and the cross-section of the heading taking the air out of the longwall is at least 8 m²
- The sum of streams of methane volume flowing out of the longwall and its goaf to the longwall

heading does not exceed 15 m³/min, and the cross-section of the heading taking the air out of the longwall is at least 6 m²

The ventilation system in "Y" form with taking the used air out along the longwall goaf (Figure 3) is very favorable in methane hazard conditions, because it causes moving the dangerous methane concentration in the goaf away from the longwall heading. However, it is less advantageous in respect of the endogenous fire hazard because of deep penetration of air in the longwall goaf.

If the calculated value of criterial methanity is less than the predicted or actual methanity, with assumed daily advance, it is necessary to apply methane drainage or take other steps, including limitation of daily advance, in order to reduce the volume of methane emission to the headings of mining area. In 2008, methane drainage was carried out in 20 mines, with the use of 14 surface and 4 underground methane drainage stations. The application of classical methane drainage methods in the ventilation system in "U" form on the body of coal is characterized by the effectiveness of methane drainage about 35%. Methane drainage of the longwall environment based on classical technologies is most often insufficient to provide the safe mining conditions. In such longwall conditions, the circumstances enforce limiting of mining advance.

In hard coal mines with high methane content the technologies based on overlying drainage of the rock mass are introduced with the effectiveness of 70–80%. Shown schematically in Figure 4 is a vertical cross-section through the deposit with the applied technology of overlying drainage, based on a specially made drainage gallery.

The rules of methane prevention set the requirements for monitoring methane hazard. The current

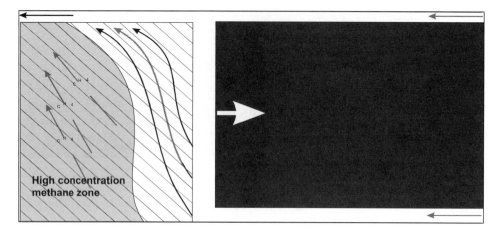

Figure 3. Distribution of methane concentration in the goaf of the longwall ventilated in the "Y"system with "refreshing" of the outlet current from the longwall and with taking the air out along the goaf

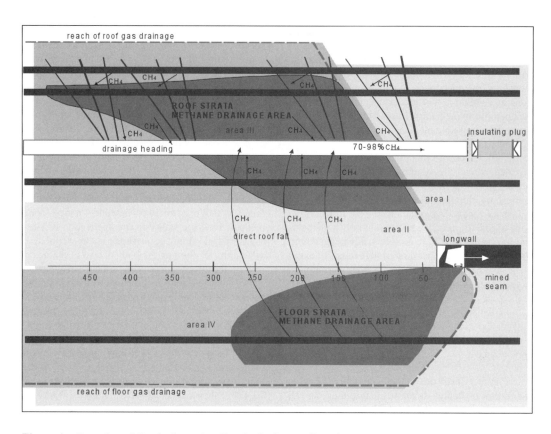

Figure 4. Location of the drainage heading in the longwall environment

state of coal mine technical equipment and solutions offered by the producers allow full completion of monitoring systems within the range presented below:

- Methane content measurement in the working at locations of possible hazard occurrence, including the hazard of high methane concentration under the longwall conveyor
- Measurements of methane content in methane-drainage pipelines
- As far as it is possible and necessary, automatic control of methane drainage networks, depending on methane concentration in pipelines and workings
- Measurement of air velocity (delivery) in the workings, and methane concentrations resulting from changes of the air flow
- Condition of closure of ventilation stoppings that influence the intensity of working ventilation, and the changes of distribution of aerodynamic potential in the longwall environment
- Measurement of other gases content (e.g., oxygen, carbon monoxide, carbon dioxide) if a possibility of hazard exists

The legally binding safety regulations impose application of an automatic methanomery with shortened repetition time of measurements, or with continuous measurement for newly-built central control stations, and in cases when methane hazard appears in the form associated with the rockburst hazard. In the regulations, the minimal conditions for the number and locations of methanometric sensors and the air flow, have been defined. For longwalls with high concentration of mining production, conducted in the conditions of high methane hazard, only complying with the requirements proves to be insufficient. For those longwalls, continuous monitoring of safety and production parameters in the aspect of their influence on the state of methane hazard, state of ventilation and methane drainage and other hazard control becomes necessary. Currently, in respect of quantity, the coal mine equipment is sufficient in terms of measuring and control equipment.

COAL DUST EXPLOSION HAZARD

The hazard of coal dust explosion belongs to principal natural hazards in hard coal mines. An appropriate safety level in that area is achieved by the application of an effective dust explosion prevention. Successively decreasing number of mining longwalls is perceived as a reduction of dust emission sources, which, in consequence, leads to limitation of the

area of coal dust explosion hazard. Nevertheless, an increase of production capabilities of longwalls results also in a significant increase of intensity of dust emission from those sources. It is estimated, that in the process of mining, transport and preparation, from 1 to 3% of the output is converted into coal dust, which, at the present level of mining production, gives the output of coal dust in the Polish coal mines in the range of 1–3 million tones a year. Dispersion of the dust in mine workings causes that the scale of dust hazard in coal mines, both in the aspect of explosion, and of miners health is large. Hence, the dust hazard is common in coal mines and the considerable improvement of this situation is not expected. Additionally, taking into account the fact that all currently mined hard coal seams in Poland are the seams threatened with coal dust explosion, there is the necessity of maintaining in each coal mine dust service teams to evaluate this hazard, and, first of all, to conduct efficient preventive actions. We all remember the tragic year 2002, when the coal dust explosion in "Jas-Mos" coal mine caused loss of 10 lives of miners, as well as the year 2006, when in consequence of coal dust and methane explosion in "Halemba" mine 23 miners died. Therefore the quality of explosion prevention actions must be among the basic tasks for the safe operation of a coal mine.

Line of Defense Against Coal Dust Explosions

1. Limitation of the coal dust formation, its removal and elimination of its ability to form a cloud
2. Limitation of an ignition source (possibility of ignition of coal dust by primary source)
3. Elimination of coal dust explosion development (propagation of coal dust explosion)
4. Limitation of coal dust explosion extent (explosion suppression)

Limitation of Coal Dust Formation, Its Removal and Elimination of Its Ability to Form a Cloud

Keeping in the working the amount of dangerous coal dust below the lower explosion limit is performed by removal of dust deposits (sweeping, washing). At long sections of workings it is difficult to be applied on account of labor intensity and necessity of transport of the removed dust. In such workings, it is necessary to perform periodical measurements of dust settlement intensity in order to prepare a schedule of removal dust deposits.

Elimination of the ability to form a cloud from the coal dust deposits is performed by washing, sprinkling or with the use of hygroscopic agents.

Washing and sprinkling as the method of defense against coal dust explosion is more and more frequently applied owing to easiness of application and low costs. In the conditions of increased climatic hazard, and consequently of increased temperature, effective application of water is made difficult, and often practically impossible, because of its fast evaporation. The dust just wetted quickly becomes dry, and the working threatened with the possibility of explosion propagation.

The application of washing or sprinkling is also hindered in the workings in which electrical equipment is installed. This method is effective when surfactant additives (wetting agents) are added to the water, especially for dusts originated from coking coals. Elimination of coal dust ability to be lifted from deposits can be widely applied in all the workings, especially for maintenance of zones protecting against the possibility of actuation and propagation of coal dust explosion, or in workings with belt conveyor transportation of the output.

In coal mines, stone dust constitute the basic measure (above 80% actions of dust prevention) applied for coal dust neutralization, i.e., protection against dust explosion propagation. The stone dust in the underground workings is applied to maintain the protective zones, as a fire extinguishing agent of explosion barriers as well as for dusting of mine workings.

The hazard of coal dust explosion exists in all workings. The degree of that hazard depends on conditions in a specific underground working. In Table 2, the data from the years 1999–2008 concerning the total quantity of consumed stone dust, taking into account the participation of stone waterproof dust have been presented.

The increased amount of consumed stone waterproof dust and its considerable participation in the total quantity of consumed stone dust results from its greater efficiency in longer time periods, and is dictated by economic reasons, as the frequency of renewed preventive actions is smaller than in the case of typical coal dust use.

Elimination of Ignition Source

Ignition sources of coal dust explosion in underground workings may result natural hazards (methane, fire, rockbursts) or from mining machinery and equipment used in underground workings. Limitation of fire, methane and rockburst hazards, together with methods and measures for detection of those hazards reduce the possibility of coal dust explosion initiation. Sometimes, even a small shortcoming may cause of a serious catastrophe.

Table 2. Amount of stone dust consumed in years 1999–2008

Years	Amount of Consumed Stone Dust (t)		Participation of Waterproof Dust (%)
	Total	Including Waterproof Dust	
1999	57 635	22 382	38,8
2000	52 896	25 455	48,1
2001	50 706	19 204	37,9
2002	79 541	32 720	41,1
2003	71 234	35 099	49,3
2004	67 687	36 180	53,4
2005	63 493	32 582	51,3
2006	59 675	29 339	49,1
2007	66 102	35 018	53,0
2008	70 520	37 185	52,7

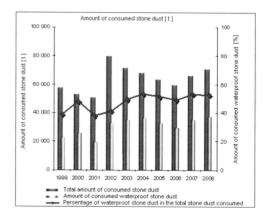

The applied machines, equipment and electric networks in the workings threatened with coal dust explosion must have appropriate levels of protection against the initiation of coal dust explosion. This responsibility results from legally binding, at present, mining regulations and directives of the European Union.

The factors that cause the initiation of coal dust explosion are: methane explosion, rocks sparking, explosive material, detonating cord, firing of several fuses together, electrical arc, open source of light, explosion of fire gases.

Elimination of Development and Propagation of Coal Dust Explosion

The protective zones are the sections of workings being safe relative to initiation and propagation of coal dust explosion. As a rule, they are maintained on a distance of at least 200 meters in all directions from the places of possible coal dust explosion initiation. The zones are made (depending on the existing humidity conditions) by dusting with a stone dust in the whole perimeter of the working until the inflammable matter content of at least 80% in methane panels, or at least 70% in non-methane panels is reached, and also by washing with water.

In the protective zones, the workings are washed with water and sprayed with stone dust in the entire perimeter, including the supports, on a distance of at least 200 meters from the locations of possible dust explosion initiation. The size of protective zones is as follows:

200 m   from places of possible initiation of coal dust explosion
25 m   from switching stations, transformer stations, rectifiers and switch rooms
5 m   from cable connections (cable boxes over 230 V AC)

Additional protective zones in methane panels are:

1. On the whole length of the working ventilated with an air duct
2. In workings classified as degree "c" of methane explosion hazard, on the sections with installed cables or electrical power conduits

Places of possible coal dust explosion initiation:

1. Places of blasting in the workings endangered with coal dust explosion hazard
2. Locations of coal extraction
3. Places of identified methane concentration of at least 1,5%
4. Places of dangerous coal dust concentration in the amount of at least 500 g/m^3 of working in the mine dust unprotected on a distance greater than 30 m, in the working where machines or electrical equipment are used
5. Fire panels
6. Coal containers
7. Stores of explosive materials
8. Zones of particular rockburst hazard in the seams of II, III and IV categories of methane hazard
9. Workings with a slope greater than 10° with a rope, road transport or with narrow-gauge railways, in which cables or electric conduits are installed

In the coal seams classified as category IV of methane hazard outside the 200-meter zones, the mining regulations require dusting of workings with stone

Table 3.   Length of protective zones in the years 1999–2008

Years	Length of Protective Zones (km)		Participation of Washed Zones (%)
	Sprayed With Stone Dust	Washed With Water	
1999	1 680,0	142,3	7,8
2000	1 317,8	118,2	8,2
2001	1 323,7	245,4	15,6
2002	1 460,0	270,7	15,6
2003	1 424,6	310,7	17,9
2004	1 394,2	322,7	18,8
2005	1 454,2	312,1	17,7
2006	1 393,2	318,9	18,6
2007	1 367,5	385,8	22,0
2008	1 468,9	386,5	20,8

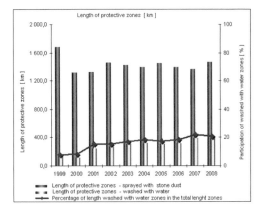

dust up to the content of at least 50% of incombustible matter or washing of workings with water to its content in the mine dust deposits at least 60% of the value required in the protective zone.

The data concerning a total length of protective zones, together with the participation of protective zones made by washing with water, in total length of protective zones in the years 1999–2008, have been presented in Table 3.

Limitation of Range-Suppression of Coal Dust Explosion, Explosion Barriers

Explosion barriers constitute the fourth—last defense line against coal dust explosions. The basic and only task of the barriers is to restrict the explosion extent i.e., suppress the explosion already developed. At present, in the Polish coal mines, generally two types of explosion barriers are used that differ in the type of fire-extinguishing agent placed on shelves of the barrier—stone dust barriers and water barriers. Depending on its location, the explosion barriers can be divided into auxiliary barriers (protecting directly places of possible coal dust explosion initiation) and main barriers (built at inlets and outlets of ventilation areas).

- Explosion barriers-auxiliary, built inside ventilation areas at a distance of 60–200 meters from the places of possible initiation of coal dust explosion

Additional barriers constructed every 200 m:

- Galleries in the methane panel ventilated by air ducts
- Galleries in the panels of II–IV category of the methane hazard ventilated by the main fan:
  - where CH4 > 0.5% and cables or electric power conduits are installed
  - where CH4 > 1.5% - methane accumulations at the roof
  - where zones of particular rockburst hazard are marked

Requirements for explosion barriers:

- Quantity of extinguishing agent in the barrier and on shelves of the barrier
  - 200 dm^3 of water or 200 kg of stone-dust— per 1 m^2 of the cross section
  - 400 dm^3 of water or 400 kg of stone-dust— per 1 m^2 of the cross section

The operation principle of explosion barriers consists in dispersing, in consequence of explosion pressure, extinguishing agent stored on the barrier over the cross-section of the working.

In Table 4, the development of the number of stone dust explosion barriers and water barriers together with the participation of water barriers in a total number of barriers in the Polish hard coal mines, in the years 2000–2008, has been presented.

Classification of Coal Seams in Terms of Coal Dust Explosion Hazard

A coal seam threatened with coal dust explosion it is such a coal seam where volatiles content in coal is greater than 10% in water- and ash-free carbon substance. At present, all mined hard coal seams in the Polish coal mining industry are threatened with coal dust explosion.

- Class A of coal dust explosion hazard—coal seams together with headings driven in the seams where coal dust is protected in a natural way

Table 4. Number of explosion-proof barriers in the years 2000–2008

| Years | Number of Explosion-Proof Barriers | | Participation of Water Barriers (%) |
	Stone-Dust Barriers	Water Barriers	
2000	2 572	257	9,1
2001	2 297	610	21,0
2002	2 116	879	29,3
2003	1 927	1 114	36,6
2004	2 010	1 175	36,9
2005	1 871	1 107	37,2
2006	1 898	1 069	36,0
2007	2 069	752	26,6
2008	2 264	625	21,6

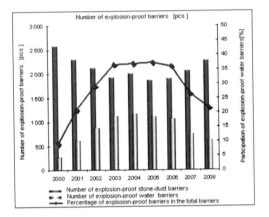

- Class B of coal dust explosion hazard—coal seams together with headings driven in the seams, that do not meet the above requirements

Classification of Mine Workings in Terms of Coal Dust Explosion Hazard

Basic Notions:

Coal dust—grains of coal passing through a sieve with the mesh size equal to 1x1 mm,

Protected mine dust—mine dust that contains:
a. At least 70% of incombustible matter in non-methane panels, or
b. At least 80% of incombustible matter in methane panels, or
c. Passing water that makes impossible coal dust explosion propagation and completely disable dispersion of the dust

Unprotected mine dust—mine dust that does not meet the above requirements.

Classification of Headings in Terms of Coal Dust Explosion Hazard

The working not threatened with a coal dust explosion:

a. Dangerous coal dust does not occur, or
b. Mine dust contains at least 80% of incombustible matter of natural origin, quantity of dangerous coal dust is smaller than 10 g/m3 of working, and intensity of dust settlement is smaller than 0,15 g/m^3 a day, or
c. Mine dust includes at least 50% of a passing water, or
d. Mine dust includes at least 80% of incombustible matter, and content of passing water is at least 30%, and neighbor headings, connected with it, are headings not threatened with coal dust explosion or were classified as class A of coal dust explosion hazard

- Class A of coal dust explosion hazard of the working in which coal dust occurs, protected in the natural way or there are no sections with mine dust unprotected in a natural way longer than 30m, at the same time the distance between these sections must not be smaller than 100m
- Class B of coal dust explosion hazard of workings that do not meet the above requirements

CONCLUSIONS

1. The methane and dust hazard in the Polish hard coal mining, in spite of decrease of hard coal production, is still very high. This results, in particular, from the increase of methane content in the deposits, together with the depth of deposition and conducting mining operations in lower and lower situated parts of coal seams as well as concentration of mining production.
2. The yearly mining production of coal mines in 2008 was 83.4 million Mg, and was smaller than the mining production in 2007, by about 3.8 million Mg.
3. In the year 2008, the participation of methane coal seams in the total output was about 66.6 million Mg, that is 79.9% of the output.
4. In the year 2008, in 20 coal mines, the methane drainage was permanently applied. A total quantity of methane collected with methane drainage pipelines increased from 268.75 million m^3/year in the year 2007 to V=274.2 million m^3/year in the year 2008.

5. From the total volume of methane 274.2 million m^3/year collected 2008, 156.5 million m^3/year was utilized, the remaining part i.e., 117.7 million m^3/year was released into the outer atmosphere. Therefore, the effectiveness of utilization of methane collected was 57.1%.
6. Collecting methane with methane drainage pipelines enables to utilize it for economical purposes. The methane-containing mixture can be burnt in the systems generating electric and thermal energy. Such activity limits the methane emission into the atmosphere, brings in profits from the sale of electric and thermal energy, positively affects safety during exploitation of methane deposits.
7. In the year 2008, a total length of maintained protective zones made by dusting with stone dust was 1 468.9 km (the highest value within last 9 years) and the number of maintained explosion barriers in the Polish coal mines at the end of the previous year was 2889 barriers.

LITERATURE

1. Krause E., Łukowicz K.: Dynamiczna prognoza metanowości bezwzględnej ścian (poradnik techniczny). Instrukcja nr 14 GIG, luty 2000r.
2. Krause E., Łukowicz K.: Zasady prowadzenia ścian w warunkach zagrożenia metanowego. Główny Instytut Górnictwa. Seria Instrukcje nr 17. Obligatoryjnie stosowana w polskich kopalniach węgla kamiennego. Katowice 2004r. s 23–35.
3. Krause E.: Technologie ujmowania metanu w polskich kopalniach węgla kamiennego—doświadczenia oraz perspektywy. Międzynarodowa sesja nt. New Trends in Coal Mine Methane Recovery and Utilization. XXI Światowy Kongres Górniczy. Kraków—Katowice, 7–11 października 2008r.
4. Dubiński J., Krause E., Cybulski K.,: Dynamic Prediction of Absolute Methane Emissions on Longwalls. Proceedings of China International Conference on Coal Mine Gas Control and Utilization. China University of Mining Technology Press. 23–24.10.2008r., Huainan City, Anhui Province, Chiny, s. 95–104.
5. Dubiński J., Krause E., Cybulski K.,: Comprehensive assessment of the methane hazard of designed and mined longwalls. International Conference on Coal Mine Gas Control and Utilization. China University of

Mining Technology Press. 23–24.10.2008r., Huainan City, Anhui Province, Chiny.

6. Krause E., Sebastian Z.: Raport roczny o stanie podstawowych zagrożeń naturalnych i technicznych w górnictwie węgla kamiennego. Rozdział 3. Zagrożenia gazowe. Praca zbiorowa pod kierunkiem prof. dr hab. inż. Władysława Konopko. Główny Instytut Górnictwa. Katowice 2008r. s. 36–58.

7. Cybulski K., Malich B.: Raport roczny o stanie podstawowych zagrożeń naturalnych i technicznych w górnictwie węgla kamiennego. Rozdział 4. Zagrożenia pyłowe. Praca zbiorowa pod kierunkiem prof. dr hab. inż. Władysława Konopko. Główny Instytut Górnictwa. Katowice 2008r.

8. Cybulski K., Koptoń H.: Zagrożenie metanowe i wybuchem pyłu węglowego w polskim górnictwie węgla kamiennego. Górnictwo węglowe Polski i Ukrainy—doświadczenia i współpraca. Konferencja naukowo—techniczna. Katowice 13–15 maja 2009r. s. 15–25.

A Centennial of Mine Explosion Prevention Research

Michael J. Sapko
Sapko Consulting Services, Finleyville, Pennsylvania, United States

Eric S. Weiss
NIOSH, Pittsburgh, Pennsylvania, United States

Marcia L. Harris
NIOSH, Pittsburgh, Pennsylvania, United States

Chi-Keung Man
NIOSH, Pittsburgh, Pennsylvania, United States

Samuel P. Harteis
NIOSH, Pittsburgh, Pennsylvania, United States

ABSTRACT: A mere 100 years ago the mining industry, scientific investigators, and a concerned public struggled with the notion that coal dust could lead to mine explosions. A succession of disasters in 1907 left the U.S. mining industry desperately seeking answers. The pioneering work of Taffanel in France and Rice in the U.S. convinced the mining industry of the dangers of coal dust. By 1911, the United States Bureau of Mines (USBM) was conducting full-scale studies of mine explosions at the experimental mine at Bruceton, Pennsylvania. While this work has dramatically improved mine safety since the early days, methane and coal dust remains a threat for miners even now.

This paper provides a brief historical overview of full-scale mine explosion research conducted primarily at the USBM, now the National Institute for Occupational Safety and Health (NIOSH). The paper will evaluate the factors common to explosion disasters over the last century and identify some of the new safety challenges created by modern mining methods. This report reviews the Federal Mine Health and Safety Acts that have been passed over the last century and discusses how explosion research and enforcement of safety regulations have led to a significant reduction in the number of fatalities and disasters.

PROVING TO THE WORLD THAT COAL DUST IS EXPLOSIVE (1900–1930)

The coal dust question in this country cannot be said to have awakened widespread interest in mining men until the terrible disasters of December 1907…While it is probable that for several years the leading mining men in the country have believed in the explosibility of coal dust without the presence of firedamp, yet until the public demonstrations were given at the testing station in Pittsburg [sic] during 1908–09…a large proportion disbelieved. These tests were so convincing…that it is exceptional to find a mining man who does not accept the explosibility of coal dust.

—George Samuel Rice, 1910

Coal had become a chief source of energy in the second half of the 1800s as underground coal mining emerged as an important industry in England and the rest of Europe. The explosion dangers of firedamp (methane), hydrogen, and other combustible gases were quickly recognized by the scientific and industrial community. It was strongly and widely believed that when a mine explosion occurred, a flammable atmosphere of firedamp must be present. Unfortunately in the early years of coal mining, few scholars considered the possibility and hazard of a dust explosion, an oversight that led to the loss of hundreds of lives.

Perhaps the first major impetus to conduct more thorough investigations into the cause and prevention of mine explosions came from the 1906 Courrières mine disaster in France. This horrific mining disaster killed 1,099 men and brought the issue of coal dust explosibility to the forefront of public attention. Soon after the Courrières disaster, many countries began extensive experimental investigations of coal dust explosions. In 1907, Taffanel began his seminal experiments in a small surface gallery at Liévin, France and within a year had reproduced and expanded the work in a full-sized gallery. Taffanel devoted much of his attention to the chemistry of dust explosions, including the effect of volatile matter on coal dust explosibility. His pioneering work would serve as the foundation for research into the newly identified danger of dust in mine explosions that continues to this day.

The early history of dust explosions in the U.S. was quite similar to European experiences in that the explosion hazard of coal dust was not widely considered in the U.S. mining industry until the U.S. suffered a number of its own disasters. The coal mine disasters of 1907, especially the Monongah explosion in West Virginia and the Darr explosion in Pennsylvania which caused a combined total of 601 deaths, led the U.S. Congress to appropriate funds for an investigation of mine explosions.

U.S. research on the explosibility of coal dust began at the Federal Geological Survey during the latter part of 1908. Preliminary tests were made in a 6.3-ft. (1.9-m.) diameter, 100-ft. (30.5-m.) long steel gallery erected in Pittsburgh. In 1910, the work was transferred to the newly created United States Bureau of Mines (USBM) in the Department of the Interior. The first director of this new bureau, Dr. Joseph Holmes, would prove to be a crusader for mine safety. Because of the continued reluctance of the mining industry to acknowledge the explosive dangers of coal dust, Holmes determined that experimental testing must be performed in a real mine so results from the tests would not only be valid, but would be accepted as conclusive. Using the newly chartered Bruceton Experimental Mine (BEM), the original USBM scientists demonstrated that coal dust alone was capable of propagating an explosion in the absence of any methane gas. Demonstrations of the hazards of coal dust within the BEM conclusively proved that many practices such as using loose coal dust in mines to pack explosives in boreholes were too hazardous to continue. Shown in Figure 1 is a demonstration of a coal dust explosion conducted in the BEM.

The widespread use of rock dusting in U.S. coal mines has been in use for nearly a century as a precautionary measure against the dust explosion hazard. George S. Rice of the USBM recommended the use of rock dusting to prevent or limit coal dust explosions as early as 1911 (Rice 1911) and referenced experiments being conducted in Great Britain. Experiments at the USBM confirmed that ignitions could not be obtained with mixtures of Pittsburgh seam coal dust and rock dust having an incombustible content of 64 percent. In 1927 (Rice, Paul, and Greenwald 1927) and again in 1937 (Mine Safety Board 1937), the USBM recommended that a total

Figure 1. Bruceton Experimental Mine coal dust explosion demonstration

incombustible content of 65 percent should be required in all bituminous coal mines.

The effectiveness of rock dusting in preventing mine explosions is illustrated by the decline in fatalities associated with mine explosions. In the U.S., mine explosions declined from 33 per year in the late 1920s to about 20 per year in the late 1960s with further decreases in later decades.

FOCUS ON EXPLOSIONS AND COMBUSTION (1930s AND 1940s)

One of the original missions of the USBM was to provide the mining industry with information on blasting materials and techniques that could be used safely in the presence of flammable mine gases and dust. This meant identifying both the useful and dangerous chemical and physical characteristics of explosives. At the USBM, work focused on identifying the processes associated with initiation, growth, and eventual extinguishment of mine explosions. Scientists discovered that weak methane explosions can initiate violent coal dust explosions and that rock dust and other quenching agents can arrest these explosions. The USBM became a vocal advocate for the use of permissible explosives and denounced the use of black powder through demonstrational and educational programs, as well as research. By listening to the requests of the mining industry and through extensive experimentation, the USBM approved an increase in the permissible charge limit which greatly improved productivity while maintaining safety. In 1941, a year after 257 miners died in four separate mine explosions, Congress passed the Coal Mine Health and Safety Act. Under that law, the government was authorized to conduct safety inspections in mines for the first time. Unfortunately,

Table 1. Partial list of the USBM/NIOSH contributions to various fields

Medicine	Virtually eliminated the risk of explosions in hospital operating rooms after extensive investigation into the spark ignition of anesthetic gases
Transportation	Supported the construction of the Holland Tunnel between New York and New Jersey and the Liberty Tunnel in Pittsburgh through safety research
	Established ventilation requirements for car tunnels to maintain carbon monoxide at safe levels
	Developed storage and transport systems for liquid natural gas, hydrogen, and chlorine
	Established methods for the safe handling of hazardous materials
Utilities	Improved burner designs capable of maintaining stable gas flames
Aviation	Provided for the safe use of fuels and fluids for conventional aircraft and jet engines
Military	Made major contributions to the formulation of new military explosives
	Contributed to the development of the trigger for the atomic bomb—the first experimental work on cylindrical implosions
NASA	Studied the effect of extraterrestrial atmospheres on the performance of explosives
	Apollo mission support and accident investigation
	Safe fuel ejection techniques at high altitude

safety regulations remained advisory during this era of mining.

TOWARD SAFER EXPLOSIVES (1950s AND 1960s)

U.S. mines were becoming safer through voluntary efforts on the part of the mining industry and safety research pioneered by the USBM. In 1951, however, the Orient No. 2 Mine in Illinois suffered a massive coal explosion killing 119 miners. One year later, an updated Federal Coal Mine Health and Safety Act was passed which called for the elimination of black powder and set requirements for rock dusting along with other regulations. During this era, scientists discovered that a small amount of sodium chloride—table salt—added to explosives would greatly reduce the occurrence of methane gas explosions. This advance in permissible explosives gave the mining industry a safe and effective alternative to black powder. Since the start of the USBM explosives testing program, there has been no mine disaster linked to proper use of permissible explosives. After decades of education and research, black powder use in mines was finally outlawed in 1966.

USBM CONTRIBUTIONS TRANSCEND COAL MINES

In addition to practical advancements in mining safety, the USBM produced an important and extensive body of information on the flammability and explosibility of numerous gases, liquids, and dusts, as well as on the toxicity of the subsequent combustion products. USBM scientists developed methods for determining flammability hazards of combustible materials. Standard tests, developed by USBM personnel, to determine auto-ignition temperatures, flammability limits, and minimum ignition energies were adopted by the American Society for Testing and Materials (now ASTM International). Fundamental and highly innovative research into the manner in which flames and explosions originate, propagate, and are quenched assured the USBM and the United States an authoritative place in the international scientific community. In addition, early research at the BEM supported the growing infrastructure of the country. Technology initially developed for the coal mining industry was extended to meet the needs of many other government agencies, including the Department of Defense (DoD), the Department of Transportation (DOT), and the National Aeronautics and Space Administration (NASA). Over the last 100 years, the USBM and later NIOSH have provided technology, research data, and analysis to various government and civilian organizations (Table 1).

THE FEDERAL COAL MINE HEALTH AND SAFETY ACT OF 1969

Despite volumes of research on proper explosives use and mine safety, regulations were simply advisory and, as such, variably implemented in actual mines. It would take the death of 78 miners in an explosion at the Consolidation Coal No. 9 Mine in Farmington, West Virginia to permanently change the attitude of the mining industry toward mine safety regulations. Congress passed the Federal Coal Mine Health and Safety Act of 1969 (Public Law 91-173), which, for the first time, provided the mandate for federal mine

Figure 2. Bruceton Experimental Mine oil shale dust explosion

safety enforcement. This act instituted procedures for developing *mandatory* standards and called for expanded health and safety research. As a consequence, the USBM conducted the research necessary to identify and eliminate coal mining hazards and to reduce the risk of health impairment, injury, or death.

This legislation also redefined and expanded the mission and responsibilities of the USBM at the Pittsburgh Research Center. To the traditional work on fires, explosions, and explosives were added the responsibilities of addressing health, safety, and productivity problems associated with mechanized mining, longwall mining, and the pursuit of deeper, less accessible resources. The Act increased the number of coal mine inspectors and the frequency of inspection. Also during this period, training programs expanded rapidly. In 1971, the Coal Mine and Safety Academy was established in Beckley, West Virginia.

Numerous techniques, devices, and systems were generated during this time of renewed national emphasis on health and safety. Rescue apparatus for application in mines became more sophisticated. Before the 1970s, a USBM-approved self-rescuer protected the wearer against carbon monoxide for about 30 minutes. However, lessons learned from the Sunshine Mine disaster on May 2, 1972 heralded the era of self-rescuer development that supplied the wearer with oxygen and protection from toxic gases. On June 21, 1981, U.S. mine operators were required (30 CFR 75.1714) to make available to each underground coal miner a self-contained self-rescuer (SCSR) that provided respiratory protection with an oxygen source for at least 1 hour. This advancement proved to be a crucial component to post-disaster survival and rescue efforts in subsequent years.

The extensive experience of the personnel at the Pittsburgh Research Center in mine fires and explosions research was applied to generating new and more quantitative knowledge on the burning behavior of a wide variety of dusts, gases, fluids, and solids, and on the relative merits of flame inhibitors, quenching agents, and high-expansion foam. Large-scale research in the BEM produced basic information on the origin, growth, and suppression of fires and explosions. Passive barriers designed to disperse a quenching agent under the action of an oncoming explosion were successfully deployed in a working mine. Improved cutting bits and directed water sprays were developed to combat the growing problem of frictional methane ignitions associated with modern, mechanized coal-cutting equipment. In the 10 years following the Coal Mine Health and Safety Act of 1969, the number of major explosions was only one-third of that in the previous 10 years, and there were two 4-year periods with no major explosions.

EXPLOSION HAZARDS IN OIL SHALE MINES (1970s)

While dependence on foreign oil is now recognized as a threat to national security, the 1973 oil embargo by the major oil exporting nations was the first time the nation became acutely aware of the scope of the problem. It was during this period that the U.S. began to develop more efficient techniques to extract oil from shale rock.

Historically, the mining operations in oil shale were regulated under the standards developed for metal and non-metal mines; however, in the late 1970s the USBM and the newly formed Mine Safety and Health Administration (MSHA) raised concerns regarding the adequacy of these regulations for oil shale mining. This surge in commercial oil shale mining and the unique dangers associated with it led the USBM to study the explosibility of oil shale dust and the effects of blasting in large diameter holes in the presence of methane. This research continued through the early 1990s (Richmond, Sapko, and Miller 1982; Weiss et al. 1996).

Even though oil shale dust had been shown to be explosible in experimental mine tests (see Figure 2), sampling of dust depositions following blasting in oil shale mines had shown that the dust generated during blasting was an order of magnitude below the concentrations required to propagate an explosion in the large headings of typical oil shale operations. However, cameras monitoring these same full-scale blasts recorded high concentration dust clouds generated by the detonation of the blast holes. The ignition

Figure 3.   Frames from a high-speed movie during an oil shale face blast showing the non-electric explosive initiation (left photo) and subsequent flame and dust generation (right photo)

Figure 4.   Coal dust ignited by explosives at the Lake Lynn Laboratory Cannon Gallery

of these dust clouds can result in localized secondary explosions near the face (Figure 3). Further, the presence of methane in many of the deep oil shale formations posed a significant hazard to underground blasting operations. Tests within the BEM had shown that even a small amount of methane can significantly reduce the lower explosible concentration for pre-dispersed oil shale dusts. To verify these findings, the USBM developed and installed a tube bundle gas sampling system at the White River Shale Project in Utah and at the Horse Draw mine in Colorado to measure the methane emission rates following blasting operations.

Similar to coal and oil shale dusts, sulfide ore dusts exhibit a higher degree of explosibility as particle size decreases. Also, explosibility increases as sulfur content increases. Numerous documented sulfide dust ignitions occurred following blasting operations and had resulted in personnel injuries in addition to production and equipment losses.

The USBM research in laboratory, gallery (Figure 4), BEM tests, and through numerous full-scale blasts in operating oil shale and sulfide ore mines contributed to the better understanding of the fire and explosion hazards associated with blasting in these commercial-scale underground mining operations. Through careful analysis of dust depositions, full-scale blasting studies, and the determination of methane accumulations in deep oil shale reserves, researchers at the USBM were able to demonstrate the specific dangers posed by this mining. The use of inert gelled water stemming was shown to reduce the dust generated from blasting and to also cut air blast overpressures in half. The use of low-incendive explosives also reduced the blast-induced explosion hazards in oil shale and sulfide ore mines.

FEDERAL MINE SAFETY AND HEALTH AMENDMENTS ACT OF 1977

In 1977, Congress passed the Federal Mine Safety and Health Amendments Act (Public Law 95-164). For the first time, this document provided a single piece of comprehensive legislation for coal, metal and nonmetal mining operations and extended the research mandates of prior legislation to all segments of the mining industry. The Act required four annual inspections at each underground coal mine other mines and two annual inspections at each surface mine. The Secretary of the Interior was authorized to institute civil action for relief, including injunctions, for violations or interferences by operators with enforcement of this Act. Penalties were increased for violations. Provision was made for mandatory health and safety training of miners using both initial and refresher courses. In addition, the Mining and Enforcement Safety Administration (MESA) became the Mine Safety and Health Administration (MSHA) and was later transferred to the U.S. Department of Labor, headed by an Assistant Secretary of Labor.

ACQUISITION OF LAKE LYNN LABORATORY (1980)

In keeping with the philosophy of Joseph Holmes that there is no perfect substitute for explosions research performed in a full-scale mine environment, the Lake Lynn Laboratory was acquired in 1979 and became operational in 1982. This 400-acre facility near Fairchance, Fayette County, Pennsylvania was constructed in a limestone bed adjacent to an abandoned limestone mine. Lake Lynn is a highly sophisticated laboratory that provides an unparalleled research venue for mine disaster prevention and response. The Lake Lynn Experimental Mine (LLEM) can be adapted to simulate virtually any modern coal mine geometry while tightly controlling ventilation and humidity. The surface facilities provide an isolated environment in which large-scale research and testing can be done in a realistic, yet environmentally controlled manner. For nearly 30 years, the complex has been used for numerous and significant explosion, fire, and explosives research projects as well as research efforts of national and international interest.

MINE SEALS AND STOPPINGS

In research programs dating back to the early 1930s, the USBM (Rice, Greenwald, and Howarth 1930; Mitchell 1971; Nagy 1981) conducted research on mine seal and stopping performance to address the many issues associated with the design and installation of these structures. Seals are barriers constructed in underground mines throughout the U.S. to separate abandoned workings from the active workings. Stoppings are designed to direct the flow of fresh air to the working faces and remove the contaminated air from the mine while minimizing air leakages. Seals must be resistant to specified explosion overpressures while stoppings need only resist very small overpressures (39 lb/ft^2). Until the tragic explosion in early 2006 at the Sago Mine in West Virginia, the general consensus was that seals would most likely be subjected to an explosion initiated within the active workings of the mine well away from the seal location. Therefore seals were primarily designed to prevent an explosion from entering the methane-rich sealed areas. Prior to the 1992 regulations, seal designs were based on that of a fully-mortared, solid-concrete-block seal keyed into the

ribs and floor as built and explosion tested within the BEM (Mitchell 1971).

In the early 1990s, MSHA requested that the USBM evaluate alternative seal designs to that of the solid-concrete-block seal. Over the next 15 years, many different full-scale designs of seals were evaluated within the LLEM for strength characteristics and air-leakage resistance, and those designs that met the requirements of the 1992 regulations were deemed suitable by MSHA for use in underground coal mines.

NIOSH developed and recommended a new, preferred approach for evaluating seals under static load conditions to address the limitations of the explosion test program within the LLEM entries. This method allowed for the determination of the ultimate strength of the seal by evaluating the seal to failure under a controlled static load using water pressure (Sapko, Weiss, and Harteis 2005).

Likewise, mine stopping designs have also been evaluated under USBM (Kawenski et al. 1965) and NIOSH (Weiss et al. 2008) research studies to evaluate the strength characteristics and air leakage resistance of many typical and innovative stopping designs that can perform their intended function under the requirements of the regulations. The LLEM explosion data on stoppings, coupled with the use of predictive wall strength models have assisted investigators to more accurately determine the range of explosion pressures that destroy or damage stoppings during actual coal mine explosions and can lead to the development of enhanced stopping designs.

CLOSURE OF THE USBM AND TRANSFER TO NIOSH (1995–1997)

USBM research played a fundamental role in reducing accidents, fatalities, and health-related problems in the U.S. coal mining industry by providing innovative technology and groundbreaking research. In September 1995, Congress recommended the closure of the USBM; this both surprised and angered many in the industry. The closure resulted in the dismissal of many federal employees and threatened to quiet some of the leading voices in mining research, including the experts of explosion prevention and safety. After restructuring and reorganization, the Mining Health and Safety Research Program at the Pittsburgh and Spokane Research Centers was assigned on an interim basis to the U.S. Department of Energy and, in 1996, it was permanently transferred to NIOSH. Under NIOSH, the Pittsburgh and Spokane Research "Centers" were renamed the

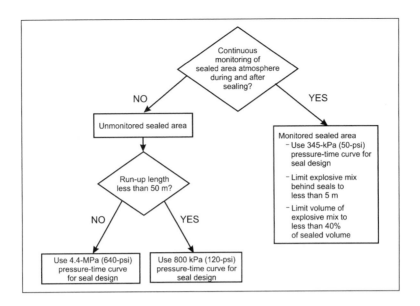

Figure 5. Flowchart for selecting design pressure-time curve for new seals

Pittsburgh Research Laboratory (PRL) and Spokane Research Laboratory (SRL), respectively. Despite a brief period of challenge and uncertainty at that time, cutting edge, life-saving research continued at both facilities.

THE MINER ACT OF 2006

Tragedies at the Sago Mine, the Aracoma Alma Mine No. 1, and Darby Mine No. 1 in early 2006 once again raised serious concerns about underground coal mine explosion safety. These events led to sweeping federal and state legislation, including the Mine Improvement and New Emergency Response (MINER) Act in 2006. Some of the major areas addressed in the Act were enhanced oxygen supply, refuge chambers, and communications and worker tracking.

Post-Sago seals research by NIOSH focused on developing explosion pressure design criteria for new seal designs (Zipf, Sapko, and Brune 2007) and retrofit techniques (Weiss and Harteis 2008) for *in situ* strengthening of thousands of existing 20-psi seals to meet the new, temporary seal strength standard of 50 psi. Zipf, Sapko, and Brune (2007) examined seal design criteria and practices used in the U.S., Europe, and Australia and then classified seals by their various applications. They then considered various kinds of explosive atmospheres that can accumulate within sealed areas and used thermodynamic calculations and simple gas explosion models

to estimate worst-case explosion pressures that could impact seals. Three design pressure-time curves were developed for the dynamic structural analyses of new seals under the conditions in which those seals may be used: (1) unmonitored seals where there is a possibility of methane-air detonation or high-pressure nonreactive shock waves and subsequent reflections behind the seal; (2) unmonitored seals with little likelihood of detonation or high-pressure nonreactive shock waves and subsequent reflections; and (3) monitored seals where the methane-air is strictly limited and controlled. The recommended design pressure-time curves for the three scenarios described above are (1) 640 psi falling back to a constant volume explosion overpressure of 120 psi, (2) a constant volume explosion overpressure of 120 psi, and (3) a 50 psi design that requires monitoring and active management of the sealed area atmosphere, respectively. Shown in Figure 5 is the proposed flowchart by NIOSH for selecting design pressure-time curve for new seals.

Subsequently, MSHA promulgated the Final Rule on "Sealing of Abandoned Areas" in April 2008 which included requirements for seal strength, design, and construction of seals. In this Final Rule, seals must either:

1. Withstand 50 psi if the sealed area is monitored and maintained inert
2. Withstand 120 psi if the sealed area is not monitored

Figure 6. Cross-section of thin 0.25 mm (0.01 inch) float coal dust layer deposited on top of a 20 mm (0.75 inch) thick layer of rock dust. This amount of float coal dust will propagate an explosion.

3. Withstand greater than 120 psi if the area is not monitored and certain conditions exist that might lead to higher explosion pressure

NEW MINING METHODS HAVE CHANGED THE NATURE OF HAZARDS

Mining of coal by explosives has diminished significantly over the years, and the dust or gas ignition hazards associated with properly used permissible explosives has essentially been eliminated. However, the increased use of continuous miner and longwall mining practices produce other hazards. These techniques tend to produce finer dust and an increased incidence in frictional ignitions of methane when cutting bits strike sandstone or pyrites. Also, as deeper seams are mined, the methane content tends to increase, as does the explosion hazard. These modern mining techniques continue to demand innovative research approaches to meet new safety challenges. Improved methods are needed to optimize mine ventilation systems to better manage flammable methane accumulations in mines. Further, improved explosion-resistant ventilation controls that continue to perform their intended function after an explosion will play a key role in future post explosion survival and rescue operations.

Even though rock dusting has been the primary means of defense against coal dust explosions for the past century, there is still room to improve rock dusting procedures and overall miner safety. The most promising method for improving the quality of rock dusting practices combines the results of theoretical and experimental work with practical hardware development, all of which have been developed

in large part by the USBM and later NIOSH. The explosibility of thin deposits of "float" coal dust (Figure 6) is well documented and in fact, the most severe experimental dust explosion conducted by the USBM involved float coal dust (Nagy 1981). There is little data regarding the rate of float dust deposition in mechanized mines, but there is no doubt that it correlates with production rates. Experiments indicate that only the top layer (2 to 4 mm) of floor dust participates in a weak to moderate float dust explosion (Sapko, Weiss, and Watson 1987).

The present practice of gathering dust samples then sending the samples to an outside laboratory does not allow for the immediate correction of potentially dangerous situations. Moreover, if the collected samples do not represent the top layer of dust this may misrepresent the actual risk. Further, the current sampling and testing methods do not take into account variations of coal particle size when assessing the local level of protection afforded by rock dusting. In 2006, NIOSH won a prestigious award (R&D Magazine's selection as one the 100 most technological significant new products of year) for the development of a hand-held Coal Dust Explosibility Meter (CDEM) shown in Figure 7, which rapidly determines the explosibility of a mixture of coal and rock dust *in situ*. The CDEM addresses a long standing problem in our underground coal mines by providing a rapid technique for evaluating the explosibility of coal and rock dust mixtures taking into account particle size in the topmost layers of mine dust. The general use of the CDEM will no doubt lead to more effective rock dusting practices and greatly improved mine safety (Sapko and Verakis 2006).

Over the past 2 years, NIOSH in cooperation with MSHA conducted a comprehensive survey on the coal dust particle size in numerous mines representing 10 of the 11 MSHA coal mine safety districts

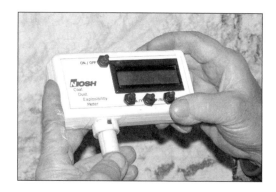

Figure 7. Coal Dust Explosibility Meter

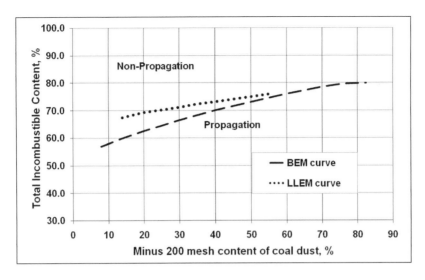

Figure 8.   Effect of particle size of coal dust on the explosibility of Pittsburgh seam bituminous coal

(not anthracite). The data revealed that coal dust in the intake airways of mines today is much finer than that on which the existing rock dusting regulations were based (Sapko, Cashdollar, and Green 2007). Full-scale explosion tests within the LLEM determined that a higher total incombustible content than the current 65% regulation is necessary to render this finer dust inert. Figure 8 shows the effect of coal dust particle size on the explosibility of Pittsburgh seam bituminous coal; finer dust requires more incombustible to inert. The primary difference between the BEM and LLEM data curves in Figure 8 is that explosions in the larger entries of the LLEM (which are typical of mines today) behave more adiabatic than in the smaller BEM entries (Richmond, Liebman, and Miller 1975). NIOSH has recommended that the total incombustible content for intake airway be 80% (Cashdollar et al. 2009).

The next hundred years of mine explosion research will hopefully see an even further decline in mine explosion disasters. While full-scale tests are an invaluable source of information for testing and validating concepts, future work will also likely focus on computer modeling techniques. Development of computer hydrocode and advanced reactive flow models for reconstructing gas and dust explosions will provide the ability to completely reconstruct complicated explosion dynamics. These tools will allow government and independent investigators to unravel complicated blast patterns and chemical reactions of participating gas and coal dust. These models may even facilitate the development of new techniques for explosion mitigation.

SUMMARY

Mining disasters have been the major driving force of mine explosion research. From the seminal work of Taffanel after the Courrières disaster in France in 1906 to the Sago Mine disaster in the U.S. in 2006, mine explosion researchers have responded to these incidents by conducting research and making recommendations that improve the safety of underground personnel. The USBM was and NIOSH continues to be at the forefront of proactive mine explosion research, producing many of the innovations and breakthroughs in the field over the last century. Just as early researchers proved that coal dust represents a significant explosion hazard that is separate from methane gas, NIOSH continues to make major contributions to mine safety in its world-class research laboratory at Lake Lynn Laboratory and in Pittsburgh and Spokane. Some of the early U.S. investigators who studied explosion hazards are listed in the Bibliography. A more extensive bibliography of early investigators that studied coal dust explosions is presented in USBM Bulletins 20 and 167 (Rice 1911; Rice et al. 1922).

After reviewing the past century of mining practices in the U.S., we observe that the major milestones in explosions safety have come from legislation and enforcement supported by extensive research efforts. Each of the Federal Coal Mine Health and Safety Acts and major amendments were passed soon after a major mining disaster. Mine operators and miners must continue to maintain their vigilance through compliance with the mandates of these Acts and regulations to mitigate the explosion hazards. Notwithstanding the effort devoted to prevention, coal mine gas and dust ignitions and explosions continue to occur with unsettling frequency. The increased utilization of high production, mechanized mining methods in deeper and gassier coal seams place additional demands on those responsible for mine safety, particularly in the area of explosion prevention. Despite the vast amount of knowledge gathered and painfully learned over the past century, much more can be discovered through high quality forensic investigations of mine disasters and by identifying ways to prevent future occurrences. Compliance and enforcement along with the development and implementation of affordable, practical, and useful technologies can eliminate deadly mine explosions in the next century and beyond.

REFERENCES

Cashdollar, K.L., Sapko, M.J., Weiss, E.S., Harris, M.L., Man, C.K., and Harteis, S.P. 2009. *Recent Coal Dust Particle Size Survey and the Implications for Mine Explosions.* Pittsburgh, PA: U.S. Department of Health and Human Services, National Institute for Occupational Safety and Health, RI *in press.*

Federal Coal Mine Health and Safety Act of 1969, United States Public Law 91-173.

Kawenski, E.M., Mitchell, D.W., Bercik, G.R., and Frances, A. 1965. *Stoppings for Ventilating Coal Mines.* Pittsburgh, PA: U.S. Department of the Interior, Bureau of Mines, RI 6710.

Mine Improvement and New Emergency Response Act of 2006 ("MINER Act"), United States Public Law 109-236.

Mitchell, D.W. 1971. *Explosion-Proof Bulkheads: Present Practices.* Pittsburgh, PA: U.S. Department of the Interior, Bureau of Mines, RI 7581.

Nagy, J., 1981. *The Explosion Hazard in Mining.* Washington, D.C.: U.S. Department of Labor, Mine Safety and Health Administration, Informational Report 1119.

Rice, G.S., Greenwald, H.P., Howarth, H.C., and Avins, S. 1931. *Concrete Stoppings in Coal Mines for Resisting Explosions: Detailed Tests of Typical Stoppings and Strength of Coal as a Buttress.* Pittsburgh, PA: U.S. Department of Commerce, Bureau of Mines, Bulletin 345.

Rice, G.S. 1911. *The Explosibility of Coal Dust.* Washington, D.C.: U.S. Bureau of Mines, Bulletin 20.

Rice, G.S., Paul, J.W., and Greenwald, H.P. 1927. *Coal-Dust Explosion Tests in the Experimental Mine, 1919–24, Inclusive.* Washington, D.C.: U.S. Bureau of Mines, Bulletin 268.

Rice, G.S., Greenwald, H.P., and Howarth, H.C. 1930. *Explosion Tests of Pittsburgh Coal Dust in the Experimental Mine, 1925 to 1932, Inclusive.* Washington, D.C.: U.S. Bureau of Mines, Bulletin 369.

Richmond, J.K., Liebman, I., and Miller, L.F. 1975. *Effect of Rock Dust on Explosibility of Coal Dust.* Pittsburgh, PA: U.S. Department of the Interior, Bureau of Mines, RI 8077.

Richmond, J.K., Sapko, M.J., and Miller, L.F. 1982. *Fire and Explosion Properties of Oil Shale.* Pittsburgh, PA: U.S. Department of the Interior, Bureau of Mines, RI 8726.

Sapko, M.J., Cashdollar, K.L., and Green, G.M. 2007. Coal dust particle size survey of U.S. mines. *J Loss Prev Process Ind,* 20(1):616–620.

Sapko, M.J., Weiss, E.S., and Watson, R.W. 1987. *Explosibility of Float Coal Dust Distributed Over a Coal-Rock Dust Substratum.* In Proceedings of the 22nd International Conference of Safety in Mines Research Institutes (Beijing, China, November 2–6, 1987). Beijing, China: China Coal Industry Publishing House, pp. 459–468.

Sapko, M.J., and Verakis, H. 2006. *Technical evaluation of the coal dust explosibility meter.* In Transactions of the 2006 Society for Mining, Metallurgy and Exploration (SME) Annual Meeting, St. Louis, MO, March 26–29, 2006.

Sapko, M.J., Weiss, E.S., and Harteis S.P. 2005. *Methods for Evaluating Explosion Resistant Structures.* In Proceedings of the 8th International Mine Ventilation Congress, Brisbane, Australia, July 6–8, 2005.

Weiss, E.S., Cashdollar, K.L., Sapko, M.J., and Bazala, E.M. 1996. *Secondary Explosion Hazards During Blasting in Oil Shale and Sulfide Ore Mines.* Pittsburgh, PA: U.S. Department of Energy, Pittsburgh Research Center, ISSN 1066-5552, RI 9632.

Weiss, E.S., and Harteis, S.P. 2008. *Strengthening Existing 20-psi Mine Ventilation Seals with Carbon Fiber-Reinforced Polymer Reinforcement.* Pittsburgh, PA: U.S. Department of Health and Human Services, National Institute for Occupational Safety and Health, DHHS (NIOSH) Publication No. 2008-106, RI 9673.

Weiss, E.S., Cashdollar, K.L., Harteis, S.P., Shemon, G.J., Beiter, D.A., and Urosek, J.E. 2008. *Explosion Effects on Mine Ventilation Stoppings.* Pittsburgh, PA: U.S. Department of Health and Human Services, National Institute for Occupational Safety and Health, DHHS (NIOSH) Publication No. 2009-102, RI 9676.

Zipf, R.K., Sapko, M.J., and Brune, J.F. 2007. *Explosion Pressure Design Criteria for New Seals in U.S. Coal Mines.* Pittsburgh, PA: U.S. Department of Health and Human Services, National Institute for Occupational Safety and Health, IC 9500.

BIBLIOGRAPHY

Dobroski, H.Jr, Stephan, C.R., and Conti, R.S. 1996. *Historical Summary of Coal Mine Explosions in the United States, 1981–94.* Washington, D.C.: U.S. Bureau of Mines, Information Circular 9440.

Humphrey, H.B. 1960. *Historical Summary of Coal-Mine Explosions in the United States, 1810–1958.* Washington, D.C.: U.S. Bureau of Mines, Bulletin 586.

Nagy, J. 1981. *The Explosion Hazard in Mining.* Washington, D.C.: U.S. Department of Labor, Mine Safety and Health Administration, Informational Report 1119.

Rice, G.S., and Greenwald, H.P. 1929. *Coal-Dust Explosibility Factors Indicated by Experimental Mine Investigations, 1911 to 1929.* Washington, D.C.: U.S. Bureau of Mines, Technical Paper 464.

Rice, G.S., Greenwald, H.P., and Howarth, H.C. 1933. *The Experimental Mine of the United States Bureau of Mines.* Washington, D.C.: U.S. Bureau of Mines, Information Circular 6755.

Rice, G.S., Jones, L.M., Clement, J.K., and Egy, W.L. 1913. *First Series of Coal Dust Explosion Tests in the Experimental Mine.* Washington, D.C.: U.S. Bureau of Mines, Bulletin 56.

Rice, G.S., Jones, L.M., Egy, W.L., and Greenwald, H.P. 1922. *Coal Dust Explosions Tests in Experimental Mine 1913 to 1918, Inclusive.* Washington, D.C.: U.S. Bureau of Mines, Bulletin 167.

Richmond, J.K., Price, G.C., Sapko, M.J. and Kawenski, E.M. 1983. *Historical Summary of Coal Mine Explosions in the United States, 1959–81.* Washington, D.C.: U.S. Bureau of Mines, Information Circular 8909.

Sapko, M.J., Greninger, N.B. and Watson, R.W. 1989. *Review Paper: Prevention and Suppression of Coal Mine Explosions.* In Proceeding of 23rd International Conference of Safety in Mines Research Institutes, Washington, D.C., September 11–15, 1989.

Verakis, H.C. and Nagy, J. 1987. A Brief History of Dust Explosions. In *Industrial Dust Explosions*, American Society for Testing and Materials, ASTM STP 958. Edited by K.L. Cashdollar and M. Hertzberg. Philadelphia, PA: ASTM, pp.342–350.

Development of Mine Gas Management Technology and Practices in Australia Following the Moura Mine Explosions

Paul Harrison
Simtars, Ipswich, Queensland, Australia

Stewart Bell
Queensland Mines and Energy, Brisbane, Queensland, Australia

Darren Brady
Simtars, Ipswich, Queensland, Australia

ABSTRACT: The government inquiries following the explosions at the Moura No. 4 and Moura No. 2 underground coal mines in 1986 and 1994 respectively, where 23 miners lost their lives made a number of recommendations about gas management in underground coal mines.

Simtars (the Safety in Mines Testing and Research Station) was tasked with improving the mining industry's understanding of gas management, improving gas management practices and assisting industry to implement better gas monitoring technology. Once embarked on this journey it was soon realised that there was a dearth of tools available to miners in this area.

This paper describes the path travelled to establish in Australian coal mines arguably the best gas management practices in underground coal mining in the world today. It details how the Moura disasters contributed to the initial and ongoing development of monitoring technology, inertisation techniques and management practices.

BACKGROUND

The arguably best gas management practices observed throughout the underground coal mining industry in Queensland have come at considerable cost. They follow recommendations from inquiries into fatal mine explosions and the subsequent legislative changes, advances in technology, intensive training programs and adoption by industry of standard practices relating to gas management.

On *July 16, 1986*, 12 miners died as a result of an explosion at the Moura No. 4 Mine. The Inquiry into this explosion found that a roof fall in the goaf pushed a mixture of methane, air and coal dust into the working area (Warden's Inquiry 1987). An explosive atmosphere developed and was ignited creating a violent explosion which caused extensive damage throughout the section. The Inquiry considered that a flame safety lamp, although properly assembled, was the most likely source of ignition. A number of recommendations were made by the members of the Inquiry that lead to significant changes to gas monitoring within the Queensland industry. These included prohibiting flame safety lamps from use in underground coal mines in Queensland and that a chromatograph with support equipment be purchased for each Mines Rescue Station in central and northern Queensland to accurately and expeditiously determine the explosibility of a mine air sample.

Less than ten years later eleven miners died after an explosion at the Moura No. 2 Mine on *August 7, 1994*. The Inquiry found that the explosion originated in the just sealed 512 Panel of the mine after a heating in that panel went unrecognised and ineffectively treated (Warden's Inquiry 1995). The heating progressed to become the ignition source for the methane that accumulated in the sealed panel. The Inquiry made recommendations to prevent the future re-occurrence of a similar accident. The Inquiry also identified a number of areas requiring investigation and improvement to assist in securing the safety of those employed in the underground coal mining industry.

These two inquiries and the advances and legislative changes resulting from their recommendations have significantly shaped the gas management practices that are standard throughout the industry today. The Moura No. 2 report clearly identified that, although there would be early efforts in improvement, there was a need for ongoing work to prevent another disaster otherwise history would repeat itself and in about ten years time there could possibly be another disaster.

MOURA NO. 4 INQUIRY

A number of recommendations were made by the members of the Inquiry. The report itself identifies the most important of these as being that flame safety lamps be prohibited from use in underground coal mines in Queensland subject to limited exceptions.

The recommendations from the Inquiry related to mine gas management included (Warden's Inquiry 1987):

Flame Safety Lamp

The first recommendations made in the report from the Warden's Inquiry into the Moura No. 4 explosion was that Simtars should continue research to establish the ignition hazards associated with approved flame safety lamps in explosive mixtures of methane and coal dust. This research continued with Simtars demonstrating that under certain conditions a fully assembled flame safety lamp could be an ignition source of a methane/coal dust contaminated atmosphere.

It was also recommended that legislation (Coal Mining Act 1925) be amended to prohibit the use of flame safety lamps in underground coal mines in Queensland and that all statutory underground officials be required to carry instruments for the detection of methane and oxygen deficiency.

The withdrawal of the iconic safety lamp from use underground in Queensland obviously had a significant impact on personal gas detection techniques. As a result of the withdrawal, flame safety lamps were replaced with portable electronic gas detectors which have advanced considerably over the past twenty years and are now relied upon for the measurement of multiple gases including carbon monoxide, further assisting gas management. These instruments are also capable of data logging and individual instruments can have specific alarm set points including time weighted exposures, something not possible with a lamp.

Gas Analysis Equipment

The Inquiry also called for mines rescue stations to be equipped with gas chromatographs to accurately and expeditiously determine the explosibility of mine air samples. This was because the time required for transporting a gas chromatograph and skilled operator to a remote mine site in case of an emergency, as was the practice, was less than adequate. Following the detection of the event at the Moura No. 4 Mine it was some 11 hours before the first qualitative analysis of mine air was available. Little was it known at the time that this would result in the development by Simtars of the Computer Assisted Mine Gas Analysis

System (Camgas), used throughout the industry today and equipping mines with immediate chromatographic gas analysis capabilities.

Emergency Procedures

No emergency procedure existed which would automatically alert and call up experts and equipment. This and the absence of other procedures for use in emergencies led to identification of the need for mine managers to prepare detailed emergency procedures to be used to assist in identification of the occurrence, evacuation of employees, notification of officials and statutory requirements. Also the need for duty sheets for persons assigned specific tasks, and lines of delegation in the case of absence of statutory officials. The involvement of outside services such as police, state emergency services, and hospitals, and rapid transportation of persons and equipment, medical officers and specialist assistance was also recommended.

Inertisation

The benefit of inertisation during emergency situations was clear and a recommendation was made for the Chief Inspector of Coal Mines to appoint a committee to investigate all aspects of the control of mine fires, post-explosion conditions and heatings by inertisation. A task group was formed and investigated and reported on techniques available around the world.

Continuous Gas Monitoring

The requirements for continuous gas monitoring were also outlined in the recommendations. It was recommended that an approved continuous monitoring system, capable of automatically determining the composition of the mine atmosphere at pre-determined points in the airway system be required at all underground coal mines in Queensland. The minimum standard for the system was outlined as:

i.   The ability to accurately determine the concentration of methane and carbon monoxide and to record the results at the surface
ii. Monitoring points in all return airways where methane content has exceeded 0.5%
iii. Monitoring points in return airways from waste or goaf areas where spontaneous combustion may develop
iv. Monitoring points on main conveyor roadways which are not frequently travelled by men
v.   A continuous recording system at the surface which provides an historical record of gas levels at each monitoring point

vi. An auxiliary power supply which enables continuity of operation in the event of a power interruption

vii. The positioning of monitoring points should be determined by the Manager, approved by the Inspector of Mines and recorded in a book kept for this purpose. No person should be permitted to move the position of a monitoring point except with the permission of the Manager

These recommendations paved the way for the gas monitoring systems and protocols in use throughout Australia today.

FLAME SAFETY LAMP INTRINSIC SAFETY RESEARCH

Following the recommendations made in the Warden's report from the Moura No.4 Inquiry, Simtars continued research into the intrinsic safety of the flame safety lamp in atmospheres of methane and coal dust as part of a National Energy Research, Development and Demonstration Program (NERDDP) funded project.

During this project it was determined that a bonneted flame safety lamp in proper working order could provide an ignition source for a coal mine explosion (Figure 1). It was identified that during normal ventilation such an ignition would not occur, but was possible following a major roof fall in either room and pillar or longwall workings. Ignition didn't occur in every test but was frequent enough to present an unacceptable risk. Ignitions occurred with gas velocities between 13.1 m/s and 23.6 m/s with the lamp in the vertical position. At the time of the ignitions the methane concentration was between 4.1 and 5.8% (Golledge 1991).

CAMGAS

Following problems experienced with supposedly portable gas chromatographs and the delay in obtaining results for gas samples collected, the Moura No. 4 Warden's report recommended mines rescue stations in Queensland be equipped with a gas chromatograph (Warden's Inquiry 1987). Differing opinions were expressed at the Moura No. 4 Inquiry as to the ability required to maintain and operate a gas chromatograph. The practical solution to this problem developed by Simtars was that gas chromatographs were installed at each of the underground mines and linked back to a "master" chromatograph (Figure 2) at a central laboratory staffed by gas chemists, rather than at rescue stations. (Brady, Harrison and Bell 2009). This allowed expert gas chemists based at the

Figure 1.   Flame safety lamp igniting methane

Simtars facility at Redbank near Brisbane to oversee chromatograph operation at each mine site remotely, thereby reducing the expertise required at the mine site and assisting in the ongoing operability at the mine site of this complex and often temperamental piece of analytical equipment. Effectively Camgas provided each mine site with an expert gas chemist 24/7 and provided rapid and accurate analytical support during a mine emergency.

The first Camgas system was installed at Cook Colliery in 1989. In the twenty years since, improvements in technology have reduced sample analysis times and improved sensitivity and the instrument in current use (Figure 3) was conceived by NASA for gas analysis in space. This decreased analysis time has resulted in mines conducting more routine chromatographic gas analysis for assessment of the underground environment. As such, the role of the gas chromatograph has shifted from just gas analysis during emergencies to routine assessment of the underground environment to assist in avoiding the disasters that they were originally intended to assess.

Camgas was developed by Simtars to provide a complete analytical solution to gas chromatographic analysis of gas samples at coal mines. With the Camgas system, gas chromatographs are tested and optimal methods developed by gas chemists prior to installation on the mine site. Following commissioning onsite, mine personnel are trained in the operation and ongoing maintenance of the gas chromatograph by experienced gas chemists. The mine site operators are then able to access 24 hour assistance from these gas chemists should they encounter any problems.

Mines recognise that gas chromatography is the only technique available on a mine site which provides complete analysis of gas composition by accurately quantifying all components of a gas mixture, essential when assessing the underground

Figure 2.   Original Camgas system

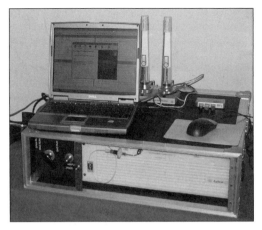

Figure 3.   Current Camgas system

atmosphere during an emergency (Brady 2008). Gas chromatographs have been responsible for identifying the onset of spontaneous combustion events and from this early detection the mine has been able to control and mitigate any potential mine fires.

One of the main reasons for the ongoing success of Camgas systems has been the implementation of processes which make their operation and the interpretation of the data they produce as user friendly as possible. A lot of this can be attributed to the software written by Simtars dedicating the system to mine gas analysis by mine site personnel rather than chemists. Integrating the decision support software Segas Professional with the gas chromatograph operating software has ensured that the results generated by the chromatograph are able to be easily used for interpretation and assessment of the underground environment by relatively unskilled operators.

MOURA NO. 2 INQUIRY

Within a 20 year period there were three mining disasters in the Moura district at a cost of 36 lives, Moura No. 2 being the third. The Moura No. 2 report detailed concerns that although initially there would be early activity to prevent a reoccurrence, that without fundamental and permanent change in the current approaches and attitudes in the coal industry, over time industry would lose sight of the lessons of the past and within a decade we could again witness another disaster (Warden's Inquiry 1995).

The Inquiry reported that it recognised and supported the need for a major revision of the existing Coal Mining Act (Coal Mining Act 1925) and the regulations with fundamentally different philosophies and approaches in both its formulation and implementation. This resulted in Queensland introducing risk management based legislation (Coal Mine Safety and Health Act 1999 and Coal Mine Safety and Health Regulation 2001). Between the release of the Warden's report and revised legislation, many Approved Standards were released to cover identified gaps in industry practice as well as existing legislation.

Although not a recommendation, the report identified that reliance on incubation period as a primary, if not the sole, determinant of the likelihood of spontaneous combustion led to some false sense of security and likely resulted in a failure to take precautions and be sufficiently alert to other indicators of spontaneous combustion and that this may have contributed to the failure to detect the developing heating through some complacency based on incubation period expectations.

The Inquiry report (Warden's Inquiry 1995) stated, "It must now be obvious that reliance on the concept of an incubation period is not an adequate defence in the face of the many other factors likely to influence the likelihood of a heating." One of the major recommendations made that has changed the approach to gas management was the introduction of safety management plans based on detailed risk/hazard analysis to cater for key risk areas.

The recommendations from this Inquiry had a dramatic and lasting effect on gas management practices in the industry today. The recommendations related to mine gas management are detailed below.

Spontaneous Combustion Management

The report identified that the lack of a specific and durable system for the management of the spontaneous combustion risk contributed to the ultimate outcome at Moura No. 2. It was therefore recommended that all mines be required to develop and implement a spontaneous combustion management plan to provide effective long term control of that risk. This management system was to be based on an assessment of the spontaneous combustion risk present at the mine and required reassessment of that risk from time to time and modification of the system, if required. The system needed to be reviewed for adequacy both on a regular basis and as a result of defined events or any significant change in operating conditions. The management plan was specified as requiring effective means for the gathering of information related to spontaneous combustion with an emphasis on early detection and evaluation, including but not limited to appropriate gas monitoring, personal observation and reporting processes. Also a defined means for evaluating spontaneous combustion related information and decision processes covering both the evaluation of that information and resulting actions. It was noted that at Moura there was no system to trigger the bringing together of people to consider the overall picture.

Mine Safety Management Plans

Several areas in addition to spontaneous combustion were also identified as risk areas requiring structured management. It was recommended that mines be required to put in place mine safety management plans based on detailed risk/hazard analysis to cater for key risk areas. The areas identified included but were not limited to:

- Ventilation
- Gas management
- Methane drainage
- Emergency evacuation
- Strata control

Safety management plans are developed to formalise processes used by mines to address principal hazards and other related matters to ensure the safety of mine personnel. They set out the action, controls and procedures and are intended to provide consistency (independent of personnel) in the way hazards are controlled while catering for changes in conditions through regular reviews of their operation and adequacy. The requirement for safety management plans to control principal hazards at a mine was included in the updated legislation (Coal Mine Safety and Health Act 1999 and Coal Mine Safety and Health Regulation 2001).

Gas Monitoring System Protocols

The warden's Inquiry recommended that mines be required to develop and implement protocols for the setting, re-setting and the noting and acceptance of alarm conditions raised by any gas monitoring system in use at the mine covering:

- Who is authorised to set or change alarm levels and the recording of those settings or changes
- Who is responsible for the acknowledgment of alarms and recording of acknowledgments
- Who is responsible for communicating the occurrence of alarms and initiating action as a result of those alarms
- How the actions of responsible persons, and the identity of those persons, are to be recorded

There was no set protocol for management of tube bundle alarms at Moura No. 2 and the recommendation for such a protocol followed considerable evidence surrounding the less than desirable practice at Moura No. 2 for the setting and acceptance of alarms. It was also identified that mines need to schedule gas monitoring system testing to occur before critical times when the system may be required. It was also noted that consideration should be given to making gas alarms readily distinguishable from other alarms.

The software issues including alarm acknowledgment and setting and distinguishable alarms have been addressed in Simtars gas monitoring controlling software, Safegas, with procedural requirements addressed in the reviewed legislation (Coal Mine Safety and Health Act 1999 and Coal Mine Safety and Health Regulation 2001).

Inertisation

It was noted in the Warden's report that had Moura No. 2 decided to inert the atmosphere inside 512 Panel that there was no suitable equipment available anywhere in Queensland. It was therefore recommended that the work previously undertaken as a result of the Moura No. 4 Inquiry be reviewed and the most appropriate method of inertisation determined for Queensland mines.

Further to that recommendation was that the Queensland Government obtain such a system to be maintained and operated by the Queensland Mines Rescue Service to service all the mines in Queensland on a fee for service basis. This ultimately led to the purchase of two GAG jet engines by the Queensland Government. These units were then donated to the Queensland Mines Rescue Service who are now responsible for the maintenance and operation of the equipment.

Statutory Positions

Evidence heard at the Inquiry lead to the recommendation that a process be introduced to ensure that holders of statutory certificates maintain a sound knowledge base on, and keep abreast of, technical developments in coal mining and most particularly those relevant to coal mine safety.

It was also recommended that the statutory position of ventilation officer be established and person appointed must have demonstrated competencies appropriate to the duties and responsibilities of the position. The position was determined as being required to be directly responsible to the mine manager for the planning, design and implementation of the mine ventilation system and for the establishment of effective standards of ventilation for the mine, methods for its control and protection, monitoring of performance, reporting procedures, maintenance of ventilation records and plans, and emergency action plans. The ventilation officer is now a statutory position requiring appropriate competencies.

Sealing—Designs and Procedures

It was noted in the report that the sealing of any area in a gassy mine should never be considered a routine or trivial event. The Inquiry raised the question on the adequacy of the then current designs of seals and sealing practices. It believed that current legislation (Coal Mining Act 1925) concerning seals needed to be comprehensively reviewed and that the review take into account the best available technology and practice as assessed world-wide. The Inquiry also believed it necessary to set minimum standards and requirements for the design, installation and maintenance of seals and for the maintenance, control and management of sealed areas.

In respect of the design, installation and maintenance of seals, the Inquiry recommended that:

- The location of final seals be subject to approval by the District Inspector of Mines
- It be a requirement that seals be constructed using only materials that have been approved for the purpose by the Chief Inspector of Coal Mines
- The Chief Inspector of Coal Mines should determine and then apply requirements appropriate for the design and installation of seals and for their long term stability

In respect of the control and management of sealed areas, the Inquiry recommended that minimum requirements provide for:

- The continuous and effective sampling and monitoring of the atmosphere in a sealed area including a minimum number of sampling points and suitable location(s)
- Means whereby the pressure difference between the inside and outside surfaces of seals can be measured
- The effective ventilation of the outside surfaces of seals
- Regular inspection and periodic auditing on the long term performance of seals and sealed areas

The Inquiry recommended that it be a requirement that no part of a mine be sealed without the prior written approval of the District Inspector of Mines (other than in an emergency, whereupon the inspector must be informed as soon as practicable thereafter) and that the mine manager be required to submit a formal proposal to seal an area of the mine to the inspector that should include:

- A complete specification for the seals proposed including the location, method of construction, and the materials to be used
- A risk assessment (involving mine management, the district inspector and the district union inspector) of the potential hazards introduced by sealing (or by not sealing) the area
- The gas monitoring system, location of the gas monitoring points proposed and arrangements for post-sealing ventilation of the seals
- A management plan for the sealing operation including the long term security of the seals and the arrangements for evacuation of the mine during and after sealing

The report also included the statement "Persons should not be allowed to remain in or enter a mine following a sealing without the manager first having obtained the written consent of the District Inspector of Mines."

The updated legislation (Coal Mine Safety and Health Act 1999 and Coal Mine Safety and Health Regulation 2001) detailed requirements for sealing and the need for the mine manager to give at least 30 days notice of the proposed sealing to a mines inspector and an industry, or site, safety and health representative. Also included was, that without an inspector's written consent, a person must not enter or remain in an underground mine after the mine, or part of it, has been sealed.

Withdrawal of Persons

There were no protocols at Moura No. 2 for the withdrawal of persons from the mine in response to potential dangers. This left consideration of questions of withdrawal to those officials who happened to be on duty at any particular time. It was therefore recommended that mines be required to develop and implement protocols, as a statutory requirement, for the withdrawal of persons when conditions warranted such action.

Research into Spontaneous Combustion

The Inquiry identified that there was a great deal of basic scientific and technical knowledge already available on the subject of spontaneous combustion of coal; on its causes, its detection and methods of dealing with it, but that much of this information was not widely known nor readily available to mine operators. It therefore recommended that funds be made immediately available to undertake an exhaustive international literature and data search, to critically review the literature and data and to prepare a comprehensive state-of-the-art report on the subject of spontaneous combustion in coal mines.

In response to this recommendation, Simtars prepared a reference text "Spontaneous Combustion in Australian Underground Coal Mines" (Cliff, Rowlands and Sleeman 1996) for the underground coal mining industry for reference material on spontaneous combustion which relates to Australian mining practices and conditions.

Simtars has continued research into spontaneous combustion using small scale, medium scale and large scale test vessels with a focus on early detection.

Training

The Inquiry identified that a lack of appreciation of the fundamentals of spontaneous combustion contributed to the disaster at Moura No. 2. To address this issue it was recommended that there was a basic need for all members of the coal mining industry in Queensland to improve their knowledge with regard to the fundamentals of spontaneous combustion and the underground mining problems associated therewith. In particular it was recommended that all employees be effectively trained to:

- Recognise indicators of specific mine hazards, such as spontaneous combustion, and their control
- Become sufficiently familiar with mine gases, and associated risks

Since this recommendation, the mining industry in Queensland introduced a requirement for all workers to undergo a Generic Induction prior to undertaking

work at minesites. One of the components of this training for underground coal mines is spontaneous combustion and mine gases.

Simtars alone has delivered training on mine gases and spontaneous combustion to thousands of coal mine workers since this recommendation was made.

CONTINUOUS MONITORING SYSTEMS

Monitoring on its own is never going to prevent a mine fire or put it out if it starts. What it does do is provide a means of identifying a problem early and taking appropriate controlling actions. The earlier a problem is identified the better the chance of successfully dealing with it. The best chance of getting an early warning is by continuous monitoring. Queensland mines still have spontaneous combustion activity but these have been identified early enough to take effective control.

Each underground coal mine in Queensland utilizes real time and tube bundle continuous monitoring systems. The industry has realised that both techniques are needed for effective monitoring of the mine's atmosphere. Each technique has advantages and disadvantages which must be known to those using the systems and those using the results to interpret the status of the underground environment. The real time systems are used for real time warning, essential for incidents such as belt fires. Tube bundle systems (Figure 4) suit long term trending used for identification of the onset of spontaneous combustion or for the determination of explosibility during the routine sealing of worked areas.

These automated monitoring systems are programmed to alarm for gas concentrations, gas ratios and explosibility. These alarms are then used to initiate predetermined actions to take control of the situation and to prevent compromising the safety of workers and the loss of resources. Dedicated user friendly software packages have been developed to assist in the interpretation of the large volume of results generated.

CONTINUOUS MONITORING SOFTWARE

Simtars has been developing controlling and decision support interpretative software on an ongoing basis for the past twenty years. The hardware component of the monitoring system is controlled and data displayed through Safegas software. Segas Professional is decision support software for the trending, analysis and interpretation of gas data and determination of the explosibility of gas samples. Segas Professional

Figure 4.   Typical tube bundle surface installation

is tightly integrated with Safegas and allows operators to readily convert complex gas data into easily understood graphs and visual interpretive tools. This meets the requirements of Queensland legislation for mines to display values and trends from their gas monitoring systems at the surface of the mine where the record can be easily accessed by coal mine workers and in a way that the record can be easily read and understood by the workers (Coal Mine Safety and Health Act 1999 and Coal Mine Safety and Health Regulation 2001).

The mine plan indicating monitoring locations is central to the Safegas display (Figure 5). This displayed mine plan helps operators to visualise and gain a better understanding of the underground environment. Current gas composition is displayed on the main Safegas screen. Each monitoring location can be clicked on to display the history/trending for that point. Segas Professional allows the operator to view the gas data in a variety of different ways including time graphs (Figure 6), explosibility diagrams and trends (Figure 7) and statistical trends. The different views allow the operator to analyse the data from a variety of perspectives, with the aim

being to gain the best understanding of the underground atmosphere.

Safegas allows four individual alarm thresholds (two levels of high and two levels of low) to be set for each parameter per monitoring location. Any value which exceeds the threshold, in either direction, will trigger Safegas to raise an alarm. Safegas uses a combination of colour and sound to attract the operator's attention and will repeat the alarm until the situation has been acknowledged. If an acknowledged alarm is of a serious nature then the alarm acknowledgement will expire after an administrator defined period of time, and the alarm will be retriggered to alert the operator that the alarm condition is still present.

The Safegas system also monitors the status of the hardware and will raise an alarm if a fault is detected. An audit trail of all operator actions, alarms logs and gas readings is also kept. Safegas has been developed with multi-level user access, controlled by password, preventing unauthorised acknowledgement of alarms or changes to any operational parameters.

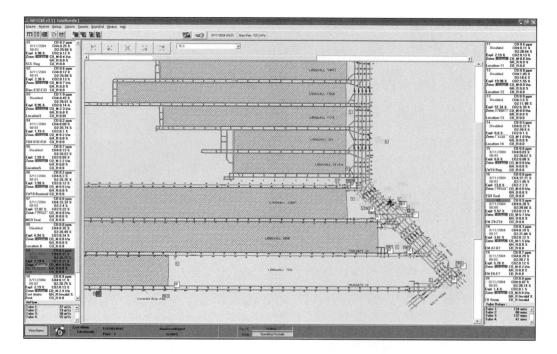

Figure 5.   Safegas main screen

TRIGGER ACTION RESPONSE PLANS

Individual mines determine gas levels that should not be exceeded during normal operations. Often these numbers are based on historical data collected from the monitoring systems. To handle the large volumes of measurements made, gas monitoring software has been developed that will automatically trigger a visual and audible alarm if one of the preset levels is exceeded. Alarm set points can be different for every sample point. These alarms activate what are known as Trigger Action Response Plans (TARPs) that have predetermined procedures to follow. These procedures have been formulated to ensure that appropriate actions are taken to safeguard of the wellbeing of workers and maintain control of the mine. These TARPs make up part of the mine's safety management plans.

Alarms can be generated from absolute concentrations, gas ratios or explosibility. They are most useful as an early warning, not as an alert to an emergency or a need to evacuate the mine. When alerted early enough the mine will have time to take remedial action to rectify the problem. The frequency and scope of monitoring is often included in a TARP to ensure that the situation is not escalating or that the control measures are effective.

INERTISATION

Simtars was involved in two inertisation research projects in 1997. These were aimed at providing a solution to the industry following the realisation in the Moura No. 2 Wardens Inquiry that even if there had been a desire to use inertisation there was no means available in Queensland.

The two trials involved different techniques, one a high flow device (typically 20 m³/s) and the other a low flow device (typically 0.5 m³/s). Both trials received financial support from the Australian Coal Association Research Program (ACARP). The high flow trial involved the use of a GAG-3A jet engine system, similar to that used in Eastern Europe for decades, in which aviation fuel is burnt, consuming oxygen and producing carbon dioxide and water. To ensure that the oxygen is reduced to the low concentrations required to be effective, the system is fitted with an afterburner for complete combustion of fuel.

Bell et al (1997) demonstrated that the GAG could be operated underground to inert a panel or small localised area or connected to a portal seal to inert a large area of the mine. The effectiveness was demonstrated by gas measurements and the ability to inert a small coal fire deliberately lit in a small steel sled. Since this project the Queensland Government purchased two GAG units which are now under the control of the Queensland Mines Rescue Service.

354

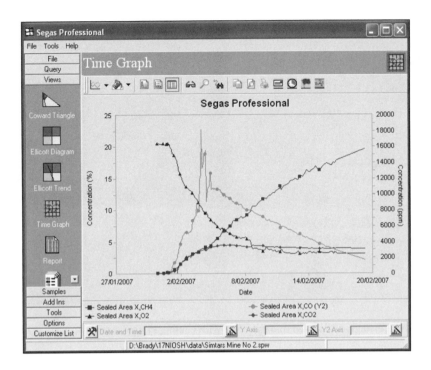

Figure 6.   Segas Professional gas trends

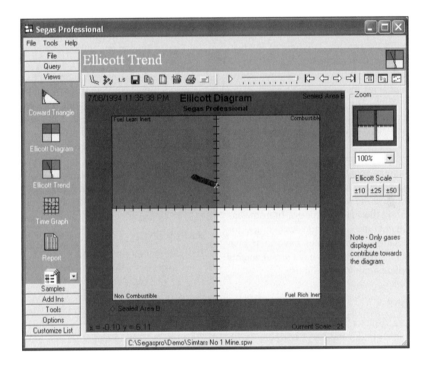

Figure 7.   Segas Professional explosibility trend

Figure 8. GAG jet engine deployed underground with cooling system fitted

There is also a legislative requirement for mines to have a safe means of docking the GAG to the mine from the surface.

The low flow project was conducted at Cook Colliery in central Queensland in association with Mr John Brady of Statutory Management Services (Brady 1997). Simtars was contracted to perform gas monitoring during the project. The type of inertisation used in the low flow project was a Tomlinson Boiler. This technique utilises a diesel fired steam generating boiler with the combustion gases being the desired product. The project showed that the boiler was capable of successfully inerting a large open goaf. It was also shown to suppress the release of seam gas.

The boiler has become a regularly used tool by coal mines to combat spontaneous combustion activity by reducing the oxygen concentration, as well as to avoid or minimise the period of time that a panel is explosive during the sealing process. Industry has also utilised membrane separation techniques for low flow inertisation. Many mines have their own low flow units on site or at least have arrangements for their immediate deployment and use.

CONCLUSIONS

- The Moura disasters have shaped where the Queensland mining industry is today with regards to its world class mine gas management and legislation..
- The Australian mining industry has developed, sourced and implemented available technology to assist with mine gas management..
- Australian mines develop and implement safety management plans to control hazards (including gas management and spontaneous combustion) that exist in the mine.
- Simtars has made a significant contribution to the mine gas monitoring systems used throughout the Australian underground coal mining industry and elsewhere around the world, and considered to be amongst the best in the world.
- Inertisation is proactively used in the Australian coal mining industry for mine gas management.
- Ongoing vigilance is required. We must never forget the lessons learnt from past disasters and make the same mistakes again.

REFERENCES

Bell, S., Humphreys, D., Harrison, P., Hester, C. and Torlach, J. 1997. *A report on the Demonstration of the GAG-3A Jet Inertisation Device at the Collinsville No 2 Coal Mine on 11, 16, 17 and 18 April 1997.* Australian Coal Association Research Program No: C6019, Safety in Mines Testing and Research Station (Simtars), Redbank.

Brady, D. M. 2008. The role of gas monitoring in the prevention and treatment of mine fires. *Proceedings from 12th U.S./North American Mine Ventilation Symposium*, Reno.

Brady, D., Harrison, P. and Bell, S. 2009. 20 Years of Onsite Gas Chromatographs At Queensland Underground Coalmines. *Proceedings from Ninth International Ventilation Symposium*, Delhi.

Brady, J. 1997. *Sealing, Monitoring and Low Flow Goaf Inertisation*. Australian Coal Association Research Program No. C6002. Cook Resource Mining, JP Brady Statutory Management Services, Rockhampton.

Cliff, D., Rowlands, D. and Sleeman, J. 1996 *Spontaneous Combustion in Australian Underground Coal Mines*. Safety in Mines Testing and Research Station, Reprinted 2004, Redbank.

Coal Mining Act 1925 (Repealed 1999). Office of the Queensland Parliamentary Counsel, Brisbane.

Coal Mining Safety and Health Act 1999. Office of the Queensland Parliamentary Counsel, Reprinted as in force on 4 September 2009. Reprint No. 3B, Brisbane.

Coal Mining Safety and Health Regulation 2001. Office of the Queensland Parliamentary Counsel, Reprinted as in force on 1 July 2009. Reprint No. 3C, Brisbane.

Golledge, P. 1991. *Research on the Intrinsic Safety of the Flame Safety Lamp in Atmospheres of Methane and Coal Dust*. National Energy Research Development and Demonstration Program (NERDDP) End of Grant Report QSR9004. Safety in Mines Testing and Research Station (Simtars), Redbank.

Warden's Inquiry. 1987. *Report on an Accident at Moura No. 4 Underground Mine on Wednesday, 16th July. 1986*. Rockhampton.

Warden's Inquiry. 1995. *Report on an Accident at Moura No. 2 Underground Mine on Sunday, 7 August 1994*. Brisbane.

Large Scale Tests U.S. BoM and CMI Collaboration

Kazimierz Lebecki

Central Mining Institute, Katowice, Poland

ABSTRACT: Paper describes the results of common efforts of the U.S. Bureau of Mines and the Central Mining Institute in Poland on enhancement of explosion safety in underground coal mines. Investigation had been concentrated on the coal dust explosion suppression. Coal dust explosions research were focused on:

- Performance of solar panels as the detectors in the triggered barrier system
- Possible use of triggered barriers to suppress weak explosion
- Course of the explosion, especially weak one within the system of parallel entries

These studies had been financed by grants from the Maria Sklodowska-Curie—Joint Fund II. Results achieved were crucial for further work on explosion suppression in Polish coal mines.

INTRODUCTION

In 1990 the Maria Sklodowska—Curie Fund II recommended its activity in Poland. The Central Mining Institute (CMI) received grant funds for 10 research projects—two of them devoted strictly to the explosion safety in underground coal mines. From the US side the partner was the US Bureau of Mines' Pittsburgh Research Laboratories which traditionally addressed problems of fire and explosion from the Bureau/s inception in 1910. From CMI's side the principal partner was Experimental Mine "Barbara," whose underground experimental facilities served as the research site. This author served as the principal investigator for both projects listed below:

- Mine-worthy triggered barrier system to suppress gas and dust explosion in underground coal mines
- Control of Dust Explosion in Multiple Entry Gate Systems

The results achieved are presented below.

Concept of Triggered Barrier

Explosion barriers are widely used in European coal mining where the extraction of the coal is by the longwall system. Their operational idea is simple—in case of stone dust barriers several shelves with stone dust are suspended under the roof. When explosion occurs, a pressure wave moving ahead of the flame front blows off the stone dust forming a cloud of incombustible stone dust/air mixture, extinguishing and thus interrupting flame propagation. It is observed that the pressure wave is only weakened

and propagates until its energy fades, The action of water barriers is similar. Such a barrier is called a passive one, because its action requires energy supplied by the explosion itself. The passive barrier is effective if placed at a distance of 60 or more meters from the ignition source, therefore at least 60 meters of the working is affected by flame and pressure wave. To suppress the explosion rather close to the likely ignition source a triggered barrier system was tested in Western Europe (Goffart, 1984; Michelis, 1985), US (Greninger et al) and Poland (Lebecki et al., 2000). There are three basic elements of the triggered barrier:

- Detector or transducer which reacts to selected explosion-specific phenomenon as pressure wave. temperature rise, radiation from the flame front
- Operational element which in response to an impulse received from detector, puts the barrier into action and disperses extinguishing agent
- Extinguishing agent in form of powder or liquid.

In our common experiments we have used a solar panel (Cortese et al., 1991) being a set of photovoltaic cells as a flame detector, giving electric energy to ignite the detonator. As the operational element for dispersing water a short piece of detonating cord was used. The extinguishing agent was water.

EXPERIMENTS WITH TRIGGERED BARRIER SYSTEM

Experiments were performed in the underground galleries of the Experimental Mine "Barbara" with two types of triggered barriers both actuated by a solar panel detector. It consists of 48 photo elements, the size of the whole panel is 69.8 × 36.8 cm, and weight 2.95 kg. It reacts on the radiation from methane and coal dust flames. In the panel's electrical circuit a pressure switch was included in such a manner that only the simultaneous presence of flame radiation and overpressure of 5.0–5.7 kPa could generate an output signal for the combined detector. The solar detector was investigated for its reaction to a methane and coal dust flame. In the laboratory experiments it was confirmed that (1) solar panel has a fairly good sensitivity to the radiation emitted by dust particles, (2) its sensitivity drops with the incombustible content in the dust cloud, and (3) its sensitivity grows with the finesse of coal dust. The influence of humidity on and the effect of dust covering panel surface was investigated showing that (1) the part of the flame radiation useful for the detection is less absorbed than that for other wavelengths and (2) humidity with associated water condensation on the panel surface can diminish electrical efficiency even by 50%.

Tests With Triggered Barriers in Underground Entries

Two types of barrier were tested:

Type A—six barrel-shaped plastic containers, each filed with 30 liter of water—the inside of each of the containers was weakened from one side by a few cuts just to control water dispersion direction. Before filling the containers with water a plastic bag was inserted into the container to keep it watertight. In each container a section of 0.5m of detonating cord was inserted to disperse water. The barrier consisted of six barrels, one end of detonating cord was connected to the detonator ; detonators were connected in series with solar panel. The containers were hung on the entry's sidewalls at the distance 40 meter from the ignition source, three on each side with one meter distance between vessels. Total amount of water in all tests was about 25kg/m² of the entry cross section. In all tests coal dust was used with finesse 50% below 200-mesh and volatile content 38%. The dust was deployed along a length of 120 m.

The primary explosion was accomplished (initiated) by using 25 m³ of a stoichiometric methane–air mixture ignited with a detonator. In all tests the flame velocity and static pressure were measured. Tests results are summarized in the Table 1.

Type B—known as Belgian Triggered Barrier, a plastic cylinder 2 meter long, 25cm of diameter, filled with polyurethane foam saturated with water. The volume of the cylindrical container is 110 liters. Water is dispersed by detonating cord placed inside

Table 1.   Test results with A type barrier

Test No.	P_{max} (bar)	V_f (m/s)	T_d (ms)	T_w (ms)	Flame Range (m)
I	0.28	100	42	158	40
II	0.31	99	65	138	40
III	0.32	101	39	133	40

Table 2.   Test results with B type barrier

Test No.	P_{max} (bar)	V_f (m/s)	T_d (ms)	T_w (ms)	Flame Range (m)
I	0.325	87	32	199	40
II	0.375	141	31	112	40
III	0.355	103	31	165	40
IV	0.360	179	26	96	40

P_{max}—maximum static pressure at the barrier position
V_f—flame velocity between 30 and 40 m from ignition source (in front of barrier)
T_d—delay of the detector, time lapse between flame arrival at the detector position and the beginning of dispersing of extinguished agent
T_w—time of the barrier actuation, time lapse between the beginning of dispersing of extinguished agent and flame arrival at the barrier position
Flame range—the position of the last flame sensor marking flame passage

the cylinder along its axis. Barrier system is built of containers interconnected by Nonel, a detonating cord with 25 mg of explosive per meter. Tests were performed with three containers suspended parallel under the roof (test I and II), two containers placed parallel at the side walls (test III) and one container under the roof at the center of the entry (test IV). Test results are presented in Table 2.

All test results were positive—in every case explosion did not propagate behind the barrier.

COURSE OF DUST EXPLOSION IN MULTIPLE ENTRY GATE SYSTEM

Both American and Polish mining industries have a common interest in dealing with explosion fighting in parallel interconnected entries. In American room and pillar mining practice entries are connected by cross-cuts, of which the first is open, the second closed by the ventilation barrier, and the third closed by a dam (seal) made of blocks. Polish mining practices do not use a configuration identical with the American one—but there are parallel entries connected by cross–cuts. Thus there is common interest

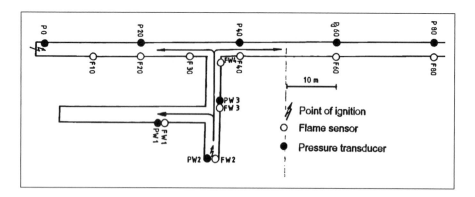

Figure 1. Plan of the test site

Table 3. Characteristics of primary explosions

Test Number	Explosible Mixture	Mixture Volume (m³)	Pressure Pulse (kPa s)	Average Pressure (bar)
1	9.5% CH₄	40	2.40	0.201
2	9.5% CH₄ + 2kg/m³ coal dust	18.8	4.02	0.335
3	9.5% CH₄	13.9	1.82	14.0
4	10m of coal dust, 0.5kg/m³	75.0	3.22	0.248

Table 4. Test results with weak explosions

Test Number	Vd (m/s) flame Velocity in Main Gallery	Maximum Pressure in the Drift (kPa)	Maximum Pressure in Parallel Gallery (kPa)	Maximum Pressure in the Main Gallery Face (kPa)	Maximum Pressure in the Main Gallery (kPa)
5	67.5	80.0	80.0	80.0	67.5
6	45.0	45.0	50.0	50.0	37.5
7	17.5	20.0	17.5	17.5	15.0
8	40.0	40.0	37.5	37.5	30.0

in investigating the course of an explosion in multiple entries and ways of suppressing it. In such a system barriers should be deployed as close as possible to the likely ignition point. So, the triggered barriers are the most suitable in this case which is characterized by weak explosion with low pressure and slow flame movement. [Passive barriers require sizeable explosion pressure to disperse the water or stone dust and therefore need to be located at least 60 meters or more from the likely ignition source.]

The course of weak explosion was investigated in the underground galleries of the Experimental Mine "Barbara." The plan of the site is shown on Figure 1.

Four tests were performed with primary explosion. For each explosion test the measurements

included the average static pressure and pressure pulse as

$$\int_{t1}^{t2} p\,dt .$$

The results are shown in Table 3. There were three distinctive path of the flame (Figure 1):

• From the ignition source, along the drift and main gallery towards the open end
• From the ignition source, along the drift and main gallery towards the closed end
• From the ignition source towards the parallel gallery

Summary of tests results is given in Table 4.

V_d value is average flame velocity in the section of main gallery between 40 and 60 meter from the closed end. This is the velocity with which flame leaves test site. Because of existence of blind gallery, the influence of reflected pressure wave could be expected. The analysis of pressure distribution at the gallery cross-road revealed that flame propagation is not affected by abnormal pressure distribution. Three tests with triggered barrier Type A and with solar detector were performed. Two kinds of primary explosion (initiating explosion) were used:

- 10mPII- 10m long gallery section with fine pure coal dust distributed uniformly on the floor, ignited by the 750 gram of gun powder from cannon
- 10mPII as V_d above, enforced with $20m^3$ of stoichiometric methane-air mixture (hybrid mixture)

In all experiments dust with properties mentioned in the previous section was distributed in short galleries and in main one up to 80m. Test results are presented graphically showing separately the flame and pressure distributions. The delay between beginning of barrier action and flame arrival to the barrier position, marked by τ, is the most important parameter, deciding the barrier efficiency. Courses of the two explosions with 10mPII primary explosion show in both cases that flame velocity between the detector and barrier was about 40 m/s and the delay times were respectively 110 ms and 112 ms. Flame was detected at F40 station only, just around the corner of the main gallery. In the test with stronger primary explosion—the respective flame velocity between detector and barrier was almost the same as previously—42.5 m/s. but the delay was considerably shorter—85 ms. The arrangement is not enough to fill the whole entry cross section with water and consequently the flame passed the longer path reaching station at 600 m in the main entry. The flame was extinguished further from the initiating source than in the previous cases. It happened probably because the flame radiation was less intense and then the energy needed to fire the detonator was reached later. The course of pressure shows that barrier efficiency is the same; however in the third test the highest pressure growth, up to 33kPa was observed. It was due to the longer burning of the coal dust, resulting in pushing the flame into parallel entry which was not observed in previous tests

FINAL REMARKS

Investigations performed in the Central Mining Institute proved that triggered barrier can be effectively used to suppress explosion of methane and

coal dust. These investigations have again demonstrated the potential of the CMI and its possibility to test explosions in the large-scale for various mining configurations. The presented results are rather historical in point of view of techniques used during the experiments. But the practical idea associated with using a triggered barrier is still worth pursuing. It would be reasonable to return to the ideas associated with triggered barriers since new technological advances mainly in solar cell design and fabrication have taken place.

Prior to the use of the solar panel coupled to a pressure switch—detectors for use with triggered barriers were massive, expensive, required a power supply, required frequent maintenance checks, and suffered from reliability problems. The same was with containers, which were not enough tight. A quality pressure switch coupled with solar detector is needed—because explosion is marked by pressure rise and flame radiation. The triggered barriers may represent supplementary measures to deal with otherwise difficult areas to protect against explosions. They can augment passive barriers at roadway intersections (where additional paths dilute the pressure for putting passive barriers into play). Triggered barriers can be used to protect areas plagued with high rates of float coal dust deposition. Since the overwhelming majority of ignitions take place near the working face, the use of triggered barriers well within 60 meters from the working face can significantly shorten the zone or range of damage or danger zone in the event an explosion takes place. There is definite merit in pursuing triggered barrier research and developing a prototype one for deployment in an operating coal mine. More work is needed to optimize the dispersal of extinguishing agent to combat an explosion flame in wide roadways. More work is merited in developing an effective deployment strategy for their use in mining. All coal producing countries are concerned with protection against gas and dust explosions in coal mining operations. Some of these mines might benefit from the use of triggered barriers. Some major coal mining countries have to deal with strategies to combat gas and coal dust explosions. Separate from worker safety matters the cost-benefit of the use of triggered barriers merits consideration.

Triggered barriers would also have definite application for use in protecting against grain dust explosions—as water would not introduce a toxic substance into the food materials. Triggered barrier ideas have found application for use in fire/explosion suppression systems within aircraft in flight. Triggered barriers would have application to the chemical processing industry to suppress explosions—in such situations there are excellent alternatives to the solar panel detector and also various options for the use of extinguishing agents that can chemically alter the combustion process.

REFERENCES

Cortese R.A. Sapko M.J. Flame-Powered Trigger Device for Activating Explosion Suppression Barrier, BUMines RI 9365, 1991.

Greninger N.B., Cortese R.A., Weiss E.S. Suppression of Dust Explosions Using Triggered Barriers in Wide Rectangular Entries; Proceedings of the 26th International Conference of Safety in Mines Research Institutes, ed. Central Mining Institute, Katowice, Poland 1995.

Goffart P.R. La detection des Explosions en Mine au Moyen du Systeme Belge d'arret-barrage a Declenchement; Symposium "La Belgique Face aux Nouvelles Techniques Charbonnieres" 1984.

Lebecki K., Sliz J., Dyduch Z., Cybulski K. The possible use of Solar Panels as flame Detector of Triggered Barriers; Journal of Loss Prevention in the Process Industries vol.13, 311–318, 2000

Michelis J. Application of the Triggered Barriers Type "Tremonia" as a Stationary and Mobile Explosion Barrier, 21st International Conference of Safety in Mines Research Institutes, ed. Balkema, Sydney 1985.

Underground Coal Mine Disasters 1900–2010:
Events, Responses, and a Look to the Future

Michael J. Brnich, Jr.
NIOSH, Pittsburgh, Pennsylvania, United States

Kathleen M. Kowalski-Trakofker
NIOSH, Pittsburgh, Pennsylvania, United States

ABSTRACT: This paper captures almost 110 years of history of underground coal mine disasters in the United States. The deadly disasters of the first ten years of the twentieth century led to the U.S. Congress founding the U.S. Bureau of Mines (USBM) in 1910. The authors examine the changing trends in mine disasters including the frequency of fatalities, causal types, the responses to those disasters and most importantly, the growing body of research on human behavior in mine emergencies. Emphasis is on the future—integrating the research on human behavior in disasters into the mining industry. This research includes the integration of the judgment decision-making process, communication, leadership in escape, expectations training, incident command center issues including fatigue, shifts and leadership, plus issues concerning the introduction of refuge chambers into U.S. mines. The authors suggest that a key factor in meeting the goal of increasing successful mine escape and rescue while decreasing fatalities and injuries lies in the field of social-psychological research and human behavior interventions.

INTRODUCTION

Mine disasters have been a focal point among mine operators, safety and health personnel, and miners, as well as mine safety and health researchers in the United States for decades. Hundreds of disasters, resulting in thousands of mine worker deaths, have occurred in mines since 1900.

Because most of these catastrophes have occurred in underground coal mines, the authors have chosen to focus on these disasters in this paper. The authors examine and discuss the history of coal mine disasters in the U.S. with emphasis on changing trends in disasters, responses to the disasters, the growing body of research on human behavior in mine emergencies, and implications for future integration of research on human behavior into the mining industry.

The Mine Safety and Health Administration (MSHA) classifies disasters by number of fatalities and by cause. MSHA defines a "disaster." as an incident with five or more fatalities. In this paper, the authors do not differentiate "incident." from "disaster," and discuss incidents with too few fatalities to be classified as "disasters." by MSHA. Causes are grouped into six categories: (1) explosion; (2) fire; (3) haulage (transportation of personnel, materials, or equipment); (4) ground fall/bump (fall of roof rock or an outward, violent burst of a pillar); (5) inundation (the sudden inrush of water or toxic gases from old workings); and (6) other (MSHA 2006a).

From 1900–2006, 11,606 underground coal mine workers died in 513 U.S. underground coal mining disasters, with most disasters resulting from explosions (Kowalski-Trakofler, et al.,2009a). In 2007, 9 additional workers died in the Crandall Canyon disaster (Gates, et al. 2007a), bringing the total to 11,615 miners killed in 514 disasters. Table 1 summarizes the number of disasters by category and the related number of fatalities for the period.

The major coal mine disasters are presented within three time periods: *1900–1909,* preceding the founding of the Bureau of Mines; *1910–1969,* a period of significant decrease in underground coal mine disasters; and *1970–present,* a period when human behavior and psycho-social factors came into play. Industry and congressional responses to these events are also presented in these timeframes.

THE EVENTS: MAJOR COAL MINE DISASTERS

The Period 1900 Through 1909

The period 1900 through 1909 was the deadliest decade in U.S. underground coal mining, and led to the legislation that founded the Bureau of Mines with the express mandate of reducing fatalities in the mining industry. From 1900 through 1909, 3,660 miners perished in a total of 133 mine disasters. Sixteen major mine disasters during this time period killed 2,070 miners. These 16 events are summarized in Table 2.

Table 1. Number of underground coal mine worker fatalities by type of disaster, 1900 through 2008

Type incident	Number of Events	Number of Fatalities
Explosion	420	10,390
Fire	35	727
Haulage	21	145
Ground fall/Bump	14	92
Inundation	7	62
Other	17	199

Table 2. Major U.S. underground coal mine disasters, 1900–1909

Year	Mine	Type Disaster	No. Killed
1902	Fraterville	Explosion	184
1902	Rolling Mill	Explosion	112
1903	Hanna No. 1	Explosion	169
1904	Harwick	Explosion	117
1905	Virginia City	Explosion	112
1907	Stuart	Explosion	84
1907*	Naomi	Explosion	34
1907*	Monongah Nos. 6 & 8	Explosion	362
1907*	Yolande	Explosion	57
1907*	Darr	Explosion	239
1907*	Bernal	Explosion	11
1908	Hanna No. 1	Explosion	59
1908	Rachel and Agnes	Explosion	154
1908	Lick Branch	Explosion	50
1909	Lick Branch	Explosion	67
1909	Cherry	Fire	259

*Occurred in December, 1907

December 1907, known as "Bloody December," is the deadliest on record for the U.S. underground coal mining industry. That month 703 miners died in 5 mine explosions.

The Monongah Nos. 6 & 8 disaster, the Darr disaster, and the Cherry disaster remain 3 of the deadliest events in U.S. coal mines.

The Period 1910 Through 1969

After the founding of the Bureau of Mines, a number of major events were still to come. Many of the events lead to changes in U.S. mine safety and health regulations. Overall, 1910–1969 was a period of significant decrease in disasters.

Stag Canyon No. 2 Mine—1913
An October 1913 explosion rocked the Stag Canyon No. 2 mine in New Mexico, killing 263 workers in the mine. Fourteen miners escaped from an unaffected part of the mine; 9 others were rescued from the bottom of the air shaft. The explosion was caused by detonating an overcharged shot in a dusty pillar section (MSHA, 1998a).

Eccles Nos. 5 and 6 Mines—1914
A methane explosion occurred in April 1914 at the Eccles Nos. 5 and 6 mines in West Virginia. There were 246 miners working when the explosion occurred. All 172 miners working in the No. 5 Mine were killed. Afterdamp killed 8 miners in the No. 6 mine, working in the coal seam above. Sixty-six miners escaped from the No. 6 mine (Ibid).

Castle Gate No. 2 Mine—1924
A series of three explosions struck the Castle Gate No. 2 mine in Colorado in March 1924. The disaster claimed the lives of 172 workers. The initial explosion occurred when a mine examiner attempted to relight his key-locked safety lamp from his carbide cap lamp and ignited methane (Ibid).

Mather No. 1 Mine—1928
An explosion in May 1928 at the Mather No. 1 mine in Pennsylvania killed 195 of the 279 workers underground. Methane accumulated in a working section, possibly caused by an open mandoor. The gas was ignited by an arc from a battery locomotive working in the area (Ibid).

After the Mather mine explosion, there were no major mine disasters for 12 years, until 226 miners died in three major disasters between January and July 1940. Two other major disasters, in 1942 and 1943, killed 130 miners (Ibid).

Centralia Mine—1947
An explosion in the Centralia No. 5 mine in Illinois killed 111 miners. Eight miners were rescued and 24 managed to escape after the explosion. The disaster is believed to have been caused by either an insufficiently charged shot or a blown-out shot stemmed with coal dust (Ibid).

Orient No. 2 Mine—1951
An explosion in December 1951 at the Orient No. 2 mine in Illinois killed 119 of the 256 miners working underground. A main ventilation door was left open, relieving the ventilation pressure on old abandoned panels. It is believed methane came out of abandoned panels and traveled to an active section where it ignited, possibly due to an arc from electrical equipment (Ibid).

Farmington No. 9—1968
The frequency and severity of underground coal mine disasters continued to decline in the 1950s and 1960s. Many mine safety practitioners believed the day of major disasters had come to an end, until a major explosion tore through the Farmington No. 9 mine in West Virginia on November 20, 1968. At the time of the explosion, 99 miners were working underground; only 21 managed to escape. With no hope of finding survivors, the mine was sealed on November 30, 1968 (MSHA 1989). The mine was recovered and between September 1969 and April 1978, the bodies of 59 victims were removed. The mine was permanently sealed in November 1978, leaving 19 victims entombed. The cause of the explosion was never determined. However, the event would have far reaching effects that would forever shape U.S. mine safety and health (MSHA 1998a).

The Period 1970 Through 2008
While the frequency and severity of underground coal mining disasters along with the number of miner deaths decreased substantially during 1970–2005, several significant disasters did occur that had further impact on mine safety and health in the U.S. Before 2005, the total number of mining disasters had decreased from a high of 20 in 1909 to an average of one every 4 years during 1985–2005 (Kowalski-Trakofler, et al. 2009a). However in 2006, disasters at the Sago mine in West Virginia and the Darby mine in Kentucky claimed a total of 17 lives. The 2007 Crandall Canyon disaster killed 9 workers. No mine disasters occurred in 2008.

Scotia Mine—1976
In March 1976, two explosions occurred on separate days at the Scotia Mine in Kentucky. The first occurred on March 9, killing 15 of the 116 miners working underground. The second explosion took place on March 11, during recovery operations. It killed 11 of 13 workers underground, including several inspectors. Investigators believed the first explosion occurred when an accumulation of methane in a track entry was ignited by arcing created in the controller of a locomotive. The second explosion is also believed to have occurred when methane was ignited by an arc from electrical equipment (MSHA 1998b).

Following the Scotia mine disaster, a handful of other disasters occurred in the U.S. Table 3 summarizes these events. After the South Mountain mine explosion in 1992, there were no major coal mine disasters until 2001.

Table 3. Underground coal mine disasters, 1977–1999

Year	Mine	Type Disaster	No. Killed
1977	Porter Tunnel	Inundation	9
1980	Ferrell No. 17	Explosion	5
1981	Dutch Creek No. 1	Explosion	15
1981	Adkins Coal No. 11	Explosion	8
1981	Grundy Mining No. 21	Explosion	13
1982	RFH Coal No. 1	Explosion	7
1983	Clinchfield No. 1	Explosion	8
1984	Wilberg	Fire	27
1986	Loveridge No. 22	Other*	5
1989	William Station	Explosion	10
1992	South Mountain	Explosion	8

*Stock pile collapse

Jim Walter Resources No.5 Mine—2001

In September 2001, two separate mine explosions occurred at the Jim Walter Resources No. 5 Mine in Alabama, killing 13 miners. The first explosion occurred after a roof fall at a scoop battery charging station. The fall damaged a scoop battery and ventilation controls, and an arc flash from the damaged scoop battery ignited methane. The explosion damaged critical ventilation controls and injured four miners who were working in the affected section. Three of the miners escaped while the fourth was left behind because of the seriousness of his injuries.

The second explosion occurred as 12 miners made their way to rescue the miner left behind. This explosion was most likely caused when a signal light system ignited methane in the track entry. At least 12 miners were killed by the second explosion. It is not known if the 13th miner, the one left behind after the first explosion, died from his initial injuries or as a result of the second explosion (McKinney, et al. 2002).

Sago Mine—2006

An explosion in January 2006 at the Sago Mine in West Virginia killed 12 miners. A crew of 12 miners entered the mine shortly before the explosion, and proceeded toward their section. A second crew of 16 miners entered the mine shortly after the first crew and traveled toward their section. After the first crew reached their section, an underground methane explosion occurred approximately two miles from the entrance, near a worked out, sealed area of the mine. The explosion instantly killed one miner, a mine examiner working alone in the vicinity of the blast.

After the explosion, the 12 miners of the first crew were unable to escape because of smoke and dust. The crew returned to their section, barricaded, and awaited rescue. Eleven of the 12 died from carbon monoxide before rescuers reached them. The second crew was outby the explosion and was able to escape the mine. Investigators believe energy from a lightening strike was transferred onto an abandoned length of damaged electrical cable in the sealed area, igniting methane that accumulated in the abandoned workings (Gates, et al. 2007b).

Alma No. 1 Mine—2006

Two miners died in a mine fire at the Alma No. 1 Mine in West Virginia in January 2006. A total of twenty-nine miners were working underground when the fire broke out at the longwall section belt take-up. Because of the ventilation arrangement, the longwall crew was able to evacuate the section in fresh air through a set of cut-through entries that brought air into the longwall panel from the mains. Another crew, working inby the fire, proceeded to evacuate. Partway out of the mine, this crew abandoned their mantrip because of heavy smoke. For unknown reasons, two miners from this crew became separated from the rest of the group and perished. The bodies of the two missing miners were found two days after the fire began (Murray, et al. 2007).

Darby No.1 Mine Explosion—2006

In May 2006, an explosion occurred at Darby No.1 Mine in Kentucky, killing 5 miners. At the end of the afternoon shift, a group of 5 miners headed outside on the mantrip. Two afternoon shift miners stayed behind to cut roof straps near a ventilation seal in the return airway. At the same time, the midnight shift crew was entering the mine. Shortly after the afternoon shift crew reached the outside, an explosion occurred in the mine. The two miners performing the cutting work died in the explosion, and three miners from the entering midnight shift crew died while trying to escape. One miner survived and was able to travel part of the way towards the mine entrance wearing his self-contained self-rescuer (SCSR). He was later rescued. The explosion was the result of improper construction of the mine seals and inappropriate use of cutting/welding equipment in the return airways (Light, et al. 2007).

Crandall Canyon—2007

The most recent mining disaster in the United States occurred in August 2007, when a major coal bump or bounce occurred on the Main West pillar section at Crandall Canyon mine, trapping and killing 6 miners. Underground rescue and recovery work began immediately but was suspended 10 days later, when a second major bump/bounce killed 3 rescue workers. All rescue/recovery work was suspended on August 31, 2007. In all, 9 miners died as a result of the disaster (Gates, et al. 2007a).

THE RESPONSE: SYNOPSIS OF PERTINENT MINE SAFETY LEGISLATION

For years mine operators, federal and state mine safety agencies, and researchers have looked at numerous aspects of mine disasters. The most intense efforts in this arena have occurred following major mine emergency incidents. The response to these disasters have taken a variety of forms, with the most impactful being legislation adopted at the federal and/or state level. As Nieto and Duerksen (2008) point out, this legislative response is cyclical with each event. Pennsylvania adopted the first significant mine safety legislation in 1870. The first federal mine safety regulations were passed by Congress in 1891 (MSHA, 1998a).

U.S. Bureau of Mines Established

As mentioned earlier, 16 significant mine disasters occurred in the U.S. from 1900 through 1909. These events, including five disasters in December 1907, caused citizens and lawmakers to focus on coal mining and the dangers it presented to workers. After intense public pressure, Congress passed Public Law 61-79 in 1910. This law created the United States Bureau of Mines, an agency whose primary mission was mine safety research and investigation (MSHA, 1998a). The goal was to mitigate underground coal mine disasters through application of research.

Public Law 77-49—1941

After the establishment of the Bureau of Mines, additional coal mine disasters continued to occur. These included the disasters at the Stag Canyon No. 2 mine, 1913; Eccles Nos. 5 and 6 mines, 1914; Castle Gate, 1924; and Mather mine, 1928. As mentioned previously, several additional major disasters occurred in the early 1940s.

These disasters, coupled with the overall high death rate among underground coal miners, led Congress to pass Public Law 77-49 in 1941. This law gave Federal mine inspectors right of entry to carry out annual mine safety inspections and investigations.

Public Law 80-326—1947

The March 1947 Centralia mine explosion led Congress to enact Public Law 82-326. This new law, which expired one year after enactment, authorized the crafting of the first Code of Federal Regulations for bituminous and lignite coal mine safety. However, there were no provisions in the law for enforcement of the regulations (Ibid).

Public Law 82-522—1952

The December 1951 Orient mine explosion, along with fallout from the 1947 Centralia tragedy prompted Congress to pass Public Law 82-522, the 1952 Federal Coal Mine Safety Act (1952 Act). The law emphasized prevention of major mine explosions. It mandated regular annual inspections of mines and provided the Bureau of Mines with limited enforcement power. Although anthracite coal mines were covered by the new law, surface mines and all mines employing fewer than 15 workers were exempt (Ibid).

Public Law 89-376—1966

Despite strengthened safety regulations, a number of disasters occurred following the passage of the 1952 Act. In 1966, Congress passed Public Law 89-376, extending provisions of the 1952 Act to all underground coal mines. The new law permitted mine inspectors to issue withdrawal orders in the event of an unwarrantable failure on the part of a mine operator to comply with regulations (Ibid). Withdrawal orders require operators to withdraw all workers from the affected area until the hazard has been mitigated.

Public Law 91-173—1969

The provisions of the 1952 Act, strengthened in 1966, still did not stop major disasters. The U.S. underground coal industry was rocked by the 1968 Farmington disaster. The Farmington event led to Public Law 91-173, the Federal Coal Mine Health and Safety Act of 1969 (1969 Act). This new law was the most sweeping and comprehensive mine safety legislation ever in the U.S. and forever changed the face of coal mine health and safety. The law included both surface and underground coal mines, and mandated two annual inspections at surface mines and four per year for underground mines. Enforcement powers were increased, safety standards strengthened, and new health standards adopted. The law established fines for violations and criminal penalties for willful violations of the law and set forth basic safety and health training requirements (Ibid).

Public Law 95-164—1977

After the 1976 Scotia mine disaster, Congress enacted Public Law 95-164, the Federal Mine Safety and Health Act in 1977. This legislation moved mine safety and health enforcement to the Mine Safety and Health Administration and placed the agency under the Department of Labor. The new law combined coal and metal/nonmetal mine safety and health regulations and contained enforcement provisions similar to those of the 1969 Act. This law strengthened miners' rights, established the Federal Mine Safety and Health Review Commission, and strengthened training requirements first set forth in the 1969 act (Ibid).

Public Law 109-236—2006

Following the 1977 Federal Mine Safety and Health Act, only a handful of additional mine disasters occurred in the U.S. However, in the first five months of 2006, the Sago and Darby mine explosions and the Alma No. 1 fire all occurred, and brought mine safety and health to the forefront again. Following these disasters, Congress passed Public Law 109-236, the Mine Improvement and New Emergency Response Act of 2006 (MINER Act). The MINER Act contained provisions for improving safety, health, preparedness, and emergency response at U.S. mines. Among other elements, the MINER Act requires operators to develop and maintain emergency preparedness and response plans to reduce delay and improve response quality. The Act also requires mine operators to provide additional SCSRs, escapeway lifelines, wireless two-way communication and tracking, and refuge alternatives. It also mandates care for the families of trapped miners (MINER Act 2006).

BEHAVIORAL RESEARCH IN MINE DISASTERS

Until relatively recently, human behavior in mine emergencies was largely ignored as an area of research. Following many major mine disasters, legislation was enacted to improve mine safety and health, focusing on prevention through research interventions, mandated inspections, regulatory compliance, and eventually safety and health training. However, until NIOSH embarked in research aimed at assessing miners' behavior in mine emergencies in the late 1980s, the effect of human behavior on the outcome of major events remained unstudied. Since that time, such behavioral research has proved to have far-reaching implications, especially with respect to training for mine escape or disaster management. In their report, the National Mining Association's Mine Safety Technology and Training Commission (2006) reported on the importance of behavior research in mine emergency response and the need for continued efforts in this area.

Worker Behavior in Mine Emergencies

After a mine emergency occurs, the event is investigated by both the state mining agency and MSHA. These investigations focus on the elements of the incidents in an attempt to identify root causes, actions taken, and to recommend ways to prevent such events in the future. Before 1988, no one had studied mine emergency events from the human behavior perspective.

From 1988 through 1990, NIOSH researchers interviewed 48 miners who escaped three different mine fires. Miners discussed their actions and thoughts from the time they first became aware that there might be a problem in their mine until they reached safety. The interviews resulted in more than 2,000 pages of data. In analyzing miners' testimony, researchers discovered an array of decision

variables related to various aspects of individual and group behavior. Researchers constructed a model of workers' behavior escaping a mine fire (Vaught, et al. 2000). These human behavior elements play an important role in miners' ability to respond to a mine emergency.

Judgment and Decision Making

There is limited literature on the process of judgment and decision-making under stress (Kowalski-Trakofler, et al. 2003). The study by Vaught, et al. (2000) was the first one to examine the judgment and decision making process within the context of a mine emergency. Based on analysis of miners' testimony, researchers discovered that the miners underwent a complex decision making process as they escaped. The researchers were able to construct a model of the judgment and decision making process (Figure 1).

Escaping miners go through a multi-step process in judgment and decision making. This process is ongoing and continues from when they first perceive there is a problem until they reach safety. Initially, miners are presented with the nominal problem. In the case of the 48 miners interviewed, this was a mine fire. As miners begin perceiving the problem, background problems and contextual issues factor in. Background problems include information such as knowledge of the fire location or the smell of smoke. Contextual issues include miners initially framing the problem as a normally occurring event, such as smelling smoke from bonds being welded at rail joints.

Once the miners perceived the problem, they entered a diagnosis or analysis phase. Stress from a variety of sources, including information uncertainty, affected their ability to effectively analyze the situation. After analyzing the situation, miners assessed options available for responding to the circumstances. Then, they selected an option and executed their decision. In some instances, miners made choices and executed decisions only to find that they made the wrong choice. They would then be required to re-evaluate the situation, perhaps through further diagnosis, and then make new decisions about courses of action. This judgment and decision making process continued throughout the entire escape.

Researchers identified several important points about judgment and decision making. First, miners tended not to perceive the problem accurately at first. Often they tried to place the problem within the context of normal activities, rather than as an emergency situation. Second, the diagnosis made by escapees was affected by the nature of the warning message they received. Third, miners' perceived options and choices were impacted most by their overall knowledge of the mine and the quality of information available.

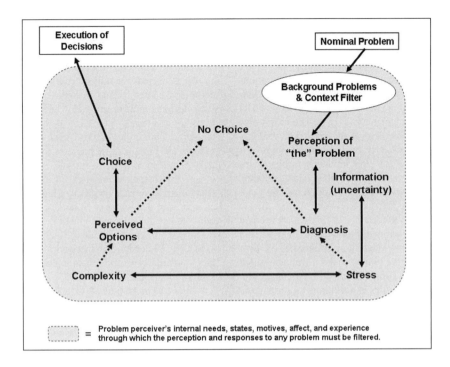

Figure 1. Model of judgment and decision making

Researchers discovered other factors affecting judgment and decision making in a mine emergency such as a fire. These include wayfinding and cognitive (mental) mapping, which are two important components in escape; leadership, which played a significant role in escape from the three mine fires (Kowalski, et al. 1994); fire warning and information uncertainty; and group dynamics (Vaught, et al., 2000).

Leadership
Researchers investigated the emergence of leadership in escape from the three underground mine fires (Kowalski, et al. 1994). Attributes of leaders who facilitated escapes were identified and compared, as were instances of lack of leadership. The structure of the group leadership before the crisis was compared with the reality of the leadership during the crisis. The day-to-day leader was not necessarily the one who lead the miners out of the burning mine. Some groups experienced leadership breakdown during their escape, while others witnessed the emergence of leadership.

Researchers identified six consensus characteristics of leaders. The leaders (1) were aware and knowledgeable, (2) were decisive, yet flexible, (3) were open to input from others, (4) had a calming influence and gained followers' confidence, (5) were

logical decision-makers, and (6) and allowed leadership to develop naturally.

Refuge Chambers
After the incidents of 2006, attention turned to the use of refuge chambers in underground coal mines. The 2006 MINER Act mandated that every coal mine in the country must have such chambers in place by 2009. The use of these chambers has raised a number of psycho-social questions beginning with when, and if, to go into a chamber. Escape is always the first option. NIOSH initiated a project to "develop engineering guidelines associated with the location, construction, and general application of various refuge alternatives." (Ounanian 2007, p.1). The first phase of the project evaluated the impact that refuge chambers could have made in the outcome in coal mine explosions, fire emergencies, and mine inundations in which fatalities occurred from 1970 through 2006. The potential effect of refuge chambers on both survivors and fatalities was estimated. Thirty-eight disasters were studied and the results of the event tree analysis indicated that the availability of "refuge stations (chambers) would have had a positive impact on the outcomes of 12 disasters or 32%. The total number of miners that would have been positively impacted was 83 (19%) of the 429 underground and impacted (miners) by these accidents. A

total of 74 (29%) of the 252 fatalities would have been positively impacted and potentially would have survived the accident. These numbers are based on the assumptions that were made during the analysis of the mine disaster reports. These assumptions are conservative and based on sound understanding of coal mining environments, operations and procedures." (Ounanian 2007, p.3). The report also specified that psychological factors would strongly influence miners when using a refuge chamber.

At the time of the 2006 incidents, refuge chambers were not available. If they had been available, they may have saved the lives of 12 of the Sago miners and three of the Kentucky Darby miners (Ounanian 2007). Recently, NIOSH has developed and tested two training products to help miners understand and make decisions about refuge chambers. The first helps miners make a decision about whether to enter a refuge chamber in an emergency, and the second provides expectations training in the form of information about what behaviors emotions and physical reactions a miner might expect in a chamber over a period of time (Vaught, et al. 2009; Margolis, et al. 2009).

Command Center Issues
Human behavior issues in mine emergency command centers have been assessed to some extent, although more research is needed. The behavioral sciences can contribute to knowledge of productive command center behaviors such as leadership, limiting fatigue, decision-making under stress, setting appropriate shifts, and maintaining nutrition for command center and rescue personnel. Additional topics that could be addressed include command center organization, data overload, and integration of psychosocial support.

THE FUTURE: BEHAVIORAL HEALTH ASPECTS OF MINE DISASTER

In the 21st century, the U.S. mining industry can look back over 100 plus years of disaster history and note several trends that stand out. Initially, the industry focused on controlling fires and explosions in underground coal mines to mitigate disasters. Gradually, researchers expanded their inquiries to broader engineering control solutions for prevention. It is only in the past 25 years that research has slowly moved into studying human behavior in mine safety and health, beginning with disaster behavior in escape and rescue.

Psychological issues can play a major part in a successful escape. An individual's state-of-mind affects his/her cognition, emotional, behavioral, and physical state. Miners escaping from an underground coal mine, in fear for their lives, suffer considerable stress and the resultant symptoms. Decision makers in the command center are also subject to stress.

It is normal for miners, rescue personnel, and command center leaders to experience these reactions during a mine emergency. The immediate reactions are commonly referred to as the *fight or flight response*. This is the survival instinct that all individuals are born with; it prepares the individual to fight or to run in order to survive. In a mine emergency, this response prepares the body to escape. Psychologists refer to an incident like an emergency escape from a mine as a "traumatic incident." and research has documented some of the normal human reactions to traumatic incidents, both short term and long term. Understanding these normal reactions can facilitate orderly escape behavior and good judgment and decision-making.

Short-term responses to a traumatic incident usually include strong reactions, especially emotional reactions. These emotions are normal reactions to an extra-ordinary situation. Individuals are different and report some or none of the following symptoms: rapid heartbeat, dry mouth, sweaty palms, increased anxiety, fear, sweating profusely, feeling confused, feeling overwhelmed, shallow breathing, nausea, disorientation and poor problem solving. Longer term symptoms might include a continuation of some of the more immediate responses plus nonspecific aches and pains, nightmares, disorientation, poor concentration, guilt, irritability, depression, sense of failure, intense anger, withdrawal, change in sexual functioning, or change in eating and alcohol habits. A few larger mines in the U.S. have contracted with disaster trained mental health professionals to assist with operation planning and emergency response. NIOSH has found in focus groups that stakeholders are increasingly aware of the importance of incorporating psychological responses in training and after-event support (Kowalski et al. 2009b, in press).

Psychologically, preparation is the most important predictor of reactions in emergencies. Regular drills, equipment testing and donning, and expectations training can help to mitigate symptoms. Few programs in the mining industry address expectations training from a human behavior perspective. One good example of sound expectations training was developed after Sago. NIOSH researchers conducted a study investigating the human response issues related to wearing an SCSR, with the goal of developing training for miners on what they could expect from their units during an escape. Subjects included miners who had experience wearing SCSRs, manufacturers, and researchers. Results identified nine key areas of concern: (1) starting the unit, (2) unit heat, (3) induction of coughing, (4) unit taste, (5) difficulty in breathing while wearing the unit, (6) quality of the air supplied, (7) nose clips, (8) goggles, and (9) the behavior of the breathing bag (Kowalski-Trakofler, et al. 2008). These findings resulted in a training component published in MSHA's SCSR expectations training module (MSHA 2006b).

CONCLUSION

The authors presented the major coal mine disasters within three time periods: 1900–1909 which preceded the founding of the U.S. Bureau of Mines, 1910–1969, a period of significant decrease in fires and explosions in underground coal mines, and 1970–present, a period when the importance of human behavior and psycho-social factors were recognized. It is only recently in the history of mine disasters that behavioral health, psychology, and the behavioral sciences in general have impacted mine safety and health. The authors suggest that 2010 and beyond could be the era of the integration of aspects of human behavior with the technical expertise important to miner safety and health, especially in escape and rescue.

REFERENCES

Gates, R.A., Gauna, M., Morley, T.A., O'Donnell, J.R., Smith, G.E., Watkins, T.R., Weaver, C.A., Zelanko, J.C. 2007a. *Report of Investigation Underground Coal Mine Fatal Underground Coal Burst Accidents, August 6 and 16, 2007, Crandall Canyon Mine, Genwal Resources Inc., Huntington, Emery County, Utah, ID No. 42-01715*. Arlington, VA: U.S. Department of Labor, Mine Safety and Health Administration.

Gates, R.A., Phillips, R.L., Urosek, J.E., Stephan, C.R., Stoltz, R.T., Swentosky, D.J., Harris, G.W., O'Donnell, J.R., Dresch, R.A. 2007b. *Report of Investigation: Fatal Underground Coal Mine Explosion, January 2, 2006, Sago Mine, Wolf Run Mining Company, Tallmansville, Upshur County, West Virginia, ID No. 46-08791*. Arlington, VA: U.S. Department of Labor, Mine Safety and Health Administration.

Kowalski, K.M., Mallett, L.G., Brnich, M.J. 1994. *Emergence of Leadership in a Crisis: A Study of Group Escapes from Fire in Underground Coal Mines*. Pittsburgh, PA: U.S. Department of the Interior, Bureau of Mines IC 9385.

Kowalski-Trakofler, K.M., Vaught, C., Scharf, T. 2003. Judgment and decision-making under stress: an overview for emergency managers. *Int J Emergency Management* 1(3):278–289.

Kowalski-Trakofler, K.M., Vaught, C., Brnich, M.J. 2008. Expectations training for miners using self-contained self-rescuers in escapes from underground coal mines. *J Occup Environmental Hygiene* Oct 5(10): 671–677.

Kowalski-Trakofler, K.M., Alexander, D.A., Brnich, M.J, McWilliams, L. 2009a. *Underground Coal Mining Disasters and Fatalities—United States, 1900–1906*. In: MMWR, Vol. 57, No. 51/52, January 2, 2009, pp. 1,379–1,383.

Kowalski-Trakofler, K.M., Brnich, M.J., Vaught, C., Jansky, J.J. 2009b. A Study of First Moments in Underground Mine Emergency Response. Submitted to the *International J. Emergency Management*, in press.

Light, T.E., Herndon, R.C., Guley, A.R., Cook, G.L., Odum, MA., Bates, R.M., Schroeder, M.E., Campbell, C.D., Pruitt, M.E. 2007. *Report of Investigation: Fatal Underground Coal Mine Explosion, May 20, 2006, Darby No. 1 Mine, Kentucky Darby LLC, Holmes Mill, Harlan County, Kentucky, ID No. 15-18185*. Arlington, VA: U.S. Department of Labor, Mine Safety and Health Administration.

Margolis, K.A., Kowalski-Trakofler, K.M., Kingsley-Westerman, C.Y. 2009. *Refuge Chamber Expectations Training*. Pittsburgh, PA: U.S. Department of Health and Human Services, Centers for Disease Control and Prevention, National Institute for Occupational Safety and Health, DHHS, *in press*.

McKinney, R., Crocco, W., Stricklin, K.G., Murray, K.A., Blankenship, S.T., Davidson, R.D., Urosek, J.E., Stephan, C.R., Beiter, D.A. 2002. *Report of Investigation Fatal Underground Coal Mine Explosions, September 23, 2001, No. 5 Mine Jim Walter Resources, Inc., Brookwood, Tuscaloosa County, Alabama I.D. No. 01-01322*. Arlington, VA: U.S. Department of Labor, Mine Safety and Health Administration.

Mine Improvement and New Emergency Response Act of 2006 (MINER Act), Pub. L. No. 108-236 (S 2803) (June 15, 2006).

Mine Safety Technology and Training Commission 2006. Improving Mine Safety Technology and Training: Establishing U.S. global leadership. National Mining Association, Dec. 2006, 193 pp. http://www.coalminingsafety.org/documents/msttc_report.pdf. Accessed September 2009.

MSHA. 1989. *Informational Report of Investigation—Underground Coal Mine Explosion and Fire, Consol No. 9 Mine, Mountaineer Coal Company, Division of Consolidation Coal Co., Farmington, Marion County, West Virginia, November 20, 1968.* Arlington, VA: U.S. Department of Labor, Mine Safety and Health Administration.

MSHA. 1998a. *Historical Summary of Mine Disasters in the United States. Volume I: Coal Mines, 1810–1958.* Beckley, West Virginia: US Department of Labor, Mine Safety and Health Administration, National Mine Health and Safety Academy.

MSHA. 1998b. *Historical Summary of Mine Disasters in the United States, Volume II: Coal Mines, 1958–1998.* Beckley, West Virginia: US Department of Labor, Mine Safety and Health Administration, National Mine Health and Safety Academy.

MSHA. 2006a. Accident, illness and injury, and employment self-extracting files (part 50 data). http://www.msha.gov/stats/part50/p50y2k/p50y2k.htm. Accessed September 2009.

MSHA. 2006b. *A Comprehensive Guide to the Inspection, Care and Use of Self-Contained Self-Rescuers, DVD 013.* Beckley, West Virginia: US Department of Labor, Mine Safety and Health Administration, National Mine Health and Safety Academy.

Murray, K.A., Pogue, C.W., Stahlhut, R.W., Finnie, M.G., Webb, A.A., Burke, A.L., Beiter, D.A., Francart, W.J., Tjernlund, D.M., Waggett, J.N. 2007. *Report of Investigation Fatal Underground Coal Mine Fire, January 19, 2006, Aracoma Alma Mine #1, Aracoma Coal Company, Inc. Stollings, Logan County, West Virginia I.D. No. 46-08801.* Arlington, VA: U.S. Department of Labor, Mine Safety and Health Administration.

Nieto, A., Duerksen, A. 2008. The effects of mine safety legislation on mining technology in the USA. *Int. J Mining and Mineral Engineering* 1(1): 95–103.

Ounanian, D. 2007. *Refuge Alternatives in Underground Coal Mines, Phase I Final Report.* NIOSH Office of Mine Safety and Health Research Contract No. 200-2007-20276. Waltham, MA: Foster-Miller, Inc.

Vaught, C., Brnich, M.J., Mallett, L.G., Cole, H.P., Wiehagen, W.J., Conti, R.S., Kowalski, K.M., Litton, C.D. 2000. *Behavioral and Organizational Dimensions of Underground Mine Fires.* Pittsburgh, PA: U.S. Department of Health and Human Services, Centers for Disease Control and Prevention, National Institute for Occupational Safety and Health, DHHS (NIOSH) Publication No. 2000-126, IC 9450.

Vaught, C., Hall, E., Klein, K. 2009. *Harry's Hard Choices: Mine Refuge Chamber Training, Instructors Guide.* Pittsburgh, PA: U.S. Department of Health and Human Services, Centers for Disease Control and Prevention, National Institute for Occupational Safety and Health, DHHS (NIOSH) Publication No. 2009-122, IC 9511.

MSHA Mine Emergency Operations—Then and Now

John E. Urosek
MSHA, Pittsburgh, Pennsylvania, United States

Virgil F. Brown
MSHA, Beckley, West Virginia, United States

ABSTRACT: The United States Bureau of Mines, the predecessor agency to the Mine Safety and Health Administration (MSHA), was a founder and a leader in mine emergency response and mine rescue. MSHA continues this role and constantly evaluates and makes changes to Mine Emergency Operations (MEO) to improve its response to emergency situations. Some recent changes include the establishment of three mine rescue stations for MSHA's Coal Mine Safety and Health (CMS&H) at Price, Utah; Beckley, West Virginia; and Pittsburgh, Pennsylvania. Each station has MSHA's CMS&H Mine Emergency Unit (MEU) personnel report to and train at least 4 times each year. Mine rescue team trucks are also located at two of these sites. The Pittsburgh Station is the National Headquarters for Mine Emergency Operations. It is the home of the Mine Emergency Technology Team (METT). The METT provides expertise in gas analysis and ventilation. The robotics response team and seismic location system is also in Pittsburgh. The mine rescue team for MSHA's Metal and Nonmetal Mine Safety and Health (M&NM) is stationed at the National Mine Health and Safety Academy in Beckley and is under the direction of the administrator for M&NM. This paper will look back at the early years of mine emergency response, provide an overview of mine emergency operations since 1969, and discuss current mine emergency operations and procedures.

THE EARLY YEARS OF MINE EMERGENCY RESPONSE

The Bureau of Mines was established in 1910 in response to the growing number of fires and explosions occurring in underground coal mines. The Bureau of Mines was in the United States Department of the Interior. The headquarters was located in Washington, DC. The "Mining Experiment Station" was located in Pittsburgh. Dr. Joseph A. Holmes was the first Director of the Bureau of Mines. The Bureau of Mine's initial duties included the treatment of ores and other minerals, the use of explosives and electricity in mines, and the prevention of mine accidents.

Mine rescue was another concern for the newly formed Bureau of Mines. In October 1911, the first National Mine Safety Demonstration, the processor to the mine rescue contest, was held. About 15,000 persons, chiefly miners, attended the event. William Taft, the President of the United States at the time, attended the event. Figure 1 shows a picture of an early mine rescue contest in Pittsburgh at the former Forbes Field, which was the home of the Pittsburgh Baseball Club.

In addition to sponsoring mine rescue contests to improve emergency response, the Bureau of Miners obtained four railroad cars from the Pullman Company and converted them to movable safety and rescue stations. One of these railroad cars was stationed in Pittsburgh and the others toured the nation's

Figure 1. Mine rescue contest in Pittsburgh in 1911

coalfields training miners in mine rescue, first aid, and miner's safety. In 1915, Congress authorized the establishment of 7 mine safety stations and 10 mining experimental stations. Seven of the stations were equipped with mine rescue trucks. There were 10 railroad cars, of which six were new steel cars. These railroad cars did not always remain at their stations but moved from one mining camp to another. When a serious explosion or fire occurred at a mine, the car would be moved to the mine by a special locomotive or by the first available train. Figure 2 is a picture of one of the mine rescue railroad cars.

Figure 2.   Mine rescue railroad car

By 1919, the nation was divided into nine mining districts. District A was the Northern Appalachian District with headquarters in Pittsburgh. There were 3 mine rescue stations in the district. They were located in Pittsburgh; Wilkes-Barre, PA; and Norton, VA. Mine rescue trucks were located in Pittsburgh and Norton. Mine rescue railroad cars were located in Pittsburgh (No. 3) and Huntington, WV (No. 8). District B was the Southern Appalachian District with headquarters in Birmingham, AL. There were 2 mine rescue stations in the district, which were located in Birmingham and Knoxville, TN. A mine rescue truck was located in Knoxville. District C was the Eastern Interior District with headquarters in Vincennes, IN. There were 2 mine rescue stations in the district, which were located in Vincennes and Evansville, IN. A mine rescue truck was located in Vincennes. Mine rescue railroad cars were located in Terre Haute, IN (No. 9) and Des Moines, IA (No. 7). District D was the Lake Superior District with headquarters in Minneapolis, MN. A mine rescue railroad car was located in Ironwood, MI (No. 10). District E was the Southwestern District with headquarters in McAlester, OK. There was 1 mine rescue station in the district, which was located in McAlester. A mine rescue truck was also located in McAlester. A mine rescue railroad car was located in Pittsburg, KS (No. 4). District F was the Rocky Mountain District with headquarters in Denver, CO. A mine rescue railroad car was located in Raton, NM (No. 2). District G was the Northern Rocky Mountain District with headquarters in Salt Lake City, UT. A mine rescue railroad car was located in Rock Springs, WY (No. 11). District H was the Northern Pacific District with headquarters in Seattle, WA. There was 1 mine rescue station in the district, located in Seattle. A mine rescue railroad car was located in Butte, MT (No. 5). District I was

the Southern Pacific District with headquarters in Berkeley, CA. There was 1 mine rescue station in the district, which was located in Berkeley. A mine rescue truck was also located in Berkeley. A mine rescue railroad car was located in Reno, NV (No. 1).

In 1921, the number of experimental stations had increased to 13 with 12 field offices. In the following years, the number of explosions and fires began to decrease due, at least partially, to the work being done by the Bureau of Mines. By 1939, only 2 railroad cars were in service. The field headquarters remained in Pittsburgh. The Knoxville; Evansville; and Berkeley mine rescue stations were closed. New mine rescue stations were established at Juneau, AK; Denver; Duluth, MN; Jellico, TN; and Phoenix, AZ.

A typical mine rescue station in 1939 traditionally had the following equipment: 10 self contained breathing apparatus; 10 extra oxygen bottles; 2 extra reducing valves; extra parts for repairing apparatus; 5 compressed oxygen cylinders; 6 flame safety lamps; 12 flashlights, cap lamps, or hand lamps; 1 charging rack; 1 oxygen pump; 1,000-ft. lifeline; 2 canaries and a portable cage; 1 hoolamite carbon monoxide detector; 1 pyrotannic-acid carbon monoxide detector; 1 Orsat apparatus for measuring gases; 100 extra regenerators; and other miscellaneous equipment.

AN OVERVIEW OF MINE EMERGENCY OPERATIONS SINCE 1969

On November 20, 1968, an explosion occurred at the Consolidation Coal Company's No. 9 Mine in Farmington, WV. Of the 99 men that were underground at the time, only 21 escaped. Ten days later on November 30, 1968 mine rescuers were forced to seal the mine at the surface because of uncontrolled fires and repeated explosions. The tragic loss of 78 miners prompted Congress to pass the Federal Coal Mine Health and Safety Act of 1969. Figure 3 is a picture of smoke pouring from the Lewellen Shaft of the No. 9 Mine.

On December 3, 1968, the Director of the Bureau of Mines asked the National Academy of Engineering to conduct a study on post-mine disaster survival and rescue. On February 25, 1969, the National Academy of Engineering submitted a proposal to the Bureau of Mines. The proposal was accepted on March 5, 1969. The final report was released in March of 1970. The report contained various recommendations to the Bureau of Mines. These recommendations included developing and acquiring a seismic system capable of locating a trapped miner. In March 1970, Westinghouse Electric Corporation was awarded a contract to carry out the recommendations by the National Academy of Engineering.

Figure 3. Smoke pours from the Lewellen Shaft of the No. 9 Mine

In 1971, the MEO group was established in Pittsburgh at the Pittsburgh Technical Support Center (PTSC) of the Bureau of Mines. This group was charged with coordinating the activities of the Westinghouse Electric Corporation contract during the operation of the seismic location system, a borehole camera, a command system, a large hole drilling system and the associated logistics. The equipment was originally staged in Charleston, WV. MEO worked closely with the other technical personnel at the PTSC who provided expertise in ventilation and gas analysis. In 1972, a second staging area for MEO equipment was established in Salt Lake City.

In 1972, CMS&H of the Bureau of Mines established the MEU. The MEU members reported to the CMS&H District Managers in Pittsburgh and Morgantown. The MEU trained in mine rescue techniques and were apparatus wearers. There were eight MEU members from Pennsylvania and eight from West Virginia. The MEU supervisor and trainer were also from West Virginia. They trained at the Pittsburgh Station with the McCaa Breathing Apparatus.* Later, in 1974, they began using the Dräger BG-174A breathing apparatus. The MEU remained under the direction of CMS&H until 1999.

In 1973, an administrative procedure created the Mine Enforcement and Safety Administration (MESA) as a new agency in the Department of the Interior. MESA assumed the enforcement

---

* The use of any manufacturer's name in this paper does not constitute endorsement by the Mine Safety and Health Administration.

functions formerly carried out by the Bureau of Mines. Technical Support was also a part of the new agency. The Pittsburgh Safety and Health Technology Center (PSHTC) was formed. MEO became part of the PSHTC. In 1975, the staging area in Charleston was closed and the equipment transferred to a facility in Aliquippa, PA. The staging area in Salt Lake City was reduced to a quarterly maintenance cycle. The command vehicle from Salt Lake City was transferred from MEO to the Denver Technical Support Center.

AN OVERVIEW OF MINE EMERGENCY OPERATIONS SINCE 1977

When Congress passed the Federal Mine Safety and Health Act of 1977, MESA became the Mine Safety and Health Administration and was transferred from the Department of the Interior to the Department of Labor. The Salt Lake City facility was closed and the equipment, as well as the large diameter drilling equipment in Aliquippa, was transferred to other agencies. In 1991, MSHA ended the contract for MEO with Westinghouse Electric Corporation and reached an agreement with the Montgomery County Fire Department to maintain, house, and operate the seismic location system, the borehole camera, and other related equipment if needed. The command vehicle was moved to MSHA's National Mine Health and Safety Academy in Beckley. MEO personnel were transferred to other functions. The facility in Aliquippa was closed.

In 1992, MSHA re-established MEO in the Pittsburgh Safety and Health Technology Center. MSHA re-acquired the seismic location system, the borehole camera, and other related equipment from the Montgomery County Fire Department. The METT was formalized at about this time. It was composed of engineers, chemists, and various other experts in mine gases and ventilation from the Ventilation Division and the Physical and Toxic Agents Division of the PSHTC.

In 1994, the MEU began to use the Dräger BG-4 breathing apparatus. The Dräger 174a was eventually phased out. In 1994, a new division within CMS&H was established for mine emergencies. The control of the MEU was transferred from the District Managers in Pittsburgh and Morgantown to the new division located in Beckley. The mine rescue station was still maintained in Pittsburgh. It continued to be staffed by members from Pennsylvania and West Virginia. In 1998, a mine rescue station was established in Beckley. The unit members came from Virginia, Southern West Virginia, and Kentucky.

Figure 4.   Picture of the Sago Mine

In 1999, the control of the MEU was transferred from CMS&H to the MEO Group in the PSHTC of Technical Support. MEO and MEU were now in the same group with the same leader. Mine rescue stations were located in Pittsburgh and Beckley. A new command vehicle for the western United States was purchased in 1998 and located under the control of CMS&H, District 9 in Price. In 2005, CMS&H District 9 formed a small mine rescue station in Price. The team members were under the control of the District 9 Manager. Also in 2005, Program Evaluation and Information Resources (PEIR) began to respond as an MEO resource. They provided satellite communications and improved computer and communication technology.

In January 2006, an explosion occurred in a sealed area of the Sago Mine of the Wolf Run Mining Company resulting in 12 fatalities. Figure 4 is a picture of the Sago Mine. There were 29 miners underground at the time of the accident. One miner died shortly after the explosion. Sixteen miners escaped successfully. Twelve miners attempted to evacuate but were unsuccessful and barricaded themselves deeper in the mine. The mine rescue efforts eventually reached the barricade after 41 hours. Only one of these miners survived.

In January 2006, a fire occurred in the Aracoma Alma Mine No. 1 of the Aracoma Coal Company resulting in 2 fatalities. There were 29 miners underground at the time of the accident. Two miners working underground died as they tried to escape.

In May 2006, an explosion occurred in a sealed area of the Darby Mine No. 1 of Kentucky Darby LLC resulting in 5 fatalities. There were 6 miners underground at the time of the accident. Two miners died from the forces of the explosion. Three of the four other miners died as they tried to escape.

In June 2006, Congress passed the Mine Improvement and New Emergency Response Act

of 2006 (MINER Act). In 2007, the agency created two new positions. One was known as the Mine Emergency Response Coordinator (MERC). This position was designed to coordinate all of MSHA's emergency response capabilities. MEO/Coal MEU personnel reported directly to the MERC during a mine emergency. The second position was known as the Chief of Scientific Development. It was developed for discovering and bringing to fruition new technologies for MEO, mine rescue, mine recovery, and for interacting with research entities and other governmental agencies, such as NIOSH. New technologies included mine rescue communication systems, trapped miners location systems, drilling systems and robotics. In 2008, the Western Mine Rescue Station and members became the responsibility of the MERC.

CURRENT MINE EMERGENCY OPERATIONS EQUIPMENT

MSHA is currently prepared to respond to various types of mine emergency situations. This effort is coordinated by the MERC who is located at the National Mine Emergency Headquarters in Pittsburgh at the PSHTC. Equipment is located in one of the three mine rescue stations for CMS&H. The stations are located in Pittsburgh; Beckley; and Price.

1.   Command vehicles are located at the Price and Beckley Stations. These vehicles serve as mobile on-site offices. They are equipped with a fax and copy machine. Telephones, either satellite or land-line, can be installed. A new command vehicle is being constructed for the Pittsburgh Station. In addition to serving as a mobile office, this new command vehicle is to contain communications and computer equipment to enhance command center capabilities. Figure 5 is a picture of the command vehicle located at the Beckley Station.

Figure 5.   Command vehicle located at the Beckley Station

Figure 6.   Mobile monitoring vehicle located at the Pittsburgh Station

Figure 7.   Mobile laboratory located at the Pittsburgh Station

2.   The Ventilation Division of the PSHTC maintains portable infra-red gas detection equipment at the Pittsburgh Station. They have a monitoring vehicle at the Pittsburgh Station. They also provide expertise in gas analysis. They can analyze for methane, oxygen, carbon dioxide, and carbon monoxide. They can collect a gas sample from over 5,000 feet away. Figure 6 is a picture of the Mobile Monitoring Vehicle located at the Pittsburgh Station.

3.   The Physical and Toxic Agents Division of the PSHTC maintains chromatographic equipment and a Mobile Laboratory at the Pittsburgh Station and a one man satellite office in Denver. They provide expert chromatographic analysis of mine fire gases. CMS&H also maintains some chromatographic capabilities for fire gases at offices in Mt. Hope, WV. They can analyze for methane, ethane, acetylene, oxygen, nitrogen, hydrogen, carbon dioxide, carbon monoxide and other low level hydrocarbons. Figure 7 is a picture of the Mobile Laboratory located at the Pittsburgh Station.

4.   The Seismic Location System is located at the Pittsburgh Station. The system uses seven geophone sub-arrays, each having 7 geophones that are installed on the surface above the mine. The system is capable of detecting and locating miners to depths of 1,500 feet. A portable unit is also maintained at Pittsburgh. The portable is designed to be easily transported into a mine. Figure 8 is a picture of the Seismic Vehicle located at the Pittsburgh Station.

5.   A Borehole Camera Truck is located at the Pittsburgh Station. The camera system is maintained permissible. Each of the two cameras have a light source, a remote focus, a zoom lens, an iris, and have a 360 degree rotation. The camera can be used in a 6" cased borehole up to 1,500 feet deep. Figure 9 is a picture of the Borehole Camera Truck located at the Pittsburgh Station.

Figure 8.   Seismic vehicle located at the Pittsburgh Station

Figure 9.   Borehole camera truck located at the Pittsburgh Station

6.   MSHA's two robots are located in the Pittsburgh Station. One robot is permissible and can currently travel into the mine about 1,500 feet. This is a battery operated, modified Remotec Andros Wolverine model. The Wolverine Robot is equipped with three cameras and utilizes a remote display and has a continuous monitor for mine gases including methane, oxygen, and carbon monoxide. The second robot is made by iRobot. The iRobot is not permissible and can travel only a few hundred feet. Figure 10 is a

Figure 10. Wolverine robot located at the Pittsburgh Station

picture of the Wolverine Robot located at the Pittsburgh Station.

7. MSHA responds with other Federal Agencies in the event of other types of national disasters, such as earthquakes, hurricanes, etc. An urban search and rescue (USAR) trailer is maintained at the Beckley Station. This trailer is used to provide a base of operations including housing for first responders. The trailer can also be used as a mobile field office. Figure 11 is a picture of the USAR/Field Office Trailer located at the Beckley Station.

8. PEIR maintains the satellite and communications systems, one system at the Beckley Station and another in Denver. Satellite equipment is used to provide communications from the disaster site to MSHA headquarters. The satellite equipment is transported to the mine in a trailer that also serves as the computer based hub for the on-site command and control center. Figure 12 is a picture of the PEIR Trailer located at the Beckley Station.

9. MSHA also maintains other Emergency Response Equipment. Caches of infrared emergency equipment, gases for chromatographs, and other sampling equipment are located throughout the country at strategic locations including: Vacaville, CA; Price; Denver; Benton, IL; Birmingham; Green River, WY; and Norton. Additionally, first response handheld gas detection and sampling equipment is maintained in every CMS&H field office. This equipment includes an oxygen detector, a 0–100% methane detector and a 1–10,000 ppm carbon monoxide detector.

CURRENT MINE EMERGENCY UNIT EQUIPMENT

There are three mine rescue stations for CMS&H. Each station houses various mine rescue equipment.

There are approximately 30 members on the CMS&H MEU and they practice four days each quarter. The stations are located in Pittsburgh, Beckley, and Price. Trained mine emergency unit members from CMS&H report to and train at the individual stations. The M&NM mine rescue station is located at the National Mine Health and Safety Academy in Beckley.

Each CMS&H mine rescue station is equipped with at least the following equipment:

• Twelve Dräger BG-4—4-hour closed circuit breathing apparatus
• Twelve spare oxygen bottles
• Spare parts for the Dräger BG-4
• Two Dräger Universal Test Sets (Rz-25 testers)
• One Dräger Test-it 6100
• A six hour supply of compressed oxygen
• One Masterline Oxygen Booster Pump Model 7000A-2, MDM-4 or similar
• Two National Mine Service Air Mask Dryer DM6 Drymask or equivalent
• Six—18kg containers of Dragersorb 400 and one 50-pack filter mats
• One freezer w/ice forms for Dräger BG-4's
• Twelve Wheat cap lamps and charging rack
• Ten Industrial Scientific MX 6 Ibrid Gas Detectors for methane, (0–5%, 0–100%) oxygen

Figure 11. USAR/field office trailer located at the Beckley Station

Figure 12. PEIR trailer located at the Beckley Station

(0–30%), and carbon monoxide (0–9,999ppm) and one DS2 docking station

- One Dräger CMS multigas analyzer
- One MSA Evolution 5200 HD Thermal Imaging Unit or equivalent
- One Pro/pack Portable Multipurpose Foam System
- Twelve Motorola (HT1000 or HT750) or Kenwood TK-390 IS hand held radios and chargers
- One Con-space CSI 2100 unit with reel
- Twelve CSE SR-100 SCSR's
- Ten Ocenco M20 SCSR's
- Ten PMA-2008 pocket anemometers
- Magnahelic pressure gauges
- Sampling bags and pumps
- One F-100 Performance Fog Generator or similar
- Three Dräger Carevent Resuscitators for Dräger BG-4's
- One BioPak 240R 4-hour Breathing Apparatus with service/test kits
- One BioPak 240S 4-hour Breathing Apparatus with service/test kits
- One stretcher
- Six complete sets of fire turnout gear

Additional equipment is located at the Beckley Station. This additional equipment includes: airlift bags, a hydraulic Jaws of Life unit, Two Hughes Supply Company Model HSC RDR-O2 Radio Repeaters, one 3M Heat gun, one MSA Ultralite II pressure demand 1 hour air mask, twelve Level A Hazmat Suits, and one National Foam, Inc. NF HI-EX-4000 foam generator.

A mine rescue team truck is located at the Beckley Station. This truck contains the equipment needed by MSHA's MEU to respond to an emergency situation. This truck contains the necessary support equipment that is used to bench test the breathing apparatus prior to use, including an oxygen pump for refilling the oxygen bottles in the apparatus. The truck has its own on-board generator for power and a water supply for cleaning the breathing apparatus after use. Figure 13 is a picture of the mine rescue team truck located at the Beckley Station. A second rescue team truck is located in the Price Station. The second truck contains similar equipment to the one at the Beckley Station with the exception of an onboard water supply.

CURRENT OPERATIONAL PROTOCOL

MSHA has established mine emergency response guidelines. The MERC has certain responsibilities that are outlined in MSHA Headquarters Mine

Figure 13. Mine rescue team truck located at the Beckley Station

Emergency Response Guidelines Handbook. Promptly after learning of a mine accident where rescue and recovery operations are necessary, the district office will either notify the administrator's office who will contact the MERC or the district office will notify the MERC directly. The district or administrators office will provide preliminary information about the mine emergency.

Immediately upon learning of a mine emergency from the coal district or the administrator's office, the MERC, based on discussions, will notify the CMS&H MEU to deploy to the emergency. For M&NM, the Metal and Nonmetal District will notify the Administrator's Office and based on discussions, the administrator's office will notify the M&NM Mine Emergency Unit to deploy to the emergency.

Additionally, the MERC will contact PEIR who will make personnel available to provide support for the assembly and configuration of MSHA's satellite communications equipment. The District or Administrator's Office will consult directly with the MERC to determine the specific mine emergency equipment and/or resources that are needed, then prioritize the MEU and equipment requests for the MERC accordingly.

The CMS&H MEU members receive training regularly in donning and using breathing apparatus and serve as apparatus wearers during emergencies. The MEU can set up air sampling stations underground, brief and debrief mine rescue teams, monitor rescue team activities, travel with rescue teams, and observe general rescue and recovery procedures. Generally, eleven MEU members are deployed during a mine emergency to provide around-the-clock operations. Two members serve as supervisors on the surface working 12 hour shifts. These two members also work directly with the command center. Three members work each 8 hour shift with one member traveling underground with the exploring team, one with the underground back-up team, and one with the surface back-up team.

The MERC may notify the METT depending on the situation. The METT generally consists of employees of the PSHTC in the Ventilation Division, the Physical and Toxic Agents Division, and Mine Emergency Operations. The Ventilation Division provides expert gas analysis and interpretation of the results. The Physical and Toxic Agents Division operates the gas chromatographs. MEO operates seismic, camera, and command and control systems. The METT will be placed on alert and deployed to the emergency site by the MERC or Administrator's Office.

REFERENCES

Forbes, J.J., Owings, C.W., Miller, A.U., Griffith, F.E., Westfield, James, Jr., Quenon, E.E., and Williams, M.L. 1939. *Central Mine Rescue Stations.* Miners Circular 39. United States Department of the Interior, Bureau of Mines.

Tuchman, Robert J., and Brinkley, Ruth F. *A History of the Bureau of Mines Pittsburgh Research Center* Technical Publication. United States Department of the Interior, Bureau of Mines.

Powell, Fred Wilbur. 1922. *The Bureau of Mines, Its History, Activities, and Organizations.* Institute for Government Research.

Gates, R., Phillips, R., Urosek, J., Stephan, C., Stoltz, R., Swentosky, D., Harris, G., O'Donnell, J. Jr., and Dresch, R.. 2007. *Report of Fatal Underground Coal Mine Explosion—Sago Mine, Wolf Run Mining Company.* Mine Safety and Health Administration.

Murray, K., Pogue, C., Stahlhut, R., Finnie, M., Webb, A., Burke, A., Beiter, D., Francart, W., Tjernlund, D., and Waggett, J. 2007. *Report of Fatal Underground Coal Mine Fire—Aracoma Alma Mine, Aracoma Coal Company.* Mine Safety and Health Administration.

Light, T., Herndon, R., Guley, A., Jr., Cook, G. Sr., Odum, M., Bates, R., Schroeder, M., Campbell, C., and Pruitt, M. 2007. *Report of Fatal Underground Coal Mine Explosion—Darby Mine No. 1, Kentucky Darby LLC.* Mine Safety and Health Administration.

Coal and Rescue History, United States Mine Rescue Association. www.usmra.com/photos/united-states/powerpoint.htm Accessed September 2009.

History of MSHA, United States Mine Rescue Association www.usmra.com/photos/united-states/powerpoint.htm. Accessed September 2009.

Kravitz, Dr. J. H. *An Examination of Major Mine Disasters in the United States and a Historical Summary of MSHA's Mine Emergency Operation Program.* www.msha.gov/s&hinfo/techrpt/meo/majormin.pdf Accessed September 2009.

The Use of Self-Contained Self Rescuers (SCSRs) During Recent Mine Accidents in the U.S. and Subsequent Regulatory Improvements for SCSRs

Jeffery H. Kravitz
MSHA, Pittsburgh, Pennsylvania, United States

John H. Gibson
MSHA, Pittsburgh, Pennsylvania, United States

ABSTRACT: Several major mine accidents have occurred in the United States since the beginning of 2006. Self-Contained Self-Rescuers (SCSRs) were used by escaping miners in all events involving explosions and fires. Mine Safety and Health Administration (MSHA) and National Institute for Occupational Safety and Health (NIOSH) investigations regarding the testing of the SCSRs involved in the accidents provide insight, due to the thoroughness of the procedures and the results of the investigations. The subsequent regulatory improvements are seminal, and are required by MSHA's Emergency Mine Evacuation Final Rule and the Miner Act of 2006.

This paper gives a historical overview regarding the development of SCSRs in the United States and summarizes major accidents where SCSRs were used. The results of the MSHA and NIOSH investigations are discussed, and the new MSHA SCSR regulations are delineated. Recommendations for improvements to SCSRs are also discussed.

BACKGROUND

SCSRs were first introduced into mines in the U.S. in 1981. These were known as the first generation SCSRs. They were large and heavy, and several mine operators required miners to wear these units throughout their entire shifts underground. Many miners complained of back problems due to wearing these units. Some miners carried the units in pouches or other carrying cases, and did not take very good care of their SCSRs. Many problems were reported with some of these SCSRs due to neglect.

The first major mine disaster involving the use of SCSRs was the Wilberg Mine fire on December 19, 1984. The mine was located in Emery County, Utah, near Price, Utah. Twenty seven miners lost their lives during this incident, and only one miner on the affected section escaped using his SCSR. According to the final MSHA accident investigation report, "No apparent attempt was made by the miners in 5th right panel to obtain a SCSR after the first notification of the fire and prior to smoke arriving on the section." Also, after retrieving their SCSRs, some of the miners carried them for a distance before donning them. Had the miners immediately gone to the stored SCSRs when notified of the fire and donned the SCSRs, they would have greatly increased their chances to escape from the fire and exit the mine.

The MSHA investigators reviewed each miner's activities after they were warned of the fire. Their final report (Huntley, et al. 1984) found ``the actions of the victims in obtaining and using self-rescue devices indicate many were not sufficiently instructed to be considered adequately trained in the use of the self-rescue devices." Each of the 27 miners had a filter self-rescuer (FSR). Four miners wearing FSRs walked past stored SCSRs in their attempt to escape. They died from lack of oxygen. Three other miners attempted escape with only an FSR and were overcome by carbon monoxide. Six miners first attempted to use their FSRs and then switched to their SCSRs. Four apparently died due to improper donning, removing the mouthpiece, or switching from the FSR to the SCSR. One miner used three SCSRs and almost made it to fresh air; however, he removed his SCSR prematurely. The rest of the deceased miners had not attempted to use either the FSRs or SCSRs. According to the report, MSHA believed that better training, along with a better location of stored SCSRs could have resulted in a different outcome. Improper training of donning and transferring from one device to another, as well as the use of FSRs in such an environment, contributed to the severity of the disaster.

Based on the findings at the Wilberg mine, MSHA issued an Emergency Temporary Standard (ETS) in June 1987. The 1987 ETS required that all training in the use of SCSRs include complete donning procedures. This training was required for any person going underground in coal mines for the first time and as part of regularly scheduled annual refresher training required by 30 Code of Federal Regulations (CFR) Part 48.

Figure 1. The Wilberg Mine Disaster Memorial in Castledale, Utah

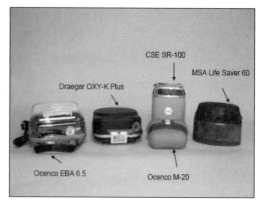

Figure 2. Approved SCSRs

Among the victims was Nannett Wheeler, the first woman to die in a Utah mine since women entered mining in 1973. Figure 1 is a picture of the Wilberg Mine Disaster Memorial in Castledale, Utah.

The MSHA investigation of the Wilberg Disaster indicated that miners were poorly trained in the use of their SCSRs. This outcome led to improved MSHA training regulations for SCSRs.

In another incident, during the escape from a fire at the Mathies Mine in 1990, seven of 18 miners removed their SCSR mouthpieces in order to talk or get more comfortable during the escape. Only seven miners donned their SCSRs at the first sign of smoke. One miner took his nose-clip off during the escape. Another miner claimed that he could not get enough oxygen from his SCSR (Kovac, Kravitz, et al. 1991). If any of these persons had encountered a toxic atmosphere at the point when they removed their protection, the miners may not have escaped successfully.

Also, in November 1998, during the escape from the Willow Creek Mine fire, the two miners that used SCSRs had difficulty starting the oxygen flow of their devices and removed the mouthpieces prior to reaching the main fresh airway. If the carbon monoxide in the mine atmosphere had been higher, the miners that removed their mouthpieces may not have escaped successfully. In situations where miners remove the mouthpiece prematurely, additional training will increase knowledge and reinforce that miners should continue to use the self-rescue device until escaping into fresh air (Kravitz, 1991).

APPROVED SCSRS

There are currently five models of SCSRs that are used in coal mines in the U.S. today that are approved by both MSHA and NIOSH (Figure 2). They are the Ocenco EBA 6.5, the Dräger OXY-K PLUS, the Dräger OXY-K PLUS S, the CSE

SR-100, and the Ocenco M-20. The OXY-K-PLUS and OXY-K-PLUS S are similar with the exception of having different opening mechanisms. All SCSRs except the M-20 are approved for 60 minutes duration. The M-20 is approved for 10 minutes. The MSA Life-Saver 60 was previously approved, but removed from service voluntarily by MSA due to lack of market interest and problems with their chlorate candle starter.

The Ocenco EBA 6.5 is a first generation SCSR, while all the others are second generation. Second generation SCSRs are also known as Belt-Wearable or Person-Wearable SCSRs (PW-SCSRs).

RECENT MAJOR MINE ACCIDENTS

Since 2006, there have been four major mine accidents requiring the use of SCSRs. These are: (1) the Sago Mine explosion, (2) the Aracoma Alma #1 Mine fire, (3) the Darby Mine explosion, and (4) the Crandall Canyon Mine Disaster.

Sago

The explosion at the Sago mine occurred on January 2, 2006, and originated behind a recently sealed area that was located near the mouth to the last section (2 Left) where 12 miners were working (Figure 3). The CSE SR-100 (Figures 4 and 5) was used at this mine. The miners on this section donned their SCSRs, but when four miners reported that they were having difficulty with their SCSRs, all 12 barricaded in the #3 face area and awaited rescue. The MSHA investigation concluded that eleven of these miners succumbed to carbon monoxide poisoning, and one miner was rescued. One outby miner was overcome by the explosion and gases and did not have a chance to don his SCSR. Thirteen other miners who were working on an outby section escaped.

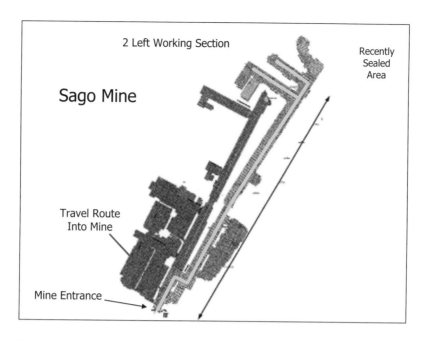

Figure 3.   Map of the Sago Mine

Only seven of the thirteen used their SCSRs during the escape. Twenty-four SCSRs were recovered and examined by MSHA and NIOSH.

MSHA and NIOSH conducted a joint investigation to: (1) inspect and catalog the condition of each SCSR used during the disaster, and (2) evaluate the life support performance of each apparatus. They performed the following: (1) measured the residual life support performance of each SCSR, (2) assessed the condition of the chemical beds, and (3) documented the condition of each component with a visual record (Figure 6).

The results from the SCSR investigation indicated that each SCSR had all critical parts intact and functioning. All SCSRs produced oxygen. The oxygen generating chemicals were used to varying degrees in each unit.

Aracoma Alma #1

A fire occurred in the Aracoma Alma #1 Mine on January 19, 2006 in the belt entry, about 3 miles inby the portal on 2 Section (Figure 7).

Ten miners escaped successfully wearing their CSE SR-100 SCSRs. Two miners took different paths and died due to carbon monoxide poisoning. The conclusions from the MSHA investigation were:

*   The fire occurred as a result of frictional heating when the longwall belt became misaligned

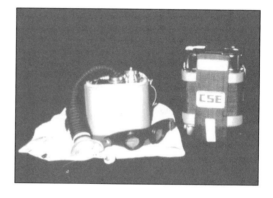

Figure 4.   The CSE SR-100 Deployed

in the 9 Headgate longwall belt takeup storage unit
*   This frictional heating ignited accumulated combustible materials and the belt
*   The lack of a fire suppression system allowed the fire to spread
*   The lack of water compromised fire fighting activities
*   The lack of separation between the No. 7 Belt entry and the primary escapeway for 2 Section allowed smoke and CO to inundate the primary escapeway

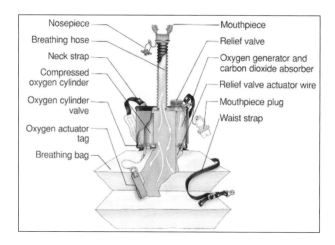

Figure 5.   The SR-100 illustrated

Exhibit #	Location	SN	MFR Date	Notes	Start-up Oxygen Activated?	Spent KO₂	Did the SCSR produce Oxygen?
57P	2nd Left Barricade	57517	12/97	Breathing tube has taken a set but is pliable and open. Cap missing from relief valve. Breathing bag has goggle impression or stain on bag.	YES	20%	YES

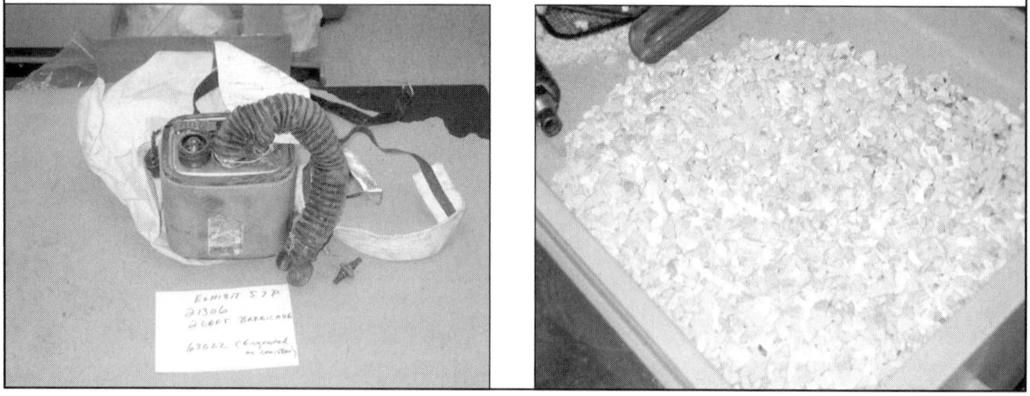

Figure 6.   Example of catalog and visual record (The material on the right is unused KO_2)

The results from the SCSR investigation showed that all escaping miners used a percentage of the oxygen producing chemicals which roughly correlated with the amount of time that they used their SCSRs. One recovered SCSR was found in water and could not be evaluated. The other SCSR from a victim was found to have approximately 85% of oxygen generating chemicals expended.

Darby Mine #1

An explosion occurred in the Darby Mine #1 on May 20, 2006. After arriving at the #3 return seal, miners began cutting a roof strap at the seal with an acetylene torch. At some point during this cutting operation, an explosive mixture of methane and air was ignited behind the seal, causing an explosion and killing the two miners instantaneously. Persons on

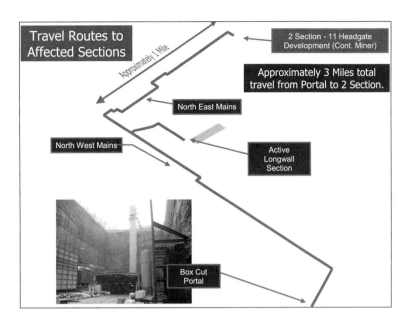

Figure 7.   Travel routes to affected sections at Aracoma Alma #1 Mine

Table 1.   Results from the Joint MSHA-NIOSH SCSR Investigation.

Exhibit No.	Locations of SCSR	Serial No.	MFR. Date	Dust Shield Cracked	Canister Dented	Notes	Start-up Oxygen Activated	Spent KO₂	Evidence of Oxygen Production
D-01	Outside	93609	07/02	Yes	Yes	Bottom bushing out of place (survivor)	Yes	45%	Yes
D-02	In Mine	84784	06/01	No	Yes	No comments	Yes	40%	Yes
D-03	In Mine	89962	01/02	Yes	No	Name on dust cover	Yes	75%	Yes
D-04	In Mine	105936	07/04	Yes	No	Goggles have smoke residue	Yes	15%	Yes
D-05	In Mine	84698	06/01	Yes	N/A	Unopened SCSR. Pulled freely from pouch, bottom moisture indicator dented into lid, unit open to atmosphere dents			

the surface observed dust and debris coming out the mine portals. Four other miners, working on another section inby the explosion attempted to escape using their CSE SR-100 SCSRs. They donned their SCSRs and used a mantrip to travel out the primary escapeway. The escape was blocked by a damaged overcast, and they resumed their escape on foot in heavy smoke. Three of these miners died due to Carbon Monoxide (CO) poisoning and the fourth made it to fresh air and was brought out by rescuers; however, he also had a significant amount of CO in his blood. The results from the joint MSHA-NIOSH SCSR investigation is shown in Table 1.

Dozer Entrapment

In 2007 a dozer operator was buried alive under 30–40 feet of coal while he was working on a clean coal pile (Music 2007). The accident occurred at Premier Elkhorn Coal's Burke Branch tipple in Dorton, KY while the dozer was working over a void created by a coal feeder. The hidden void collapsed, causing the dozer to be buried. The dozer cab was equipped with two SR-100 SCSRs. When the operator felt that he was running out of oxygen, he donned a SCSR and used part of a second SCSR. He was buried for about 2 ½ hours and was rescued uninjured.

Crandall Canyon

On August 6, 2007, six miners in the Crandall Canyon Mine were killed in a catastrophic coal outburst when pillars failed and violently ejected coal into the mine entries over a half-mile area. Ten days later, two mine employees and an MSHA inspector perished in a coal outburst during rescue efforts.

- The operator provided CSE SR-100, SCSRs for use as required by 30 CFR Part 75 requirements. The Emergency Response Plan (ERP) defined storage and reliability requirements for the units. At the time of the accident, all SCSR units were located in the areas stipulated in the ERP.
- Following the accident, five miners initiated a rescue attempt into the South Barrier section.
- Two miners used Dräger, 30-minute, self-contained breathing apparatus units (SCBAs), two used the CSE SR-100, and one did not don any type of unit.
- Two miners were trained to use the SCBAs as members of the mine fire brigade. Two other miners donned the SCSRs to cope with dust in the atmosphere. These miners stated that the SCSR units activated properly, and performed as expected without incident. One miner did not don any type of breathing apparatus. During their attempt to advance into the section, the miners retreated after encountering low levels of oxygen and adverse ground conditions. The lowest oxygen level detected was 16.0%.

RULEMAKING

MSHA promulgated several new regulations following the mine disasters in 2006 including the Emergency Mine Evacuation rule, and Congress passed the MINER ACT of 2006. Rulemaking included the following provisions:

- The Emergency Mine Evacuation rule and the MINER Act require that each miner have available two hours of breathable air.
- The Emergency Mine Evacuation rule includes a provision that an additional SCSR shall be stored at fixed underground work locations and at 30-minute travel time intervals along travel ways for miners without fixed work locations.
- Mine operators are required to provide additional SCSRs in both the primary and alternate escapeways at 30-minute travel time distances.
- Mine operators are required to provide additional SCSRs on mantrips.

- Each miner is required to have hands-on training in the complete donning and transferring from each type of SCSR device used at the mine *quarterly*, including: inserting the mouthpiece, and putting on the nose clip and goggles, and must be given comprehensive instruction on the use, care, and maintenance of all SCSRs used at the mine.
- Training should also emphasize the importance of:
  1. Recognizing when the SCSR is not functioning properly and demonstrating how to initiate and reinitiate the starting sequence (manual starting)
  2. Not removing the mouthpiece, even to communicate, until the miner reaches fresh air
  3. Proper use of the SCSR by controlling breathing and physical exertion
- Once each year, miners are required to have training on realistic SCSR training devices that produce heat and breathing resistance to airflow.
- Miners are also required to participate in annual expectations training, that includes donning and transferring SCSRs in smoke, simulated smoke, or an equivalent environment.

Also, in 2006, MSHA issued Program Policy Letter P06-V-10 which states that:

> The Emergency Response Plan (ERP) [as required by the MINER ACT] should contain a provision adopting the manufacturer's recommendations for SCSR maintenance, routine examinations, storage, and retirement. The ERP should also address SCSR performance by specifying a schedule for opening, initiating the breathing cycle, and establishing operational reliability for a representative number of SCSR units on an annual basis. Units at the end of their service life, if available, may be used for this purpose. The ERP should also provide for replacement of retired SCSRs with technologically advanced SCSRs as they become commercially available and are approved for use in mines.

New Technology

The MINER ACT includes a provision for "... introducing new self-rescuer technology, such as units with interchangeable air or oxygen cylinders not requiring doffing to replenish airflow and units with supplies of greater than 60 minutes, as they are approved by the Administration and become available on the market."

NIOSH and MSHA have sponsored several workshops to identify new SCSR technology, and NIOSH has funded a research contract with one contractor for the development of a "dockable" and "hybrid" SCSR. The dockable SCSR would achieve the MINER ACT goal of having an SCSR which does not require doffing, i.e., removal of the mouth-piece or nose clip. The manufacturer has submitted the final design for NIOSH approval.

Self-Contained Breathing Apparatus (SCBAs) which can be quickly filled, without removal, have been used in mines in Australia for years. The use of SCBAs with quick-fill stations has been approved for one coal mine in New Mexico, and other mine operators are considering use of this technology. SCSRs are worn by miners and are provided on man-trips. SCBA caches are located on working sections, and quick-fill stations are located at 30 minute travel intervals along escapeways. This new technology fulfills the requirements of the MINER ACT and the Mine Evacuation regulations. It also offers miners the use of a full face-piece respirator instead of just having a mouth-piece and nose-clip; this allows miners to communicate more readily. The SCBA also supplies cooler air, with very low breathing resistance (Figure 8).

Figure 8.   SCBA with Quick Fill System

use, care, and maintenance of SCSRs. Reliability has been improved by requiring mine operators to test a representative sample of their SCSRs annually.

New, improved SCSR technology is on the horizon, pursuant to requirements in the MINER ACT. MSHA and NIOSH will continually seek new technology and improvements for SCSRs. An SCBA Quick Fill System has been introduced, and has potential to be an effective alternative to a 100 percent SCSR deployment in coal mines, meeting MINER ACT requirements.

SUMMARY

Since SCSRs were introduced into mines in the U.S. in 1981, they have saved several lives; however, there have been several usage problems stemming from a lack of quality training, insufficient training, removal of the mouth-piece to talk, and other factors. New MSHA regulations have increased the availability of SCSRs for miners facing a mine escape. The increased frequency and improved quality of SCSR training in realistic scenarios has made the industry more aware of the proper procedures for the

REFERENCES

Huntley, et al. 1984. Final Report of Investigation, Wilberg Mine Fire

Kovac, J., Kravitz, J.H., et al. 1991. Investigation of Respirators Used at the Mathies Mine Fire, MSHA Special Report, February 1991

Kravitz, J.H. 1991. in MSHA Report of Investigation, Mine Fire, Willow Creek Mine

Music 2007, Appalachian News-Express, "Pike Miner Survives Being Buried Alive."

A Century of Bureau of Mines/NIOSH Explosives Research

Richard J. Mainiero
NIOSH, Pittsburgh, Pennsylvania, United States

Harry C. Verakis
MSHA, Triadelphia, West Virginia, United States

ABSTRACT: The U.S. Bureau of Mines (USBM) was created by an Act of Congress on July 1, 1910. The USBM strove to eliminate the use of black powder in underground coal mines and replace it with safer permissible explosives. The testing and approval of permissible explosives, blasting units, and blasting practices evolved and are now codified in the Code of Federal Regulations Title 30, Part 7, Subpart C and Parts 15 and 75. In subsequent years explosives and blasting safety research expanded from Pittsburgh, PA to include USBM labs in Minneapolis, MN, Denver, CO, Spokane, WA and College Park, MD. Research grew from studies of newer explosives and blasting technology for underground coal mines to include applications for all types of mining. During wartime the USBM applied it's expertise to studies and research that supported national defense. Following World War II, the USBM continued explosives research for industrial applications. Advances were made on vibration from blasting, evaluations of the state-of-art in safer explosives, electronic detonators and blasting devices, and research that supported changes in regulations for blasting in the mining industry.

INTRODUCTION

The history of explosives and blasting research at the U.S. Bureau of Mines (USBM) spans many years, many locations, and wide-ranging applications. This work has been documented in hundreds of publications over the years but there has been no publication that summarizes the totality of the USBM's explosives and blasting research. The authors' goal is to summarize this work.

U.S. BUREAU OF MINES EXPLOSIVES RESEARCH—THE EARLY YEARS

The early history of the U.S. Bureau of Mines (USBM) was closely related to the development of permissible explosives. In the USA, the first blasting-related explosion disaster occurred in the Coulterville coal mine in Illinois in 1883 and resulted in 10 fatalities. It was initiated by a blown-out black blasting powder shot and propagated by coal dust. The worst coal mine explosion disaster in the USA in the 1800s occurred in the Laurel Mine in Virginia on March 13, 1884 and resulted in 112 fatalities. The cause indicated for this disaster was the use of large quantities of black blasting powder to blast coal off-the-solid (Humphrey 1960).

Due to the disastrous coal mine explosions during the 1800s, actions were taken by various states in the USA to enact laws and requirements for mine safety improvements. In 1872, Illinois was the first state to enact a bituminous coal mine safety law. Over the next 45 years, 24 other states enacted mine safety laws. Observance of the laws fell short and often needed revisions because of the subsequent occurrence of coal mine explosion disasters. The view that coal dust without the addition of methane was explosive was not generally accepted by industry. Coal mine explosions continued to occur into the 1900s with more disastrous results.

During the ten-year period, 1901–1910, there were 111 major explosion disasters. A major disaster was defined as resulting in 5 or more fatalities. The 111 major explosion disasters resulted in a total of 3, 316 fatalities. This 10-year period was catastrophic for the US coal mining industry. The details of these disastrous explosions pointed to the use of black powder in shooting overburdened and overloaded blast holes. The flame from blasting the overburdened and overloaded holes was a primary ignition source that caused the explosion of accumulations of methane and coal dust. Blasting constituted about 44 percent of the 111 ignition sources in these disasters (Humphrey 1960). Although the major causes of coal mine explosions were known, particularly since studies were made by Britain and other European countries, many in the U.S. coal mining industry were not convinced. Black blasting powder and dynamite continued to be used and could not be made safe for blasting in underground coal mines. An obvious safety hazard is illustrated in Figure 1 which shows an early miner with an open light on his hat while forming a cartridge of black powder. Instances such as illustrated in Figure 1 continued on a daily basis. It took numerous coal mine explosion

**Figure 1.   Miner making black powder
cartridge for blasting coal**

of the 1900s, greatly aroused public attention. The attention brought about action through federal agencies to find means to control the occurrence of disastrous coal mine explosions and to reduce the loss of life and injuries. The study of explosives as a cause of mine explosions began in 1908 as part of the Technologic Branch in the U.S. Department of Interior's Geologic Survey (USGS). The Technologic Branch of the USGS had been created in 1907 to test fuels and structural materials but its mission expanded in 1908 to include the study of mine explosions in response to the series of disastrous coal mine explosions in December, 1907 (Powell 1922). An act of congress appropriated $150,000 "for the protection of lives of miners in the territories and in the District of Alaska, and for conducting investigations as to the causes of mine explosions with a view to increasing safety in mining." Work began with examinations of explosives, conducted chiefly at the Geological Survey's laboratory in Pittsburgh, PA. This work passed to the Bureau of Mines following its creation in 1910. The initial goal of this work was to replace the black powder used for blasting in underground coal mines with safer explosives. These were termed permissible explosives and they would produce flame of short duration with relatively low temperature.

disasters to force the changes needed for improvements in blasting safety.

The worst coal mine explosion disaster in the USA occurred on December 6, 1907 in the Monongah Number 6 and 8 coal mines in Monongah, West Virginia. The explosion resulted in a reported 362 fatalities (Humphrey 1960). However, a recent book indicates that close to 500 men and boys lost their lives in the Monongah coal mine explosion (McAteer 2007). Although the exact cause of the Monongah coal mine explosion was not determined, several theories suggested an electric arc, open lights, or a blown-out shot from blasting with black powder may have been the cause. The second worst coal mine explosion in the USA occurred thirteen days later, on December 19, 1907, at the Darr Mine, Jacobs Creek, Pennsylvania. When the explosion occurred, an awful rumbling was followed by a loud report and a concussion shook the nearby buildings. The underground coal mine explosion was felt within a radius of several miles. This explosion caused 239 fatalities; no miners escaped the explosion. The cause of this explosion was indicated as a blown-out shot or open lights (Humphrey 1960).

The increasing frequency of mine explosions and the tremendous loss of human lives, especially during the latter part of the 1800s and the first decade

Early explosives work by the Bureau of Mines covered a wide range of topics including:

*   Publishing information describing the characteristics and chemistry of various explosives (Hall 1912; Hall 1915)
*   Promoting the use of permissible explosives (Rutledge, Hall 1912)
*   Reporting on tests employed to evaluate the safety of explosives (Hall, Howell 1913)
*   Promoting the safe and effective use of explosives (Rutledge, Hall 1912; Hall, Howell 1913; Munroe, Hall 1911)
*   Testing of explosives for the U.S. Army and Navy departments prior to and during World War 1 (U.S. Bureau of Mines 1919)

Over the next half century, new and different types of explosives were developed and introduced for commercial use. Prior to 1981, the majority of permissible explosives were nitroglycerin-sensitized granular, semi-gel, or gelled-typed explosives (Verakis, Uraco 1992). However, water-gel explosives accounted for nearly 50 percent of the commercial market within a decade of first being approved in 1970. Water-gels generally consist of about 10 to 20 percent water and contain no nitroglycerin sensitizer. Subsequent to the introduction of water-gels, new emulsion-type explosives were approved and introduced for use in coal mines in 1981. These explosives were in the form of water-in-oil emulsions and utilized artificial microspheres made from glass or other material for sensitizers. The development of water-gel and emulsion explosives improved product safety and made it possible to encompass a wide range of mine blasting requirements.

PERMISSIBLE EXPLOSIVES

The use of permissible explosives has significantly reduced blasting accidents in coal mines. Enactment of legislation, such as the Federal Coal Mine Health and Safety Act of 1969 and enforcement of regulations has also improved blasting safety. In the 1980s, MSHA began a major revision of its regulations for blasting in coal mines. In 1988, MSHA published final rules for the approval and use of permissible explosives and blasting safety requirements in coal mines. The approval requirements for permissible explosives addressed testing and evaluation in which some test changes and additions were made based on technical studies (Verakis, Uraco 1992). The requirements included tests for sheathed explosives, which represented a new technology and permitted unconfined or "open shooting." (See Figure 2). The

sheathed explosive basically consists of a permissible explosive surrounded by an inert material. The inert material serves to suppress the flame from the detonation of the permissible explosive. The regulations also provided increased safety protection for miners by including qualifications for blasters, multiple-shot blasting and the use of sheathed explosives (Mainiero 1990).

Today, black blasting powder, which was the primary blasting agent a century ago, peaking in 1917 at about 125,645 metric tons (277 million pounds), is no longer permitted to be taken into underground coal mines and has been replaced by permissible explosives (ISEE 1998). However, the use of permissible explosives has declined significantly from peak sales in 1947 of about 60,328 metric tons (133 million pounds) to about 998 metric tons (2.2 million pounds) in 2007. The 998 metric tons (2.2 million pounds) of permissible explosives represents about 0.03 percent of the total amount of commercial explosives consumed in the United Sates in 2007 (USGS 2008). As a point of interest, about 12,247 metric tons (27 million pounds) of permissible explosives were consumed in 1912 versus about 12,247 metric tons (27 million pounds) consumed in 1988 when the steep decline began to occur in the use of permissible explosives in underground coal mines (ISEE 1998). Over the past several decades

Figure 2.   A miner and an explosives company technical representative preparing to break rock using the commercial version of the sheathed explosive charge, the Rock-Buster

continuous and longwall mining have become the dominant methods of extracting coal in underground coal mines; therefore, the demand for permissible explosives has declined.

USBM LABS

Explosives Research Lab (ERL)

The Bureau of Mines explosives program in Bruceton, PA made significant contributions to the war effort during World War II. In the summer of 1940 the National Defense Research Committee (NDRC) was looking for a location to set up a centralized laboratory for testing explosives. They wanted to place this lab in "…the hands of staff which was already familiar with the subject." (Kistiakowsky, Connor 1948). The USBM was selected to operate the laboratory because they had many years of experience testing commercial explosives at their facilities in Bruceton, PA.

Following acceptance by USBM the lab, designated the Explosives Research Laboratory (ERL), was set up in Bruceton in the fall of 1940. The USBM explosives scientists contributed generously to operation of the lab but their lack of experience with military explosives and their need to continue working on USBM activities created a need to recruit additional personnel. Initially, recruiting was only moderately successful due to the difficulties of recruiting personnel under the civil service system and only three scientists had been recruited by January, 1941. The NRDC looked for ways around this bottleneck and decided to contract out operation of the Bruceton ERL to Carnegie Institute of Technology (CIT). Under this arrangement research at ERL would continue under the direction of USBM scientists but CIT would recruit and pay new personnel.\

ERL grew quickly and by the end of the war 90% of the personnel in Bruceton were CIT employees. Growth of ERL activities may be judged from the number of personnel in Bruceton. In January, 1941 there were 5 ERL personnel and the number grew to 162 by 1945 (Kistiakowsky, Connor 1948). During the war, research at the lab included development of shaped charges for anti-armor applications (Walters, 2008), development and production of propellants, and work on cylindrical implosions. Following the war the number of personnel in Bruceton dropped significantly. CIT continued to operate ERL at a reduced level of effort, while USBM research and testing in Bruceton continued as before. During the 1960s CIT discontinued work in Bruceton as ERL was shut down.

Explosives research in Bruceton contributed in a significant way to the successful development of the atomic bomb. During 1943 scientists pursued two techniques for the creation of a nuclear chain reaction. A good chain reaction required that a critical mass of Uranium 235 or Plutonium be created quickly, in a time frame of milliseconds. If the critical mass was assembled too slowly the chain reaction would not release the desired amount of energy. In the gun method a subcritical mass of nuclear material would be fired into another subcritical mass thereby creating a critical mass. Another alternative was to collapse and compress a sphere of nuclear material to the point that it represented a critical mass, resulting in a chain reaction.

In 1943 the gun method for forming a critical mass was the first choice because gun technology was well developed and no one knew if a spherical implosion would work. At this time scientists envisioned gun type bombs employing both Uranium 235 and Plutonium. There was some uncertainty whether the technique would function for Plutonium since the timing of the critical mass assembly was more critical than that required for Uranium 235; if assembly of the critical mass of fissionable material didn't occur quickly enough the chain reaction would begin before the critical mass was in its optimum configuration.

As research progressed it became apparent that a gun type Plutonium bomb would require Plutonium of a purity previously not obtained. An implosion type bomb would function with a lower purity Plutonium and a smaller quantity would be required.

One member of the Los Alamos community, Dr. Neddermeyer, remained optimistic about the possibility of creating an implosion type bomb. In 1943 Dr. Neddermeyer visited ERL in Bruceton. At that time the lab was operated by Dr. Kistiakowsky of Harvard, a renowned explosives scientist. Experiments were conducted in which explosives were formed around steel pipe and initiated at multiple points. These shots demonstrated that a detonation could uniformly compress the pipe, thus demonstrating that an implosion-type atomic bomb was feasible (Hoddeson, Henriksen, Meade, Westfall 1993).

From this beginning, research continued in Bruceton and Los Alamos, resulting in sophisticated explosive lenses that made the implosion type bomb possible. An explosive lens was constructed of two explosives of different detonation velocities that would bend a detonation wave much the same way as a glass lens bends light. A new explosion formulation, Baritol, was developed in Bruceton to serve as the slower detonation velocity explosive in explosives lenses. This development was also fortuitous because the scientists learned during 1944 that Plutonium would not function in a gun type bomb.

Eastern Experimental Station/Applied Physics Lab

The Eastern Experimental Station in College Park, MD opened in 1937. The lab was set up to provide a facility where selected experiments could be conducted under the direct supervision of the USBM's headquarters staff. During and after World War II research into the physics of blasting was carried out here by USBM scientists. In 1942 air blast instrumentation was developed and used to study glass breakage by detonating cartridges of dynamite at various distances (Ireland 1942).

Following World War II the study of blasts in rock began. The research is well summarized by the following excerpt from the introduction of a paper published in 1949 (Obert 1949),

> This report describes the development of a dynamic strain gage and a companion amplifier and recording cameras—an apparatus designed to pick up and record the strain waves produced in rock by a nearby explosion. Also included in the report is a discussion of the relationships connecting the strain with the displacement, velocity, and acceleration for both plane and spherical waves?

The following year a group of USBM scientists from the Eastern Experimental Station used these strain gauges to measure the strain in rock surrounding a blast hole. This was a simulation of shooting oil and gas wells to increase permeability in rock. Following each shot the permeability of the rock surrounding the borehole was determined by pouring in water and timing how fast the borehole emptied.

Over the following two decades researchers at the USBM Eastern Experimental Station (later renamed the Applied Physics Laboratory) continued to develop instrumentation for measuring ground vibrations from blasting (Duvall 1964), study how detonating explosives in blastholes break rock (Atchison, Duvall, Pugliese 1964), evaluate the application of photography to blasting studies (Petkof, Atchison, Duvall 1961), and study damage to homes caused by blast vibrations (Nicholls, Johnson, Duvall 1971). This was the beginning of blasting science as we know it today. In the mid 1960s the blasting research at the USBM's facilities in College Park, Maryland was discontinued and the work was moved to other USBM laboratories.

Twin Cities Research Center (TCRC), 1959–1996

In 1959 the USBM opened a mining research center in Minneapolis, MN. In 1972 the lab was renamed the Twin Cities Research Center (TCRC). During the 1960s the lab began research on explosives cratering (Johnson, Fischer, 1963), factors affecting blasting performance (Dick, 1970), vibration produced by blasting operations, and effects of blast vibrations on the roof and ribs in underground mines (Snodgrass, Siskind 1974). This work was intended to develop a better understanding of the interaction between detonating explosives and rock so that more efficient blasting methods could be developed, as well as protecting homes around a mine from vibrations produced by a blast.

A study was conducted to combine data collected in TCRC research with data from earlier USBM studies to produce two landmark publications on the effects of blasting-related ground vibration and airblast on residential structures (Siskind, Stagg, Kopp, Dowding, 1980; Stachura, Siskind, Engler 1981). During the 1980s additional research was conducted to develop a better understanding of the effects of blast vibrations on structures (Stagg, Siskind, Stevens, Dowding 1984), evaluate techniques for measuring ground vibrations (Siskind, Stagg 1985), and study the effects of underground workings on the blast vibrations from surface mines (Siskind, Crum, Otterness, Kopp 1989).

Millisecond delay blasting was studied as a way to yield more stable highwalls (Stachura, Fletcher, and Peltier 1986), and a technique for controlling air blast and ground vibrations (Kopp, Siskind 1986).

In 1982 TCRC published the USBM Information Circular 8925, "Explosives and Blasting Procedures Manual." which provided industry with a comprehensive review of current safe, productive blasting practices (Dick, Fletcher, D'Andrea 1982). During its first year of publication, more than 5000 copies of IC8925 were distributed.

TCRC research and publications on airblast and ground vibrations from blasting rank as one of the top USBM contributions to the science of blasting safety. Data generated in the TCRC studies formed the basis for the Office of Surface Mining and Reclamation's (OSMRE) air blast and ground vibration limits published in their 1983 regulations (U.S. Government Printing Office, 1982; U.S. Government Printing Office, 1983). In addition, a majority of the U.S. jurisdictions base their regulations on TCRC data (Eltschlager, 2003; Seismic Regulation Subcommittee, 2000). The publications generally referenced are Bulletin 656 published in 1971 and Report of Investigation 8507 published in 1980 (Eltschlager, 2003; Seismic Regulation Subcommittee, 2000).

Figure 3. A USBM technician prepares to conduct a test that would eventually become Test 8(d) in the UN Test Manual. If the 45.3 kg (100 lb) of blasting agent loaded explodes when subjected to the fire it will not be approved for bulk transport.

USBM Pittsburgh Research Center (PRC) — NIOSH Pittsburgh Research Lab (PRL)

As mentioned earlier, PRC in Bruceton, PA has been involved in explosives and blasting research since the creation of the USBM in 1910. Evaluating permissible explosives and blasting practices was a significant part of the PRC's activities for the entire life of USBM. The early years and work on permissible explosives are described above. This section will discuss PRC's explosives research other than that conducted as a part of ERL and for permissible explosives safety.

Following closure of the Explosives Research Lab in the 1960s, PRC continued to conduct explosives research for the military and other government entities under Interagency Agreements. This work included studies of the effects of hypervelocity projectiles on metallic target plates, a study of the factors affecting the energy transferred from detonating explosives to coupled metal plates, a study of detonation initiation phenomena in liquid explosives, the destruction of earthen tunnels used by the Viet Cong (Tuchman, Brinkley 1990), the development of a test to evaluate the explosive reactivity of solid wastes (Bajpayee, Mainiero 1988), and others.

In 1967 PRC's Research Director, Dr. Robert Van Dolah, was awarded the Nitro-Nobel Gold Medal for the theory he developed to explain the accidental initiation of liquid explosives (Person 1999; Tuchman, Brinkley 1990). In 1972 PRC published guidelines for the safe use of ammonium nitrate/fuel oil (ANFO) blasting agents (Damon, Mason, Hanna, Forshey 1972). Beginning about 1980 Richard W. Watson, Research Supervisor for PRC's Explosives Section, participated in a United Nations effort to develop an international classification system for explosives and other hazardous materials. The outcome of this work was the development of the U.N. Recommendations on the Transportation of Dangerous Goods and a manual containing procedures and tests to be used in evaluating the safety of new explosives and explosives devices, and assign them to a hazard class and division.

This work involved close work with the U.S. Department of Transportation (USDOT), the agency tasked with coordinating U.S. activities in this international effort (See Figure 3). These close ties lead to PRC's designation as one of two U.S. laboratories authorized to test explosives pursuant to classification by USDOT during the 1980s. This activity continued until 1997 when it was discontinued following closure of the USBM and PRC's transfer to the National Institute for Occupational Safety and Health (NIOSH).

From 1997 to 2005 PRL conducted a research program looking at all aspects of toxic gases produced by blasting. The effort began with the development of a technique for measuring toxic gases produced by blasting agents. The research studied factors controlling the quantity of toxic gases produced and resulted in recommendations surface mines could employ to minimize production of nitrogen oxide by blasts (Sapko, Mainiero, Rowland, Zlochower 2002). Later the work expanded to include toxic fumes produced by high explosives. This work was initiated by concerns that carbon monoxide could migrate from a blast into nearby homes or other confined spaces (Harris 2005). Due to staffing and budget shortfalls, the NIOSH explosives program in Pittsburgh was discontinued in 2005 (Sharpe 2009).

The problem of air blast and noise produced by explosives research at the USBM's Bruceton facility had always been a concern. This became especially true during World War II when the explosives research effort expanded significantly with the opening of ERL. Initially residents of neighboring communities were tolerant of the noise because the research was in support of national defense. However, as the population of the surrounding area increased the public became less tolerant of the noise. During the early 1960s a study was initiated to determine how the noise from research at PRC might be reduced. This study led to enclosing the bombproofs, closing the end of the permissible gallery tube, and monitoring weather reports. Research and testing would be halted on days when a weather inversion would reflect noise down on the neighbors (Van Dolah, Gibson, Hanna 1964). Limitations on explosives research at PRC led to the relocation of large scale explosives research to the USBM's Lake Lynn Laboratory following its opening in 1982.

USBM Spokane Research Center (SRC) — NIOSH Spokane Research Lab (SRL)

The USBM's Spokane Research Center (SRC) began research on blasting safety relatively recently with one exception. During the 1960s researchers at SRC needed to measure differential movement within a borehole as a way to understand the movement of rock formations in and around ore deposits before, during, and after mining operations. The plan was to install an anchor at one end of a drill hole and measure the movement of wire extending from that point to the collar. Wire was threaded through a bolt from the collar to an anchor at the other end of the borehole. Standard expansion-type anchors could not be used because rotating the bolt to tighten the anchor would twist and tangle the wire. To alleviate this problem, SRC developed an explosive-expansion center-hole anchor which did not require turning of the bolt; an explosive

charge in the anchor would expand the anchor, locking it firmly in place (Parsons, Osen 1963).

In 2006 the Spokane Research Lab began a 5-year research effort to develop techniques for evaluating damage to a mine's roof and ribs caused by the blasting process and develop blasting techniques that would produce more stable roof and ribs. The program recognized that fall of ground was a significant hazard in underground mines and a leading cause of fatalities. This hazard is increased if the standard drill and blast process leaves the roof and ribs significantly fractured and unstable. The goal of the program is the perfection of perimeter blasting techniques that remove the desired rock but do not produce fractures into the remaining rock (Johnson, Hustrulid, Iverson 2009). The program continues as of the writing of this paper.

CONCLUSION

For the last century the USBM and NIOSH have contributed to explosives and blasting science far more than would be expected for an organization of its size. Prior to World War II the USBM was considered the preeminent explosives research organization, which led to it selection to head up the Department of Defense Explosives Research Lab where explosives research was conducted for application to weapons systems and development of explosive components for the first implosion type atomic bomb. The USBM's reputation continued long after World War II. During the mid sixties the military explosives work was moved to other facilities which could accommodate large-scale explosives testing. The USBM, and later NIOSH, have continued with mining-related explosives and blasting research to the present time.

REFERENCES

Atchison, T.C., Duvall, W.I., and Pugliese, J.M. 1964. *Effect of Decoupling on Explosion-Generated Strain Pulses in Rock.* Report of Investigations 6333. U.S. Bureau of Mines.

Bajpayee, T.S., and Mainiero, R.J. 1988. *Methods of Evaluating Explosive Reactivity of Explosive Contaminated Solid Waste Substances.* In Proceedings of the 4th Mini-Symposium on Explosives and Blasting Research, Feb., pp. 137–145.

Damon, G. H., Mason, C. M., Hanna, N. E., and Forshey, D. R. 1972. *Safety Recommendations for Ammonium Nitrate-Based Blasting Agents.* Report of Investigations 8746. U.S. Bureau of Mines.

Dick, R.A. 1970. *Effects of Type of Cut, Delay, and Explosive on Underground Blasting in Frozen Gravel.* Report of Investigations 7356. U.S. Bureau of Mines.

Dick, R.A., Fletcher, L.R., D'Andrea, D.V. 1983. *Explosives and Blasting Procedures Manual,* Information Circular 8925. U.S. Bureau of Mines.

Duvall, W.I. 1964. *Design Requirements for Instrumentation to Record Vibrations Produced by Blasting.* Report of Investigations 5487, U.S. Bureau of Mines.

Eltschlager, K.K. 1983. *Blast Vibration and Seismograph Section 2002.* Proceedings of the 29th Annual Conference on Explosives and Blasting Technique. International Society of Explosives Engineers. Nashville, TN. February 2–5, 2003.

Hall, C., and Howell, S.P. 1913. *Tests of Permissible Explosives.* Bulletin 66. U.S. Bureau of Mines.

Hall, C., and Howell, S.P. 1915. *A Primer on Explosives for Metal Miners and Quarrymen.* Bulletin 80. U.S. Bureau of Mines.

Hall, C., Snelling, W.O., Howell, S.P., and Burrell, G.A. 1912. *Investigations of Explosives Used in Coal Mines.* Bulletin 15. U.S. Bureau of Mines.

Harris, M.L., Sapko, M.J., Mainiero, R. J. 2005. *Field Studies of Carbon Monoxide Migration from Blasting.* Proceedings of the 31st Annual Conference on Explosives and Blasting Technique. International Society of Explosives Engineers. Orlando, FL. February 6–9, 2005.

Hoddeson, L., Henriksen, P.W., Meade, R.A., and Westfall, C. 1993. *Critical Assembly: A Technical History of Los Alamos during the Oppenheimer Years, 1943–1945.* University of Cambridge, United Kingdom.

Humphrey, H.B. 1960. *Historical Summary of Coal-Mine Explosions in the United States, 1810–1958.* Bulletin 586. U.S. Bureau of Mines.

International Society of Explosives Engineers (ISEE). 1998. *Blasters' Handbook.* 17th Edition, Cleveland, OH.

Ireland, A.T. 1942. *Design of air-blast meter and calibrating equipment.* Technical Paper 635. U.S. Bureau of Mines.

Johnson, J.B., and Fischer, R.L. 1963. *Effects of Mechanical Properties of Material on Cratering: A Laboratory Study.* Report of Investigations 6188, U.S. Bureau of Mines.

Johnson, J.C., Hustrulid, W.A. and Iverson, S.R. 2009. *Extent of Damage Associated with the Passage of the Compressive Stress Wave Generated by Blasting.* Presented at the 43rd U.S. Rock Mechanics Symposium, American Rock Mechanics Association. June 28–July 1. Ashville, NC.

Kistiakowsky, G.B., and Connor, R. 1948. *Chemistry: A History of the Chemistry Components of the National Defense Research Committee 1940–1946. Chapter III: The History of Division 8 Central Laboratories.* Little, Brown and Company. Boston. pp. 26–29.

Kopp, J.W. and Siskind, D.E. 1986. *Effects of Millisecond-Delay Intervals on Vibration and Airblast from Surface Coal Mine Blasting.* Report of Investigations 9026. U.S. Bureau of Mines.

Mainiero, R.J., Santis, L.D. 1999. *Use of Sheathed Explosive Charges on Longwalls.* Report of Investigations 9294. U.S. Bureau of Mines.

McAteer, D. 2007. *Monongah: The Tragic Story of the 1907 Monongah Mine Disaster.* West Virginia University Press. 332 pp.

Munroe, C.E., and Hall, C. 1911. *A Primer on Explosives for Coal Miners.* Bulletin 17. U.S. Bureau of Mines.

Nicholls, H.R., Johnson, C.F., and Duvall W.I. 1971. *Blasting Vibrations and their Effects on Structures.* Bulletin 656. U.S. Bureau of Mines.

Obert, L.O., and Duval, W.I. 1949. *Gage and Recording Equipment for Measuring Dynamic Strain in Rock.* Report of Investigation 4581. U.S. Bureau of Mines.

Parsons, E.W. and Osen, L. 1963. *Explosive-Expansion Center-Hole Anchor.* Report of Investigations 6704. U.S. Bureau of Mines.

Petkof, B., Atchison, T.C, and Duvall, W.I. 1961. *Photographic Observation of Quarry Blasting.* Report of Investigations 5849. U.S. Bureau of Mines.

Powell, F.W. 1922. *The Bureau of Mines Its History, Activities, and Organizations, Service Monographs of the United States Government No. 3.* The Institute for Government Research, Washington, DC, D. Appleton and Company, New York.

Rutledge, J.J., and Hall, C. 1912. *Use of Permissible Explosives.* Bulletin 10. U.S. Bureau of Mines.

Sapko, M.J., Rowland J.H., Mainiero, R.J., Zlochower I. A. 2002. *Chemical and Physical Factors that Influence N0x Production During Blasting: Exploratory Study.* Proceedings of the 28th Annual Conference on Explosives and Blasting Technique. International Society of Explosives Engineers. Las Vegas, NV. February 10–13. pp. 317–330.

Seismic Regulation Subcommittee. *Summary Report of the Seismic Regulation Subcommittee.* The Journal of Explosives Engineering. International Society of Explosives Engineers. Vol. 17. No 5. September/October 2000.

Sharpe, D. 2009. NIOSH Research Program Undergoing Change. Sharpe's Point, July 30, 2009.

Siskind, D.E., Crum, S.V., Otterness, R.E., and Kopp, J.W. 1989. *Comparative Study of Blasting Vibrations from Indiana Surface coal Mines.* Report of Investigations 9226. U.S. Bureau of Mines.

Siskind, D.E., and Stagg, M.S. 1985. *Blast Vibration Measurements Near and on Structure Foundations.* Report of Investigations 8969. U.S. Bureau of Mines.

Siskind, D.E., Stagg, M.S., Kopp, J.W., and Dowding, C.H. 1980. *Structure Response and Damage Produced by Ground Vibration from Surface Mine Blasting.* Report of Investigations 8507. U.S. Bureau of Mines.

Snodgrass, J.J., and Siskind, D.E. 1974. *Vibrations from Underground Blasting.* Report of Investigations 7937, U.S. Bureau of Mines.

Stachura, V.J., Fletcher, L.R., and Peltier, M.A. 1986. *Delayed Blasting Test to Improve Highwall Stability-A Final Report.* Report of Investigations 9008. U.S. Bureau of Mines.

Stachura, V.J., Siskind, D.E., and Engler, A.J. 1981. *Airblast Instrumentation and Measurement Techniques for Surface Mine Blasting.* Report of Investigations 8508, U.S. Bureau of Mines.

Stagg, M.S., Siskind, D.E., Stevens, M.G., and Dowding, C.H. 1984. *Effects of Repeated Blasting on a Wood-Frame House.* Report of Investigations 8896. U.S. Bureau of Mines.

Tuchman, R.J. and Brinkley, R.F., 1990. *A History of the Bureau of Mines Pittsburgh Research Center.* U.S. Bureau of Mines, January.

U.S. Geologic Survey (USGS). 2008. Explosives. *Minerals Yearbook-September.*

U.S. Bureau of Mines. 1919. *Explosives and Miscellaneous Investigations.* Bulletin 178D. U.S. Bureau of Mines.

U.S. Government Printing Office. *Part V, Department of the Interior, Office of Surface Mining and Reclamation and Enforcement, Permanent Regulatory Program; Use of Explosives and Training, Examination and Certification of Blasters*; Proposed Regulations. Federal Register, Wednesday, May 24, 1982, pp. 12,760–12,784.

U.S. Government Printing Office. *Part III, Department of the Interior, Office of Surface Mining Reclamation and Enforcement, Surface Coal Mining and Reclamation Operations; Initial and Permanent Regulatory Programs; Use of Explosives.* Federal Register. March 8, 1983. pp. 9,788–9,811.

Van Dolah, R.W., Gibson, F.C., and Hanna, N.E. 1964. *Abatement of Noise from Explosives Testing.* Report of Investigations 6351. U.S. Bureau of Mines.

Verakis, H.C., and Uraco, J.L. 1992. *The approval and use of explosives in coal mines.* In Proceedings of the 18th Annual Conference on Explosives and Blasting Technique, International Society of Explosives Engineers. Orlando, FL. January 19–23. 9 pp.

Verakis, H.C. and Lobb, T. 2009, *Mine Blasting Safety: A Century in Review.* In Proceedings of the 35th Annual Conference on Explosives and Blasting Technique, International Society of Explosives Engineers. Denver. CO. February 8–11. 10 pp.

Walters, W. 2008. *A brief history of shaped charges.* In Proceedings of the 24th International Symposium on Ballistics, vol. 1. New Orleans, LA, September 22–26, pp. 3–10.

A Summary of U.S. Mine Fire Research

Alex C. Smith

NIOSH, Pittsburgh, Pennsylvania, United States

Edward D. Thimons

NIOSH, Pittsburgh, Pennsylvania, United States

ABSTRACT: Since 1910, the U.S. Bureau of Mines and NIOSH has conducted research to eliminate fires in underground mines. Early studies concentrated on the investigation of the causes of mine fires and recommendations for preventing fires, spontaneous combustion, and mine fire rescue, all of which served to improve safety in mines. In the 1950s, research characterized the flammability of gases, dusts, and vapors. The 1960s brought an emphasis on mine fire prevention, the hazards of combustible materials used in mines, and mine fire extinguishment methods. Results from these studies formed the basis for fire prevention regulations contained in the 1969 Coal Mine Health and Safety Act. The 1969 Act also greatly expanded the Bureau of Mines' role in coal and metal/nonmetal mine fire research. This led to a significant reduction in the severity and incidence of mine fires. The MINER ACT of 2006 placed a renewed emphasis on research in the areas of mine fire prevention, detection, control, and extinguishment. This paper summarizes the highlights of the Bureau of Mines and NIOSH fire research program since 1910 and describes current research to reduce or eliminate the hazards of fires in mines.

INTRODUCTION

Mine fires have always been a serious hazard in mining. One of the contributing factors in the call for the establishment of the U.S. Bureau of Mines was the Cherry Mine fire in Illinois in 1909 (Keenan 1963). This disaster took the lives of 259 miners. In the years preceding the Bureau of Mines, there were an average of about 500 fires per year. Fires continue to be a serious hazard to the health and safety of our Nation's mining workforce.

From 1990 through 2001, there were 1,060 fires, resulting in 560 injuries and 6 fatalities at U.S. mine operations (DeRosa 2004). As recently as 2006, a fire in an underground coal mine in West Virginia claimed two lives. From the Bureau of Mines inception in 1910, through today under NIOSH, the United States has conducted a rigorous research program to reduce the fire hazards in mining.

Through the years, the emphasis has focused on making mines safer through improved fire prevention, detection, control, and extinguishment methods. There have been many tangential benefits from this research, including improvements in fire safety in the transportation, defense, and space industries, to name a few. This paper will describe the progress made in mine fire safety through the history of the Bureau of Mines and NIOSH.

THE EARLY YEARS

The genesis of the U.S. Bureau of Mines is well documented in other papers and books (Kirk 1994).

Briefly, the Bureau's scope included "investigations into the methods of mining, especially in relation to the safety of miners, and the appliances best adapted to prevent accidents; the possible improvement of the conditions under which mining operations are carried on; the treatment of ores and other mineral substances; the use of explosives and electricity; the prevention of accidents; and other inquiries and technological investigations pertinent to said industries." (U.S. Public Law 179). Initially, the Bureau conducted a country-wide study of mine fires and mine fire hazards, including mine accident investigations, to find the cause of fires, means to prevent fires, and to determine the oxidation rate of coal in mine air. Another early interest in fire research was in the spontaneous heating during coal storage. Large amounts of coal were stockpiled by the government for heating of government buildings and powering naval boats. Though largely for economic reasons, this research laid the groundwork for spontaneous combustion prevention in underground coal mines.

The early investigations of the causes of coal mine fires showed the importance of proper equipment and methods for preventing and fighting coal mine fires. The first causes were identified as the careless handling of open lights, short circuits in electrical equipment, improper fuel storage practices, the use of long-flame explosives, the lighting of accumulations of gas, and spontaneous heating. Note the open flame cap lamp on the miner in Figure 1. Early recommendations included the use of permissible explosives, the use of electrical lamps

Figure 1. Miner with open flame cap lamp for illumination

for illumination, better ventilation in the working areas, the guarding of trolley wires, fireproofing of shaft stations, and the use of wire rope to replace hemp rope (Rice 1912). Prior to 1941, the Federal government was prohibited from inspecting mines or enforcing recommendations. After the Cherry Mine disaster, the state of Illinois drafted the most complete code for fire protection in mining at the time. This code included the need for adequate fire-fighting equipment, including hose lines, water tanks and hydrants, portable extinguishers, telephone and alarm apparatus, breathing apparatus, ventilation fans, and mine fire escape plans.

A critical component after a mine fire or explosion in the early 1900s was evaluating the mine atmosphere in order to re-enter safely for rescue of miners and for post-accident investigation. Gas analysis at the time was rudimentary and facilities for analyzing gas samples were far removed from mine sites. Research was completed in 1913 on a gas detector for determining inflammable gases (Burrell and Siebert 1913). By 1921, The Bureau of Mines equipped railroad cars, shown in Figure 2, with the latest technology in mine rescue apparatus, first aid equipment, and gas analysis equipment and stationed them in 10 mining districts throughout the country. These cars were used for first aid and mine rescue training during normal circumstances, and immediately deployed to the mines during emergencies.

By 1921, the Bureau's role in fires was still primarily as an accident investigations branch with no dedicated fire research program. As a result of World War I, several new areas of interest emerged. A program on metal dust fires studied the fire hazards of shipping metal dusts, including aluminum and zinc. The use of gas masks, developed in large part by the Bureau for the Army during the war (Farrow 1920), was examined for use in fighting fires. Of course, the filters removed smoke particles, but did not remove carbon monoxide or provide oxygen. This led to research on carbon monoxide filters and to the development of carbon monoxide detectors. It was found that carbon tetrachloride and foamite fire extinguishers, typical of the time, produced hydrochloric acid and phosgene (Fieldner and Katz 1921). Studies continued on spontaneous combustion during coal storage (Hood 1922). It was determined that coal characteristics such as coal type, moisture content, and pile size contributed to the spontaneous combustion hazard. Other research findings during the 1920s included the determination that rock dust could readily extinguish a coal fire when the miner was able to fight the fire directly (Howarth and Greenwald 1927), and that a hot surface could ignite methane (Leitch et al. 1925).

Flammability Laboratory

In 1924, a cooperative arrangement was started between the Safety in Mines Research Board of Great Britain and the Bureau of Mines to deal with the prevention or abatement of accidents in mines. Part of this program included the determination of the limits of flammability and explosibilty of gases and dusts encountered in mines and the mineral industries. Dr. Coward of England and Dr. G.W. Jones of the

Figure 2. Bureau of Mines mine rescue and training car

Bureau of Mines led this program. Dr. Coward stated that "this knowledge is of fundamental importance in the study and prevention of mine fires and explosions." In 1929, a flammability laboratory was set up in Pittsburgh, PA to determine the inflammable limits of methane and of the distillation products of coal in air. A gas flammability apparatus is shown in Figure 3. The laboratory also studied other natural and manufactured gases used as anesthetics, refrigerants, fumigants, insecticides, and solvents. Initial studies were completed on ethylene, ethyl ether, ethylene dichloride, vinyl chloride, carbon disulfide, pentane, benzol, acetone, furfural, and ethanol for the purpose of preventing gas fires and explosions in the metallurgical, petroleum, gas-manufacturing, and related industries. By 1952, the laboratory had completed studies of 155 dusts and gases. In 1952, a comprehensive report of all data was published (Coward and Jones 1952).

The 1930s

In 1931, there were 21 fires and 1 fatality due to fires. Overall, there were 1430 mining fatalities. With the Great Depression affecting the whole country, and the government not immune, Director Turner did his best to fend off budget cuts. Director Turner, in his 1932 report to the Secretary of Commerce, compared the 1931 statistics to the annual average of 2409 fatalities in the 25 years prior, and stated "Various factors are undoubtedly responsible for the excellent safety record in 1931, but doubtless the activities of the Bureau of Mines had a vital influence." (Turner 1932) In 1934, the Bureau's budget was cut to one-half of the 1928 funding level, and coal mine fire and coal dust explosion research was limited to just a few explosion demonstrations.

In 1936, a new laboratory investigation was started to study the causes, behavior, and control of mine fires. The Bureau began collecting extensive data on the atmospheric composition of gases during mine fires. The Bureau's gas laboratory is shown in Figure 4. A chemical method for locating fires in mines based on gas data (Ash and Felegy 1948), and a study of the mechanism of flame propagation in mine entries was initiated (Godbert and Greenwald 1936). Improvements were also made in the use of ventilation to eliminate gas accumulation hazards.

Mine Investigations and Inspections

Prior to this time, the Federal government had jurisdiction to investigate mine accidents only for the purpose of determining the cause and means to prevent the accidents, and to assist in any recovery operations. In 1931, the Director of the Bureau of Mines,

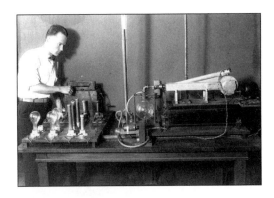

Figure 3.   A gas flammability apparatus

Scott Turner, stated in his Annual Report to the Secretary of the Commerce, that "fires are generally preventable, and that data from investigations are confidential. So there is lack of knowledge/appreciation of dangerous practices and conditions." Turner stated that he was "going to ask Congress to allow publication with conclusions and recommendations so that they can be put into effect." (Turner 1931)

In 1938, reports on recent mine fires (Harrington and Fene 1938) and recent lab research (Fieldner 1938) were published. Both reports emphasised recommendations to improve safety in mines. In 1941, Congress approved the Coal Mine Inspection Act, a year after 257 miners died in four separate explosions. The 1941 Act now allowed the Federal government to inspect mines for the purpose of making recommendations to improve safety.

Initially, there were 107 Federal mine inspectors and over 14,000 coal mines. In 1946, agreement was reached between the John L. Lewis of the United Mine Workers Union and Julius A. Krug, the Secretary of the Interior, to make compliance with Federal recommendations mandatory based on the 1946 Federal Mine Safety Code.In the 1947 Annual Report to the Secretary, Bureau of Mines Director Sayers stated that "thousands of safety improvements were made in U.S. mines as a result of the Krug-Lewis agreement." (Sayers, 1947).

World War II

During the months prior to the United States' involvement in World War II, the Bureau of Mines diverted its efforts into gearing up for war. There was an abnormal demand for metals and minerals, particularly to make steel. During the war years, Bureau scientists worked to supply the Nation with mineral commodities. The Bureau's fire scientists tested western coals for making coke. They studied

Figure 4. Early Bureau of Mines gas laboratory

the flammability of metallic dusts used in bombs and methods for extinguishing burning powders. They also conducted training in the extinguishment of incendiary bombs. Because of the need for synthetic rubber, the Bureau conducted flammability studies of rubber-making chemicals, such as butadiene, styrene, acrylonitrile, chlorobenzene, and trichloroehtylene to ensure safety in the rubber-making process. Research was also conducted on the detection of atmospheric contaminants that might be encountered on the battlefield. This work had significant mining application after the war in the improvement of mine gas detection devices.

POST-WAR YEARS

The post-war years featured a much larger emphasis on mine safety research as a whole, including fire research. The 1946 Federal Mine Safety Code giving the Bureau of Mines the right to inspect and enforce mandatory compliance with the Code, combined with the technological advances from the war effort, led to a marked increase in safety research. A research study was conducted on the efficacy of various materials and methods for extinguishing coal mine fires (Howarth et al. 1949). In 1947, the Bureau was working on the development of diesel equipment in underground mines as a means to eliminate electrical hazards associated with electrical trolley lines. The first approval of a permissible diesel locomotive for underground use occurred in 1951. In addition,

the Bureau conducted classified research through the 1940s for the Navy on gas-indicating devices, which led to improvements in mine fire detection.

The Bureau of Mines celebrated its 40th anniversary in 1950 by reorganizing into three divisions; Minerals, Fuels and Explosives, and Health and Safety. In 1950, there were still 593 fatalities in the mining industry. The Bureau decided that they would investigate all fatalities with a view to determining preventive measures. Prior to 1950, they had only investigated major disasters, defined as having more than 5 fatalities. In 1951, there were five major coal mine disasters, including the Orient Mine disaster in Illinois, which killed 119 and was the worst disaster since 1928 (Keenan 1963). This led to the 1952 Federal Coal Mine Safety Act, and authorized Federal inspectors to act to prevent disasters in all mines with greater than 15 underground workers.

In the early 1950s, the Bureau conducted extensive studies on the fire resistance of conveyor belting, hydraulic fluids and electrical cables to determine their fire hazards. An important milestone was reached in 1955 when the Bureau prescribed a schedule of acceptance requirements for fire resistant conveyor belting (Schedule 28). By 1958, 60 conveyor belts were accepted as fire resistant. Flame resistance tests were also conducted on 20 electrical cables and on fire resistant hydraulic fluids, and the first testing on frictional ignition of conveyor belts was completed. In 1958, suppression tests of high expansion foam plugs to extinguish large-scale coal and mining machinery fires were started at Bruceton, Pa. Figure 5 shows a coal mine fire in the Bruceton Experimental Mine just prior to extinguishment. By 1960, the Bureau had completed six extinguishment tests of 15 ton coal fires and demonstrated the results to the coal industry. This resulted in the employment of portable foam generators at mines throughout the United States (Nagy et al. 1961).

Fundamental Combustion Research

An integral part of the Bureau of Mines fire research program throughout the 1940s and 1950s was the fundamental combustion research program. In 1948, researchers began a study of the mechanism of ignition, propagation, and stabilization of flame and flame quenching. The research contributed significantly to the fundamental knowledge of combustion, fluid dynamics, kinetics, and thermodynamics. Among the accomplishments were an understanding of laminar and turbulent flames; kinetic studies of the oxidation of various hydrocarbons, important in understanding the theory of combustion; an experimental technique for measuring burning velocities of turbulent flames at low pressures, which was useful

Figure 5. A 15-ton coal fire prior to extinguishment with high-expansion foam

in the design of jet engine combustors and rockets; and burning velocity measurement at above-atmosphere pressures. An instrument to measure turbulent flame structure is shown in Figure 6. The results of this research also had direct application to the understanding of coal mine fire and explosion behavior. In 1951, Bernard Lewis and Guenther von Elbe of the Bureau of Mines published the seminal book "Combustion, Flames and Explosions of Gases." (Lewis and von Elbe 1951). This book had two subsequent editions, the latest in 1987, and is still widely used as a graduate textbook for combustion science. In 1954, Lewis was instrumental in founding the Combustion Institute, in Pittsburgh, Pa. This international organization is an educational, nonprofit scientific society whose purpose is to promote and disseminate research in combustion science. The organization publishes the journal, Combustion and Flame, and holds the International Combustion Symposium every two years.

1960s

Experiments continued in the early 1960s on coal mine fire control. The efficacy of various fire extinguishing agents and application methods of these agents were examined on simulated mining machines (Mitchell et al. 1961). These studies included water and alkali metal salts or wetting agents, water fog, and the remote application of high expansion foam. Experiments and field trials were conducted with rigid polyurethane foam for sealing mine openings, controlling ventilation, and combating mine fires (Mitchell et al. 1964). This investigation also required testing of the flame-retarding and toxic properties of polyurethane Mitchell et al. 1966). The first ventilation studies using the Bureau's fluid

network analyzer for contaminant spread were completed in 1962. That year also included the first evaluation of health and safety, including fire hazards, in metal/nonmetal mines.

The mid-1960s saw the introduction of longwall mining into the United States, and the need to evaluate the effect of this new technology on fire hazards. The first investigations into the effect of air velocity on a burning conveyor belt were started in 1964. Also included in this study was the determination of types of contaminant gases produced by the fires for detection (Mitchell et al. 1967). With the increased mechanization, new fluids were introduced into the mines. A special report was published on the ignitibility and flammability of more than 80 lubricants, hydraulic fluids, and engine oils used in mines. Auto-ignition temperatures were measured for these fluids at pressures up to 15,000 psig (Zabetakis 1965). In 1968, a new approach to the concept of the extinguishment of fires was started to gain a better knowledge and understanding of the mechanism of inerting in the prevention of fires. This program was fundamental in nature, and made significant advances in the mechanism of flame inhibition (Biordi et al. 1973).

The Bureau of Mines also contributed significantly to the space program in the 1960s.

Figure 6. An apparatus to measure turbulent flame structure

Figure 7. A large surface vehicle fire

Researchers were asked by NASA to develop solutions to the problem of extinguishing fires in space vehicles, venting to low pressures, and the flammability characteristics of a variety of rocket propellant combinations, including studies in high vacuum for vehicle attitude control. The Bureau was also asked to assist in the investigation of the tragic Apollo 1 fire, studying the fires in oxygen-rich environments. The results of the Bureau's investigation showed that a treatment for flame proofing space fabrics may actually enhance the ignitibility or flammability of the material (Kutcha et al. 1969).

FEDERAL COAL MINE HEALTH AND SAFETY ACT OF 1969

In 1968, a series of explosions at the Farmington No. 9 Mine in Mannington, WV killed 78 miners. This disaster prompted a renewed demand for improved safety measures in coal mining and led to the Federal Coal Mine Health and Safety Act of 1969 (Public Law 91-173). Congress said in the preamble that "the first priority in mining must be the health and safety of its most precious resource—the miners." The legislation directed the Bureau to conduct research and development needed to eliminate coal mining hazards and reduce the risks of health impairment, injury, or death. The 1969 Act led to the development of mandatory safety standards in 1970. At that time, the Bureau health and safety program was reorganized to strengthen enforcement, and a Health and Safety Technical Support Center was created for approvals and testing of mine equipment and materials, including fire resistance. In 1973, the enforcement function and Technical Support Center was moved into the newly created Mine Enforcement Safety Administration (MESA), the precursor to the Mine Safety and Health Administration (MSHA).

1970s

The new law greatly expanded the Bureau's role in coal mine fire research. In 1971, contract research was initiated to develop methods of remote sealing, fire detection, automatic fire suppression on mining equipment, and to evaluate suppression systems in conveyor belt entries. The results of these studies led to many of the in-mine detection and suppression systems still used today. In-house research continued on studying the reaction mechanism of flame inhibition, the early detection of smoke and fire in underground mines, and the efficacy of various extinguishing agents to fight coal mine fires.

In 1972, the Sunshine Mine fire at a silver mine in Kellogg, Idaho killed 91 miners. This led to an increase in fire prevention research in metal and nonmetal mines. Research was initiated to evaluate and develop improved detection and suppression systems for deep shaft metal and nonmetal mines, the use of fire-protective coatings on shaft liners, the use of inflatable stoppings in these mines to control ventilation, and the development of fire suppression systems for mobile mining equipment, both underground and on the surface. Figure 7 shows the devestating effects of a large mobile equipment fire.

Research continued at a fast pace throughout the 1970s, concentrating on the detection, control, and extinguishment of mine fires. In the area of mine fire detection, spontaneous combustion studies were made to understand the low temperature oxidation mechanism for early detection. This also included the development of tube bundle systems for sampling worked-out areas, and the evaluation of various gas indices to understand the state of a mine fire. These indices are routinely used today at mine fires as guides for rescue and re-entry. The first generation of submicron smoke and metal oxide detectors were developed. Contract research evaluated the adequacy and equivalency of fire detection and suppression systems for belt entries. Research was also conducted in-house and through contracts to understand the spread of fire contaminants through the mine ventilation system and the effect of fire size and ventilation on contaminant spread. One of the products of this research was the MFIRE mine fire ventilation code, still in use today (Greuer 1977).

For the remainder of the decade, the Bureau's research program began to de-emphasize small-scale studies and shifted to large or full-scale in-mine investigations. The Bureau entered into a cooperative agreement with MESA with respect to development and evaluation of mining regulations. Under this agreement, the Bureau conducted evaluations of the Schedule 2G criteria for conveyor belt flammability and Schedule 30 criteria for flammability of hydraulic fluids, and evaluated the vulnerability of bulkheads to fires. A full-scale in-mine evaluation of a second generation CO monitoring system was also completed (Litton 1979).

1980s

Fire research in the 1980s continued on developing new or improved fire test standards for mine combustibles, such as plastic fluid containers, rigid and flexible ventilation ducting, and polyurethane sealants, based on large-scale experiments. In addition, a comprehensive program was completed to quantify the toxicity loads of mine combustibles (Egan 1990). In 1980, the Lake Lynn Laboratory (LLL), a state-of-the art mining research facility located near Uniontown, Pa, began operation (Triesbsch and Sapko 1990). In 1984, a fire gallery was constructed at the Lake Lynn Laboratory, which allowed full-scale flammability testing of mine conveyor belting. Figure 8 shows a full-scale conveyor belt flammability experiment in the fire gallery.

These tests clearly demonstrated the inadequacy of the small-scale Federal acceptance test for fire-resistant belting and led to the development of a new laboratory-scale test method (Lazzara and Perzak 1989). Automatic sprinkler and dry chemical suppression systems in conveyor belt entries were evaluated in the fire gallery to determine the effect of ventilation on their effectiveness (Smith et al). The integrity of steel and concrete block stoppings, used to separate the airways carrying air to and from the working face, was evaluated in large-scale fire tests (Ng et al. 1988). Spontaneous combustion research investigated variables that contribute to the self-heating process and developed a laboratory test method to predict the potential for spontaneous combustion (Smith and Lazzara 1987).

Research continued on the development of rapid and reliable fire and product of combustion sensors. The Bureau patented and commercially licensed a sensitive smoke detector for underground mines (U.S. Patent 4,053,776). A prototype model of a diesel-discriminating smoke detector was developed and evaluated in mines. The location and placement of sensors is critical in the performance of a detection system. Mine studies were conducted to determine the effect of crosscuts, air velocity, and placement in the entry on detector response. Guidelines for spacing of carbon monoxide and smoke detectors in conveyor belt entries were published (Litton et al. 1991). The Bureau also assisted MSHA in the establishment of regulations for metal and nonmetal

Figure 8. A conveyor belt fire experiment in fire gallery

mines. Research studies were completed on the flame propagation, rate of heat release, and products of combustion of timber-lined shaft fires. Deep shaft fire suppression systems were evaluated and in-mine validation of the MFIRE ventilation code for toxic fume spread was completed.

1990s

The fire research program of the 1990s was centered on the prevention, detection, and suppression of mine fires. The Bureau began a series of very successful Open Industry Briefings on Mine Fire Preparedness. These 2-½ day briefings were conducted at the Lake Lynn facility, and were tailored to mine safety and training personnel, rescue and fire brigade team members, and supervisors and shift foremen. Participants heard the latest in fire research from Bureau researchers and viewed large-scale demonstrations. The demonstrations included full-scale conveyor belt and ventilation tubing fires, the response of fire detection systems to small coal fires, and the extinguishing of liquid fuel fires using various fire-fighting equipment (Conti 1994).

In 1995, the Bureau developed a computer program to assess the spontaneous combustion risk of an underground mining operation, based on the coal's spontaneous combustion potential and the geologic and mining conditions encountered in the mining of the coal, and the mining practices employed (Smith et al. 1995). This program is used by mine operators to determine the best in-mine ventilation methods to control spontaneous combustion. Detection studies focused on strategies for sensor deployment and understanding the effect of a fire on the ventilation patterns and smoke rollback. Research continued on the improvement of sensor technologies, particularly

Figure 9. "Smart" sensor detection system installed at mine

the development of the diesel-discriminating detector. Studies were also conducted to determine the feasibility of a remote optical methane detector.

THE MOVE TO NIOSH

In September 1995, Congress recommended the closure of the Bureau of Mines. Through the efforts of many stakeholders, the function of mining health and safety was retained by the government, first under the Department of Energy. In 1996, the Office for Mine Safety and Health (OMSH) was established under the National Institute for Occupational Safety and Health (NIOSH). The OMSH immediately began the formulation of a Mining Strategic Plan to guide health and safety research activities, consistent with the needs of stakeholders and the mission outlined by the U.S. Congress. The OMSH relied heavily on stakeholder insight and empirical statistical data to establish the most efficient and effective Strategic Plan.

The plan set goals in the area of fire prevention to reduce fires in underground coal mines associated with conveyors and haulage systems, fires associated with flame cutting and welding, and the establishment of performance standards and location strategies for mine fire detection systems. Research was completed to evaluate the application of new and improved fire suppression systems, such as water mist suppression systems, to protect underground mine haulage ways and stationary equipment, such as compressors and electrical substations, and diesel fuel storage areas (Smith and Lazzara 1998). Methods for the remote sealing and remote suppression of mine fires were developed (Trevits et al. 2007).

Because of the issue of using belt air at the working face, experiments were completed to determine the effect of air velocity on fire suppression system performance (Rowland et al. 2009). The root causes of flame cutting and welding fires in both underground and surface mining were identified and guidelines for safe flame cutting and welding operations were developed (Monaghan et al. 2006).

The goal of reducing noise exposure in mining led to an increased use of noise abatement materials in both aboveground and underground equipment cabs. The flammability hazards of different types of noise materials were identified and guidelines for use were published (Litton et al. 2003).

Research was also conducted to determine the intrinsic causes of large surface vehicle and equipment fires and new or improved fire protection systems for these vehicles and equipment were developed (DeRosa and Litton 2007). A "smart." sensor system utilizing neural network methodology was developed. The system has the capability to detect the fire, discriminate the fire combustible source in a diesel background, and provide a mine fire location strategy in an underground mine. This system was successfully demonstrated at an underground coal mining operation (Franks et al. 2008). The sensor installation at the mine is shown in Figure 9.

THE MINER ACT

On January 19, 2006 an underground mine conveyor belt fire occurred at a West Virginia coal mine, fatally injuring two miners. This accident, along with two other mine tragedies in 2006, which resulted in 17 additional fatalities, led to The Mine Improvement and New Emergency Response Act of 2006 (MINER ACT). This legislation amended the Mine Safety Health Administration Act of 1977 to improve the safety of the mines and mining. Section 11 of the Miner Act required that a Technical Study Panel (The Panel) be formed to provide recommendations on the use of belt air and new technology that may be available for increasing fire safety. These recommendations include the use of recent technological advances applied to the current state of the art for fire-resistant conveyor belt materials, belt fire suppression systems, atmospheric monitoring systems, and escapeway integrity to improve miner safety.

In 2007, the fire research program was re-evaluated and re-directed to address the recommendations of the Act. A project was started to reduce the hazards of underground coal mine fires by applying recent technological advances in the areas of fire-resistant conveyor belt materials, belt fire suppression systems, atmospheric monitoring systems, and computer codes for predicting and assessing in real-time the impact of fire on the mine ventilation system and the spread of fire contaminants throughout the mine. The project includes tasks to address some of the major issues concerning fires in underground coal mines; flammability of conveyor belts, detection, suppression systems, fire modeling of conveyor belts, fire modeling of contaminant spread, and fire risk assessment. This project is currently ongoing.

Research is being conducted to compare the results of full-scale conveyor belt flammability tests to the recently developed reliable small-scale test to determine the effect of high ventilation velocities on the flammability hazards of conveyor belts. Research is planned to define sensor location and response time in conveyor belt entries with wider belts and high air velocities to ensure adequate warning time to mine personnel. Experiments are being conducted to evaluate novel suppression systems currently available in the United States and in other countries in full-scale tests to determine if improvements can be made, such as changes in spacing of sprinklers or dry powder nozzles, to improve their effectiveness. Mine Fire Risk Assessment software aids are being developed utilizing a set of potential hazards that must be considered. The MFIRE ventilation code is undergoing modernization to increase its capabilities for smoke management during a mine fire emergency.

Spontaneous combustion continues to pose a hazard for U.S. underground coal mines, particularly in western mines where the coal is generally of lower rank. The risk of an explosion ignited by a spontaneous combustion fire is also present in those mines with appreciable levels of accumulated methane. In fact, three of the mine fires from the reported period 1990–1999 resulted in subsequent methane explosions. A new project is using CFD modeling technique to develop new methods to prevent, detect, control and suppress the spontaneous heating in mines (Yuan and Smith 2007).

SUMMARY

Mine fire research has been an integral part of the Bureau of Mines and NIOSH mission for 100 years. There have been many major advances in mine fire safety that have led to a significant decrease in the occurrence and injury and fatality rates over this time. However, the potential for disaster remains, so we must not rest on our laurels. With the use of new technology to mine coal and other minerals, and the mining of deeper and gassier mines, the hazards will continue to evolve. Our research must also continue to evolve to meet these challenges to continue to provide the safest and healthiest workplaces for our miners.

ACKNOWLEDGMENTS

We would like to acknowledge all of the many Bureau of Mines, Mine Safety and Health Administration, and NIOSH scientists and engineers, technical staff, and others who have worked to reduce the hazards and risks of mine fires over the last 100 years. We would also like to acknowledge the help of our Librarian, Kathleen Strabyla, for helping to gather the historical documents needed for this paper.

REFERENCES

Ash, S. H. and Felegy, E. W. 1948. *Analyses of Complex Mixtures of Gases. Application to Control and Extinguish Fires and To Prevent Explosions in Mines, Tunnels, and Hazardous Industrial Processes.* Washington, D.C.: U.S. Bureau of Mines, Bulletin 471.

Biordi, J. C., Lazzara, C. P., and Papp, J. F. 1973, *Chemical Flame Inhibition Using Molecular Beam Mass Spectrometry.* U.S. Bureau of Mines, Report of Investigation 7723.

Burrell, G. A. and Siebert, F. M. 1913. *Apparatus for Gas Analysis Laboratories at Coal Mines.* U.S. Bureau of Mines, Technical Paper 14.

Conti, R. S. 1994. *Fire Fighting Resources and Fire Preparedness for Underground Coal Mines.* U.S. Bureau of Mines, Information Circular 9410.

Coward, H. F. and Jones, G. W. 1952. *Limits of Flammability of Gases and Vapors.* Washington, D.C.: U.S. Bureau of Mines, Bulletin 503.

DeRosa, M. I. 2004. *Analysis of Mine Fires for All U.S. Underground and Surface Coal Mining Categories, 1990–1999.* Pittsburgh, PA: U.S. Department of Health and Human Services, National Institute for Occupational Safety and Health, IC 9470.

DeRosa, M. I. 2004. *Analysis of Mine Fires for All U.S. Metal/Nonmetal Mining Categories, 1990–2001.* Pittsburgh, PA: U.S. Department of Health and Human Services, National Institute for Occupational Safety and Health, IC 9476.

DeRosa, M. I. and Litton, C. D. 2007. *Rapid Detection and Suppression of Mining Equipment Cab Fires.* Fire Technol 2007 Aug; 17:1–11.

Egan, M. R. 1990. *Summary of Combustion Products From Mine Materials: Their Relevance to Mine Fire Detection.* U.S. Bureau of Mines, Information Circular 9272.

Farrow, E. S. 1923. *Gas Warfare.* New York, E. P. Dutton & Company.

Federal Mine Safety Code for Bituminous-Coal and Lignite Mines of the United States 1946. Washington, D.C.: U.S. Bureau of Mines.

Fieldner, A. C. and Katz, S. H. 1921. *Gases produced in the Use of Carbon Tetrachloride and Foamite Fire Extinguishers in Mines.* U.S. Bureau of Mines, Report of Investigation 2262.

Fieldner, A. C. 1938. *Annual Report of Research and Technologic Work on Coal, Fiscal Year 1937.* Washington, D.C.: U.S. Bureau of Mines, Information Circular 6992.

Franks, R. A., Friel, G. F., Edwards, J. C, and Smith, A. C. 2008. *In-Mine Evaluation of Smart Mine Fire Sensor.* Trans Soc Min Metal Explor 2008 Dec; 324:81–88.

Godbert, A. L. and Greenwald, H. P. 1936. *Laboratory Studies of the Inflammability of Coal Dusts: Effect of Fineness of Coal and Inert Dusts on the Imflammability of Coal Dusts.* Washington, D.C.: U.S. Bureau of Mines, Bulletin 389.

Greuer, R. E. 1977. *Study of Mine Fires and Mine Ventilation. Part I, Computer Simulation of Ventilation Systems Under the Influence of Mine Fires.* U.S. Department of the Interior, Bureau of Mines, Contract No. S0241032, OFR 115(1)-78, NTIS No. PB288231.

Harrington, D. and Fene, W. J. 1938. *Coal-Mine Explosions and Coal- and Metal-Mine Fires in the United States During the Fiscal Year Ended June 30, 1937.* Washington, D.C.: U.S. Bureau of Mines, Information Circular 6986.

Hood O. P. 1922. *Factors in the Spontaneous Combustion of Coal.* Washington, D.C.: U.S. Bureau of Mines, Technical Paper 311.

Howarth, H. C. and Greenwald, H. P. 1927. *Tests with Rock Dust for Extinguishing Fire.* U.S. Bureau of Mines, Report of Investigation 2801.

Howarth, H. C., Nagy, J., and Hartmann, I. 1949. *Experiments on the Control of Coal Mine Fires.* Proc. 39th Annual Convention, Mine Inspectors' Inst. America, June, 1949.

Keenan, C. M. 1963. *Historical Documentation of Major Coal-Mine Disasters in the United States Not Classified as Explosions of Gas or Dust: 1846–1962.* Washington, D.C.: U.S. Bureau of Mines, Bulletin 616.

Kirk, W. S. 1994. *The History of the U.S. Bureau of Mines.* U.S. Department of Interior, U.S Bureau of Mines, Minerals Yearbook, v. 1.

Kuchta, J. M., Furno, A. L., and Martindill, G. H. 1969. *Flammability of Fabrics and Other Materials in Oxygen-Enriched Atmospheres: Ignition Temperature and Flame Spread Rates.* Fire Technology, v. 5.

Lazzara, C. P. and Perzak, F. J. 1989. *Conveyor Belt Flammability Tests: Comparison of Large-Scale Gallery and Laboratory-Scale Tunnel Results.* Proc. 23rd Int. Conf. of Safety in Mines Research Institutes, Washington, D.C.

Leitch, R. D., Hooker, A. B., and Yant, W. P. 1925. *The Ignition of Firedamp by Exposed Filaments of Electric Mine-Lamp Bulbs.* U.S. Bureau of Mines, Report of Investigation 2674.

Lewis, B. and von Elbe, G. 1951. *Combustion, Flames and Explosions of Gases.* New York, Academic Press, Inc.

Litton, C. D. 1979. *Fire Detection in Underground Mines.* U.S. Bureau of Mines, Information Circular 6992.

Litton, C. D., Lazzara, C. P., and Perzak, F. J. 1991. Fire Detection for Conveyor Belt Entries. Report of Investigation 8370.

Litton, C.D., Mura, K. E., Thomas, R.A., and Verakis, H.C. 2003. *Flammability of Noise Abatement Materials Used in Cabs of Mobile Mining Equipment.* Proc: Fire & Materials 2003. London: Interscience Communications, 2003 Feb; :297–306.

Mitchell, D. W., Murphy, E. M., Nagy, J. and Christofel, F. P. 1961. *Practical Aspects of Controlling an Underground Fire on a Mining Machine.* U.S. Bureau of Mines, Report of Investigation 5846.

Mitchell, D. W., Nagy, J., and Murphy, E. M. 1964. *Rigid Foam for Mines.* U.S. Bureau of Mines, Report of Investigation 6366.

Mitchell, D. W., Murphy, E. M, and Nagy, J. 1966. Fire Hazard of Urethane Foam in Mines. U.S. Bureau of Mines, Report of Investigation 6837.

Mitchell, D. W., Murphy, E. M., Smith, A. F., and Polack, S. P. 1967. *Fire Hazard of Conveyor Belts.* U.S. Bureau of Mines, Report of Investigation 7053.

Monaghan, W. D., Brüne, J.F., and Smith, A.C. 2006. *Determining the Root Causes of Flame Cutting and Welding Fires in Underground U.S. Coal Mines.* Mining Media, Inc., Denver, CO: Proc. National Coal Show, Pittsburgh, Pennsylvania.

Nagy, J., Murphy, E. M., and Mitchell, D. W. 1960. *Controlling Mine Fires With High-Expansion Foam.* U.S. Bureau of Mines, Report of Investigation 5632.

Ng, D. L., Lazzara, C. P., Mura, K. E., and Luster, C. K. 1988. *Fire Endurance of Mine Stoppings.* U.S. Bureau of Mines, Report of Investigation 9207.

Paul, J. W. 1912. *Mine Fires and How to Fight Them.* Washington, D.C.: U.S. Bureau of Mines, Miner's Circular 10.

Rice, G. S. 1912 *Mine Fires, A Preliminary Study.* Washington, D.C.: U.S. Bureau of Mines, Technical Paper 24.

Rowland, J. H., Verakis, H. C., Hockenberry, M. A., and Smith, A.C. 2009. *Effect of Air Velocity on Conveyor Belt Fire Suppression Systems.* SME Preprint No. 09135. Littleton, CO: SME.

Sayers R. R. 1947. *Annual Report of the Director of the Bureau of Mines to the Secretary of the Interior, for the fiscal Year ended June 30, 1947.* Washington, D.C.: U.S. Bureau of Mines, DAR 1947.

Schedule 28 1955. *Fire-Resistant Conveyor Belts: Tests For Permissibility; Fees.* Approved 1955; amended 1957. Federal Register, v. 20, No. 220, Nov. 10, 1955, v. 22, No. 238, Dec. 10, 1957.

Smith, A. C. and Lazzara, C. P. 1987. *Spontaneous Combustion Studies of U.S. Coals.* U.S. Bureau of Mines, Report of Investigation 9079.

Smith A. C., Pro R. W. and Lazzara C. P. 1995. *Performance of Automatic Sprinkler Systems for Extinguishing Incipient and Propagating Conveyor Belt Fires Under Ventilated Conditions.* U.S. Bureau of Mines, Report of Investigation 9538.

Smith A.C., Rumancik W.P., and Lazzara C.P. 1995. *SPONCOM—A Computer Program for the Prediction of the Spontaneous Combustion Potential of an Underground Coal Mine.* Proc. 26th Int.Conf. on Safety in Mines Research Institutes.v. 4. Katowice, Poland.

Smith A.C. and Lazzara C.P. 1998. *Water Mist Suppression of Fires in Underground Diesel Fuel Storage Areas.* Annual Conference on Fire Research—Book of Abstracts. Gaithersburg, MD: U.S. Department of Commerce, Technology Administration, National Institute of Standards and Technology.

Staff, Bureau of Mines 1988. *Recent Developments in Metal and Nonmetal Mine Fire Protection.* U.S. Bureau of Mines, Information Circular 9206.

Trevits, M.A., Smith, A.C., and Brüne J.F. 2007. *Remote Mine Fire Suppression Technology.* Proc. 32nd Int. Conf. of Safety in Mines Research Institutes, September 2007, Beijing, China. Beijing, China: National Center for International Exchange & Cooperation on Work Safety (SAWS), 306–312.

Triesbsch, G. and Sapko, M. J. 1990. *Lake Lynn Laboratory: A State-of-the-Art Mining Research Laboratory.* Proc. Int. Symp. On Unique Underground Structures, Denver, CO.

Turner, S. 1931 *Annual Report of the Director of the Bureau of Mines to the Secretary of Commerce, for the Fiscal Year ended June 30, 1931.* Washington, D.C.: U.S. Bureau of Mines, DAR 1931.

Turner, S. 1932 *Annual Report of the Director of the Bureau of Mines to the Secretary of Commerce, for the Fiscal Year ended June 30, 1932.* Washington, D.C.: U.S. Bureau of Mines, DAR 1932.

Yuan, L and Smith, A. C. 2007. *Computational Fluid Dynamics Modeling of Spontaneous Heating in Longwall Gob Areas.* Trans Soc Min Metal Explor 2007 Dec; 322:37–44

Zabetakis, M. G. 1965. *Flammability Characteristics of Combustible Gases and Vapors.* U.S. Bureau of Mines, Bulletin 627.

Proactive Strategies to Prevent Fires and Explosions in Longwall Mines

Rao Balusu

CSIRO Exploration and Mining, Kenmore, Queensland, Australia

Ting Ren

CSIRO Exploration and Mining, Kenmore, Queensland, Australia

Kelvin Schiefelbein

Anglo Coal Australia Pty Ltd, Brisbane, Australia

Paul O'Grady

Anglo Coal Australia Pty Ltd, Brisbane, Australia

Tim Harvey

Anglo Coal Australia Pty Ltd, Brisbane, Australia

ABSTRACT: An extensive research programme has been undertaken in Australia in recent years to reduce the risk of fires and explosions in longwall mines. The research projects have integrated extensive field studies and computational fluid dynamics (CFD) models of goaf gas flow behaviour to develop optimum proactive strategies to prevent and control spontaneous combustion and fires in goaf areas of longwall mines. The research has developed a fundamental understanding of gas flow behaviour in longwall goafs and provided a detailed understanding of the effects of various geological, ventilation, operational and inertisation design parameters on goaf gas distribution. Field studies conducted at a number of underground coalmines in Australia have shown that the proactive strategies were successful in reducing oxygen ingress into the longwall goaf areas and in reducing the risk of fires and explosions in longwall mines. The results of various research investigations and guidelines for proactive strategies are presented in this paper.

INTRODUCTION

The risk of fires and explosions is a major safety hazard in underground coal mines with severe consequences. A number of new underground coal mines developed in Australia in recent years during the 1990s and early 2000s have a moderate to high susceptibility to spontaneous combustion (sponcom). The coal seams in these new areas are generally thick and the risk of sponcom, heatings, fires and explosions increases significantly during longwall extraction due to the large quantities of broken coal left in the goaf and its subsequent exposure to high oxygen concentration levels. This poses a major risk to the safety of the people, longwall face equipment and economic viability of the modern and highly capital intensive coal mines. There was an urgent need to develop and implement proactive technologies and strategies to reduce the risk of sponcom and heatings development in goaf areas in order to prevent fires and explosions in longwall mines.

To address this critical safety issue in underground coal mines, a major research programme has been undertaken by CSIRO in collaboration with Australian coal mining companies under the ACARP research program. While the earlier projects in Australia focussed on development of inert gas generators (Bell et al., 1998), this research concentrated on development of proactive technologies and strategies. The main objective of the research programme was to develop and demonstrate effective proactive strategies to reduce the risk of heatings and fires in longwall panels, particularly under modern longwall mining conditions in Australia that are characterised by high face ventilation quantities in the range of 50 to 80 m³/s and thick coal seam mining conditions.

The research has adopted an integrated approach involving (i) detailed monitoring and analysis of goaf gas distribution in longwall panels for improved understanding of oxygen ingress patterns into the goaf areas; (ii) extensive Computational Fluid Dynamics (CFD) modelling to provide a fundamental understanding of the gas flow behaviour in longwall goafs and to develop a detailed understanding of the effect of various geological, ventilation, operational and inertisation design parameters on goaf gas distribution; (iii) development of optimum proactive strategies for prevention of heatings and fires in longwall panels; and (iv) detailed field

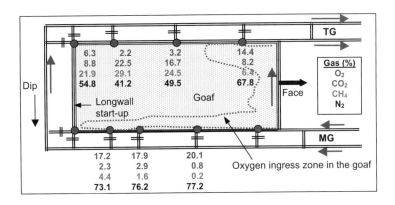

Figure 1. Typical gas distribution in the longwall goaf at one of the mines

investigations and demonstration studies in longwall mines to evaluate the effectiveness of various proactive strategies. A brief summary of the results of the research work are presented in this paper.

FIRES AND EXPLOSIONS ISSUES IN LONGWALL MINES

The longwall mines in Australia typically employ a two gateroad layout for development and extraction of longwall panels. The longwall faces are 250 to 300 m wide with panel lengths in the range of 2,000 to 3,000 m. The barrier pillar width ranges from 30 to 45 m, with cut-through spaced normally at 100 m intervals. A typical layout of the longwall panel in Australia is shown in Figure 1. The total airflow rates used in longwall panels varies from 50 to 150 m^3/s depending on the goaf gas emission flow rates. It is to be noted here that most of the mines use bleederless ventilation systems to manage sponcom in the longwall goaf areas. The longwall extraction process also involves construction of seals in all the cut-throughs around the goaf during face retreat operations.

The traditional practice of gas monitoring in longwall extraction panels was to monitor the return air for CO and other gases to detect any heatings or fires in the panel, similar to the practice used in bord and pillar extraction panels. However, it was identified during the early stages of this research that in order to investigate the heatings and fires issues in longwall panels, it is very important to characterise the oxygen ingress patterns in longwall goafs under different mining conditions. The longwall panel shown here from one of the mines has access around the perimeter of the panel and therefore has enabled extensive monitoring of gas distribution in the goaf on both maingate (MG) and tailgate (TG) sides of the panel. Tube-bundle monitoring system was installed

at the mine to continuously monitor goaf gas distribution at varying distances behind the longwall face.

A snapshot of the typical goaf gas distribution behind the longwall face at one of the mines is presented in Figure 1. The field data show that intake air ingress on the maingate side of the panel was very high, with the oxygen level at more than 17% even at 400 m behind the longwall face. Oxygen ingress on the tailgate appears to extend only up to 100 m behind the face due to higher goaf gas emissions in the panel. Goaf gas monitoring results from several other longwall panels in different mines also revealed similar trends, with high oxygen ingress on the intake side of the panel. This level of high oxygen ingress into the goaf could lead to heatings and fires in the goaf if the working coal seam or nearby coal seams in the caving zone have moderate to high propensity to sponcom. Heatings and fires in longwall goafs could lead to explosions under favourable conditions. It is crucial to adopt some proactive strategies to reduce the length of the oxygen ingress zone in the goaf to prevent the risk of heatings, fires and explosions in longwall mines.

Cliff (2005) highlighted that given the difficulty in detecting an active heating in longwall goafs, we should focus on preventing a heating from occurring. Cliff pointed out that early response should be triggered by detection of high oxygen in the areas of the goaf where it should not be. Although high oxygen does not immediately cause trouble, it will be the catalyst if this condition remains in place for any length of time. Remedial actions such as proactive inertisation to reduce the oxygen supply into the goaf can avert a heating.

In order to develop effective proactive strategies, it is critical to obtain a fundamental understanding of the goaf gas flow patterns under different mining conditions and to investigate the effect of various mining and operational parameters on goaf gas distribution. A brief summary of the modelling investigations carried out to investigate the effect of various parameters and the results of field studies are presented in the following sections.

FUNDAMENTAL UNDERSTANDING OF GOAF GAS FLOW PATTERNS

Extensive computational fluid dynamics (CFD) modelling investigations have been carried out to obtain a fundamental understanding of the goaf gas flow and distribution patterns in the longwall goaf areas. The studies involved innovative CFD modelling, calibration and validation of initial base models using data obtained from field studies, extensive parametric studies to investigate the effect of various parameters and development of optimum proactive strategies.

Oxygen Ingress Patterns into the Longwall Goaf

A number of models have been developed to represent various longwall panel geometries under different mining and gas emission conditions based on the preliminary field investigations carried out at the field mine sites. For each mine site, the base model was 1 km in length along the panel, 250 m in width and 80 m in height to cover the immediate high porosity caving regions in the goaf. Goaf gas emission was varied between 100 l/s and 3,000 l/s to represent different goaf gas environments. A commercial CFD code, FLUENT, has been used for development of these 3D longwall goaf gas flow models. A number of user defined functions/subroutines were written to represent the goaf permeability and different gas emissions scenarios, which were then combined with the main FLUENT program to carry out the simulations. Goaf gas data obtained from field measurements were used to calibrate and validate the initial base models.

The results of the base case simulations for a typical longwall panel in one of the mines with face air quantity around 50m³/s are presented in Figure 2, showing the oxygen gas distribution in the goaf. In this base case, intake to the panel was through the maingate (MG), which was at 20 m lower elevation compared to the tailgate return airway. The goaf gas emission rate in this base case was around 600 l/s. In the colour coding scale shown along with the simulation results in Figure 2 and other figures, 0.21 represents 21% oxygen, i.e., fresh air composition.

Results of the simulations indicate that oxygen ingress into the goaf was high, due to low goaf gas emissions in the panel. Oxygen level in the goaf on the maingate side was over 14% at 350 m behind the face and over 10% even at 600 m behind the longwall face. Analysis of the results indicated that intake airflow and ventilation pressures have a major influence on gas distribution up to 50 to 150 m behind the face, and beyond that goaf gases' buoyancy and panel geometry seems to play a major role on goaf gas distribution. Simulation results presented in Figure 2 tallied well with the results of field goaf gas monitoring studies carried out at that mine site. Longwall goaf gas distribution may vary significantly at different mines based on panel geometry, ventilation, gas emission rates and other site specific conditions.

Effect of Various Parameters on Oxygen Ingress into the Goaf

A number of parametric studies have been carried out to investigate the effect of changes in geological features, mining conditions and operational parameters on gas flow patterns in longwall goafs. A few of the parametric studies include: (i) the effect of various goaf gas emission rates and composition;

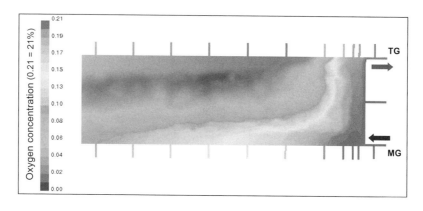

Figure 2.   Oxygen distribution in the longwall goaf—base model results for a typical mine

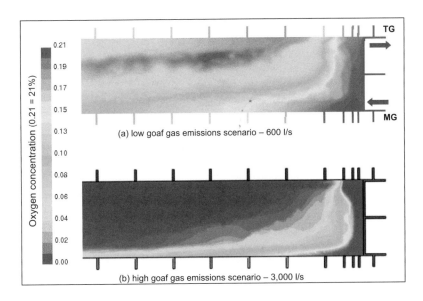

Figure 3.   Effect of goaf gas emissions on oxygen distribution in the longwall goaf

(ii) the effect of bleeder ventilation; (iii) the effect of changes in seam gradients; (iv) the effect of changes in face airflow; (v) the effect of various ventilation layouts; (vi) the effect of seal leakages; (vii) the effect of increased permeability on collapsed gateroads due to increased support density; (viii) the effect of geological features such as dykes; (ix) the effect of various goaf gas drainage strategies; and (x) the effect of various inertisation strategies.

The results of one of the typical parametric studies are presented in Figure 3, showing the effect of goaf gas emissions on oxygen distribution in the goaf. Analysis of the results show that oxygen ingress into the goaf was very high in the case of lower goaf gas emission flow rates of around 600 l/s, with around 10% oxygen at 600 m behind the face (Figure 3a). The results indicate that gas emission flow rate has a major effect on goaf gas distribution and the oxygen ingress into the goaf was generally lower with high goaf gas emissions of around 3,000 l/s. However, the oxygen ingress along the narrow collapsed maingate area was still high with around 10% oxygen at 300 m behind the face (Figure 3b). The results indicate that high goaf gas emissions alone would not be enough to completely inertise the goaf in most cases.

The parametric studies with bleeder ventilation indicated that the oxygen ingress into the goaf would be very high, with over 10% concentration even at 1,000 m behind the face. The modelling results with gas drainage through seals at the back of the goaf at lower gas concentrations also indicated similar high oxygen ingress patterns in the goaf. The results also indicated that leakage through the seals located at deeper sections of the goaf would lead to increased oxygen ingress into the goaf. The results indicate that all these factors should be considered carefully while developing proactive strategies for prevention of heatings and fires in longwall mines.

The research studies indicated that in addition to the above parameters, a number of other geological, mining and operational parameters would have a significant effect on goaf gas distribution and oxygen ingress into the longwall goaf (Balusu et al., 2002, 2004). For example, increasing the face airflow or increased permeability along the gateroads would significantly increase oxygen ingress into the longwall goaf. Detailed analysis of the CFD modelling results and field goaf gas monitoring results from various mines revealed that oxygen ingress depends on a number of factors including intake airflow, ventilation layout, pressure differentials, seal condition and leakages, caving characteristics, high permeability areas adjacent to the gateroads, goaf gas emission rates, seam gas composition, panel geometry, seam gradients and dip direction.

Heating Gas Flow Patterns in Longwall Goafs

CFD modelling studies have also been carried out to simulate the behaviour of heatings and related CO and H_2 gases flow patterns in the longwall goafs. The predicted gas distribution patterns of CO produced during early stages of heating development are shown in Figure 4. The results indicate that CO produced from heating on the MG side would disperse over a wide

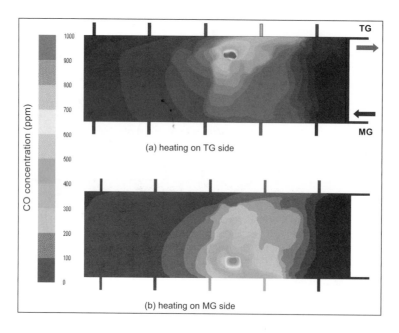

(a) heating on TG side

(b) heating on MG side

Figure 4. **CO gas dispersion patterns in the goaf—at early stages of heatings development**

area in the goaf (Figure 4b) before reporting to return airflow. The results show that CO readings in the goaf would reach 100s ppm before a significant change in CO levels in the TG return airway. The results shown in Figure 4a indicate that CO produced from heatings on the TG side would also disperse over a wide area in the goaf, although the dispersion area would be significantly smaller compared to the heatings on the MG side. The simulation results indicate that heating gases CO and H_2 would also disperse to the opposite side of the goaf as shown in Figure 4.

The simulations indicated that the dispersion patterns of CO and H_2 gases in the goaf are significantly different, particularly at advanced stages of heatings with H_2 gas dispersing towards higher elevations in the goaf and CO closely following goaf gas flow patterns. The results indicate that goaf gas monitoring is the best way to detect heatings or fires at early stage of development, as substantially higher readings can be measured in the goaf long before noticing any significant increase in CO readings in the TG return airway. The results of these simulations coupled with field data would be helpful in proper interpretation of goaf gas data during early stages of heatings development and in estimating the location of heatings in the goafs. The results of these studies are also useful for development of appropriate control strategies to prevent or manage heatings and fires in goaf areas.

Inertisation Simulations

Extensive parametric studies involving changes in inert gas injection locations, inert gas flow rates and different inertisation strategies were also carried out to investigate their effect on goaf inertisation in active longwall panels (Balusu et al., 2006). The results of one of the typical inertisation simulations are presented in Figure 5, showing the effect of inert gas injection at two different locations behind the longwall face. A comparison of the results of investigations with no inertisation and inert gas injection at 30 m and 200 m behind the face on maingate side is shown in Figure 5. In these simulations, inert gas was injected at the flow rate of 0.5 m³/s.

Comparison of (a) and (b) in Figure 5 indicates that inert gas injection at 30 m behind the face at a flow rate of 0.5 m³/s had negligible effect on oxygen ingress patterns into the goaf, with oxygen levels well over 13% at 300 m behind the face. The simulation results also indicated that increasing the inert gas injection rate to 1.0 m³/s at 30 m behind the face had only a minor effect on goaf gas distribution, with oxygen level around 10% at 300 m behind the face. These results indicated that inert gas injection just behind the face is not an effective strategy for goaf inertisation.

The results presented in Figure 5c indicates that inert gas injection at 200 m behind the face resulted in substantial reduction in oxygen ingress into the goaf, with oxygen levels reducing below 5% within 150 m behind the longwall face. Although the

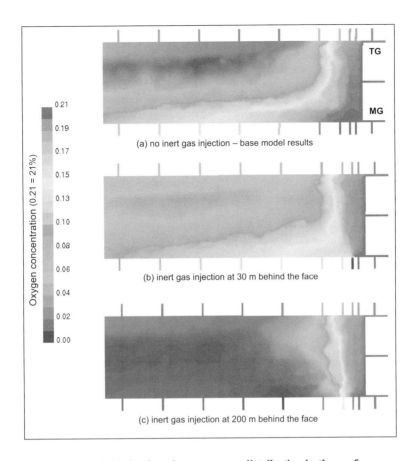

Oxygen concentration (0.21 = 21%)

0.21
0.19
0.17
0.15
0.13
0.10
0.08
0.06
0.04
0.02
0.00

TG

MG

(a) no inert gas injection – base model results

(b) inert gas injection at 30 m behind the face

(c) inert gas injection at 200 m behind the face

Figure 5.   Effect of inert gas injection location on oxygen distribution in the goaf

optimum location for inert gas injection in a long-wall panel depends on site specific parameters, the modelling simulations with various inertisation locations indicated that inert gas injection at around 200 to 400 m behind the face would be far more effective than inert gas injection close to the face.

FIELD STUDIES

Extensive field investigations were carried out over the years to obtain a better understanding of the longwall goaf airflow patterns, goaf gas and heating gases flow behaviour and the effect of various gas and sponcom control technologies and strategies. Tube bundle gas monitoring systems were installed in several longwall mines and used for extensive goaf gas monitoring during the field investigations. In addition to the continuous tube bundle data, weekly gas samples were also collected from all the cut-through seals around the goaf and were analysed using gas chromatograph. The results of a few typical field studies are presented in this section.

Monitoring of Heatings Development

During the course of research over the years, a number of heatings incidents have occurred in goaf areas of longwall panels. The data collected during one of the heating incidents on MG side of the longwall goaf is presented in Figure 6.

The results show that during the early stages of heating development (Figure 6a), the CO gas reading started increasing in the vicinity of the heating in the longwall goaf. The CO gas reading reached over 250 ppm, but the H_2 reading was only around 5 ppm. There was no change in CO readings at the tailgate longwall return gas monitoring location. The gas readings during the next stage of heating development are presented in Figure 6b. The time lag between these two sets of readings was around 3 weeks. The results show that CO and H_2 gases have spread to a wider area in the goaf and reached over 600 ppm CO and 250 ppm H_2 at MG seal locations in the vicinity of heating in the goaf. However, there

413

was no significant change in the gas concentration readings at the tailgate return monitoring location.

The data collected during other heating incidents on both MG and TG side of the longwall goaf have also shown similar trends, with CO and H_2 readings increasing significantly in the vicinity of heating without recording any major change in gas concentration levels at the tailgate return air monitoring location. The CO and H_2 gas dispersion patterns measured during these heating incidents in the field were similar to the CFD modelling patterns shown in Figure 4.

The goaf gas monitoring systems have detected the heatings in goaf areas during their development stages and assisted in successful control of the goaf heatings without completely sealing the longwall panels. The data obtained from the field studies highlight the importance of detailed goaf gas monitoring at cut-through seal locations and identify this strategy as one of the critical proactive strategies to prevent fires and explosions in longwall mines.

Proactive Inertisation Studies

Optimum inertisation strategies for active longwall panels have been developed based on the analysis of the modelling investigation results and review of the oxygen ingress patterns in the field at the mine sites. These proactive inertisation strategies have then been implemented at the mine sites to investigate their performance in the field, particularly with respect to their effect on oxygen ingress patterns in the goaf. The results of one such typical field study are presented in Figure 7.

The longwall face at this mine had progressed for about 450 m from the panel start-up line before the start of field trials of proactive inertisation. The normal goaf gas distribution in the panel showed that oxygen ingress into the goaf was very high with oxygen levels over 17% at 400 m behind the face. The proactive inertisation strategy for this panel consisted of injecting inert gas into the goaf through the cut-through seals near the start-up area, at the rate of around 0.15 to 0.5 m^3/s. The gas distribution in the longwall goaf before and after injection of inert gas is shown in Figure 7.

Field measurements show that this proactive inertisation strategy was highly successful in the field and substantially reduced oxygen ingress into the longwall goaf on both sides of the panel. Field data in Figure 7b show that the oxygen level was below 6% at the second cut-through behind the face, i.e., 200 m behind the face. As can be seen from comparison of Figures 7a and 7b, inert gas injection had also resulted in reduced oxygen ingress on the tailgate side of the panel. Field studies at other mine sites also demonstrated that the proactive inertisation strategies were highly successful in converting the general goaf environment into an inert atmosphere.

The results of the field studies carried out in a number of longwall panels at several mines have

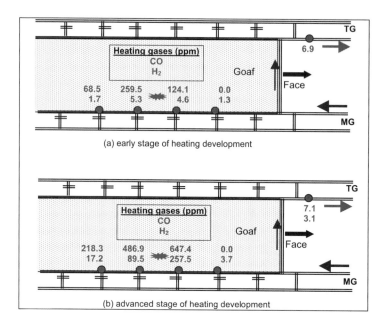

Figure 6. Goaf gas readings during different stages of heating development on MG side

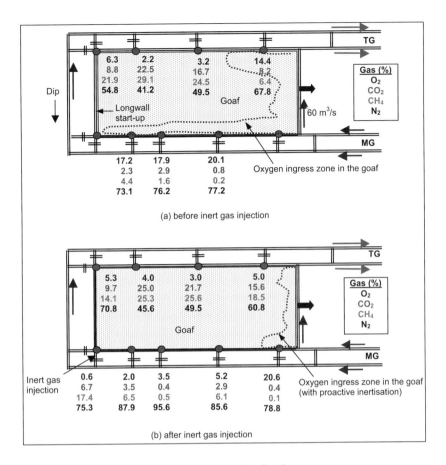

Figure 7. **Effect of proactive inertisation on goaf gas distribution**

provided large data sets and invaluable information on gas behaviour in longwall goafs. The field results were utilised in combination with modelling results to develop various proactive strategies for prevention of heatings and fires in longwall mines.

OPTIMUM PROACTIVE STRATEGIES

The longwall goaf gas distribution measurements at the mine sites showed that air ingress into the goaf was very high on the maingate side of the panels that could lead to heatings and fires development in the longwall goafs. It is recommended that the following proactive strategies may be introduced to prevent fires and explosions in longwall mines:

- Goaf gas monitoring
- Proactive inertisation
- Preparedness for fires control and emergency situations

- Longwall panel design optimisation to reduce crushing of barrier pillars
- Ventilation layout and design changes to reduce oxygen ingress into the goaf areas
- Optimisation of goaf gas drainage systems to minimise the risk of sponcom

The specific guidelines for key proactive strategies are presented in the following sections.

Goaf Gas Monitoring for Early Detection of Heatings

It is recommended that an extensive goaf gas monitoring plan should be implemented as one of the critical proactive strategies to detect goaf heatings at very early stages in order to prevent fires and explosions in longwall mines. A schematic of the longwall panel showing recommended locations of tune bundle monitoring points for continuous monitoring

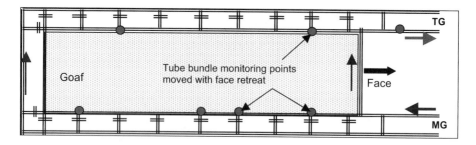

Figure 8. A schematic of goaf gas monitoring—locations of tube bundle monitoring points

of goaf gas conditions is presented in Figure 8. The monitoring points positioned immediately behind the face are to be moved forward with face retreat. The exact location of these continuous monitoring points depends on the depth of air ingress zones into the goaf areas at any specific longwall panel.

It is recommended that in addition to the continuous goaf gas monitoring, weekly bag samples should be collected from all or at least alternate cut-through seals around the working longwall panel goaf and analysed with gas chromatograph for detailed analysis of all gases including O_2, CH_4, CO_2, CO, H_2 and higher hydrocarbon gas concentrations. It is also recommended to monitor the adjacent sealed goaf for major changes in gas concentrations during face retreat operations, particularly near the suspected heating areas in the adjacent old goaf.

Proactive Intervention

It is recommended that proactive strategies and/or proactive inertisation be introduced in the case of the following scenarios to reduce the risk of heatings and fires in the longwall goafs:

- Longwall panels extraction in the coal seams with very high spontaneous combustion potential (and during extraction of any specific areas with high sponcom potential)
- Extraction in areas with severe geological disturbances, where it is expected that longwall retreat will be very slow for a period of weeks/months
- Longwall stoppage for prolonged periods—due to tailgate/face collapse or other reasons
- Abnormal CO readings behind the face or steady rise in CO readings at a number of seals in the active longwall goaf
- When the longwall face is retreating past suspected heating areas in the adjacent goafs
- Abnormal CO readings or other indications of heatings in adjacent sealed goafs (in this case, proactive inertisation should be introduced in the adjacent sealed goaf)

Proactive Inertisation Strategies to Reduce Oxygen Ingress into the Goaf

If the mining conditions at any specific site require proactive inertisation to be implemented, the recommended guidelines for proactive inertisation strategy are:

a. Inert gas should be injected into the goaf at 200 to 400 m behind the face, or inbye side of a suspected heating location in the goaf
b. Inert gas flow rate of around 0.5 m^3/s is recommended for most cases
c. Inert gas should be injected on intake side of the goaf in most cases
d. Inert gas may be injected through goaf holes (preferable, if goaf holes are available) or through cut-throughs on the intake side of the goaf
e. Inert gas may need to be injected on both sides of the goaf if some heating is suspected on return side of the goaf
f. Onsite nitrogen generation units or boiler gas are some of the feasible options for proactive inertisation into longwall goafs
g. Inert gas injection to be continued until face resumes normal production in case of prolonged stoppages or until face has retreated for more than 300 to 500 m past the suspected heating location, in the case of advanced heatings

Fire Control and Emergency Preparedness

The recommended guidelines for fire control and emergency preparedness strategy are:

a. All the seals should have sampling pipes installed near the roof and the sampling end of the pipes should be extended and supported up to the goaf edge.
b. The sampling points in some of the seals (in around 6 to 8 seals) should be connected to the tube bundle monitoring system for continuous goaf gas monitoring. Weekly gas bag samples

should also be collected from all or at least alternate seals around the goaf.

c. Inertisation pipes should be installed in at least alternate seals and should be connected through an inertisation pipeline to a cased borehole drilled from the surface.

d. The mine should have inert gas generators installed on the surface or the capability and the facilities to install and operate the units immediately, if required.

In addition to the proactive strategies discussed above, it is also very important to appropriately design the longwall panels, ventilation systems and goaf gas drainage systems to reduce the risk of sponcom, heatings and fires in longwall goafs. Specifically, the ventilation systems should be designed to reduce oxygen ingress into the longwall goaf, taking into consideration the panel layout and gas emissions. The goaf gas drainage system design and operation should ensure that gas drainage does not result in excessive oxygen ingress into the goaf area. The reduction in oxygen ingress into the goaf would significantly reduce the risk of sponcom and heatings development and consequently reduces the risk of fires and explosions in longwall mines.

CONCLUSIONS

The risk of heatings, fires and explosions is a major safety hazard in coal mines with severe consequences including risk to the safety of people, longwall face equipment and economic viability of the modern highly capital intensive longwall mines. It is critical to obtain a fundamental understanding of the gas flow patterns in longwall goafs and to investigate the effect of various parameters on goaf gas distribution in order to develop optimum proactive strategies.

The goaf gas distribution measurements at the mine sites showed that air ingress into the goaf areas was very high and was contributing to the development of heatings and fires, particularly in sponcom prone mines during slow retreat periods. Analysis of the modelling results and field data from various mines revealed that oxygen ingress depends on a number of factors including intake airflow, ventilation layout, pressure differentials, seal condition and leakages, caving characteristics, high permeability areas adjacent to the gateroads, goaf gas emission rates, seam gas composition, panel geometry, seam gradients and dip direction.

The results of field studies in combination with modelling results were utilised to develop a number of proactive strategies for prevention of heatings and fires in longwall mines. Technology transfer to the industry through implementation of the newly developed techniques and strategies at the mine sites has been one of the highlights of the research. The field studies at mine sites in Australia have shown that the proactive strategies were successful in reducing oxygen ingress into the goaf areas and in reducing the risk of fires and explosions in the longwall mines.

ACKNOWLEDGMENTS

This paper presents a brief summary of the work carried out under various ACARP projects and field studies at different mine sites in Australia. The authors would like to record their grateful thanks to the 'Australian Coal Association' and ACARP administration for funding this research over the years. The authors wish to thank the management and staff of the participating mines for their in-kind support, field data and permission to carry out the field investigations. We wish to express our special appreciation to Mr Bruce Robertson of Anglo Coal Australia Pty Ltd for his advice and valuable contribution to the research.

REFERENCES

Balusu, R., Deguchi, G., Holland, R., Moreby, R., Xue, S., Wendt, M., and Mallett, C. 2002. Goaf gas flow mechanics and development of gas and sponcom control strategies at a highly gassy coal mine. *Coal and Safety Journal*, No 20, March, pp. 17–34.

Balusu, R., Tuffs, N., Peace, R., Harvey, T., Xue, S., and Ishikawa, H. 2004. Optimisation of goaf gas drainage and control strategies—ACARP project C10017, *CSIRO Exploration and Mining Report 1186F*, June, 136 pp.

Balusu, R., Ren, T. X., and Humphries, P. 2006. Proactive inertisation strategies and technology development—ACARP project C12020, *CSIRO Exploration and Mining Report P2006/26*, January, 94 pp.

Bell, S., Cliff, D., Harrison, P., and Hester, C. 1998. Recent developments in coal mine inertisation in Australia, *Proceedings of the Coal 1998—1st Australasian Coal Operators' Conference*, 18–20 February, Wollongong, Australia, pp. 701–717.

Cliff, D. 2005. The ability of current gas monitoring techniques to adequately detect spontaneous combustion, *Proceedings of the Coal 2005—6th Australasian Coal Operators' Conference*, 26–28 April, Brisbane, Australia, pp. 219–223.

Development of Methods for Determining the Evacuation Routes in Case of Fire in Polish Underground Mines

Andrzej Strumiński
KGHM Cuprum OBR, Wroclaw, Poland

Barbara Madeja-Strumińska
Wroclaw University of Technology, Wroclaw, Poland

ABSTRACT: Numerous mine disasters caused by underground fires, mainly due to obstruct crew evacuation from threaten sites, are known in mining practice. One of the most efficient ways to rescue people during the fire in coal or ore mines is their evacuation from hazardous places. Currently, different methods for determining the escape routes are used. Some of those methods, especially the ones providing the miner with high level of safety, are presented in the paper.

INTRODUCTION

Present day Polish underground hard coal and copper mines are mechanized to a great extend, which ties with, among others, the possibility of occurrence of spontaneous fires. In most cases, the fires occur without any warning signs and spread very quickly thus presenting high safety risks for underground mine crews, especially, where vast and complex mine workings exist.

Very dangerous for mine crew are also the spontaneous fires in gobs from where, during low barometric pressure conditions, blackdamp, toxic or explosive gases may be discharged into the active working. Especially unsafe are gobs in methane bearing beds, which are also highly susceptible to self-ignition, creating highly dangerous conditions.

Numerous mine disasters caused by underground fires, mainly due to difficulties encountered during crew evacuation from threaten sites, are known in mining practice. One of the most efficient ways to rescue people during the fire in coal or metalliferous mines is their immediate evacuation from hazardous areas. Thus, the Polish mine regulations (Decree of Minister of Economy. 2002) say that in every underground mine the evacuation routes, called escape routes, must be specified.

In Poland, many scientists and mine specialists are investigating methods to determine safety escape routes (Badura, H., Sułkowski, J. 1994; Badura, H., Biernacki, K., Sułkowski, J., Żur, K. 1996; Ćwięk, B. 1987; Dziurzyński, W. 1988; Firganek, B., Marszałek, E. 1988; Kozdrój, M., Mola, J. 1996; Litwiniszyn, J. 1947; Pałka, T. 2001; Strumiński, A., Madeja-Strumińska, B. 2005; Strumiński, A., Madeja-Strumińska, B. 2007; Tracz, J., Wilczek, K. 1987; Trutwin, W. 1968; Turek, W., Hansen, S., Parchański, J. 2001; Walkiewicz, W. 1983).

This paper presents some methods for establishing the escape routes in Polish underground mines, with special consideration given to high level of personnel safety during the evacuation from dangerous areas.

FACTORS INFLUENCING EVACUATION ROUTES SELECTION

Major hazard resulting from underground mine fire is the spread of gases and fumes in mine workings. Gases and fumes are usually not diluted to safe levels and may catch the crew unawares causing intoxication or suffocation, even at a considerable distance from the fire center. Additionally, the combustible gases occur in the fire smokes which, in certain conditions, may explode (Budryk, W. 1961; Bystroń, H. 1977; Litwiniszyn, J. 1947; Strumiński, A. 1996). To determine the risk of gas explosion created during fires, Le Chatelier's index is used. The fire gases are considered explosive if the Le Chatellier's index is not less than 0.6 while oxygen concentration is higher then the minimum value, which is dependent on the concentration of combustible components (Budryk, W. 1961; Bystroń, H. 1977; Strumiński, A. 1996).

The temperature of fire gasses highly influences the safety of the crew being evacuated. High temperature would impede crew's evacuation from the hazardous areas as the temperature of fire gases at the fire site often exceeds 1000°C and sometimes reaches even 1500°C. However, as the distance from the fire center increases, the temperature of fire gases decreases.

The temperature of fire gasses may be determined based on the energy balance of combustion process or certain empirical relationships (Bystroń, H. 1977; Strumiński, A. 1996). According to (Badura, H.,

Biernacki, K., Sułkowski, J., Żur, K. 1996), the maximum time τ_{max}(min) that the mining personnel may remain in the area with high temperature level of fire gases T_g(°C) may be calculated from:

$$\tau_{max} = 1812\exp\left(-0,046T_g\right) \qquad (1)$$

For example, for gases' temperature of T_g=120°C, the maximum time is τ_{max}=10 min.

It has been shown that during the initial phase of underground fire development, the temperature of fire gases generally does not present risks to personnel's life. However, in some cases, high temperature of fire gases, especially due to exogenous fires, may make the evacuation from the dangerous areas impossible.

Appearance of fire smokes (gases) in mine workings during the underground fire limits the visibility in those workings, to lesser or higher degree, which can make it difficult, if not impossible, to evacuate the personnel. According to (Pałka, T. 2001), poor visibility due to fire smokes results in the decrease in crews' evacuation speed from 30 to 50% in comparison to full visibility conditions, with the crew wearing respiratory protective equipment (escape apparatus).

In Polish underground mines, the fire evacuation routes are established during the planning phase (Strumiński, A. 1996; Strumiński, A., Madeja-Strumińska, B. 2005; Strumiński, A., Madeja-Strumińska, B. 2007). An escape route is defined as a system of series-connected, unobstructed mine workings from the area threatened by fire smokes to the nearest node of ventilation network with smoke- and/or gas-free air flow or to the mine surface. When several routes are available for the evacuation, usually the escape route with the shortest traversability time to a smoke/gas-free area is selected.

During the selection of fire escape routes various factors are considered including: individual characteristics of human body, type of possible fire and phase of its development, type and condition of mine workings forming the escape route, complexity of ventilation network, type of facilities protecting the air-passages used at a given mine, time of their operation and level of miners' training.

Individual's physical and mental characteristics have a great impact on the speed of crew relocation, especially along the smoke-filled workings. Individual workers have different physical efficiency and resistance to toxic substances (Badura, H., Biernacki, K., Sułkowski, J., Żur, K. 1996). Moreover, under stressful conditions, which are usually caused during underground fires, some miners may experience mental impairment, causing panic among the crew being evacuated. That is why, in general, the withdrawal speed for a group of people is slower than the speed of relocation for an individual person; the speed of evacuation of a group is always determined by the speed of the slowest person.

The type of an underground fire and the level of its development highly influenced the temperature, density and toxicity of the fumes. Those factors, in turn, determine the speed at which the crew proceeds along the escape routes. For instance, during the developed fire of conveyor belts or diesel oil, the visibility in some workings may be minimal and, consequently, the speed of crew's movement will be very low. Selection of escape routes is highly dependent on the types of mine workings, their inclination and presence of equipment with the ease of passage being the main concern. Therefore, workings with high inclination, slippery floors, equipped with numerous machines blocking the way, low, narrow and with mould filled with water or sludge are avoided (Kozdrój, M., Mola, J. 1996).

Complexity of ventilation networks and especially presence of numerous diagonal or dependent air currents may create disturbances in the flow of fire gases, which may have detrimental effect on the escaping crew. However, if underground mine crew is frequently and systematically trained in emergency evacuation and self-rescue rules, the evacuation process would be more efficient (Strumiński, A., Madeja-Strumińska, B. 2007).

The analyzes of numerous fire evacuations show that visibility is a crucial component influencing the crew evacuation in underground mines. Experiments conducted (Badura, H., Biernacki, K., Sułkowski, J., Żur, K. 1996). to determine the impact of visibility on the crew evacuation showed, among others, that the crew evacuation when the working are smoke-filled is very difficult and, in many cases, when escape routes are very long, may exceed the so called available time i.e., the duration for which the breathing apparatus is designed to afford protection.

CHARACTERISTIC OF SOME METHODS FOR FIRE ESCAPE ROUTES

In the analysis of hazard resulting from mine workings smokiness, the knowledge of air propagation through mine ventilation network during underground fire is extremely important. In Poland, this issue has been under intensive investigation since the second half of twentieth century after a series of underground disasters due to fires worldwide, including Polish mines (Budryk, W. 1961; Bystroń, H. 1977; Strumiński, A. 1996; Strumiński, A., Madeja-Strumińska, B. 2007).

The impact of an underground fire on air propagation through the ventilation network were discussed, for example, in (Budryk, W. 1961; Dziurzyński, W. 1988; Litwiniszyn, J. 1947; Struminski, A. 1996; Trutwin, W. 1968). The issue of mine crew evacuation from dangerous areas during underground fires was also analyzed by many scientists (Badura, H., Biernacki, K., Sułkowski, J., Żur, K. 1996; Budryk, W. 1961; Bystroń, H. 1977; Ćwięk, B. 1987; Litwiniszyn, J. 1947; Pałka, T. 2001; Struminski, A., Madeja-Struminska, B. 2005; Struminski, A., Madeja-Struminska, B. 2007; Tracz, J., Wilczek, K. 1987; Trutwin, W. 1968; Walkiewicz, W. 1983).

In the sixties, the phenomenon of fumes propagation in mine ventilation networks during underground fires was presented as a random process in studies (Litwiniszyn, J. 1947; Trutwin, W. 1968), among others. In those studies, it was assumed that the fire formation site is known. However, mine practice shows that, generally, during the initial phase of fire development, the location of the fire within the ventilation network is unknown (Struminski, A. 1996; Struminski, A., Madeja-Struminska, B. 2005; Struminski, A., Madeja-Struminska, B. 2007). Thus, to improve the process of fire localization, various types of sensors are used to record physical and chemical parameters of the airflow, which enables early detection of underground fires. The precise placement of these sensors within the mines' ventilation networks enabling relatively accurate determination of possible sources of fumes, i.e., formation of fire centers, were presented, among others, in (Turek, W., Hansen, S., Parchański, J. 2001).

Numerous publications provide information related to methods for determining the escape routes in terms of optimal time of mine crew withdrawal from the fire hazard sites (Badura, H., Sułkowski, J. 1994; Firganek, B., Marszałek, E. 1988; Kozdrój, M., Mola, J. 1996; Pałka, T. 2001; Struminski, A., Madeja-Struminska, B. 2005; Struminski, A., Madeja-Struminska, B. 2007; Walkiewicz, W. 1983).

In one of the first studies, the total time needed to traverse the escape route is calculated as a sum of times needed to pass each horizontal, upward and downhill section of the escape route according to the formula:

$$\Theta(\alpha,z) = \sum_{n=1}^{n_p} \frac{L_{pn}}{w_p} + \sum_{n=1}^{n_w} \frac{L_{\alpha wn}}{w_{\alpha wn}} + \sum_{n=1}^{n_u} \frac{L_{\alpha un}}{w_{\alpha un}} \qquad (2)$$

where:

$\Theta(\alpha,z)$ time necessary to traverse the escape route from the work place to the nearest node of the

ventilation network (crossing), where stream of smoke-free air flows, min,

L_{pn}, $L_{\alpha wn}$, $L_{\alpha un}$ lengths of working sections being a part of escape route, horizontal, upwards and downhill, respectively, m,

w_p, $w_{\alpha wn}$, $w_{\alpha un}$ average speeds of crew relocation, wearing rescue apparatus, in smoke-filled workings, horizontal, upwards and downhill, respectively, m³/min.

Speeds used to calculate the time $\Theta(\alpha,z)$ i.e., w_p, $w_{\alpha wn}$, $w_{\alpha un}$ are calculated from suitable formulas, tables or charts (Badura, H., Biernacki, K., Sułkowski, J., Żur, K. 1996; Kozdrój, M., Mola, J. 1996; Pałka, T. 2001; Walkiewicz, W. 1983) or are determined experimentally.

Time $\Theta(\alpha,z)$ for traversing the escape route must meet the following criteria:

$$\Theta(\alpha,z) \le \tau_{au} \qquad (3)$$

where τ_{au}(min) means the time of operation of the breathing apparatus. In Polish mines, this time is usually 60 min.

The study (Badura, H., Biernacki, K., Sułkowski, J., Żur, K. 1996) presents the method for determination the escape routes from the smoke zone. In order to evaluate the fire formation in any working, the arbitrary established fire hazard index is used. The index is determined basing on the numerous factors describing the fire hazard. The factors have ascribed the specific scores. The number from 0 to 100 is a measure of hazard.

In this method, the crew escape time is, similarly as in the previous method, dependent on the crew speed. This time is corrected for so called concentrated (such as ventilation barriers) and distributed (such as flooded workings) obstructions.

If multiple evacuation routes are available, a decision matrix is constructed and advantages of each route are defined; an escape route which has the highest number of advantages can then be selected i.e., an optimum escape route. An advantage value is calculated as a difference between the critical time and time needed to pass along the specific escape route, where critical time is a maximum time of operation of the personal protection equipment. If the the escape routes are established according to the rules given in (Firganek, B., Marszałek, E. 1988), the target function, defined as total length of workings being part of escape routes with certain weighting parameters, must be minimized. The defined target function depends on the directions and velocities of airflow as well as propagation and anticipated changes to the concentration of fumes. It also takes into consideration the effort required by the crew to pass through the individual workings, hazards resulting from fire location, size of area covered by fire, fire propagation rate and changes in air temperature.

All these parameters are expressed by special indexes defined based on experiments and practical knowledge. In addition, to accurately implement this technique within the computer model, detailed knowledge of fire formation site is necessary; information which is usually not available in practice.

In Polish hard coal and copper mines, the escape routes are often determined based on the knowledge and experience of mine specialists or *a priori* imposed by the local ventilation network structure (Strumiński, A., Madeja-Strumińska, B. 2007). Generally, it is assumed that the time to traverse the escape route in case of smoke-filled workings is usually 30% longer than in smoke-free working. Thus, for smoke-filled workings, the evacuation time is calculated from the relationship:

$$\tau_i(\alpha,z) = 1,30 \frac{L_i}{w_i(\alpha)} \tag{4}$$

where:

$\tau_i(\alpha,z)$—time of passage along i section of smoke-filled evacuation route, min,

L_i—length of i section of this route, m,

$w_i(\alpha)$—speed (m/min) of passing i section of smoke-free working defined experimentally or from the following formulas (Pałka, T. 2001):

• uphill

$$w(\alpha) = 63,42\exp(-0,040\alpha) + 1,96 \tag{5}$$

• downhill

$$w(\alpha) = 79,61\exp(-0,018\alpha) - 11,99 \tag{6}$$

In formulas (4), (5) and (6), α means the angle of inclination of mine workings expressed in grades.

Adding the times needed to traverse the individual sections of the entire evacuation route (Formula 4), the total time necessary for passing this route, which includes smoke-filled sections, is calculated as:

$$\Theta(\alpha,z) = 1,30 \sum_{i=1}^{n} \frac{L_i}{w_i(\alpha)} \tag{7}$$

If the total time necessary for traversing the escape route under consideration $\Theta(\alpha,z)$ is shorter or equal to the time (τ_{au}) of the operation of the protective rescue equipment i.e., when relationship (3) applies, it is recognized that the selected route may be used for crew evacuation.

Theoretical calculations of the time necessary for traversing the escape routes by the mine crew not always corresponds with practice as variety of factors influence the speed of crew's passage and, especially, as mentioned previously, the fire smoke has the highest impact on the crew's evacuation speed. Therefore, a research project sponsored by the Polish Ministry of Science and Higher Education undertaken in 2005–2007, presented a method which allows the determination of evacuation routes based on the probabilities of mine workings, which form mine ventilation network, being filled with smoke (Strumiński, A., Madeja-Strumińska, B. 2005; Strumiński, A., Madeja-Strumińska, B. 2007). In this method, it is assumed that every mine working within this network may be a source of fire smokes (gases). The probability of the formation or existance of the smokes source in any working of ventilation network may be calculated from the relationship:

$$P(Z) = \frac{1}{B} \tag{8}$$

where B is a number of workings forming the considered ventilation network.

If an underground fire occurs and its location is unknown, the probability of a specific working being smoke-filled i.e., to which the fumes may flow from other workings, is equal to the sum of probabilities of those workings being smoke-filled, namely:

$$P_i(Z) = \sum_{r=1}^{R} P_r(Z) \tag{9}$$

For example, if we are interested in the probability of working 12–13 being filled with smoke (Figure 1) when data on fire location is limited, it is evident that air will flow to working 12–13 from workings: 1–2, 2–3, 3–4, 3–10, 4–10, 10–11, 11–12, and 12–13. Therefore, if in any of these workings, the fire fumes are present or the fire starts, working 12–13 will be filled with smoke.

Probability of fumes occuring in any working of considered ventilation network according to Formula (8) is P(Z)=1/40=0.025. Because the probability of fumes occurring in 12–13 air split is the sum of probabilities off all mentioned workings being smoke-filled, according to the relationship (9), that is:

$$P_{12-13}(Z) = \sum_{r=1}^{9} P(Z) = \frac{9}{40} = 0.225 \tag{10}$$

In order to define the optimal, in terms of absence of fumes, evacuation route, all D_u routes for withdrawing the workers from certain mine areas are determined based on the following formula:

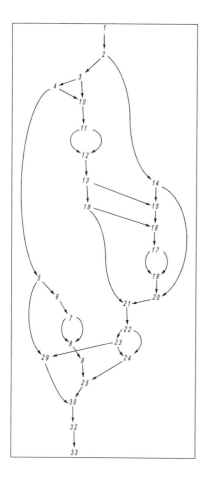

Figure 1. Fragment of simplified PS ventilation network

$$D_u = D\left[\left(d_1 - w_1\right),...,\left(d_n - w_n\right)\right] \qquad (11)$$

where:
$(d_1 - w_1),...,(d_n - w_n)$—mine workings used for mine crew evacuation,
d_i, w_i—initial and final node of i working.

Subsequent sections of evacuation route have different probabilities of being smoke-free. If the probability for the entire escape route is to be as high as possible then, due to the sequential nature of crew evacuation process then the probability of each section along this route must be as high as possible.

Probability of being smoke-free $P_\beta(\bar{Z})$ for a specific β working, which is part of ventilation network, is determined using the relationship:

$$P_\beta(\bar{Z}) = 1 - P_\beta(Z) \qquad (12)$$

To define the probability of entire evacuation route being smoke-free, the following formula is used:

$$P\left[D_u(\bar{Z})\right] = \prod\left[P_\beta(\bar{Z})\right]_{D_u} \qquad (13)$$

The optimum escape route is the one for which the probability of being smoke-free is the highest i.e., the route:

$$D_{uop} = \max_{D_u}\prod\left[P_\beta(\bar{Z})\right]_{D_u} \qquad (14)$$

If, due to the structure of ventilation network, components of the evacuation route must have a high probability of being smoke-filled, they should be equipped with the equipment that can facilitate as quick as possible relocation of mine crew towards safe and smoke-free workings (for example safety wires, reflective strips, infrared receivers and others) (Strumiński, A., Madeja-Strumińska, B. 2007; Turek, W., Hansen, S., Parchański, J. 2001)

SUMMARY

Underground mining of mineral resources in Poland, especially hard coal and copper ore, is carried out in difficult and hazardous conditions. One of the biggest hazards is the occurrence of underground fires, which many times resulted in mine disasters.

Current methods of determining the escape routes in hard coal and copper ore mines, especially the method which calculates the possibility of mine workings being smoke-filled, substantially improve safe crew evacuation from sites threatened by fire.

Currently, a new method of mine crew evacuation when the fire location is unknown is being developed (Strumiński, A., Madeja-Strumińska, B. 2007), which takes into account the presence of smoke due to existence of fire source as well as airflow disturbance.

REFERENCES

Badura, H., Sułkowski, J. 1994. Methods of determining escape routes for fire threatened crew. *WUG, Bezpieczeństwo Pracy i Ochrona Środowiska w Górnictwie*, no. 4/12.

Badura, H., Biernacki, K., Sułkowski, J., Żur, K. 1996. Factors deciding on the velocity of mine crew withdrawal along the smoked workings during fire in mine. *Przegląd Górniczy*, no. 6.

Budryk, W. 1961. *Mines ventilation. Part I, Ventilation of workings*, WGH, Katowice.

Bystroń, H. 1977. Phase of fire, combustion and gases explosion during active and passive extinguishing. *Archiwum Górnictwa*, vol. 42, no.1.

Ćwięk, B. 1987. Rules of self-rescue for underground workers during withdrawal from danger area. *Wiadomości Górnicze*, no. 2–3.

Dziurzyński, W. 1988. Predicting the ventilation process in underground mine during fire. *Studia Rozprawy Monografie*, vol. 56, IGSMiE PAN.

Firganek, B., Marszałek, E. 1988. System of computer supporting the mine fire action conductor. *Mechanizacja i Automatyzacja Górnictwa*, vol.5.

Kozdrój, M., Mola, J. 1996. Probabilistic method of determining the escape time of miners group in stenosed workings. *Przegląd Górniczy*, no.1.

Litwiniszyn, J. 1947. *Principles of mines ventilation. Underground Fires*. Biuro Wydawnictw Technicznych CZPW, Katowice.

Pałka, T. 2001. *Methods of determining escape routes in case underground fire*. PhD thesis. Instytut Mechaniki Górotworu PAN, Kraków.

Decree of Minister of Economy dated 28.06.2002 concerning work safety and hygiene and running the mine as well as special fire protection equipment in underground mines (Dz. U. Nr 139).

Struminski, A. 1996. *Fires fighting in underground mines*. Wyd. Śląsk, Katowice.

Struminski, A., Madeja-Strumińska, B. 2005. Evaluation of smokiness possibility along evacuation routes during underground fire basing on structural table of J. Litwinisz. *Archives of Mining Science* 50, Special Issue.

Struminski, A., Madeja-Strumińska, B. 2007. Methods for increasing the safety of mine crews evacuation during underground fire, *Cuprum* vol. 3 (44).

Tracz, J., Wilczek, K. 1987. Problems of optimum placing smoke sensors in ventilation network of mine. *Archiwum Górnictwa*, vol.23, no.2.

Trutwin, W. 1968. About proposal on certain measure of workers danger in mine ventilation network. *Archiwum Górnictwa*, vol.13, no.2.

Turek, W., Hansen, S., Parchański, J. 2001. Withdrawing the crew from underground mine and its protection during fire hazard following example of one Australia mine. *Wiadomości Górnicze*, no. 5.

Walkiewicz, W. 1983. Method of evaluating the escape routes in coal mines. *Bezpieczeństwo Pracy w Górnictwie*, no.1.

History of Research in Mine Refrigeration, Cooling, and Ventilation in South Africa

Steven J. Bluhm
BBE Consulting, Johannesburg, South Africa

Wynand M. Marx
BBE Consulting, Johannesburg, South Africa

Russell Ramsden
BBE Consulting, Johannesburg, South Africa

Frank H. von Glehn
BBE Consulting, Johannesburg, South Africa

ABSTRACT: Formal research started in South Africa in 1901 and, in the 1960s, the Chamber of Mines Research Organisation was formally established and derivative institutes continue actively today. Initial problems related mainly to silicosis, but with the extreme depth and heat of the gold mines (and latterly the platinum mines) South African mining research became renowned in the field of refrigeration, cooling, and ventilation in deep hot mines. During this period there often was close collaboration with the USBM and other international research institutes.

This paper examines the history of mine cooling and ventilation research in South Africa and highlights the major developments that have enabled mining to be carried out safely and productively in extremely deep and hot environments. From the earliest research the aim of the research has been to provide a healthy and safe working environment in the most cost effective manner.

THE EARLY YEARS

Emphasis on Dust

The Witwatersrand Chamber of Mines (later named Transvaal Chamber of Mines and subsequently Chamber of Miners of South Africa) was founded in 1887 to create a body that could publish authoritative information on the Witwatersrand gold discoveries (Lang, 1986). One of the earliest research efforts was concerned with the adverse effect of dust upon the health of the workforce in gold mines. In 1901 a committee was established by the Mine Manager's Association and the Transvaal Medical Society to investigate the high mortality rate among rock drillers in the Witwatersrand gold mines (Low 1994). The report showed that between October 1899 and January 1902, 85 out of 100 fatalities were as a result of Miner's Phthisis (pneumoconiosis), largely due to the high quartz content of the host rock. In addition, at least a quarter of underground workers were suspected of having the disease. A government commission released a report in 1903 with recommendations including wet drilling and the need for planned and controlled ventilation. The importance of ventilation aspects was recognised and the first paper presented to the Chemical, Metallurgical and Mining Society of

South Africa (the forerunner of the Southern African Institute of Mining and Metallurgy, SAIMM) in 1903 was entitled "The ventilation of deep levels." In 1914 the Chamber of Mines established a dust department, which introduced systematic sampling of all underground working sections. Initially dust sampling was carried out using sugar tubes. It was recognised that there was a need to develop instrumentation that would allow many dust samples to be taken throughout the many areas of mines so that strategies for addressing the adverse effects of dust could be developed. In 1916 the konimeter was developed by RN Kotze (the Government Mining Engineer) and this instrument became the industry standard and was still in extensive use in the 1980s when personal gravimetric sampling was introduced. Research into alternatives to the konimeter that performed satisfactorily in the harsh underground conditions and development of sampling procedures still continues (Gardiner, 1987).

An important milestone in mine ventilation development was the formation of the Mine Ventilation Society of South Africa (MVS) in 1944. The MVS provided a Monthly Bulletin, later to be renamed the Journal of the MVS in which developments and innovations in the ventilation field were

424

recorded and discussed. One of the early presidents of the MVS, AWT Barenbrug (employed at the New Consolidated Gold Fields Research Bureau), and a recognised pioneer in the development of mine ventilation published a paper on economic aspects of mine ventilation (Barenbrug 1948). He examined the relationship between air temperature and work output and compared the cost of increased ventilation quantities against the cost of introducing refrigeration. Barenbrug developed nomograms and tables which assisted ventilation engineers in various aspects of planning. This paper established the principles for providing a healthy and safe thermal environment in a cost effective manner.

Other significant developments that were taking place during these early years were concerned with noxious gases encountered in the Witwatesrand gold mines from blasting fumes and a filter for removing blasting fumes from the ventilation air was developed, (Rabson, 1955).

In 1962 WS Rapson was appointed Research Advisor to the Chamber of Mines. He recognised that there was a need to create a centre for mining research excellence that could satisfy the gold mining industry needs, and in 1964 he recommended the establishment of COMRO which would carry out mining research on a collaborative basis on behalf of the gold mining members of the Chamber of Mines. Prior to this the Chamber controlled three main laboratories: the Dust and Ventilation Laboratory, the Applied Physiology Laboratory and the Biological and Chemical Research Laboratory. The Applied

Physiology Laboratory had been established in 1949 by BFJ Schonland and was taken over by the Chamber of Mines in 1951. The new organisation COMRO also included the Mining Research Laboratory, the Physical Sciences Laboratory and an Environmental Services Division. The objective of the combined organisation was to build on earlier research to offset the costs of mining at increasing depths and temperatures and to improve working conditions underground. COMRO was extended to include the Collieries Research Laboratory in 1966 after the Coalbrook colliery roof collapse in which 435 miners perished. MDG Salamon was appointed Research Advisor in 1974, and in 1986 H Wagner took over.

Research into ventilation and cooling issues was carried out by the Environmental Engineering Laboratory initially under Whillier, and then in the late 1970s to the early 1990s by Sheer and Bluhm. Work on heat stress was carried out at the Industrial Hygiene Laboratory initially under Wyndham, and then from the mid 1970s by Kielblock. The focus of the work was mainly aimed at the deep, hot gold mines in South Africa with very little work being done for the emerging platinum industry (at that stage not being a full member of COMRO). Because of the particular geothermal properties in the Bushveld Igneous Complex, the platinum mines would soon reach virgin rock temperatures, similar to the deep gold mines, but at significantly shallower depths (see Figure 1).

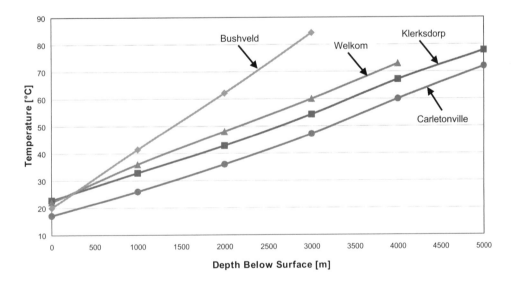

Figure 1.   Geothermal gradients in South African mining regions

Heat Stress at Increasing Depths of Mining

As mines became deeper the temperatures within the workings increased and this lead to many heat related incidents. The mining industry adopted a multi disciplined approach to this problem namely examining the causes of the heat related incidents and engineering methods to reduce the temperatures within the workings.

Research was conducted from 1930 onwards into the cause and numbers of heat stroke fatalities in individuals working at high wet-bulb temperatures (Wyndham 1965). A large number of cases occurred in the first four days of commencing underground work and this lead to the development of the Chamber of Mines acclimatization procedures (physical exercise in a controlled hot environment).

The research showed that recruits varied greatly in their tolerance to heat and this resulted in varying acclimatization times (between 14 days for heat intolerant recruits and 7 days for 'normal' recruits) being applied depending on the rise in oral temperature after one hour's work at a saturated temperature of 35°C (95°F). Wyndham concluded that the risks of heat stroke increased significantly at wet-bulb temperatures above 32°C (90°F). He proposed that wet-bulb temperatures in all working places should be reduced to 28.9°C (84°F) and below and that no working place should exceed 31°C (88°F).

These limits were the earliest recommendations for maximum temperatures within the South African mining industry. Since the publication of these limits additional work has been aimed at relating wet-bulb temperatures with productivity and occurrences of accidents. At wet-bulb temperatures above 33°C there is a tendency for a 'don't care' mental attitude to develop, productivity falls off drastically, accidents tend to become more frequent (for example, Smith, 1984), and the danger of heat stroke increases rapidly. Whillier, Director of the Environmental Engineering Laboratory of COMRO, called this the difference between heaven (28°C wet-bulb) and hell (33°C wet-bulb) (Whillier, 1976). These limits set the targets for ventilation and refrigeration engineers (see for example, Barcza and Lambrechts, 1961) and became the focus of research of the Environmental Engineering Laboratory of COMRO in the 1980s (see for example, Ramsden et al., 1987).

Although the benefits of the formal acclimatization programme were recognised, it was not popular with workers and continued research in the 1980s led to the development of the Heat Tolerance Screening tests (HTS). These screening tests defined recruits as heat tolerant or heat intolerant (Kielblock and Schutte, 1991). Heat tolerant individuals were allowed to work progressively in deeper sections as they acclimatized naturally. Those classified as heat intolerant were allocated to surface tasks. At wet-bulb temperatures below 27.5°C fit men were shown to be able to undertake all mining tasks in still air with no danger to their health and without the need for acclimatization.

COMRO RESEARCH IN THE LATER PERIOD UP TO 1990

In parallel with research work that was being carried out on human physiology, engineering research was carried out on providing an acceptable thermal environment in the most cost effective manner. To be able to control temperatures in underground workings it was necessary to understand the sources of heat and to be able to predict the influence of these sources at various depths and inlet conditions.

Two approaches were adopted for predicting stope wet bulb temperatures at increased depths. Lambrechts (1950) tabulated stope wet-bulb temperatures and suggested correlations to assist with practical temperature predictions. Graphical heat load prediction methods were developed by Barenbrug (1967), Whillier (1982) and Stroh (Hemp, 1982).

Goch and Patterson (1940) carried out a theoretical analysis of the steady state solution to the Fourier equations of rock heat conduction in a long cylindrical excavation. They developed a table which enabled the heat flow into an excavation to be estimated. With the advent of digital computers in the 1960s Starfield (1965) carried out finite difference and finite element analyses of the heat flow into a cross-section of rectangular tunnel. Starfield and Dickson (1967) used this information to calculate the air temperature along the full length of a tunnel using iterative methods. Hemp in 1995 used Starfield's basic data to obtain correlations between the numerical solution for a rectangular tunnel and the analytical solutions for the circular tunnel of Goch and Patterson. He also included the effect of varying wetness of the tunnel floor. This solution has been used as the basis for mine-wide temperature predictions in software such as Environ (von Glehn et al., 1987) and VUMA (Bluhm et al., 2001). In the 1980s COMRO carried out detailed field trials to verify prediction algorithms, see for example, Alexander et al. (1986) and Mathews et al. (1987).

From 1962 until 1994 the Chamber of Mines produced the "Annual Ventilation Returns" for gold mines, which listed various statistics such as working place wet-bulb temperatures vs. depth, and air quantity circulating in a mine as a function of rock production. A typical graph for the rise in installed

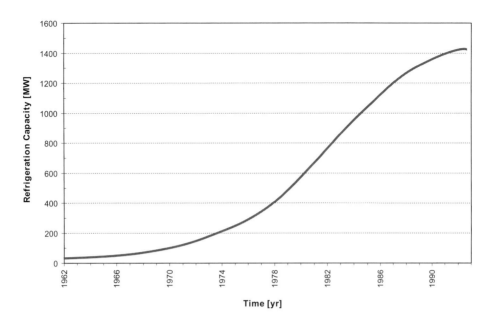

Figure 2.   Rise in refrigeration in South African gold mines

refrigeration is shown in Figure 2 (Denysschen and Mackay, 1994).

Reduction of Mine Heat Loads

During this period the work on ventilation and cooling at COMRO concentrated first on the reduction of mine heat loads at their source and then secondly on development of new technologies to ventilate and cool mines. This was also an era where Personal Computers became freely available to ventilation practitioners and ground breaking simulation software that could predict aerodynamic and thermodynamic interaction of the entire underground ventilation circuit was developed (Environ). These programs could quantify the benefits of changes to mine layouts to reduce mine heat loads, concentrated mining, backfill and insulation of mining intake airways and controlled recirculation (Burton et al., 1984).

Although the work on reducing mine heat loads clearly showed the overall benefits of concentrated mining and increased rates of mining through mechanisation using large diesel fleets it created new problems regarding heat and diesel pollutants in localised areas.

Refrigeration and Cooling Distribution

To combat high temperatures underground, refrigeration had been introduced into Witwatersrand gold mines since the 1930s. Refrigeration produces cold water and effective and practical ways of cooling mine air continually invited extensive research and development of direct and indirect water-to-air heat transfer equipment, such as cooling coils and spray chambers, (Bluhm, 1990). However, there is a limit to the cooling that can be applied to the air and in 1976 Whillier proposed chilling the water used for dust suppression and drilling as additional cooling (Whillier, 1976 and van der Walt and Whillier, 1976). This allowed cooling to be more effectively distributed to the zones where it was needed, i.e. the working face and remote developments, and in any case it was shown that the unless the service water is cooled to below the design wet-bulb temperature, it would act as a heat load and make it difficult to achieve lower design temperatures. The distribution of large quantities of water underground led to research into energy recovery devices, such as turbines, hydro lift devices (three chamber pipe feeders) to minimise the temperature rise of this water.

Other work by COMRO during this period resulted in the development of a software package to simulate the performance of water chilling installations (Bailey-McEwan and Penman, 1987). The use of ice was also examined to reduce the quantity of water required for cooling, (Sheer, 1983). The main advantage of using ice rather than water for distributing cooling from surface is due to its high latent heat which leads to smaller mass flow rates of return water being pumped to surface (approximately one fifth of the corresponding amount of water is required). Various methods for conveying ice from surface to underground melting dams were examined and guidelines were produced for distributing particulate ice from surface to underground reliably through piping systems (Sheer, et al., 1986).

THE PERIOD AFTER COMRO

In the late 1980s there was a cut-back in funding for formal research and a reduction in COMRO's research programme followed. In 1992 COMRO was taken over by the CSIR, the government statutory research council, and was renamed Miningtek. The early directors Garrett and Fritz set the scene for Gürtunca who headed Miningtek from 1996 till 2003. Miningtek research was divided into three Programmes focussing on Rock Mechanics, Environmental Engineering and Mining systems. The Environmental Engineering Programme was the centre for refrigeration, cooling and ventilation research under the leadership of Schutte, Smith, Marx, and Willis. Following this period Miningtek was reformed as part of greater strategic transformation of the CSIR. During the transition period Miningtek was headed by Vogt and the Environmental Engineering Programme by van Zyl.

Because of the release from the auspices of the Chamber of Mines, research was now extended to include problems encountered in the platinum mines. Research was carried out for individual mines/groups on an elective basis and in 1998 the collaborative DEEPMINE research programme was launched under Gürtunca's leadership and involved Mine Owners, Labour and Government. Research teams included CSIR, universities, expert individuals and mine owner specialists. The objective of DEEPMINE was to develop expertise and technology to mine gold safely and profitably at ultra-depth (3 to 5 km) in the Witwatersrand basin of South Africa (Bluhm et al., 2000; Funnell et al. 2000; Funnell et al. 2001). The challenges included worker physiology (Barotrauma, cognitive ability, etc.) at these extreme depths and engineering challenges to ventilate and cool such mines. Research work included both an investigation part to understand the challenges better and the development of new technologies to enable mining. For instance, in-stope cooling would be a given at these depths and DEEPMINE facilitated the development of a lightweight, quiet and efficient in-stope cooler. After a planned life of three years, the successful DEEPMINE research programme led to another collaborative research programme namely FUTUREMINE.

FUTUREMINE focused on improved productivity and reduced mining costs, and the initiative was sub-divided into four broad streams of technological expertise: rock engineering and seismicity; mining engineering and ore-body management; software and communications and ventilation, cooling and refrigeration. Broadly, research in mine refrigeration, cooling and ventilation was directed in areas related to the recirculation of air, cyclical use of ventilation and cooling systems, air scrubbing technology, optimization of chilled water reticulation including the use of ice, minimizing heat losses in underground dams, tunnel insulation, obtaining direct cooling at the work-face, improving underground refrigeration system through dry air-cooling and water heat rejection systems. Safety issues included provision of ventilation and refrigeration in emergency situations such as fan and refrigeration plant stoppages as a result of power failures, and a number of software simulation programs to be used in specific applications such as predicting ventilation and heat flow transients and real-time monitoring of underground environmental conditions (Biffi et al., 2005).

Based on the success of DEEPMINE, the coal mining industry and Miningtek formed COALTECH 2020. COALTECH 2020 was strongly focused on productivity and environmental research and to date little work has been done on ventilation issues. PLATMINE is a collaborative research programme formed by the platinum mining industry. Refrigeration, cooling and ventilation research for this programme has been limited and mainly determined the applicability of the gold mine research, specifically DEEPMINE and FUTUREMINE research.

None of the above collaborative research programmes concentrated on mine health and safety since this was adequately addressed by another research body. The Safety in Mines Research Advisory Committee, known as SIMRAC, was established with the principal objective of advising the statutory Mine Health and Safety Council on the determination of the safety risk on mines and the need for research into safety on mines based on the safety risk. In 2003 a partnership of the South African Government, Industry and Labour committed to Safety, Dust and Noise Milestones to be achieved by 2013. With regard to dust, by the end of 2008, 95% of individual measurements would not exceed the legal limit of 0.1 mg/m³ and by 2013 no new cases of silicosis would be recorded in individuals employed from 2008 onwards. Although recent SIMRAC work is focused on achieving these goals, some work has also been carried out into ventilation aspects of diesel pollutants.

THE CURRENT SITUATION

South African energy cost escalations and the short-age in electricity generating capacity, encouraged major electricity users in South Africa to re-evaluate their power use. This led to a search for energy effi-cient technology and power saving strategies specifi-cally by South African mines. Research in this regard focused on main fans and refrigeration systems. However it should be noted that very little collabora-tive research and development work in mine refrig-eration, cooling and ventilation is currently being conducted in South Africa. Most of the work is being done to address mine specific problems.

Mine refrigeration, cooling and ventilation sys-tems are electrical energy intensive and lend them-selves to improved energy efficiency because of the cyclic operation of mining and the diurnal changes in surface temperatures. Opportunities exist in reducing main fan absorbed power by reducing fan speed, pre-rotation of inlet air stream, or damping the fan outlet. Pre-rotation of inlet air via Radial Vane Controllers has proved to be a most practical and cost effective option, especially when retrofitting existing main fans, and this technology has been implemented on many South African mines with great success (Pooe et al. 2009.).

Investigations into the use of ice and/or water thermal storage and cyclical control of ventilation and cooling systems to reduce energy consumption have also resulted in significant savings with several ice and water thermal storage projects being imple-mented in South Africa (Marx et al., 2008, Wilson et al., 2003 and Wilson et al., 2005).

Other refrigeration research efforts investigated the potential combination of solar energy and refrig-eration, the comparison between hard ice and vac-uum ice, and the status of water vapour refrigeration technology and equipment.

In addition, recent work has again examined the effect of wet-bulb temperature on accidents and pro-ductivity in South African platinum mines (Marx et al., 2009). The work also aimed to establish an optimum design temperature in the context of modern labour requirements and current engineering/equipment costs and realistic power cost projections. From the base-line evaluation, the conclusion was that the optimum design wet-bulb temperatures should be about 25.6°C for mines with average depth around 900 m, but this will increase to, for example, 28.5°C for deeper mines with average depth approaching 1700 m. The fact that the optimum design tempera-ture increases clearly demonstrates the increased difficulty of building and operating refrigeration equipment in the deeper mines (see Figure 3).

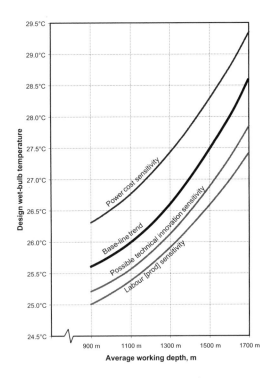

Figure 3. Optimum design wet-bulb tempera-ture for base-line and possible future trends

CONCLUSION

Most of the research carried out in mine refrigera-tion, cooling and ventilation over the past 100 years is applied in practice today. In South Africa mining is taking place at Virgin Rock Temperatures of up to 70°C. Simulation software and equipment, meth-ods and systems born out of research (especially DEEPMINE) are used and implemented daily. For example, one of the deep gold mine projects in South Africa will have 120 MW of refrigeration installed, including 100 kg/s of ice being send down the mine and about 1300 l/s of chilled water circulated through the mining sections. An additional challenge is the shortage of electricity forcing designs to be optimal from an energy efficiency point of view.

It is also evident that this research is spreading to other parts of the world as refrigeration is becom-ing necessary in mines in Central and North Africa, North and South America, Australasia and the East.

Research in mine cooling and ventilation is not being carried out with the same intensity as in the past. It is the role of governments, mine owners and researchers to continually review the require-ments and goalposts in mine refrigeration, cooling and ventilation and to ensure that the necessary knowledge and skill bases are protected into the future.

REFERENCES

Alexander N.A., Bluhm S.J., March T.C.W., Bottomley P., von Glehn F.H., 1986. The measurement of the heat load in a deep level stope in the Klerksdorp Gold Field, Journal of the Mine Ventilation Society South Africa, Vol 39, No 10.

Bailey-McEwan M., Penman J.C., 1987. Interactive computer program for simulating the performance of water chilling installations on mines, APCOM 87 Symposium, Johannesburg.

Barcza N. and Lambrechts J. de V., 1961. Ventilation and air conditioning practice in South African Gold mines, Trans Seventh Commonwealth Mining and Metallurgical Congress, SAIMM, Johannesburg.

Barenbrug AWT, 1967. The battle of the B.T.U.s, Journal of the Mine Ventilation Society of South Africa, Vol. 20.

Barenbrug A.W.T., 1948. Economy in mine ventilation, Bulletin of the Mine Ventilation Society of South Africa, Vol 1, March.

Biffi M., du Plessis J., and Marx W., 2005. The FutureMine Collaborative Research Initiative: Making Research Work, Proceedings, 8th International Mine Ventilation Congress (Ed. A D S Gillies), Brisbane QLD, (The Australasian Institute of Mining and Metallurgy, Carlton).

Bluhm S.J., 1990. Thermal performance of direct-contact water-air heat exchangers, PhD Thesis, University of the Witwatersrand.

Bluhm S.J. et al., 2000. Generation and distribution of refrigeration for ultra-deep mining: new challenges and insights. Proceedings: Frigair 2000, Gallagher Estate, South African Institute of Refrigeration and Airconditioning, Johannesburg.

Bluhm S.J., von Glehn F.H., Marx W.M. and Biffi M., 2001. VUMA mine ventilation software, Journal of the Mine Ventilation Society of South Africa, Vol 54, No 3.

Burton R.C., Plenderleith W., Stewart J.M., Pretorious B. and Holding W., 1984. Recirculation of air in the ventilation and cooling of deep gold mines. Proceedings, Third International Mine Ventilation Congress, Harrogate, England, The Institution of Mining Engineers.

Denysshen C.J and Mackay J.G., 1994. Annual ventilation returns for the period 1992–1993, CSIR-Miningtek.

Funnell R.C., Bluhm S.J., Kempson W.J., Sheer T.J. and Joughin N.C., 2000. Examination of cooling effects in stopes using cold hydropower water. Proceedings ASHRAE-FRIGAIR Congress, SAIRAC, Johannesburg.

Funnell R.C., Bluhm, S.J. and Sheer T.J., 2001. Optimisation of cooling resources in deep stopes. Proceedings 7th International Mine Ventilation Congress, EMAG, Krakow.

Gardiner L.R., 1987. Feasibility of personal gravimetric dust sampling for health hazard assessment in the South African mining industry, Proceedings Mine Safety and Health Congress, Chamber of Mines of South Africa, Johannesburg.

Goch D.C. and Patterson H.S., 1940. The heat flow into tunnels, J Chem. Met. Min. Soc of SA, Vol. 1.

Hemp R., 1982. Sources of heat in mines, in Environmental Engineering in South African Mines, Burrows J. (ed), The Mine Ventilation Society of South Africa.

Hemp R., 1995. Air temperature increases in airways, Journal of the Mine Ventilation Society of South Africa, Vol 38.

Kielblock A.J. and Schutte P.C., 1991. Heat stress Management, A comprehensive guideline, Chamber of Mines of South Africa Research Organization, Johannesburg.

Lambrechts J. de V., 1950. The estimation of ventilation air temperatures in deep mines, Journal of the Chemical and Metallurgical Society of South Africa, Vol 50.

Lang J., 1986. Men, Mines and the Challenge of Conflict, Bullion Johannesburg, Ball Publishers, Johannesburg.

Low M., 1994. History of the MVS, Mine Ventilation Society of South Africa, Johannesburg.

Matthews M.K., McCreadie H.N. and March T.C.W., 1987. The measurement of heat flow in a backfilled stope, Journal of the Mine Ventilation Society of South Africa, Vol 40, No 11.

Marx, W.M., von Glehn F.H. and Wilson R.W., 2008. Design of energy efficient mine ventilation and cooling systems MVS Symposium 2008, The Mine Ventilation Society of South Africa, Johannesburg.

Marx W.M. ;et al., Identification of optimum underground working temperatures, Anglo Platinum Personal Communication, 2009.

Pooe T., Marx W., du Plessis J., Janse van Rensburg J., van den Berg L., 2009. Main fan energy management—actual savings achieved, Proceedings 9th International Mine Ventilation Congress, New Delhi.

Rabson S.R., 1955. Investigation into the elimination of nitrous fumes, Journal of the Mine Ventilation Society of South Africa, Vol 8, No. 5.

Ramsden R., Bluhm S.J., Sheer T.J. and Burton R.C., 1987. Research into the control of the underground thermal environment in South African gold mines, Proceedings Mine Safety and Health Congress, Chamber of Mines of South Africa, Johannesburg.

Sheer T.J., 1983. Investigations into the use of ice for cooling deep mines, Proceedings: Frigair '83 Conference, ASHRAE/SAIRAC, Pretoria.

Sheer T.J., del Castillo D., Csatary C.J.L., 1986. Unconventional systems for removing heat from mines, Proceedings Frigair '86, ASHRAE/SAIRAC, Pretoria.

Smith O., 1984. Effects of a cooler underground environment on safety, labour and labour productivity, Proceedings, Third International Mine Ventilation Congress, Harrogate, England, The Institution of Mining Engineers.

Starfield A.M., 1965. A theoretical study of heat flow in rock surrounding underground excavations, PhD thesis, University of the Witwatersrand, Johannesburg.

Starfield A.M. and Dickson A.J., 1967. A study of heat transfer and moisture pick-up in mine airways. South African Institute of Mining and Metallurgy, Vol 68.

Van der Walt J. and Whillier A., 1978. Considerations in the design of integrated systems for distributing refrigeration in deep mines, Journal of the South African Institute of Mining and Metallurgy, Vol 79, December.

Von Glehn F.H., Wernick B.J., Chorosz C. and Bluhm S.J., 1987. Environ—a computer program for the simulation of cooling and ventilation systems on South African mines. Proceedings 12th International Symposium on the Application of Computers and Mathematics in the Mineral Industries, SAIMM, Vol 1, Johannesburg.

Whillier A., 1976. Machine water, the forgotten factor in mine cooling, Journal of the Mine Ventilation Society of South Africa, Vol 29, June.

Whillier A., 1982. Present day trends on cooling practices on South African gold mines, Paper presented at the 12th CMMI Congress, Johannesburg.

Wilson R.W., Bluhm S.J. and Smit H.J., Surface bulk air cooler concepts, producing cold air and utilising ice thermal storage. Proceedings: Managing The Basics, CSIR Conference Centre, The Mine Ventilation Society of South Africa, Johannesburg, 2003.

Wilson R.W., Bluhm S.J. and von Glehn F.H., Thermal storage and cyclical control of mine cooling systems, Journal of the Mining Ventilation Society South Africa, Vol. 58, No. 4, October/December 2005.

Wyndham C.H., 1965. A survey of the causal factors in heat stroke and of their prevention in the gold mining industry, Journal of the South African Institute of Mining and Metallurgy, Vol. 66, No. 11.

A Summary of USBM/NIOSH Respirable Dust Control Research for Coal Mining

Jay F. Colinet
NIOSH, Pittsburgh, Pennsylvania, United States

James P. Rider
NIOSH, Pittsburgh, Pennsylvania, United States

John A. Organiscak
NIOSH, Pittsburgh, Pennsylvania, United States

Jeff Listak
NIOSH, Pittsburgh, Pennsylvania, United States

Gregory Chekan
NIOSH, Pittsburgh, Pennsylvania, United States

ABSTRACT: Workers in the coal mining industry can be exposed to elevated levels of respirable coal and silica dust, which can lead to the development of Coal Workers' Pneumoconiosis or silicosis. These lung diseases can be disabling or fatal depending upon the dose of dust exposure that is received. Research into the development of dust control technologies for coal mining operations was initiated in the United States Bureau of Mines and continues today within the National Institute for Occupational Safety and Health. A summary of the historical developments of this dust control research and its impact on the coal mining industry will be provided.

INTRODUCTION

Inhalation of excessive levels of respirable coal mine dust can lead to the development of Coal Workers' Pneumoconiosis (CWP), commonly called Black Lung Disease, a chronic disease of the lungs. Likewise, excessive inhalation of respirable silica dust can result in the lung disease called silicosis. Both of these lung diseases can be disabling or fatal. In coal mining occupations, dust exposure typically occurs for 10 years or more before signs of the disease can be detected. Once contracted, there is no cure for these lung diseases so the goal is to prevent the disease by limiting the dust exposure of mine workers through the implementation of effective dust control technologies.

In the United States (US), the passage of the Federal Coal Mine Health and Safety Act of 1969 (1969 Act): established respirable dust standards to limit exposure to coal and silica dust, specified dust sampling requirements for assessing compliance with the dust standards, created a voluntary x-ray surveillance program for underground coal miners to detect CWP in the industry, and provided a benefits program for those miners that contracted CWP.

The 1969 Act established the respirable coal mine dust standard of 2 mg/m³ over an 8-hour shift, as long as the dust sample contains 5% silica

or less. If silica is greater than 5%, a reduced dust standard is calculated with the formula of 10 ÷ (% silica). Consequently, if a sample contains 10% silica, a reduced dust standard of 1 mg/m³ is enforced. Enforcement of a reduced dust standard is designed to limit exposure to respirable silica dust to a maximum of 100 µg/m³. The Mine Safety and Health Administration (MSHA) and mine operators are required to collect personal dust samples on a periodic basis to assess compliance with these dust standards. Researchers at the US Bureau of Mines (USBM) and now the National Institute for Occupational Safety and Health (NIOSH) routinely analyze this sampling data to identify high-risk occupations. Research is then directed at developing control technologies to reduce the dust exposures for these occupations.

The Division of Respirable Disease Studies of NIOSH administers the x-ray surveillance program for underground coal miners. Approximately every five years, underground coal mine workers can volunteer to have a chest x-ray taken and analyzed for signs of lung disease. Figure 1 illustrates the prevalence rate of CWP since the inception of the x-ray program (NIOSH 2008) and shows that significant reductions in CWP have been realized since the passage of the 1969 Act. From 1970–1975, nearly 33%

of examined miners with 25 or more years of experience were diagnosed with CWP. With a recent upturn in prevalence, approximately 8% of miners with 25 or more years of experience that were examined from 2000–2006 were diagnosed with CWP.

Although significant reductions in CWP have been realized, the human and financial toll of CWP on the coal mining industry has been staggering. From 1970 through 2004, CWP has been the direct or contributing cause of 69,377 deaths of underground coal mine workers (NIOSH 2008). In addition, from 1980–2005, over $39,000,000,000 in CWP benefits have been paid to underground coal miners and their families (NIOSH 2008).

With the passage of the 1969 Act, a major reorientation and expansion of the USBM's research efforts were realized (Tuchman and Brinkley 1990). Health research with a significant focus upon dust control was added to traditional USBM research related to fires, explosions, and explosives. Research was conducted to address various sources of dust and examine a variety of control technologies. This paper will summarize a number of control technologies that were developed through USBM/NIOSH research that have been shown to be successful in reducing dust levels and have had a positive impact on the dust control efforts of the coal industry.

LONGWALL DUST CONTROLS

Longwall mining was introduced in the US in the 1960s (Arentzen 1970) and initially was not very productive. Average longwall production was 490 tons per shift during compliance sampling by MSHA inspectors in 1970 (Niewiadomski 2009). Longwalls continued to improve in performance and by the mid 1980s, 112 longwalls were operating (Sprouls 1986) and averaged 1500 tons per shift during MSHA compliance sampling (Niewiadomski 2009). Continuing increases in the power of longwall equipment, improvements in equipment availability and mining practices, and increases in panel size have allowed longwall mining to substantially increase in production. Despite having 46 active longwall faces in 2008 (Fiscor 2009), longwall mining accounted for 50% of the total underground coal production in the US (EIA 2009). Average shift production during MSHA compliance sampling increased to 5200 tons per shift in 2008 (Niewiadomski 2009). Unfortunately, higher production generates greater quantities of respirable dust (Webster, Chiaretta, and Behling 1990).

Shearer Dust Control

On average, the shearer accounts for approximately 50% of the dust generated on the longwall face

(Colinet, Spencer, and Jankowski 1997) and shearer operators can have some of the highest dust exposures of underground coal mine workers. Water sprays can be used to suppress dust before it becomes airborne, "capture" dust out of the airstream, or act as small fans by inducing airfow. Various water spray types and spray locations were evaluated on shearer drums to identify the most effective combination. Pick face flushing sprays operated at 100 psi or less were shown to be the most effective for dust control (Shirey, Colinet, and Kost 1985).

One of the most successful uses of water sprays was a directional spray system for the shearer, commonly called the shearer-clearer (Jayaraman, Jankowski, and Kissell 1985). This spray system utilizes the air moving capabilities of the water sprays to direct shearer-generated dust away from the shearer operators. The sprays create a split in the ventilating air coming down the longwall face so that dust from the shearer is held near the longwall face. A physical barrier provided by a "splitter arm" mounted on the headgate side of the machine also helps develop this split in the face airflow. These directional sprays, which are mounted on the splitter arm and the shearer body, are directed at an angle with the ventilating air so that they move air without forcing dust into the walkway. Figure 2 combines a schematic illustrating the principle of the shearer clearer spray system with a photo of the sprays operating on a shearer at an underground face. Recent surveys on 10 longwall faces across the US revealed that all operations were using some variation of this directional spray system (Rider and Colinet 2007).

Headgate Dust Control

In the past, the stageloader-crusher unit was a significant source of dust liberation (Colinet and Jankowski

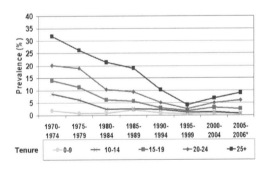

Figure 1. Trends in CWP prevalence by tenure among examinees employed at underground coal mines, National Coal Workers' X-ray Surveillance Program

1997) and this dust had the potential to expose all workers on the longwall face. Improvements in dust control for the stageloader-crusher include the use of high volume/low pressure water sprays to suppress dust before it becomes airborne and completely enclosing the stageloader-crusher to prevent generated dust from being entrained into the airstream (USBM 1985a). Research indicated that the quantity of water applied at the crusher was more important than the pressure at which it was applied. High pressure sprays could force dust out of the crusher unit, allowing it to become entrained. Likewise, by completely enclosing the crusher and stageloader, the dust would be isolated from the ventilating air and not contaminate intake air to the face. Figure 3 illustrates these controls for the stageloader-crusher.

Ventilation can improve dust control through two methods. Increasing the quantity of air can further dilute generated dust, while increasing air velocity can more quickly move dust clouds away from the breathing zone of workers. Therefore, it is desirable to utilize all available ventilating air on the longwall face. Unfortunately, because of roof bolting in the headgate entry, the gob area behind the shields at the headgate may stay open longer than other areas of the gob. This open area allows intake air coming up the headgate entry to flow into the gob and away from the face. The installation of a section of line brattice between the first shield and the rib (gob curtain), as shown in Figure 4, can increase air velocity down the face by 35% (Jankowski, Jayaraman, and Potts 1993). During recent longwall surveys, several mines were found to extend this curtain along the first five to ten shields to maximize the quantity of intake air that travels down the face (Rider and Colinet 2007).

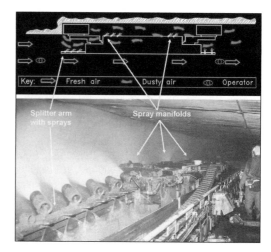

Figure 2. Schematic of shearer-clearer water spray system (above) and sprays installed on a shearer (below)

CONTINUOUS MINER DUST CONTROLS

Cutting Techniques

One of the most effective methods of reducing worker exposure to respirable dust is by minimizing the quantity of dust generated. Research has shown that this can be accomplished by utilizing optimized cutting techniques along with optimally designed bits. Undercutting of roof rock can reduce the amount of respirable dust generated, particularly dust potentially containing high quantities of silica (USBM 1985b). Likewise, research on various bit designs showed that dust generation can be minimized when the bit is designed with a large carbide insert and

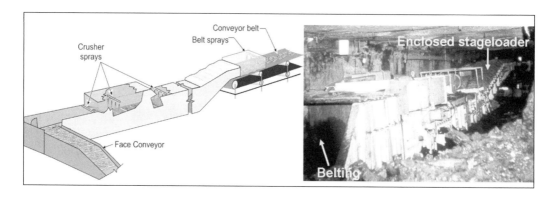

Figure 3. Schematic of sprays and enclosed stageloader-crusher (left) and actual installation (right)

Figure 4. Schematic of gob curtain location (left) and photo of gob curtain installed (right)

smooth transition to the bit body (Organiscak, Khair, and Ahmad 1996). Finally, regardless of cutting technique or bit selection, a key to minimizing dust generation during cutting is to routinely inspect the bits and replace worn bits. This is especially critical when cutting rock since silica becomes a concern and bits tend to wear more quickly.

Water Spray Systems

As early as 1939, water sprays were shown to be a significant improvement in reducing airborne dust when compared to dry cutting (Kobrick 1970). In the 1980s, USBM research focused upon improving the performance of water sprays by researching a number of variables that could impact dust control and included nozzle type, nozzle location, spray quantity, and spray pressure. Results from this research indicated that different nozzles were better suited for different applications. In general, it was found that sprays located close to the cutting bits and operated at pressures less than 100 psi were effective in suppressing dust at the cutter head without creating turbulence to force dust to roll back toward the miner operator (Jayaraman, Kissell, and Schroeder 1984). In addition, large-orifice sprays operated at low pressure were mounted on the underside of the cutting boom to wet the coal as it is being loaded by the gathering arms, while also minimizing turbulence (Foster-Miller, Inc. 1986).

More recently, NIOSH research has shown that the air moving capabilities of sprays can be utilized to help confine dust in the face area (Goodman 2000). Blocking sprays were mounted on each side of the miner and sprayed at an angle away from the miner body to induce airflow toward the face and prevent roll back along the sides of the machine.

For spray systems to be effective, the nozzles must remain open and free of debris that can cause clogging. Dirt and rust particles in the water line are the primary cause of clogging. A simple, non-clogging water filtration system was designed and offers improved protection for spray nozzles (Divers 1976).

Flooded-bed Scrubbers

A flooded-bed scrubber is a fan-powered dust collector mounted on the continuous miner and is capable of removing over 90% of the respirable dust from the air that is drawn into the collector (Colinet 1997). In addition, the scrubber helps move air toward the cutting face, which has enabled mine operators to receive approval from MSHA to take cuts greater than 20 feet in depth. The combination of these operating characteristics has led to scrubbers being used today on the majority of continuous mining machines.

Flooded-bed scrubbers capture dust-laden air through inlets located near the cutting boom of the miner, carry this air through ductwork to the rear of the miner, and pass the air through a filter panel that is wetted with water sprays, as shown in the schematic on the left in Figure 5. As dust particles impact and travel through the filter panel, they mix with water droplets and are removed from the airstream by a mist eliminator. The cleaned air is discharged from the scrubber back into the mine environment. The density and type of media used in the filter panel influence the dust collection efficiency and air-moving capacity of the scrubber. NIOSH conducted research to quantify the dust collection efficiency of multiple scrubber filters and their impact on scrubber airflow (Colinet and Jankowski 2000). Filters ranged in efficiency from nearly 75% for less dense filters to over 90% for the densest filter, as shown in the graph on the right side of Figure 5. This research also showed that the overall performance of a flooded-bed scrubber depends on the collection efficiency of

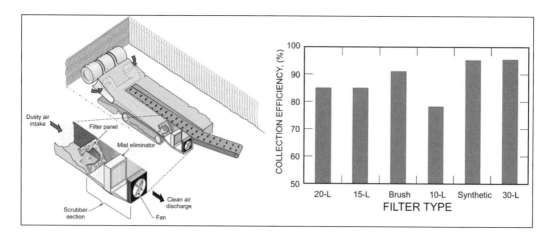

Figure 5. Schematic of flooded-bed scrubber (left) and collection efficiency for different filters (right)

the filter panel and the capture efficiency, which is the percentage of face air drawn into the scrubber. Both factors are impacted by filter selection.

ROOF BOLTER DUST CONTROLS

Roof bolter operators can have high silica dust exposures since their job requires them to drill through roof rock that may contain large quantities of silica. The primary means of controlling dust from roof bolting in the US is through the use of dry vacuum collection systems installed on the bolters. When operating properly, this system has been shown to be quite effective in capturing drill-generated dust (Beck and Goodman 2008). However, roof bolter operators can be exposed to several dust sources which include: dust not captured by the collection system and liberated at the drill hole, collected drill dust that passes through the collection system and is discharged back into the mine air, dust liberated while maintaining the collection system, and dust generated by the continuous miner when the bolter is downwind of the miner. It should also be noted that the mining cycle should be structured to eliminate or minimize the need of the bolter to work downwind of the miner.

Dry Dust Collectors

A vacuum pump pulls air and drill cuttings out of the drill hole, through the drill steel and hoses, and into a dust collector box. The collector box is equipped with cyclones and a cartridge filter to remove dust from the airstream before discharging the cleaned air back into the mine atmosphere. However, to operate effectively, all components of the system must be maintained on a periodic basis and this maintenance

can expose the roof bolter operators to high dust levels.

Dust must be emptied from the collector box and typically, this is done by opening the door on the box and pulling the dust out onto the mine floor, left photo in Figure 6. Although relatively short in duration, a high concentration dust plume can be created and the collector dust can soil the clothes of the worker cleaning the box. Sampling of this activity has shown that dust levels up to 14 mg/m³ were measured (Goodman and Organiscak 2002). If this technique is used, care must be taken so that the bolting machine in positioned in a entry with good air velocity. The mine worker should be positioned so that any dust generated by the emptying of the box is quickly carried away from the worker's breathing zone.

NIOSH recently evaluated a collector bag that was introduced to reduce dust exposure from the dust box cleaning. A retrofit kit is installed in the collector box that allows a bag to be inserted to capture the majority of the dust entering the collector box, right photo in Figure 6. The use of dust collector bags allows workers to easily remove the contained dust from the main compartment, which reduces dust exposure and cleaning time (Listak and Beck 2007). The bag can be deposited against the rib so that equipment running in the entry does not disturb the dust. Also, it was found that the cartridge filter can remain in operation for a much longer time period due to reduced dust loading. The popularity of this control continues to grow throughout the industry.

Drill Bits

Two common types of drill bits have been available for use on roof bolting machines. Shank type bits

Figure 6.   Miner cleaning dust collector box (left) and retrofitted collector bag filled with dust (right)

were mounted on the drill steel and dust from drilling was captured through a hole located in the drill steel. A second type of bit, the "dust hog" design, had a hole in the body of the bit for capture of the drill dust. This design placed the capture point closer to the dust generating bit tip, which was particularly beneficial in the first few inches of drilling. Testing was conducted that showed the dust hog bit to be at least 67% more effective in dust capture for the first 12 inches of the drill hole (Colinet, Shirey, and Kost 1985). The majority of bolting operations are using the dust hog style of bit.

SURFACE DRILL DUST CONTROLS

At surface coal mines, the greatest dust threat to mine workers is typically associated with silica dust exposure which occurs from the drilling and removal of the rock overlying the coal seam. Historic MSHA sampling results show that the surface drill operators are at greatest risk of overexposure to respirable silica dust. Bull dozer operators are typically the second highest occupation with silica overexposures. Research has focused upon developing controls for surface drills and also, improving the protection afforded by enclosed cabs on mobile equipment.

Controlling Surface Drill Dust

Dry dust collection systems are the most common type of dust control incorporated into surface drills by the original equipment manufacturers in the US because of their ability to be operated in freezing temperatures. The typical dry dust collection system is shown in Figure 7 and is comprised of a self-cleaning (compressed air back-pulsing of filters) dry dust collector sucking the dusty air from underneath the shrouded drill deck located over the hole. The

greatest quantity of drill dust is generated by compressed air (bailing airflow) flushing the drill cuttings from the hole and blowing the dust into the ambient air.

Typically, conveyor belting or other rugged material is used to construct a deck shroud that encloses the drill table. As shown in Figure 7, this shroud is designed to confine dust within the shroud for capture by the dry collector. Research has shown that improved dust capture is realized when the gap between the shroud and the ground is minimized. This can be accomplished by better vertical positioning of the drill table shroud by the operator to minimize the ground-to-shroud gap (Organiscak and Page 1999). Also, the ground-to-shroud gap can be reduced by using a flexible shroud design that can be mechanically raised and lowered to the ground via cables and hydraulic actuators. An adjustable height shroud design maintains a better seal with uneven ground and was found to keep dust emissions next to the shroud below 0.5 mg/m^3 at several drill operations (Organiscak and Page 2005).

Two other parameters that can minimize dust escape from the shroud are an optimized collector-to-bailing-airflow ratio and a shrouded collector discharge. Research examined the relationship between the quantity of air sucked from under the shroud by the collector and the amount of bailing air used to flush the hole. Dust emissions were significantly decreased around the shroud when a collector-to-bailing airflow ratio at or above 3:1 was achieved, even in the absence of a tight shroud to ground seal (Organiscak and Page 2005). After the collector accumulates dust, this dust is dropped from a discharge port that can be several feet above the ground. With high wind velocities, this dust can be entrained into the air and expose the drill operator or nearby

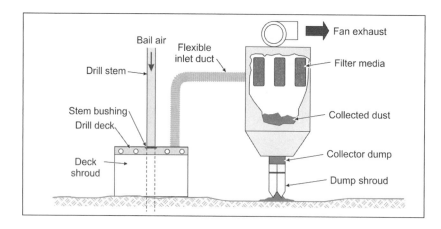

Figure 7.   Typical dry dust collection system used on surface drills

workers. Research showed that this dust can be minimized by installing a simple shroud on the collector dump (Reed, Organiscak, and Page 2004) that
extends to the ground as shown in Figure 7. These
shrouds can be fabricated by wrapping brattice cloth
around the perimeter of the collector discharge port
and securing it with a hose clamp in a few minutes.

Enclosed Cabs

An enclosed cab filtration system is the primary engineering control for reducing exposure to airborne
respirable dust for mobile equipment operators at
surface mines. Enclosed cabs with heating, ventilation, and air conditioning (HVAC) systems are typically integrated into the drills and mobile equipment
to protect the operator from the outside environment.
Air filtration is often part of the HVAC system as an
engineering control for airborne dusts. When new,
these enclosed cabs and filtration systems can offer
acceptable protection to the equipment operator.
However, as the equipment ages, the effectiveness
of the filtration systems and the integrity of the cab
degrades and may not offer suitable protection.

A number of research efforts were completed
to demonstrate the potential of retrofitting filtration and pressurization systems onto enclosed cabs.
These research efforts identified several factors that
are critical in order to control dust levels inside the
cab to less than 10% of levels observed outside of
the cabs. Significant improvements were achieved in
the protection provided by older cabs in field studies
where positive cab pressures at or above 0.07 inches
of water gauge were achieved (Cecala et al. 2003).
The positive pressure prevents dust from outside of
the cab from being blown into the cab. Improvements
in cab integrity were achieved by sealing cracks and

gaps with silicone and sealing doors with closed cell
foam tape.

The filters used for the intake air circuit must
be 95% efficient or higher in removing respirable
sized dusts (Organiscak et al. 2004). Also, a recirculation circuit substantially improved the protection
afforded by enclosed cabs. Laboratory experiments
showed an order of magnitude increase in cab protection factors when using an 85% to 95% efficient filter as compared to no recirculation filter (Organiscak
and Cecala 2007). This testing also showed that the
time for the interior cab concentration to decrease
and reach stability after the door has been opened
and closed was cut by more that half when using a
recirculation filter.

DISCUSSION

Prior to the 1969 Act, mandatory dust sampling was
not required by the federal government and dust
exposures and CWP rates were high. In April of
1968, the USBM initiated a study to measure the dust
exposure of numerous occupations in underground
coal mines (Doyle 1970). Table 1 lists the dust concentrations obtained with a gravimetric sampler for a
number of occupations. The range and mean concentrations are the actual concentrations that were measured, while the MRE equivalent column of the 1968
data shows the mean dust levels after being multiplied by 1.38. These data then become comparable to
today's MSHA compliance sampling results, which
use the same 1.38 MRE equivalency factor (Tomb
et al. 1973).

Table 1 also shows mean dust levels for 2008
from the MSHA inspector compliance sampling
database for these same occupations. All occupations
show substantial reductions in MRE equivalent dust

Table 1. Comparison of dust levels from 1968 and 2008 for select occupations

| | USBM Dust Samples Collected in 1968 | | | | | MSHA 2008 Samples | |
Occupation	Number of Mines	Number of Samples	Range, mg/m³	Mean, mg/m³	MRE Equiv* mg/m³	Number of Samples	Mean, mg/m³
Continuous miner operator	21	178	0.02–21.44	4.08	5.63	3,830	0.89
Cutting machine oper.	15	98	0.71–15.42	3.69	5.09	92	1.14
Loading machine oper.	18	97	0.25–39.56	3.75	5.18	139	0.32
Roof bolter operator	25	296	0.09–38.50	2.46	3.39	5,147	0.68

* MSHA calculates a MRE equivalent concentration by multiplying the gravimetric dust concentration by 1.38.

exposures, with miner and bolter operators having reductions of 84% and 80%, respectively. Despite the recent upturn shown in Figure 1, CWP prevalence has also been substantially reduced from levels identified in 1970. The significant reductions in dust exposure and disease have been realized through the combined efforts of: the USBM/NIOSH developing control technologies, MSHA promulgating and enforcing dust standards, mine operators implementing dust controls, and mine workers effectively utilizing these controls.

However, the recent upturn in CWP indicates that more work needs to be done. A comment by Mr. James R. Garvey at the Symposium on Respirable Coal Mine Dust (USBM 1970), which was held on November 3–4, 1969, sheds some guidance on reducing dust exposure among miners:

> The only effective way in which compliance with the standards can be achieved is by controlling the exposure of the miner. He must learn how to do his job with a minimum of exposure to high dust levels, and he must know how to take full advantage of whatever means are available, such as water sprays, ventilation, etc., to keep dust away from his breathing zone. This will not be easy, because the respirable dust with which we are concerned is invisible. The miner must learn how to avoid it without being able to see it or otherwise detect it.

Mr. Garvey's comments on control technologies are still valid today in that multiple controls must be utilized to control dust levels and NIOSH continues to support a strong dust control research program. His last comment provides an opportunity to mention the results of NIOSH research in the area of instrumentation that offers a significant advantage to today's miner in efforts to further prevent CWP.

The Personal Dust Monitor (PDM) is a real-time dust sampler that provides the wearer with a running average of their respirable dust exposure from the beginning of the shift (Volkwein et al. 2006). The PDM also displays the percent of the allowable dust standard that has been reached. When miners are wearing this sampling instrument, they will have the ability to monitor their dust exposure while still having the opportunity to influence their subsequent exposure over the remainder of the shift. Ideally, improvements in dust controls can be made that will prevent overexposures from occurring. The PDM is approved for use in underground coal mines by MSHA and became commercially available in July of 2009. Utilization of the PDM along with application of improved control technologies should result in reduced dust exposures for mine workers and further reductions in lung disease among mine workers.

REFERENCES

Arentzen, R.F. 1970. Machine design: state of the art. In Proceedings of the Symposium on Respirable Coal Mine Dust, Washington, DC, November 3–4, 1969, USBM Information Circular 8458, pp. 178–183.

Beck, T.W., and Goodman, G.V.R. 2008. Evaluation of dust exposures associated with mist drilling technology for roof bolters. Mining Engineering, Vol. 60, No. 12, pp. 35–39.

Cecala, A.B., Organiscak, J.A., Heitbrink, W.A., Zimmer, J.A., Fisher, T., Gresh, R.E., and Ashley, J.D. 2003. Reducing enclosed cab drill operator's respirable dust exposure at surface coal operation with a retrofitted filtration and pressurization system. SME Transactions 2003, Vol. 314, Littleton, CO: SME.

Colinet, J.F., Shirey, G.A., and Kost, J.A. 1985. Control of respirable quartz on continuous mining sections. USBM Contract J0338033, NTIS Publication No. PB86-179546.

Colinet, J.F., Spencer, E.R., and Jankowski, R.A. 1997. Status of dust control technology on U.S. longwalls. In Proceedings of the 6th International Mine Ventilation Congress, Pittsburgh, PA, May 17–22, pp. 345–351.

Colinet, J.F., and Jankowski, R.A. 1997. Ventilation practices on longwall faces in the United States. In Proceedings of the 5th International Symposium on Ventilation for Contaminant Control, Ottawa, Ontario, Canada, September 14–17.

Colinet, J.F. 1997. Exposure to silica dust on continuous mining operations using flooded-bed scrubbers. NIOSH Hazard ID. DHHS (NIOSH) Publication No. 97-147.

Colinet, J.F., and Jankowski, R.A. 2000. Silica collection concerns when using flooded-bed scrubbers. Mining Engineering, Vol. 52, No.4, pp. 49–54.

Divers, E.F. 1976. Nonclogging water spray system for continuous mining machines: installation and operating guidelines. U.S. Department of the Interior, Bureau of Mines, Information Circular 8727. NTIS No. PB 265-934.

Doyle, H.N. 1970. Dust concentrations in the mines. In Proceedings of the Symposium on Respirable Coal Mine Dust, Washington, DC, November 3–4, 1969, USBM Information Circular 8458, pp. 27–33.

EIA. 2009. Annual Coal Report 2008. Energy Information Administration, U.S. Department of Energy, Washington, DC, Report No.: DOE/EIA 0584 (2008), http://www.eia.doe.gov/cneaf/coal/page/acr/table3.pdf

Fiscor, S.J. 2009. Total number of longwall faces drops below 50. Coal Age, Vol. 114, No. 2, pp. 24–25.

Foster-Miller, Inc. 1986. Development of optimal water spray systems for ventilation of coal mine working faces. U.S. Bureau of Mines, Contract Report No. J0113010, 292 pp.

Goodman, G.V.R. 2000. Using water sprays to improve performance of a flooded-bed scrubber. Appl. Occup. Env. Hygiene, Vol. 15, No.7, pp. 550–560.

Goodman, G.V.R., and Organiscak, J.A. 2002. An evaluation of methods for controlling silica dust exposures on roof bolters. SME Transactions, Vol. 312, Littleton, CO: SME.

Jankowski, R.A., Jayaraman, N.I., and Potts, J.D. 1993. Update on ventilation for longwall mine dust control. U.S. Bureau of Mines, Information Circular 9366, NTIS No. PB94-112190.

Jayaraman, N.I., Kissell, F.N., and Schroeder, W. 1984. Modify spray heads to reduce dust rollback on miners. Coal Age, Vol. 89, No.6, pp. 56–57.

Jayaraman N.I., Jankowski, R.A., Kissell, F.N. 1985. Improved shearer-clearer system for double-drum shearers on longwall faces. U.S. Bureau of Mines, Report of Investigations RI 8963, 11 p.

Kobrick, T. 1970. Water as a control method, state of the art sprays and wetting agents. In Proceedings of the Symposium on Respirable Coal Mine Dust, Washington, DC, November 3–4, 1969, USBM Information Circular 8458, pp. 123–132.

Listak, J.M., and Beck, T.W. 2007. Evaluation of dust collector bags for reducing dust exposure of roof bolter operators. NIOSH Technology News, No. 2007–119.

Niewiadomski, G. 2009. Summary of valid longwall "DO" samples by calendar year. Mine Safety and Health Administration, personal communication.

NIOSH. 2008. The Work-Related Lung Disease Surveillance Report, 2007. DHHS (NIOSH) Publication No. 2008-143a, September, 2008.

Organiscak J.A., Khair A.W., and Ahmad M. 1996. Studies of bit wear and respirable dust generation. SME Transactions, Vol. 298, Littleton, CO:SME.

Organiscak, J.A., and Page, S.J. 1999. Field assessment of control techniques and long-term dust variability for surface coal mine rock drills and bulldozers. International Journal of Surface Mining, Reclamation, & Environment, Vol. 13, No. 4, pp. 165–172.

Organiscak, J.A., Cecala, A.B., Thimons, E.D., Heitbrink, W.A., Schmitz, M., and Ahrenholtz., E 2004. NIOSH/industry collaborative efforts show improved mining equipment cab dust protection. SME Transactions 2003, Vol. 314, Littleton, CO: SME.

Organiscak, J.A., and Page, S.J. 2005. Improve drill dust collector capture through better shroud and inlet configurations. DHHS (NIOSH) Publication No. 2006-108, NIOSH Technology News, No. 512, October, 2005.

Organiscak, J.A., and Cecala, A.B. 2007. Recirculation filter is key to improving dust control in enclosed cabs. NIOSH Technology News, No. 528, October.

Reed, W.R., Organiscak, J.A., and Page, S.J. 2004. New approach controls dust at the collector dump point. Engineering and Mining Journal, July, pp. 29–31.

Rider, J.P., and Colinet, J.F. 2007. Current dust control practices on US longwalls. Longwall USA, Pittsburgh, PA, Proceedings # 3564, June 4–6.

Shirey, G.A., Colinet, J.F., and Kost, J.A. 1985. Dust control handbook for longwall mining operations. USBM Contract J0348000, NTIS Publication No. PB86-178159.

Sprouls, M.W. 1986. Longwall census '86. Coal Mining, Vol. 23, No. 2, pp. 32–33.

Tomb, T.F., Treaftis, H.N., Mundell, R.L., and Parobeck, P.S. 1973. Comparison of respirable dust concentrations measured with MRE and modified personal gravimetric sampling equipment. U.S. Department of Interior, Bureau of Mines, Report of Investigation 7772.

Tuchman, R.J. and Brinkley, R.F. 1990. A history of the Bureau of Mines Pittsburgh Research Center. Pittsburgh, PA, US Department of the Interior

USBM 1970. Proceedings of the Symposium on Respirable Coal Mine Dust, Washington, DC, November 3–4, 1969, Information Circular 8458, p253.

USBM 1985a. Improved stageloader dust control in longwall mining operations. Department of Interior, Bureau of Mines, Technology News No. 224, August.

USBM 1985b. How twelve continuous miner sections keep dust levels at 0.5 mg/m3 or less. Department of Interior, Bureau of Mines, Technology News No. 220, July.

Volkwein, J.C., Vinson, R.P., Page, S.J., McWilliams, L.J., Joy, G.J., Mischler, S.E., and Tuchman, D.P. 2006. Laboratory and field performance of a continuously measuring personal respirable dust monitor. DHHS (NIOSH) Publication No. 2006-145. Report of Investigation 9669, September.

Webster, J.B., Chiaretta, C.W., and Behling, J. 1990. Dust control in high productivity mines. SME Preprint No. 90-82, Littleton, CO: SME.

An Overview of USBM/NIOSH Technology to Reduce Silica Dust in Metal/Nonmetal Mines and Mills

Andrew B. Cecala
NIOSH, Pittsburgh, Pennsylvania, United States

John A. Organiscak
NIOSH, Pittsburgh, Pennsylvania, United States

Gregory J. Chekan
NIOSH, Pittsburgh, Pennsylvania, United States

Jeanne A. Zimmer
NIOSH, Pittsburgh, Pennsylvania, United States

James P. Rider
NIOSH, Pittsburgh, Pennsylvania, United States

Jay F. Colinet
NIOSH, Pittsburgh, Pennsylvania, United States

ABSTRACT: Inhalation of respirable silica dust can lead to silicosis, a disabling and potentially fatal lung disease. For the metal/nonmetal industry, silica exposures occur during the mining and processing of the ore. A focus of the U.S. Bureau of Mines (USBM) and now the National Institute for Occupational Safety and Health (NIOSH) has been to conduct research to lower silica concentrations in metal/nonmetal operations. This report reviews various control techniques that have been developed at the USBM/NIOSH over the past 30 years and have been demonstrated to be successful at lowering respirable silica levels at metal/nonmetal mines and mills.

INTRODUCTION

In 1977, the Mine Safety and Health Administration implemented the Federal Mine Safety & Health Act which states "the first priority and concern of all in the coal or other mining industry must be the health and safety of its most precious resource—the miner" (MSHA 2009a). Obviously, this Act impacted the health and safety research program of the USBM, and subsequently NIOSH. One of these research programs was to lower the respirable dust exposure of mine workers, and as it specifically relates to this paper, silica dust in metal/nonmetal mines and mills.

The U.S. metal/nonmetal mining industry is composed of a wide range of different types of mineral operations and processes. In 2007, there were 12,270 mines (256 underground/12,014 surface) in the metal, nonmetal, stone, and sand and gravel industries, employing 149,180 miners (MSHA 2009b).

Throughout the metal/nonmetal mining cycle, ore is mined underground or in open pits/quarries. The ore is then processed in a mineral processing facility where it goes through a series of crushing, grinding, cleaning, drying, and product sizing sequences as it is processed into a marketable commodity. Because these operations are highly mechanized, they are able to process high tonnages of ore. This in turn generates large quantities of dust, often creating elevated levels of respirable silica dust, which is liberated into the work environment and exposes the miners that work at these operations.

The serious health effects from overexposure to respirable dust from these mining operations has been known for many years. When one considers that the ore being processed within the M/NM mining industry often contains crystalline silica, the health effects to the miners are even greater (NIOSH 2002; Hessel et al. 1988). This is the reason that the Mine Safety and Health Administration (MSHA) mandates a reduced respirable dust standard when silica is present (MSHA CFR 2008).

The purpose of this document is to review control techniques that have been developed over the past 30 years by the USBM and the NIOSH since the implementation of the 1977 Safety and Health Act. Many different engineering control techniques developed in this research program have improved the health of miners by lowering respirable silica dust exposure at U.S. metal/nonmetal mines and mills.

DUAL BAG NOZZLE SYSTEM

The dual bag nozzle system was designed to reduce major dust sources of the bag filling process, and thus, lower the dust exposure of the bag operator. A number of dust sources must be controlled to achieve this goal. The primary dust sources from the fill nozzle and bag valve area are product blowback and product spewing from the fill nozzle area. Product blowback occurs as excess pressure builds inside the bag during bag filling and is then relieved by air and product exiting the bag around the fill nozzle, creating a considerable amount of dust. As the bag is ejected from the filling machine, a "rooster tail" of product is thrown from the bag valve and fill nozzle. The rooster tail occurs because the bag is pressurized as it leaves the machine, releasing dust into the air and contaminating the outside of the bag. These contaminated bags then become a major dust source for the bag stacker and any other individuals handling the bags.

Figure 1 depicts the components of the dual-bag nozzle system. The dual-bag nozzle device uses a two-nozzle arrangement with an improved bag clamp to control the dust sources. The two-nozzle arrangment uses an inner nozzle to fill the bag and an outer nozzle to relieve excess pressure from the bag after it has been filled. Depressurizing of the bag is accomplished once filling is completed with the aid of an eductor, which uses the venturi principle to exhaust excess air from the bag at approximately 50 ft³/min. A pinch valve is then used to open and close the bag exhaust. The bag is slightly overfilled and held in place until the exhaust system depressurizes the bag. After a few seconds, the bag clamp opens and the bag falls from the fill station. The exhaust system continues to operate as the bag falls away, cleaning the bag valve area. The exhausted material is then recycled back into the system.

The other key component of the system is an improved bag clamp. This bag clamp makes direct contact with approximately 60 percent of the nozzle, thus reducing the amount of product blowback during bag filling. A controlled amount of blowback is necessary so the bag does not rupture, but this occurs at the bottom of the nozzle, minimizing dust contamination to the outside of the bag.

Several field evaluations were performed to determine the effectiveness of the dual-bag nozzle system (Cecala, Volkwein, and Thimons 1984). During one study, there was an 83 percent reduction in the bag operator's dust exposure with the dual-bag

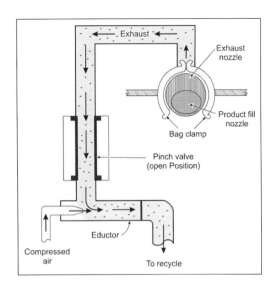

Figure 1.   Dual bag nozzle system design

nozzle system. Further, a 90 percent reduction was measured in the hopper below the fill station, indicating a substantial reduction in product blowback during bag filling.

The use of the improved bag clamp allows for a significant decrease in the amount of dust and product on the outside of the bag. This resulted in a 90 percent reduction in the bag stacker's dust exposure while bags were being loaded into an enclosed vehicle during the testing of this system.

The dual-bag nozzle system is mainly recommended for operations with three- and four-fill nozzle bag machines because there is a slight decrease in production due to the time needed to depressurize the bags after filling is completed. The system can be used on a one- or two-nozzle machine, but this decreases the production rate even further because the bag operator must wait on each individual bag instead of a cycle of bags.

OVERHEAD AIR SUPPLY ISLAND SYSTEM

The overhead air supply island system (OASIS) was developed to provide an envelope of clean, filtered air to a worker at a stationary location (Volkwein, Engel, and Raether 1986). One of the main advantages of the OASIS is that it is suspended over a worker and operates independently of any processing equipment, as shown in Figure 2. Mill air is drawn into the unit and passes through a primary HEPA cartridge filter. After the air exits the primary filter, it passes through an optional heating or cooling chamber, which can be incorporated in the unit if temperature control is desired. The air then flows through a distribution manifold, which also serves as a secondary filter, and finally exits the unit. The resulting filtered air flows down over the worker at an average velocity of roughly 375 ft/min, which normally keeps any mill air from entering this clean air core. The system can detect a filter overload based upon an increase in pressure and automatically self-clean the HEPA filter using one of various cleaning techniques.

During an evaluation on the OASIS at two different sites, the operator's respirable dust exposure was reduced by 82 and 98 percent, as compared to when the unit was not being operated. At both of these operations, the dust concentration within the clean filtered air of the OASIS remained under 0.04 mg/m^3.

An additional benefit provided by the OASIS is that the filtering system provides some general cleaning and reduces the overall air quality at the structure. During one of the evaluations, there was a 12 percent reduction in respirable dust levels in the entire mill building where the OASIS was installed. The volume of clean air delivered by the OASIS is variable based upon the size of the unit, but it is normally in the range of 6000 to 10,000 cfm. The OASIS is generic in design and can be fabricated and installed in-house or through any local engineering company that handles ventilation and dust control systems.

BAG AND BELT CLEANING DEVICE

The bag and belt cleaner device (B&BCD) is designed to reduce the amount of dust escaping from bags as they travel from the bag loading station to the stacking/palletizing process. It is intended to be self-supporting so that it can be incorporated anywhere along the conveyor belt line. This device reduces the dust exposure of all workers in and around the conveying process, as well as, anyone handling the bags once they are filled at the loading station. This system should be applicable to any mineral processing operation that loads product into 50- to 100-lb paper bags.

It is more logical to place the cleaner in close proximity to the loading station to eliminate the dust hazard as quickly as possible. Also, it is important to have the B&BCD unit under sufficient negative pressure to insure that any dust removed from the bags of product or from the conveyor belt, does not leak from the unit.

A prototype B&BCD, which was designed and built by the USBM, was 10 ft long and used a combination of brushes and air jets to clean all sides of

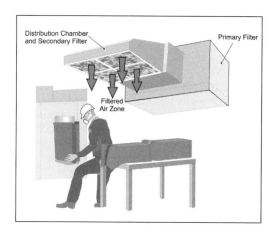

Figure 2. OASIS positioned over bag operator to deliver clean filtered air down over work area

the bags, as shown in Figure 3. As the bag travelled through the B&BCD, it entered through a curtain made of clear, heavy-duty flexible plastic stripping and into the device. Inside, a stationary brush on a swing arm started the cleaning process on the front and top of each bag. The bag would go through a second set of plastic stripping and into the main chamber section, where it travelled under a rotating circular brush that further cleaned the top of the bag. The sides were cleaned by a stationary brush located on each side of the bag. An air jet was located at the end of these brushes to provide for additional cleaning. The side with the bag valve used a higher volume (velocity) air jet to provide for the maximum amount of cleaning. After passing through the air jets, the bag travelled over a rotating circular brush located beneath the bag which cleaned the bottom of the bag. The bag exited the device by traveling through another air lock chamber with flexible plastic stripping.

A chain conveyor was used for the entire length of the device in the prototype design to allow product removed from the bags during cleaning to fall into a hopper. Product collected in this hopper was then recycled back into the process. Once exiting the B&BCD, both the bags of product and the conveyor belt should be essentially dust-free.

An evaluation of the B&BCD was conducted at two mineral processing plants. The device showed very favorable results in lowering respirable dust levels. The most relevant reduction data for this testing was the amount of dust removed from the surface of the bags, which ranged from 78 to 90 percent (Cecala, Timko, and Prokop 1995).

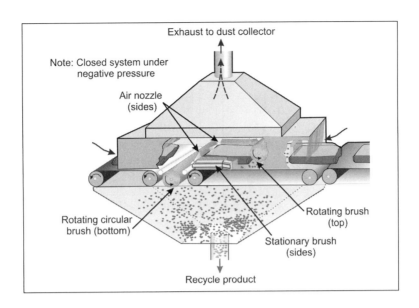

Figure 3.   Bag and Belt Cleaner Device

BAG STACKING

There are many different types of semi-automated systems that assist workers by performing some tasks in the bag loading and stacking process. Two of these systems will be discussed in this section. In the first system which was designed by the USBM, a worker performs the bag stacking task manually, but is assisted by using a hydraulic lift table. This lift table allows the height for stacking the bags to remain constant throughout the entire pallet loading cycle. The bag loading height is set to approximately knuckle-high for the worker, which is the most ergonomic loading height, as shown in Figure 4. A push-pull ventilation system is used on either side of this pallet to capture the dust liberated during the bag stacking (palletizing) process (Cecala and Covelli 1991).

In the second system, the worker slides the bag of product on an air table one layer at a time, but the actual stacking of the bags onto the pallet is performed automatically. Since bag injuries are a major lost-time injury for bag stackers, this design significantly reduces the stress by not requiring them to manually lift any bags of product. One problem with an air slide device is that it can cause dust to be blown from the bags of product into the worker's breathing zone. A NIOSH study (Cecala et al. 2000) revealed that the OASIS system worked very effectively along with this air slide device. In addition to the OASIS system, an exhaust hood was used next to the air slide area to capture the dust blown from the bag, as seen

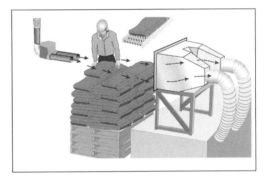

Figure 4.   Semi-automated bag palletizing system to ergonomically improve bag stacking process using a push-pull ventilation to capture the dust generated

in Figure 5. This exhaust system is then tied into the facilities local exhaust ventilation (LEV) system.

CLEANING DUST FROM SOILED WORK CLOTHING

A significant source of respirable dust exposure to workers at mineral processing operations can come from contaminated work clothing (NIOSH 2002). It must be noted also that once a worker's clothing becomes contaminated, it is a continual source of dust exposure until the cloths are changed or cleaned.

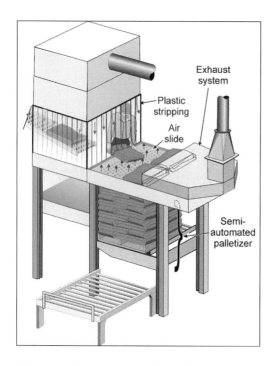

Figure 5. Semi-automated bag palletizing system where a worker transfers the bags onto air slide and then positions each layer of bags on the pallet. Both OASIS and LEV systems are used to lower the worker's dust exposure during this bag palletizing process.

A cooperative research effort between Unimin Corporation and NIOSH led to the development of a clothes cleaning system. This clothes cleaning system consists of four major components: (1) a cleaning booth, (2) an air reservoir, (3) an air spray manifold, and, (4) an exhaust ventilation system. Figure 6 represents the design of the clothes cleaning system.

The clothes cleaning booth is 48 by 42 inches which provides the worker with sufficient space to rotate in front of the air spray manifold to perform the cleaning process. An air reservoir is necessary to supply the required air volume to the air nozzles used in the spray manifold. The size requirement can be either a 120- or 240-gallon for the reservoir, depending on the number of workers needing to clean their clothes in sequence. If multiple individuals will be using the booth one after another, then the 240-gallon reservoir should be used. This reservoir should be pressurized to at least 150 psi and it should be located close to the cleaning booth and be hard-piped to the air spray manifold located in the booth. A pressure regulator must be installed immediately before

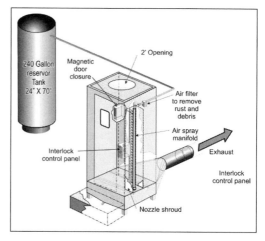

Figure 6.   Design of clothes cleaning system

the air spray manifold to regulate the nozzle pressure to a maximum of 30 psi.

Intake air enters the booth through a 2-foot opening in the roof, and then flows down through the enclosure before exiting through an air plenum on the bottom back wall of the booth. As this intake air flows through the booth, it entrains dust removed from the worker's clothing during the cleaning process and forces it down towards the air plenum and away from the worker's breathing zone. The exhausted dust-laden air then travels from this air plenum at the base of the booth to the exhaust ventilation system.

The air spray manifold is composed of 26 spray nozzles spaced 2 inches apart. The bottom nozzle is located 6 inches from the floor and is a circular designed nozzle for cleaning the worker's boots. This nozzle is used in conjunction with an adjustable ball-type fitting so that it can be directed downward. The other 25 air spray nozzles are flat-fan sprays, which lab testing proved to be the most effective for cleaning at close distances. The air spray nozzles deliver slightly less than 500 cfm of air.

In order to perform the clothes cleaning process, a worker dons personal protective equipment, enters the cleaning booth, activates a start button and rotates in front of the air spray manifold while dust is blown from the clothing via forced air. After a short time period (18 seconds), the air spray manifold is electronically de-activated and the worker can exit the booth with significantly cleaner work clothing.

For effective dust removal, it is critical that the cleaning booth be ventilated under negative pressure at all times so as to not allow any dust liberated from the clothing to escape from the booth and into the

Table 1. Amount of dust remaining on coveralls after cleaning and cleaning times for 100% cotton and polyester/cotton blend coveralls

Cleaning Method	100% Cotton		Polyester/Cotton Blend	
	Dust Remaining on Coveralls, grams	Cleaning Time, seconds	Dust Remaining on Coveralls, grams	Cleaning Time, seconds
Vacuuming	63.1	398	45.5	346
Air Hose	68.8	183	48.4	173
Clothes Cleaning Booth	42.3	17	21.9	18

work environment. In the design stage of this technology, testing validated that an exhaust volume of 2,000 cfm was sufficient to maintain a negative pressure throughout the entire clothes cleaning cycle (Cecala et al. 2008).

During development, the new clothes cleaning technology was compared to both the MSHA-approved vacuuming approach and a single handheld compressed air hose. In this testing, several 100% cotton and cotton/polyester blend type coveralls were soiled with limestone dust before the worker entered the clothes cleaning booth. Table 1 shows the cleaning times and effectiveness of the three different techniques. Clearly, the use of the clothes cleaning system was much more effective than the other two approaches, and required only a fraction of the time. Table 1 also shows that the polyester/cotton blend coveralls were cleaned more effectively than the 100% cotton type. Figure 7 depicts a worker before and after entering the clothes cleaning booth.

TOTAL STRUCTURE VENTILATION DESIGN

The first strategy to lower dust exposures in any structure is to have an effective primary dust control plan that captures major dust sources at their point of origin, before the dust is allowed to liberate out into the plant and contaminate workers. Engineering controls discussed in this manuscript are effective at reducing and capturing the dust generated, but they do not address the continual buildup of dust from background sources, such as:

- Product residue on walls, beams, and other equipment which becomes airborne from plant vibration
- Product that accumulates on walkways, steps, and access areas, and which may be released as workers move through the plant and by plant vibration
- Leakage or falling material from chutes, beltways, and dust collectors

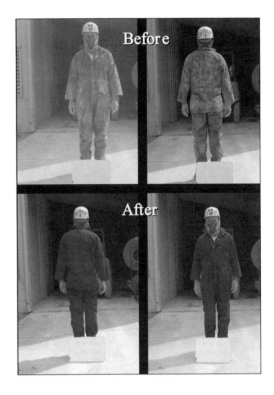

Figure 7. Test subject wearing polyester/cotton blend coveralls before and after using clothes cleaning booth

- Lids and covers of screens that are damaged or when they are removed for inspection and cleaning
- Product released because of imperfect housekeeping practices

Since most mineral processing structures can be considered closed systems, the background dust sources can cause dust concentrations to continually increase as the day or shift progresses inside these closed buildings. One method that can be used to control these background dust sources is a total structure ventilation system.

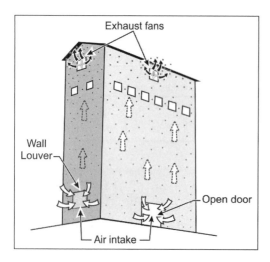

Figure 8. Basic design of total structure ventilation system

The basic principle behind the total structure ventilation design is to use clean outside air to sweep up through a building to clear and remove the dust-laden air. This upward airflow is achieved by placing exhaust fans at or near the top of the structure and away from plant personnel working both inside and outside the structure. The size and number of exhaust fans is determined based upon the initial respirable dust concentration and the total volume of the structure. It must be noted that all the processing equipment within a mill generates heat and this produces a thermodynamic "chimney effect" that works in conjunction with the total structure ventilation design (Cecala and Mucha 1991). Figure 8 shows the concept of the total structure ventilation design. To be effective, the total structure ventilation design must meet three criteria: (1) provide a clean makeup air supply, (2) an effective upward airflow pattern, and (3) a competent shell structure.

Intake air needs to be brought in at the base of the structure by strategically located wall louvers or open plant doors. It is very important that this air be free from outside dust sources such as bulk loading, high traffic areas, etc., which could cause dust-laden air to be drawn into the structure and could increase respirable dust levels. Through the use of wall louvers or closing doors, the intake air locations could be changed based upon the outside dust conditions.

The system should provide an effective upward airflow pattern that ventilates the entire structure by sweeping the major dust sources and work areas. Strategic positioning of both exhaust fans (high in the walls or roof) and makeup air intakes (at the

base) create the most effective airflow pattern for purging the entire mill.

A competent outer shell of the structure is necessary because the ventilation system draws the makeup air through the points of least resistance. Therefore, the structure's outer shell must be free of open or broken windows, holes, cracks, and openings, especially in the vicinity of the exhaust fans.

A normal range of airflow for a total structure ventilation design would be in the 10–35 air changes per hour (acph) level. During the development of the total structure ventilation design concept, two different field studies were performed in an effort to document its effectiveness. In the first study, with a 10 acph ventilation system, a 40 percent reduction in respirable dust concentrations was achieved throughout the entire structure. In the second study, the total structure ventilation system was capable of providing both 17 and 34 acph. Average respirable dust reductions throughout the entire structure ranged from 47 to 74 percent for the two exhaust volumes, respectively, (Cecala, Klinowski, and Thimons 1995).

OPERATOR BOOTHS/CONTROL ROOMS/ ENCLOSED CABS

Many times at mineral processing plants, workers will be located in an operator's booth, control room, or enclosed cab to give them a safe work area and to isolate them from dust sources. If these areas are properly designed, they can provide very good air quality to the worker. On the other hand, if these enclosed areas are not properly designed and maintained, the air quality can deteriorate to unacceptable and unsafe levels.

The most effective technique for reducing operators' exposure to airborne dust in booths/control rooms/enclosed cabs at mineral processing operations is with filtration and pressurization systems. The most effective filtration and pressurization systems have the heating and air conditioning (HVAC) components tied in as an integral part of the system.

NIOSH conducted a controlled laboratory study to evaluate the key factors necessary for achieving an effective enclosure filtration and pressurization system (Organiscak and Cecala 2008). In the course of the laboratory study, the significance of the filtration system parameters were evaluated and the following mathematical model was developed:

$$PF = \frac{C_o}{C_i} = \frac{Q_I + Q_R \eta_R}{Q_I(1 - \eta_I + l\eta_I) + Q_W}$$

(Note: The above equation is dimensionless, therefore air quantities used must be in equivalent units. Also, filter efficiencies and intake air leakage must be fractional values (not percentage values).

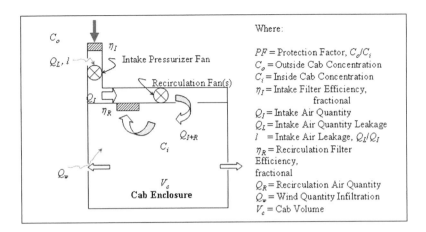

Where:

PF = Protection Factor, C_o/C_i
C_o = Outside Cab Concentration
C_i = Inside Cab Concentration
η_I = Intake Filter Efficiency, fractional
Q_I = Intake Air Quantity
Q_L = Intake Air Quantity Leakage
l = Intake Air Leakage, Q_L/Q_I
η_R = Recirculation Filter Efficiency, fractional
Q_R = Recirculation Air Quantity
Q_w = Wind Quantity Infiltration
V_c = Cab Volume

Figure 9. Key factors for effective enclosure filtration and pressurization system

This model (see Figure 9) was formulated from a basic time-dependent mass balance model of airborne substances within a controlled volume with steady state conditions. It determines the protection factor in terms of intake air filter efficiency, intake air quantity, intake air leakage, recirculation filter efficiency, recirculation filter quantity, and outside wind quantity infiltration into the cab. The equation allows for a comparison of how changes in the various parameters and components in the system impact the protection factor. The wind quantity infiltration (Q_w) can be assumed to be zero if the cab pressure exceeds the wind velocity. By using this equation, operations have the ability to determine the desired parameters necessary to systematically achieve a desired protection factor in an operator's booth, control room, or enclosed cab to improve the air quality to safe levels, and to ultimately protect their workers.

The results of this laboratory study indicate that intake filter efficiency and the use of a recirculation filter had the greatest impact on improving the air quality. When considering the use of an intake air filter, the addition of the recirculation component significantly improved the air quality due to the repeated filtration of the cab's interior air. The addition of an intake pressurizer fan to the filtration system increased both intake airflow and cab pressure significantly. The cab air quality was also affected by intake filter loading and air leakage.

Through this laboratory and numerous field studies, the following items were identified as key components to an effective system:

• **Ensure booth/control room/cab integrity.** Effective protection factors were realized in various field studies when pressures between 0.01

and 0.40 inches of water gauge were achieved because of good enclosure integrity. These pressures correspond to wind velocity equivalents of 4.5 and 29 miles per hour, respectively, and prevent wind penetration (Cecala et al. 2009).

• **Use high efficiency filters on intake air.** Only intake filters with an efficiency of 95% or greater were used during field studies. Laboratory experiments showed an order of magnitude increase in protection factors when using a 99% efficient filter versus a 38% efficient filter on respirable sized particles (Organiscak and Cecala, 2009a).

• **Use an efficient recirculation filter.** All the field evaluations used recirculation filters that were 95% or greater on respirable sized dusts. Laboratory experiments showed a 10-fold increase in protection factors when using an 85% to 94.9% efficient filter on respirable sized dusts as compared to using no recirculation filter. Laboratory testing also showed that the time needed for the interior to stabilize after the door was closed was reduced by more than 50% when using the recirculation filter (Organiscak and Cecala, 2009b).

• **Minimize interior dust sources.** Good housekeeping practices are needed to keep enclosure interiors clean and eliminate any dust sources. One field study showed a significant increase in dust levels (0.03 mg/m^3 to 0.26 mg/m^3) when a floor heater was used. The fan from the floor heater stirred up dust lying on the cab floor (Cecala et al. 2005).

• **Keep doors closed.** In a study on an enclosed cab of a surface drill, the operator's dust exposure averaged 0.09 mg/m^3 inside the cab with the door closed and 0.81 mg/m^3 when the door was briefly opened to add drill steels. This procedure was performed after drilling stopped and the visible dust dissipated, it nevertheless produced a nine-fold increase in dust concentrations inside the cab each time a drill steel was added (Cecala et al. 2007).

As determined in both laboratory and field studies, there are two critical factors for an effective filtration and pressurization system. First, the filtration system needs to be comprised of an effective outside intake (make-up) air, and a recirculation air component. Second, it is also critical to establish and maintain cab integrity in order to achieve an acceptable level of positive cab pressurization. Without positive pressure in the cab, both the effectiveness of the system and the air quality are greatly compromised.

PRIMARY CRUSHING

The primary dump and crushing operation is a significant dust source at metal/nonmetal operations. As previously mentioned, a very effective method to minimize the crusher operator's respirable dust exposure is to place the operator in a control booth with a filtration and pressurization system. This is not always possible and the dust generated during the dumping and crushing process does expose and contaminate secondary workers in and around this area. A number of techniques have been studied in an attempt to minimize the dust generated and liberated during this mining process in both underground and surface operations.

In a study performed at an underground facility, a comparative evauation was performed to determine if a portable diesel-powered propeller fan could perform more efficiently for dust dilution and transport than an axial fan for localized ventilation. The results of this testing indicated the propeller fan reduced both respirable and silica dust by 20 pct at the dumping/crushing facility as compared to the vane-axial fan (Chekan, Colinet, and Grau 2006).

In another study, water spray suppression systems were evaluated to lower respirable dust generated at the primary dumping location. The general rule of thumb is to add one percent of moisture to the product, which is based on the weight of ore processed compared to the weight of the water added (Quilliam 1974). From this point, the percentage can be adjusted based upon the improvement gained from additional moisture versus any consequences from adding too much water. It must be noted that the amount of moisture that can be added at primary dump locations is normally not as sensitive as in later stages of the mineral processing cycle, and thus, higher rates or percentages can usually be tried. One important feature with a primary dump application is to only activate the water sprays during the actual dump cycle through the use of a photo cell or a mechanical switching device.

Enclosures for primary dump application normally require a custom design and are usually dependant on the type and size of dump vehicles being used. In some cases, walls can be constructed around the primary dump location to form an enclosure. The walls can be either stationary or removable, based on maintenance requirements. In some cases for a removable enclosure, a breathable tarp fabric material, similar to the material used on over-the-road haul trucks, can simply be laid over to seal the top of the enclosure.

Overlapping plastic can also be used to provide for an effective seal over the front of the enclosure, as shown in Figure 10. This plastic stripping is not damaged when contacted by the bucket of the front-end loader or the bed of the haul truck during dumping and remains very durable over time.

When using an enclosure, it is also possible to incorporate a LEV system to filter the dust-laden air from the hopper area. This would be most applicable when the primary dump is at a location where the dust could enter an adjoining structure or impact outside miners. The enclosure helps to contain the dust cloud that billows up from the hopper during the dumping process, but the dust remains airborne unless it is suppressed or removed. An LEV can be an effective technique incorporated to remove and filter this dust if it is properly designed and sized to the hopper.

Another dust source at primary dump operations is the dust that is allowed to liberate or to roll back under the dumping mechanism. A tire stop or jersey barrier should be positioned at the most forward point of dumping for the primary hopper. To the back side of this tire stop, a water spray manifold should be attached to knock down and force the dust that would otherwise roll back under the dumping mechanism to remain in the hopper. Additionally, a shield should be placed over this water spray manifold to protect it from damage from falling ore. Finally, a system should also be incorporated that allows the water sprays to only be activated during the actual dumping process, as previously discussed.

CONCLUSION:

The USBM and NIOSH has and continues to perform research to reduce the dust exposure to miners in the U.S. metal/nonmetal mining industry. This

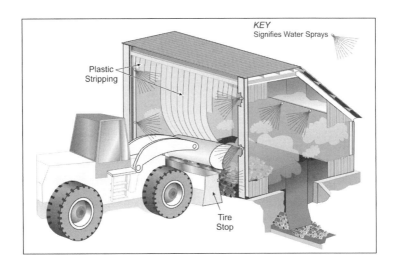

Figure 10. Flexible plastic stripping used to hold dust inside enclosure and water sprays used to knock down airborne dust

manuscript briefly highlighted some of the engineering control technologies that have been developed through this research effort at the Pittsburgh Research Laboratory over the past 30 years. Through the implementation of these control technologies, it is our hope that the likelihood of miners developing silicosis and/or other respiratory diseases in this industry would be eliminated.

REFERENCES

Cecala, A.B., Organiscak, J.A., Zimmer, J.A., Hillis, M.S., and Moredock, D. 2009. Maximizing Air Quality Inside Enclosed Cab With Uni-Directional Filtration And Pressurization System. SME Preprint No. 09-099. Denver, CO: SME.

Cecala, A.B., Pollock, D.E., Zimmer, J.A., O'Brien, A.D., and Fox, W.F. 2008. Reducing Dust Exposure From Contaminated Work Clothing With A Stand-Alone Cleaning System. Proceedings of 12th US/North American Mine Ventilation Symposium—Wallace, ed. Omnipress. ISMN 978-0-615-2009-5, pp. 637–643.

Cecala, A.B., Organiscak, J.A., Zimmer, J.A., Moredock, D., and Hillis, M.S. 2007. Closing the door to dust when adding drill steels. Rock Products. October. pp. 29–32.

Cecala, A.B., Organiscak, J.A., Zimmer, J.A., Heitbrink, W.A., Moyer, E.S., Schmitz, M., Ahrenholtz, E.,Coppock, C.C., and Andrews, E.H. 2005. Reducing enclosed cab drill operator's respirable dust exposure with effective filtration and pressurization techniques. *J. of Occ. and Env. Hyg.*, Vol. 2, No. 1, January. pp. 51–63.

Cecala, A.B., Zimmer, J.A., Smith, B., and Viles, S. 2000. Improved dust control for bag handlers. Rock Products. April. Vol. 103, No. 4, pp. 46–49.

Cecala, A.B., Timko, R.J., and Prokop, A.D. 1995. *Reducing Respirable Dust Levels During Bag Conveying and Stacking Using Bag and Belt Cleaner Device*. U.S. Bureau of Mines Report of Investigations RI 9596. December. 20 p.

Cecala, A.B., Klinowski, G.W., and Thimons, E.D. 1995. Reducing respirable dust concentrations at mineral processing facilities using total mill ventilation system. Mining Engineering. June. Vol. 47, No. 6, pp. 575–576.

Cecala, A.B., and Covelli, A. 1991. Automation to control silica dust during pallet loading. Mining Engineering. December. Vol. 43, No. 12, pp. 1440–1443.

Cecala, A.B., and Mucha, R. 1991. General ventilation reduces mill dust concentrations. Pit and Quarry. July. Vol. 84, No. 1, pp. 48–53.

Cecala, A.B., Volkwein, J.C., and Thimons, E.D. 1984. *New Bag Nozzle to Reduce Dust from Fluidized Air Bag Machine*. U.S. Bureau of Mines Report of Investigations RI 8886. Pittsburgh, PA.

Chekan, G.J., Colinet, J.F., and Grau, R.H. 2006. Impact of Fan Type for Reducing Respirable Dust in an Underground Limestone Crushing Facility. Proceedings of the 11th North American/Ninth US Ventilation Symposium, University Park, PA, June 5–7. pp. 203–210.

Hessel, P.A., Sluis-Cremer, G.K., Hnizdo, E., Faure, M.H., Thomas, R.G., and Wiles, F.J. 1988. Progression of silicosis in relation to silica dust exposure. *Ann. Occ. Hyg.* 32 (Suppl 1):689–696.

MSHA CFR. 2008. (Mine Safety and Health Administration Code of Federal Regulations), 30 CFR 57.5005. *Exposure Limits for Airborne Contaminants. Air Quality—Surface and Underground.* June 1, Edition.

MSHA. 2009a. Federal Mine Safety and Health Act of 1977; Public Law 91-173. http://www.msha.gov/REGS/ACT/ACT1.HTM. Accessed October 2009.

MSHA. 2009b. Number of Metal/Nonmetal Operations in the United States. MSHA website: http://www.msha.gov/STATS/PART50/WQ/1978/wq78mn01.asp. Accessed September 2009.

NIOSH. 2002. Health effects of occupational exposure to respirable crystalline silica. NIOSH Hazard Review, DHHS (NIOSH) Publication No. 2002-129. http://www.cdc.gov/niosh/docs/2002-129/02-129a.html. Accessed September 2009.

Organiscak, J.A., and Cecala, A.B. 2009a. Doing the math—the effectivness of enclosed-cab air-cleaning methods can be spelled out in mathematical equations. Rock Products. October. Vol. 112, No. 10.

Organiscak, J.A., and Cecala, A.B. 2009b. Laboratory investigation of enclosed cab filtration system performance factors. Mining Engineering. January. Vol. 61, No. 1, pp. 48–54.

Organiscak, J.A., and Cecala, A.B. 2008. *Key Design Factors of Enclosed Cab Dust Filtration Systems.* NIOSH Report of Investigation 9677. November. 43 pp.

Quilliam, J.H. 1974. Sources and Methods of Control of Dust. The Ventilation of South African Gold Mines. Yeoville, Republic of South Africa: The Mine Ventilation Socity of South Africa.

Volkwein, J.C., Engel, M.R., and Raether, T.D. 1986. Get away from dust with clean air from overhead air supply island (OASIS). Rock Products. July.

A Review of the USBM/NIOSH Diesel Research Program

Steven Mischler
NIOSH, Pittsburgh, Pennsylvania, United States

Aleksandar Bugarski
NIOSH, Pittsburgh, Pennsylvania, United States

James Noll
NIOSH, Pittsburgh, Pennsylvania, United States

Larry Patts
NIOSH, Pittsburgh, Pennsylvania, United States

Emanuele Cauda
NIOSH, Pittsburgh, Pennsylvania, United States

ABSTRACT: Since the implementation of diesel equipment in underground mines, the US Bureau of Mines (USBM) and more recently the National Institute for Occupational Safety and Health (NIOSH) have conducted research focused upon protecting the safety and health of miners using this equipment. These research initiatives ranged from initially evaluating the use of diesel equipment and the health and safety implications of this usage, to our current research evaluating control technologies and strategies to aid the mining industry in reducing miners' exposure to diesel emissions. This paper will summarize the diesel research history of the USBM and discuss more recent successes of NIOSH diesel research that have improved the health and safety conditions for underground miners.

INTRODUCTION

The use of diesel-powered equipment by the underground mining community has continuously increased since its inception in a German coal mine in 1927. In the United States approximately 3,000 pieces of diesel equipment were being operated in underground coal mines by 1995 (MSHA 2001a) and over 4000 units operating in underground metal/nonmetal mines (M/NM) by 2001 (MSHA 2001b).

Through the many decades of diesel equipment usage in underground mines, research into the many health and safety aspects surrounding diesel equipment was first conducted at the Unites States Bureau of Mines (USBM) and then at the National Institute for Occupational Safety and Health (NIOSH). The initial interest in the use of diesel engines in underground mines resulted from the hazards associated with trolley-locomotive haulage systems in coal mines, and research into acceptable alternatives. Most of this research was completed in the 1930s and early 1940s and the first diesel locomotive was delivered to a coal mine in the Appalachian region in late 1945. After 1945 and through the 1950s, USBM diesel research focused on methods to maintain a safe working environment underground when internal-combustion engines were being operated. This included research on ventilation of exhaust and evaluating the explosion hazards underground. Throughout the 1960s and 1970s, the USBM research continued evaluating testing procedures for permissibility in coal mines but the focus started to shift towards the measurement and control of emissions from diesel equipment. Characterizing the emissions for diesel equipment remained the focus of the USBM until it became part of NIOSH in 1996. The control of diesel emissions became important in the late 1990s as the Mine Safety and Health Administration (MSHA) initiated rulemaking to limit the exposure to diesel emissions.

In January of 2001, MSHA promulgated rules setting compliance standards for both underground coal and M/NM mineworkers. The underground coal rule (MSHA 2001a) controls the exposure of the miners by limiting the emission rate from newly introduced and existing diesel-powered equipment. Engine emission rates are calculated using the emission rate determined through MSHA engine certification and the MSHA-accepted particulate matter (PM) reduction factor for the emission control technology being implemented. An engine/control system must be verified to meet an emission rate of 2.5 g/hr for permissible and heavy-duty or <5 g/hr for outby light-duty equipment. MSHA determined that due to the absence of an accurate method for sampling diesel particulate matter (DPM) in underground coal mines, a performance rule was not feasible, thus air sampling, to measure DPM in the workplace is not required under this rule.

The underground metal/nonmetal rule (MSHA 2001b) requires the mine operators to limit a miner's personal exposure to DPM to an average eight-hour equivalent concentration of 160 ug/m^3 total carbon, as measured by NIOSH Analytical Method 5040 (NIOSH 1999). This limit went into effect on May 20, 2008. As industry works to achieve compliance with these regulatory limits, mine operators are looking for feasible methods for reducing DPM concentrations in their mines. In addition, the industry needs methods to accurately measure DPM to ensure that the control measures they adopt are working successfully.

Recent NIOSH research supports the use of an integrated approach to controlling DPM. This integrated approach adopts a holistic strategy of integrating all departments of a mine including management, production, maintenance, and safety.

This paper will present an overview of the USBM/NIOSH research from it's initiation in 1936 until the present and emphasize the successful NIOSH research programs which helped with the implementation of diesel equipment into underground mining. Finally this paper will present our current research on the integrated approach and review the strategies being used by US mines to reduce the concentrations of DPM emitted from the tailpipe.

BUREAU OF MINES AND NIOSH DIESEL RESEARCH

The Bureau of Mines diesel research began in the 1930s and continued until 1996, at which time the mining research was moved to NIOSH. The initial USBM Report of Investigation (RI) 3320 discussing the use of diesel equipment in mines was published

in 1936 and titled "Diesel Mine Locomotives-Development and Use in European Coal Mines." This RI was issued in "response to numerous requests for information as to the status of internal-combustion mine locomotives of the Diesel type that use heavy or non-volatile oil for fuel, with special reference to their safe use in coal mines" (USBM 1936). At this time popular interest in the diesel engine was a result of the successful application of diesel engines on railway locomotives at a low cost of fuel per mile. Additionally diesel locomotives had been used successful in Europe since 1927 and resulted in reduced haulage cost and greater safety in contrast with electric and other forms of haulage.

The main reason for the delay of the implementation of diesel into US underground mines was a prejudice against all internal-combustion engines for underground mine-use, based on hazards with the use of gasoline mine locomotives. Gasoline locomotives which were introduced into underground mines in 1912, and were found to result in dangerous levels of carbon monoxide and to be a fire and explosion hazard due to the highly volatile fuel. In 1932 the director of the USBM issued a recommendatory decision (No. 19) stating "if the ventilation requirements are met and other precautions are taken it appears reasonably safe to employ diesel locomotives that use a heavy nonvolatile fuel oil in tunnel construction or in mines"(Mine Safety Board 1933).

Beginning in 1910, coal mine companies were using the underground trolley-locomotive system for hauling coal and this was credited with the rapid increase in production of coal in the US. However, the USBM research suggested that the trolley system was hazardous and it continued to look for alternative haulage methods. This research goal was stated in the annual report of research and technologies work on coal for the fiscal year of 1942 (USBM 1943). This Information Circular (IC) stated that "at the outset, the USBM research into diesel locomotives underground were directed toward assisting the development of a flameproof diesel locomotive that might be safer than the exceedingly hazardous electric-trolley locomotive." The need for a safer haulage system was further presented in IC 7328, "Hazards of the Trolley-Locomotive Haulage System in Coal Mines," published in 1945.

The objective of this IC was to discuss the hazards of the bare trolley wire and the "imminent, constant, universal danger of contact, gas- or dust-ignition, and fire hazards along every foot of bare wire from the mouth of the mine to the farthest recess of the working to which it extends" and to present suggestions for eliminating these hazards. The use of diesel engines was recommended in this IC because "when carefully constructed and operated, they have proved to be economical, efficient, and safe for use in the driving of some large tunnels in the United States" (USBM 1945). IC 7673, published in 1953, furthered this discussion and answered questions including; 1) Why is the Bureau of Mines interested in the development of diesel-powered haulage, and 2) What factors must be considered to insure safety in the use of diesel engines in mines (USBM 1953). This document lists the problems with a haulage system using bare trolley wires, and states "30 to 40 fatalities occur each year as a result of electrocution by contact with bare trolley wires and in 1943 and 1944, 101 men were killed and extensive property damage resulted from mine fires started by falls of roof grounding bare trolley wires" (USBM 1953).

During the 1940s, USBM was also conducting studies on the use of ventilation to eliminate the health hazards associated with diesel emissions. Figure 1 (top) presents the outside of the diesel gallery used for this research and Figure 1 (bottom) is a picture of the control room containing the dynamometer. This research gallery was used to (1) determine the condition under which a diesel engine will ignite an inflammable atmosphere, (2) develop methods for testing the effectiveness of flame-protective devices to prevent such ignitions and (3) investigate if the health hazards resulting from diesel emissions can be eliminated with adequate ventilation (USBM 1943).

The USBM also initiated field studies on diesel locomotives and the effects of diesel emissions in mine air, starting in the late 1940s. The results from one initial study were published in Report of Investigation (RI) 4032 entitled, "Diesel Engines Underground, Use of Diesel Locomotives in Construction of the Delaware Aqueduct: Effect of Exhaust Gases upon Quality of Tunnel Air" (USBM 1947). This RI discusses the effect on the quality of tunnel air resulting from the operation of diesel locomotives. This and other field studies conclude that under normal operating conditions, the carbon dioxide, carbon monoxide and oxides of nitrogen content of the air did not exceed limits generally considered permissible in the air of working places, nor was the oxygen content depleted to any significant extent. Furthermore, these field studies provided practical information on the use of diesel-powered haulage equipment underground.

This early research resulted in the publication of Schedule 22, "Procedures for Testing Diesel Mine Locomotives for Permissibility and Recommendations on the Use of Diesel Locomotives Underground" in 1944 (USBM 1944) and schedule 24 "Procedure for Testing Mobile Diesel Powered Equipment for Noncoal Mines" published in 1949

Figure 1. USBM diesel testing facility, 1943

(USBM 1949). Following these publications, the USBM approved the first diesel engine for underground use, under schedule 24, in 1951 and by 1959 USBM had approved 28 vehicles from 13 manufacturers. In 1960, the USBM published RI 5616 entitled "Safety with Mobile Diesel-Powered Equipment Underground." This document summarized the research results that illustrated the need for the approval requirements set in Schedules 22 and 24. In addition RI 5616 summarized the important practical precautions needed for the safe use of diesel equipment in underground mines and tunnels.

In 1963, the USBM published IC 8183 which listed all of the equipment approved for underground use under Schedule 24, and included 59 equipment approvals and 24 certifications of subassemblies (USBM 1963). Through these publications and this early research the USBM greatly advanced the use of diesel equipment in underground mines, by establishing protocols for the safe operation and methods for reducing health hazards associated with the diesel emissions. In addition, this early research showed that diesel-powered haulage equipment could be made safer than electric trolley locomotives and should replace these systems. The extensive utilization of diesel-powered equipment was made possible through USBM research and has resulted in the existence of almost 5000 diesel-powered units operating today in underground metal/nonmetal mines in the U.S.

As stated previously, USBM research on safety issues associated with the use of diesel-powered equipment in coal mines continued throughout the 1960s and 1970s. During this period, the USBM continued research on the feasibility of using diesel equipment in coal mines. Under the Federal Coal Mine Health and Safety Act of 1969, the USBM was required to investigate safer sources of power, and this initiated the development of new and revised standards under which diesel-powered equipment was made permissible for, and used in, the nation's coal mines. The USBM issued the first experimental permit to Marin County Coal Corp in 1971 (USBM 1973). By 1977, there were 175 diesel vehicles being used in 30 underground coal mines and by 1995 there were over 3000 pieces of diesel powered equipment operating in coal mines (MSHA 2001a). The use of diesel powered equipment in coal, made possible due to USBM research, resulted in improved safety conditions for underground coal miners as well as increases in coal production capability.

By the 1980s, the overall goal of the USBM research continued to focus on reducing the exhaust emissions through ventilation and minimizing the risk of fires and explosions caused by the use of diesel-powered equipment in underground mines. However, during this decade the emphasis of the research program shifted toward the development of instrumentation and monitoring strategies to measure in-mine concentrations of exhaust pollutants (especially particles), and towards the testing of integrated control systems to reduce exhaust pollutants and minimize safety hazards (USBM 1987).

The initial USBM research on measuring diesel emissions focused on using these measurements to calculate a recommended ventilation rate that, under any engine operating condition, would dilute the exhaust to a safe hygienic concentration (USBM 1947). This research objective continued for two decades and in 1968, the USBM published RI 7074 entitled "Diesel Exhaust Contamination of Tunnel Air," which used this approach to develop calculation methods to be applied when designing the volume of ventilation for diesel haulage in proposed mines and tunnels. However, RI 7310, published in 1969, is an example of the shifting approach to exhaust measurements. The research presented in this RI is the first to use exhaust measurements in an attempt to reduce emissions based on integrated control technology. This RI evaluates the use of cetane improvers and their effect on nitrogen oxide

emissions. This report concluded that no significant effects on normal exhaust products (carbon dioxide, carbon monoxide, methane or oxygen) were caused by these additives (USBM 1969), and is the first RI presenting data evaluating engine controls on exhaust pollutants. In 1975, the USBM published a paper on the NO_x emissions from diesel engines and their control and measurement (Marshall 1975). The research presented in this paper focused on analytical methods for NO_x determination and control of NO_x from diesel engines. This work was consistent with the change in research objectives during this time period.

Initially, the evaluation of engines controls involved measuring only exhaust gases. However, in the mid-1970s the USBM also started to measure particle mass concentration in the diesel exhaust. RI 8141r reported results from research on measurements of mass sized distribution, concentration, and mass output of particulates in diesel exhaust from two types of engines, using three types of fuel and operated in five different modes (USBM 1976). This early particle research was motivated by establishing the effect of diesel particulate on measurements of coal mine dust. With the increase in usage of diesel equipment in coal mines during this time, a common worry was that the diesel particulate would cause the coal mines to exceed the coal dust standards established in 1969. This RI concluded that diesel particulates could add to the respirable dust in coal mines and cause added difficulty in meeting the 2.0 mg/cu m respirable dust standard. It further concluded that attempts should be made to minimize the mass output of particulates from diesel engines.

Throughout the 1980s and 90s, research continued on investigating methods for both control and measurement of particulates and gases emitted from diesel engines. IC 9141 (USBM 1987) presented results from investigations on a number of controls and measurement methods including; 1) monitoring and measurements of in-mine aerosols from diesel emissions, 2) effects of diesel engine maintenance on emissions 3) effects of fuel additive on diesel soot emissions, and 4) development and effectiveness of ceramic diesel particulate filters. RI 9571 reported the results of a field study to evaluate catalyzed diesel particulate filters at two underground metal mines (USBM 1995).

During the mid-1990s, the health and safety research of the USBM was moved to NIOSH, and the research objectives remained focused on evaluating measurement and control options for DPM. By this time, MSHA was seeking information about regulating DPM as a result of health studies associating DPM exposure with long-term health effects.

Ultimately, the concerns about the health effects of DPM exposure resulted in the issuing of regulations by MSHA governing the emission rates for diesel equipment used in coal mines (MSHA 2001a) and environmental compliance levels for DPM in US metal and nonmetal mines (MSHA 2001b).

The promulgation of these regulations motivated the industry to turn to NIOSH for assistance with the evaluation of control technology that would allow them to meet the new regulations. NIOSH published IC 9462 entitled "Review of Technology Available to the Underground Mining Industry for Control of Diesel Emissions" (NIOSH 2002). In 2003 and 2004, NIOSH completed two extensive field studies, in collaboration with industry, using an isolated zone in the Stillwater Mine in Nye, Montana. These studies evaluated the newest available technology, including diesel particulate filter systems (DPFs), disposable filter elements (DFEs), diesel oxidation catalysts (DOCs), and alternative fuels (biodiesel), for DPM removal efficiency and implementation feasibility, and resulted in the publications RI 9667 (NIOSH 2006a) and RI 9668 (NIOSH 2006b).

During this period, NIOSH also evaluated monitoring methods to measure DPM. In 1999, NIOSH published Analytical Method 5040 (NIOSH 1999), which is the basis for the regulatory measurements of DPM in metal/nonmetal mines. NIOSH continued evaluating methods of DPM measurements and created and tested a DPM sampling cassette (Noll, 2005). Recently, NIOSH research has resulted in the development of a continuous DPM monitor (Noll 2008). This pre-commercial prototype monitor measures in real-time the elemental carbon concentrations in mine air and is being tested in several mines to evaluate the effectiveness of DPM controls and ensure that the miners' exposures remain below the regulated levels. NIOSH is also investigating using the continuous EC monitor and other methods to measure DPM in the tailpipe of diesel equipment. In 2005, NIOSH published results on a differential pressure monitor to measure DPM in the tailpipe (Mischler, 2005). Tailpipe measurements have proven to be very useful to the mining industry for implementing engine maintenance programs and as a check on the effectiveness of diesel control measures.

Future NIOSH diesel research will continue to evaluate the efficiencies and effectiveness of new DPM control technologies. In addition, over the last several years NIOSH has initiated research focused on characterizing the chemical, physical and toxicological properties of DPM emitted from both mechanically controlled and electronically controlled engines over several different duty cycles. Figure 2a presents a picture of a modern electronically controlled engine and its associated dynamometer to be used in this research. Figure 2b presents a picture of the modern DPM monitoring equipment being used to collect DPM measurements.

The remaining section of this paper will discuss an overall strategy for controlling DPM, which builds upon NIOSH research evaluating efficiencies and feasibilities of control technologies used to reduce the exposure of miners to DPM. Currently, NIOSH is providing assistance to mines implementing this approach and completing sampling studies in mines that have adopted this approach to evaluate it's effectiveness.

Figure 2a. Electronically controlled modern diesel engine and dynamometer

Figure 2b. Modern DPM sampling equipment

CURRENT WORK ON CONTROLLING DPM

Using the results from past NIOSH research and in collaboration with our stakeholders, NIOSH is currently promoting an integrated approach to DPM control. The integrated approach is a strategy which systematically resolves issues associated with diesel emissions by integrating solutions, from each of the independent departments, throughout the mine as a whole. NIOSH has found that DPM exposure control is complicated but manageable when the proper people and proper systems are put into place and allowed to succeed. With this approach, mines learn how to properly understand the issues to be solved, tackle each issue systematically and thoroughly, research and try several different solutions, and finally implement the solutions that are most feasible for that mine. The integrated approach can be separated into five basic steps shown below:

1. Appoint a management supported leader or DPM champion
2. Get control over engine-out emissions (replace older/dirty engines, establish an emission's based maintenance program)
3. Look for administrative controls and gain an understanding of equipment inventory and usage
4. Ensure adequate ventilation
5. Examine additional control options including aftertreatment systems and alternative fuels

Finally, at each step in the process it is necessary to have a DPM measurement program that allows for documentation of the efficacy of each step in reducing the DPM concentration in the mine.

Management Supported Leader

The management supported leader under the integrated approach serves as the focal point for the program and NIOSH has labeled this person the DPM champion. Successful management of worker exposure to diesel exhaust requires that the DPM champion has adequate expertise, the authority to manage and coordinate efforts within and throughout various mine organizational structures, and the respect and cooperation of the workers. The DPM champion must have the expertise to explain the intricacies of the diesel regulations as well as an understanding of the many control options to be used. Although many mines attempt to place a mechanic in this role, since they have the greatest knowledge on the workings of the diesel equipment, NIOSH has found that often a mechanic does not have the authority to institute all the necessary changes required in the integrated approach. Because it is the DPM champion's

responsibility to bring together all persons whose actions can influence the emissions of DPM, NIOSH has found that a person currently holding authority in the organization is the best choice. The DPM champion must have the ability to assemble all the people with influence over the departments affecting DPM emissions, lead the discussion on the DPM problems and their potential solutions, and work cooperatively with each department to reach solutions to address the ultimate goal of controlling DPM in the most cost-effective and reliable manner.

Control Over Engine-out Emissions

Gaining control over the engine-out DPM emissions is the important starting point for the integrated approach. A reduction in the amount of DPM emitted by the engine is a reduction in the amount of DPM which must be otherwise controlled using alternative control strategies. In many instances, reducing these engine-out emissions may be the only step necessary to meet DPM compliance standards. The two main approaches that can be used to reduce engine-out emissions are maintenance and administrative controls.

Maintenance

Since the introduction of diesels into underground mining, the need for a good maintenance program has always been recognized. However, this recognition has not always resulted in the adoption and disciplined implementation of proper maintenance practices being applied to every diesel engine or vehicle in operation in underground mines. Nevertheless, it is extremely important to realize that the very first step on the path to reducing worker exposures to DPM is to implement an effective diesel vehicle/engine maintenance protocol and apply it to every diesel unit that operates underground. The early work in this area was performed by the U.S. Bureau of Mines (USBM, 1987). The University of Minnesota's Center for Diesel Research (Spears 1997) developed procedures for using tailpipe gas measurements as a diagnostic for engine maintenance. A comprehensive study on the relationship between diesel engine maintenance and tailpipe emissions was completed by McGinn (2000) under a research effort by the Diesel Emissions Evaluation Program (DEEP). McGinn developed a maintenance auditing procedure (McGinn et al. 2000) and guidelines (McGinn 1999), which were implemented in a hardrock mine with demonstrable results.

There are several important reasons to provide the best possible engine maintenance when considering or implementing control technology. The first reason is that the lowest emissions resulting from the application of any control technology are obtained when starting with the lowest possible engine-out emissions. The second reason is that ventilation requirements and PM emission rates determined through MSHA's engine certification process were obtained using a properly tuned, well-maintained (new) engine. It is important that the engines in the field have emission characteristics no worse than those of the certified engine so that calculations that use the MSHA ventilation and PM emission rates to ensure safe levels of toxic gases, to estimate workplace diesel particulate levels, or to compare particulate emission rates among engines are valid. In addition, excessive emissions from poorly maintained engines may jeopardize the performance of aftertreatment technologies. For example, excessive emissions of the ash caused by burning crankcase oil might result in clogging and premature failure of a diesel particulate filter. In summary, the very first step in applying control technology to reduce workplace exposures to diesel exhaust is to implement an effective maintenance program and closely monitor its effectiveness.

Administrative Controls

Effective administration of equipment usage can greatly assist in reducing the DPM concentrations in underground mines and is an important step in the integrated approach. To begin this process it is first necessary to evaluate the equipment inventory and equipment usage. Once the equipment usage has been optimized there are several other administrative controls that may reduce emissions and DPM concentrations and these include (1) minimize engine idling, (2) avoid lugging engines (low RPM high load) and (3) maintain clean fuel and lube oil. Finally traffic control has been shown to be effective in reducing the DPM concentrations in work areas. This type of control would entail routing traffic away from a miners work area, routing haul trucks in return air especially when ascending ramps while loaded and limiting horsepower in a work area based on available ventilation.

Aftertreatment Technologies

Aftertreatment technologies including diesel oxidation catalysts (DOC's), diesel particulate filters (DPF's) and disposable filter elements (DFE's) are very effective in removing DPM from the exhaust of diesel-powered equipment (Johnson 2008). These aftertreatment systems are placed in the exhaust system just after the combustion chambers and are gradually becoming more utilized for underground mining applications (NIOSH 2006a). Figure 3 shows a schematic of the basic concept of filtering where

engine emissions (blue arrows) enter at one end of the filter and then must physically flow through the filter wall, which removes the DPM, before exiting the filter. Figure 4 presents a picture of both the inlet and outlet of a filter showing that the black diesel soot is only seen on the inlet side of the filter. Disposable and non-disposable filters have been successful in efficiently removing DPM on mining equipment.

NIOSH has completed extensive research on the effectiveness of DPF's, DFE's and DOC's to control DPM (NIOSH 2006a). The results presented in Table 1 are from a DFE and three types of DPF systems representative of the systems currently available to the underground mining industry to curtail DPM emissions from diesel-powered equipment. The DPF systems with Cordierite and silicon carbide (SiC) wall flow monoliths are listed by MSHA as being 85 and 87 percent efficient, respectively, in the removal of total DPM mass (MSHA 2007). The Sintered Metal DPF system was recently added to the MSHA list as 81 percent efficient. The DFE was also shown to be over 80 percent efficient. As shown in Table 1 NIOSH results agree well with the MSHA efficiencies.

When compared with DPFs and DFEs, the DOC produced relatively modest effects on aerosol mass concentrations. The effects of the tested DOC on total mass concentrations were found to be influenced strongly by engine operating mode. However, the most substantial reductions of 42 percent in mass concentration occurred when the engine was operated under the highest work load scenario (intermediate speed and 100% load).

Biodiesel

Although DPF systems are gradually becoming more utilized for controlling DPM emissions from underground mining vehicles (NIOSH 2006a, NIOSH 2006b) their acceptance has been hindered by their

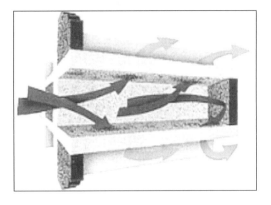

Figure 3.   Basic concept of filtering

Figure 4: Entrapment of soot at inlet of filter

Table 1.   Removal efficiencies of diesel filters

Cordierite DPF	Silicon Carbide DPF	Sintered Metal DPF	DFE
> 87%	> 90%	> 91%	> 80%

relative complexity, implementation issues and expense. Changing the fuel supply from petroleum diesel to biodiesel blends is considered by a number of underground mine operators to be a viable alternative for controlling DPM emissions. The major advantages of biodiesel over petroleum-based diesel fuels are greater cetane number, absence of aromatics and sulfur, and high oxygen content. In addition, biodiesel was shown (Williams 2006; Boehman 2005) to lower the balance point temperature of passively regenerated DPFs and therefore, could potentially facilitate regeneration and the implementation of DPF systems in underground mines.

In NIOSH testing, biodiesel fuels reduced total mass concentrations for all engine operating modes. With few exceptions, reductions in total mass concentrations rose with an increase in biodiesel fraction with 100% biodiesel reducing the DPM concentrations by slightly greater than 50%.

CONCLUSION

Throughout the 70-plus years of diesel research at USBM/NIOSH, the focus has remained to assist the mining industry to safely use diesel equipment underground. Initially this research was directed towards using diesel haulage as a safer replacement for the electric-trolley system. This initial research focused on reducing the fire and explosion hazards from the internal-combustion engine as well as using ventilation to dilute and remove the diesel exhaust.

Eventually this focus shifted towards evaluating methods to control diesel emissions with integrated systems, such as DPFs or fuel additives, and establishing methods to accurately measure these gaseous and particulate emissions. Through this research, NIOSH promotes the integrated approach to controlling DPM, which is being used by the mining industry to control the exposure of miners to diesel exhaust. Current research focuses on characterizing the chemical, physical and toxicological properties of the diesel emissions and is directed towards eliminating the components which create the greatest health concerns. The direct result of the USMB/NIOSH diesel research is an extensive and safe use of diesel equipment in today's underground mines and the current and future projects will continue to assist the mining industry with issues related to the safe use of diesel equipment as this equipment continues to change in the future.

REFERENCES

Boehman, A.L., Song, J., Alam, M. 2005. Impact of biodiesel blending on diesel soot and the regeneration of particulate filters, Energy and Fuels, 19, 1,857–1,864.

Johnson, T.V. 2008 Diesel emission control in review, SAE Paper No. 2008-01-0069.

Gillies, S., H.W. Wu. 2009. Personal communication

Marshall, W.F., 1975. NO_x Emissions from Diesel Engines—Control and Measurement Techniques. AICHE Sympoium Series; Vol. 71. No. 148. pp. 114–119. 1975.

McGinn, S. 2000. The relationship between diesel engine maintenance and exhaust emissions—final report, http://www.deep.org/reports/mtce_report.pdf.

McGinn, S., 1999. Maintenance guidelines and best practices for diesel engines. http://www.deep.org/reports/mtce_guidelines.pdf.

Mine Safety Board, Recommendations of the United States Bureau of Mines on Certain Questions of Safety as of February 3, 1933: Inf. Circ. 6732, 1933.

Mine Safety and Health Administration. 2001a. 30 CFR 72. *Diesel particulate matter exposure of underground coal miners; final rule.* 66 Fed. Reg. 5526 and corrections 66 Fed. Reg. 27864.

Mine Safety and Health Administration. 2001b. 30 CFR 57. Diesel *particulate matter exposure of underground metal and nonmetal miners; final rule.* 66 Fed. Reg. 5706 and corrections 66 Fed. Reg. 35518.

Mischler S.E., Volkwein J.C. 2005. Differential Pressure as a Measure of Particulate Matter Emissions from Diesel Engines. *Environ Sci Technol* 2005 Apr; 39(7):2,255–2,261.

NIOSH, 1999, National Institute for Safety and Health. *Elemental Carbon (Diesel Particulate): Method 5040, Issue 3 (Interim).* In NIOSH Manual of Analytical Methods, 4th rev. ed. http://www.cdc.gov/niosh/nmam/pdfs/5040f3.pdf.

NIOSH. 2006a. Effectiveness of selected diesel particulate matter control technologies for underground mining applications: Isolated zone study, 2003, U.S. Department of Health and Human Services, DHHS (NIOSH) Publication No. 2006-126, Report of Investigations 9667.

NIOSH. 2006b. Effectiveness of selected diesel particulate matter control technologies for underground mining applications: Isolated zone study, 2004, U.S. Department of Health and Human Services, DHHS (NIOSH) Pub. No. 2006-138, Report of Investigations 9668.

Noll, J. D., Timko, R.J., McWilliams, L., Hall, P., Haney, R. 2005. Sampling Results of the Improved SKC Diesel Particulate Matter Cassette. *Journal of Occupational and Environmental Hygiene,* vol. 2, pp. 29–37.

Noll, J.D. 2008. Reducing exposures to diesel particulate matter (DPM) using direct reading instruments. *NIOSH Direct-Reading Exposure Assessment Methods Workshop,* November 13–14, 2008, Washington D.C.

Pope, A.C.; Burnett, R.T.; Thun, M.J.; Calle, E.E.; Krewski, D.; Ito, K.; Thuston, G.D. 2002. Lung cancer, cardiopulmonary mortality and long-term exposure to fine particulate air pollution. J. Amer. Med. Ass., 287(9), 1,132–1,141.

Spears, M.W. 1997. An emissions-assisted maintenance procedure for diesel-powered equipment. Minneapolis, MN: University of Minnesota, Center for Diesel Research. NIOSH contract No. USDI/1432 CO369004. [http://www.cdc.gov/niosh/mining/eamp/eamp.html].

USBM. 1936. United States Bureau of Mines Report of Investigations. *Diesel Mine Locomotives—Development and Use in European Coal Mines.* R.I. 3320. Department of the Interior—Bureau of Mines. 1936.

USBM. 1943. United States Bureau of Mines Information Circular. Annual Report of Research and Technological Work on Coal, Fiscal Year 1942. IC 7241. Department of the Interior—Bureau of Mines. May 1943.

USBM. 1944. United States Bureau of Mines. Procedures for Testing Diesel Mine Locomotives for Permissibility and Recommendations on the Use of Diesel Locomotives Underground. Schedule 22. Department of the Interior—Bureau of Mines. July 1944. Amended 1955.

USBM. 1945. United States Bureau of Mines Information Circular. Hazards of the Trolley-Locomotive Haulage System in Coal Mines. IC 7328. Department of the Interior—Bureau of Mines. July 1945.

USBM. 1947. United States Bureau of Mines Report of Investigations. Diesel Engines Underground, Use of Diesel Locomotives in Construction of the Delaware Aqueduct: Effect of Exhaust Gases Upon Quality of Tunnel Air. R.I. 4032. Department of the Interior—Bureau of Mines. 1947.

USBM. 1949. United States Bureau of Mines. Procedure for Testing Mobile Diesel Powered Equipment for Noncoal Mines. Schedule 24. Department of the Interior—Bureau of Mines. July 1949. Amended 1955

USBM. 1953. United States Bureau of Mines Information Circular. The Development of Permissible Requirements for Safe Underground Deisel Haulage. IC 7673. Department of the Interior—Bureau of Mines. December 1953.

USBM. 1960. United States Bureau of Mines Report of Investigations. Safety with Mobile Diesel-Powered Equipment Underground. RI 5616. Department of the Interior—Bureau of Mines. 1960.

USBM. 1963. United States Bureau of Mines Information Circular. Hazards of the Trolley-Locomotive Haulage System in Coal Mines. IC 8183. Department of the Interior—Bureau of Mines. July 1963.

USBM. 1968. United States Bureau of Mines Report of Investigations. Diesel Exhaust Contamination of Tunnel Air. RI 7074. Department of the Interior—Bureau of Mines. February 1968.

USBM. 1969. United States Bureau of Mines Report of Investigations. Effect of Cetane Improvers in the Fuel on Nitrogen Oxides Concentrations in Diesel Exhaust Gas. RI 7310. Department of the Interior—Bureau of Mines. October 1969.

USBM. 1973. United States Bureau of Mines Information Circular. Proceedings of the Symposium on the Use of Diesel-Powered Equipment in Underground Mining; Pittsburgh, PA, January 30–31, 1973. IC 8666. Department of the Interior—Bureau of Mines. January 1973.

USBM. 1976. United States Bureau of Mines Report of Investigations. Size Distribution and Mass Output of Particulates from Diesel Engine Exhausts. RI 8141r. Department of the Interior—Bureau of Mines. February 1976.

USBM. 1987. United States Bureau of Mines Information Circular. Diesels in Underground Mines; Proceedings: Bureau of Mines Technology Transfer Seminar, Louisville, KY, April 21, 1987, and Denver, CO, April 23, 1987. IC 9141. Department of the Interior—Bureau of Mines. April 1987.

USBM. 1995. United States Bureau of Mines Report of Investigations. In-mine evaluation of Catalyzed Diesel Particulate filters at two underground metal mines. RI 9571. Department of the Interior—Bureau of Mines. 1995.

Williams A. McCormick, R.L., Hayes, R., and Ireland, J., 2006, Biodiesel effects on diesel particle filter performance. National Renewable Energy Laboratory, Milestone Report NREL/TP-540-39606.

462

Simultaneous Ventilation, Power, and Cooling Supply with Compressed Air

George Danko
Department of Mining Engineering, University of Nevada, Reno, Nevada, United States

Davood Bahrami
Department of Mining Engineering, University of Nevada, Reno, Nevada, United States

Rajeev Gunda
Department of Mining Engineering, University of Nevada, Reno, Nevada, United States

ABSTRACT: Ancient knowledge about compressed air usage in underground mines is re-visited. Compressed air as power supply alone has lost popularity due to a relatively low power efficiency of the compressor and the air motor. The issue merits a new investigation due to two important features of compressed air power supply. First, the air from the air motor can be kept un-contaminated, representing fresh air supply to the working area of the power equipment. In this regard, a compressed air power supply line from the surface represents auxiliary ventilation with fresh air delivery. Second, the expansion work in the air motor absorbs and removes the same amount of heat from the working area. In this regard, the compressed air supply power represents a cooling unit with the same cooling power as of the air motor. A numerical example is provided to demonstrate the potential advantage of using combined compressed air power supply and ventilation system for an underground coal mine.

A new, coupled, ventilation, heat and moisture transport model is configured, using the qualified MULTIFLUX (MF) code. A published ventilation example for a UK coal mine is chosen as a baseline case for benchmarking the conventional ventilation solution for comparison with the new, compressed air application. The MF model is benchmarked against the baseline model results. Excellent agreement is found between the MF model and the measured temperature and humidity distributions. The mining example is then reconfigured, by increasing the depth and temperatures. Ventilation of the new task demands cooling, which is solved in two different ways for comparison. First, conventional spot air chiller is used. Second, compressed air powered auxiliary ventilation with spontaneous air cooling is applied. The example shows that compressed air usage for power may eliminate the need for refrigeration.

INTRODUCTION

A ventilation example is chosen for a development end in a mine. It is important to focus the attention on these zones for several reasons. First, the work force is concentrated in or around the production zone which constitutes the working environment. Second, a production zone is the place where the highest sources of dust, heat, moisture, hazardous gas, and noise are encountered. The emission from the freshly exposed rock and minerals imposes the highest demand on mitigation by ventilation for fresh, cool, and clean air supply. Third, mine development often results in dead ends, accessed by single-entry long drifts starting at the existing ventilation and transportation drifts. The delivery of fresh air in such a drift can only be provided by auxiliary ventilation.

Ventilation in a single-entry drift in a modern mine with a high advance rate and high virgin rock temperature is a challenging task. Earlier simplified numerical simulation models were published for short tunnel drifts by Voss (1980); Kertikov (1997);

Ross et al., (1997). Longwall coal faces were modeled by Longson and Tuck (1985), Gupta et al., (1993), and Lowndes et al., (2005).

The present paper applies a new, coupled thermal-hydrologic ventilation model, implemented in MULTIFLUX (MF) (Danko, 2008). The MF model was developed for detailed studies of heat, moisture, and air flow problems in high-level, radioactive waste storage facilities in deep geologic formations. Such storage facilities encounter high temperature and strong thermal-hydrologic interactions between the drifts and the host rock mass. It is advantageous to apply the MF model and software to mine ventilation to increase the accuracy of the simulation relative to simpler models (Danko and Bahrami, 2008). In the present single-entry drift analysis, MF provides the necessary simulation model to comply with the need for the fast-changing, transient conduction in the rock wall, complex heat transport interactions between the airflows in the auxiliary air duct and the drift, viscous dissipation, and compression-expansion work in the variable-pressure air flow.

In this paper, the MF model is first bench-marked against published results for an underground coal mine (Lowndes et al., 2005), discussed as baseline Case 1 in Section 2. The Case 1 task does not require any cooling. A more challenging task is created as Case 2 with an increased depth as well as higher air and virgin rock temperatures relative to those in Case 1. The MF model is applied to the Case 2 study in Section 3.1 for predicting the working conditions and refrigeration demand. The task in Case 2 requires air cooling to provide acceptable working conditions at the face. A conventional cooling solution is assumed in the Case 3 study, using a spot cooler at the intake of the forcing auxiliary ventilation duct, are described in Section 3.2. The necessary cooling power is determined using the MF model based on the criterion of effective temperature at the face. Effective temperature is calculated based on the empirical formula in literature (Tuck, et al., 1997), given in the Appendix. The Case 3 study is the reference for a comparison when cooling is needed. A different, new solution is presented in Case 4, in Section 4, applying compressed air for power, cooling, and additional fresh air delivery. The new solution is discussed and compared with the conventional one in Section 5. The conclusions of the study are summarized in Section 6.

THERMAL-MOISTURE-AIR FLOW MODEL FOR A SINGLE-ENTRY DRIFT WITH MF

MF Model Description

MF Version 5.0 is used to evaluate the air flow, temperature, and humidity distributions. The MF model and software is well documented since it became a qualified code for studying hot, subsurface nuclear waste storage facilities (Danko, 2008). Only a brief summary of the MF model is provided here.

The Thermal Model of the Strata

The strata around the drift is represented by a series of NTCF (Numerical Transport Code Functionalization) surrogate models along the drift length. The NTCF surrogate models are in matrix equation form (Danko, 2006):

$$qh = qh^c + hh \cdot \left(T - T^c\right) \tag{1}$$

The hh dynamic admittance matrix is identified based on Eq. (1) by fitting qh to simulation data from a numerical, detailed heat conduction model for the strata. In the present model, TOUGH 2 (Pruess et al., 1999) was used for generating input data for the

NTCF surrogate model. The NTCF model identification method, a U.S. patent (Danko, 2009), follows the technique previously described by Danko (2006). The model for each drift section perfectly reproduces qh^c, the central output fluxes from TOUGH2, for $T=T^c$, the central input boundary conditions, that are included in the pre-selected set of boundary conditions. Other T input variations can produce outputs from the NTCF model for qh without actually re-running TOUGH2. For the current model, 984 NTCF models are generated from only a few TOUGH2 runs by scaling, following the technique published by Danko and Bahrami (2008).

The moisture transport of the strata may be modeled with another NTCF matrix equation similar to Eq. (1) in MF. The underlying model of the porous and fractured strata is TOUGH2, a thermal-hydro-logic model itself, that may account for vapor and water inflow. However, the MF model configuration in the current example applies a moisture transport element based on the wetness factor for moisture evaporation from the drift surface.

CFD Model for Heat, Moisture, and Air Flow Transport in the Drift

The energy balance equation in the CFD model of MF is used in a simplified form, as follows, for an x-directional flow with v_i velocity in a flow channel of cross section dy by dz (and with no convective heat transport in y and z directions while considering the x-directional flow):

$$\rho c \frac{\partial T}{\partial t} + \rho c v_i \frac{\partial T}{\partial x} = \rho c a \frac{\partial^2 T}{\partial x^2} + \rho c a \frac{\partial^2 T}{\partial y^2} + \rho c a \frac{\partial^2 T}{\partial z^2} + \dot{q}_h \tag{2}$$

In Eq. (2), ρ and c are density and specific heat of moist air, respectively; a is the molecular or eddy thermal diffusivity for laminar or turbulent flow; and $\dot{q}_h$ is the latent heat source or sink for condensation or evaporation. The second and the third terms on the right-hand-side of Eq. (2) represent heat conduction (or effective heat conduction) in the y and z directions, normal to the x axis of the flow channel; these terms are substituted with expressions for transport connections using heat transport coefficients for flow channels bounded by solid walls. Eq. (2) is discretized and solved numerically and simultaneously along all flow channels for the temperature field T in MF.

The simplified moisture transport convection-diffusion equation in the CFD model of MF is similar to Eq. (2) as follows:

$$\frac{\partial \rho_v}{\partial t} + v_i \frac{\partial \rho_v}{\partial x} = D \frac{\partial^2 \rho_v}{\partial x^2} + D \frac{\partial^2 \rho_v}{\partial y^2} + D \frac{\partial^2 \rho_v}{\partial z^2} \qquad (3)$$
$$+ \dot{q}_{cm} + \dot{q}_{sm}$$

In Eq. (3), ρ_v is the partial density of water vapor; D is the molecular or eddy diffusivity for vapor, calculated from the thermal diffusivity, a; $\dot{q}_{cm}$ is the moisture source or sink due to condensation or evaporation; and $\dot{q}_{sm}$ is the vapor flux the air flow distribution in superheated steam form in high-temperature applications.

The air flow field is also modeled numerically. The Navier-Stokes momentum balance equation for 3D flow of the bulk air-moisture mixture is used as follows, following Welty et al. (1984):

$$\rho \left(\frac{\partial v_x}{\partial t} + \mathbf{v} \cdot \nabla v_x \right) = \rho g_x - \frac{\partial Pb}{\partial x} + F_x \qquad (4a)$$

$$\rho \left(\frac{\partial v_y}{\partial t} + \mathbf{v} \cdot \nabla v_y \right) = \rho g_y - \frac{\partial Pb}{\partial y} + F_y \qquad (4b)$$

$$\rho \left(\frac{\partial v_z}{\partial t} + \mathbf{v} \cdot \nabla v_z \right) = \rho g_z - \frac{\partial Pb}{\partial z} + F_z \qquad (4c)$$

The viscous terms in Eqs. (4a–c) can be expressed with the viscous normal-stress $(\sigma)_v$ and shear-stress (τ) components as:

$$F_x = \frac{\partial (\sigma_{xx})_v}{\partial x} + \frac{\partial \tau_{yx}}{\partial y} + \frac{\partial \tau_{zx}}{\partial z} \qquad (5a)$$

$$F_y = \frac{\partial \tau_{xy}}{\partial x} + \frac{\partial (\sigma_{yy})_v}{\partial y} + \frac{\partial \tau_{zy}}{\partial z} \qquad (5b)$$

$$F_x = \frac{\partial \tau_{xz}}{\partial x} + \frac{\partial \tau_{yz}}{\partial y} + \frac{\partial (\sigma_{zz})_v}{\partial z} \qquad (5c)$$

The viscous force terms in Eqs. (4a–5c) are integrated along the grid lines of the flow channels and expressed as a function of the convective air flow components in the air space. The lumped-parameter CFD model approach allows for reducing the number of discretization elements in the computational domain. MF allows for defining connections between lumped volumes, applying direct air flow, as well as heat and moisture transport relations between them. The current, lumped-parameter CFD model in the drift applies 3000 nodes for the heat, and the same number of nodes for the moisture transport as well as for air flow transport.

Coupled Solution Between the NTCF and CFD Model-Elements

The NTCF (approximating the rock mass response) and CFD models are coupled on the rock-air interface by MF until the heat and moisture fluxes are balanced at the common surface temperature and partial vapor pressure at each surface node and time instant. Two inner iteration loops are used to balance the in-rock and in-drift transport processes on the rock-air interface:

1. Heat flow balance iteration between the NTCF and airway CFD models for each time division
2. Moisture flow balance iteration between the NTCF and airway CFD models for each time division

An additional outer iteration is also performed:

3. Air flow field re-calculation with balanced thermal and moisture transport

The nested iterations continue until full convergence is achieved.

MF Model Benchmarking Against Published Field Data—Case 1 Study

The benchmarking exercise applies a carefully configured MF Model. The inputs for the simulation are: drift geometry, ventilation air delivery, and input temperature and humidity for the given mining method. The mining production rate and work schedule also follows the published study. Some important input data for the simulation are reverse-engineered with the use of the MF model, most notably the power load factors for the road header, scraper, forcing and exhaust fans, and the wetness factors at the face and drift walls. The powered units are specified only by their nominal power in the reference study (Lowndes et al., 2005), whereas the thermal and climatic model requires actual power dissipation input data. The power load factors together with the wetness factors are determined by trial and error until good match is achieved between the MF model output results and the published data for dry-bulb and wet-bulb temperatures. The thermal input power of the forcing auxiliary fan is reprocessed from the measured temperature rise across the fan, given in Table A1 of the Appendix. The process water application rate used for dust suppression at the production face is back-calculated from measured dry bulb and wet bulb temperature variations between Stations 3 and 5, given in Table A3 of the Appendix.

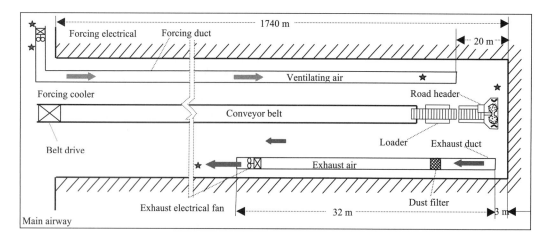

Figure 1. Single entry drift with auxiliary ventilation and cooling after Lowndes et al. (2005)

Mine Layout and Input Data

A development drift studied and published by Lowndes et al., (2005) is used as a reference in the present ventilation example for two practical reasons: (1) the results are readily available and well documented; and (2) in situ measurement results are also included that can be used for benchmarking comparisons with our model results. The schematic plan of the single entry drift is shown in Figure 1.

The schematic shows a single entry drift with 2 auxiliary ducts. The forcing duct brings in fresh air to the face/ head end. The exhaust duct removes the air with an electrical fan, and filters it with a dust filter. A road header is used for mining, accompanied by a loader which transports coal to a motor-driven conveyor belt that carries away the load.

The geometry, properties of the rock, power inputs to fans and machinery as well as ventilation air flow rates and production rate all follow those of the reference case. The input parameters and properties are given in Table 1. The reference working temperatures are determined from the published measurement results of Lowndes et al., (2005). The recorded temperatures from measurement Sites 1 through 6 were digitized and averaged over an entire week, the 33rd in mining operation, and provided weekly reference values for comparison. The results are given in Table 2.

Comparison Between Simulated and Measured Temperatures

The Case 1 dry and wet bulb temperature variations along the flows in the forcing duct and drift from the MF model are shown in Figure 2. The reference temperatures from Table 2 at six stations are also plotted in Figure 2. As shown, excellent agreement

is obtained between the simulated and reference dry and wet bulb temperatures. Figure 2 also shows the effective, equivalent temperature, T_E, relevant to thermal stress, calculated from T_d, T_w and the air velocity (see equations in Table A4 of the Appendix).

At the working face, the temperature field is more detailed from the MF model than from measurements, showing variations in Figure 2 along flow paths due to heating from strata and machinery, as well as cooling by water evaporation. A close-up enlargement of this area is shown in Figure 3. Evaporation is the sole source of cooling since the incoming and exiting ventilation air is nearly at the same temperature, which is apparent from the curves of the dry bulb temperature variation. Figure 4 depicts the velocity field in the coarse grid used in the MF air flow model around the working face. Table 3 summarizes the temperature results and shows the comparison between MF results and the reference values.

Discussion of the Benchmarking Results

The MF model has responded to input load factor and wetness factor adjustments and shows an excellent match with the reference data in both dry and wet bulb temperatures. The adjustments by manual trial-and-error gave power load factors of 0.31, 0.1 and 0.21 for the cutter and scraper power pack, the conveyor loader and the exhaust duct fan, respectively. The wetness factor at the face area need to be set to 1.21 to account for process water evaporation, indicating that the entire surface is fully wet and either (1) the real transport surface is larger by 21% than the boundary geometry due to roughness and tortuosity; or (2) the moisture transport coefficient in

466

Table 1. Geometry and main input data (Lowndes et al., 2005; McPherson, 2003)

Road header length	12.2 m
Drift length	1740 m
Drift cross sectional area	19.85 m² (height: 5.7 m × 3.5 m)
Depth at drift entry	743 m
Depth at drift development area	710 m
Virgin Rock Temperature (VRT)	30.4°C
Thermophysical properties of the rock strata	Thermal conductivity: 2.175 W/K
	Thermal diffusivity: 1.286×10^{-6} m²/s
Forcing Duct and Fan	
Diameter	1.066 m (for 1st 400 m)
	0.900 m (for next 1,340 m)
Relative surface roughness (k/D_h)	0.05 (Mcpherson, 1993)
Forcing fan utilization	100% (7 days @ 24 hours/day per week)
Air flow rate at intake	6 m³/s @ P = 108.88kPa; T_{d1} = 25.46°C; T_{w1} = 18.71°C
	(averaged values for 33rd week from Lowndes et al., (2005))
Thermal dissipation of forcing fan	74.62 kW (from calorimetric calculations; see Table A2 of Appendix)
Forcing fan nominal power	90 kW
Exhaust Duct and Fan	
Diameter	0.600 m
Relative surface roughness (k/D_h)	0.0003 (Mcpherson, 1993)
Exhaust fan utilization	67.86% (4 days @ 24 hours/day, 1 day @ 18 hours/day per week)
Nominal power	56 kW
Road header nominal power	142 kW(Power pack: Cutter & Scraper)
Rate of advance (average)	54 m per week (from 10.8 m/day advance rate for 5 days/week
	(Lowndes et al., 2005))
Age of drift	33 weeks (Drift length/weekly advance rate)

Table 2. Averaged reference temperatures (at the 33rd week)

Measurement Site	Dry Bulb Temperature (°C)	Wet Bulb Temperature (°C)
Site 1	T_{d1} = 25.46	T_{w1} = 17.81 + 0.91*
Site 2	T_{d2} = 35.14	T_{w2} = 22.76
Site 3	T_{d3} = 31.5	T_{w3} = 21.58
Site 4	T_{d4} = 31.65	T_{w4} = 23.57
Site 5	T_{d5} = 30.99	T_{w5} = 23.8
Site 6	T_{d6} = 31.73	T_{w6} = 25.01

* Note: Adjustment for possible measurement error, suggested by Lowndes et al., (2005)

reality is 21% higher than that in the MF model; or a combined effect, sharing the 21% between reasons (1) and (2). The duct area from the working zone required a wetness factor of only 0.03 to account for the increase in the moisture content between Stations 5 and 6.

An exceptionally good match with measurement data is achieved in both dry and wet bulb temperatures, shown in Figure 2 and summarized in Table 3. The largest discrepancy is 0.52°C (1.69%) in T_d at Site 5; and –0.91°C (–4.24%) in T_w at Site 3, the exit point of the forcing duct. The difference

indicates perhaps an uncertainty in the measurement reference data for T_w. Indeed, such uncertainty is indicated by the fact that the wet bulb input temperature at Site 1, directly affecting Site 3, was suspected and adjusted upward by 0.91°C by Lowndes et al (2005) in their simulation model.

After reaching the matched temperature results shown in Figures 2 and 3 as well as in Table 3, the MF model can be used to determine ventilation parameters, such as total, barometric pressure variations and the hydraulic power consumption of the forcing and exhaust fans, moisture content variations

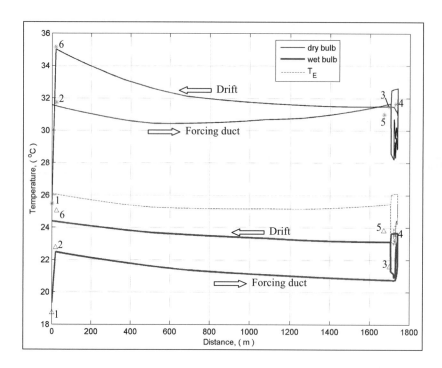

Figure 2. Dry and wet bulb temperature variations along the flows in the forcing and exhaust ducts and drift in Case 1

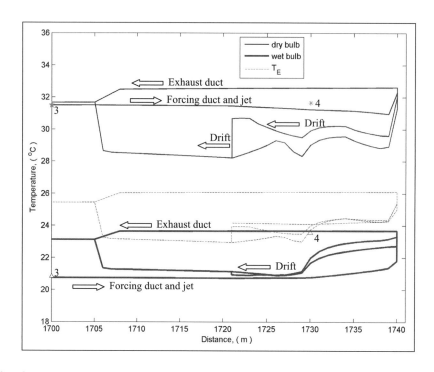

Figure 3. Close-up enlargement of the temperature fields in the working area in Case 1

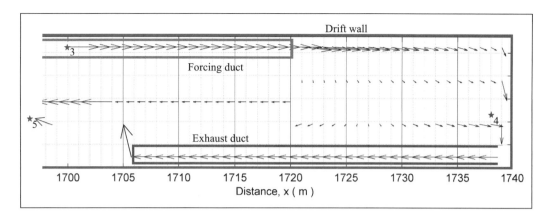

Figure 4. Velocity field in the working area from the MF model

Table 3. Comparison between measurement and MF results for Case 1

| | Measurement | | MULTIFLUX Prediction | | | | | |
Measurement Station	Dry Bulb Temperature (°C)	Wet Bulb Temperature (°C)	Dry Bulb Temperature (°C)	Error (°C)	Relative Error (%)	Wet Bulb Temperature (°C)	Error (°C)	Relative Error (%)
Site 1	25.46	18.72	25.48	0.02	0.08	19.31	0.59	3.14
Site 2	35.14	22.76	34.95	−0.19	−0.53	22.46	−0.30	−1.30
Site 3	31.50	21.58	31.24	−0.26	−0.81	20.67	−0.91	−4.24
Site 4	31.65	23.57	31.50	−0.15	−0.49	23.07	−0.50	−2.11
Site 5	30.99	23.80	31.51	0.52	1.69	23.54	−0.26	−1.11
Site 6	31.73	25.01	31.44	−0.29	−0.90	24.58	−0.44	−1.74

in the air flow, and other interesting parameters such as the air flow and temperature distribution fields around the working area at any location of interest.

The total forcing fan power from the MF model is 115 kW, which indicates an overload for the fan motor, which is only 90 kW in nominal power. This overload may explain the excessive power loss across the booster fan and a low overall efficiency of 35.1%, albeit a somewhat higher value from the MF model than the 30% reported by Lowndes et al. (2005) from their model. The evaporation water flow rate from the MF model at the working face is 0.024 kg/s. This value compares excellently with 0.025 kg/s, determined from measurements and shown in Table A3 of the Appendix.

In summary, the Case 1 study results qualifies the MF model against measurement results without any concern for discrepancy, making it ready to predict Cases 2, 3 and 4, different from the baseline case. The Case 1 study also shows that working environment with $T_E = 25.2°C$ is perfectly adequate for a full work load without any cooling needs.

REFRIGERATION DEMAND PREDICTION WITH THE MF MODEL FOR HOTTER CONDITIONS THAN THOSE IN CASE 1

A more challenging ventilation task as Case 2 is presented by lowering the depth to 1210 m at the drift entry, and increasing the virgin rock temperature to 45°C. The air intake temperature and relative humidity were also both increased due to the greater depth, to 27.46°C and 60%, respectively, giving a wet bulb temperature of 21.62°C. The air flow rate was still kept constant at 6 m³/s. Other input parameters, such as power and wetness factors and fan efficiencies are kept unchanged.

Mining Environmental Prediction at an Increased Depth—Case 2 Study

The simulation results are shown in Figure 5 for the dry and wet bulb temperatures along the air flow for Case 2. The T_E temperature rises to 29.91°C, violating the acceptable limit of 28°C for a full work load (Lowndes et al., 2004). This environment requires

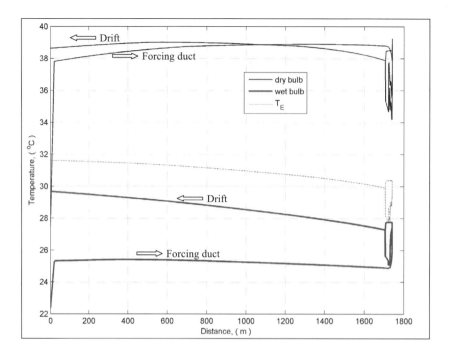

Figure 5. Dry and wet bulb temperature variations along the flows in the forcing and exhaust ducts and drift in Case 2

either air climatization by cooling or the reduction of heat sources.

Effective Temperature Reduction at the Working Face With Conventional Ventilation and Cooling—Case 3 Study

A spot cooler is applied as a conventional solution approach right after the forcing fan, shown as an option in Figure 1. The cooling power of the coil is manually adjusted by trial-and-error in the MF model. After a few trials, a cooling power of 115 kW was reached which produced the output temperature results shown in Figure 6 with T_E= 28.05°C at the face, a very close value to the required 28°C. At this point, it should be considered quite coincidental that the required cooling power equals the electrical input power of the forcing fan for the arbitrarily chosen input parameters (intake air and virgin rock temperatures) in Case 3.

COOLING AIR AND POWER DELIVERY BY COMPRESSED AIR SUPPLY

Concept Description

A concept is analyzed for powering the auxiliary fan motor by compressed air. The benefit of using compressed air power for air cooling is recognized in the literature. A survey of heat sources and cooling loads was conducted in seven underground mines in 1973 reported by Hartman et al. (1997). At the time, the total heat removal shared by compressed air, air conditioning, and ventilation was 5%, 25%, and 70%, respectively. The power flow from input P_{in} to the fan hydraulic (useful) power, P_h and fan heat dissipation, Q_{hF} is shown for both electrical and compressed air power supplies in Figure 7. The fan hydraulic efficiency, η_F, and motor efficiency, η_M, are separated from each other since the same fan blades may be mounted on either an electrical or a compressed air motor of different efficiencies. The electric motor and fan efficiencies are 90% and 39% in the numerical model, resulting in the overall efficiency of $\eta_M \cdot \eta_F = 35.1\%$ in Cases 1 through 3.

For the air motor, the thermodynamic conversion equation between isothermal expansion power, P_{IE} which equals the input power to the air motor, P_{in}, and, thermal power Q_{hT}, is as follows:

$$P_{IE} = P_{in} = -Q_{hT} \qquad (7)$$

where the isothermal expansion power due to pressure change from P_{b2} or P_{b1} is:

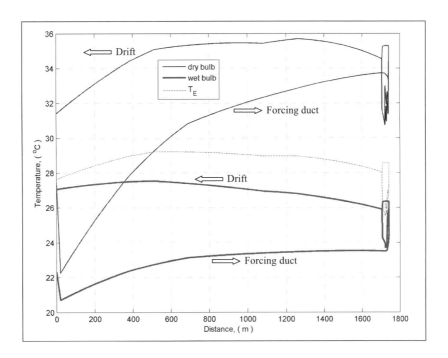

Figure 6. Dry and wet bulb temperature variations along the flows in the forcing and exhaust ducts and drift in Case 3

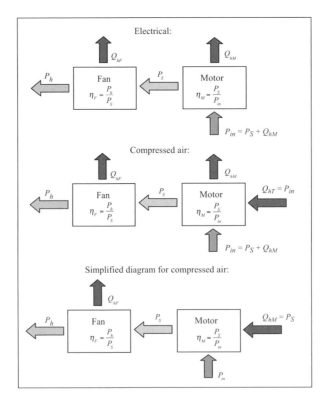

Figure 7. Power flow charts for electrical and compressed air driven fans

Table 4. Input data for the compressed air motors and forcing fans in the MF model in Case 4

Compressed air supply data	Joint Air Motors	Separate Air Motors With Two Fans
Input pressure, P_{b2} [MPa]	0.6 12	0.612
Exhaust Pressure, P_{b1} [MPa]	0.1	0.1
Air quantity to motor, Q_a [m³/s]	0.75	0.75/2+0.75/2
Compressed air power, P_{CA}, [kW] of motor @ $\eta_M = 70\%$	147.8	129.6/2+ 129.6/2
Shaft power to fan, P_s [kW]	103.6	90.7/2+ 90.7/2
Cooling power of air motor [kW]	103.6	90.7/2+ 90.7/2
Auxiliary fans data		
Air quantity through fan [m³/s]	6.0	5.25/2 + 5.25/2
Pressure rise through fan [kPa]	6.73	6.08/2 + 6.08
Hydraulic efficiency of fan [%]	39	39
Fan heat dissipation due to hydraulic loss [kW]	63.2	58.8/2 + 58.8/2
Total shaft power input for fan [kW]	103.6	129.6/2 + 129.6/2

$$P_{IE} = \frac{Q_a \cdot P_{b1}}{\rho_1} \ln\left(\frac{P_{b2}}{P_{b1}}\right) \qquad (8)$$

The thermodynamic conversion heat removal from the environment while delivering the P_{in} power in compressed air form is the basis for the air cooling principle. An air motor can deliver only a fraction of P_{IE} on the shaft, P_S, due to its mechanical and fluid power losses as well as conversion irreversibility, expressed by a combined motor efficiency, η_M:

$$P_S = P_{IE} \cdot \eta_M \qquad (9)$$

A portion of the input power, $P_{in} - P_S = P_{in(1 - \eta_M)}$ returns immediately back to the system as heat. The net cooling power is therefore reduced to $Q_{hM} = P_{in} - (P_{in} - P_S) = P_S$, that is, to the useful shaft power of the air motor. Since $P_S = \eta_M \cdot P_{in}$ is only a fraction of the input power, both the net output as well as the cooling capacity are hampered by the efficiency of the air motor. The η_M value for a large, 60 hp compressed air motor currently available in the market is 35%, although higher efficiencies can be achieved for smaller motors, up to 75% (Hydraulics&Pneumatics, 2009). An efficiency value of 70% is assumed in the current study, as a feasible value if the air motor industry is challenged with a strong demand.

Solution Example with Compressed Air-Driven Booster Fans—Case 4 Study

Two 45kW compressed air motors are used instead of one nominal 90kW electrical forcing fan motor in Case 4. The two motors can be used together or separately at two different locations. The input data for the MF model simulations are identical to those used in Case 3, except for the forcing fan driving power and heat dissipation which both have

different components, most notably, the spot cooler heat removal being displaced by compressed air expansion power. The input data are given in Table 4 for the air motors. No credit is applied in this case to injecting the exhaust of compressed air into the duct and reducing by the same amount of the air flow through the fan. The temperature results from the MF model for Case 4 with two air motors used together at the intake location are shown in Figure 8. As shown, the results agree exactly with those in Figure 6 for Case 3.

It may be advantageous to separate the two air motors, and use one at the intake, and one in the middle of the length at 870m into the drift, shown in Figure 9. The distributed application splits the total temperature decrease in the duct into two portions. The pressure distribution is also more balanced, reducing leakage, which is a fact in auxiliary ventilation, but not accounted for here for brevity.

Another advantage of the separation of the air motor locations is that lower air quantity (by the exhaust compressed air quantity of the second air motor which injects into the duct at 850 m) is carried through the first fan and along a significant length. Credit can also be claimed for power savings, since each air powered fan carries less air quantity by their own used air, injected downstream to the duct. These flow rate reductions reduce the total shaft power to the fans from 103.6 kW to 90.7 kW which was determined by a few steps of iteration. The simulation results for this case are shown in Figure 10.

Comparison Between Conventional, Refrigerated, and Compressed Air-Cooled Solutions

The temperature results for Case 3 (in Figure 6) and Case 4 with joint motors and one fan (in Figure 8) are identical, allowing for easy comparison of the

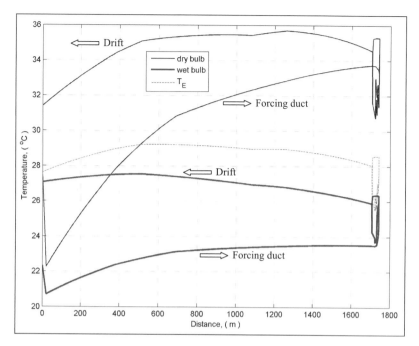

Figure 8. Dry and wet bulb temperature variations along the flows in the forcing and exhaust ducts and drift in Case 4, using both air motors at the intake fan

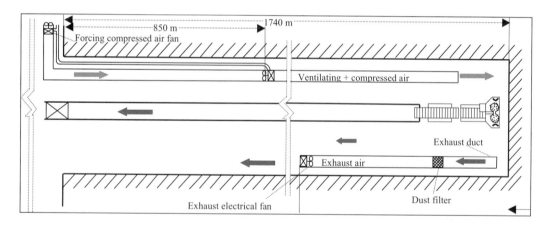

Figure 9. Single-entry drift with compressed air ventilation and cooling, using two air motors and fans

two different climatization solutions based on input power and equipment requirements. Case 4 with separate air motors and credit taken for air quantity reduction by injecting the air power exhaust air into the auxiliary duct, tips the balance in favor of the compressed air power solutions. The input power requirements for Cases 3 and 4 with joint and separate air motors are compared in Table 4.

As shown, the total power demand at the motor is the same for both cases assuming a performance index of 3.5 for the refrigerator in Case 3, and an air motor efficiency of 70%, for joint air motors. This agreement is due to a careful selection of the depth and input data set, given in Section 3 for Case 2 in order to make an easy comparison of the power requirement. The two different solutions in this case are identical regarding power and working face temperatures.

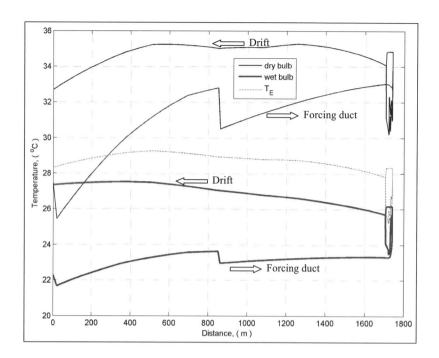

Figure 10. Dry and wet bulb temperature variations along the flows in the forcing and exhaust ducts and drift in Case 4, using two auxiliary fans, and taking credit for air quantity reduction through the fans

Table 5. Input power requirement for Cases 3 and 4

	Case 3	Case 4 Joint motors, one fan	Case 4 Separated motors, total
Fan motor electrical input [kW]	115	—	—
Spot cooler capacity [kW]	115	—	—
Spot cooler's electrical input power [kW] @ 3.5 performance index	32.8	—	—
Compressed air cooling capacity @ η_M = 70% [kW]	—	103.6	90.7
Compressed air power input [kW]	—	147.8	129.6
Total power demand [kW]	147.8	147.8	129.6

Many differences remain, such as (1) installation and maintenance of a 115 kW (32.7 tons refrigeration) spot cooler at the entrance of the drift in Case 3, versus a 147.8 kW air compressor either underground or on the surface; (2) distribution losses for electrical versus compressed air; (3) compressor inefficiency versus the handling of the coolant fluid or air for the spot chiller which rejects 115 kW heat; and many others. These considerations are all quite outside the scope of the paper, but will be the deciding factors since the total power demand is identical.

The balance becomes positive in favor of the compressed air power solution if the two fans are separated, with lower power requirement, given in the second column for Case 4 in Table 5. In this case, 12.3% less total input power is needed than that in Case 3. In addition, the effective temperature is reduced to 27.2°C, shown in Figure 10.

DISCUSSION OF THE SOLUTION USING COMPRESSED AIR FOR POWER, COOLING, AND FRESH AIR DELIVERY

As shown in Point 4.3, compressed air power can compete with electrical power due to its cooling capacity which may eliminate the need for an additional air cooler. For a break-even competition in power consumption alone, a minimum air motor efficiency of around 70% is needed. This defines a demand for the compressed air equipment industry to consider. The expansion ratio in the air motors must be increased from the currently favored 20% to around 80%, which may require multi-stage air motors with interheaters. A new generation of air motors may revolutionize mine climatization, allowing the economic use of compressed air underground instead of operating refrigerators in auxiliary ventilation systems.

Many obstacles will remain and need further studies, such as noise and control of cooling power which is, as shown, linked directly to shaft power, P_s. What appears an advantageous link in some cases (a longer drift needs higher cooling power) may become an obstacle for others. For example, what happens if spot cooling is needed without the need for shaft power? The solution in the example suggests that a promising application field is compressed air powered auxiliary fans which act as thermo-dynamic chillers with a cooling capacity of the shaft power. As shown, particularly attractive is the injection of the exhaust compressed air into the air duct downstream from the blades. This way, higher air flow rate is propelled for the cost of lower flow rate by the exhaust air quantity of the air motor.

CONCLUSIONS

1. A new mine ventilation, heat, and moisture simulation model is applied to a complex air flow, heat, and vapor transport problem in a single-entry drift, using MF. The MF model shows exceptional performance in matching mine monitored data from a publication.

2. The problem used in the example describes a real work plan in an operating mine with sparse in situ data, and missing information about basic parameters, such as operating power load factors. In spite of all drawbacks, the MF model matches very well the measurement data and reveals other missing data elements. The identified power load factors and surface wetness factors are within a realistic range. Only a few default values in the transport model-elements in MF needed adjustment. Based on this positive experience, it is quite realistic to assume that a MF model can be made to self-adjust to monitored data. There was no need to adjust the basic strata properties, such as rock thermal conductivity, or diffusivity in the example, quite the contrary to what the reference work (Lowndes et al., 2005) had to do to match their model with the mine data.

3. The feasibility of using the cooling effect of compressed air expansion power is demonstrated with numerical simulation, and with a new, benchmarked MF model for a challenging situation. The solution shows that compressed air power supply to a forcing fan can completely eliminate the need for refrigeration at identical total electrical power consumption.

4. Many challenges remain to make the suggested solution a reality for routine mining application. The critical element is the availability of air motors at or above 70% in power efficiency.

5. Multi-stage air motors with interheaters are seen as a unique combination of the heating need for the expanding compressed air stream (and prevent it from freezing the water content) with the cooling need of the hot ventilating air in the air duct. Consequently, auxiliary booster fans are suggested with multi-stage, multi-impeller air motors connected with interheater air lines. Figure 11 is a schematic diagram of the suggested solution for the air motor industry.

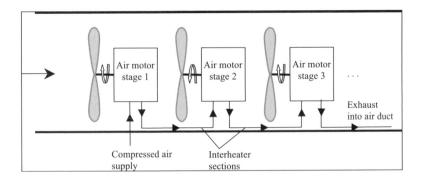

Figure 11.   Schematic diagram of multi-stage, multi-impeller air motors in a booster fan application

ACKNOWLEDGMENT

Support from NIOSH is appreciated for the studies as part of the research project "Safety, Health, and Ventilation Cost Benefit Optimization with Simulation and Control," grant number 200-2009-30157. The assistance of mining engineering students Barret Loehden and Raymond Grymko in preparing the manuscript is thankfully recognized. The editorial comments of Professor Mousset-Jones is thankfully appreciated.

APPENDIX

Table A1.   Conversion between dry and wet bulb temperatures

Energy balance for the wet bulb temperature sensor (McPherson, 2003, p.506):

$$L(X_S - X) = (1 + X)c_{pm}(T_d - T_w) \tag{A1}$$

From Equation A1:

$$T_w = T_d - \frac{L}{c_{pm}}\left(\frac{X_S - X}{1 + X}\right) \tag{A2}$$

Table A2.   Thermal power of forcing fan

$c_{pm} = 1.0217$ kJ/kg°C
$\rho = 1.257$ kg/m³
$Q_{av} = 6$ m³/s
$Q_a = Q_v * \rho = 7.543$ kg/s
$Q_h = c_{pm} \cdot Q_{am} \cdot (T_1 - T_2)$
where, $T_1 - T_2 = 35.14 - 25.46 = 9.68$°C
$Q_L = 74.6$ kW (Thermal dissipation across forcing fan)

Table A3.   Process water added

Site 3:
The moisture content calculated for average temperature over the week:
$X_3 = 1.08 \times 10^{-2}$
$\omega_3 = 1.07 \times 10^{-2}$
Site 5:
The vapor content calculated for average temperature over the week:
$X_5 = 1.42 \times 10^{-2}$
$\omega_5 = 1.40 \times 10^{-2}$
$Q_{am} = 7.577$ kg/s
$Q_d = 7.504$ kg/s
The moisture content at Site 5 = Water added from mine + Dust suppression water added
Dust suppression water added $= (\omega_5 - \omega_3)Q_{am}$
$\qquad = (X_5 - X_3)Q_d = 2.51 \times 10^{-2}$ kg/s

Table A4.   Effective temperature calculation

The Basic effective temperature can be calculated using the following equation for velocities in the range of 0.5–3.5 m/s (Tuck et al., 2003, p. 46):

$$T_E = \left(\frac{4(4.12 - x_1) + x_2}{1.65176}\right) + 20 \tag{A3}$$

where

$$x_1 = \frac{\left[8.33\left(17x_3 - (x_3 - 1.35)(T_w - 20)\right)\right]}{\left[(x_3 - 1.35)(T_d - T_w) + 141.6\right]} \tag{A4}$$

$$x_2 = \frac{\left[17(T_d - T_w) \cdot x_3 + 8.33 \cdot (T_w - 20)\right]}{\left[(x_3 - 1.35) \cdot (T_d - T_w) + 141.6\right]} \tag{A5}$$

$$x_3 = 5.27 + 1.3v - 1.15 \cdot \exp \cdot (-2v) \tag{A6}$$

$T_E = 28$°C effective is a commonly applied design condition applied to mine workings.

Nomenclature

T_d = Dry bulb temperature [°C]
T_w = Wet bulb temperature [°C]
Q_{av} = Volumetric flow rate of air [m³/s]
Q_a = Mass flow rate of air [kg/s]
Q_v = Mass flow rate of vapor [kg/s]
Q_d = Mass flow rate of dry air [kg/s]
ρ = Density of air [kg/m³]
Q_h = Thermal dissipation of forcing fan [kJ]
c_{pa} = Specific heat of air at constant pressure [kJ/kg °C]
T_i = Dry bulb temperature at Site i [°C] for (i = 1 to 6)
P_b = Barometric (total, static) pressure [Pa]
P_v = Partial vapor pressure [Pa]
P_{vsw} = Saturation pressure at wet bulb temperature [Pa]
L = Latent heat of vaporization at $(X_S + X)/2$ and $(T_d + T_w)/2$ [J]
R_d/R_v = Ratio of gas constants of air and vapor = 0.622
ΔT = Wet bulb depression = $T_d - T_w$ [°C]
X = Actual water content
X_S = Saturated water content at t_w
c_{pm} = Mixed air-vapor specific heat at $(X_S + X)/2$ and $(T_d + T_w)/2$ [kJ/kg °C]
ω = Vapor mass fraction = $X/(1 + X)$
v = Air velocity [m/s]
T_E = Effective temperature [°C], see Table A4 of the Appendix
P_h = Hydraulic power [kW]
P_S = Shaft power [kW]
Q_{hF} = Fan power loss by heat dissipation [kW]
Q_{hM} = Motor power loss by heat dissipation [kW]
P_{in} = Input power [kW]
Q_{hT} = Thermal conversion heat removal [kW]
η_M = Motor efficiency
η_F = Fan hydraulic efficiency

REFERENCES

Danko, G., 2008. *MULTIFLUX V5.0 Software Qualification Documents*. Software Tracking Number: 1002-5.0-00, Software Management Office, Berkeley National Laboratory.

Danko, G. and Bahrami, D., 2008. *Application of MULTIFLUX for air, heat and moisture flow simulations*. 12th U.S./North American Mine Ventilation Symposium, Reno, NV, June 9–11, 2008, pp. 267–274.

Danko, G., 2006. *Functional or Operator Representation of Numerical Heat and Mass Transport Models*. ASME J. of Heat Transfer, Vol. 128, pp. 162–175.

Danko, G., 2009. *Multiphase physical transport modeling method and modeling system*. US Patent No 7,610,183.

Hartman, H.L., Mutmansky, J.M., Ramani, R.V., and Wang, Y.J., 1997. Mine Ventilation and Air Conditioning. John Wiley and Sons, Inc., 3rd Ed., p. 646.

Gupta, M.L., Panigrahi, D.C., Banerjee, S.P., 1993. *Heat flow studies in longwall faces in India*. In: R. Bhaskar, ed., Proceedings of 6th U.S. Mine Ventilation Symposium, SME, Littleton, CO, pp. 421–427.

Hydraulics & Pneumatics-Turbine Motors. 2009. *http://www.HydraulicsPneumatics.com/200/ TechZone/FluidPowerAcces/Article/True/6422/ TechZone-FluidPowerAcces.* Accessed October 2009.

Kertikov, V., 1997. *Air temperature and humidity in dead-end headings with auxiliary ventilation*. Proceedings of 6th International Mine Ventilation Congress, SME, Littleton, CO, USA, pp. 269–276.

Longson, I., Tuck, M.A., 1985. *The computer simulation of mine climate on a longwall coal face*. In: P. Mousset-Jones, ed., Proceedings of 2nd U.S. Mine Ventilation Symposium, A.A.Balkema, Rotterdam, pp. 439–448.

Lowndes, I.S., Yang, Z.Y., Jobling, S., and Yates, C., 2005. *A parametric analysis of a tunnel climatic prediction and planning model*. Tunnelling and Underground Space Technology 21(2006), pp. 520–532.

Lowndes, I.S., Crossley, A.J., and Yang, Z.Y., 2004. *The ventilation and climate modelling of rapid development tunnel drivages*. Tunnelling and Underground Space Technology 19 (2004), pp. 139–150.

McPherson, M.J., eds. 1993. *Subsurface Ventilation and Environmental Engineering*. London: Chapman & Hall, pp. 138–139, 506.

Pruess, K., Oldenberg, C., and Moridis, G., 1999. *TOUGH2 User's Guide, Version 2.0.* Report LBNL-43134, Lawrence Berkeley National Laboratory, Earth Sciences Division, Berkeley, California.

Ross, A.J., Tuck, M.A., Stokes, M.R., Lowndes, I.S., 1997. *Computer simulation of climatic conditions in rapid development drivages*. Proceedings of 6th International Mine Ventilation Congress, SME, Littleton, CO, pp. 283–288.

Tuck, M.A., Stokes, M.R., Lowndes, I.S., 1997. *Mine climate control options in underground working zones*. Final Report on ECSC research Project No. 7220 - AC/006.

Voss, J., 1980. *Mine Climate in Mechanised Drivages and Its Determination by Advanced Calculation*. Methane, Climate, Ventilation in the Coalmines of the European Communities, Vol. 1. Colliery Guardian, pp. 285–305.

Welty, J. R., Wicks, C.E., and Wilson, R.E., eds. 1984. *Fundamentals of Momentum, Heat, and Mass Transfer*. 3rd Ed., John Wiley and Sons, pp. 126–127.

Historical Development of Technologies for Controlling Methane in Underground Coal Mines

Charles D. Taylor
NIOSH, Pittsburgh, Pennsylvania, United States

C. Özgen Karacan
NIOSH, Pittsburgh, Pennsylvania, United States

ABSTRACT: For 100 years, a major emphasis has been placed on controlling methane in underground coal mines in order to improve worker health and safety. This emphasis and related research, as well as legislative actions, have led to new mining equipment and methods which have changed methane control practices in coal mines. The U.S. Bureau of Mines (USBM), and more recently the National Institute for Occupational Safety and Health (NIOSH), have been an integral part of many of these developments which have improved mining health and safety, either by initiating research programs or by collaborating with the U.S. mining industry. Two primary goals of this research have been to reduce underground methane explosions by controlling methane emissions from the coal bed and the gob, and diluting methane gas that has been liberated in the face areas. The decreased number of fatalities due to methane explosions clearly shows the success of this program. This paper discusses several milestone events that show the progress of this methane control research.

INTRODUCTION

Most explosions in coal mines occur when an explosive methane-air mixture is present. Since its inception in 1910, the U.S. Bureau of Mines and the National Institute for Occupational Safety and Health have conducted research to help prevent methane explosions. Some of this methane research has been in direct response to underground explosions. Other research has been the result of applying new technologies to reduce the likelihood of underground explosions (for example, degasification with directional drilling). This paper examines several of the major technologies that the USBM and NIOSH developed for controlling methane in underground coal mines. The technologies are grouped and discussed according to the decade(s) of their development.

BACKGROUND

In 1907, following the explosions in four mines that killed over 600 miners, the United States coal mining industry gained an international reputation as "...being the most backward of the civilized nations" and was characterized as "...callous where financial interests have weighed against human lives."(Colliery Guardian 1908). Responding to the large loss of life in the mines, the United States public demanded that safety in U.S. coal mines be improved. In 1910, the existing U.S. Geological Survey department was used to form the basis for the U.S. Bureau of Mines.

CHRONOLOGICAL DEVELOPMENT OF METHANE CONTROL TECHNOLOGIES

1910 to 1950

In the early 1900s, it was a common belief that most mine explosions were the result of unsafe practices that occurred throughout the mines. Early USBM research, therefore, emphasized education of the mining industry. It was assumed that a more educated mining community would act in a way that would result in safer mines.

In 1910, the Bureau of Mines opened an experimental mine in Bruceton, PA. From 1914 to 1925, the Bureau focused almost exclusively on the causes of explosions and ways to prevent them. Explosion demonstrations conducted at Bruceton for the public did not explain why methane ignitions occurred, but graphically showed the effects of igniting dust or methane gas in a coal mine (Figure 1).

Early Ventilation Studies

Only limited ventilation research was conducted in the early 1900s because there was a common belief that explosive methane-air mixtures would accumulate in face areas regardless of the quantity of air provided by the ventilation system. The emphasis of the research was on removing sources of methane ignitions, such as explosives and open flame lights, rather than improving ventilation.

The quantities of air normally provided in mines were considered adequate (Harrington and Denny 1938). As long as coal was loaded by hand, relatively small amounts of fresh air were needed to

Figure 1. Early explosion test at the U.S. Bureau of Mines

Figure 2. Mining conditions at a work place

dilute the quantities of methane released at the face. Health requirements of the workers and animals, not methane control, usually determined the amount of air provided to the working places (Miller 1940) (Figure 2). In the 1930s and 1940s, airflow quantities increased as more mines began to use mechanical fans for ventilation.

Ignition Sources

Permissible Explosives. The USBM played an important role in the development and testing of permissible explosives which, when used properly, would not ignite methane. As mines began to use permissible explosives, the number of ignitions decreased; however, at some mines with less gassy seams, people continued to think permissible explosives were not necessary. Opposition to use of the permissible explosives also came from miners who said they were too expensive and broke coal into pieces that were too small. That is, the miners who were paid for lump coal saw their incomes decrease when they used permissible explosives.

Open Flame Lights. Use of open flames for underground illumination was known to be hazardous, but there were few alternatives until the Bureau helped to develop the electric cap lamp (also known as a "closed light"). The Bureau issued the first permit for an electric light in 1914 (Fieldner 1950), but the use of open flame lights continued in many less gassy mines where they were considered safe. Initially, mines were not required to use the electric lights. Mines were often identified as either "open light" or "closed light" depending on whether they were considered non-gassy or gassy, respectively.

Some miners preferred open lights because they provided more light than early electric lights, and the electric lights had heavy batteries that often leaked

Figure 3. Thomas Edison with early electric cap lamp

electrolyte (Figure 3). Some workers opposed the use of electric cap lamps because they believed ventilation would be neglected in mines that used electric lights. There were reports that some companies tried to force workers to mine in gassy areas of the mine while wearing the electric lights.

Early Coal Bed and Gob Degasification Studies

The potential benefits of using boreholes to drain gas from coal in advance of mining were recognized in the early 1900s, especially when mining deeper and

Figure 4.   Early continuous mining machine

Figure 5.   Miner with safety lamp

more gassy seams (Diamond 1994). However, the quantities of methane removed by drilling short horizontal holes in the mine ribs were small (Lawall and Morris 1934). From the early 1930s until 1950, some mines tried methane drainage technologies developed in Europe, with some success. Two methods used were drilling holes in broken sandstone above gassy coal seams (Coal Age 1938), and using cross-measure boreholes to drain the gas and move it to surface (Diamond 1994).

1950 to 1969

In the 1950s and 1960s, the introduction of new mining equipment and mining methods had a big effect on how ventilation was used to control methane concentrations.

Continuous Mining Machines
One USBM researcher noted that the continuous mining machine (Figure 4) revolutionized coal mining (Stahl 1958). Methane liberation rates during continuous mining were much higher than during conventional mining, due to the faster coal extraction rates. Methane concentrations during continuous mining sometimes exceeded 5% (the lower explosive limit for methane gas) at distances as far as 20 feet from the face (Stahl 1958). Ventilation practices available at that time were inadequate for controlling methane during continuous mining.

Face Airflow Control
Although the increased use of mechanical fans resulted in more airflow underground, much of the air was lost before it could reach the working faces due to leakage through doors and curtains. Tests examined ways to improve delivery of air to the face.

- *Curtains—* Most curtains were made of jute fabric. Air leakage increased with increasing porosity. Less porous materials, such as plastics, along with better installation techniques that allowed less leakage between the curtain and roof were recommended (Dalzell 1966).
- *Tubing—* When permissible auxiliary fans became available for underground use, studies examined factors affecting the delivery of air to the face using fans with attached tubing. Tubing materials and construction techniques were tested to determine how airflow shock losses could be reduced and airflow quantities increased (Peluso 1968).
- *Blowing versus exhausting ventilation—* Studies using comparable blowing and exhausting face ventilation systems with curtains measured methane levels near the mining face for a range of operating conditions. The blowing systems were on average more than 5 times more effective for diluting methane in the face area (Luxner 1969).

Methane Monitors
Until the 1950s, the flame safety lamp (Figure 5) was the primary means used for detecting methane in mines (Ilsley 1933). The effectiveness of the ventilation system was determined more by observing operating conditions than by measuring gas levels. In 1958, the Bureau of Mines initiated programs to develop a methane detector system that would provide continuous monitoring of methane at the mining face (James 1959).

Eventually all mining machines had to be equipped with these machine-mounted methane monitors [(30 CFR § 75.342(a)(1)]. Data from these monitors became the basis for evaluating ventilation effectiveness for methane control and compliance with methane standards.

Improved Methane Drainage

- *Hydraulic stimulation*—The efficiency of vertical degasification wells was improved by using hydraulic stimulation (Spindler and Poundstone 1960). In one coalbed, there was more than a 16-fold increase in gas production. In another study, the methane content in the main return air currents after stimulation was reduced more than 86% (Diamond 1994; Merritts et al. 1963).
- *Methane drainage on longwall panels*—Use of vertical boreholes reduced methane concentrations in the returns while decreasing the level of ventilation required for methane control (Elder 1969).

1970 to 1980

1969 Coal Mine Health and Safety Act

The Federal Coal Mine Health and Safety Act of 1969 set standards for methane and dust control. Several research projects in the 1970s examined ways to improve ventilation and reduce airborne dust levels to the new required levels. Most of the studies concluded that exhaust ventilation was needed for adequate dust control (Kingery 1969; Mundell 1977). Earlier work had shown that blowing ventilation was better for methane control (Luxner 1969) (Figure 6), and some mines wanted to continue using blowing ventilation systems. These mines began cooperative studies with the Bureau to develop a machine-mounted dust scrubber (Tomb 1973). The flooded bed machine-mounted scrubber was shown to lower dust levels when using blowing ventilation. Initially, there were concerns that scrubber recirculation would cause methane levels to increase, but tests showed that concentrations did not increase as long as the scrubber exhaust did not reduce intake air flow to the face.

Developments in Coal Bed Methane Drainage

In the 1970s, studies examined techniques for removing methane from longwall gob areas.

- Use of an exhauster with a slotted vertical pipe allowed faster mining rates and lower ventilation airflow in the mine (Moore and Zabetakis 1972).
- Isolating panels during degasification improved gas removal by 70% (Findlay et al. 1973).
- The best levels for bit thrust and rotational speed for rotary drilling of holes were investigated

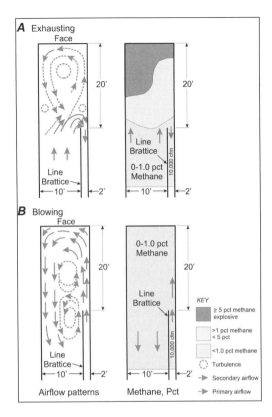

Figure 6. Data comparing blowing and exhausting face ventilation

to find the optimum combination to stay in the coalbed and drill ahead of coal production (Cervik et al. 1975).

- Horizontal boreholes drilled from the bottom of the shaft in virgin coal reduced methane emissions at the working face by 50% (Diamond, 1994).
- Hydraulic stimulation had only minimal effects on improving degasification of a longwall panel (Maksimovic et al. 1977), but boreholes stimulated with foam and proppant sands significantly reduced methane emissions (Steidle 1978). The stimulation treatment did not cause damage to roof or floor.
- Water infusion techniques developed by the Bureau of Mines were used to move gas away from face areas and into the return airways. Use of this technique reduced gas emissions in the more permeable coal seams by almost 90% (Cervik 1977).

1980 to 1990

By 1980, the Bureau of Mines had demonstrated effective face ventilation techniques for controlling both methane and respirable dust on continuous mining operations. Maximum curtain or tubing setback distances were usually 10 ft for exhausting and 20 ft for blowing ventilation systems. Most mining cuts were 20 ft deep and the machine operator sat at the rear of the machine. In the 1980s more machines were equipped for remote control operation.

Remote Control

As more mining machines were equipped with remote controls to remove the operator from the immediate face area, cutting depths in the 1980s increased to 35 ft and greater. Studies examined whether available face ventilation practices were adequate for controlling methane levels during deep cut mining (cut depths greater than 20 ft).

- The benefits of using directional water sprays for improving face airflow (spray fan system) had been demonstrated earlier by Kissell (1979). With exhausting ventilation and a modified spray fan system, cutting depths could be increased to 40 ft with no significant increases in methane levels (Ruggieri 1987).
- When using a machine-mounted scrubber with blowing ventilation, cutting depths to 50 ft could be attained without increasing methane levels (Volkwein 1986).

Most mines employing deep cut mining used scrubbers with blowing ventilation or a spray fan system with exhausting ventilation.

Longwall Methane Control

Longwall mining does not require curtain or tubing to ventilate the face. However, airflow must be maintained along the entire face for good methane control (Cecala 1986). Studies showed that:

- The highest methane levels were measured during cutting at the headgate. Placement of a curtain at the headgate diverted needed air flow away from the gob and down the panline (Cecala 1986).
- A modified water spray system on the shearer (shearer clearer) improved air flow over the mining machine by reducing eddy zones where methane could accumulate (Cecala 1986).

Frictional Ignition Suppression

Past research had eliminated most ignition sources from coal mines, but the number of ignitions at the

Figure 7. Methane frictional ignition test

mining face increased following the introduction of continuous mining machines. These frictional ignitions occurred when the metal machine bit in the rotating cutting drum became hot enough, usually during cutting in sandstone rock, to deposit a streak of hot metal on the rock surface. When the hot metal contacted a flammable methane-air mixture, an ignition could occur (Courtney 1990).

Laboratory tests designed to simulate cutting in hard rock (Figure 7) showed that frictional ignitions could be reduced by changing bits more frequently and/or using bits with larger carbide inserts to reduce contact between the steel bit shank and the rock. Ignitions could also be reduced by cooling the rock surface with water directed from a nozzle mounted directly behind the bit (Courtney 1990).

Coal Seam Methane Drainage Research

- The Bureau of Mines developed a "direct method" for determining coal bed gas content (Diamond and Levine 1981) to make preliminary estimates of ventilation requirements and to determine if methane drainage would be beneficial (Diamond et al. 1986).
- A comprehensive report (Diamond 1982) detailed geological factors important during the drilling of degasification boreholes in order to achieve optimum methane control was written. Techniques for drilling boreholes from the surface to intercept coal beds were demonstrated (Oyler and Diamond 1982).
- Progress was made on developing numerical models for methane control in coal mines (Schwerer et al. 1984). The two-dimensional mathematical model and accompanying numerical model were used to predict methane influx into the coal mines under a variety of mining conditions.

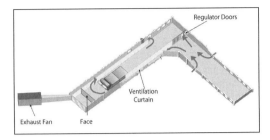

Figure 8.   NIOSH ventilation gallery

1990 to 2000

In 1996 certain research programs of the Bureau of Mines were transferred to NIOSH. The emphasis of most research during this decade was on improving the effectiveness of available ventilation techniques and providing data needed by the mining industry for better implementation of existing technologies. A ventilation test gallery (Figure 8) was specially designed to evaluate factors affecting face ventilation and methane monitoring techniques.

Airflow Patterns and Methane Distributions
Earlier Bureau tests showed that mapping airflow patterns and methane concentration near the face helped during the design of face ventilation systems (Luxner 1969). A surface test facility was used to further study airflow patterns and methane distributions in the area between the ventilation curtain or tubing and the face. Study results showed that:

- The narrower the entry, the more likely flow to the face will be disrupted and methane levels will increase (Taylor 2005).
- Operation of machine-mounted water sprays and scrubber can increase face airflow by a factor of 5 (Thimons 1999).
- A spray fan system that uses water sprays directed toward the return air side of the face reduced methane levels by increasing airflow across the face. If the water sprays are directed straight toward the face methane levels are also reduced but concentrations can increase in some areas near the face, especially if intake airflows are low (Taylor 1997; Chilton 2006).

Instrument location on continuous mining machines. In the ventilation gallery, methane measurements made at the face and at multiple locations on the machine were compared. Researchers determined that methane concentrations should be measured on the return side of the mining machine, and that if alternative methane sampling locations further

from the face are selected, action levels should be reduced to assure protection to worker is not diminished (Taylor 2001).

Instrument location on a roof bolting machine. Based on results from tests conducted in the ventilation gallery, Taylor (1999) proposed an alternative procedure for methane monitoring during roof bolting of deep cuts. The procedure calls for methane to be monitored continuously using a bolter mounted methane monitor, while a hand held detector with extendable probe is used to periodically "sweep" the area 16 feet inby the last permanent row of roof supports.

Methane Control Research
Studies concluded that:

- About 90% of the gas from longwall gobs comes from the overlying coalbeds (Diamond et al. 1992).
- On longwall panels, 77% more gas came from holes drilled at the end of the panels and at panel margins where strata were in tension compared to holes drilled in the center of the panel. Drilling gob vent holes in panel margins has become standard practice in most U.S. mines (Diamond et al. 1994).
- Methane emission rates increase as the length of the longwall face increases, but not necessarily in the same proportion (Diamond and Garcia 1999).

2000 to 2010

Emphasis was placed on improving the evaluation and use of methane monitors. Ultrasonic anemometers were used to plot airflow in face areas in order to evaluate factors affecting methane control at the face. Computer modeling was used to study the flow of methane gas through rock strata and after its liberation at the mine face

Methanometer Response Time
Methane concentration rise and fall quickly during mining. In order for instruments to read accurately they must response quickly to these changes. Researchers developed test procedures and equipment for measuring methanometer response times. Using the new procedures it determined that the design of the dust caps used to protect the sensor heads was the major factor affecting response times. Response data was presented for both machine-mounted monitors (Taylor 2004) and hand held detectors (Chilton 2005). Tests compared methane response times of instruments using infrared and heat of combustion sensors. Again, dust cap design was

the biggest factor affecting differences in response times. In the ventilation gallery where face airflow conditions were simulated, it was shown that when exposed to the same gas concentrations instruments with faster response times showed larger peaks and dips in concentration (Taylor 2008).

Methane Measurements Distant from the Working Face

Although regular methane monitoring is required near working faces, much less methane sampling occurs in areas distant from the face. Small person wearable methane monitors were evaluated. Equipped with aural, visual and vibratory alarms, the instruments provided alarms to warn workers when methane levels exceed preset limits (Chilton 2005).

Measuring Airflow Speed and Direction

Researchers developed techniques for using ultrasonic anemometers to measure airflow direction and speed in the ventilation gallery (Taylor 2005). The anemometers were used to measure air flow direction and speed inby the ventilation curtain. Flow profiles drawn for different size empty entries showed that less air reaches the face when the entry is narrower. The same type of flow pattern can occur at the face when machine-mounted water sprays or scrubbers are not used. Data obtained using the anemometers showed an increases in face airflow when the scrubber was used (Taylor 2006).

Development of Computer Models for Face Airflow and Methane Distribution

Airflow and methane data obtained in the ventilation gallery was compared to data generated by a computer simulation model. The flow and the methane concentration data from the model were in good agreement with the empirical results.

Methane Control and Degasification Research

Besides the field work and observations, a special emphasis was placed on integrating analytical and numerical reservoir simulation techniques. Research in this area, based on the mining industry's need to respond to changing mining conditions and economic drivers, produced several outputs and recommendations. Study results showed that:

- The methane emission increase due to a face width increase can be controlled by increasing ventilation quantities and reducing shearer transit times (Schatzel et al. 2006; Krog et al. 2006), increasing the number of gob gas ventholes, and drilling boreholes on the headgate side close to the start up of the panel. These holes should be operated continuously (Karacan et al. 2005).

- Horizontal in-seam degasification boreholes can be effectively used for methane drainage of the coal bed prior to the start of longwall mining. The patterns of the boreholes can be optimized by using numerical reservoir simulations. In general, boreholes drilled perpendicular to face cleats produce more methane per borehole length and more effectively degasify the coal bed. Extension of methane drainage duration prior to mining start is also an effective way of decreasing methane emissions during mining. When this is not possible, the boreholes should continue producing until the mining face reaches their location (Karacan et al. 2007a).

- Entry development processes can be modeled by reservoir simulations. Methane emissions into the entries during their development can be controlled by using shielding boreholes drilled parallel to the entries. These boreholes can serve both degasification and shielding purposes (Karacan et al. 2007a; Karacan 2007b).

- Methane emissions during continuous miner operation are a function of mining parameters (Karacan 2007b) and coal bed reservoir parameters (Karacan 2008a). Optimum methane control strategies can be developed based on these two considerations.

- Geological and structural characteristics of the coal bed are important to decide where to drill the boreholes (Karacan et al. 2008) and how they can affect the borehole productivity (Karacan and Goodman 2008).

- Expert systems can be used in determining and optimizing longwall emissions (Karacan 2008b), degasification system selection (Karacan 2009a) and forecasting the performance of gob gas ventholes (Karacan 2009b).

- Longwall mining changes the hydraulic conductivity of the overlying strata. Hydraulic conductivity of the fractured zone affects the control of methane using gob gas ventholes. Hydraulic conductivities can be determined using slug testing methods (Karacan and Goodman 2009). It was determined that the hydraulic conductivities of gob gas ventholes vary based on the overburden thickness, the rate of the longwall face advance, and their locations over the panel (Karacan and Goodman 2009).

SUMMARY AND CONCLUSIONS

Research programs to prevent methane ignitions began with the establishment of the U.S. Bureau of Mines and remain an important part of the current NIOSH mining research program. The substantial reduction in the number of injuries and deaths due to mine explosions is probably the best indicator of the program's success. Although the potential for explosions resulting from methane in mines still exists today, with knowledge gained from Bureau and NIOSH research it is now possible to work underground much more safely than previously. Research into better ways to utilize ventilation for the control of methane gas by underground ventilation and extraction prior to mining continues.

References provided with this paper give more information about past Bureau of Mines/NIOSH methane control research. Additional sources of information include:

- "Handbook for Methane Control in Mining" (Kissell 2006).
- "Guidelines for the Prediction and Control of Methane Emissions on Longwalls" (Schatzel et al. 2008).

REFERENCES

Cecala, A.B, et al. 1986, *Cut longwall ignitions by lowering methane concentrations at shearers.* Coal Age.

Cervik, J., Fields, H.H and Aul, G.N. 1975. *Rotary drilling holes in coal beds for degasification.* Bureau of Mines RI 8097.

Cervik, J., et al. 1977. *Water infusion of coal beds for methane and dust control.* U.S. Bureau of Mines RI 8241.

Chilton, J. E., et al. 2005. *Evaluation of person wearable methane monitors.* Intl. Conf. of Safety in Mines Res Inst., Brisbane, Australia.

Chilton, J.E., et al. 2006. *Effect of water sprays on airflow movement and methane dilution at the working face.* 11th US/North American Mine Ventilation Symposium, University Park, PA.

Anon. 1938. *Fans in parallel and core-drilling gas bleeding feature Pocahontas fuel improvements.* Coal Age, v. 43.

Colliery Guardian. January 14,1908.

Courtney, W.G. 1990. *Frictional ignition with coal mining bits,* Bureau of Mines IC 9251.

Dalzell, R.W. 1966. *Face ventilation in underground bituminous coal mines, performance characteristics of common jute line brattice.* Bureau of Mines RI 6725.

Diamond, W.P. and Levine, J.R. 1981. *Direct method determination of the gas content of coal: procedures and results.* Bureau of Mines RI 8515.

Diamond, W.P. 1982. *Site-specific and regional geologic considerations for coal bed gas drainage.* Bureau of Mines IC 8898.

Diamond, W.P., et al. 1988. *Results of direct-method determination of the gas content of U.S. coal beds.* Bureau of Mines IC 9067.

Diamond, W.P., et al. 1992. *Determining the source of longwall gob gas: Lower Kittanning coal bed, Cambria County, PA.* Bureau of Mines RI 9430.

Diamond, P.W. 1994. *Methane control for underground coal mines.* Bureau of Mines IC 9395.

Diamond, W.P., Jeran, P.J. and Trevits, M.A. 1994. *Evaluation of Alternative placement of longwall gob gas ventholes for optimum performance.* Bureau of Mines RI 9500.

Diamond, W.P. and Garcia, F. 1999. *Analysis and prediction of longwall methane emissions: a case study in the Pocahontas No. 3 coal bed, VA.* NIOSH RI 9649.

Elder, C.H. 1969. *Use of vertical boreholes for assisting ventilation of longwall gob areas.* Bureau of Mines TPR 13.

Fieldner, A.C. 1950. *Achievements in mine safety research and problems yet to be solved.* Bureau of Mines IC 7573.

Findlay, C., et al. 1973. *Methane control by isolation of a major coal panel-Pittsburgh coal bed.* Bureau of Mines RI 7790.

Harrington, D., and Denny, E.H. 1938. *Some observations on coal-mine fans and coal-mine ventilation.* Bureau of Mines IC 7032.

Ilsley, L.C., and Hooker, A.B. 1933. *Evolution of methane detecting devices for coal mines.* Bureau of Mines IC 6733.

James R. S. 1959. *A continuous methane monitoring system at the working face.* Mining Congress Journal.

Lawall, C.E and Morris, L.M. 1934. *Occurrence and flow of gas in the Pocahontas No. 4 coal bed in southern West Virginia.* Trans. AIME.

Karacan, C.Ö., Diamond, W.P, Esterhuizen, G.S and Schatzel, S.J. 2005. *Numerical analysis of the impact of longwall panel width on methane emissions and performance of gob gas ventholes.* International Coalbed Methane Symposium, Tuscaloosa, Alabama.

Karacan, C.Ö., Diamond, W.P. and Schatzel, S.J. 2007a. Numerical analysis of the influence of in-seam horizontal methane drainage boreholes on longwall face emission rates. *International Journal of Coal Geology* 72(1):15.

Karacan, C.Ö. 2007b. Development and application of reservoir models and artificial neural networks for optimizing ventilation air requirements in development mining of coal seams. *International Journal of Coal Geology* 72:221.

Karacan, C.Ö. 2008a. Evaluation of the relative importance of coal bed reservoir parameters for prediction of methane inflow rates during mining of longwall development entries. *Computers and Geosciences* 34(9):1093.

Karacan, C.Ö. 2008b. Modeling and prediction of ventilation methane emissions of U.S. longwall mines using supervised artificial neural networks. *International Journal of Coal Geology* 73:371.

Karacan, C.Ö and Goodman GVR. 2008. *Coal bed discontinuity effects on the production of degasification boreholes and on emissions during longwall mining.* SPE Eastern Regional/AAPG Eastern Section Joint Meeting, Pittsburgh, Pennsylvania.

Karacan, C.Ö, Ulery, J.P. and Goodman, G. 2008. A numerical evaluation on the effects of impermeable faults on degasification efficiency and methane emissions during underground coal mining. *International Journal of Coal Geology* 75(4):195.

Karacan, C.Ö. and Goodman, G. 2009. Hydraulic conductivity and influencing factors in longwall overburden determined by using slug tests in gob gas ventholes. *International Journal of Rock Mechanics and Mining Sciences* 46(7):1162.

Karacan, C.Ö. 2009a. Degasification system selection for U.S. longwall mines using an expert classification system. *Computers and Geosciences* 35:515.

Karacan, C.Ö. 2009b. Forecasting gob gas venthole production performances using intelligent computing methods for optimum methane control in longwall coal mines. *International Journal of Coal Geology* 79:131.

Kingery, D.S. et al. 1969. *Studies on the control of respirable coal mine dust by ventilation.* Bureau of Mines TPR 19.

Kissell, F.N. 1979. *Improved face ventilation by spray jet systems.* Second Annual Mining Institute, University of Alabama.

Kissell, F.N. 2006. *Handbook of methane control in mining.* NIOSH Information circular 9486.

Krog, R., et al. 2006. *Predicting methane emissions from longer longwall faces by analysis of emission contributors.* 11th U.S./North American Mine Ventilation Symp., University Park, PA.

Maksimovic, S.D., et al. 1977. *Hydraulic stimulation of a surface borehole for gob degasification.* Bureau of Mines RI 8228.

Merritts, W.M., et al. 1963. *Removing methane (degasification) from the Pocahontas No. 4 coal bed in southern West Virginia.* Bureau of Mines RI 6326.

Miller, A.U. 1940. Some *information for miners about coal-mine ventilation.* Bureau of Mines IC 7137.

Moore, T.D. and Zabetakis, M.G. 1972. *Effect of surface borehole on longwall gob degasification (Pocahontas No. 3 coal bed).* Bureau of Mines RI 7657.

Mundell, R.L. 1977. *Blowing versus exhausting face ventilation for respirable dust control on continuous mining sections.* MESA IR 1059.

Oyler, D.C. and Diamond, W.P. 1982. *Drilling a horizontal coal bed methane drainage system from directional surface borehole.* Bureau of Mines RI 8640.

Peluso, R.G. 1968. *Face ventilation in underground bituminous coal mines, airflow characteristics of flexible spiral-reinforced ventilation tubing.* Bureau of Mines RI 7085.

Ruggieri, S.K. et al. 1987. *Deep cut mine face ventilation. Bureau of Mines*, Final Contract Report HO348039

Schatzel, S.J., et al. 2006. *Prediction of longwall methane emissions and associated consequences of increasing longwall face length: a case study in the Pittsburgh coal bed.* 11th U.S./North American Mine Ventilation Symp., University Park, PA.

Schatzel, S.J., Karacan, C.Ö., Krog, R.B., Esterhuizen, G.S., Goodman, G.V.R. 2008. *Guidelines for the prediction and control of methane emissions on longwalls.* NIOSH Information Circular 9502.

Schwerer, F.C., et al. 1984. *Methane modeling—predicting the inflow of methane gas into coal mines.* Bureau of Mines research contract report No. J0333952.

Spindler, M.L. and Poundstone, W.N. 1960. *Experimental work in the degasification of the Pittsburgh coal seam by horizontal and vertical drilling.* AIME, Preprint 60F106.

Stahl, R.W. 1958. *Auxiliary ventilation of continuous miner places.* Bureau of Mines RI 5414.

Steidl, P.F. 1978. *Foam stimulation to enhance production from degasification wells in the Pittsburgh coal bed.* Bureau of Mines RI 8286.

Taylor, C.D., Rider, J.P., and Thimons, E.D. 1997. *Impact of unbalanced intake and scrubber flows on methane concentrations.* 6th International Mine Ventilation Congress, Pittsburgh, PA.

Taylor, C.D., Thimons, E.D., and Zimmer, J.A. 2001. *Factors affecting the location of methanometers on mining equipment.* 7th International Mine Ventilation Congress, Krakow, Poland.

Taylor, C.D., Chilton, J. E., and Zimmer, J.A., 2004. *Use of a test box to measure response times for machine-mounted monitors.* 10th US/NA Vent Symp, Anchorage, Alaska.

Taylor, C.D., Timko, R.J., Thimons, E.D., and Mal, T. 2005. *Using ultrasonic anemometers to evaluate factors affecting face ventilation effectiveness.* SME Annual Meeting, Salt Lake City, Utah. Preprint 05-80.

Taylor, C.D., Chilton, J.E., Hall, E. and Timko, R.J., 2006. *Effect of scrubber operation on airflow and methane patterns at the mining face,* 11th US/NA Mine Vent Symp, University Park PA.

Taylor, C.D., Chilton, J.E., and Martikainen, A.L. 2008. *Use of infrared sensors for monitoring methane in underground mines.* 12th U.S./North American Mine Ventilation Symp., Reno, NV.

Thimons, E.D., et al.1999. *Evaluating the ventilation of a 40-ft two-pass extended cut.* NIOSH Report of Investigations, No. 9648.

Tomb, T.F., et al.1973. *Evaluation of a machine-mounted dust collector,* Bureau of Mines, RI 7788.

Volkwein, J.C., and Thimons, E.D. 1986. *Extended advance of continuous miner successfully ventilated with a scrubber in a blowing section.* SME Annual Meeting Preprint 86-308.

Section 5. Training

Global Mining Engineering Education: Past, Present, and Future

Michael Karmis
Virginia Tech, Blacksburg, Virginia, United States

Bruce Hebblewhite
University of New South Wales, Sydney, Australia

Hans de Ruiter
Technical University of Delft, Delft, The Netherlands

Malcolm Scoble
University of British Columbia, Vancouver, British Columbia, Canada

Mario Cedrón
Universidad Catolica, Lima, Peru

Huw Phillips
University of the Witwatersrand, Johannesburg, South Africa

ABSTRACT: Mining engineering education across the globe has been fundamentally restructured in the past 10 years to follow trends in engineering education, foster national and international collaborations, assume interdisciplinary approaches and satisfy lifelong learning objectives. Reinventing the mining academic infrastructure has not always followed a smooth path and has challenged many traditional concepts and ideas on the number, size and location of mining schools, the balance between depth and breadth of the curriculum, the relationship between government/state support and industry commitment, the emphasis on undergraduate education versus post-graduate studies and the bond between research and education.

This paper presents a historical review of global mining engineering education in major parts of the world, and a discussion on the dynamics of change that will continue to shape educational and research objectives and goals into the future in those countries and regions. The paper includes global as well as regional coverage and highlights a number of developments and initiatives as "best practices" in minerals education and research around the world.

HISTORY AND STATUS OF MINING ENGINEERING EDUCATION

The global minerals industry of the 21st century is a very different industry than that which existed even a decade ago. It is different in structure, in its contribution to both global well-being and that of individual nations, and in its competitive posture in the world. Indeed, the minerals industry is different from the standpoint of technology, environmental awareness and employee and community consciousness. These significant changes in industry necessitate the rethinking and restructuring of mining and minerals education. This new mining industry seeks not just to employ the specialist mining engineer, but desires to employ a professional who can function as engineer and manager, who can understand labor problems and global finances, who has the ability to communicate with fellow professionals as well as with the public and media, and who is sensitive to environmental and safety issues. In short, industry is looking to the universities to educate mining engineers who "can get it all together."

Minerals education, globally, has been in a crisis for almost three decades. Many historic mining schools have terminated their minerals programs and those that have so far survived, in both developed and developing countries, are under severe pressure and scrutiny. Numerous factors have contributed to this decline. At the undergraduate level, recruitment and retention of students has been a matter of serious concern. At the post-graduate level, the absence of research funding by industry and government is a serious threat to the sustainability of post-graduate programs in minerals education. Recruitment, retention and development of academic staff have also been impacted, affecting the future of the "professorate." Demographics suggest that a significant portion of the staff in minerals programs is at the senior level, fast approaching retirements, posing serious succession and continuity issues. Finally, the limited recognition of the minerals engineering field by the broad research and academic communities is a serious issue of morale amongst the ranks and also affects the way that university administration evaluates, respects and supports such programs.

This environment has forced minerals programs to become more innovative, seek new partnerships, develop joint educational and research efforts and expand to new areas and disciplines. The following is a brief review of the history and status of mining engineering education in various countries/regions.

Australia

Australia is endowed with diverse and abundant mining resources that have been mined since at least 1788. The thriving minerals industry in Australia not only fulfils much of the nation's energy needs, but exports a significant amount of energy as well. The industry is dominated by several multinational companies, some of which are based in Australia.

One of the biggest challenges facing the industry in Australia, as in other parts of the world, is the continued availability of skilled staff resources, and in particular, professional mining engineers. Education of professional mining engineers has been undertaken in Australia for over one hundred and fifty years. Many programs have opened and closed; currently four year professionally accredited mining engineering undergraduate degree programs operate at just five universities: Curtin, Adelaide, Wollongong, New South Wales and Queensland.

Two key features continue to characterize Australian mining education. First, demand for graduate mining engineers has always been greater than the supply, as industry has grown and the industry population aged. Cyclic trends are a natural part of the minerals industry, with education providers inevitably operating in a counter-cyclical mode due to the lag in student recruitment and four years of education. Current national demand is estimated at present to be approximately 220–250 B.S. graduates per year, whereas national supply in 2008 was only 163, and not all of those graduates entered mainstream industry positions.

Second, there is a chronic shortage of current or potential future mining academics, meaning that there is a tendency for the limited staff pool to move from one institution to another, without the academic pool growing, so the shortages continue. Opening new mining programs only exacerbates the problem.

At the end of the twentieth century, Australia realized that "more of the same" in conventional mining education was an unacceptable option. A number of major issues were identified by the industry in Australia at the time, resulting in the publication of a landmark document, "Back from the Brink," in 1998 (Minerals Council of Australia 1998). This work reviewed the three major professional disciplines serving the industry—mining engineering, geosciences and metallurgy—and led to the formation of the Minerals Tertiary Education Council (MTEC), in 1999. MTEC was established and funded by industry to develop initiatives in these three disciplines, initially for a three year period but subsequently beyond, commencing in 2000. An overview of the initial MTEC initiatives in mining engineering has been documented by Hebblewhite (2006). The subsequent major new initiative, Mining Education Australia (MEA), is discussed later in this paper.

Canada

In response to declining enrolments and sustainability threats to mining engineering departments, Canadian mining universities and technology schools formed an alliance, the Canadian Mining Education Council (CMEC) in 2000, aiming to network the schools with industry, government and other organizations. The Mining Industry Training and Adjustment Council—Canada (MITAC-Canada) was established in 1997 to address the various human resource issues that had been identified and to develop an ongoing human resource strategy for the Canadian Mining Industry. The Canadian federal government has established a Sector Study project in 2003 to assess the current situation with respect to human resource demands and gaps in the minerals industry and to conduct research into emerging trends and future human resources and skill needs.

In Canada it is estimated that demand for graduates in minerals and metals is about twice the annual university output in these disciplines. The industry is committed to a robust hiring program, motivated in part by the findings that the industry could lose up to 40 percent of the existing workforce in the next ten years (MITAC Annual Report 2005). Recent data from educational surveys in that report indicates that enrolment numbers are expected to increase for each of the next ten years.

A recent Canadian initiative has created the Canadian Mining Innovation Council (CMIC), which brings together collaboration between industry, government, universities and the Canadian Institute of Mining and Metallurgy. CMIC is collaborating with the Mining Industry Human Resources Council, for example, in strategies for development of the Canadian mining industry's Highly Qualified People (HQP). HQP are those with the technical, business and social skills on which the operational integrity and leadership of the mining industry is dependent. Their skills, competence and motivation result from a combination of educational and training experiences, ranging from secondary through tertiary education, employment training and experience and lifelong learning. Five action areas have been identified by the CMIC HQP working group, as follows: early actions towards countering the impacts of commodity cycles on human resources; improved industry-academia engagement; fostering co-operative mining education; strengthening the HQP research base; mapping Canada's mining HQP stocks and flows today and into the future; and identifying best practices in HQP attraction, retention and development.

European Union

Despite the decline in direct mineral activities in Europe, the mining and processing supply industries play a significant role in the European Union and its member states. In addition, the E.U. is a substantial consumer and processor of imported minerals. For these reasons, minerals engineering is important in the E.U. and it is vital to maintain the very high level of knowledge and skill in minerals engineering which exists in the E.U., through strong and well balanced education programs.

At the end of the eighties it became clear that innovation was necessary to continue offering high quality mining courses. In order to optimize and enhance the quality of education at Delft University of Technology in the Netherlands, it was decided in 1990 to send mining students during their final year to Imperial College's Royal School of Mines in London. This was the first initiative of this kind in Europe and led in 1996 to the establishment the European Mining Course, a joint curriculum during the final year, organized by RWTH Aachen, TU Delft, Helsinki University of Technology and Imperial College. These universities were primarily chosen because of existing cooperative links and similarity in structure, culture and desire to internationalize. Each partner concentrated on subjects in which it already excelled, thus a joint, high quality, curriculum was realized.

The success of the EMC led in 1998 to the European Mineral Engineering Course (EMEC), which emphasizes mineral processing, metallurgy and recycling. In 2003, a third program, the European Geotechnical and Environmental Course (EGEC), started, emphasizing geotechnical and environmental subjects. Since the organizational considerations are the same in all cases, they now are offered as three options of one program, the European Mining, Minerals and Environmental Program (EMMEP).

There are also exchange opportunities between students of the organizing universities and a number of associated universities. Currently, the following six European Universities are now involved in EMMEP: RWTH Aachen (Germany), Delft University of Technology (the Netherlands), University of Exeter (UK), TKK Helsinki (Finland), Miskolc University of Technology (Hungary) and Wroclaw University of Technology (Poland).

In 1999 the positive results of the EMC and EMEC raised the interest of industry. In order to formalize relations between the universities and industry, the Federation of European Mineral Programs (FEMP) was founded in Delft in December, 1999. The main goals were to strengthen the ties between industry and academia, while creating possibilities for other (European) universities to participate. FEMP, whose board of directors includes representatives from academia and industry, oversees and carries out EMMEP.

Latin America

Latin America comprises the territory between the Rio Grande in the border between Mexico and the USA down to Tierra del Fuego in southern Argentina and Chile, plus an array of small republics in the Caribbean. The region is divided into 22 countries, has over 400 million inhabitants; Spanish and Portuguese are the most common languages. This subcontinent ranks first in the world's production of several minerals, accounting for 45% of the production of copper, 40% of silver and 35% of zinc, comes second in tin and lead and third in other nonmetal minerals like lithium and emeralds.

Mining has been a traditional occupation ever since the arrival of the Spanish and Portuguese conquerors. Gold and silver and to a certain extent copper and tin were the main metals exploited during pre-Columbian times. Upon the arrival of the Spanish conquerors, silver and gold became the main targets of the colonial mining industry, with Mexico and Peru being the centers of this activity. In 1790, the Spanish king sent two missions of mining experts to set up the first mining schools in the new world, one in Mexico and the other in Peru. The Palacio de Mineria at Mexico City was the site of the first mining school, but the one established in Peru lasted only a very short time. In republican times, the Mexican school became the Department of Mines of the Universidad Autonoma de Mexico. Since then, 45 mining schools have been established in Latin America.

Unlike what happens in other parts of the world, Latin America has a surplus of mining students as compared to its needs, having eight thousand undergraduate mining students and one thousand annual graduates in mining engineering in a region that needs no more than 250 new mining engineers per year. Only two (Catholic Univerisities of Chile and Peru) schools are private and almost half of the 45 schools belong to the Aismin (Association of Iberoamerican Mining schools), which meets annually and has similar objectives as those of the Society of Mining Professors. Most of these schools concentrate on teaching as resources for research are very scarce, although some of the schools have worldwide reputation and agreements established with mining schools in other parts of the world and with industry.

With globalization, increasing numbers of multinational mining companies arriving to the region and the need of qualified mining engineers familiar with state of the art technology, it is expected that Latin American mining schools will improve their academic standards, establish more links with their first world counterparts, incorporate distance learning, establish alliances with industry and become a source of first class mining engineers not only for the region but also for the world.

South Africa

The discovery of vast mineral resources has in large part spurred the economic development of South Africa from a largely agrarian society to a modern, industrial nation over the last 150 years. The opening up of the diamond fields at Kimberly in 1870, the discovery of gold on the Witwatersrand in 1886 and the later development of the vast platinum group metals deposit in the Bushfield complex contributed to the creation of a significant mining industry and an economy still heavily reliant on the export of minerals. In response to the technical and managerial needs of industry, a School of Mines was established in Kimberley in 1896. In 1904, this institution relocated to Johannesburg and was the founding School of the University of the Witwatersrand in 1922. This School has continued to educate mining engineers continuously since its formation and was joined in this task by the University of Pretoria's Department of Mining Engineering upon its formation in 1962.

No review of mining education in South Africa would be complete without comment on the distortions and anomalies created by apartheid, the government policy demanding total racial segregation, which was in place from 1948 to 1994. In particular, the liberal spirit that dominated the policies of the University of the Witwatersrand since the beginning of apartheid inevitably influenced the School of Mining Engineering in carrying out its duties. For decades the School advocated the extension of its educational activities to all students, irrespective

of color or gender. In spite of this advocacy, the recruitment of non-white students to the mining engineering program was gradual and only began in the 1980s, when political conditions permitted. The South African mining industry supported it through the financial support of bursaries and bridging schemes for students from disadvantaged educational backgrounds.

Over the past 15 years enrolments in the four mining engineering programs in South Africa have grown, mainly due to interest from the black community, who see mining engineering as an attractive and long-term career opportunity.

United States

The workforce in the mining industry has been declining for more than 20 years, as of 2004. Caused by, in part, the consolidation of mining companies and the increasing size of mines, this overall decrease in the workforce has also affected technical personnel, such as mining engineers. Funding for academic research and education, both from industry and government, has markedly declined in the same period. As a result, the number of U.S. programs in the field has dropped from a high of 25 in 1984 to 15 in 2007 (Figure 1). Limited educational research funding also threatens the sustainability of minerals engineering programs and severely restricts the ability to recruit, retain and develop students and faculty.

Furthermore, the average age of working mining engineers has increased and the number of expected retirements in the next decade is substantial. According to 2004 SME data, 90% of mining engineers were over the age of 50. The required number of new mining engineers for the nation, assuming new openings and expected retirements, has been estimated by industry and other sources to be about 350 mining engineers per year. Currently, after reaching a low output of less than 100 mining engineers per year in the early 2000s, the graduates are in a respectable range of over 150 per year with expectations (looking at positive enrolment trends) that this gap will close in a few years.

The expected massive retirement is applicable not only to practicing mining engineers, but also to mining engineering faculty (Figure 2). Assuming attrition and retirements, about 35 mining engineering academics have or will retire in the period 2004–2012.

The increasing specialization required for training a mining engineer and the need to broaden the student's background to include knowledge of social, environmental and sustainability concepts, when added to the ever-increasing sophistication of technology that is essential in the mining industry

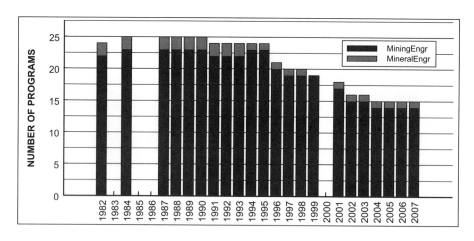

Figure 1.　Mining and minerals engineering programs in past 25 years (McCarter 2007)

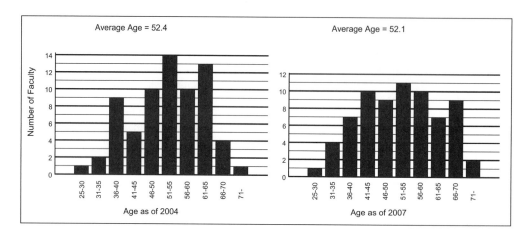

Figure 2.　Age distributions for faculty in the 13 mining engineering programs (McCarter 2007)

today, makes it more expensive than ever to educate a mining engineer. This is at a time when government budgets for academic funding are severely reduced and industry has shown limited interest in providing significant resources towards such academic efforts.

Finally, increasing numbers of mining students (McCarter 2007) are less interested in faculty positions, potentially contributing to future problems filling the positions created by retirement of aging faculty. All of these data suggest that U.S. mining educators must apply innovative methods to redress these shortages. Approaches to solving these problems involve increased collaborations with industry, joint partnerships with other universities and increased participation in interdisciplinary programs.

Mining education in the U.S. can be broadened in two ways. First, mining education and research should seek collaboration with areas of related competence and expertise, such as carbon capture and storage, water issues, energy and environmental issues, enhanced coalbed methane and carbon management. Collaborations in these areas will have the added benefit of improving the image of the industry, and these are lucrative potential sources of funding as well. Second, structural impediments in forging national and international collaborations should be addressed. This will require faculty and administrators working jointly to achieve course transference, meet accreditation requirements, resolve tuition/fee imbalances and provide work experiences and internship opportunities.

HISTORY AND CURRENT ROLE OF THE SOCIETY OF MINING PROFESSORS

The Society of Mining Sciences, or Societät der Bergbaukunde, was established in Schemnitz by

494

Ignaz Von Born, more than two hundred years ago, in 1786, as the first international professional society. Membership included many great scientists of the day, including Sir Humphrey Davy. The Society developed branches and offices in 15 countries and, at its peak, listed 154 members. Within a few years of the death of Von Born in 1791, the Society was disbanded.

Even without an official Society, the links between mining professionals in Europe continued to be strong. In 1990, after changes in European laws requiring EU member-countries to recognize degrees of every member country, 34 mining professors from 20 countries founded a new Societät der Bergbaukunde with the English name, Society of Mining Professors (SOMP). Professor Gunther Fettweis proposed the formation of the Society during the 150th anniversary of his home institution, the Montanuniversität Leoben. The SOMP initially had approximately 70 members, representing 39 mining faculties (or colleges) in 27 countries.

The modern Society quickly agreed on the contents of a standard curriculum for mining engineering degrees. To establish the standard curriculum, a comparison was made of the actual curricula of member universities. A database of these curricula was compiled that currently includes information on 109 curricula from 96 universities in 49 countries.

One of the Society's most important roles since its inception has been networking, and the Annual General Meeting facilitates that goal. At several meetings, 60–70 percent of all members have attended the three to four day meetings and discussions. Invaluable contacts are made and many important initiatives have been undertaken as a result of these meetings. The extensive geographical spread of the membership gives the society a true worldwide character, helping it to communicate easily about educational and research matters.

The initiatives for the development of the European Mining Course (EMC), the European Mineral Engineering Course (EMEC), and the European Geotechnical and Environmental Course (EGEC) led to the combined European Mining, Minerals and Environmental Program (EMMEP) program. Delft University of Technology spearheaded the development of these programs with SOMP also playing a key role in forging partnerships.

Research was the theme of the 1998 Annual General Meeting in Italy, which was jointly organized with Eurominerals and Euromines, the professional organizations representing professional institutions and the minerals industry in Europe. The Society has also undertaken several initiatives to improve the image of the discipline and serve as the global representative of mining engineering in academia. For example, in 1999 after the "Minetime 99" conference in Düsseldorf, Germany, the Society co-organized an international press conference aimed at improving the image of mining and emphasizing the positive societal contributions of the minerals sector.

At the 2003 Annual General Meeting, in Milos, Greece, the SOMP co-sponsored, with many other organizations, including SME, the Declaration for Sustainable Development, known as the "Milos Declaration" (Karmis, et al 2005) and endorsed by the major professional and scientific international groups. This is a landmark statement of the role, responsibilities and opportunities of scientific and professional societies in the transition to sustainable development.

Also at the 2003 Annual General Meeting a committee was established with the task of providing a platform for an expanded, global role of the Society, including a revitalized mission and new goals and priorities. The committee also addressed constitutional changes. It recommended that the Society should be a vibrant global Society, representing the majority of minerals academics, and should make a significant contribution to a sustainable future. According to the statement, the main goal of the Society should be to guarantee the scientific, technical, academic and professional knowledge required to ensure a sustainable supply of minerals for mankind.

The Society actively seeks to enhance the funding base for minerals-related research with governments and industry. The Society also actively promotes coordination of research grants and the undertaking of major research initiatives. The fact remains, however, that minerals research today requires interdisciplinary approaches and talents and it is, or it should be, driven by the enabling technologies that are fuelling the high technology revolution. To this end, the Society promotes interdisciplinary research and joint research initiatives amongst member institutions, and plays an important role in assisting colleagues in emerging countries. Finally, the Society is engaged in communication programs focusing on fostering pride in the minerals industry and protecting and enhancing its image

Over the last two decades, Society membership has increased to 180 members representing 70 mining schools in 40 countries. SOMP has acted as a catalyst and platform for the development of innovative programs, new partnerships and joint educational and research efforts, and for the expansion into new areas and disciplines. It has also fostered information exchange and dissemination of "best practices" in minerals education and research around the world.

RESTRUCTURING THE MINING ENGINEERING CURRICULUM

Curriculum and research emphasis have evolved in response to technological, industrial and societal change, including advances in mining practice; discoveries of new mining regions and types of mineral deposit; economies of scale and technology breakthroughs; and sadly, mine safety events and disasters. This evolution can be broadly viewed in six eras, as follows:

- The pre-60s curriculum was dominated by geology, mineralogy, engineering science, surveying, ventilation, drilling and blasting, underground environmental and safety, and basic electro-mechanical applications. The curriculum and research were characterized by an empirical approach and reliance on rules of thumb, graphical strengths, and work experience.
- The 60s saw the emergence of economics, management, rock mechanics, operations research, ore estimation, and exploration science. Doctoral programs increased in number, and international perspectives were recognized.
- The 70s saw the emergence of computer-based applications, monitoring tools and techniques, instrumentation, maintenance, environmental science, and waste management.
- The 80s saw the emergence of geomechanics, computer simulation, mineral process control, and financial evaluations; legislative issues were increasingly important.
- The 90s saw the emergence of environmental assessment, concern with public image, managing and business perspectives, automation and robotics, and risk assessment
- The 21st century has seen the emergence of ethics, corporate social responsibility, sustainability, CO_2 and global warming, water and energy conservation, entrepreneurship and innovation. It has witnessed the start of a new era of significant collaboration and networking in education and research.

The expectations of the capabilities of the future mining engineer, on the part of both industry and society, are shifting in response to the need to accommodate the dynamics of technology advances as well as changing real world issues and the paradigm of sustainable mining (Scoble and Laurence 2008). Mining engineering curricula in Australia evolved, at least in part, in response to the industry's needs for mine managers. In most jurisdictions in Australia a mine manager's statutory certificate of competency can only be granted to an applicant with a degree in mining engineering (Laurence and Galvin 2006). It is no longer adequate, however, to simply educate mining engineers in how to design and operate mines and mills safely. There remains a necessity for a strong focus on these skills, but this potentially risks the exclusion of other knowledge and skill sets which are becoming increasingly relevant. Accommodating this broader paradigm in mining education raises the need to stretch out into social sciences, ethics and law. This may be a difficult transition for some who still only recognize the past narrower definition of a mining engineer.

Figure 3 attempts to capture four domains of responsibility and capability that the mining engineer will need to accommodate in future mining, which in itself is seen to relate to design and planning and operational and business management.

The new paradigm of mining education recognizes that mining engineers, mine designers and mine managers will be called upon ever more frequently to serve as communicators with the public and media. The old paradigm of one industry CEO or public relations manager speaking for industry perspectives is no longer viable in this age of 24-hour new and weblogs. This will require students to acquire interpersonal and written and oral communication skills. Courses in professional development, management, psychology, technical writing and public speaking already exist on most campuses, and those educational resources can be readily marshaled in unique and innovative collaborations with mining engineering faculty to educate mining students at all levels.

Another educational issue for future development relates to the optimum, modern learning experience. This applies, for example, to resolving the role of different options in educational components and concepts: e.g., design projects, problem-based learning, laboratories, internet, teleconference, and lifelong learning.

BEST PRACTICES

Mining Education Australia (MEA)

Mining Education Australia (MEA) is a national initiative to create a standardized curriculum for the last two years of an undergraduate degree program in mining engineering. According to the vision statement for the MEA, the goal was to create a virtual national mining school, "one Program and one School delivering a world class program of undergraduate education in mining engineering by integrating and coordinating the resources of Australia's premier mining universities."

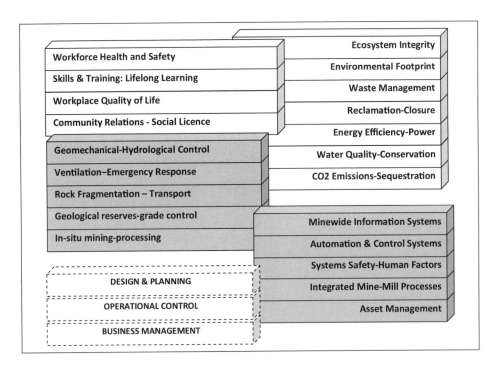

Figure 3. Domains of responsibility for a mining engineer (Scoble 2008)

The founding universities were: Curtin University of Technology (WA School of Mines); the University of New South Wales; and The University of Queensland. A fourth university, the University of Adelaide, joined MEA in October, 2008, and commenced teaching the MEA curriculum in March, 2009. In 2008 these four MEA universities accounted for almost 90 percent, or 260 of the 290, newly enrolled mining engineering students across Australia. MEA produced 123 mining graduates from the first cohort of students to complete the MEA third and fourth year curriculum in 2008.

MEA provides students with a far more comprehensive and higher quality education than was available at any of the individual universities previously, as a result of access to a much broader range of faculty experience and expertise, as well as a totally new, state of the art curriculum, designed by the combined staff in conjunction with industry, with funding provided by the MCA and the federal government. The curriculum has been developed along a series of program themes or streams, as illustrated by Scoble (2008), shown in Figure 4.

Key features of MEA are:

- An unincorporated joint venture between the university members, with a parallel funding

agreement contract between the MCA and MEA members

- Funding managed through a small MEA Administration Unit at The University of New South Wales (2006–2009), with responsibilities for management of course development, all course and related resource data repository, marketing and promotion, MEA staff development programs and financial management
- A Governing Board comprising 50% industry and 50% university representation
- Comprehensive team of over 30 mining academics across the country
- Innovative delivery methods
- Financial incentives based on participation and performance

Financial support from the MCA provided initial funding for the establishment of MEA, and continues to provide sustaining support to fund the MEA Administration Unit. It also provides for a flow of funds, through MEA, back out to the member institutions, for support of mining educational initiatives. MEA is also looking forward to broader international collaborative opportunities in the future, with a range of operational models available for discussion with prospective partners. Since the mining industry is an

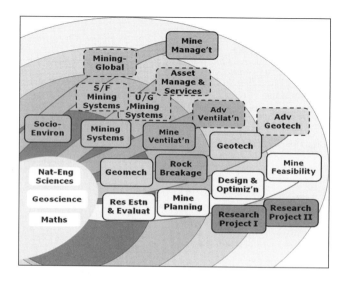

Figure 4. MEA curriculum themes (Scoble 2008)

international industry, the mining education sector must also adopt an international or global focus, and MEA provides a means to achieve this on the international education stage.

Erasmus Mundus Program (EMMEP)

The Erasmus Mundus Minerals and Environmental Programme (EMMEP) is a two-year Masters-level course of study to educate new talent, future managers and leaders in the European and worldwide minerals industry. Emphasizing important European perspectives while maintaining an international focus, EMMEP is taught at multiple locations in Europe and leads to a double M.Sc. degree in the field of Minerals and Environmental Engineering. Coordinated at TU Delft, the program accepts 15–20 non-European Union students per year in addition to up to 40 European students.

Combining the areas of excellence of six leading European universities, EMMEP provides students with knowledge of important issues concerning sustainability and the environment, and with the research skills necessary for continuous professional development either at the Ph.D. level or in researching special topics required by industry. It features small class sizes and the courses are conducted in English. Other language courses are offered in addition to the program. The consortium offers a unique curriculum with mineral resource and environmental courses in a combination which is not found in any single European country or institution.

The study plan comprises four semesters. Two semesters are reserved for a ten month joint curriculum at four participating universities, and the remaining two semesters are spent at the two universities awarding the double degree. Formally initiated in October 2007, the first class included 18 non-EU students, all supported by 2 × 21,000 Euros (2 years) scholarships.

The participating universities also have strong links with the European Technology Platform on Sustainable Mineral Resources (SMR) and the Society of Mining Professors. Both organizations promote joint research activities and recognize the need for a good international education program. They therefore strongly support the EMMEP program.

Capacity Building and Global Mining Issues and Challenges in South Africa

Four institutions in South Africa provide tertiary mining engineering education. All are at historically high enrolment levels in 2009, for several reasons:

- The transformation of the country to a democracy in 1994 also led to significant political, social and economic changes. This coupled with high and on-going levels of crime has seen the emigration of many skilled professionals, leaving the mining industry with the urgent task of rebuilding its skills base.
- Following the 1994 election there has been need to transform the racial composition of the ownership and management of the mining industry. In 2004 a formal agreement between the industry and government (the Mining Charter) set targets of ten percent female employment,

498

40 percent female or black management by 2009 and 26 percent ownership by historically disadvantage South African citizens by 2014.

- The re-entry of South Africa into regional relationships has meant that South African universities now educate many mining engineers for the whole of Sub-Saharan African, particularly at post-graduate level.

Of the four universities offering mining programs, both the University of Witwatersrand and the University of Pretoria are traditional universities offering fully accredited professional undergraduate degrees and a full range of postgraduate qualifications. The University of Johannesburg offers technologist qualifications at bachelor and diploma levels and the University of South Africa provides distance learning for the diploma level.

Currently, there are approximately 1,260 undergraduates registered in mining engineering programs in South Africa and 350 students registered for postgraduate studies. Because of the national imperative for capacity building, this high level of enrolment has forced all four institutions to curtail their research activities and concentrate on undergraduate teaching and course-work for Master's programs. Since the global skills shortage of mining industry professionals is currently partially obscured by effects of the recession, it is important to remember that an upturn in economic activity will again expose the problem of skills, nowhere more significantly than in Africa.

Sustainability, Interdisciplinarity and the Curriculum

A number of universities are attempting to integrate sustainability into their curricula, and although strategies vary, most follow one of the following two strategies for accommodating sustainability. The first involves modifying the contemporary curriculum by:

- Introducing new sustainability courses into undergraduate mining programs in topics such as sustainable development, environmental stewardship, social as well as environmental impact assessment, community resilience, history and culture of indigenous people, and corporate social responsibility
- Developing courses in key topics such as climate change, carbon management, clean coal, energy efficiency, and water conservation
- Relating existing courses to the concept of sustainable development

The second strategy for integrating sustainability into mining research degrees is to foster innovative

collaborations with other departments. Students in other disciplines can be 'inducted into mining research,' building on their strengths and motivation to learn about mining. Recognizing that the postgraduate degree carries no qualification towards professional engineering status, these students, drawn from such diverse areas as sociology, science and technology studies, history and environmental studies, will readily see the value of exploring the mining sciences for their rich applications as case studies in current science and technology to important social issues. Joint research project supervisory committees representing key disciplines for interdisciplinary research focused on mining application can be fostered. Most universities offering mining engineering education are broad-based and include relevant disciplines and interested faculty, but it is also possible to build such collaborative, interdisciplinary efforts among students and faculty at other universities.

Life Long Learning—Teaching and Learning Methods

The objectives of 'educating' a mining engineer will only be fully realized through life-long learning. The education of mining engineers does not stop after four years at university. However, both during the initial graduate education period and through subsequent postgraduate and continuing professional development (CPD) programs, universities must consider the full range of alternative delivery options, and also capitalize on the rapid changes and developments in technology, in order to deliver educational material more effectively and to a wider and often remotely located audience around the world.

A number of universities, particularly in Australia and Canada, have pioneered teaching and learning alternatives and methods, using advanced technologies, as summarized below.

- Block teaching—key course modules can be effectively taught in block-mode, over anything from half day workshops to periods of three or four days full-time. Such sessions work well where there is a combination of lecturing plus student project work, presentations and discussion forums.
- Keynote lectures/learning sessions by subject matter experts—use of key expertise from anywhere in the world, can be delivered to different universities, either as a keynote platform for a particular learning module, or simply to offset a skills shortage at a certain university.
- Interactive teaching classrooms, where classes from across the country, or from around the world, can be connected with both video and

audio capability for staff and students alike. This type of system also enables international connections as well as remote site location single student engagement through a facility such as Skype.

- Where online interactive teaching is not possible due to facilities or timetabling, key lectures can be recorded as podcasts for distribution to either other site lecturers, or direct to students. (A database of such lectures could be established from all over the world, representing content provided by both academic and industry experts and mining case studies).
- Project and problem-based student learning—both individually and group work.
- Fully interactive, independent web-based learning management platforms, independent of particular university systems, for managing course teaching resources and student access etc.
- Distance-based learning modules—either provided on web, or via CD resource material—to provide full distance-based on-line learning.
- Use of Virtual Reality (VR) teaching modules for group or individual learning of complex mining systems and concepts; high hazard scenarios; or field-oriented learning where it may be difficult to expose students in the field to such environments.
- Staff exchange initiatives—for periods from just a few weeks to a full semester, possibly across non-teaching periods between different hemispheres (e.g., a southern hemisphere academic could exchange to teach in the northern hemisphere during his/her summer non-teaching period, and vice-versa).
- Greater use of industry guest lecturers/adjunct appointments, for either face to face, or interactive lectures—within an institution, a network such as MEA, or across the globe.

CONCLUSION

The "University of the 21st Century" is an institution for teaching and evaluating ways of behavior, based on new ideas about human beings, knowledge and practice. Moreover, it is not just a physical place, but a network of resources—faculty, students, laboratories, libraries and classrooms—all linked together not only physically but electronically. At the same time, the global mining industry is focusing on new technologies, environmental impacts, sustainability practices and social and corporate responsibility. Under this environment, mining engineering education must continue to develop by improving learning and teaching methods, introducing sustainable development

principles, developing national and international partnerships and focusing on career-long learning.

As Georgius Agricola stated in *De Re Metallica*, in 1556, "the art (of mining) is one of the most ancient, the most necessary and the most profitable to mankind." While this comment is still true almost 500 years later, the social context that the mining industry operates has changed dramatically. Mining educators must respond to that context to ensure the supply of graduates needed to meet global minerals and energy demands.

REFERENCES

Agricola, G. 1556. *De Re Metallica.* Translated by H. Hoover. New York: Dover Publications, 1950.

Hebblewhite B. 2006. Mining engineering education initiatives in Australia. *Mining Engineering*, 58(2):31–37.

Karmis, M., Shaw, C.T., and de Ruiter, H. 2005. Global minerals education and the Society of Mining Professors/Societät Der Bergbaukunde: A vision for the future and a plan of action. Annual Meeting of SME, Salt Lake City, Utah, February 28–March 2, 2005. SME Preprint Number 05-122.

Laurence, D.C., and Galvin, J.M. 2006. Educating future mine managers—maintaining the gene pool. *Proc. 1st International Mine Management Conference*, Melbourne, AusIMM.

McCarter, M.K. 2007. Mining engineering: Demand and supply. PowerPoint presentation to the Report Committee. National Commission on Energy Policy Coal Study. Sept. 17, 2007.

Minerals Council of Australia. 1998. Back from the brink: Reshaping minerals tertiary education. Discussion paper. http://www.minerals.org.au/__data/assets/pdf_file/0015/8007/backfrom thebrink_exec_summ.pdf. Accessed October 2009.

Mining Industry Training and Adjustment Council (MITAC). 2005. *Prospecting the Future: Meeting the Human Resources Challenges in Canada's Minerals and Metals Sector.* 3pp. http://www.canadianminingnews.com/mining%20hr%20release.pdf. Accessed October 2009.

Scoble, M. 2008. Mining Education Australia (MEA) mining engineering degree program—External review: assessment, recommendations & conclusions, Independent Report to MEA Executive, June, 2008 (unpublished).

Scoble, M., and Laurence, D. 2008. Future mining engineers: Educational development strategy. *Proc. Int. Conf. Future Mining*, Australian Inst. Min. Metall., Sydney.

A Review of NIOSH and U.S. Bureau of Mines Research to Improve Miners' Health and Safety Training

Robert H. Peters
NIOSH, Pittsburgh, Pennsylvania, United States

Charles Vaught
NIOSH, Pittsburgh, Pennsylvania, United States

Launa Mallett
NIOSH, Pittsburgh, Pennsylvania, United States

ABSTRACT: This paper reviews NIOSH and US Bureau of Mines research to improve miners' health and safety training. The program began in the late 1970s—after Federal Mine Safety legislation had been passed requiring all mine operators to provide formal safety and health training to miners on a regular basis. These regulations offered unique opportunities for conducting research to improve miners' training. Initial efforts to provide safety and health training were often found lacking in several respects. This paper describes how miners' safety and health training has improved since the 1970s, and identifies areas where further improvements are needed.

INTRODUCTION

This paper presents highlights of a federal government research program to improve miners' safety and health (S&H) training that has existed for more than 30 years. The program began in the US Interior Department's Bureau of Mines in the 1970s. In 1996, the Bureau of Mines was closed, and the mine safety and health research function was transferred to the Centers for Disease Control's National Institute for Occupational Safety and Health (NIOSH). NIOSH continued the Bureau's mine safety training research program, without interruption, to the present. Throughout the remainder of this paper, the mine safety training research carried out by the Bureau of Mines and NIOSH will simply be referred to as "government research." The information presented in this paper is organized into the following major sections: (1) overview of three decades of mine safety training research; (2) examples of major research accomplishments; (3) future mine safety training research needs; and (4) concluding remarks.

OVERVIEW OF THREE DECADES OF MINE SAFETY TRAINING RESEARCH

The scope of this paper is limited to mine safety training research that has been conducted or funded by the federal government since regulations were passed requiring mine operators to provide such training.

The Federal Coal Mine Health and Safety Act of 1969 (Public Law 91-173) contained specific language regarding safety training for coal miners (United States Public Laws 91st Congress—First Session 1969). The Federal Mine Health and Safety Act of 1977 (Public Law 95-164) expanded these requirements to metal and nonmetal mining, and mandated training for all individuals on mine property who met the definition of "miner" (United States Public Laws 95th Congress—Second Session 1977).

These acts significantly increased the funding available for mine S&H research by the US Bureau of Mines. The government's mine S&H training research program was implemented during the mid 1970s. Like the 1969 Mine Act, it started with a focus on the coal industry. In the 1970s the coal mining industry was establishing or strengthening training programs to meet the new legal requirements. Training research was "directed toward assisting the mining community in structuring, formalizing, and evaluating training investments" (Pittsburgh Research Center Staff 1981). After 1978, it was directed at the training requirements found in Title 30, Part 48 of the US Code of Federal Regulations (CFR 2009).

Some of this initial work was conducted by government researchers, but the majority was completed by non-government contractors. Researchers gathered baseline information about the state of mine training, developed guidelines for conducting and evaluating training, and created instructional materials about basic mining tasks and activities. Details of this work can be found in Informational Circular Report 8858, "Mine Safety Education and Training Seminar Proceedings" (Pittsburgh Research Center Staff 1981).

Along with the development of user-ready training materials, these contract projects covered topics such as (1) assessing the impact of training on accident rates, (2) methods to improve instruction in the classroom and on-the-job, and (3) the development of training simulators. Deficiencies identified during the research of the 1970s pointed the way for future work in areas such as training evaluation and the need for professional development for mine trainers.

At the end of the 1970s and early in the 1980s research on training for non-routine situations was added to the program. The Federal Mine Safety and Health Act of 1977 required mine rescue teams for all underground mines. The government supported training for the team members through a contract to design instructional materials (Pittsburgh Research Center Staff 1981). Training materials were designed for both small and large metal-nonmetal and coal mining operations. These training exercises included problem-solving activities, which became a component of many future training research and development projects.

During the 1980s, government researchers expanded their focus on training to prepare miners for non-routine emergency situations. During this same time period, the employment of new miners significantly declined, lessening the demand to make advances in new miner training about routine mining hazards. In addition, disasters at the Clinchfield Coal Company (1983) and Wilberg (1986) brought the reality of mine disasters to the forefront. Attempts to address the deficiencies identified in post-event analyses of those disasters led to training research and development in the areas of emergency decision-making and use of personal protective equipment—most notably the donning procedure for Self-Contained Self-Rescuers (SCSR).

Along with these two disasters, three coal mine fires occurred between 1988 and 1990 which had the potential to have caused major loss of life. Government researchers conducted in-depth interviews with 43 survivors of those events. The information gained through these interviews led to significant advances in knowledge about human behavior during escape from mine fires. The results from this investigation led to many publications including several sets of training materials and a book (Vaught, Brnich, Mallett, Cole, Wiehagen,

Conti, Kowalski, and Litton 2000). These materials have been used extensively, and have served as the basis for changes in mine emergency response training regulations.

In the years between 1992 and 2001, there were no major mine disasters in the United States. The economics of the 1980s and 1990s had created an environment where few new miners entered the industry. Mine employees were primarily experienced workers who planned to stay in their jobs for some time. The training research conducted during the 1990s reflected this relatively stable period. Research was conducted to improve training on both routinely and non-routinely used skills. For example, topics included emergency warning and escape strategies, as well as the hazards of roof falls and the need to stay away from unsupported roof (Peters 1992). Also, work directed at a growing population of contract workers and of workers at small mines was started.

As the year 2000 approached, concern grew regarding the training needs of the future mining workforce. The baby-boomer, experienced workforce began thinking about retirement. Mine trainers saw a population of increasingly older workers caused by the hiring gap in the 1980s and 1990s and began to think about the potential problems associated with bringing in large numbers of new employees within a short period of time. Because the complexity of mining equipment and miners' jobs had increased dramatically over the past 40 years, these new miners would need a substantial amount of training to get them up to speed quickly.

Government training researchers started addressing the needs of this new generation of miners in two ways. Questions regarding appropriate training methodologies for them were explored and support for the trainers who would be preparing them was developed. The generation of miners who began their career shortly after the new millennium was expected to be better educated and more technologically savvy than those who had started during the previous hiring boom of the mid 1970s. Many safety professionals believed that these two cohorts of miners would require different approaches to training. The trainers who would be teaching them would need to understand (1) how to use new training delivery technologies, and (2) would need to employ new styles of instruction. Details of the research conducted to help trainers address these changes can be found in three NIOSH Information Circular Reports (Kowalski-Trakofler, Vaught, Mallett, Brnich, Reinke, Steiner, Wiehagen, and Rethi 2004; Peters 2002; Vaught and Mallett 2008).

In 2006 and 2007, mine disasters again caught the attention of the nation. The deaths at Sago, Darby, and Crandall Canyon were seen as a call to action by many. Analyses of these events identified problems that needed, in part, training solutions. In response, training research and development has recently been conducted in the areas of (1) switching between SCSR units, (2) the use of refuge alternatives, (3) communication during emergencies, (4) emergency mine evacuation, and (5) decision-making during mine emergencies. But NIOSH's training researchers are not focused exclusively on improving emergency response training. Significant research and development is also underway to help trainers make the best use of new computer based technologies in the classroom and in other training venues. Computer-based technologies hold much potential for training both new and experienced miners about routine and non-routine workplace hazards (Mallett and Orr 2008).

EXAMPLES OF MAJOR TRAINING RESEARCH ACCOMPLISHMENTS

The government's mine safety training research program can be divided into two major categories: (1) training miners how to respond to mine emergencies, and (2) training miners to recognize and respond to more routinely encountered hazards. This latter focus has been especially important for new miners because, without proper training, they will not understand the types of hazards that are inherent to their new work environment. Because miners do not routinely utilize the knowledge, skills and abilities they would need to respond to mine emergencies, there are some important differences in the training strategies that must be used in order to maintain their emergency response competencies at an acceptable level. The next two sections present examples of the research performed to improve miners' training on each of these two important areas of mine safety.

Improving Miners' Ability to Respond to Mine Emergencies

Unlike skills such as the ability to recognize and avoid hazards, which are employed in the workplace every day, there is another set of skills miners need that they may seldom (if ever) employ in the course of their career. These skills come into play if they should get caught up in problems involving fires, explosions, or other catastrophic events. In such situations they need to be able to gather information, analyze risks, communicate with each other, and make informed judgments and decisions. If they can do that, their actions may prevent or

limit some of the worst consequences of the problem they are facing (Cole, Berger, Vaught, Haley, Lacefield, Wasielewski, and Mallett 1988). If they are lacking these non-routine skills, however, manageable problems may quickly get out of hand. University of Kentucky (UK) researchers, working on a Bureau of Mines contract, identified two critical non-routine skill domains that needed to be addressed through training and assessment: (1) self rescue, which requires task training to proficiency on devices needed to cope with the situation, particularly SCSRs; and (2) escape, which requires, among other things, training in the area of judgment and decision making.

Self-Contained Self-Rescuers

The SCSR studies began in the summer of 1985 when UK project staff interviewed 50 mine safety experts about SCSR usage. The key finding from these interviews was the fact that most underground coal miners never actually put the apparatus on. Instead, procedures for donning the unit were covered in annual refresher classes and were typically explained by a trainer who stood before the class and demonstrated the steps involved (Cole et al. 1988).

SCSR donning is a motor task, and the only effective way to teach a motor task is to have the trainee learn by doing (Schmidt 1988). The researchers decided to develop one universal hands-on donning task. They used a procedure in which the miner would kneel, loop the neck strap, activate the oxygen, insert the mouthpiece, put on the nose clips, put on the goggles, complete tying the straps, replace his or her cap, and be ready to travel (Vaught, Brnich, Wiehagen, Cole, and Kellner 1993).

The "3+3" donning method had some compelling features other than just the fact that it would be easy to teach and learn. First, it could be applied to all apparatuses so that instruction would be the same for each one. Second, because of this, every donning performance regardless of the unit being used could be evaluated the same way. Third, due to the donning procedure's simplicity, it should be possible to develop a simple evaluation instrument. A "connect the dots" evaluation form was developed. This device could show sequencing errors and actions done incorrectly so that the evaluator could see, in graphic detail, how each person had performed.

A second feature of motor tasks is that people forget over time (Hagman and Rose 1983; Johnson 1981). To explore this aspect of motor tasks a team of government researchers worked with a coal mine in the west. They trained the mine's workforce to the point that each miner was able to perform a perfect donning sequence on the day he or she was trained.

Then these trained miners were divided into two groups (a control group and an experimental group) from which samples would be taken during the ensuing year. These samples were drawn without replacement every three months. The control group received no additional training during the year. At nine months no one sampled from this group was able to don his or her SCSR proficiently. The experimental group had SCSR practice and evaluation added as part of their escapeway travel (every 45 days) and fire drills (every 90 days). At the end of the year 65 percent of those sampled from the experimental group were still able to perform a perfect sequence. While this study was only one in a series, it is the one that cut to the heart of the matter: workers need to be well trained initially and then given the opportunity to practice regularly. Additionally, it could be argued that the best place to do this is in the workplace.

In 1993 the research team published recommendations from lessons they had learned from their studies: (1) only one procedure should be taught; (2) training should be hands-on, with evaluation and feedback; (3) training ought to be conducted in-mine to minimize the interruption of production; (4) hands-on practice should be scheduled as part of fire drills and other emergency preparedness routines; (5) training models with easily cleaned and replaceable mouthpieces ought to be used; (6) distributed mental rehearsals could be provided between hands-on practice sessions; and (7) trainers should sample their workforce periodically and do spot evaluations in order to keep track of proficiency levels, with remediation given as needed (Vaught et al. 1993). Many of these recommendations have now been put into practice in the industry (US Department of Labor 2006).

Judgment and Decision Making

Having established judgment and decision making as a critical non-routine skill, the UK project team turned their attention to how it could best be taught and assessed. They sat in on many annual refresher classes in several states as observers, and sometimes as participants (Cole, Wasielewski, Lineberry, Wala, Mallett, Heley, Lacefield, and Berger 1988). They found that it was when there was an opportunity to discuss and resolve actual or realistic problems, often presented in story form (such as one trainer's account of a multiple fatality which had happened in Virginia) that they paid the most attention. This was not too surprising because problem solving is an integral part of living, and problems often are posed in story form. The use of storytelling (or narrative) has been explored in the literature and shown to be the primary way in which people make sense out

of things that occur in their daily lives (Clandinin and Connelly 2000; Cole 1997; Czarniawska 1997; Cullen and Fein 2005).

A good simulation presents an interesting and believable story that includes the teaching points the developers want to get across. In developing a simulation the researchers began with a problem, often taken from accident reports or from interview accounts by people who had been on the scene of a mine emergency. They would then turn the problem into a story that unfolded over time and had predicaments that the person working the problem would be required to deal with. As the story went along, there would be decision points that presented the trainee with an array of choice alternatives. Alternatives at each decision point in an exercise would consist of good choices and poor choices, but all of them plausible. For several reasons, one being their low cost and ease of use in miner training, the project team decided to develop paper and pencil simulations.

Once a simulation had been developed an instructor's guide and student workbook with an answer sheet were prepared. The workbook presented the story's decision points one page at a time. At each decision point in the story the trainee would make one or more selections among the choice alternatives on the answer sheet. The choices were printed in invisible ink, which he or she made visible with a special developing pen. Once developed, there was a response to each choice that gave feedback as to whether the choice was a good or poor one, and explained the reasons why. In this way the person working through the story received immediate corrective feedback where needed, and positive reinforcement when earned. The simulation, a form of programmed learning, proved to be an effective teaching device. In addition, however, the developed choices were a visible record of how the trainee performed at each choice point. So, it was also an embedded test (Cole 1994). The data from these tests could be examined to indicate where people were having problems and where remediation would be advisable. In addition, a trainer could get an idea of how effective his or her instruction was in any particular area. Therefore, in addition to being an effective instruction method, the simulations were inconspicuous measures that could be used for many positive purposes.

In all, there were 65 latent image simulations constructed by the UK staff, dealing with a variety of problems. They were field tested with a national sample of roughly 4,000 miners and found to be effective and reliable teaching and assessment instruments. There have been more than 400,000 copies of the answer sheets distributed in the US and abroad (Cole, Wiehagen, Vaught, and Mills 2001). And, simulations based on stories are still being developed—now by researchers at the NIOSH Office of Mine Safety and Health Research. Many of the interactive mine safety training simulations developed by government researchers have been converted to electronic delivery format by the Mine Safety & Health Administration, see *http://www.msha.gov/interactivetraining.htm*).

MERITS: A Training Simulation for Emergency Command Center Personnel

Near the conclusion of the UK contract work, government researchers began to focus their attention on training to improve the abilities of command center personnel to make good decisions at the outset of potentially disastrous situations. A computer-based training simulation titled "Mine Emergency Response Interactive Training Simulation" (MERITS) was developed and tested (Mallett 2002).

MERITS is an extension of previous work that compiled a knowledge base relevant to mine disaster management via interviews, analysis of literature, reviews of past events, and observations of mock mine disasters. The MERITS simulation is used to train emergency command center personnel to effectively manage an underground mine fire. Modeled after training simulations developed for other industries (e.g., nuclear, chemical, etc.), MERITS simulates underground and surface events related to the disaster. It exposes the user to events that typically occur during a mine emergency, such as lack of information and miscommunication. It also presents trainees with issues that must be addressed, such as briefing news media and victims' families, ordering supplies, interfacing with government enforcement agencies, and housing mine rescue teams.

MERITS allows individuals who are typically present in a command center to practice information gathering, situation assessment, decision-making, and coordination skills without risk to personnel or property. The outcome of the scenario is determined by the users' decisions and their emergency response plans. Field tests have been conducted with several groups of mining officials. Evaluations conducted at the completion of the simulation suggest that MERITS is a very beneficial experience for those who may someday need to manage the response to mine emergency.

Improving Miners' Ability to Respond to Routine Hazards

It is vital that miners be able to recognize and respond to common workplace hazards. Examples of two lines of research studies to improve miners' training in this domain are provided: (1) visual hazard recognition, and (2) story telling videos.

Visual Hazard Recognition Studies

The ability to perceive hazards is, perhaps, more difficult to achieve in underground mining than in other types of work areas because the work environment is confined, dark, and constantly changing due to the mining process. Workers must be alert and continuously cognizant of their surroundings. The information needed to recognize hazards is often available in the form of visual cues found throughout the workplace. Government researchers have conducted a series of studies on the effectiveness of novel approaches to improving miners' abilities to *visually* recognize mining hazards. Two examples of these studies are described below.

Stereoscopic images. In order to realistically portray certain types of hazardous conditions in the workplace, it is very desirable to add cues that convey information about depth and/or distance. For example, cracks that indicate the potential for a structural failure are often almost impossible to perceive in conventional slides and photographs. However, stereoscopic (3-D) slides have been found to be quite effective at illustrating such features. Two research studies have found that 3-D images are superior to 2-D for the purpose of training underground miners to recognize various geological conditions that are known precursors of mine roof failure (Barrett, Wiehagen, and Peters 1988; Blignaut 1979).

Although there appears to be no data on the question, stereoscopic photography may also be effective at portraying a variety of other potentially hazardous conditions in the workplace, such as the danger of falling from elevated work areas. During the 1980s and 90s government researchers created several training modules that involve using sets of 3-D slides to portray hazards at a variety of mining operations including underground coal and limestone mines, and surface aggregates mines. These training packages include 3-D slides, a student problem booklet and an instructor's guide (see "View-Master Reel" Training Exercises at http://www.cdc.gov/niosh/mining/products/#Training%20Packages).

Degraded images. Government researchers have conducted studies on the effectiveness of using degraded versus highlighted illustrations of hazards commonly found in underground mines. Highlighted examples of hazards are easy to detect. They are clear, close-up, unobstructed views. In contrast, degraded pictures of hazards are more difficult to detect. They are partially hidden, poorly illuminated, or were photographed from an eccentric angle, or through hazy or dusty conditions. During the 1970s, experiments were conducted on the use of degraded images to train military aircraft personnel to correctly identify ground targets (Cockrell 1979).

Researchers found that, in contrast to those trained with highlighted examples, those who were trained with degraded examples performed significantly better on a subsequent target recognition test.

Underground miners work in an environment where the hazards they need to be able to detect are often degraded—they are difficult to see because they are partially hidden, poorly illuminated, and covered with dust. Government researchers decided to conduct an experiment similar to Cockrell's to determine whether degraded images are better than highlighted images for the purpose of teaching miners to recognize mining hazards (Kowalski-Trakofler and Barrett 2003).

The researchers took three sets of pictures of hazards found in underground coal mines. One set showed highlighted views of 35 hazardous conditions. These slides were used to train one group of miners. A second set of slides showed degraded views of the same types of hazardous conditions. These slides were used to train a second group of miners. Participants in both groups viewed each slide for five seconds. Then, the instructor reviewed the slides and discussed the hazards depicted in each scene.

After the training phase, participants in both groups were asked to view a third set of slides. This last set of slides illustrated 36 hazards of the same type that were shown during the earlier training phase. Participants viewed each slide for five seconds and then reported any hazards to the experimenter. The dependent measure in this experiment was the number of hazards the participant correctly reported. It was found that participants trained with degraded images were able to identify a significantly higher number of hazards than those who had been trained with the highlighted set of hazards. Based on these findings, a "Degraded image hazard recognition" training program was created. The training module includes degraded image slides, a student problem booklet and an instructor's guide.

Story Telling Videos

A second example of government research to improve training on routine hazards involves using narrative accounts of accidents based on interviews with miners. During the past 20 years, researchers videotaped several miners as they told about various types of mining accidents, e.g., roof collapses, fires, electric arc flashes, etc. These stories are told in a very compelling fashion. They are a valuable tool for imparting historical information about mine safety to the next generation of miners. As previously mentioned, storytelling (or narrative) has been explored in the literature and shown to be the primary way in which people make sense out of their world. Along with each video, the training module includes an instructor's guide. These guides contain: questions the instructor could use to encourage group discussions about key training points, a picture or diagram of the accident scene, and additional background information about the safety topics covered in the video (see videos at http://www.cdc.gov/niosh/mining/products/#Videos).

FUTURE MINE TRAINING RESEARCH NEEDS

Although many types of improvements to miners' safety training are still needed, it appears particularly important for future research to focus on two objectives: (1) utilizing methods that improve realism and increase trainee engagement (participation or involvement), and (2) developing methods for evaluating miners' competencies.

Research on Improvements to Training Methods

Research on the effectiveness of occupational safety training methods suggests that such training is more effective when the training methods are highly realistic and engaging (Burke, Sarpy, Smith-Crowe, Chan-Serafin, Salvador, and Islam 2006; Cohen 2004; Robson, Stephenson, Schulte, Chan, Bielecky, Wang, Heidotting, Irvin, Eggerth, Peters, Clarke, Cullen, Boldt, Rotunda, and Grubb 2009). To facilitate the transfer of training to the job, it is important that practice drills and simulations be as realistic as possible. Training should reflect actual conditions on the job as closely as possible. Such training builds miners' confidence and enables them to respond appropriately to emergency situations and hazardous conditions. Great strides have been made in improving the realism of miners' safety training over the past 40 years, but there is definitely room for further improvements.

Research also strongly suggests that higher levels of engagement in occupational H&S training are positively associated with knowledge acquisition and reduction in accidents, injuries and illnesses. Low engagement H&S training typically employs oral, written or multimedia presentations of information by an expert source, but requires little or no active participation by the learner, other than attentiveness. During low engagement learning sessions, miners often do not have an active cognitive or behavioral role that can be clearly documented.

With high engagement training methods, the trainee has a much more active role in the learning process. The trainee engages in significant cognitive and behavioral interaction with the material, and has many opportunities to ask questions of experts/ instructors and engage in focused discussion with other trainees. High engagement training methods frequently provide trainees with opportunities to discover new cognitive strategies related to problem-solving and decision-making. Participants are often involved in hands-on practice of the behaviors to be learned. Examples can range from table-top exercises conducted in a classroom setting, to mine emergency escape and rescue training within a real or simulated mine.

According to Burke (2006), behavioral modeling training is also an effective way to increase engagement in S&H training. Behavioral modeling involves trainee observation of a role model, followed by trainee modeling or practice, and feedback designed to modify behavior.

Several additional computer simulations, such as MERITS, need to be created to provide mine managers and responsible persons with the opportunity to practice handling a wide variety of mine emergency situations. The more experience people gain through participating in such role playing simulations, the better prepared they will be to handle real-world events. Virtual reality technologies also appear to hold much potential for improving the realism and engagement of miners' safety training (Mallett et al. 2008).

Research on Competency Evaluation Methods

For the most part, US mine H&S training regulations simply require miners to attend training classes for a prescribed number of hours. Other than donning their SCSR properly, regulations do not require that miners demonstrate their mastery of specific H&S competencies. Definitions and standards concerning what constitutes "successful" completion of mandated mine H&S training need to be developed, along with detailed and valid instruments, checklists and procedures for measuring individual miners' competencies. The Mine Safety Technology and Training Commission recently cited the lack of suitable methods for evaluating competencies as a particularly important deficiency in the area of miners' emergency response training (Mine Safety Technology and Training Commission 2006).

CONCLUSION

Collectively, our nation's miners sit through millions of hours of mandated S&H training each year, and mining companies spend millions of dollars to provide this training. Unless effective training materials and methods are used, miners are unlikely to learn what they need to know to actually help them reduce their risk of suffering occupational injury and illness.

Many miners sit through the same training lectures and films year after year in order to fulfill the requirements of the law. In these situations, their "training" ends up being a very unfortunate waste of time and resources, i.e., a wasted opportunity. On the other hand, when training is done well, it is extremely valuable and worthwhile to both the miners and their employer.

With help from academic institutions, mining companies, MSHA, and other providers of miners' training, government researchers have produced over 90 training modules on a wide variety of mine S&H topics. However, the primary emphasis of the government's training research program is not upon the production of training materials per se, but rather upon finding better *processes and methods* of training. Most of these training modules were developed in the course of research studies to determine the feasibility and effectiveness of using innovative new methods of presenting occupational S&H information to miners. Although the technologies available for developing and delivering miner training continue to evolve and improve, the basic principles behind the types of training that government researchers strive to produce remain constant.

Several major improvements have taken place in the way mine S&H training is conceptualized and practiced since the federal mine training regulations were instituted almost 40 years ago. The government's training research program is but one of several important forces that are responsible for bringing about the following improvements to miners' training:

1. Greater emphasis upon learning that requires collaboration and active problem solving.
2. Greater integration of miners' practical knowledge and experience with the mandatory S&H information they are required to receive annually.
3. Greater realism in training scenarios and greater fidelity of visual illustrations.
4. Greater use of training materials that are thoroughly authenticated and field tested. All materials should be authenticated with subject matter experts and thoroughly field tested with multiple groups of miners before releasing them to mine trainers.

The goal of the government's mine training research program has been—and will continue to be—providing assistance to the mining industry with appropriate use of new theories, methods, and technologies as they take their training programs into the future.

REFERENCES

Barrett, E, W Wiehagen, and R Peters. 1988. Application of Stereoscopic (3-D) Slides to Roof and Rib Hazard Recognition Training. U.S. Department of the Interior, Bureau of Mines, Information Circular Report 9210.

Blignaut, C. 1979. The perception of hazard: The contribution of signal detection to hazard perception. *Ergonomics* 22:1,177–1,183.

Burke, M, S Sarpy, K Smith-Crowe, S Chan-Serafin, R Salvador, and G Islam. 2006. Relative effectiveness of worker safety and health training methods. *American Journal of Public Health* 96:315–324.

CFR. 2009. Code of Federal Regulations. Washington DC: U.S. Government Printing Office, Office of the Federal Register.

Clandinin, D, and F Connelly. 2000. *Narrative Inquiry: Experience and Story in Qualitative Research*. San Francisco: Jossey Bass.

Cockrell, J. 1979. Effective Training for Target Identification Under Degraded Conditions. U.S. Army Research Institute for the Behavioral and Social Sciences.

Cohen, A. 2004. Report from the 1999 National Conference on Workplace Safety & Health Training: Putting the Pieces Together & Planning for the Challenges Ahead. Cincinnati, Ohio: U.S. Department of Health and Human Services, CDC, NIOSH Publication No. 2004-132.

Cole, H. 1994. Embedded performance measures as teaching and assessment devices. *Occupational Medicine: State of the Art Reviews* 9 (2):261–281.

———. 1997. Stories to live by: A narrative approach to health behavior research and injury prevention. In *Handbook of Health Behavior IV: Relevance for Professionals and Issues for the Future*, edited by Edited by D.S. Gochman. New York, NY: Plenum Press.

Cole, H, P Berger, C Vaught, J Haley, W Lacefield, R Wasielewski, and L Mallett. 1988. *Measuring Mine Health and Safety Skills*. Pittsburgh, PA: US Department of Interior, Bureau of Mines, Contract Technical Report No. 2.

Cole, H, R Wasielewski, GT Lineberry, A Wala, L Mallett, J Heley, W Lacefield, and P Berger. 1988. Miner and trainer responses to simulated mine emergency problems. Paper read at Proceedings: Mine Safety Education and Training Seminar, Information Circular Report 9185, at Pittsburgh, PA.

Cole, H, W Wiehagen, C Vaught, and B Mills. 2001. Use of Simulation Exercises for Safety Training in the U.S. Mining Industry. U.S. Department of Health and Human Services, CDC, Information Circular Report 9459, NIOSH Publication 2001-141.

Cullen, E, and A Fein. 2005. Tell Me a Story: Why Stories are Essential to Effective Safety Training. U.S. Department of Health and Human Services, CDC, Report of Investigation 9664, NIOSH Publication No. 2005-152.

Czarniawska, B. 1997. *Narrating the Organization: Dramas of Institutional Identity*. Chicago, IL: University of Chicago Press.

Hagman, J, and A Rose. 1983. Retention of military tasks: A review. *Human Factors* 25 (2):199–213.

Johnson, S. 1981. Effect of training device on retention and transfer of a procedural task. *Human Factors* 23 (3):257–272.

Kowalski-Trakofler, K, and E Barrett. 2003. The concept of degraded images applied to hazard recognition training in mining for reduction of lost-time injuries. *Journal of Safety Research* 34:515–525.

Kowalski-Trakofler, K, C Vaught, L Mallett, M Brnich, D Reinke, L Steiner, W Wiehagen, and L Rethi. 2004. Safety and Health Training for an Evolving Workforce: An Overview From the Mining Industry. U.S. Department of Health and Human Services, CDC, Information Circular Report 9474, NIOSH Publication 2004-155.

Mallett, L. 2002. NIOSH Releases New Computer-Based Training Exercise Called MERITS. U.S. Department of Health and Human Services, CDC, NIOSH Technology News No. 496.

Mallett, L, and T Orr. 2008. Working in the Classroom - A Vision of Miner Training in the 21st Century. In *The International Future Mining Conference & Exhibition*. Sydney, Australia.

Mine Safety Technology and Training Commission. 2006. Improving Mine Safety Technology and Training: Establishing U.S. Global Leadership. National Mining Association.

Peters, RH. 1992. Miners' Views About Why People Go Under Unsupported Roof and How To Stop Them. U.S. Department of the Interior, Bureau of Mines, Information Circular Report 9300.

———. 2002. Strategies For Improving Miners' Training. U.S. Department of Health and Human Services, CDC, Information Circular Report 9463, NIOSH Publication 2002-156.

Pittsburgh Research Center Staff. 1981. Mine Safety Education and Training Seminar Proceedings. Pittsburgh, PA: Department of Interior, U.S. Bureau of Mines, Information Circular Report 8858.

Robson, L, C Stephenson, P Schulte, S Chan, A Bielecky, A Wang, T Heidotting, E Irvin, D Eggerth, R Peters, J Clarke, K Cullen, L Boldt, C Rotunda, and P Grubb. 2009. Systematic review of the effectiveness of training & education programs for the protection of workers. Toronto Canada and Cincinnati Ohio: Institute for Work & Health and National Institute for Occupational Safety and Health.

Schmidt, R. 1988. *Motor Control and Learning: A Behavioral Emphasis*. Champaign, IL: Human Kinetics Publishers.

United States Public Laws 91st Congress—First Session. 1969. Federal Coal Mine Health and Safety Act, Public Law 91-173. Washington D.C.: U.S. Government Printing Office.

United States Public Laws 95th Congress—Second Session. 1977. Federal Mine Safety & Health Act of 1977, Public Law 91-173, as amended by Public Law 95-164. Washington D.C.: US Government Printing Office.

US Department of Labor. 2006. *Mine emergency evacuation: Final rule*. Federal Register, US Government Printing Office.

Vaught, C, M Brnich, L Mallett, H Cole, W Wiehagen, R Conti, K Kowalski, and C Litton. 2000. Behavioral and Organizational Dimensions of Underground Mine Fires. U.S. Department of Health and Human Services, CDC, Information Circular Report 9450, NIOSH Publication 2000-126.

Vaught, C, M Brnich, W Wiehagen, H Cole, and H Kellner. 1993. *An Overview of Research on Self-Contained Self-Rescuer Training*. Pittsburgh, PA: US Department of the Interior, Bureau of Mines, Bulletin 695.

Vaught, C, and L Mallett. 2008. Guidelines for the Development of a New Miner Training Curriculum. U.S. Department of Health and Human Services, CDC, Information Circular Report 9499, NIOSH Publication 2008-105.

Training Using Virtual Environments: The Problems of Organisational Knowledge Creation

Damian Schofield

State University of New York (SUNY), New York, United States

ABSTRACT: Virtual Reality (VR) simulators represent a powerful tool for training humans to perform tasks which are otherwise expensive or dangerous to duplicate in the real world. The idea is not new. This technology has been successfully extended and utilised in the mining industry in a range of applications (Bise 1997; Denby and Schofield 1999a; Schofield 2005; Henning et al. 2002; Kaiser et al. 2005).

There are a number of lessons that can be learned from other industries that have effectively utilised virtual technology for a number of years. A number of organisational knowledge models have been used in their creation such as the SECI model (Socialisation, Externalisation, Combination and Internalisation) developed by Nonaka et al. who have extensively published on its use (Nonaka et al. 1998; Nonaka et al. 2000). However, recently an increasing number of authors have raised issues with these theories, many reporting that the empirical basis of the models are unsatisfactory and thus that the models themselves are flawed.

The work in this paper follows on from work undertaken by the author on improving knowledge organisation when creating virtual reconstructions in other thematic disciplines (Hussin et al. 2004; Schofield et al. 2005). This paper will introduce knowledge organisation theory in relation to examples of virtual simulations developed within the mining industry for safety training purposes.

INTRODUCTION

Inevitably the future will be digital. The continuing digital revolution has had an enormous impact on the way training is undertaken and safety information disseminated in a range of industries. A wide range of digital media is currently being used, to varying degrees in training centres around the world. In many industries around the world technology can be slow to become fully accepted. It is fair to say that, in general, acceptance of digital training media in many commercial mining organisations can often lag behind the technological development (Schofield et al. 2002).

Advanced three-dimensional computer graphics and virtual environment technology, similar to that used by the film and computer games industry have been used to generate interactive learning environments allows mining personnel to perform a range of simulated training. Virtual teaching and training applications from a range of industries (flight, surgery and driving simulators to name a few) have proved the value of this technology, however it is currently not as widely utilised in the mining sector (Schofield et al. 2001).

Early attempts at 'virtual' mining simulators, particularly those which tried to apply three-dimensional computer graphics based technology, were often constrained by lack of realism detail in their graphical interfaces and crude level of simulation (Bise 1997; Denby and Schofield 1999b). However, it was also noted that even given these limitations, these virtual environments had the potential to allow users to experience situations which would not readily exist within the real world, e.g., to see 'into' a chemical reaction or to cause a major catastrophe through their actions (Filigenzi et al. 2000; Schofield et al. 2001).

The author's experience of building a range of interactive, virtual reality based engineering simulators has demonstrated the enormous benefits of using this type of learning in a mining environment, and also highlighted a few of the potential problems (Tromp and Schofield 2004; Schofield et al. 2005; Vasak et al. 2008). Many of these early virtual reality mining simulators were built either by university departments or by large mining corporations (Schofield et al. 2001). To develop high quality virtual simulations requires teams of experienced software/media developers, many of whom often have no a priori knowledge of the mining industry and environment. Hence there is a requirement for specialist, context specific knowledge to be effectively and accurately transferred to the software/media developers and then passed into any software training tools created. Many training simulator projects fail or do not achieve their full potential due to poor organisational knowledge management models and inefficient transfer techniques (Hussin et al. 2004; Schofield et al. 2005).

TECHNOLOGY

It is important to realise that the use of simulator based computer-generated training media in the mining industry is only the current manifestation of a long history of visual training systems (Denby and Schofield 1999b; Schofield et al 2001). However, computer animations and interactive virtual simulations are unparalleled in their capabilities for presenting complex information. The use of such enabling visualisation technology can affect the manner in which data is assimilated and correlated by the viewer; in many instances, it can potentially help make safety policies and procedures more relevant and easier to understand (Vasak et al 2008). Hence it is crucial to ensure that relevant, accurate and understandable knowledge is being transferred into the simulators being developed (Hussin et al 2004).

At this point, it is perhaps worth defining and describing the technologies under discussion in this paper. Visual media displays and digital simulation systems cover a wide variety of technologies. This paper focuses on computer-generated imagery, particularly computer graphics. Computer graphics in this context refers to a suite of software applications that can be used to produce outputs such as rendered images and animations. Rendering is the process of generating a digital image from a virtual model, by means of computer software. The term may be thought of by analogy with an 'artist's rendering' of a scene (Schofield, 2007).

Computer graphics systems utilise numerical three-dimensional models of real world objects to create artificial virtual environments. Based on scene survey data, objects such as equipment, vehicles, human figures, mine environment details and other relevant evidence items can be accurately positioned and precisely scaled within the artificial three-dimensional environment. The scene objects can then be texture mapped with relevant photographic images to produce a credible lifelike appearance.

Computer technology can be employed to build an animation from one these virtual environments showing the movement of these scene objects (usually controlled using a physics model). This animation is usually rendered frame-by-frame (as a series of still images). These frames, when played back in quick succession, create an experience of space, motion and time. Popular cultural examples of this technique include recent animated movies such as Shrek and Beowulf. These two recent, popular films demonstrate two distinct animation and representation styles. The first, Shrek by Dreamworks Animation (IMDB 2009a) demonstrates on a cartoon-like, abstract approach to present its narrative. The second, Beowulf by Imagemovers (IMDB 2009b) relies on a more realistic representational

form. In an accident reconstruction or safety training environment the term often used to describe evidence presented in this format is *scientific animation*.

Virtual reality involves interactive, real-time, three-dimensional graphical environments that respond to user input and action, such as moving around in the virtual world or operating virtual equipment. An important aspect of such a simulation system is its underlying processes, simulations, behaviour and reactions, and the way a user can interact with objects within the virtual world. A virtual reality user could, for example, sit in a virtual vehicle and drive it. Accurately comparing the 'real world' view of the driver of a vehicle involved in a mining accident with the field of view of a camera in a virtual reconstruction is a complex issue.

Popular cultural examples of virtual reality interactive simulations include recent three-dimensional computer games such as Unreal Tournament (Epic Games 2009) and Grand Theft Auto (Rockstar Games 2009). These popular computer game titles provide a good example of distinct viewing configurations through various game playing styles. Unreal Tournament belongs to a genre known as the First Person Shooter (FPS), distinguished by a first person perspective (egocentric) that renders the game world from the visual perspective of the player character. Grand Theft Auto is a Third Person Shooter (TPS), this is a genre of video game in which an avatar of the player character is seen at a distance from a number of different possible perspective angles (exocentric).

In any training simulation (as in any computer game), the choice of the viewing perspective may have an enormous impact on the way an image is interpreted by the viewer. Changing the viewing perspective can potentially alter which 'character' in a simulation that a viewer identifies with, or aligns themselves with (Bryce and Rutter, 2002). In some cases it may be possible to show views of an incident from the viewpoints of the multiple parties involved (Noond et al., 2002). In a training context the term often used to describe simulations presented in this format is *virtual simulation* or *virtual reconstruction*.

At first glance, these graphical simulations may be seen as potentially useful in many mining training situations, and they are often treated like any other form of digital media regarding their application. However, perhaps this specific form of digital media warrants special care and attention due to its inherently persuasive and engaging nature, and the undue reliance that the viewer may place on the information presented through such a powerful visualisation medium (Schofield and Goodwin 2007).

KNOWLEDGE

Knowledge is broader, deeper, and richer than data or information. Unlike data and information, knowledge usually contains judgment. In any mining company, staff will speak of a "knowledgeable individual," and mean someone with a thorough, informed, and reliable grasp of a subject. Knowledge is often a fluid mix of framed experience, values, contextual information, and expert insight that provides a framework for evaluating and incorporating new experiences and information (Davenport and Prusak 2000a). In mining organisations, it often becomes embedded not only in documents or repositories but also in staff and in organisational routines, processes, practices, and of course, also in training procedures and systems.

Knowledge transfer activities take place within and between humans, often facilitated by technology. Better knowledge can lead, for example, to improvements in mine safety and production (Druker, 1999). Mine workers can use knowledge to make wiser decisions about their safety and the safety of others. Of course, since knowledge and decisions usually reside in people's heads, it can often be difficult to trace the path of knowledge in an organisation; for example from the expert through the training system to the action performed by the employee in the mine (Hussin et al. 2004).

In many training systems, knowledge is usually transferred using a structured approach, in a virtual reality based simulation it is transferred using a structured media approach. It is important to understand exactly what is being communicated to the learners by this media (Schofield et al. 2005).

Knowledge of any mining operation or process develops over time, through experience that includes what is absorbed from courses, books, mentors, informal learning and training systems. Experience refers to what we have done and what has happened to us in the past. Experts in the mining industry are commonly defined as people with deep knowledge of a subject who have been tested and trained by experience. One of the prime benefits of such experience is that it provides a historical perspective from which to view and understand new situations and events. Knowledge born of experience recognises familiar patterns and can make connections between what is happening now and what happened in the past (Davenport and Prusak 2000b). One can posit the hypothesis that virtual training simulators are unique compared to other forms of training as they allow the ability for a trainee to actually gain 'experience' of an unfamiliar setting or situation (Schofield 2007)

Experience changes ideas about what *should* happen into knowledge of what *does* happen (Davenport and Prusak 2000a). A good example of effective knowledge management and transfer comes from the U.S. Army's Center for Army Lessons Learned (CALL) which uses examples of real situations experienced close up to train army personnel. The centre focuses on lessons learned on the ground, rather than from theory or generalisation (Psotka et al. 1998).

Knowing what to expect and what to do in military situations can be literally a life-or-death matter. Army personnel need to know what really works and what doesn't. Experts from CALL take part in real military operations as learning observers and disseminate the knowledge they gather through photos, video tapes, briefings, and simulations (Weber et al. 2001).

We could make a similar distinction between how mining actually happens and how it is taught in mining engineering programs at universities around the globe. Managers of mining companies usually recognise the importance of real-life knowledge, experienced mining personnel exchange personal experiences and talk about "life at the coal-face." In other words, they share the detail and meaning of real experiences because they understand that knowledge of the everyday, complex, often messy reality of mining is generally more valuable than academic theories about it.

Training of mine personnel is a complex task; no training simulation can teach everything one needs to know about every situation that will be faced in the dangerous and dynamic mine environments. Trainees must be able to generalise using the knowledge that has been given to them. It is often tempting to look for simple answers to complex problems and deal with uncertainties and gaps in knowledge by pretending they don't exist. However, it is important to remember that improving knowledge usually leads to better decisions (Grefenstette et al. 2004).

Most of us have met mining "experts" whose knowledge seems to consist of stock responses and who offer the same old answer to any new question: every problem looks like a nail to a person who has only a single conceptual hammer in his toolbox (Davenport and Prusak 2000a). One could argue that the expertise of these experts ceases to be real knowledge when it refuses to examine itself and evolve. It becomes opinion or dogma instead (Jenson and Meckling 1995). Evolution is particularly important in this context as the application of new technology in the design and development of mines greatly changes how experiences are learnt. As is the case in many industries that rely on technology, two decades of 'hands on' experience can sometimes actually be an impediment to knowledge generation.

Most people in the mining industry work each day using rules of thumb: flexible guides to action that have developed through trial and error and over long experience and observation. These rules of thumb (or heuristics) are shortcuts to solutions to new problems that resemble problems previously solved by experienced workers. Those with knowledge see known patterns in new situations and can respond appropriately. A problem arises when the rules of thumb do not evolve with the technological changes that are taking place in the mining workplace.

Roger Schank, a computer scientist at Northwestern University, calls these internalised responses "scripts" (Schank and Abelson 1977). Like film scripts (or computer program codes), these scripts are efficient guides to complex situations. These scripts are the core elements that need to be externalised by mining experts and passed to software/media developers to be encapsulated into virtual training simulators (Schofield et al. 2005).

These scripts are, however, notoriously elusive and difficult to capture as knowledge can be deeply embedded. The skill of an experienced Load Haul Dump (LHD) unit driver provides an example of this kind of intuitive knowledge. They *know* how to drive, rapidly accomplishing a series of complex actions without having to think about them, as a beginner would. The veteran driver also develops an intuitive sense of what to expect in the mine environment. Hundreds of hours of driving have led them to *know* that another LHD is likely to pull out of a cross-cut or switch into reverse without warning. Experience has made them aware of minute signs that a novice driver would almost certainly miss and that may be too subtle to easily verbalise into scripts.

It may seem odd to be describing these vague, intellectual concepts in a discussion of knowledge in mining organisations. Many people assume that these organisations are objective and neutral; their purpose is to generate a product. In fact, the individual miner's knowledge has a powerful impact on organisational knowledge. Mining companies are, after all, made up of people whose knowledge inescapably influences their thoughts and actions (Bhatt 2001; Gold 2001).

People in mining organisations have always sought, used, and valued knowledge, at least implicitly. Mining companies hire for experience more often than for intelligence or education because they understand the value of knowledge that has been developed and proven over time. Miners faced with making difficult decisions are much more likely to go to people they respect and avail themselves of their knowledge than they are to look for information in databases. However, explicitly recognising knowledge as a corporate asset is a relatively new idea. There is a constant need to improve the level of individual knowledge through training. However, the complex process of transferring information from the mining expert to the trainee using simulation technology is still not very well understood (Quinn 2005; Schofield 2007).

ORGANISATIONAL KNOWLEDGE MODELS

Many mining organisations developing training simulations follow a model of integrated research through knowledge mobilisation. Most follow this model unconsciously, being unaware of the explicit process. The most famous version of this model is the SECI model (Socialisation, Externalisation, Combination and Internalisation) developed by Nonaka et al. who have extensively published on its use (Nonaka et al. 1998; Nonaka et al. 2000).

It is fair to say that Nonaka's theory of organisational knowledge creation centring on this SECI model is one of the most widely cited theories in knowledge management. Many authors report on its use, in particular among technology development projects in large organisations. However, recently an increasing number of authors have raised issues with this theory, many reporting that the empirical basis of the model is unsatisfactory and thus that the model itself is flawed (Gourlay 2003; Suzuki and Toyama 2004; Okada 2005; Vorakulpipat and Rezgui 2008).

Applying the SECI model to the development of a virtual training simulation, four modes of knowledge transfer within a organisation can be identified and these are listed below (Nonaka et al. 1998).

1. **Socialisation (Entry Phase)**—*tacit to tacit*
   - Literature Synthesis, Strategic Direction, Expert Interrogation, Specification
2. **Externalisation (Directed Research)**—*tacit to explicit*
   - Demonstration Projects, Implementation Plans, Pilot Studies, Data Reviews and Analysis
3. **Combination (Mobilisation)**—*explicit to explicit*
   - Integrated Development, Presentation and Dissemination, Advocacy and Marketing
4. **Internalisation (Sustainability)**—*explicit to tacit*
   - Evidence Based Practice, Policy Development, Evaluation and Self Management

A number of the problems raised with the model are concerned with the nature of *"ba,"* where *"ba"* is a Japanese word with no exact translation; it refers to a shared place or context for human interaction. *"ba"* is a crucial concept in the development of virtual simulators as by their very definition these are usually 'shared places where human-computer interaction takes place'. In the context of the SECI model *"ba"* is of particular importance in the establishment of Socialisation (where people develop an understanding of the basic situation, i.e., what they are learning) and Externalisation (where they are making explicit the information, i.e., creating the training simulations for knowledge transfer) (Gourlay 2003; Suzuki and Toyama 2004).

Since learning through safety training, even in simulators, is often defined as a social process its effects in the epistemological dimension are on the individual. The second ontological dimension depicts the passage of knowledge from individual to an organisation/group level. The SECI model predicts that an individual's personal knowledge may be amplified and crystallised as part of the knowledge network of the organisation to which they belong (Nonaka et al. 1998; Nonaka et al. 2000). In the generic mining training simulations under consideration here, the SECI model can also be considered on both an epistemological dimension (individual experience and effectiveness of the training simulations developed) and an ontological dimension (strategic, staged schedule of the development process at an organisational level). However, problems with the SECI model have been identified in all four of the modes of knowledge conversion and creation. If this model is to continue to be utilised then these concerns need to be addressed.

The SECI model claims that knowledge conversion begins with the tacit acquisition of tacit knowledge by people who do not have it from people who do; a process Nonaka and his colleagues named Socialisation (Nonaka et al. 1998; Nonaka et al. 2000). In the case of training simulation development which is under consideration in this paper the Socialisation stage will involve the gathering of the information from mining experts. However, research has repeatedly shown that many groups studied do not understand how particular goals are realised or tasks achieved, i.e., that many people have a tacit understanding of how to perform in the mining workplace but find it difficult to 'tacitly' pass on the good practice guideline/rules that make them successful (Collins 1994; Collins 2004 and Engestrom 1999).

The main problem with the Externalisation stage was originally stated by Polanyi who claimed "we can have a tacit foreknowledge of yet undiscovered things" (Polyani 1966; Polyani 1969). The phenomenon Polyani described, that one person draws inferences from another's ideas, can be explained with reference to individuals operating in an organisational context. This can be seen as an expansion of the problem described above, where people who can do something are not able to fully describe (explicitly) how they do it (Collins 1994; Collins 2004 and Engestrom 1999). A number of researchers are now concluding that the process of Externalisation is more complex than that described in Nonaka's SECI model flawed (Gourlay 2003; Okada 2005; Vorakulpipat and Rezgui 2008).

The Combination stage involves the process of "systemising concepts into a knowledge system," which happens when people synthesise different sources of explicit knowledge into, for example, a report or training system (Nonaka et al. 1998; Nonaka et al. 2000). Combination thus involves many kinds of activities as knowledge is 'embodied' into some form of virtual training simulator. As Adler has noted, pursuit of Combination appears to require "an important dose of tacit knowledge" (Adler 2000). Thus Combination appears to also involve aspects of the Socialization and Externalization stages.

The final stage in the SECI process is labelled Internalisation which is described as "a process of embodying explicit knowledge into tacit knowledge." Nonaka has described this process as being closely related to "the traditional notion of learning" and to "learning by doing" (Nonaka et al. 1998; Nonaka et al. 2000). In a similar manner to the problems with the Externalisation process, many researchers have also highlighted the fact that the Internalisation process is confusing appearing to involve many different activities from the other areas of the process. There is also dispute about the terms used such as "embodying of explicit knowledge" and "traditional notion of learning." Many researchers do however seem to agree that the Internalisation process involves an individual acquiring a subjective sense of meaning from the explicit knowledge which can then be fed back, perhaps tacitly, at an organisational level (Gourlay 2003; Suzuki and Toyama 2004; Okada 2005; Vorakulpipat and Rezgui 2008).

While much of the above discourse concentrates on the SECI model as it affects the individual, the same reasoning can be extended to the application of the model as a template for organisational knowledge flow. Hence, it is imperative that developers of training systems who (consciously or unconsciously) are relying on the SECI structure are able to constrain and guide their progress carefully examine the implications of this model. There is a need to constantly evaluate its potential advantages and disadvantages, create pedagogically sound learning content, ensure the efficacy of the learning process and fully assess the training tool's impact on the workforce.

CONCLUSION

The ever progressing pace of technological development has created new infrastructures for knowledge exchange and transfer and opened up new knowledge management opportunities within organisations. The use of advanced technology now allows the development of virtual training systems. These provide novel and far reaching possibilities as knowledge enablers among mining workforces. One must appreciate that these systems are only the conduits and storage mechanisms for knowledge exchange.

A virtual training system does not create knowledge and also cannot guarantee to promote knowledge generation or knowledge sharing if the organisational knowledge management structure does not favour those activities. This paper has discussed the SECI process, one of the most popular organisational knowledge models and highlighted some of the issues with the model. Hopefully this discussion will allow developers of these virtual training systems to think about some of the knowledge transfer issues within their own organisations as they build these training systems.

The medium turns out not to be the message, although many advocates of virtual training system would have you believe that it can act as a universal panacea. In fact the use of a virtual training system does not even guarantee that there will be a message.

REFERENCES

Adler, P.S., 2000. Comment on I. Nonaka: Managing Innovation as an Organizational Knowledge Creation Process. In *Technology Management and Corporate Strategies: A Tricontinental Perspective*. Edited by *J.* Allouche, and G. Pogorel, (Amsterdam, Holland: Elsevier).

Bhatt, G.D, 2001. Knowledge Management in Organizations: Examining the Interaction between Technologies, Techniques and People, *Journal of Knowledge Management*, 5(1): 68–75.

Bise, C.J., 1997. Virtual Reality: Emerging Technology for Training of Miners, *Mining Engineering*, 49(1).

Bryce, J. and Rutter, J., 2002. Spectacle of the Deathmatch: Character and Narrative in First-Person Shooters. In *ScreenPlay: Cinema/Videogames/Interfaces.* Edited by G. King, and T. Krzywinska, London: Wallflower.

Collins, H.M., 1994. The TEA Set: tacit Knowledge and Scientific Networks, *Journal of Science Studies*, 4: 165–186.

Collins, H.M., 2001. Tacit Knowledge, Trust and the Q of Sapphire, *Journal of Social Studies of Science*, 31(1): 71–85.

Davenport, T.H. and Prusak, L., 2000a. Working Knowledge: How Organisations Manage What They Know, *ACM Ubiquity*, 2000(2).

Davenport, T.H. and Prusak, L., 2000b. *Working Knowledge*, Cambridge, MA: Harvard Business Press.

Denby, B. and Schofield, D., 1999a. Advanced Computer Techniques: Developments for the Minerals Industry Towards the New Millennium, Keynote Paper, *Proceedings of the 1999 International Symposium on Mining Science and Technology 1999*, (Beijing, China).

Denby, B. and Schofield, D., 1999b. The Role of Virtual Reality in the Safety Training of Mine Personnel, *Mining Engineering*, October 1999: 59–64.

Drucker, P.F, 1999. Knowledge-Worker Productivity: The Biggest Challenge, *California Management Review*, XLI(2): 79–94.

Engestrom, Y., 1999. Innovative Learning in Work Teams: Analyzing Cycles of Knowledge Creation in Practice. In *Perspectives on Activity Theory*. Edited by Y. Engestrom, R. Miettinen, and R.L. Punamaki, (Cambridge, UK: Cambridge University Press).

Epic Games, 2009. Epic Games Website, http://www.epicgames.com. Accessed September 2009.

Filigenzi, M.T., Orr, T.J. and Ruff, T.M., 2000. Virtual Reality for Mine Safety Training, *Journal of Applied Occupational and Environmental Hygiene*, 15(6): 465–469.

Gold, A.H., 2001. Knowledge Management: An Organisational Capabilities Perspective, *Journal of Management and Information Systems*, 18 (1): 185–214.

Gourlay, S., 2003. The SECI Model of Knowledge Creation: Some Empirical Shortcomings, *Proceedings of the 4th Conference on Knowledge Management*, (Oxford, UK).

Grefenstette, J.J., Ramsey, C.L. and Schultz, A.C., 2004. Learning Sequential Decision Rules using Simulation Models and Competition, *Machine Learning*, 5(4): 335–381.

Henning, J., Kaiser, P.K., Cotesta, L. and Dasys, A., 2002. Innovations in Mine Planning and Design Utilizing Collaborative Immersive Virtual Reality (CIVR), *Proceedings of the 104th CIM Annual General Meeting 2002,* (Vancouver, Canada).

Hussin, N., Schofield, D. and Shalaby, M.T., 2004. Visualising Information: Evidence Analysis for Computer-Generated Animation (CGA), *Proceedings of 8th International Conference on Information Visualisation IV04*, (London, UK).

IMDB, 2009a. Shrek, http://www.imdb.com/title/tt0126029/. Accessed September 2009.

IMDB, 2009b. Beowulf, http://www.imdb.com/title/tt0442933/. Accessed September 2009.

Jensen, M.C. and Meckling, W.H., 1995. Specific and General Knowledge and Organizational Structure, *Journal of Applied Corporate Finance*, 8(2).

Kaiser, P.K., Vasak, P. and Suorineni, F.T., 2005. New Dimensions in the Interpretation of Seismic Data with 3D Virtual Reality Visualization in Burst-Prone Mines. Keynote Address, *Proceedings of the 6th International Symposium on Rockburst and Seismicity in Mines 2005*, (Perth, Australia).

Nonaka, I., Reinmoeller, P. and Senoo, D., 1998. The 'ART' of Knowledge: Systems to Capitalize on Market Knowledge, *Journal of European Management*, 16(6): 673–684.

Nonaka, I., Toyama, R. and Konno, N., 2000. SECI, BA and Leadership: A Unified Model of Dynamic Knowledge Creation, *Journal of Long Range Planning*, 33(1): 5–34.

Okada, T., 2005. An Attempt to Prevent Operational Incidents by Applying the SECI Model, *Proceedings of the 1st World Congress of the International Federation for Systems Research*, (Kobe, Japan).

Polyani, M., 1966. *The Tacit Dimension*, (London, UK: Routledge and Kegan Paul).

Polyani, M., 1969. Knowing and Being. In *Knowing and Being: Essays*. Edited by M. Greene, (London, UK: Routledge and Kegan Paul).

Psotka, J., Massey, L.D. and Mutter, S.A., eds. 1998. *Intelligent Tutoring Systems: Lessons Learned*, Philadelphia, PA: Lawrence Erlbaum.

Quinn, C.N., 2005. *Engaging Learning: Designing e-Learning Simulation Games*. Misenheimer, NC: Pfeiffer Press.

Rockstar Games, 2009. Rockstar Games Website, http://www.rockstargames.com. Accessed September 2009.

Schank, R.C. and Abelson, R., 1977. *Scripts, Plans, Goals, and Understanding*, Hillsdale, NJ: Earlbaum Assoc.

Schofield, D., Denby, B. and Hollands, R., 2001. Mine Safety in the Twenty-First Century: The Application of Computer Graphics and Virtual Reality. In *Mine Health and Safety Management*, 1st ed. Edited by M. Karmis, Littleton, CO: SME.

Schofield, D., Noond, J. and Burton, A., 2002. Reconstructing Accidents: Simulating Accidents Using Virtual Reality, *Proceedings of APCOM XXX Symposium 2002*, (Phoenix, Arizona).

Schofield, D., 2005. Improving Safety through Virtual Learning: Transferring Good Practice from Other Industries, *Proceedings of Reducing Hazards on Operational Processes of Machines and Equipment Conference*, (Zakopane, Poland).

Schofield, D., Hussin, N. and Shalaby, M.T., 2005. A Methodology for the Evidence Analysis for Computer-Generated Animation (CGA), *Proceedings of 9th International Conference on Information Visualisation IV05*, (London, UK).

Schofield, D., 2007. Animating and Interacting with Graphical Evidence: Bringing Courtrooms to Life with Virtual Reconstructions, *Proceedings of IEEE Conference on Computer Graphics, Imaging and Visualisation*, (Bangkok, Thailand).

Schofield, D. and Goodwin, L., 2007. Using Graphical Technology to Present Evidence. In *Electronic Evidence*. Edited by S. Mason, London, UK: Lexis-Nexis.

Suzuki, Y. and Toyama, R., 2004. A Self-Evaluating Method of SECI Process in Knowledge Management, *Proceedings of the Engineering Management Conference, IEEE*, (Singapore).

Tromp, J. and Schofield, D., 2004. Practical Experiences of Building Virtual Reality Systems, *Proceedings of Designing and Evaluating Virtual Reality Systems Symposium*, (University of Nottingham, UK).

Vasak, P., Dasys, A., Malek, F. and Thibodeau, D., 2008. Extracting More Value from Complex Monitoring Data—Seismic Excavation Hazard Maps, *Proceedings of the Strategic and Tactical Approaches in Mining Conference*, (Quebec City, Canada).

Vorakulpipat, C. and Rezgui, Y., 2008. Value Creation: The Future of Knowledge Management, *The Knowledge Engineering Review*, 23: 283–294.

Weber, R., Aha, D.W. and Becerra-Fernandez, I., 2001. Intelligent Lesson Learned Systems, *Expert Systems with Applications*, 20(1): 17–34.

Training for Tomorrow—Application of Virtual Reality Training for the Mining Industry

Bruce Dowsett

Coal Services Pty Limited, Sidney, Australia

BACKGROUND

Coal Services Pty Limited (CSPL) operates from NSW Australia, serving the coal industry both within Australia and internationally. CSPL operates under the Coal Industry Act 2001 and is jointly owned by the Mining & Energy Division of the CFMEU and the NSW Minerals Council (the coal owners). CSPL is solely owned by the coal industry of NSW. Originally established as the Joint Coal Board (JCB) in 1946, the organization's vision is to be a recognised leader in the provision of worker's compensation, health, safety, mines rescue services and occupational training. CSPL's Mission, to enable customers to improve their outcomes by providing quality services and expert solutions, led to the formation of our key business units—Coal Services Health, Coal Mines Insurance, Mines Rescue, Occupational Hygiene and Coal Mines Technical Services. Many of the services provided by these business units have expanded to all states of Australia and internationally. Virtual Reality Training sits within the Mines Rescue Service to support the CSPL mandate to provide workers quality training in Mines Rescue, Health and Safety.

The Australian coal industry is striving to achieve "Zero Harm"—people not suffering any injuries at work. However the constant mining challenges of nature, large equipment in restricted environments, changing workforce experience, processes and systems continually improving, etc make quality training even more important. The coal industry has a concern and desire to continue to improve OHS as accidents are still occurring. To this end Coal Services have established a Health & Safety Trust, funded again by the NSW coal industry, to encourage innovations in this important area.

CONCEPT—CREATION—REALITY

Technology developments of Simulators were initially seen as a short-term solution to the mining industry's skill crisis in Australia and possibly other countries. Improved technology, graphics and computer power, accelerated by the rapid development of the Gaming industry, brought Virtual Reality to the attention of the mining community at the turn of the 20th century.

During 1999–2000, Coal Services (then JCB) engaged and funded the University of NSW's School of Mining to explore VR for simulation capability for the coal industry.

In 2002—A 'proof of concept' VR Flat Screen was installed at the Newcastle Mines Rescue Station in NSW with three basic mining modules being trialed—Self Escape, Rib Stability, and Open Cut Truck Inspection. Favorable responses from industry stakeholders encouraged further development.

From 2003 to 2007—The Health & Safety Trust supported further research and development with the UNSW to establish new hardware, VR Curved Screens, VR Spheres and 360 degree 3D theatres. In addition VR software module development was expanded and improved. Industry stakeholders were consulted and endorsed the VR developments. The concept was established, proven and accepted as having the potential to provide the industry with a "Step Change" in mine safety training.

In 2007—CSPL, led by the Executive Director Mr Ron Land, made a major commitment to build four VR Training centres at the four district Mines Rescue Training facilities of NSW (in regions south of Sydney, west of Sydney, in the Hunter Valley and Newcastle). The first VR Training Centre was officially opened at the Newcastle Mines Rescue Station in December 2007.

In 2008—Further VR software development continued with five more modules developed – including Hazard Awareness, Isolation, Deputies Inspection, Spon Comb and Gas Outburst. These modules were established with Coal Services expert advisers guiding the software developers. A second VR Training Centre was opened in Wollongong (south of Sydney).

In 2009—VR Training Centres at Lithgow (West of Sydney) and Singleton (Hunter Valley) were officially opened. R&D with the UNSW ceased and a commercial VR Developer, VR Space, was engaged to accelerate and further the virtual mine model development. Complete VR Underground and Open Cut mine models containing a variety of scenarios were completed.

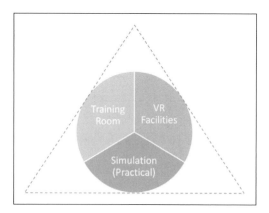

Figure. 1.   NSW mines rescue training triangle

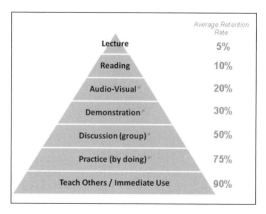

Figure 2.   Learning retention pyramid
(Adapted from National Training Laboratories
USA 2009)

TRAINING "STEP CHANGE"

An Australian capability for developing VR interactive training environments has been achieved,
addressing key industry issues and providing a clear
direction for ongoing R&D projects. The process of
data collection, model building and transfer between
CSPL VR Team and the developers was essential to
providing a quality virtual mine environment. High
resolution VR images were found to be beneficial,
easily recognised and accepted by mine workers as
"looking like our mine." Trainees responded better
to this realistic environment and the new scenarios
apply directly to national training competencies.
Most importantly industry experienced professional
Trainers who know the topics have been found to be
essential for quality VR interaction.

Trainees identify the hazards, risks and controls, even experience the consequences if things go
wrong, all whilst in a safe environment. Just like the
real world work environment, the trainee's response
to the scenario situations will affect outcomes.
Trainers can demonstrate and directly explain the
issues associated with the mining hazards in a controlled and safe learning environment. This cannot
be achieved in the real mine environment when hazardous equipment is in operation or when there are
irrespirable atmospheres. A complete training strategy has been developed, based on the MRS Training
Centre's "Training Triangle." As can be seen in
Figure 1, The Training Triangle encompasses three
different teaching methodologies: (1) Classroom
(2) Simulator (3) VR Platform.

Figure 2 attempts to show the average retention rate from a variety of teaching methods and as
shown, the application of VR to these training methods allows traditional and teaming techniques. Use of

the virtual environment applies to the Audio–Visual,
Demonstration, Discussion and Practice by Doing
strategies listed in Figure 2. This provides the adult
learner with more control over their learning experience and allows them to apply their new knowledge
soon after it is formed and is therefore more likely to
contribute to a change in behavior.

Interactivity is experienced by the Trainees by
providing group discussion, navigating their own
way through the virtual environments, witnessing
consequences of decisions made (both good and
bad), responding to challenges confronting them, and
by being fully immersed in the environment whereby
they can observe subtle signs and changing conditions. Adult learning principles are applied to the
training modules. There are hand held "Responders"
for the Trainees to indicate their answers to challenges. Answers are confidential to the group, however the Trainer has a hand held Tablet that provides
a range of information to him, including Trainees
answers. This feedback to the trainer is invaluable—
indicating which Trainees are not keeping pace with
the training concepts and allowing the trainer to
guide Trainees to the right choices if required. These
answers are recorded electronically for inclusion in
the Competency Assessments that accompany the
training modules undertaken.

The VR facilities are highly innovative teaching
tools that will have an impact on the way new knowledge is presented and imparted to training course
participants. Encompassing traditional methods
(classroom) with VR platforms, a "Step Change" has
been achieved. Research has been conducted in many
fields regarding the perceived benefits of quality VR
Training but CSPL has established a research project

by analysis experts associated with the University of Queensland to track the many perceived benefits over time. Research supports the opinion that VR interaction does provide a Trainee with extended learning retention from modules experienced.

CURRENT RESULTS

There has been wide industry interest and support by Underground coal mine operators and Open Cut coal mine operators across NSW and interstate.

Currently there are approximately 2,000 people per month experiencing the VR Training platforms, across the four centres in NSW. Trainee groups include: Mines Rescue Brigadesmen, new employees to the coal industry, experienced miners (undertaking refresher training), contractors (undertaking mine industry inductions), mine officials (nationally accredited statutory training programs), and fire fighting teams.

As a registered training organization (RTO), Mines Rescue also provides a wide variety of mine specific training as requested by individual mining companies, with training programs developing continuously. The aim has been to integrate VR into as many programs as possible for the benefit of the learner and in response to the ongoing support

the coal industry continues to show for the use and development of this powerful training tool.

WHAT DOES THE FUTURE HOLD FOR VR TRAINING?

VR Training will develop further quality mine models for both Underground and Open Cut operations that are fully functional, flexible and interactive, based on consultation with industry stakeholders. There is also wide interest internationally in the CSPL VR Training model being considered by the USA, China, Europe, New Zealand and South Africa.

Other industry interest has been forthcoming from fields such as the Military, Police and Anti-terrorist Police, Electricity Suppliers, Ore Mines, Rescue Organisations, Nuclear Energy Suppliers, and Oil & Gas suppliers. The VR Training model is transferable and can be applied to any industry with major hazards.

People are an organisation's greatest asset, deserving the best quality training available. Coal Services Pty Limited VR Training is rapidly becoming the "Go To" organization for industries desiring a step change in "Training for Tomorrow."

REFERENCES

National Training Laboratories USA 2009.

Virtual Environments for Surface Mining Powered Haulage Training

Ryan P. McMahan
Virginia Tech, Blacksburg, Virginia, United States

Steven Schafrik
Virginia Tech, Blacksburg, Virginia, United States

Doug A. Bowman
Virginia Tech, Blacksburg, Virginia, United States

Michael Karmis
Virginia Tech, Blacksburg, Virginia, United States

ABSTRACT: This project investigated technologies and intervention strategies for reducing injuries and fatalities from powered haulage equipment in the metal/non-metal mining industry through improved worker training using virtual environments (VEs). The investigation focused on two primary equipments: haul trucks and conveyor systems. The research team developed VE applications to improve worker training for each type of equipment. The effectiveness and cost-benefit ratio of these VE training tools were empirically evaluated. Based on the results, the VE training tools were found beneficial for training pre-shift inspections of haul trucks and for training safety and operational procedures for conveyor systems.

INTRODUCTION

The integration of personnel and machines to transport tons of material over long distances is a dangerous formula, with haulage trucks and conveyor systems being key components. Haul trucks are widely used in the majority of large open pit mines and some large tunnel underground mines, moving material from the working face to processing operations. The operators of these vehicles must be highly aware of any hazards that may exist in their surroundings. Conveyor systems are large, energy-intensive pieces of equipment, generally used to move materials from one point to another within a mine site that is not constantly changing. Conveyors at mine sites are highly regulated and covered under parts 46 to 48 of the Mine Health and Safety law enforced by the Mine Safety and Health Administration (MSHA). Haul truck and conveyor accidents, with other powered haulage accidents, accounted for 37.2 percent of surface mining fatalities and 24.5 percent of underground mining fatalities between 2001 and 2005, in the United States (NIOSH 2008).

Based on the analysis of the fatal investigation reports of MSHA from 1995 through 2007 (2008), inadequate or insufficient training was considered the precursor to many of the causes identified in powered haulage accidents. In a mining operation, ignorance of the full consequences of an action may lead to unsafe work practices and in turn to accidents. While the mining industry has been advancing machine technology to improve health and safety, these advances do not take the individual worker and potentially insufficient training into account.

Typically the training requirements outlined by MSHA are met through classroom training, with information presented via videotapes and PowerPoint presentations by a training professional. Oral, written, and practical tests are used to demonstrate training completion, but these methods are passive in nature as the trainees observe more than participate. According to Burke et al. (2006), engaging training methods are the most effective for knowledge acquisition and retention of material. Kowalski and Vaught (2002) have also found that adult learners, such as new miners, are "task-centered." and "solution-driven." Hence, the traditional methods for training new miners may not be the most effective training solutions; however, alternative methods of training are available.

One alternative to traditional methods of training miners is to use virtual reality (VR) or virtual environments (VEs). Virtual environments are synthetic three-dimensional spaces, which are seen from a first-person point of view, and which are under the real-time control of a user (Bowman et al. 2005). Given this definition, many different types of systems can be considered VEs, including 3D computer games, desktop animations, and web-based 3D communities. Particular VEs that create the illusion that the virtual world surrounds the user spatially are referred to as immersive virtual environments. Immersive VEs can lead to the sensation of presence, the feeling of "being there," where the virtual world replaces the physical world as the user's reality.

Outside of mining, VEs have already been used successfully for phobia training (Rothbaum et al. 1995), military training (Durlach and Mavor 1995), and medical training (Gutiérrez et al. 2007). In the mining field, VEs in the form of simulators (see Figure 1) have also been successfully used for training. Immersive Technologies and Fifth Dimension Technologies both commercially offer simulators for training operators of equipment such as haul trucks, bulldozers, and shovels. In addition to commercial simulators, other forms of VEs have been investigated for improving safety through accident simulation (Delabbio et al. 2007), helping identify hazards (Ruff 2001; Orr et al. 2003), and simulating health and safety activities (Stothard et al. 2004). The AIMS Research Unit has also developed SafeVR, a revolutionary tool that may be used to create VE training applications quickly (AIMS Research 2008).

Due to the ability to safely simulate real life events in a digital environment that might otherwise be too dangerous or expensive to create, VEs have been used with great success (Haller et al. 1999). Using VEs for training offers valuable interactive training environments that allow trainees to take charge of their surroundings and learn at their own pace. Another advantage of using VE technology is that trainees can be placed in virtual situations that range from normal to extraordinary, allowing them to learn from the virtual experience what they have not yet encountered in real life. In consideration of these advantages, the goal of this research was to develop training techniques using VE technology to reduce the dangers of working around haul trucks and conveyor systems.

IDENTIFICATION OF TRAINING NEEDS

In order to develop training VEs for reducing injuries and fatalities from powered haulage equipment through improved worker training, this research began with the identification of training needs pertaining to haul trucks and conveyor systems. These forms of powered haulage were chosen because the manner of their operation and maintenance has high implications for worker safety; they require teaching multiple skills to personnel and the execution of complex tasks. The research team worked with industrial and regulatory partners to identify specific training needs for these equipments that were high priority, based on accident reports, field observations, and interviews. These needs were kept separated, related

Figure 1. An example of a simulator used for training haul truck operation

to haul trucks and conveyor systems, as were the rest of the methodology and procedures.

Haul Truck Training Needs

Considering the number of accidents involving haul trucks, and that haul trucks are a common piece of mining equipment, more safety measures are needed to ensure worker safety around this type of equipment. Improper maintenance and inspection combined with inadequate training comprised over half of all cited violations in the accidents analyzed involving haul trucks. Problems not detected during maintenance and pre-shift inspections can lead to costly mechanical failures and worker injuries. Inadequate training and ignorance of consequences of actions can lead to unsafe work practices, which result in injuries. Improving pre-shift inspection training was an obvious corrective measure to focus on and was expected to reduce the impacts of poor maintenance and inadequate training.

Haul truck pre-shift inspections include both a safety inspection and a maintenance inspection by the operator before using a haul truck (Miller 2007). During these pre-operational checks, the operator walks around the vehicle to inspect certain parts of the haul truck and to identify any possible defects or mechanical failures. Ideally, the operator should use a checklist for the pre-shift inspection to avoid missing an inspection point and possible defects. If the operator detects any problems with the haul truck during the pre-shift inspection, the problem is to be reported to one of the mining operation's mechanics. In the case of a reported defect, the operator should be assigned a new haul truck, and the mechanic is responsible for repairing the defect using proper maintenance and safety procedures. These inspections can avert costly mechanical failures and, more importantly, dangerous accidents.

Based on training and materials received, specific inspection points and preventable failures were identified for inclusion in developing training VEs for pre-shift inspections of haul trucks. These points, detectable defects, and possible failures can be found in an earlier publication (McMahan et al. 2008).

Conveyor System Training Needs

Analysis of the accidents involving conveyor systems showed that additional measures are necessary to ensure the safety of miners working around conveyors. Improper maintenance, improper operational procedures, and unsafe work conditions were the most common causes of accidents around conveyor systems. These causes are directly related to a lack of knowledge of safety procedures and hazardous conditions. Considering these factors, the research team decided to focus on identifying hazards, proper safety procedures, and regulations for conveyor system training needs.

Conveyor system pre-operational checks incorporate both safety procedures and maintenance issues as a precursor to operational tasks. During these inspections, the operator walks around the conveyor system to identify potential hazards and maintenance issues. Once the operator has determined that the conveyor system is free of hazards and obvious mechanical problems, a lockout, tag-out procedure is required to start the system and begin operation. Once operation has begun, the operator is then responsible for identifying any mechanical issues that arise in order to properly shutdown the system to avoid further damage and even potential injuries. These aspects of the tasks required before and during the operation of conveyors were the focus of the research team's efforts to improve miner training. The specifics of these tasks were covered in an earlier publication (Lucas et al. 2008).

DESIGN OF VE TRAINING APPLICATIONS

Based on the training needs identified for both haul trucks and conveyor systems, three major considerations were taken into account for using training VEs. The first consideration was the presentation of information and how to effectively present training to the trainee. Because training VEs can simulate unlimited access to expensive or unavailable equipment, the research team decided the best method for presenting training information was to present the information while the trainee was virtually near the relevant equipment. The second consideration was

the assessment of the trainee's retention of information while still in the training process. The benefit of using VEs for training in regard to assessment is that the training can be assessed in a simulation of real life situations. The third consideration was the reinforcement of training importance. If the importance of information is explained or demonstrated, the trainee is more likely to remember the information presented, in order to avoid dire consequences. Therefore, the research team decided to use "negative." simulations to emphasize the importance of the training content.

Haul Truck Applications Design

Based on the three major considerations, three correlating phases of training were designed for the haul truck training VEs: the virtual tour, the virtual inspection, and the shift simulation.

The first phase of the training approach, the virtual tour, introduces the information required to perform an adequate pre-shift inspection of a haul truck to the trainee. To effectively present this training information, the trainee is guided around a haul truck on a tour of the various inspection points identified in the training needs. At each inspection point, information windows are used to identify parts by name and to explain defects to look for with corresponding proper actions to take. Figure 2 shows an example from the virtual tour of a hydraulic hose inspection point and the information window associated with it.

The second phase of the haul truck training, the virtual inspection, assesses the trainee's retention of the information presented during the virtual tour phase. During this phase of the training, the trainee self-navigates around a haul truck to the various inspection points. At each inspection point, the trainee looks for defects and notifies the system if a defect is detected. Notification occurs by selecting the defective part and then selecting a corrective

Figure 2. Example hydraulic hose inspection point and associated information window

action option from a newly activated window. When the trainee finishes virtually inspecting the haul truck, the trainee indicates that the inspection is complete by selecting a completion box located near the haul truck. All of the trainee's actions are graded and reported to the trainee after the virtual inspection to heighten retention levels.

The third phase of training, called the shift simulation, was designed to reinforce the importance of pre-shift inspections of haul trucks. During the shift simulation, the trainee is shown a simulation of the work shift following the virtual inspection. If the trainee missed any defects during the virtual inspection, a haul truck accident or mechanical failure related to a missed defect is animated to stress the catastrophic consequences of failure. If the trainee properly detected all defects during the virtual inspection, the shift simulation ends with the worker driving safely home after working. This animation is a positive reinforcement of the importance of the training.

Conveyor System Applications Design

Similar to the design of the haul truck training VEs, for the conveyor system training VEs, the research team designed an instructional tour to present training information related to conveyor systems and a virtual shift to assess the trainee's retention and to reinforce the importance of training.

The instructional tour was designed to take the trainee around a conveyor system along an automated path. This allows the trainee to become familiar with different areas of the conveyor system without needing to navigate manually. This also ensures the trainee reviews all of the training information provided at designed stops or stations. Each station contains a series of flashing colored spheres, or hot-points, that correlate to information about maintenance problems, safety procedures, hazard awareness, and conveyor components. Figure 3 shows an example of a hot-point related to the lockout, tag-out procedures for turning off the conveyor system. When the trainee has reviewed all of the hot-points at a station, the instructional tour continues to the next station. The research team designed six stations with training points pertaining to the specific training needs identified previously.

The virtual shift was designed to assess the trainee's retention of information after the instructional tour and to reinforce the importance of proper safety procedures. During the virtual shift, the trainee is allowed to navigate freely around a conveyor system while completing objectives. The first objective is to complete a pre-operational check to ensure the work area is clear of hazards and that the conveyor runs properly before operation. The second

objective is to properly start-up the conveyor system, including turning on breakers, sounding the alarm, and turning the lever on to each conveyor belt. The third objective is to ensure the belts are running properly, which should lead to the discovery that a damaged idler needs to be fixed. The trainee must follow the proper lockout, tag-out procedure in order to initiate the repair and complete the final objective. This sequence of objectives was designed to assess the trainee's retention of information. If the trainee makes a mistake during an objective, a simulation of an accident or failure is used to reinforce the importance of the training.

DEVELOPMENT OF VE TRAINING APPLICATIONS

With the training VEs designed, the research team decided to utilize two different development processes to provide insight to better approaches for future development of training applications. The

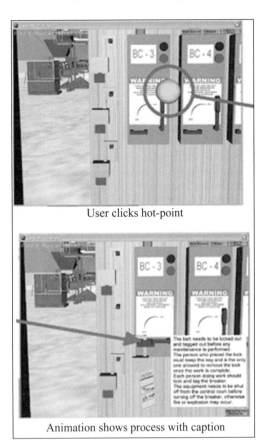

User clicks hot-point

Animation shows process with caption

Figure 3.   Hot-point example: lockout and tag-out

Figure 4. The CAVE version of a virtual environment for training haul truck pre-shift inspections

Figure 5. The CAVE version of the instructional tour virtual environment for conveyor systems

models and animations of the haul truck training VEs were created using Studio™. This same modeling and animation program was used to create Virtual Reality Modeling Language (VRML) files to use as desktop versions of the haul truck training VEs. In order to create more immersive versions of the training VEs, DIVERSE (Kelso et al. 2002), an application programming interface for VEs, was used to create versions of the training VEs for a CAVE™ (Cruz-Neira et al. 1993), a room-sized visualization system. See Figure 4 for an example of these immersive training VEs. For the conveyor system training VEs, the models needed were created using Autodesk® 3ds Max®. Right Hemisphere's Deep Creator™ was then used to create executable desktop versions for Microsoft Windows®. DIVERSE was used again for creating CAVE™ versions of the conveyor system training VEs (see Figure 5).

EVALUATION OF TRAINING VIRTUAL ENVIRONMENTS

After designing and developing the haul truck and conveyor training VEs, the research team set out to evaluate the effectiveness of these training tools to determine any potential benefits of using immersive technologies and to help determine the cost-benefit ratio of developing desktop and immersive versions of such training tools. In order to determine such benefits, the retention of information by a trainee after using the training VEs would have to be evaluated to determine the effectiveness of each set of VEs. Gutiérrez et al. (2007) conducted an experiment with a similar goal by utilizing a learning evaluation method based on Pathfinder-based representations

of knowledge to quantify the effectiveness of two versions of a medical training VE.

The learning evaluation method employed by Gutiérrez et al. uses a knowledge assessment test consisting of weighting the relatedness of pairs of concepts critical to the training. By administering this related-pairs test before and after the intended training, information on how the trainee correlates critical concepts is established both before and after training. These weighted correlations can then be symbolized by networks, with nodes representing the range of critical concepts and the length of links representing the weight of correlations. A Pathfinder algorithm is then used to prune the network down to a minimal set of links required to keep each node or concept within the network. This pruned network of links is referred to as a knowledge structure. The knowledge structure of the trainee is then compared to the knowledge structure of a predetermined domain expert with a link-by-link comparison. The percentage of links in common is regarded as a measurement of the trainee's knowledge on a scale of 0 to 1. By obtaining a measurement of the trainee's knowledge before and after using the training VEs, the difference between the measurements can be used as an indicator of the retention of information by the trainee and the effectiveness of the training. A more detailed explanation of the Pathfinder-based representations of knowledge is provided by Johnson et al. (1994).

An issue of evaluating learning using Pathfinder-based representations of knowledge is that the Pathfinder algorithm necessitates that all of the nodes or concepts be connected to each other by some weighted link. One solution is to ask a relatedness question about every possible pair of concepts, but this requires approximately N^2 questions, where N is the number of concepts. This was obviously not feasible for evaluating the VE training tools developed by the research team due to the number of concepts incorporated into the training VEs.

Another solution to the issue of having connected nodes is to assign the lowest possible weight to links without corresponding relatedness questions. The problem with this solution is that it can be counterproductive to capturing domain knowledge with related-pairs questions. For instance, assume there are two concepts considered "absolutely not related." by the domain expert. In the expert's knowledge network, the link between these concepts would be weighted 7 using the 7-point scale described by Johnson et al. (1994). Now assume the same two concepts are rated as "most likely not related." by a trainee. In the trainee's knowledge network, the link between these concepts would be weighted 6. When the Pathfinder algorithm traverses the nodes of the expert's knowledge network, the link between the two unrelated concepts will be removed assuming there is already an indirect path between the two concepts. When the Pathfinder algorithm traverses the nodes of the trainee's knowledge network, the same link will remain since the link's weight of 6 will be greater than any indirect paths weighted 7. Now, the rest of the trainee's knowledge network may change drastically compared to the expert's due to this "most likely not related." link weighing more than links without corresponding relatedness questions. This randomness in the pruning of the network defeats the very purpose of using related-pairs questions to determine knowledge.

The research team decided to avoid these issues by leaving the Pathfinder algorithm step out and modifying the graph similarity computation. Instead of pruning the knowledge networks created from the related-pairs test with the Pathfinder algorithm, the research team modified the graph similarity computation to account for the relative weight of nodes. To demonstrate this modification, consider the following example.

At node A in the expert's knowledge network, there are three links with neighboring nodes B, C, and D. The three links correlate to the three relatedness questions pertaining to the concept of node A. The weight of the links between A and the other nodes are 7, 1, and 1, relative to B, C, and D (i.e., the weight of the link between A and D is 1). In the trainee's knowledge network, at node A the same three links weigh 6, 6, and 4, respectively. Now when graph similarity is computed the relative weight of each node is counted towards similarity. For instance, the expert and trainee share a common link between

A and B because 6 is relatively closer to 7 than 1. The expert and trainee do not share a common link between A and C for the same reason. The expert and trainee also do not share a common link between A and D because 4 is relatively as close to 7 as to 1.

With these modifications to the learning evaluation method, the research team was able to account for the relatedness information captured by the knowledge assessment tests without asking thousands of questions to build a knowledge network for the Pathfinder algorithm.

Haul Truck Training Evaluation

For evaluating the effectiveness of both the desktop and immersive versions of the haul truck training VEs, the research team used three types of relatedness questions to compose the knowledge assessment test. The first type of relatedness questions involved the image of a haul truck part to inspect and the name of a haul truck part. If the name matched the part in the image, the two concepts should have been considered highly related. The second type of relatedness questions involved the image of a haul truck part to inspect and the description of a potential defect. If the potential defect described a defect that should be looked for in regard to the pictured part, the two concepts should have been considered highly related. The third type of relatedness questions involved the name of a part and the description of a potential defect. Again if the potential defect described a defect that should be looked for in regard to the named part, the two concepts should have been considered highly related. Otherwise, any pairs of concepts not highly related should have been considered definitely not related.

A total of 29 trainees recruited from around Virginia Tech's campus participated in the evaluation of the haul truck training VEs and a PowerPoint presentation that was evaluated as a traditional tool. Nine of the trainees were given the knowledge assessment test, received training using the desktop VRML versions of the haul truck VEs, and then given the knowledge assessment test for a second time. Ten of the trainees were given the knowledge assessment test, received training using the CAVE™ versions of the haul truck VEs, and finally given a second knowledge assessment test. The other ten trainees were given the knowledge assessment test, allowed to use the PowerPoint presentation for self-paced training, and then given a second knowledge assessment test.

Overall, the training tools and the PowerPoint presentation were found to significantly increase a trainee's knowledge as expected. The desktop versions significantly increased a trainee's knowledge (p = 0.0041) by an average of 32.14 percent. The CAVE™ versions significantly increased knowledge (p = 0.0012) by an average of 28.57 percent. The traditional PowerPoint presentation also significantly increased a trainee's knowledge (p = 0.0045) by an average of 23.04 percent. Because trainees with significant prior knowledge can only improve by a limited number of percentage points, the research team also calculated what percentage of possible new knowledge was retained by dividing the difference between the pre-training and post-training scores by the difference between the pre-training score and a perfect score. Based on these calculations, the desktop versions afforded learning 48.42 percent of possible new knowledge, while the CAVE™ and PowerPoint versions afforded 43.47 percent and 45.31 percent, respectively.

In comparing the desktop, CAVE™, and PowerPoint versions, the research team found that there was not a significant difference between the three versions in regard to the average increase of knowledge (p = 0.7170) and in regard to affording learning the most possible new knowledge (p = 0.8809). Because there was not a significant difference in the increases of knowledge between the versions, and because the CAVE™ versions were only enabled by immersive technologies costing thousands of dollars, the difference in the cost-benefit ratios of these versions was clear. Based on these facts, the cost-benefit ratio of developing desktop VEs was much more attractive than the cost-benefit ratio of developing immersive CAVE™ versions for these haul truck training tools. A similar difference can be identified in the cost-benefit ratios of the desktop VEs compared to the traditional PowerPoint presentation when the time required to create the 3D models and animations is considered.

Conveyor System Training Evaluation

For evaluating the effectiveness of both the desktop and immersive conveyor system training VEs, the research team again used three types of relatedness questions to compose a knowledge assessment test. The first type of relatedness questions involved the image of protective gear and the name of such protective gear. If the name matched the protective gear in the image, the two concepts should have been considered the same. The second type of questions involved the image of a conveyor system part and the name of a conveyor system part. Again, if the name matched the part in the image, the two concepts should have been considered the same. The third type of questions involved the image of a hazard and the description of a hazard. If the hazard description

matched the picture of a hazard, the two concepts should have been considered the same. Otherwise, any pairs of concepts not considered the same should have been considered definitely not the same.

A total of 12 trainees recruited from around Virginia Tech's campus participated in the evaluation of the conveyor system training tools. Six of the trainees were given the conveyor knowledge assessment test prior to training with the Deep Creator versions of the conveyor system training VEs and again afterwards. The other six trainees were given the knowledge assessment test prior to training with the CAVE™ versions of the conveyor VEs and again afterwards.

Overall, the conveyor training VEs were found to significantly increase a trainee's knowledge as expected. The desktop versions significantly increased a trainee's knowledge (p = 0.0010) by an average of 35.83 percent. The CAVE™ versions significantly increased knowledge (p = 0.0194) by an average of 26.67 percent. Again accounting for trainees with significant prior knowledge, the research team determined what percentage of possible new knowledge was retained by dividing the difference between the pre-training and post-training scores by the difference between the pre-training score and a perfect score. The desktop versions of the conveyor system training VEs afforded learning 72.43 percent of possible new knowledge, while the CAVE™ versions afforded 41.84 percent.

In comparing the desktop and CAVE™ versions, the research team found that there was not a significant difference between the versions in regard to the average increase of knowledge but there was a trend (p = 0.1339). In regard to helping learn the most possible new knowledge, there was a significant difference (p = 0.0461) with the desktop versions affording better training than the CAVE™ versions of the training VEs. This shows that the desktop versions provided better training for retaining as much new information as possible. Based on these results, the cost-benefit ratio of the desktop versions of the conveyor system training VEs is better than the cost-benefit ratio of the immersive versions, which use immersive technologies totaling thousands of dollars.

CONCLUSIONS

Based on the results from the empirical evaluations, the training VEs developed for this research are beneficial for training pre-shift inspections of haul trucks and for training safety and operational procedures for working around conveyor systems. The haul truck training VEs significantly increased trainee knowledge by an average 30.26 percent while the conveyor system training VEs significantly increased trainee knowledge by an average 32.08 percent. These results demonstrate the importance of the three major considerations taken during the design process of these tools—presentation of information, assessment of retention, and reinforcement of training importance.

In addition to validating the effectiveness of the training VEs developed, the research team demonstrated that immersive technologies are not necessary to provide effective training. In fact, for the conveyor system training VEs, a trend of the desktop version providing more effective training can be seen, because the desktop version increased training knowledge by an average of 35.83 percent while the immersive version only increased training knowledge by an average of 26.67 percent. These trends show that desktop VEs may have a more attractive cost-benefit ratio for training purposes than expensive immersive VEs, depending on the training content and application.

ACKNOWLEDGMENTS

The information and developments presented in this paper were funded by a grant from NIOSH. The authors would also like to thank the many companies and research participants that contributed to this study.

REFERENCES

AIMS Research. 2008. AIMS research introduces SafeVR. www.nottingham.ac.uk/aims/VRSite/HTML/SafeVR.html. Accessed May 2008.

Bowman, D.A., Kruijff, E., LaViola, J.J. Jr., and Poupyrev, I. 2005. *3D User Interfaces: Theory and Practice*, Pearson Education.

Burke, M.J., Sarpy, S.A., Smith-Crowe, K., Chan-Serafin, S., Salvador, R.O., and Islam, G. 2006. Relative effectiveness of worker safety and health training methods. *American Journal of Public Health*. 96(2):315.

Cruz-Neira, C., Sandin, D.J., and DeFanti, T.A. 1993. Surround-screen projection-based virtual reality: the design and implementation of the CAVE. *Proc of 20th Annual Conference on Computer Graphics and Interactive Techniques*. 135.

Delabbio, F.C., Dunn, P., Iturregui, L., and Hitchcock, S. 2007. The application of 3D CAD visualization and virtual reality in the mining & mineral processing industry. www.hatch.ca/Mining_Mineral_Processing/Articles/3D_CAD_Visualization_Virtual_reality_in_MMP.pdf. Accessed May 2008.

Durlach, N., and Mavor, A. 1995. *Virtual reality: scientific and technological challenges*. National Academy Press.

Gutiérrez, F., Pierce, J., Vergara, V.M., Coulter, R., Saland, L., Caudell, T.P., Goldsmith, T.E., and Alverson, D.C. 2007. The effect of degree of immersion upon learning performance in virtual reality simulations for medical education. *Studies in Health Technology and Informatics*. 125:155.

Haller, M., Kurka, G., Volkert, J., and Wagner, R. 1999. omVR—a safety training system for a virtual refinery. *Topical Workshop on Virtual Reality and Advanced Human-Robot Systems*. 10:291.

Johnson, P.J., Goldsmith, T.E., and Teague, K.W. 1994. Locus of Predictive Advantage in Pathfinder-Based Representations of Classroom Knowledge. *Journal of Educational Psychology*. 86:617.

Kelso, J., Arsenault, L., Satterfield, S., and Kriz, R. 2002. DIVERSE: a framework for building extensible and reconfigurable device independent virtual environments. *Proc of IEEE Virtual Reality*.

Kowalski, K.M., and Vaught, C. 2002. *Principles of adult learning: application for mine trainers*. Strategies for Improving Miners' Training, (ed. R Peters), US Department of Health and Human Services, Public Health Service, Centers for Disease Control and Prevention, National Institute for Occupational Safety and Health, DHHS (NIOSH) Publication No 2002-156, Information Circular 9463, 2002 September:3.

Lucas, J., McMahan, R., Engle, R., Bowman, D., Thabet, W., Schafrik, S., and Karmis, M. 2008. Improving Mining Health and Safety through Conveyor System Training in a Virtual Environment, *International Future Mining Conference & Exhibition*. Sydney, Australia. 161.

McMahan, R., Bowman, D., Schafrik, S., and Karmis, M. 2008. Virtual Environment Training for Preshift Inspections of Haul Trucks to Improve Mining Safety, *International Future Mining Conference & Exhibition*, Sydney, Australia. 167.

Miller, T.L. 2007. Making haul truck maintenance routine. *Aggregates Manager*. 12(12):26.

Mine Safety and Health Administration (MSHA). 2008. Fatal alert bulletins, fatalgrams and fatal investigation reports. www.msha.gov/FATALS/FAB.HTM. Accessed May 2008.

National Institute for Occupational Safety and Health (NIOSH). 2008. Mining fatalities. www.cdc.gov/niosh/mining/statistics/fatalities.htm. Accessed May 2008.

Orr, T.J., Fligenzi, M.T., and Ruff, T.M. 2003. *Desktop Virtual Reality Miner Training Simulator*. Center for Disease Control and Prevention.

Rothbaum, B.O., Hodges, L.F., Kooper, R., Opdyke, D., Williford, J.S., and North, M. 1995. Effectiveness of computer-generated (virtual reality) graded exposure in the treatment of acrophobia. *American Journal of Psychiatry*. 152(4):626.

Ruff, T.M. 2001. *Miner training simulator: user's guide and scripting language/documentation*. US Department of Health and Human Services, Public Health Service, Centers for Disease Control and Prevention, National Institute for Occupational Safety and Health, DHHS (NIOSH) Publication No 2001-136.

Stothard, P.M., Galvin, J.M., Fowler, J.C.W. 2004. Development, demonstration, and implementation of a virtual reality simulation capability for coal mining operations. *Proc of ICCR Conference*, Beijing, China.

A Century of Mining Visualisation: Moving from 2D to 5D

Damian Schofield
State University of New York (SUNY), New York, United States

Andrew Dasys
MIRARCO Mining Innovation, Sudbury, Ontario, Canada

Pavel Vasak
MIRARCO Mining Innovation, Sudbury, Ontario, Canada

ABSTRACT: Historically, static images such as diagrams and charts have been used to explain complex information associated with the mining industry. We now live in a digital culture that is dominated with images whose value may be simultaneously over-determined and indeterminate, whose layers of significance can only be teased apart with difficulty.

Computer graphics and simulation technology advances rapidly and the media literate public, who watch computer generated movies and play real-time simulated games in virtual worlds expect to see this technology used in their workplace. This paper will review the evolution of the visualisation systems and virtual training simulators used in the mining industry, focusing on some of the latest technology and examples available as applied to mine evacuation simulations and to the analysis of complex geometric and time domain information to identify hazards within mining operations.

INTRODUCTION

Mine rescue training began in the United States in 1910, the year the U.S. Bureau of Mines was created. The training efforts evolved into local and regional competitions and, a year later on October 30, 1911, the first national mine-safety demonstration was held at Forbes Field, in Pittsburgh. Approximately 15,000 persons attended the demonstration. The field exhibits were witnessed by President William H. Taft who presented Red Cross medals and first aid packages to rescue team captains, and many National and State government officials (Brnich et al. 2002).

It was a devastating fire at the Hollinger Mine in Timmins in 1928 that gave birth to the mine rescue program in Canada. Thirty-nine miners died in the fire, and the commission investigating it later recommended a service to ensure that trained and equipped crews would be available to respond to such emergencies in the future. The Ontario mine rescue program was up and running by the next year (Chadbaum 2004).

By its very nature, underground mining can be a hazardous activity in addition to accidents that may occur as in any other industrial activity the very nature of mining requires that additional analysis be undertaken to identify hazards and thus reduce worker risk. The history of all countries where mining has taken place is unfortunately punctuated by major disasters. Even in recent years there have been numerous incidents world-wide which re-enforce the need for continuing assessment of the hazards inherent in mining and the vigilance to identify changes that can, and do take place in the underground environment (Brenkley et al. 1999; Cole et al. 1988).

There remains a significant working lifetime probability that a workforce will have to evacuate a mine under emergency circumstances. In a survey of seven US mines, Vaught et al. (1996) identified that an average of 21% of miners had, at some time, donned a self-contained self-rescuer or filter self-rescuer in an emergency.

Over the years, as mines have generally become safer and major disasters fewer. This decrease can be attributed to better design and a greater corporate focus on developing both prevention and training programmes to insure that workers are ready to respond to emergencies.

The recent economic downturn has also meant that many areas have a lower density of mining operations and hence less available experienced emergency responders. This decrease in response expertise could obviously have serious ramifications in the event of a future mine disaster. Managers who are responsible for day-to-day operations may suddenly be called upon to act as command centre leaders with little or no previous experience in that role. Their lack of knowledge and skills could put in danger the workers who were present when the emergency occurred and also those called upon to respond (Brnich et al. 2002). It would be reckless not to acknowledge that the potential for disaster still exists and that the protection of miners in such circumstances must remain a priority (Mallet et al. 1993; Van Wyk 2006).

There is an emerging recognition, world-wide, that major incidents have profound and lasting repercussions on all stakeholders involved with mining operations. This is shifting the emphasis of safety management to a proactive, risk-based approach, where hazard assessment, effective emergency preparedness measures and multiple training simulations are key elements of reducing workplace and business risks (Jones et al. 1999). In highly mature mining jurisdictions such as Canada, Australia and the United States it is not uncommon for companies to target and attain zero loss hours of production. Reaching these goals within reasonable budgets require that training be focused, effective and timely.

EMERGENCY TRAINING

Historically, emergency response has sometimes been given a low priority in training planning because catastrophic events occur infrequently. While there were usually extensive training requirements for mine rescue team members, in many countries there was no mandated training for command centre leaders. In the past many managers have had little or no experience in dealing with large-scale mine emergencies. Since no one wanted to believe a disaster could happen, day-today job pressures made taking time for emergency response training seem like a luxury for another day (Cole et al. 1998).

Although Command centre training is now available through a number of organisations, such as the Mine Safety and Health Administration (MSHA) in the USA and the Mines and Aggregates Safety and Health Association (MASHA) in Canada, frequently training is in the form of a full-scale simulation and presents the complexity of a major response to the trainees. It gives trainees the needed view of the overall response, but is resource-intensive to conduct (Brnich et al. 2002).

Mine rescue contests are still conducted on a regular basis to sharpen skills and test the knowledge of miners who may one day be called upon to respond to a real mine emergency. The contests usually require teams each to solve a hypothetical rescue problem while being timed and observed by judges according to precise rules. The simulated problem often involves trapped miners who have to be found and rescued. Mine safety experts evaluate each team as they work through their rescue problem in a simulated mine environment. Teams are rated on adherence to safety procedures and how quickly they complete their task (Conti and Chasko 2001).

One form of enhanced command centre training is in the form of a mock mine disaster. Mock disasters are role play exercises designed to present a realistic mine emergency scenario. They use actual mine facilities and involve mine personnel in their assigned roles at the operation. However, staging these events requires significant time and the devotion of considerable resources from the mine and other organisations. For personnel at small, remotely located operations, full-scale mock drills may not be feasible (Brnich et al. 1997; Vaught et al. 2006).

A number of computer augmented training systems have been developed to perform or assist in the process of mine rescue training. MERITS (developed by NIOSH) is a prime example of one such system. It augments existing command center preparation with a training simulation that can be delivered at any location with basic computer equipment and an Internet connection. Many, if not most, modern mine sites have such equipment (Brnich et al., 2002).

Computer generated disaster simulation systems (such as MERITS) are often built primarily as training tools and are designed to be such. However, many such systems are also valuable evaluation tools that can reveal much about what different groups of participants know and do not know, as well as strengths and weaknesses in their logical decisions when they are faced with the predicaments involved in dealing with a mine emergency.

VIRTUAL REALITY (VR) TRAINING SIMULATIONS

Modern simulation systems range from tactile systems that physically represent the real world through to purely computer generated visualisations. These computer-generated, three-dimensional, artificial worlds are commonly referred to as Virtual Environments (VE) and in many instances people are able to interact with the data and images that are presented by these computer-based visual systems (Stothard 2007; Schofield 2008).

In a mining context, a primary aim of developing virtual environments is to allow mine personnel to practice and experience mine processes that will be encountered in the day to day operations at a mine site. In today's industry, safe and efficient planning and production is fundamental to profitable mine operations and virtual environments provide an intuitive means of exploring the diverse and disparate information associated with mining processes.

It is anticipated that providing easy access to virtual environments will facilitate the avoidance of high risk situations through improved knowledge, skills and decision making in the workplace (Tromp and Schofield 2004). An advantage is that mine personnel are also able to practice seldom experienced scenarios thereby maintaining a level of preparedness and developing their knowledge and experience. In addition, routine mining operations can also be developed, practised and monitored in virtual environments.

Virtual Training simulation for the mining industry was pioneered at the University of Nottingham in the UK where low cost simulator software was developed. A number of early examples are extensively discussed by (Denby and Schofield 1999; Schofield et al. 2001). Most of this early research and development of mining VR based systems was concerned with the training of mine operators in the area of hazard identification and hazard avoidance. The systems were PC based and took advantage of the (then) new specialist 3D accelerator cards (aimed at the fledgling PC gaming market) to provide acceptable performance in terms of graphics quality and frame rates (Denby et al. 1998).

In the systems developed at the University of Nottingham VR provided an intuitive, graphical interface for non-computer literates to allow them to interact with advanced computer simulations of dynamic and potentially hazardous situations. The systems were based on computer simulations of dynamic mining scenarios which were integrated with three-dimensional models of the physical environment. Equipment and personnel were included in order to simulate specific situations (Denby and Schofield 1999). A number of the systems developed in Nottingham also included a risk assessment facility to enable the creation of safer work places. As a trainee navigated around a mine, various hazards were presented and the trainee had to identify the type of hazard, the level of risk and suggest remedial action (Denby et al. 1998; Schofield et al. 2001a).

Researchers at the National Institute for Occupational Safety and Health (NIOSH) in the U.S. also developed a number of early prototype VR applications to train surface and underground mine workers and rescue personnel in hazard recognition and evacuation routes and procedures. Using the QUAKE II and Unreal games' level editors, low cost VR systems were developed complete with vehicles, equipments and various hazards. Using the game graphic engine, the user navigates the mine identifying and avoiding the hazards (Filigenzi et al. 2000).

Another early prototype VR model was developed by CSIR in South Africa to increase mine workers' abilities to identify hazardous ground conditions and reduce occurrences of rock fall related accidents. In the model developed, trainees were required to

successfully negotiate their way around the model identifying the hazards and selecting appropriate corrective action (Squelch 2001).

Other key early work was undertaken at Laurentian University's Virtual Reality Lab (VRL) in Sudbury. This facility offers a unique data interpretation environment that combines a large format spherical stereoscopic projection system, equipped with high definition projectors that provide the highest resolution of active stereo available with the advanced features of modelling software. This creates the first VR facility geared specifically toward geometry, geology, geochemistry, geomechanics, mining and mine safety applications (Delabbio et al. 2003).

The facility, in addition to providing a Virtual environment for first and third person rescue simulations, has also been used for the design and evaluation of line of sight issues as well as developing a comprehensive 5 dimensional risk portofolio. Both the dynamic nature of the mine's changing geometry and more importantly the changing stress fields in the host rock are modelled and presented as a means of assessing the location and timing of potential hazards.

Although proprietary systems are still in use, advanced simulation systems can be built from off-the-shelf components and deployed to industry. To date, mining VR systems have placed much of their focus on representing synthetic virtual models of the mine environment. In particular, this approach has proved successful with respect to producing a large number of virtual mine training environments (Schofield et al. 2001a; Stothard et al. 2001; Squelch 2001; Kizil 2003; Schmid and Bracher 2004; Unger and Mallet 2004; Van Wyk 2006; Stothard 2007; Schmid and Winkler 2008).

VIRTUAL REALITY (VR) EMERGENCY RESPONSE SIMULATIONS

Since the terrorist attacks of 11th September 2001, governments worldwide have invested increasing amounts of resources in the writing of emergency response plans. Particularly in the United States, the federal government has created new security organisations and urged state and local governments to draw up plans. This emphasis on the written plan tends to draw attention away from the process of the event itself and the objective of achieving emergency preparedness through awareness and training (Perry and Lindell 2003; Schmorrow et al. 2008).

A number of studies assessing the training and operational practices of a group of emergency responders to disaster search and rescue duties have hypothesised that training effectiveness could be evaluated by assessing the incidence with which event characteristics are perceived as stressors. This has shown that much of the small scale focused training did not prepare them for large scale emergency response work. The implications being that larger scale scenarios are needed to train effectively. Obviously it is much easier to replicate these events using virtual simulators than to create large scale real-world training exercises (Paton 1994).

The key to successful training is creating realistic environments, into which participants get so deeply immersed that they are stimulated and able to demonstrate realistic behaviour. In the domain of emergency response it is all about decision making under time stress, in a more or less unknown situation with limited information. Important factors are real-time, real-effect and dynamic simulation that follows all actions and decisions made by the participants through open-ended scenarios.

In most of these simulations, the visuals enhance realism by exposing the trainees to representations of hazards without putting them at risk. Registration generates data that describe the course of events in the exercise. These data form a mission history, which is a time-synchronized, event-based model of the exercise that can be replayed and reviewed in a post-training review (Jenvald et al. 2004).

There are very few 'real' examples of where virtual reality technology has been used to simulate large scale mine emergencies. A number of papers have, however, discussed the use of virtual technology for this purpose (Bise 1997; Conti et al. 1999; Denby and Schofield 1999; Filigenzi et al. 2000; Schofield et al. 2001a; Delabbio et al. 2003; Kizil 2003; Schmid 2003; Unger and Mallett 2004; Stothard 2007).

Key early work to demonstrate the effectiveness of virtual reality simulators for the training of mine emergency response teams was undertaken by the Aims Research Unit at the University of Nottingham (Schofield et al. 2001) and NIOSH in the United States (Orr and Girard 2002).

One of the early examples built by the University of Nottingham is a mine training system which was developed for the Australian Government (Department of Industry and Resources in Western Australia). This multi-modal training system involved a number of digital media forms including scientific animations, rendered still images, photographs, video clips, maps/plans and text documents (such as witness statements and accident reports). The training system reconstructed a 'real' truck roll-over disaster from the accident records of the government office (see Figure 1). A full set of animations were developed which illustrated the chain of events and circumstances which led to the accident (Schofield et al. 2001a).

In 2000 a large scale virtual reality simulation of a mine emergency was created, again for the Australian Government (Department of Industry and Resources in Western Australia). The scenario was based on a real incident, where a truck caught fire in an underground roadway and flooded the mine with toxic smoke (see Figure 2). The simulation was used in a large scale, multi-agency disaster role play event which was organized during the MineSafe conference in Perth in the year 2000 (Schofield et al. 2000).

The simulation was carried out in front of a large audience, participants included mine workers, mine management, head office staff, government officials and actors playing members of the press, family members etc. The simulation was exceptionally well received and most participants (and audience members) commented on the usefulness of virtual reality based simulation (Schofield et al. 2000; Schofield et al. 2001b).

The major focus of the early NIOSH prototype project was mine escapeway training using desktop virtual reality programs. Scenarios for evacuation

Figure 1. Stills from the truck rollover training system

Figure 2.   Still from the truck fire training system

could be practiced in a three-dimensional computer simulation of the mine in a disaster situation, complete with smoke, fire, and other dangers. Mine workers could practice evacuating various mine layouts without the danger and expense associated with escapeway training in an actual underground mine. NIOSH reported that this training experience helped to reinforce the knowledge acquired during more conventional classroom instruction, and the inherent flexibility of the program allowed training materials to be tailored to meet the specific requirements of individual mines (Orr and Girard 2002).

One of the largest mine emergency simulations to be widely used by industry is the system developed by the University of New South Wales (UNSW) for Coal Services to provides critical services to the NSW coal mining industry. Miners are immersed in a three dimensional virtual mine and taught emergency response and safety procedures. They navigate via a series of simulated underground emergency drills, including unaided self escape, hazard awareness, rib and roof stability, and truck and shovel operations (Stothard, 2007).

One of the best examples of the use of this technology is NIOSH's new computer-based training program to teach miners to read mine maps, a critical skill in learning how to navigate the labyrinth of tunnels and to stay safe in underground mines (see Figure 3). The user assumes a first person role inside a virtual coal mine, with an unseen narrator giving voice commands about what to do as the game progresses.

The software is designed to be used in safety training courses that mine operators are required to provide by federal law. The software builds on the principle that young miners who have grown up with video games and other computer applications will find this kind of interactive, game-format training more engaging and more meaningful than traditional classroom lecture and instruction.

The map reading training system is provided as a free download from the NIOSH website (NIOSH 2009).

VIRTUAL REALITY TO MODEL HAZARDS AND RISK

One of the earliest applications of large scale stereoscopic visualization as applied to mining problems was pioneered by MIRARCO (Mining Innovation Rehabilitation and Applied Research Corporation) a not for profit research group owned by Laurentian University. MIRARCO's Mine Rescue Control Group Training (MRCGT) project has developed a prototype virtual reality based mine rescue simulation system. The developed system included multiple mine specific model scenarios incorporating fire and rock fall emergencies. These three dimensional mine models were based on real mine layouts which were imported into the system. While running, the simulation includes the incorporation of realistic smoke transport into the three dimensional mine model. The smoke changes in density, direction and velocity as the parameters controlling the mine ventilation environment are altered during the life of the simulation.

MIRARCO has also focused its virtual reality research on the development of software tools to add data dimensions to the existing mine geometry to provide a means of identifying and qualifying hazards. The graphical representations as shown in Figure 4 differ significantly from other training environments discussed in this paper. The first person point of view is replaced with a god view that provides a complete view of the mining geometry. Additionally the user can not only move in 3D space but can also move in time reviewing the geometric and events that have already occurred and projecting geometry into the future. This future projection

Figure 3.   Still from the map reading training system

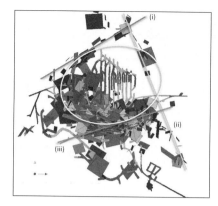

Figure 4. Long-term microseismic record on the left-hand side. Discrete seismically active planes (spatiotemporal clusters) on the right-hand side showing recognizable patterns, (i, ii, and iii) and geological discontinuities, oval zone represents the stress rotation about the excavated orebody.

is based on the mine plans that show where future workings will be located.

The use of stereoscopic projection, providing an additional dimension beyond the traditional 2D of most graphic canvases, provides researchers with an opportunity to increase the dimensions of data being explored. The time component and mine opening geometry represent 4 rather intuitive dimensions. The 5th dimension is less intuitive and represents either risk or uncertainty. An example of this uncertainty would be representing the results of conditional simulations used to define mineable reserves, which in term will define the location of the mine's tunnels and stopes. The risk dimensionality could be the graphical representation of the changing stress fields that are created by the extraction of the host rock.

Although the simplification of the graphical representation, lack of texturing and little need to deal with movement, should decrease the amount of graphical rendering power required, the need to represent much larger model views while insuring the scientific integrity of the data continues to strain the most advanced computer hardware available.

To meet these challenges most models are now being represented using ParaViewGeo a scientific visualization platform that is multi-processor capable and can deal with terabyte size datasets. Despite the technical rendering capabilities provided by gaming hardware and scientific visualization software the sheer complexity of the data requires that new data simplification algorithms be developed to identify and access areas of hazard. The left image in Figure 4 displays raw microseismic information, while a simplified view in the right pane demonstrates how data can be reduced to a series of colour

codes draped on mine geometry. This combination of accurate geometry with synthesized data is combined to create a 5 Dimensional (geometry, time and uncertainty/risk) model that can significantly improve the development of more traditional Virtual Reality training programs, by identifying areas of concern and allowing the development of advanced training scenarios.

EVALUATING SIMULATION TRAINING

Researchers have noted practical problems of the failure of training in safe work practices to sometimes result in changes to the rate of accident and injury in mining workplaces. A review of the literature in workplace training and workplace learning suggests that there has been little investigation of the relationship between how trainers train, and what trainees learn in the workplace. Analysis suggests that even though trainers and workers in the mining industry do not always work in the same organisation or geographical location, they co-participate in the ongoing construction of a community of practice that reinforces strong implicit masculine storylines. Mine workers were found to learn safety best through the experience of doing their work, while trainers report best results from safety training using simulated environments and codified practice. Understanding how mine workers learn safety in the workplace, within a community of practice, is critical to attempts to improve safety training and safety records (Somerville and Abrahamsson 2003).

With any training intervention, evaluation is critical to ensure its success. When assessing how virtual simulations prepare someone to cope in a real world situation, there is no group, including the military, that routinely conducts real-world validity studies of similar critical skills. Other groups who engage in good criterion-referenced skill training and evaluations ensure that the simulation is grounded in significant real-world problems (Brnich et al. 1997).

Based on field test results, one can be fairly sure that these simulations provide useful information and an opportunity to practice those critical cognitive skills that are needed for effective performance during emergency management. We can also be sure that little current training in this area addresses similar types of "soft" skills. It is almost certainly better to have faced these decision-making tasks initially in the context of a simulation than during an actual event. This is especially true if one's earlier learning is mainly a set of rigid protocols about how things should be done, with little awareness of how unforeseen predicaments and dilemmas cause those protocols to break down (Vaught et al. 1996).

Often simulation evaluation is assessed in terms of progress through the simulation. One should not evaluate trainees from a norm-referenced perspective. Instead, one should attempt to be criterion-referenced. Performance on this type of test is not defined by rank ordering a person's score according to the distribution of scores in some normative group, but is instead described by mastery of knowledge, skills, and strategies included in the exercise (Cole 1994).

Proper field-testing provides psychometric data on mental measurement properties of the exercise. At the same time, information is obtained to reveal skill areas in which the participants performed well and in which they performed poorly. Overall mastery levels can also be computed. This type of information can be used immediately to ensure that performance errors exhibited while working the simulation can be remediated by further instruction. Such corrective measures can help prevent these errors from occurring in actual emergencies (Brnich et al. 1997; Cole et al. 1998; Schofield et al. 2001a; Schofield 2008).

REFERENCES

Bise, C.J., 1997. Virtual Reality: Emerging Technology for Training of Miners, *Mining Engineering*, 49 (1).

Brenkley, D., Bennett S.C. and Jones, B., 1999. Enhancing Mine Emergency Response, *Proceedings of 28th International Conference on Safety in Mines Research Institutes*, (Sinaia, Romania).

Brnich, M.J., Mallett, L. and Vaught, C., 1997. Training Future Mine Emergency Responders, *Holmes Safety Association Bulletin*.

Brnich, M.J., Mallett, L.G., Reinke, D.C. and Vaught, C., 2002. Mine Emergency Response Command Center Training Using Computer Simulation, *Proceedings of the 33nd Annual Institute of Mining Health, Safety and Research*, (Blacksburg, USA).

Chadboum, J., 2004. *Under Oxygen: 75 Years of Mine Rescue in Ontario*, (Ontario, Canada: Mines and Aggregates Safety and Health Association).

Cole, H.P., 1994. Embedded Performance Measures as Teaching and Assessment Devices, *Occupational Medicine: State of the Art Reviews*, 9(26): 1–28.

Cole, H.P., Vaught, C., Wiehagen, W.J., Haley, J.V., and Brnich, M.J., 1998. Decision Making During a Simulated Mine Fire Escape, *Journal of Engineering Management*, 45(2): 153–162.

Conti, R.S., Chasko, L.L. and Cool, J., 1999. An Overview of Technology and Training Simulations for Mine Rescue Teams, *Proceedings 28th International Conference on Safety in Mines Research Institutes*, (Sinaia, Romania).

Delabbio, F.C., Dunn, P.G., Itteregui, L. and Hitchcock, S., 2003. The Application of 3D CAD Visualization and Virtual Reality in the Mining and Mineral Processing Industry, HATCH, Mining Technology Unit. http://www.hatch.ca/Mining_Mineral_Processing/Articles/3D_CAD_Visualization_Virtual_Reality_in_MMP.pdf. Accessed September 2009.

Denby, B., Schofield, D., McClarnon, D., Williams, M. and Walsha, T., 1998. Hazard Awareness Training for Mining Situations using Virtual Reality, *Proceedings of APCOM 98: Computer Applications in the Minerals Industries*, (London, UK).

Denby, B. and Schofield, D., 1999b. The Role of Virtual Reality in the Safety Training of Mine Personnel, *Mining Engineering*, October 1999: 59–64.

Filigenzi, M. T., Orr, T.J. and Ruff, T.M., 2000, Virtual Reality for Mine Safety Training, *Journal of Applied Occupational and Environmental Hygiene*, 15(6): 465–469.

Jenvald, J., Systemteknik, V., Linköping, I. and Storskiftesgatan, A.B., 2004. Simulation-Supported Live Training for Emergency Response in Hazardous Environments, *Journal of Simulation and Gaming*, 35(3): 363–377.

Jones B., Brenkley D., Burrell R.A. and Bennett, S.C., 1999. *Rescue and Emergency Support Services in Underground Coal Mines*, Report CCC/13, London, UK: IEA Coal Research.

Kizil, M., 2003. Virtual Reality Applications in the Australian Minerals Industry, *Proceedings of APCOM 98: Computer Applications in the Minerals Industries*, (Johannesburg, South Africa).

Mallett, L., Vaught, C. and Brinch, M.J., 1993. Sociotechnical Communication in an Underground Mine Fire: A Study of Warning Messages During an Emergency Evacuation, *Safety Science*, 16(5): 709–728.

NIOSH, 2009. Underground Coal Mine Map Reading Training. www.cdc.gov/niosh/mining/products/product165.htm. Accessed September 2009.

Orr, T. and Girard, J., 2002. Mine Escapeway Multiuser Training With Desktop Virtual Reality, *Proceedings of APCOM 02: Computer Applications in the Minerals Industries*, (Phoenix, USA).

Paton, D., 1994. Disaster Relief Work: An Assessment of Training Effectiveness, *Journal of Traumatic Stress*, 7(2): 275–288.

Perry, R.W. and Lindell, M.K., 2003. Preparedness for Emergency Response: Guidelines for the Emergency Planning Process, *Disasters*, 27(4): 336–350.

Schmorrow, D., Cohn, J. and Nicholson, D., 2008. *The PSI Handbook of Virtual Environments for Training and Education* [Three Volumes], Connecticut, USA: Greenwood Press.

Schofield, D., Noond, J., Goodwin, L. and Fowle, K., 2000. Recreating Reality - Using Computer Generated Forensic Animations to Reconstruct Accident Scenarios, *Proceedings of Minesafe 2000 Conference*, (Perth, Australia).

Schofield, D., Denby, B. and Hollands, R., 2001a. Mine Safety in the Twenty-First Century: The Application of Computer Graphics and Virtual Reality. In *Mine Health and Safety Management*, 1st ed. Edited by M. Karmis, Littleton, CO: SME.

Schofield, D., Noond, J., Goodwin, L. and Fowle, K., 2001b. Accident Scenarios: Using Computer Generated Forensic Animations, *Journal of Occupational Health and Safety*, 17 (2): 163–173.

Schofield, D., 2008. Do You Learn More When Your Life is in Danger? Keynote Address, *Proceedings of the 12th International Conference on Mechatronic Technology,* (Ontario, Canada).

Schmid, M., 2003. Virtually Assured—The Role of Virtual Reality in the Field of Coal Mining, *World Coal*, (11): 45–48.

Schmid, M. and Bracher, P., 2004. Use of Conceptual Models in Managing Process and Product Innovations, *Proceedings of MinExpo*, (Las Vegas, USA).

Schmid, M. and Winkler T., 2008. Interactive 3D Supports Underground Repair and Maintenance of Road Heading Machines and Shearer Loaders, *Proceedings of the 21st World Mining Congress*, (Krakow, Poland).

Somerville M. and Abrahamsson, L., 2003. Trainers and Learners Constructing a Community of Practice: Masculine Work Cultures and Learning Safety in the Mining Industry, *Studies in the Education of Adults*, 35(1): 19–34.

Squelch, A.P., 2001. Virtual Reality for Mine Safety Training in South Africa, Journal of the South African Institute of Mining and Metallurgy, July 2001: 209–216.

Stothard, P., Squelch, A., van Wyk, E., Schofield, D., Fowle, K., Caris, C., Kizil, M.S. and Schmidt, M., 2008. Taxonomy of Interactive Computer-Based Visualisation Systems and Content for the Mining Industry—Part One, *Proceedings of the Future Mining Conference*, (Sydney, Australia).

Stothard, P., 2007. Developing and Deploying and Deploying Interactive Training Simulations for the Coal Mining Industry, Proceedings of SIMTECT 2007 Conference, (Brisbane, Australia).

Tromp, J. and Schofield, D., 2004. Practical Experiences of Building Virtual Reality Systems, *Proceedings of Designing and Evaluating Virtual Reality Systems Symposium*, (University of Nottingham, UK).

Unger, R. and Mallet, L., 2004. Virtual Reality in Mine Training, *Proceedings of SME Annual Meeting and Exhibit*, Society for Mining, Metallurgy, and Exploration, (Denver, Colorado).

Van Wyk, E.A, 2006. Improving Mine Safety Training Using Interactive Simulations, *Proceedings of the ED-MEDIA World Conference on Educational Multimedia, Hypermedia and Telecommunications*, (Orlando, Florida).

Vasak, P., Dasys, A., Malek, F. and Thibodeau, D., 2008. Extracting More Value from Complex Monitoring Data—Seismic Excavation Hazard Maps, *Proceedings of the Strategic and Tactical Approaches in Mining Conference*, (Quebec City, Canada).

Vaught, C., Fotta, B., Wiehagen, W.J., Conti, R.S., and Fowkes, R.S., 1996. *A Profile of Workers' Experiences and Preparedness in Responding to Underground Mine Fires*, US Bureau of Mines, Report of Investigations RI 9584, Pittsburgh, USA.

Index

mine explosion prevention research, 335–340,
343–344
and mine illumination development, 35–40
and Polish mine safety research, 305
research to improve miners' health and safety
training, 501–509
See also National Institute for Occupational
Safety and Health

V

Ventilation
with compressed air, 463–477
and diesel emissions, 453–462
history of research in (South Africa), 424–431
total structure ventilation design, 447–448
See also Dust control; Methane

Virtual reality/virtual environments
in hazard and emergency response training,
529–536
in powered haulage training, 520–528
in training, 510–516
training software (Coal Services Pty Limited),
517–519

W

Water inflow prediction, and CSIRO, 75–77
Witwatersrand Basin goldfields (South Africa),
156–171
Women miners, 33